Annual Abstract
of Statistics

No 154
2018 Edition
Volume 1 of 2
Compiled by: Dandy Booksellers

Contacts

For information about the content of this publication, contact
Dandy Booksellers: Tel 020 7624 2993
Email: enquiries@dandybooksellers.com

Publications orders

To obtain the print version of this publication, please contact
Dandy Booksellers
Tel: 0207 624 2993
Email: dandybooksellers@btconnect.com
Fax: 0207 624 5049
Post: Unit 3&4, 31-33 Priory Park Road, London, NW6 7UP
Web: www.dandybooksellers.com

Contents

Contents

Contents

Contents

Units of measurement

Length

1 millimetre (mm)	= 0.03937 inch	
1 centimetre (cm)	= 10 millimetres	= 0.3937 inch
1 metre (m)	= 1,000 millimetres	= 1.094 yards
1 kilometre (km)	= 1,000 metres	= 0.6214 mile
1 inch (in.)		= 25.40 millimetres or 2.540 centimetres
1 foot (ft.)	= 12 inches	= 0.3048 metre
1 yard (yd.)	= 3 feet	= 0.9144 metre
1 mile	= 1,760 yards	= 1.609 kilometres

Area

1 square millimetre (mm2)		= 0.001550 square inch
1 square metre (m2)	= one million square millimetres	= 1.196 square yards
1 hectare (ha)	= 10,000 square metres	= 2.471 acres
1 square kilometre (km2)	= one million square metres	= 247.1 acres
1 square inch (sq. in.)		= 645.2 square millimetres or 6.452 square centimetres
1 square foot (sq. ft.)	= 144 square inches	= 0.09290 square metre or 929.0 square centimetres
1 square yard (sq. yd.)	= 9 square feet	= 0.8361 square metre
1 acre	= 4,840 square yards	= 4,046 square metres or 0.4047 hectare
1 square mile (sq. mile)	= 640 acres	= 2.590 square kilometres or 259.0 hectares

Volume

1 cubic centimetre (cm3)		= 0.06102 cubic inch
1 cubic decimetre (dm3)	= 1,000 cubic centimetres	= 0.03531 cubic foot
1 cubic metre (m3)	= one million cubic centimetres	= 1.308 cubic yards
1 cubic inch (cu.in.)		=16.39 cubic centimetres
1 cubic foot (cu. ft.)	= 1,728 cubic inches	= 0.02832 cubic metre or 28.32 cubic decimetres
1 cubic yard (cu. yd.)	= 27 cubic feet	= 0.7646 cubic metre

Capacity

1 litre (l)	= 1 cubic decimetre	= 0.2200 gallon
1 hectolitre (hl)	= 100 litres	= 22.00 gallons
1 pint		= 0.5682 litre
1 quart	= 2 pints	= 1.137 litres
1 gallon	= 8 pints	= 4.546 litres
1 bulk barrel	= 36 gallons (gal.)	= 1.637 hectolitres

Weight

1 gram (g)		= 0.03527 ounce avoirdupois
1 hectogram (hg)	= 100 grams	= 3.527 ounces or 0.2205 pound
1 kilogram (kg)	= 1,000 grams or 10 hectograms	= 2.205 pounds
1 tonne (t)	= 1,000 kilograms	= 1.102 short tons or 0.9842 long ton
1 ounce avoirdupois (oz.)	= 437.5 grains	= 28.35 grams
1 pound avoirdupois (lb.)	= 16 ounces	= 0.4536 kilogram
1 hundredweight (cwt.)	= 112 pounds	= 50.80 kilograms
1 short ton	= 2,000 pounds	= 907.2 kilograms or 0.9072 tonne
1 long ton (referred to as ton)	= 2,240 pounds	= 1,016 kilograms or 1.016 tonnes
1 ounce troy	= 480 grains	= 31.10 grams

Energy

British thermal unit (Btu)	= 0.2520 kilocalorie (kcal) = 1.055 kilojoule (kj)
Therm	= 105 British thermal units = 25,200 kcal = 105,506 kj
Megawatt hour (MWh)	= 106 watt hours (Wh)
Gigawatt hour (GWh)	= 106 kilowatt hours = 34,121 therms

Food and drink

Butter	23,310 litres milk	= 1 tonne butter (average)
Cheese	10,070 litres milk	= 1 tonne cheese
Condensed milk	2,550 litres milk	= 1 tonne full cream condensed milk
	2,953 litres skimmed milk	= 1 tonne skimmed condensed milk
Milk	1 million litres	= 1,030 tonnes
Milk powder	8,054 litres milk	= 1 tonne full cream milk powder
	10,740 litres skimmed milk	= 1 tonne skimmed milk powder
Eggs	17,126 eggs	= 1 tonne (approximate)
Sugar	100 tonnes sugar beet	= 92 tonnes refined sugar
	100 tonnes cane sugar	= 96 tonnes refined sugar

Shipping

Gross tonnage	= The total volume of all the enclosed spaces of a vessel, the unit of measurement being a 'ton' of 100 cubic feet.
Deadweight tonnage	= Deadweight tonnage is the total weight in tons of 2,240 lb. that a ship can legally carry, that is the total weight of cargo, bunkers, stores and crew.

Introduction

Welcome to the 2018 edition of the Annual Abstract of Statistics. This compendium draws together statistics from a wide range of official and other authoritative sources.

Dandy Booksellers have sourced and formatted these tables under instruction from various government departments/ organisations

Current data for many of the series appearing in this Annual Abstract are contained in other ONS publications, such as Economic & Labour Market Review, Population Trends, Health Statistics Quarterly and Financial Statistics. These titles can be purchased through Dandy Booksellers.

The name (and telephone number, where this is available) of the organisation providing the statistics are shown under each table. In addition, a list of Sources is given at the back of the book, which sets out the official publications or other sources to which further reference can be made.

Identification codes

The four-letter identification code at the top of each data column, or at the side of each row is the ONS reference for this series of data on their database. Please quote the relevant code if you contact them requiring any further information about the data. On some tables it is not possible to include these codes, so please quote the table number in these cases.

Definitions and classification

Time series
So far as possible annual totals are given throughout, but quarterly or monthly figures are given where these are more suitable to the type of series.

Standard Industrial Classification

A Standard Industrial Classification (SIC) was first introduced into the UK in 1948 for use in classifying business establishments and other statistical units by the type of economic activity in which they are engaged. The classification provides a framework for the collection, tabulation, presentation and analysis of data about economic activities. Its use promotes uniformity of data collected by various government departments and agencies.

Since 1948 the classification has been revised in 1958, 1968, 1980, 1992, 2003 and 2007. One of the principal objectives of the 1980 revision was to eliminate differences from the activity classification issued by the Statistical Office of the European Communities (Eurostat) and entitled 'Nomenclature générale des activités économiques dans les Communautés Européennes', usually abbreviated to NACE.

In 1990 the European Communities introduced a new statistical classification of economic activities (NACE Rev 1) by regulation. The regulation made it obligatory for the UK to introduce a new Standard Industrial Classification SIC(92), based on NACE Rev 1. UK SIC(92) was based exactly on NACE Rev 1 but, where it was thought necessary or helpful, a fifth digit was added to form subclasses of the NACE 1 four digit system. Classification systems need to be revised periodically because, over time, new products, processes and industries emerge. In January 2003 a minor revision of NACE Rev 1, known as NACE Rev 1.1, was published in the Official Journal of the European Communities.

Consequently, the UK was obliged to introduce a new Standard Industrial Classification, SIC(2003) consistent with NACE Rev 1.1. The UK took the opportunity of the 2003 revision also to update the national Subclasses. Full details are available in UK Standard Industrial Classification of Economic Activities 2003 and the Indexes to the UK Standard Industrial Classification of Economic Activities 2003. These are the most recent that are currently used. The most up to date version is the UK Standard Industrial Classification of Economic activities 2007 (SIC2007). It will be implemented in five stages and came into effect on 1 January 2008.

- For reference year 2008, the Annual Business Inquiry (parts 1 & 2) will be based on SIC 2007

- PRODCOM will also be based on SIC 2007 from reference year 2008

- Other annual outputs will be based on SIC 2007 from reference year 2009, unless otherwise determined by regulation

- Quarterly and monthly surveys will be based on SIC 2007 from the first reference period in 2010, unless otherwise determined by regulation

- National Accounts will move to SIC 2007 in September 2011

Symbols and conventions used

Change of basis
Where consecutive figures have been compiled on different bases and are not strictly comparable, a footnote is added indicating the nature of the difference.

Geographic coverage
Statistics relate mainly to the UK. Where figures relate to other areas, this is indicated on the table.

Units of measurement
The various units of measurement used are listed after the Contents.

Rounding of figures
In tables where figures have been rounded to the nearest final digit, the constituent items may not add up exactly to the total.

Symbols
The following symbols have been used throughout:

.. = not available or not applicable (also information supressed to avoid disclosure)

\- = nil or less than half the final digit shown

Office for National Statistics online:
www.ons.gov.uk
Web-based access to time series, cross-sectional data and metadata from across the Government Statistical Service (GSS), is available using the site search function from the homepage. Download many datasets, in whole or in part, or consult directory information for all GSS statistical resources, including censuses, surveys, periodicals and enquiry services. Information is posted as PDF electronic documents or in XLS and CSV formats, compatible with most spreadsheet packages.

Contact point
Dandy Booksellers welcomes any feedback on the content of the Annual Abstract, including comments on the format of the data and the selection of topics. Comments and requests for general information should be addressed to:

Dandy Booksellers
Opal Mews, Unit 3&4
31-33 Priory Park Road
London
NW6 7UP
or
enquiries@dandybooksellers.com

this page is intentionally blank

Area

Chapter 1

Area

The United Kingdom (UK) comprises Great Britain and Northern Ireland. Great Britain comprises England, Wales and Scotland.

Physical Features

The United Kingdom (UK) constitutes the greater part of the British Isles. The largest of the islands is Great Britain. The next largest comprises Northern Ireland and the Irish Republic. Western Scotland is fringed by the large island chain known as the Hebrides, and to the north east of the Scottish mainland are the Orkney and Shetland Islands. All these, along with the Isle of Wight, Anglesey and the Isles of Scilly, form part of the UK, but the Isle of Man, in the Irish Sea and the Channel Islands, between Great Britain and France are largely self-governing and are not part of the UK. With an area of about 243 000 sq km (about 94 000 sq miles), the UK is just under 1 000 km (about 600 miles) from the south coast to the extreme north of Scotland and just under 500 km (around 300 miles) across at the widest point.

- Highest mountain: Ben Nevis, in the highlands of Scotland, at 1 343 m (4 406 ft)
- Longest river: the Severn, 354 km (220 miles) long, which rises in central Wales and flows through Shrewsbury, Worcester and Gloucester in England to the Bristol Channel
- Largest lake: Lough Neagh, Northern Ireland, at 396 sq km (153 sq miles)
- Deepest lake: Loch Morar in the Highlands of Scotland, 310 m (1 017 ft) deep
- Highest waterfall: Eas a'Chual Aluinn, from Glas Bheinn, in the highlands of Scotland, with a drop of 200 m (660 ft)
- Deepest cave: Ogof Ffynnon Ddu, Wales, at 308 m (1 010 ft) deep
- Most northerly point on the British mainland: Dunnet Head, north-east Scotland
- Most southerly point on the British mainland: Lizard Point, Cornwall
- Closest point to mainland continental Europe: Dover, Kent. The Channel Tunnel, which links England and France, is a little over 50 km (31 miles) long, of which nearly 38 km (24 miles) are actually under the Channel

Area Measurements

I'd like to get area figures for all the local authorities in the UK. Where do I look?

UK Standard Area Measurements (SAM) are now available to download free of charge from the Open Geography portal in both MS Excel and CSV formats for a variety of administrative areas (countries, counties, local authority districts, electoral wards/divisions, regions, output areas, super output areas and workplace zones). The two questions below show some statistics taken from SAM 2017 (Extent of the Realm figures).

Which is the largest local authority in the UK?

- The largest anywhere in the UK is Highland (Scotland), at 2,647,291.06 hectares (ha).
- The largest in England is Northumberland, at 507,818.02 ha.
- The largest in Wales is Powys, at 519,545.61 ha.
- The largest in Northern Ireland is Fermanagh and Omagh, at 301,211.48 ha.

Which is the smallest local authority in the UK?

- The smallest anywhere in the UK is the City of London, at 314.96 ha
- The smallest in Scotland is Dundee City, at 6,222.38 ha.
- The smallest in Wales is Blaenau Gwent, at 10,872.79 ha
- The smallest in Northern Ireland is Belfast at 13,771.69 ha.

UK Standard Area Measurements (SAM)

About UK Standard Area Measurements (SAM)

The SAM product provides a definitive list of measurements for administrative, health, Census, electoral and other geographic areas in the UK. SAM will change annually for some geographies, but for others it will be 'frozen' for several years.

The land measurement figures provided are defined by topographic boundaries (coastline and inland water), where available.

Measurements are reviewed annually and include information up to the end of December.

The measurements have been produced in conjunction with the following UK government statistical organisations and independent mapping agencies: National Records of Scotland (NRS), Northern Ireland Statistics and Research Agency (NISRA), Ordnance Survey®(OS) and Land & Property Services (LPS).

The SAM User Guide explains differences in methodology and base mapping between these agencies.

Product details

If you have any queries about this dataset, please contact:

ONS Geography Customer Services
Office for National Statistics
Segensworth Road
Titchfield
Fareham
Hampshire
PO15 5RR

Tel: 01329 444 971
Email: ons.geography@ons.gov.uk

1.1 Standard Area Measurement for Local Authority Districts as at 31/12/2017

Local Authority Name	Code	Extent of the realm (km²)	Area to mean high water (km²)	Inland water (km²)	Land only (km²)
United Kingdom	**K02000001**	**248727**	**244385**	**1635**	**242750**
Great Britain	**K03000001**	**234397**	**230055**	**1098**	**228957**
England	**E92000001**	**132946**	**130462**	**152**	**130310**
Hartlepool	E06000001	98	94	0	94
Middlesbrough	E06000002	55	54	0	54
Redcar and Cleveland	E06000003	254	245	0	245
Stockton-on-Tees	E06000004	210	205	0	205
Darlington	E06000005	197	197	0	197
Halton	E06000006	90	79	0	79
Warrington	E06000007	182	181	0	181
Blackburn with Darwen	E06000008	137	137	0	137
Blackpool	E06000009	43	35	0	35
Kingston upon Hull, City of	E06000010	81	71	0	71
East Riding of Yorkshire	E06000011	2495	2406	1	2405
North East Lincolnshire	E06000012	204	192	0	192
North Lincolnshire	E06000013	876	846	0	846
York	E06000014	272	272	0	272
Derby	E06000015	78	78	0	78
Leicester	E06000016	73	73	0	73
Rutland	E06000017	394	394	12	382
Nottingham	E06000018	75	75	0	75
Herefordshire, County of	E06000019	2180	2180	0	2180
Telford and Wrekin	E06000020	290	290	0	290
Stoke-on-Trent	E06000021	93	93	0	93
Bath and North East Somerset	E06000022	351	351	5	346
Bristol, City of	E06000023	235	110	0	110
North Somerset	E06000024	391	375	1	374
South Gloucestershire	E06000025	537	497	0	497
Plymouth	E06000026	84	80	0	80
Torbay	E06000027	119	63	0	63
Bournemouth	E06000028	47	46	0	46
Poole	E06000029	75	65	0	65
Swindon	E06000030	230	230	0	230
Peterborough	E06000031	343	343	0	343
Luton	E06000032	43	43	0	43
Southend-on-Sea	E06000033	68	42	0	42
Thurrock	E06000034	184	163	0	163
Medway	E06000035	269	194	0	194
Bracknell Forest	E06000036	109	109	0	109
West Berkshire	E06000037	704	704	0	704
Reading	E06000038	40	40	0	40
Slough	E06000039	33	33	0	33
Windsor and Maidenhead	E06000040	198	198	2	197
Wokingham	E06000041	179	179	0	179
Milton Keynes	E06000042	309	309	0	309
Brighton and Hove	E06000043	85	83	0	83
Portsmouth	E06000044	60	40	0	40
Southampton	E06000045	56	50	0	50
Isle of Wight	E06000046	395	380	0	380
County Durham	E06000047	2233	2232	6	2226
Cheshire East	E06000049	1166	1166	0	1166
Cheshire West and Chester	E06000050	941	917	0	917
Shropshire	E06000051	3197	3197	0	3197
Cornwall	E06000052	3613	3550	4	3546

1.1 Standard Area Measurement for Local Authority Districts as at 31/12/2017

Local Authority Name	Code	Extent of the realm (km^2)	Area to mean high water (km^2)	Inland water (km^2)	Land only (km^2)
Isles of Scilly	E06000053	23	16	0	16
Wiltshire	E06000054	3255	3255	0	3255
Bedford	E06000055	476	476	0	476
Central Bedfordshire	E06000056	716	716	0	716
Northumberland	E06000057	5078	5026	12	5014
Aylesbury Vale	E07000004	903	903	0	903
Chiltern	E07000005	196	196	0	196
South Bucks	E07000006	141	141	0	141
Wycombe	E07000007	325	325	0	325
Cambridge	E07000008	41	41	0	41
East Cambridgeshire	E07000009	652	651	0	651
Fenland	E07000010	547	546	0	546
Huntingdonshire	E07000011	913	912	6	906
South Cambridgeshire	E07000012	902	902	0	902
Allerdale	E07000026	1321	1258	16	1242
Barrow-in-Furness	E07000027	132	78	0	78
Carlisle	E07000028	1054	1039	0	1039
Copeland	E07000029	776	738	6	732
Eden	E07000030	2156	2156	14	2142
South Lakeland	E07000031	1743	1553	19	1534
Amber Valley	E07000032	265	265	0	265
Bolsover	E07000033	160	160	0	160
Chesterfield	E07000034	66	66	0	66
Derbyshire Dales	E07000035	795	795	3	792
Erewash	E07000036	110	110	0	110
High Peak	E07000037	540	540	1	539
North East Derbyshire	E07000038	276	276	0	276
South Derbyshire	E07000039	338	338	0	338
East Devon	E07000040	824	814	0	814
Exeter	E07000041	48	47	0	47
Mid Devon	E07000042	913	913	0	913
North Devon	E07000043	1105	1086	0	1086
South Hams	E07000044	905	886	0	886
Teignbridge	E07000045	681	674	0	674
Torridge	E07000046	996	985	1	984
West Devon	E07000047	1165	1161	1	1160
Christchurch	E07000048	52	50	0	50
East Dorset	E07000049	354	354	0	354
North Dorset	E07000050	609	609	0	609
Purbeck	E07000051	428	404	0	404
West Dorset	E07000052	1087	1082	0	1082
Weymouth and Portland	E07000053	43	42	0	42
Eastbourne	E07000061	46	44	0	44
Hastings	E07000062	31	30	0	30
Lewes	E07000063	294	292	0	292
Rother	E07000064	518	512	2	509
Wealden	E07000065	836	835	2	833
Basildon	E07000066	110	110	0	110
Braintree	E07000067	612	612	0	612
Brentwood	E07000068	153	153	0	153
Castle Point	E07000069	64	45	0	45
Chelmsford	E07000070	343	342	3	339
Colchester	E07000071	347	333	4	329
Epping Forest	E07000072	339	339	0	339
Harlow	E07000073	31	31	0	31
Maldon	E07000074	427	359	0	359
Rochford	E07000075	263	169	0	169
Tendring	E07000076	366	338	0	338
Uttlesford	E07000077	641	641	0	641

1.1 Standard Area Measurement for Local Authority Districts as at 31/12/2017

Local Authority Name	Code	Extent of the realm (km^2)	Area to mean high water (km^2)	Inland water (km^2)	Land only (km^2)
Cheltenham	E07000078	47	47	0	47
Cotswold	E07000079	1165	1165	0	1165
Forest of Dean	E07000080	561	526	0	526
Gloucester	E07000081	41	41	0	41
Stroud	E07000082	476	461	0	461
Tewkesbury	E07000083	415	414	0	414
Basingstoke and Deane	E07000084	634	634	0	634
East Hampshire	E07000085	514	514	0	514
Eastleigh	E07000086	85	80	0	80
Fareham	E07000087	78	74	0	74
Gosport	E07000088	28	25	0	25
Hart	E07000089	215	215	0	215
Havant	E07000090	79	55	0	55
New Forest	E07000091	777	753	0	753
Rushmoor	E07000092	39	39	0	39
Test Valley	E07000093	628	628	0	628
Winchester	E07000094	661	661	0	661
Broxbourne	E07000095	51	51	0	51
Dacorum	E07000096	212	212	0	212
Hertsmere	E07000098	101	101	0	101
North Hertfordshire	E07000099	375	375	0	375
Three Rivers	E07000102	89	89	0	89
Watford	E07000103	21	21	0	21
Ashford	E07000105	581	581	0	581
Canterbury	E07000106	321	309	0	309
Dartford	E07000107	76	73	0	73
Dover	E07000108	321	315	0	315
Gravesham	E07000109	105	99	0	99
Maidstone	E07000110	393	393	0	393
Sevenoaks	E07000111	370	370	1	369
Shepway	E07000112	366	357	0	357
Swale	E07000113	422	374	0	374
Thanet	E07000114	112	103	0	103
Tonbridge and Malling	E07000115	241	240	0	240
Tunbridge Wells	E07000116	331	331	0	331
Burnley	E07000117	111	111	0	111
Chorley	E07000118	203	203	0	203
Fylde	E07000119	183	166	0	166
Hyndburn	E07000120	73	73	0	73
Lancaster	E07000121	654	576	0	576
Pendle	E07000122	169	169	0	169
Preston	E07000123	143	142	0	142
Ribble Valley	E07000124	584	584	1	583
Rossendale	E07000125	138	138	0	138
South Ribble	E07000126	115	113	0	113
West Lancashire	E07000127	381	347	0	347
Wyre	E07000128	329	282	0	282
Blaby	E07000129	130	130	0	130
Charnwood	E07000130	279	279	0	279
Harborough	E07000131	593	593	1	592
Hinckley and Bosworth	E07000132	297	297	0	297
Melton	E07000133	481	481	0	481
North West Leicestershire	E07000134	279	279	0	279
Oadby and Wigston	E07000135	24	24	0	24
Boston	E07000136	398	365	0	365
East Lindsey	E07000137	1831	1765	0	1765
Lincoln	E07000138	36	36	0	36
North Kesteven	E07000139	922	922	0	922
South Holland	E07000140	816	751	0	751

1.1 Standard Area Measurement for Local Authority Districts as at 31/12/2017

Local Authority Name	Code	Extent of the realm (km^2)	Area to mean high water (km^2)	Inland water (km^2)	Land only (km^2)
South Kesteven	E07000141	943	943	0	943
West Lindsey	E07000142	1158	1156	0	1156
Breckland	E07000143	1305	1305	0	1305
Broadland	E07000144	553	552	0	552
Great Yarmouth	E07000145	182	174	0	174
King's Lynn and West Norfolk	E07000146	1528	1439	0	1439
North Norfolk	E07000147	990	964	2	962
Norwich	E07000148	41	39	0	39
South Norfolk	E07000149	909	908	0	908
Corby	E07000150	80	80	0	80
Daventry	E07000151	666	666	3	663
East Northamptonshire	E07000152	510	510	0	510
Kettering	E07000153	233	233	0	233
Northampton	E07000154	81	81	0	81
South Northamptonshire	E07000155	634	634	0	634
Wellingborough	E07000156	163	163	0	163
Craven	E07000163	1179	1179	1	1177
Hambleton	E07000164	1311	1311	0	1311
Harrogate	E07000165	1309	1309	1	1308
Richmondshire	E07000166	1319	1319	0	1319
Ryedale	E07000167	1507	1507	0	1507
Scarborough	E07000168	826	816	0	816
Selby	E07000169	602	599	0	599
Ashfield	E07000170	110	110	0	110
Bassetlaw	E07000171	639	638	0	638
Broxtowe	E07000172	80	80	0	80
Gedling	E07000173	120	120	0	120
Mansfield	E07000174	77	77	0	77
Newark and Sherwood	E07000175	652	651	0	651
Rushcliffe	E07000176	409	409	0	409
Cherwell	E07000177	589	589	0	589
Oxford	E07000178	46	46	0	46
South Oxfordshire	E07000179	679	679	0	679
Vale of White Horse	E07000180	579	579	1	578
West Oxfordshire	E07000181	714	714	0	714
Mendip	E07000187	739	739	0	739
Sedgemoor	E07000188	606	564	0	564
South Somerset	E07000189	959	959	0	959
Taunton Deane	E07000190	463	462	0	462
West Somerset	E07000191	747	727	1	725
Cannock Chase	E07000192	79	79	0	79
East Staffordshire	E07000193	390	390	3	387
Lichfield	E07000194	331	331	0	331
Newcastle-under-Lyme	E07000195	211	211	0	211
South Staffordshire	E07000196	407	407	0	407
Stafford	E07000197	598	598	0	598
Staffordshire Moorlands	E07000198	576	576	0	576
Tamworth	E07000199	31	31	0	31
Babergh	E07000200	612	595	1	594
Forest Heath	E07000201	378	378	0	378
Ipswich	E07000202	40	40	0	40
Mid Suffolk	E07000203	871	871	0	871
St Edmundsbury	E07000204	657	657	0	657
Suffolk Coastal	E07000205	921	892	0	892
Waveney	E07000206	375	370	0	370
Elmbridge	E07000207	96	96	1	95
Epsom and Ewell	E07000208	34	34	0	34
Guildford	E07000209	271	271	0	271
Mole Valley	E07000210	258	258	0	258

1.1 Standard Area Measurement for Local Authority Districts as at 31/12/2017

Local Authority Name	Code	Extent of the realm (km²)	Area to mean high water (km²)	Inland water (km²)	Land only (km²)
Reigate and Banstead	E07000211	129	129	0	129
Runnymede	E07000212	78	78	0	78
Spelthorne	E07000213	51	51	6	45
Surrey Heath	E07000214	95	95	0	95
Tandridge	E07000215	248	248	0	248
Waverley	E07000216	345	345	0	345
Woking	E07000217	64	64	0	64
North Warwickshire	E07000218	284	284	0	284
Nuneaton and Bedworth	E07000219	79	79	0	79
Rugby	E07000220	354	354	2	351
Stratford-on-Avon	E07000221	978	978	0	978
Warwick	E07000222	283	283	0	283
Adur	E07000223	44	42	0	42
Arun	E07000224	224	221	0	221
Chichester	E07000225	812	786	0	786
Crawley	E07000226	45	45	0	45
Horsham	E07000227	531	530	0	530
Mid Sussex	E07000228	334	334	0	334
Worthing	E07000229	34	33	0	33
Bromsgrove	E07000234	217	217	0	217
Malvern Hills	E07000235	577	577	0	577
Redditch	E07000236	54	54	0	54
Worcester	E07000237	33	33	0	33
Wychavon	E07000238	664	664	0	664
Wyre Forest	E07000239	195	195	0	195
St Albans	E07000240	161	161	0	161
Welwyn Hatfield	E07000241	130	130	0	130
East Hertfordshire	E07000242	476	476	0	476
Stevenage	E07000243	26	26	0	26
Bolton	E08000001	140	140	0	140
Bury	E08000002	99	99	0	99
Manchester	E08000003	116	116	0	116
Oldham	E08000004	142	142	0	142
Rochdale	E08000005	158	158	0	158
Salford	E08000006	97	97	0	97
Stockport	E08000007	126	126	0	126
Tameside	E08000008	103	103	0	103
Trafford	E08000009	106	106	0	106
Wigan	E08000010	188	188	0	188
Knowsley	E08000011	87	87	0	87
Liverpool	E08000012	134	112	0	112
St. Helens	E08000013	136	136	0	136
Sefton	E08000014	205	155	0	155
Wirral	E08000015	256	157	0	157
Barnsley	E08000016	329	329	0	329
Doncaster	E08000017	569	568	0	568
Rotherham	E08000018	287	287	0	287
Sheffield	E08000019	368	368	0	368
Newcastle upon Tyne	E08000021	115	113	0	113
North Tyneside	E08000022	85	82	0	82
South Tyneside	E08000023	67	64	0	64
Sunderland	E08000024	140	137	0	137
Birmingham	E08000025	268	268	0	268
Coventry	E08000026	99	99	0	99
Dudley	E08000027	98	98	0	98
Sandwell	E08000028	86	86	0	86
Solihull	E08000029	178	178	0	178
Walsall	E08000030	104	104	0	104
Wolverhampton	E08000031	69	69	0	69

1.1 Standard Area Measurement for Local Authority Districts as at 31/12/2017

Local Authority Name	Code	Extent of the realm (km²)	Area to mean high water (km²)	Inland water (km²)	Land only (km²)
Bradford	E08000032	366	366	0	366
Calderdale	E08000033	364	364	0	364
Kirklees	E08000034	409	409	0	409
Leeds	E08000035	552	552	0	552
Wakefield	E08000036	339	339	0	339
Gateshead	E08000037	144	142	0	142
City of London	E09000001	3	3	0	3
Barking and Dagenham	E09000002	38	36	0	36
Barnet	E09000003	87	87	0	87
Bexley	E09000004	64	61	0	61
Brent	E09000005	43	43	0	43
Bromley	E09000006	150	150	0	150
Camden	E09000007	22	22	0	22
Croydon	E09000008	86	86	0	86
Ealing	E09000009	56	56	0	56
Enfield	E09000010	82	82	1	81
Greenwich	E09000011	50	47	0	47
Hackney	E09000012	19	19	0	19
Hammersmith and Fulham	E09000013	17	16	0	16
Haringey	E09000014	30	30	0	30
Harrow	E09000015	50	50	0	50
Havering	E09000016	114	112	0	112
Hillingdon	E09000017	116	116	0	116
Hounslow	E09000018	57	56	0	56
Islington	E09000019	15	15	0	15
Kensington and Chelsea	E09000020	12	12	0	12
Kingston upon Thames	E09000021	37	37	0	37
Lambeth	E09000022	27	27	0	27
Lewisham	E09000023	35	35	0	35
Merton	E09000024	38	38	0	38
Newham	E09000025	39	36	0	36
Redbridge	E09000026	56	56	0	56
Richmond upon Thames	E09000027	59	57	0	57
Southwark	E09000028	30	29	0	29
Sutton	E09000029	44	44	0	44
Tower Hamlets	E09000030	22	20	0	20
Waltham Forest	E09000031	39	39	0	39
Wandsworth	E09000032	35	34	0	34
Westminster	E09000033	22	21	0	21
Northern Ireland	**N92000002**	**14330**	**14330**	**537**	**13793**
Antrim and Newtownabbey	N09000001	728	728	157	571
Armagh City, Banbridge and Craigavon	N09000002	1437	1437	99	1338
Belfast	N09000003	138	138	0	138
Causeway Coast and Glens	N09000004	1986	1986	0	1986
Derry City and Strabane	N09000005	1251	1251	0	1251
Fermanagh and Omagh	N09000006	3012	3012	148	2864
Lisburn and Castlereagh	N09000007	510	510	5	506
Mid and East Antrim	N09000008	1061	1061	2	1059
Mid Ulster	N09000009	1957	1957	126	1831
Newry, Mourne and Down	N09000010	1682	1682	0	1682
Ards and North Down	N09000011	566	566	0	566

1.1 Standard Area Measurement for Local Authority Districts as at 31/12/2017

Local Authority Name	Code	Extent of the realm (km²)	Area to mean high water (km²)	Inland water (km²)	Land only (km²)
Scotland	**S92000003**	**80226**	**78811**	**900**	**77911**
Clackmannanshire	S12000005	164	159	0	159
Dumfries and Galloway	S12000006	6676	6438	11	6427
East Ayrshire	S12000008	1270	1270	8	1262
East Lothian	S12000010	701	679	0	679
East Renfrewshire	S12000011	174	174	0	174
Na h-Eileanan Siar	S12000013	3268	3100	41	3059
Falkirk	S12000014	315	297	0	297
Fife	S12000015	1374	1325	0	1325
Highland	S12000017	26473	26163	506	25657
Inverclyde	S12000018	174	162	2	160
Midlothian	S12000019	355	355	2	354
Moray	S12000020	2257	2238	0	2238
North Ayrshire	S12000021	904	885	0	885
Orkney Islands	S12000023	1086	1013	25	989
Perth and Kinross	S12000024	5419	5384	98	5286
Scottish Borders	S12000026	4743	4739	7	4732
Shetland Islands	S12000027	1657	1469	1	1468
South Ayrshire	S12000028	1235	1224	2	1222
South Lanarkshire	S12000029	1774	1774	2	1772
Stirling	S12000030	2255	2253	66	2187
Aberdeen City	S12000033	206	186	0	186
Aberdeenshire	S12000034	6338	6318	5	6313
Argyll and Bute	S12000035	7164	7008	99	6909
City of Edinburgh	S12000036	273	263	0	263
Renfrewshire	S12000038	269	261	0	261
West Dunbartonshire	S12000039	183	177	19	159
West Lothian	S12000040	432	429	1	428
Angus	S12000041	2203	2185	4	2182
Dundee City	S12000042	62	60	0	60
North Lanarkshire	S12000044	472	472	2	470
East Dunbartonshire	S12000045	174	174	0	174
Glasgow City	S12000046	176	175	0	175
Wales	**W92000004**	**21225**	**20782**	**46**	**20736**
Isle of Anglesey	W06000001	749	714	3	711
Gwynedd	W06000002	2622	2548	13	2535
Conwy	W06000003	1153	1130	4	1126
Denbighshire	W06000004	846	839	2	837
Flintshire	W06000005	489	437	0	437
Wrexham	W06000006	504	504	0	504
Ceredigion	W06000008	1806	1789	3	1786
Pembrokeshire	W06000009	1650	1619	0	1619
Carmarthenshire	W06000010	2439	2371	1	2370
Swansea	W06000011	421	380	0	380
Neath Port Talbot	W06000012	452	442	1	441
Bridgend	W06000013	255	251	0	251
Vale of Glamorgan	W06000014	340	331	0	331
Cardiff	W06000015	150	142	1	141
Rhondda Cynon Taf	W06000016	424	424	0	424
Caerphilly	W06000018	277	277	0	277
Blaenau Gwent	W06000019	109	109	0	109
Torfaen	W06000020	126	126	1	126
Monmouthshire	W06000021	886	850	1	849
Newport	W06000022	218	191	0	191
Powys	W06000023	5195	5195	15	5181
Merthyr Tydfil	W06000024	112	112	1	111

Source: ONS Geography Codes
Standard Area Measurement for UK Local Authority District

Parliamentary elections

Chapter 2

Parliamentary elections

This chapter covers parliamentary elections, by-elections and devolved assembly elections in the UK, Wales, Scotland and Northern Ireland.

Parliamentary elections (Table 2.1)

Information is supplied on the total electorate, average electorate and valid votes as a percentage of electorate. The number of seats by party is also listed.

Parliamentary by-elections (Table 2.2)

Information can be found on the votes recorded for each party and General Elections and subsequent by-elections between General Elections.

Devolved assembly elections (Tables 2.3 and 2.4)

Table 2.3 provides information on the devolved assembly elections in Wales and Scotland, listing information on the number of votes, percentage share and number of seats per party.

Table 2.4 provides information on the devolved assembly elections in Northern Ireland, listing information the number of seats by party and percentage share of the vote.

EU Referendum results by region (Table 2.5)
Table 2.5 provides information on the EU Referendum results by region, ranked by highest vote share for "Leave".

2.1 Parliamentary elections[1]

Thousands and percentages

		15-Oct 1964	31-Mar 1966	18-Jun 1970[1]	28-Feb 1974		10-Oct 1974	03-May 1979	09-Jun 1983	11-Jun 1987	09-Apr 1992	01-May 1997	07-Jun 2001	05-May 2005	06-May 2010	07-May 2015	08-Jun 2017
United Kingdom																	
Electorate	DZ5P	35894	35957	39615	40256	DZ6V	40256	41573	42704	43666	43719	43846	44403	44246	45597	46354	46844
Average-electors per seat	DZ5T	57	57.1	62.9	63.4	DZ6R	63.4	65.5	66.7	67.2	67.2	66.5	67.4	68.5	70.1	71.3	72.1
Valid votes counted	DZ5X	27657	27265	28345	31340	DZ6N	29189	31221	30671	32530	33614	31286	26367	27149	29688	30698	32204
As percentage of electorate	DZ63	77.1	75.8	71.5	77.9	DZ6J	72.5	75.1	71.8	74.5	76.7	71.4	59.4	61.4	65.1	66.2	68.7
England and Wales																	
Electorate	DZ5Q	31610	31695	34931	35509	DZ6W	35509	36695	37708	38568	38648	38719	39228	39266	40565	40968	41613
Average-electors per seat	DZ5U	57.8	57.9	63.9	64.3	DZ6S	64.3	66.5	67.2	68.8	68.8	68	68.9	69	70.8	71.6	72.7
Valid votes counted	DZ5Y	24384	24116	24877	27735	DZ6O	25729	27609	27082	28832	29897	27679	23243	24097	26548	27034	28732
As percentage of electorate	DZ64	77.1	76.1	71.2	78.1	DZ6K	72.5	75.2	71.8	74.8	77.5	71.5	59.3	61.4	65.4	66.0	69.0
Scotland																	
Electorate	DZ5R	3393	3360	3659	3705	DZ6X	3705	3837	3934	3995	3929	3949	3984	3840	3863	4100	3988
Average-electors per seat	DZ5V	47.8	47.3	51.5	52.2	DZ6T	52.2	54	54.6	55.5	54.6	54.8	55.3	65.1	65.5	69.5	67.6
Valid votes counted	DZ5Z	2635	2553	2688	2887	DZ6P	2758	2917	2825	2968	2931	2817	2313	2334	2466	2910	2650
As percentage of electorate	DZ65	77.6	76	73.5	77.9	DZ6L	74.5	76	71.8	74.3	74.2	71.3	58.1	60.8	63.8	71.0	66.4
Northern Ireland																	
Electorate	DZ5S	891	902	1025	1027	DZ6Y	1037	1028	1050	1090	1141	1178	1191	1140	1169	1237	1243
Average-electors per seat	DZ5W	74.2	75.2	85.4	85.6	DZ6U	86.4	85.6	61.8	64.1	67.1	65.4	66.2	63.3	65.0	68.7	69
Valid votes counted	DZ62	638	596	779	718	DZ6Q	702	696	765	730	785	791	810	718	674	718	857
As percentage of electorate	DZ66	71.7	66.1	76	69.9	DZ6M	67.7	67.7	72.9	67	68.8	67.1	68	62.9	57.6	58.1	65.4
Members of Parliament elected: (numbers)	DZV7	630	630	630	635	DZV8	635	635	650	650	651	659	659	646	650	650	650
Conservative	DZ67	303	253	330	296	DZ6D	276	339	396	375	336	165	166	198	306	331	317
Labour	DZ68	317	363	287	301	DZ6E	319	268	209	229	271	418	412	355	258	232	262
Liberal Democrat[2]	DZ69	9	12	6	14	DZ6F	13	11	23	22	20	46	52	62	57	8	12
Scottish National Party	DZ6A	–	–	1	7	DZ6G	11	2	2	3	3	6	5	6	6	56	35
Plaid Cymru	DZ6B	–	–	–	2	DZ6H	3	2	2	3	4	4	4	3	3	3	4
Other[3]	DZ6C	1	2	6	15	DZ6I	13	13	18	18	17	20	20	22	20	20	20

1 The Representation of the People Act 1969 lowered the minimum voting age from 21 to 18 years with effect from 16 February 1970.
2 Liberal before 1992. The figures for 1983 and 1987 include six and five MPs respectively who were elected for the Social Democratic Party.
3 Including the Speaker.

Source: British Electoral Facts 1832-2012
Plymouth University for the Electoral Commission: 01752 233207

General Election 12 December 2019

Thousands and Percentage

	United Kingdom	England	Wales	Scotland	Northern Ireland
Electorate	47570	39925	2319	4051	1293
Average-electors per seat	73.2	74.8	58.0	68.7	71.8
Valid votes counted	32014	26910	1544	2759	799
As percentage of electorate	67.3	67.5	66.6	68.1	61.8

Members of Parliament elected (numbers)

Conservative	365
Labour	202
Liberal Democrat	11
Scottish National Party	48
Plaid Cymru	4
Other (inc Speaker)	20
Total	650

2.2a Summary of Parliamentary by-elections in Great Britain (excluding Northern Ireland)

	Number of by-elections	Net Seat Gains and Losses						Average change in share of vote since previous election					Average turnout
		CON	LAB	LD	PC/ SNP	Other	No change	CON	LAB	LD	PC/ SNP	Other	
1945-50	50	+4	-1	...	...	-3	45	3.7%	-2.3%	-1.1%	1.4%	-1.0%	67.3%
1950-51	14	...	...	...	...	...	14	+6.8%	-2.0%	-4.6%		-0.2%	68.8%
1951-55	44	+1	-1	...	...	...	43	-0.6%	+0.3%	-0.6%	+0.6%	+0.3%	58.6%
1955-59	49	-2	+4	...	...	-2	34	-8.7%	+1.3%	+6.2%	+0.3%	+0.9%	63.5%
1959-64	61	-5	+4	+1	...	...	54	-14.1%	-2.1%	+13.7%	+1.2%	+1.5%	62.9%
1964-66	13	...	-1	+1	...	...	11	+1.3%	-1.8%	+0.5%	+0.3%	+0.4%	58.2%
1966-70	37	+11	-15	+1	+2	+1	22	+6.8%	-17.3%	+3.3%	+5.5%	+1.7%	62.1%
1970-74	30	-5	...	+5	+1	-1	20	-10.7%	-4.2%	+9.0%	+4.0%	+1.9%	56.5%
1974	1	...	...	...	...	...	1	-1.1%	-3.4%	-2.3%	...	+6.8%	25.9%
1974-79	30	+6	-6	+1	...	-1	23	+9.9%	-9.3%	-4.9%	-0.3%	+4.6%	57.5%
1979-83	17	-3	+1	+4	...	-2	11	-11.4%	-10.2%	+18.6%	+1.6%	+1.4%	56.7%
1983-87	16	-4	...	+4	...	...	11	-14.0%	+0.4%	+12.3%	+0.1%	+1.2%	63.5%
1987-92	23	-7	+3	+3	+1	...	15	-11.0%	-0.8%	-0.6%	+5.7%	+6.6%	57.4%
1992-97	17	-8	+3	+4	+1	...	9	-19.9%	+7.4%	+5.2%	+2.4%	+4.9%	52.7%
1997-2001	15	-2	+1	+2	...	-1	14	-0.6%	-11.1%	+5.0%	+3.1%	+3.6%	42.4%
2001-05	6	...	-2	+2	...	...	4	-4.2%	-19.8%	+15.8%	+1.1%	+7.0%	39.3%
2005-10	14	...	-1	+1	...	...	9	+2.6%	-10.4%	+2.0%	+4.3%	+5.3%	48.8%
2010-15	19	-3	...	...	...	+3	13	-6.4%	+5.4%	-7.6%	+1.1%	+17.9%	39.6%
2015-17	10	...	...	+1	...	...	8	-4.7%	+3.9%	+7.7%	+5.7%	-6.6%	44.1%
2017-19	4	-1	...	+1	...	...	3	-12.9%	-15.0%	+11.5%	-0.3%	+15.9%	46.6%

Sources:

1. F.W.S. Craig, Chronology of British Parliamentary By-elections 1833-1987

2. Colin Rallings and Michael Thrasher, British Electoral Facts 1832-2006

3. House of Commons Library, RP10/50 By-election results 2005-10; SN05833 By-elections since 2010 General Election; CBP 7417 By-elections since the 2015 General Election; CBP-8280 By-elections since the 2017 General Election

2.2b Parliamentary by-elections in Northern Ireland, 1974 - 2018[5]

			Change in vote share since previous election:					
Date	Constituency	Result	DUP	UUP	SF	SDLP	Other	Turnout
GENERAL ELECTION 1974 (FEB)								
None								
GENERAL ELECTION 1974 (OCT)								
None								
GENERAL ELECTION 1979								
09/04/1981	Fermanagh and South Tyrone	Anti-H Block gain from Ind Rep	...	+41.0%	...	...	+51.2%	82.4%
20/08/1981	Fermanagh and South Tyrone	Anti-H Block hold	...	-3.2%	...	...	+3.2%	87.5%
04/03/1982	Belfast South	UUP hold	+22.6%	-22.4%	...	+0.9%	+29.3%	65.7%
GENERAL ELECTION 1983[6]								
23/01/1986	East Antrim	UUP hold	...	+47.5%	...	...	-4.8%	58.9%
23/01/1986	North Antrim	DUP hold	+43.2%	...	...	...	...	53.5%
23/01/1986	South Antrim	UUP hold	...	+48.5%	...	...	...	52.2%
23/01/1986	Belfast East	DUP hold	+35.6%	...	...	...	-6.1%	63.6%
23/01/1986	Belfast North	UUP hold	...	+35.3%	...	...	+13.7%	54.7%
23/01/1986	Belfast South	UUP hold	...	+21.4%	...	...	+2.4%	56.6%
23/01/1986	North Down	UPU hold	...	...	...	...	+21.9%	60.5%
23/01/1986	South Down	UUP hold	...	+8.1%	-2.2%	+5.6%	-0.6%	73.8%
23/01/1986	Fermanagh and South Tyrone	UUP hold	...	+2.1%	-7.6%	+5.0%	+0.5%	80.4%
23/01/1986	Lagan Valley	UUP hold	...	+31.5%	...	...	+7.3%	81.4%
23/01/1986	East Londonderry	UUP hold	...	+56.0%	...	...	...	47.0%
23/01/1986	Mid Ulster	DUP hold	+16.1%	...	-2.6%	+3.0%	-0.1%	77.0%
23/01/1986	Newry and Armagh	SDLP gain from UUP	...	+0.2%	-7.7%	+8.7%	-1.2%	76.6%
23/01/1986	Strangford	UUP hold	...	+45.4%	...	...	...	55.1%
23/01/1986	Upper Bann	UUP hold	...	+23.9%	...	...	+13.8%	57.2%
GENERAL ELECTION 1987								
17/05/1990	Upper Bann	UUP hold	...	-3.5%	-1.7%	-1.6%	-4.8%	53.4%
GENERAL ELECTION 1992								
15/06/1995	North Down	UKU gain from UPU	...	...	...	...	-19.3%	38.6%
GENERAL ELECTION 1997								
21/09/2000	South Antrim	DUP gain from UUP	...	-22.2%	+3.0%	-4.7%	-5.0%	43.0%
GENERAL ELECTION 2001								
None								
GENERAL ELECTION 2005								
None								
GENERAL ELECTION 2010 (up to July 2011)								
09/06/2011	Belfast West	SF hold	-1.5%	-1.4%	-0.4%	-2.9%	+6.3%	37.4%
07/03/2013	Mid Ulster	SF hold	...	...	-5.1%	+3.1%	+1.9%	55.4%
GENERAL ELECTION 2015								
None								
GENERAL ELECTION 2017 (as at December 2018)								
03/05/2018	West Tyrone	SF hold	-3.0%	+3.1%	-4.1%	+4.9%	+1.9%	54.6%

15

2.2b Parliamentary by-elections in Northern Ireland, 1974 - 2018[5]

Notes for table 2.2b:

1. The formation of new parties in the early 1970s altered the pattern of party competition at Westminster elections. The SDLP (formed 1970) and the DUP (formed 1971) are included in Table 14b (1974-2012). Ulster Unionists are listed as Conservatives up to 1972 when they stopped taking the Conservative whip.

2. Irish Nationalist/Anti-Partitionist

3. Sinn Féin (SF) candidate T.J. Mitchell was elected as MP for Mid-Ulster at the 1955 General Election, but was in prison at the time of election and hence was disqualified from being an MP. No petition was lodged but a by-election writ was subsequently issued; in the 31 August 1955 by-election Mitchell again stood for Sinn Féin and topped the poll (therefore the by-election is here recorded as a Sinn Féin hold). However on this occasion a petition was lodged and since Mitchell was disqualified, the Conservative candidate was elected instead.

4. Prior to the by-election the seat was held by a Conservative (see footnote 1). Changes in vote share are as compared to the 1955 General Election.

* Constituency returned two MPs at previous general election. Change in vote share between general election and by-election is calculating using the total vote received by a party's candidate in the previous general election. However some electors will only voted for one candidate in the general election and of those who cast two votes, some will have voted for candidates from different parties. Additionally some parties will only have stood one candidate in a multimember seat.

5. The formation of new parties in the early 1970s altered the pattern of party competition at Westminster elections. The SDLP (formed 1970) and the DUP (formed 1971) are included in this table but not in Table 14a. The Ulster Unionist Party (UUP) took the Conservative whip at Westminster until 1972 and so in Table 14a are listed under Conservatives.

6. Multiple by-elections were held in January 1986 after fifteen unionist Members resigned their seats in protest at the Anglo-Irish Agreement.

Party descriptions:

DUP - Democratic Unionist Party	SDLP - Social Democratic and Labour Party
UUP - Ulster Unionist Party	UPU - Ulster Popular Unionist Party
SF - Sinn Fein	UKU - United Kingdom Unionist

Sources:

1. F.W.S. Craig, British Parliamentary Election Results 1918-1949
2. F.W.S. Craig, British Parliamentary Election Results 1950-1973
3. F.W.S. Craig, British Parliamentary Election Results 1974-1983
4. Colin Rallings and Michael Thrasher, British Parliamentary Election Results 1983-1997
5. House of Commons Library RP01/36, By-election results 1997-2000; RP05/34, By-election results 2001-05;
6. RP10/50, By-election results 2005-10; SN05833, By-elections since 2010 General Election, By-elections since 2015 General Election; CBP-8280 By-elections since the 2017 General Election

2.3a National Assembly for Wales elections, 1999-2016

	Number of Votes					% share				
	1999	2003	2007	2011	2016	1999	2003	2007	2011	2016
Constituency votes										
LAB	384,671	340,515	314,925	401,677	353,865	37.6%	40.0%	32.2%	42.3%	34.7%
PC	290,565	180,185	219,121	182,907	209,374	28.4%	21.2%	22.4%	19.3%	20.5%
CON	162,133	169,832	218,730	237,389	215,597	15.8%	20.0%	22.4%	25.0%	21.1%
UKIP		19,795	18,047		127,038		2.3%	1.8%		12.5%
LD	137,657	120,250	144,410	100,259	78,165	13.5%	14.1%	14.8%	10.6%	7.7%
Others	47,992	20,266	62,859	27,021	35,341	4.7%	2.4%	6.4%	2.8%	3.5%
Total	1,023,018	850,843	978,092	949,253	1,019,380	100%	100%	100%	100%	100%
Constituency seats										
LAB	27	30	24	28	27	67.5%	75.0%	60.0%	70.0%	67.5%
PC	9	5	7	5	6	22.5%	12.5%	17.5%	12.5%	15.0%
CON	1	1	5	6	6	2.5%	2.5%	12.5%	15.0%	15.0%
UKIP	0	0	0	0	0	0.0%	0.0%	0.0%	0.0%	0.0%
LD	3	3	3	1	1	7.5%	7.5%	7.5%	2.5%	2.5%
Others	0	1	1	0	0	0.0%	2.5%	2.5%	0.0%	0.0%
Total:	**40**	**40**	**40**	**40**	**40**	**100%**	**100%**	**100%**	**100%**	**100%**
Regional votes										
LAB	361,657	310,658	288,955	349,935	319,196	35.4%	36.6%	29.6%	36.9%	31.5%
PC	312,048	167,653	204,757	169,799	211,548	30.5%	19.7%	21.0%	17.9%	20.8%
CON	168,206	162,725	209,154	213,773	190,846	16.5%	19.2%	21.5%	22.5%	18.8%
UKIP		29,427	38,349	43,256	132,138		3.5%	3.9%	4.6%	13.0%
LD	128,008	108,013	114,500	76,349	65,504	12.5%	12.7%	11.7%	8.0%	6.5%
Others	51,938	71,076	119,071	95,776	95,511	5.1%	8.4%	12.2%	10.1%	9.4%
Total	1,021,857	849,552	974,786	948,888	1,014,743	100%	100%	100%	100%	100%
Regional seats										
LAB	1	0	2	2	2	5.0%	0.0%	10.0%	10.0%	10.0%
PC	8	7	8	6	6	40.0%	35.0%	40.0%	30.0%	30.0%
CON	8	10	7	8	5	40.0%	50.0%	35.0%	40.0%	25.0%
UKIP	0	0	0	0	7	0.0%	0.0%	0.0%	0.0%	35.0%
LD	3	3	3	4	0	15.0%	15.0%	15.0%	20.0%	0.0%
Others	0	0	0	0	0	0.0%	0.0%	0.0%	0.0%	0.0%
Total:	**20**	**20**	**20**	**20**	**20**	**100%**	**100%**	**100%**	**100%**	**100%**
Total seats										
LAB	28	30	26	30	29	46.7%	50.0%	43.3%	50.0%	48.3%
PC	17	12	15	11	12	28.3%	20.0%	25.0%	18.3%	20.0%
CON	9	11	12	14	11	15.0%	18.3%	20.0%	23.3%	18.3%
UKIP	0	0	0	0	7	0.0%	0.0%	0.0%	0.0%	11.7%
LD	6	6	6	5	1	10.0%	10.0%	10.0%	8.3%	1.7%
Others	0	1	1	0	0	0.0%	1.7%	1.7%	0.0%	0.0%
Total:	**60**	**60**	**60**	**60**	**60**	**100%**	**100%**	**100%**	**100%**	**100%**
Constituency turnout										
	46.4%	38.2%	43.5%	41.5%	45.5%					
Regional turnout										
	46.3%	38.1%	43.4%	41.4%	45.3%					

Sources

Colin Rallings and Michael Thrasher, *British Electoral Facts 1832-2006*

Electoral Commission, *Report on the National Assembly for Wales general election 5 May 2016*

House of Commons Library Briefing Paper CBP 7594 , *National Assembly for Wales Elections: 2016*

The next National Assembly for Wales election is expected in 2021

2.3b Scottish Parliament elections, 1999-2016

	Number of votes and seats					% Share				
	1999	2003	2007	2011	2016	1999	2003	2007	2011	2016
Constituency votes										
SNP	672,768	455,742	664,227	902,915	1,059,898	28.7%	23.8%	32.9%	45.4%	46.5%
CON	364,425	318,279	334,742	276,652	501,844	15.6%	16.6%	16.6%	13.9%	22.0%
LAB	908,346	663,585	648,374	630,461	514,261	38.8%	34.6%	32.1%	31.7%	22.6%
LD	333,179	294,347	326,232	157,714	178,238	14.2%	15.4%	16.2%	7.9%	7.8%
GRN			2,971		13,172			0.1%		0.6%
Others	63,770	184,641	43,402	21,534	11,741	2.7%	9.6%	2.2%	1.1%	0.5%
Total	**2,342,488**	**1,916,594**	**2,016,977**	**1,989,276**	**2,279,154**	**100%**	**100%**	**100%**	**100%**	**100.0%**
Constituency seats										
SNP	7	9	21	53	59	9.6%	12.3%	28.8%	72.6%	80.8%
CON	0	3	4	3	7	0.0%	4.1%	5.5%	4.1%	9.6%
LAB	53	46	37	15	3	72.6%	63.0%	50.7%	20.5%	4.1%
LD	12	13	11	2	4	16.4%	17.8%	15.1%	2.7%	5.5%
GRN										
Others	1	2	0	0	0	1.4%	2.7%	0.0%	0.0%	0.0%
Total	**73**	**73**	**73**	**73**	**73**	**100%**	**100%**	**100%**	**100%**	**100%**
Regional votes										
SNP	638,644	399,659	633,401	876,421	953,587	27.3%	20.9%	31.0%	44.0%	41.7%
CON	359,109	296,929	284,005	245,967	524,220	15.4%	15.5%	13.9%	12.4%	22.9%
LAB	786,818	561,375	595,415	523,469	435,919	33.6%	29.3%	29.2%	26.3%	19.1%
LD	290,760	225,774	230,671	103,472	119,284	12.4%	11.8%	11.3%	5.2%	5.2%
GRN	84,023	132,138	82,584	86,939	150,426	3.6%	6.9%	4.0%	4.4%	6.6%
Others	179,560	299,976	215,973	154,568	102,315	7.7%	15.7%	10.6%	7.8%	4.5%
Total	**2,338,914**	**1,915,851**	**2,042,049**	**1,990,836**	**2,285,751**	**100%**	**100%**	**100%**	**100%**	**100%**
Regional Seats										
SNP	28	18	26	16	4	50.0%	32.1%	46.4%	28.6%	7.1%
CON	18	15	13	12	24	32.1%	26.8%	23.2%	21.4%	42.9%
LAB	3	4	9	22	21	5.4%	7.1%	16.1%	39.3%	37.5%
LD	5	4	5	3	1	8.9%	7.1%	8.9%	5.4%	1.8%
GRN	1	7	2	2	6	1.8%	12.5%	3.6%	3.6%	10.7%
Others	1	8	1	1	0	1.8%	14.3%	1.8%	1.8%	0.0%
Total	**56**	**56**	**56**	**56**	**56**	**100%**	**100%**	**100%**	**100%**	**100%**
Total seats										
SNP	35	27	47	69	63	62.5%	48.2%	83.9%	123.2%	112.5%
CON	18	18	17	15	31	32.1%	32.1%	30.4%	26.8%	55.4%
LAB	56	50	46	37	24	100.0%	89.3%	82.1%	66.1%	42.9%
LD	17	17	16	5	5	30.4%	30.4%	28.6%	8.9%	8.9%
GRN	1	7	2	2	6	1.8%	12.5%	3.6%	3.6%	10.7%
Others	2	10	1	1		3.6%	17.9%	1.8%	1.8%	0.0%
Total	**129**	**129**	**129**	**129**	**129**	**100%**	**100%**	**100%**	**100%**	**100%**
Constituency Turnout										
	58.8%	49.4%	51.7%	50.4%	55.6%					
Regional Turnout										
	58.7%	49.4%	52.4%	50.4%	55.7%					

Sources

Colin Rallings and Michael Thrasher, *British Electoral Facts 1832-2006*

Electoral Commission, *Report on the Scottish Parliament election on 5 May 2011*

House of Commons Library Research Paper RP07/46, *Scottish Parliament Elections: 3 May 2007*

House of Commons Library Research Paper RP11/41, *Scottish Parliament Elections: 2011*

House of Commons Library Briefing Paper CBP-7529, Scottish Parliament Elections: 2016

The next Scottish Parliament election is expected in 2021

2.4 Northern Ireland Assembly elections: 1998-2017

	1st Pref Votes and seats won						% of votes and seats won					
	1998	2003	2007	2011	2016	2017	1998	2003	2007	2011	2016	2017
Votes												
DUP	146,917	177,944	207,721	198,436	202,567	225,413	18.1%	25.3%	30.1%	30.0%	29.2%	28.1%
Sinn Féin	142,858	162,758	180,573	178,222	166,785	224,245	17.6%	23.2%	26.2%	26.9%	24.0%	27.9%
UUP	172,225	156,931	103,145	87,531	87,302	103,314	21.3%	22.3%	14.9%	13.2%	12.6%	12.9%
SDLP	177,963	117,547	105,164	94,286	83,364	95,958	22.0%	16.7%	15.2%	14.2%	12.0%	11.9%
Alliance	52,636	25,372	36,139	50,875	48,447	72,717	6.5%	3.6%	5.2%	7.7%	7.0%	9.1%
UK Unionists	36,541	5,700	10,452	...	...	...	4.5%	0.8%	1.5%	...	...	...
PUP	20,634	8,032	3,822	1,493	5,955	5,590	2.5%	1.1%	0.6%	0.2%	0.9%	0.7%
People before Profit Alliance	...	...	774	5,438	13,761	14,100	...	...	0.1%	0.8%	2.0%	1.8%
TUV	...	...	...	16,480	23,776	20,523	...	...	...	2.5%	3.4%	2.6%
UKIP	...	...	...	4,152	10,109	1,579	...	...	...	0.6%	1.5%	0.2%
Green Party	510	2,688	11,985	6,031	18,718	18,527	0.1%	0.4%	1.7%	0.9%	2.7%	2.3%
Others	59,961	45,277	30,538	18,790	33,526	21,349	7.4%	6.4%	4.4%	2.8%	4.8%	2.7%
Total	**810,245**	**702,249**	**690,313**	**661,734**	**694,310**	**803,315**	100.0%	100.0%	100.0%	100.0%	100.0%	100.0%
Seats von												
DUP	20	30	36	38	38	28	18.5%	27.8%	33.3%	35.2%	35.2%	31.1%
Sinn Féin	18	24	28	29	28	27	16.7%	22.2%	25.9%	26.9%	25.9%	30.0%
UUP	28	27	18	16	16	10	25.9%	25.0%	16.7%	14.8%	14.8%	11.1%
SDLP	24	18	16	14	12	12	22.2%	16.7%	14.8%	13.0%	11.1%	13.3%
Alliance	6	6	7	8	8	8	5.6%	5.6%	6.5%	7.4%	7.4%	8.9%
UK Unionists	5	1	0	...	...	...	4.6%	0.9%	0.0%	...	...	...
PUP	2	1	1	0	0	0	1.9%	0.9%	0.9%	0.0%	0.0%	0.0%
People before Profit Alliance	...	...	0	0	2	1	...	...	0.0%	0.0%	1.9%	1.1%
TUV	...	...	...	1	1	1	...	...	...	0.9%	0.9%	1.1%
UKIP	...	...	...	0	0	0	...	...	...	0.0%	0.0%	0.0%
Green Party	0	0	1	1	2	2	0.0%	0.0%	0.9%	0.9%	1.9%	2.2%
Others	5	1	1	1	1	1	4.6%	0.9%	0.9%	0.9%	0.9%	1.1%
Total	**108**	**108**	**108**	**108**	**108**	**90**	100.0%	100.0%	100.0%	100.0%	100.0%	100.0%
Electorate	1,178,556	1,097,526	1,107,904	1,210,009	1,281,595	1,254,709						
Turnout	68.7%	64.0%	62.3%	54.7%	54.2%	64.0%						

Sources:
Colin Rallings and Michael Thrasher, *British Electoral Facts 1832-2006*
Electoral Office for Northern Ireland, www.eoni.org.uk

The next Northern Ireland Assembly election is expected in 2022

2.5 EU referendum results by region

Ranked by highest vote share for Leave, 23 June 2016

Counting region	Remain Votes	Remain % of valid votes	Leave Votes	Leave % of valid votes	Rejected votes	Total votes cast (incl. rejected)	Turnout (valid votes as % of electorate)	Turnup (total votes cast as % of electorate)	Electorate
West Midlands	1,207,175	40.7%	1,755,687	59.3%	2,507	2,965,370	72.0%	72.0%	4,116,572
East Midlands	1,033,036	41.2%	1,475,479	58.8%	1,981	2,510,497	74.1%	74.2%	3,384,299
North East	562,595	42.0%	778,103	58.0%	689	1,341,388	69.3%	69.3%	1,934,341
Yorkshire and the Humber (Y/H)	1,158,298	42.3%	1,580,937	57.7%	1,937	2,741,173	70.6%	70.7%	3,877,780
East of England	1,448,616	43.5%	1,880,367	56.5%	2,329	3,331,313	75.7%	75.7%	4,398,796
North West	1,699,020	46.3%	1,966,925	53.7%	2,682	3,668,628	69.9%	70.0%	5,241,568
South West and Gibraltar	1,503,019	47.4%	1,669,711	52.6%	2,179	3,174,910	76.7%	76.7%	4,138,134
Wales	772,347	47.5%	854,572	52.5%	1,135	1,628,055	71.7%	71.7%	2,270,272
South East	2,391,718	48.2%	2,567,965	51.8%	3,427	4,963,111	76.7%	76.8%	6,465,404
Northern Ireland	440,707	55.8%	349,442	44.2%	374	790,524	62.7%	62.7%	1,260,955
London	2,263,519	59.9%	1,513,232	40.1%	4,453	3,781,205	69.6%	69.7%	5,424,768
Scotland	1,661,191	62.0%	1,018,322	38.0%	1,666	2,681,180	67.2%	67.2%	3,987,112
United Kingdom	**16,141,241**	**48.1%**	**17,410,742**	**51.9%**	**25,359**	**33,577,343**	**72.2%**	**72.2%**	**46,500,001**

Source: House of Commons Library briefing paper CBP7639

http://researchbriefings.parliament.uk/ResearchBriefing/Summary/CBP-7639

International development

Chapter 3

International development

Overseas development assistance

The Department for International Development (DFID) is the UK Government Department with lead responsibility for overseas development. DFID's aim is to eliminate poverty in poorer countries through achievement of the Millennium Development Goals (MDG's). Statistics relating to international development are published on a financial year basis and on a calendar year basis. Statistics on a calendar year basis allow comparisons of aid expenditure with other donor countries. Aid flows can be measured before (gross) or after (net) deductions of repayments of principal on past loans. These tables show only the gross figures.

Aid is provided in two main ways: Bilateral funding is provided directly to partner countries while multilateral funding is provided through international organisations.

Funds can only be classified as multilateral if they are channelled through an organisation on a list in the OECD –

Development Assistance Committee (DAC) Statistical Reporting Directives – which identifies all multilateral organisations. This list also highlights some bodies that might appear to be multilateral but are actually bilateral (in particular this latter category includes some international non-governmental organisations such as the International Committee of the Red Cross and some Public-Private Partnerships). The DAC list of multilaterals is updated annually based on members nominations; organisations must be engaged in development work to be classified as multilateral aid channels although money may be classified as bilateral while a case is being made for a new multilateral organisation to be recognised.

While core funding to multilateral organisations is always classified as multilateral expenditure, additional funding channelled through multilaterals is often classified as bilateral expenditure. This would be the case in circumstances where a DFID country office transfers some money to a multilateral organisation (for example UN agency) for a particular programme in that country (or region). That is where DFID has control over what the money is being spent on and/or where it is being spent. Likewise, if DFID responds to an emergency appeal from an agency for a particular country or area, the funds will be allocated as bilateral spend to that country or region. As a result, some organisations, such as UN agencies have some of their DFID funding classified as bilateral and some as multilateral.

Bilateral assistance takes various forms:

Financial Aid – Poverty Reduction Budget Support (PRBS) – Funds provided to developing countries for them to spend in support of their expenditure programmes whose long-term objective is to reduce poverty; funds are spent using the overseas governments' own financial management, procurement and accountability systems to increase ownership and long term sustainability. PRBS can take the form of a general contribution to the overall budget – general budget support – or support with a more restricted focus which is earmarked for a specific sector – sector budget support.

Other Financial Aid – Funding of projects and programmes such as Sector Wide Programmes not classified as PRBS. Financial aid in its broader sense covers all bilateral aid expenditure other than technical cooperation and administrative costs but in SID we separately categorise this further.

Technical Co-operation – Activities designed to enhance the knowledge, intellectual skills, technical expertise or the productive capability of people in recipient countries. It also covers funding of services which contribute to the design or implementation of development projects and programmes.

This assistance is mainly delivered through research and development, the use of consultants, training (generally overseas partners visiting the UK or elsewhere for a training programme) and employment of 'other Personnel' (non-DFID experts on fixed term contracts). This latter category is growing less significant over time as existing contracted staff reach the end of their assignments.

Bilateral Aid Delivered Through a Multilateral Organisation – This category covers funding that is channelled through a multilateral organisation and DFID has control over the country, sector or theme that the funds will be spent on. For example, where a DFID country office transfers money to a multilateral organisation for a particular piece of work in that country. This also includes aid delivered through multi donor funds such as the United Nations Central Emergency Response Fund (CERF).

Bilateral Aid Delivered Through a Non-Governmental Organisation (NGO) – This category covers support to the international development work of UK and international not for profit organisations such as NGOs or Civil Society Organisations. This covers Partnership Programme Arrangements (PPAs), the Civil Society Challenge Fund and other grants.

Other Bilateral Aid – This category includes any aid not elsewhere classified such as funding to other donors for shared development purposes. More information on all of the above aid types is provided in the Glossary.

Humanitarian Assistance – Provides food, aid and other humanitarian assistance including shelter, medical care and advice in emergency situations and their aftermath. Work of the conflict pools is also included.

DFID Debt Relief – This includes sums for debt relief on DFID aid loans and cancellation of debt under the Commonwealth Debt Initiative (CDI). The non-CDI DFID debt relief is reported on the basis of the 'benefit to the recipient country'. This means that figures shown represent the money available to the country in the year in question that would otherwise have been spent on debt servicing. The CDI debt cancellation is reported on a 'lump sum' basis where all outstanding amounts on a loan are shown at the time the agreement to cancel is made.

CDC Gross Investments – **CDC Group PLC** is wholly government owned. Its investments must have a clear development objective. The net amount (that is equity purchase less equity sales) of

CDC investments in official development assistance (ODA)-eligible countries is reported as ODA and the gross amount (that is equity purchase only) is reported in GPEX.

Non-DFID Debt Relief – Comprises CDC Debt and ECGD Debt. CDC has a portfolio of loans to governments which can become eligible for debt relief under the Heavily Indebted Poor Countries (HIPC) or other debt relief deals. UK Export Finance (UKEF) is the UK's official export credit agency providing insurance for exporters against the main risks in selling overseas and guarantees to banks providing export finance. It also negotiates debt relief arrangements on commercial debt.

The Foreign and Commonwealth Office (FCO) contributes to UK GPEX in a number of ways:

The FCO Strategic Programme Fund supports a range of the UK government's international goals. Where the programme funds projects which meet the required OECD definition these projects are included in UK GPEX statistics.

The FCO supports the British Council through grant-in-aid funding. This funding goes to support a range of initiatives including building the capacity and quality of English language teaching; supporting education systems; and using cultural exchange to improve economic welfare. UK GPEX statistics include the proportion of this work which is clearly focussed on delivering economic welfare and development in ODA eligible countries.

The British Council also manages, on behalf of the FCO, the Chevening Scholarships programme, which provides funding for postgraduate students or researchers from developing countries to study in UK universities. Funding from this scheme to students from ODA eligible countries are included in UK ODA and GPEX statistics.

The FCO makes annual contributions to UN and Commonwealth organisations. A proportion of these contributions are allowed to score as ODA in line with Annex 2 of the DAC Statistical Reporting Directives.

The FCO is responsible for the UK contribution to the UN Department for Peacekeeping Operations (UNDPKO). In line with DAC rules 6 per cent of donor funding to UNDPKO is allowed to score as ODA. FCO also funds other bilateral peacekeeping missions including the Organisation for Security and Cooperation in Europe (OSCE) and the European Security and Defence Policy (ESDP) civilian missions; a proportion of which is reported as bilateral GPEX.

Conflict Stability and Security Fund (CSSF) formerly The Conflict Pool (CP) –The CSSF replaced the Conflict Pool (CP) - the Conflict Pool was a cross-departmental fund that supported a range of activities designed to reduce the number of people around the world whose lives are or might be affected by violent conflict. Unlike the CSSF, however, access to Conflict Pool funding was limited to just three Departments, namely the Foreign and Commonwealth Office (FCO), the Department for International Development (DFID) and the Ministry of Defence (MOD). As with the Conflict Pool, the CSSF combines Official Development Assistance (ODA) with non-ODA funding, enabling a wider range of responses to conflict and instability overseas .
https://publications.parliament.uk/pa/jt201617/jtselect/jtnatsec/208/20805.htm#_idTextAnchor007

Other –includes contributions from other government departments including: Department of Energy and Climate Change; Department of Health and Social Care; Department for Environment, Food & Rural Affairs; Department for Digital Culture, Media and Sport; Scottish Government;

and the Welsh Assembly Government. It also includes estimates of the UK Border Agency's costs of supporting refugees in the UK; as well as estimates of gift aid to NGOs and other official funding to NGOs.

Further details on the UK's development assistance can be found in the Department for International Developments publication Statistics on International Development which can be found on the website:

https://www.gov.uk/government/organisations/department-for-international-development

Comparisons are available in the OECD Development Assistance Committee's annual report.

3.1 DFID Gross Public Expenditure 2010/11 - 2016/17

£ thousands

	2010/11	2011/12[R]	2012/13[R]	2013/14[R]	2014/15[R]	2015/16[R]	2016/17
DFID Bilateral Programme[1]							
Poverty Reduction Budget Support	643,671	536,662	444,133	575,073	434,844	116,308	134,767
of which							
General Budget Support	*360,467*	242,290	167,343	106,600	49,698	49,706	60,000
Sector Budget Support	*283,204*	294,372	276,790	468,474	385,146	66,602	74,767
Other Financial Aid	550,728	527,292	645,648	544,700	353,506	517,955	598,823
Technical Co-operation	467,939	530,980	637,862	903,254	1,003,417	1,174,109	1,229,480
Bilateral Aid Delivered through a Multilateral Organisation[2]	1,465,789	1,426,878	1,099,909	1,519,209	1,352,995	1,430,606	1,861,934
Bilateral Aid Delivered through a NGO	626,752	750,591	750,112	1,029,242	1,058,673	1,040,066	1,149,775
Other Bilateral Aid[3]	76,009	81,372	119,220	172,786	144,298	630,877	441,279
Humanitarian Assistance	350,669	354,293	476,782	866,161	1,071,602	1,178,370	956,195
DFID Debt Relief	66,460	14,954	17,169	9,172	11,384	0	0
Total DFID Bilateral Programme	**4,248,018**	**4,223,021**	**4,190,834**	**5,619,597**	**5,430,720**	**6,088,290**	**6,372,253**
DFID Multilateral Programme							
European Commission	1,268,563	1,220,076	1,085,769	1,095,770	711,663	825,691	970,621
World Bank	926,713	1,038,568	1,025,431	1,206,227	1,407,938	1,262,581	1,220,685
UN Agencies	355,337	376,708	360,304	494,269	392,631	367,994	405,082
Other Multilateral	671,061	632,373	809,804	1,479,795	1,416,051	1,035,186	1,129,072
Total DFID Multilateral Programme	**3,221,673**	**3,267,724**	**3,281,308**	**4,276,062**	**3,928,283**	**3,491,452**	**3,725,459**
Total DFID Programme (excl. Total Operating Costs)	**7,469,691**	**7,490,745**	**7,472,142**	**9,895,659**	**9,359,003**	**9,579,742**	**10,097,712**
DFID Total Operating Costs[4]	219,457	220,352	220,220	226,743	231,973	210,093	234,898
Total DFID Programme	**7,689,149**	**7,682,191**	**7,670,916**	**10,057,810**	**9,598,783**	**9,789,834**	**10,332,610**

Source: Department for International Development (DFID)

1. Descriptions of aid types given in Technical Note 1.
2. This covers earmarked funding provided through multilateral organisations where the recipient country, region, sector, theme or specific project are known. This figure does not include all bilateral aid spent through a multilateral organisation – other types of aid such as humanitarian assistance or debt relief also include aid spent through a multilateral organisation. In total in 2016/17 £2,338 million of bilateral aid was spent through multilateral organisations. Table 5 has a more detailed breakdown of this spend.

3. Other Bilateral Aid, for example UK Aid using other governments delivery partners (for instance Sweden)
4. Includes Front Line Delivery costs and Administration spend.
R   Revised, please see overview note (published alongside these tables)

3.2 DFID Bilateral Gross Public Expenditure by Region and Country Groupings 2011/12 - 2016/17[1]

£ thousands

Region	Year	Financial Aid			Bilateral aid					DFID Debt Relief	Total DFID Bilateral Programme
		General Poverty Reduction Budget Support[R]	Sector Poverty Reduction Budget Support[R]	Other Financial Aid[R]	Bilateral aid delivered through a Multilateral[R]	Bilateral aid Delivered through an NGO[R]	Other Bilateral Aid[2,R]	Technical Co-operation[R]	Humanitarian Assistance[R]		
TOTAL ALL COUNTRIES	2011/12	242,290	294,372	527,292	1,426,878	750,591	81,372	530,980	354,293	14,954	4,223,021
	2012/13	167,343	276,790	645,648	1,099,909	750,112	119,220	637,862	476,782	17,169	4,190,834
	2013/14	106,600	468,474	544,700	1,519,209	1,029,242	172,786	903,254	866,161	9,172	5,619,597
	2014/15	49,698	385,146	353,506	1,352,995	1,058,673	144,298	1,003,417	1,071,602	11,384	5,430,720
	2015/16	49,706	66,602	517,955	1,430,606	1,040,066	630,877	1,174,109	1,178,370	0	6,088,290
	2016/17	60,000	74,767	598,823	1,861,934	1,149,775	441,279	1,229,480	956,195	0	6,372,253
Africa	2011/12	222,290	242,372	215,745	387,810	217,830	45,046	298,559	260,329	1,713	1,891,694
	2012/13	167,343	247,090	256,973	317,599	212,217	76,600	368,905	317,865	1,124	1,965,716
	2013/14	106,600	309,524	237,480	420,890	316,423	78,176	492,442	379,607	0	2,341,141
	2014/15	49,698	266,896	154,999	397,398	296,368	45,574	530,087	642,456	0	2,383,476
	2015/16	19,706	36,802	319,293	506,518	336,171	69,319	585,714	536,377	0	2,409,899
	2016/17	0	13,952	363,918	624,531	459,270	64,151	639,920	447,382	0	2,613,125
of which: *Africa: South of Sahara*	*2011/12*	*222,290*	*242,372*	*215,745*	*357,469*	*195,539*	*38,047*	*285,827*	*248,159*	*1,713*	*1,807,161*
	2012/13	*167,343*	*247,090*	*256,973*	*266,609*	*184,816*	*70,965*	*340,383*	*266,401*	*1,124*	*1,801,704*
	2013/14	*106,600*	*309,524*	*237,480*	*371,022*	*282,203*	*45,601*	*462,459*	*328,780*	*0*	*2,143,668*
	2014/15	*49,698*	*266,896*	*154,999*	*316,186*	*263,149*	*34,758*	*480,851*	*574,999*	*0*	*2,141,536*
	2015/16	*19,706*	*36,802*	*319,293*	*412,620*	*295,314*	*47,010*	*534,146*	*513,116*	*0*	*2,178,006*
	2016/17	*0*	*13,952*	*363,918*	*528,225*	*388,187*	*35,963*	*568,233*	*421,615*	*0*	*2,320,094*
Americas	2011/12	0	0	26,576	14,004	2,215	526	3,932	2,537	0	49,789
	2012/13	0	0	30,440	12,044	1,882	364	2,552	6,383	0	53,665
	2013/14	0	0	26,450	8,503	1,522	561	1,028	7,273	3,535	48,872
	2014/15	0	0	23,231	5,059	1,509	841	2,905	3,945	0	37,490
	2015/16	0	0	28,451	6,147	756	1,170	1,386	2,978	0	40,886
	2016/17	0	0	26,189	71,329	181	105	6,622	1,907	0	106,333
Asia	2011/12	20,000	52,000	282,240	415,469	136,996	15,342	144,503	89,002	9,218	1,164,770
	2012/13	0	29,700	348,605	231,597	153,867	17,503	158,589	147,785	9,781	1,097,427
	2013/14	0	158,950	271,590	363,272	212,135	54,985	208,132	439,544	0	1,708,610
	2014/15	0	118,250	166,292	307,288	202,927	63,215	208,813	391,186	0	1,457,970
	2015/16	30,000	29,800	161,854	433,177	179,786	65,657	252,597	533,911	0	1,686,782
	2016/17	60,000	60,815	200,358	579,379	229,572	37,407	217,626	450,143	0	1,835,300
Europe	2011/12	0	0	0	244	344	-14	3,230	425	378	4,607
	2012/13	0	0	6,746	782	193	12	2,305	7	0	10,046
	2013/14	0	0	6,340	66	0	12	25	129	0	6,573
	2014/15	0	0	5,915	1,389	0	500	261	6,410	0	14,475
	2015/16	0	0	5,450	3,992	0	511	4,853	17,052	0	31,858
	2016/17	0	0	4,882	99,556	1,800	800	9,170	2,740	0	118,947

3.2 DFID Bilateral Gross Public Expenditure by Region and Country Groupings 2011/12 - 2016/17[1]

£ thousands

		Financial Aid			Bilateral aid					DFID Debt Relief	Total DFID Bilateral Programme
		General Poverty Reduction Budget Support[R]	Sector Poverty Reduction Budget Support[R]	Other Financial Aid[R]	Bilateral aid delivered through a Multilateral[R]	Bilateral aid Delivered through an NGO[R]	Other Bilateral Aid[2,R]	Technical Co-operation[R]	Humanitarian Assistance[R]		
Pacific	2011/12	0	0	2,738	0	0	0	36	0	212	2,986
	2012/13	0	0	2,883	0	0	0	11	150	166	3,210
	2013/14	0	0	2,839	0	0	0	0	0	44	2,883
	2014/15	0	0	3,070	570	218	0	106	1,300	0	5,264
	2015/16	0	0	2,907	0	0	0	0	1,348	0	4,256
	2016/17	0	0	3,475	509	38	0	34	858	0	4,914
Non Region Specific	2011/12	0	0	-6	609,352	393,206	20,473	80,719	2,000	3,433	1,109,176
	2012/13[R]	0	0	0	537,888	381,952	24,740	105,500	4,591	6,097	1,060,769
	2013/14[R]	0	0	0	726,478	499,162	39,051	201,627	39,608	5,593	1,511,519
	2014/15[R]	0	0	0	641,291	557,651	34,168	261,245	26,305	11,384	1,532,045
	2015/16	0	0	0	480,773	523,353	494,221	329,559	86,703	0	1,914,609
	2016/17	0	0	0	486,630	458,914	338,815	356,109	53,166	0	1,693,634
Other groups											
Total Developing Countries[3]	2011/12	242,290	294,372	524,218	746,477	320,435	53,281	427,412	342,464	11,521	2,962,470
	2012/13	167,343	276,790	635,589	460,514	335,761	82,253	492,685	403,809	11,072	2,865,816
	2013/14	106,600	468,474	535,111	671,057	486,524	92,426	658,850	611,860	3,578	3,634,480
	2014/15	49,698	385,146	343,556	590,793	460,150	98,890	672,797	851,572	0	3,452,602
	2015/16	49,706	66,602	509,447	802,280	462,221	113,234	766,507	1,059,637	0	3,829,635
	2016/17	60,000	74,767	590,466	1,163,744	582,910	73,570	764,531	871,762	0	4,181,750
Least Developed Countries	2011/12	210,000	210,372	185,851	547,953	287,435	43,976	218,591	267,198	4,782	1,976,159
	2012/13	153,235	237,090	291,194	345,311	291,227	78,529	241,126	308,998	4,857	1,951,566
	2013/14	93,668	307,024	252,299	528,113	418,338	65,232	327,182	385,483	44	2,377,382
	2014/15	49,698	287,403	124,943	496,078	368,743	56,606	342,266	633,176	0	2,358,913
	2015/16	49,706	34,312	294,106	650,792	343,800	66,184	392,835	670,152	0	2,461,886
	2016/17	0	16,277	317,515	650,879	478,281	44,859	425,984	509,785	0	2,443,581
Commonwealth[4]	2011/12	222,290	147,539	349,455	306,863	193,542	30,273	274,349	94,825	1,305	1,620,440
	2012/13	167,343	148,290	317,442	206,468	212,259	52,367	326,083	80,472	1,291	1,512,016
	2013/14	106,600	268,824	281,091	304,835	314,681	61,085	443,086	97,907	3,578	1,881,687
	2014/15	49,698	162,496	221,943	180,079	308,892	68,811	463,529	308,257	0	1,763,706
	2015/16	49,706	66,244	290,791	236,023	324,961	80,078	505,609	228,830	0	1,782,241
	2016/17	60,000	60,252	297,578	361,839	411,792	59,886	489,867	114,458	0	1,855,673
Overseas Territories[5]	*2011/12*	*0*	*0*	*91,166*	*0*	*298*	*0*	*5,962*	*0*	*0*	*97,426*
	2012/13	*0*	*0*	*130,123*	*0*	*201*	*0*	*3,581*	*0*	*0*	*133,905*
	2013/14	*0*	*0*	*117,493*	*0*	*252*	*0*	*1,546*	*0*	*0*	*119,699*
	2014/15	*0*	*0*	*94,294*	*0*	*216*	*0*	*2,572*	*0*	*0*	*97,083*
	2015/16	*0*	*0*	*81,997*	*0*	*199*	*0*	*3,331*	*0*	*0*	*85,981*
	2016/17	*0*	*0*	*94,568*	*0*	*140*	*0*	*14,055*	*0*	*0*	*108,868*

3.2 DFID Bilateral Gross Public Expenditure by Region and Country Groupings 2011/12 - 2016/17[1]

£ thousands

	Financial Aid			Bilateral aid					DFID Debt Relief	Total DFID Bilateral Programme
	General Poverty Reduction Budget Support[R]	Sector Poverty Reduction Budget Support[R]	Other Financial Aid[R]	Bilateral aid delivered through a Multilateral[R]	Bilateral aid Delivered through an NGO[R]	Other Bilateral Aid[2,R]	Technical Co-operation[R]	Humanitarian Assistance[R]		
MENA Countries[6]										
2011/12	0	0	148	54,722	7,846	750	9,105	27,934	0	100,504
2012/13	0	0	80	54,898	18,137	178	2,327	88,280	0	163,899
2013/14	0	0	113	77,723	41,930	676	9,249	159,427	0	289,118
2014/15	0	0	338	63,005	30,550	3,649	16,747	211,522	0	325,811
2015/16	0	0	251	64,650	25,032	13,567	10,805	434,802	0	549,108
2016/17	0	0	37,053	367,506	46,077	1,692	16,795	396,723	0	865,846

Source: Department for International Development (DFID)

1. Descriptions of aid types given in Technical Note 1.
2. Other Bilateral Aid, for example UK Aid using other gover DAC List of Recipients of Official Development Assistance.
3. Developing Countries are those countries in the
4. Commonwealth countries do not include Overseas Territories. See link below for more details.
http://thecommonwealth.org/ http://thecommonwealth.org/member-countries
5. Overseas territories in GPEX 16/17 were Anguilla, Gibraltar, Montserrat, Pitcairn Island, St Helena (including Tristan da Cunha) and Turks and Caicos Islands
6. Middle East and North African Countries include - Algeria, Egypt, Iran, Iraq, Jordan, Lebanon, Libya, Occupied Palestinian Territories, Morocco, Turkey, Tunisia, Syria and Yemen.
In GPEX 16/17 there was no direct DFID bilateral spend to Algeria, Iran, Morocco and Tunisia.
R. Revised, please see overview note (published alongside these tables)

3.3 DFID Gross Public Expenditure by Input Sector Code and Delivery Channel, 2016/17

£ thousands

Input Sector Code	DFID Bilateral through a Multilateral GPEX	DFID Other Bilateral GPEX	Total DFID Bilateral GPEX
Education:			
11010 Education Poverty Reduction Budget Support	0	31,783	31,783
11020 Education Unallocable/Unspecified	34,621	14,571	49,191
11110 Education Policy and Administrative Management	8,756	44,011	52,766
11120 Facilities and Training Education	416	13,775	14,191
11130 Teacher Training	18,607	27,622	46,230
11220 Primary Education	159,603	222,635	382,239
11230 Basic Life Skills for Youth and Adults Education	1,842	28,656	30,498
11240 Pre-School	1,400	533	1,933
11320 Secondary Education	18,975	77,466	96,441
11330 Vocational Training	872	34,051	34,924
11420 Higher Education	0	38,006	38,006
11430 Advanced Technical and Managerial Training	0	7,081	7,081
Education Total	**245,093**	**540,190**	**785,282**
Health:			
12010 Health Poverty Reduction Budget Support	0	5,400	5,400
12020 Health Unallocable/Unspecified	19,372	19,139	38,511
12110 Health Policy and Administrative Management	8,566	50,672	59,239
12220 Basic Health Care	20,971	104,518	125,489
12240 Basic Nutrition	73,919	47,983	121,902
12250 Infectious Disease Control	52,205	34,629	86,834
12261 Health Education	12,502	18,354	30,857
12262 Malaria Control	32,957	54,514	87,471
12263 Tuberculosis Control	1,669	2,446	4,115
12281 Health Personnel Development	1,297	10,564	11,862
13010 Population Policy and Administrative Management	3,394	584	3,979
13021 Reproductive Health Care	22,015	29,457	51,472
13022 Maternal and Neonatal Health	39,117	83,363	122,480
13030 Family Planning	49,900	58,348	108,248
13041 HIV/AIDS including STD Prevention	3,567	19,139	22,705
13042 HIV/AIDS including STD Treatment and Care	2,680	1,758	4,438
13081 Personnel Development for Population and Reproductive Health	1,112	19,906	21,018
Health Total	**345,243**	**560,775**	**906,018**
Social Infrastructure and Services:			
16011 Social Protection	29,828	212,142	241,969
16012 Social Other	15,198	30,847	46,045
16020 Employment Policy & Admin Management	80,456	1,113	81,570
16030 Housing Policy and Admin Management	0	562	562
16040 Low-cost Housing	0	3,936	3,936
16070 Poverty Reduction Budget Support-Social infrastructure and services	0	13,214	13,214
52010 Food Aid and Food Security Programmes	30,076	22,549	52,625
Social Infrastructure and Services Total	**155,558**	**284,363**	**439,921**
Water Supply and Sanitation:			
14010 Water Resources Policy and Administrative Management	13,555	8,778	22,332
14015 Water Resources Protection	1,362	2,867	4,229
14020 Water Supply and Sanitation Large Systems	15,469	2,708	18,178
14021 Water Supply – Large Systems	1,590	3,104	4,694
14022 Sanitation - large systems, (MDG Water and Sanitation)	428	0	428
14030 Basic Drinking Water	8,778	9,111	17,889

3.3 DFID Gross Public Expenditure by Input Sector Code and Delivery Channel, 2016/17

£ thousands

Input Sector Code	DFID Bilateral through a Multilateral GPEX	DFID Other Bilateral GPEX	Total DFID Bilateral GPEX
14031 Basic drinking water supply	10,484	37,920	48,405
14032 Basic sanitation	13,535	24,242	37,777
14040 River Development	840	1,365	2,206
14050 Waste Management and Disposal	1,071	3,325	4,396
14060 Water Poverty Reduction Budget Support	0	3,000	3,000
14070 Water Unallocable/Unspecified	1,905	6,480	8,385
14081 Education and Training	5,202	1,687	6,888
Water Supply and Sanitation Total	**74,219**	**104,587**	**178,807**

Government and Civil Society:

15010 Government Poverty Reduction Budget Support	0	10,200	10,200
15020 Government Unallocated/ Unspecified	3,793	51,080	54,873
15110 Economic and Development Policy/Planning	26,181	70,747	96,928
15114 Tax policy and tax administration support	7,396	17,238	24,634
15121 Public Sector Financial Management	21,758	35,004	56,761
15122 Corruption - Public Sector Financial Management	3,866	23,668	27,533
15130 Legal and Judicial Development	8,681	16,478	25,159
15141 National Government Administration	42,743	26,052	68,795
15142 Local Government Administration	18,883	23,569	42,451
15150 Strengthening Civil Society	6,434	66,764	73,198
15152 Legislatures and political parties (Governance and Security)	390	5,963	6,353
15161 Elections	4,951	15,789	20,740
15162 Human Rights	7,742	19,502	27,243
15163 Free Flow of Information	1,189	12,257	13,445
15164 Women's Equality Organisations and Institutions	9,316	27,457	36,773
15172 Statistical Capacity Building	18,199	6,202	24,401
15173 Narcotics Control	0	398	398
15180 Ending violence against women and girls	4,496	592	5,088
15210 Security System Management and Reform	8,327	23,085	31,412
15220 Civilian Peace-Building, Conflict Prevention and Resolution	18,405	25,104	43,509
15230 Post-Conflict Peace-Building (UN)	1,822	1,747	3,569
15240 Reintegration and SALW Control	0	628	628
15250 Land Mine Clearance	0	9,870	9,870
Government and Civil Society Total	**214,572**	**489,391**	**703,963**

Economic:

Economic Infrastructure

21010 Transport Policy and Administrative Management	1,249	25,118	26,367
21021 Road Transport: Excluding Rural Feeder Roads	54,519	7,003	61,522
21022 Road Transport: Rural Feeder Roads	3,838	21,024	24,862
21031 Other Transport	8,172	21,899	30,071
22010 Communications Policy and Administrative Management	1,330	1,373	2,703
22020 Telecommunications	0	312	312
22030 Radio/Television/Print Media: Communications	36	247	283
22040 Information and Communication Technology (ICT)	840	1,175	2,015
23010 Energy Policy and Administrative Management	3,754	14,709	18,463
23020 Power Generation/Non-Renewable Sources: Energy	5,600	292	5,892
23030 Power Generation/Renewable Sources: Energy	-673	24,306	23,633
23040 Energy Access for Households, Enterprises and Communities: Energy (\	10,652	4,437	15,089
24010 Financial Policy and Administrative Management	7,196	228,112	235,308
24020 Monetary Institutions	3,840	1,831	5,671
24030 Formal Sector Financial Intermediaries	13,321	21,333	34,654

3.3 DFID Gross Public Expenditure by Input Sector Code and Delivery Channel, 2016/17

£ thousands

Input Sector Code	DFID Bilateral through a Multilateral GPEX	DFID Other Bilateral GPEX	Total DFID Bilateral GPEX
24040 Informal/Semi-Formal Financial Intermediaries	15,067	9,667	24,734
24081 Education/Training in Banking and Financial Services	594	4,151	4,746
25010 Business Support Services and Institutions	14,512	58,441	72,953
25020 Privatisation	3,243	945	4,188
Production Sectors			
31110 Agriculture Policy and Administrative Management	7,164	18,997	26,161
31120 Agricultural Development	82,442	77,222	159,664
31130 Agricultural Land Resources	16,483	17,618	34,101
31163 Livestock: Agriculture	2,470	4,874	7,344
31191 Agricultural Services	8,775	14,108	22,883
31210 Forestry Policy and Administrative Management	6,003	22,858	28,861
31220 Forestry Development	416	1,021	1,437
31310 Fishing Policy and Administrative Management	0	57	57
31320 Fishery Development	0	345	345
32110 Industrial Policy and Administrative Management	544	1,002	1,547
32120 Industrial Development	1,404	8,080	9,484
32130 Small and Medium-Sized Enterprises (SME): Development	8,701	156,884	165,585
32210 Mineral/Mining Policy and Administrative Management	2,962	9,012	11,974
32310 Construction Policy and Administrative Management	75	1,030	1,105
32350 Production Poverty Reduction Budget Support	0	4,200	4,200
33110 Trade Policy and Administrative Management	3,587	18,290	21,876
33120 Trade Facilitation	16,350	16,520	32,870
33130 Regional Trade Agreements (RTAs)	1,038	10,999	12,037
33140 Multilateral Trade Negotiations	325	3,044	3,369
33181 Trade Education/Training	2,508	1,508	4,016
33210 Tourism Policy and Administrative Management	0	1,745	1,745
43050 Non-Agricultural Alternative Development	0	1,852	1,852
Development Planning			
43020 Poverty Reduction Budget Support for Econ. Infrastructure & Dev. Planning	0	16,200	16,200
43030 Urban Development and Management	21,451	9,721	31,172
43040 Rural Development	3,197	42,311	45,508
Economic Total	**332,986**	**905,873**	**1,238,859**
Environment Protection:			
41010 Environmental Policy and Administrative Management	3,169	6,599	9,767
41031 Bio-Diversity	0	28	28
41032 Climate Change	27,524	83,301	110,826
41033 Desertification, Environment	500	446	946
41050 Flood Prevention/Control	4,429	2,554	6,983
41060 Environment: Poverty Reduction Budget Support	0	600	600
41070 Environment Unallocable/Unspecified	158	743	902
41081 Environmental Education/ Training	0	320	320
41090 Climate Change - Low Carbon Emissions	3,697	15,413	19,109
41092 Climate Change - Cross Cutting	0	0	0
41093 Climate Change - Adaptation	19,810	44,078	63,888
Environment Protection Total	**59,287**	**154,082**	**213,369**
Research:			
80010 Economic Research	8,974	27,011	35,985
80011 Education Research	1,062	20,041	21,103
80012 Health Research	15,631	68,672	84,303
80013 Water Supply and Sanitation Research	0	5,230	5,230

3.3 DFID Gross Public Expenditure by Input Sector Code and Delivery Channel, 2016/17

£ thousands

Input Sector Code	DFID Bilateral through a Multilateral GPEX	DFID Other Bilateral GPEX	Total DFID Bilateral GPEX
80014 Governance Research	1,733	15,125	16,858
80015 Social Research	2,625	11,748	14,373
80016 Humanitarian Research	4,842	10,369	15,211
80017 Renewable Natural Resources Research	1,250	2,176	3,426
80018 Environment Research	267	14,789	15,056
80019 Energy Research	1,389	8,305	9,694
80020 Agricultural Research	11,737	28,371	40,108
80021 Forestry Research	139	387	525
80022 Fishery Research	0	0	0
80023 Technological Research and Development	0	8,166	8,166
80024 Unspecified/Unallocated Research	4,197	22,316	26,513
Research Total	**53,846**	**242,706**	**296,552**
Humanitarian Assistance:			
72010 Material Relief Assistance and Services	372,754	264,361	637,116
72040 Emergency Food Aid	226,157	105,865	332,021
72050 Relief Coordination, Protection and Support Services	164,883	70,063	234,946
73010 Reconstruction Relief and Rehabilitation	14,617	24,837	39,455
74010 Disaster Prevention and Preparedness	22,612	55,546	78,158
Humanitarian Assistance Total	**801,023**	**520,672**	**1,321,695**
Non Sector Allocable:			
88801 Core Contributions to Multilateral Institutions- Global Partnerships	452	0	452
88802 Core Contributions to Multilateral Institutions - Governance & Security	0	0	0
88803 Core Contributions to Multilateral Institutions- Education	0	0	0
88804 Core Contributions to Multilateral Institutions- HIV/Aids	0	0	0
88805 Core Contributions to Multilateral Institutions - MDG Humanitarian	5,117	258	5,375
88806 Core Contributions to Multilateral Institutions- Malaria	0	0	0
88807 Core Contributions to Multilateral Institutions - MDG Other Health	9,000	0	9,000
88808 Core Contributions to Multilateral Institutions- Other MDG's	225	0	225
88809 Core Contributions to Multilateral Institutions - Poverty, Hunger & Vulner	1,072	0	1,072
88810 Core Contributions to Multilateral Institutions- Reproductive, Maternal &	1,587	0	1,587
88811 Core Contributions to Multilateral Institutions - MDG Water & Sanitation	148	0	148
88812 Core Contributions to Multilateral Institutions- MDG All	93	0	93
88813 Core Contributions to Multilateral Institutions - Multiple Pillars	0	0	0
88814 Core Contributions to Multilateral Institutions - Wealth Creation	300	0	300
88815 Core Contributions to Multilateral Institutions - Climate and Environment	512	0	512
88816 Core Contributions to Multilateral Institutions - Climate Change ICF	0	0	0
88888 Multilateral Core Contribution	3,866	0	3,866
88889 Multilateral Capacity Building and Administration	19,477	0	19,477
88890 Multilateral Institutions: Secondees to & Staffing of	317	4,787	5,104
90010 Programme Partnership Agreements	0	86,363	86,363
91020 Front Line Delivery Costs	0	2,091	2,091
92000 Support to Non-Governmental Organisations (NGOs)	153	128,600	128,753
93020 Aid to Refugees in Recipient Countries	13,215	2,924	16,139
99820 Promotion of Development Awareness	364	6,866	7,230
60010 Action Relating to Debt	0	0	0
Others			
Non Sector Allocable Total	**55,897**	**231,889**	**287,786**

1. Figues may not sum to totals due to rounding.   Source: Department for International Development (DFID)

this page is intentionally blank

Labour market

Chapter 4

Labour Market

Labour Force Survey

Background

The Labour Force Survey (LFS) is the largest regular household survey in the UK. LFS interviews are conducted continuously throughout the year. In any three-month period, nationally representative samples of approximately 110,000 people aged 16 and over in around 50,000 households are interviewed. Each household is interviewed five times, at three-monthly intervals. The initial interview is done face-to-face by an interviewer visiting the address, except for residents north of the Caledonian Canal in Scotland. The other interviews are done by telephone wherever possible. The survey asks a series of questions about respondents' personal circumstances and their labour market activity. Most questions refer to activity in the week before the interview.

The LFS collects information on a sample of the population. To convert this information to give estimates for the population, the data must be grossed. This is achieved by calculating weighting factors (often referred to simply as weights) which can be applied to each sampled individual in such a way that the weighted-up results match estimates or projections of the total population in terms of age distribution, sex, and region of residence. There is a considerable amount of ongoing research to improve methodologies. Whenever methodologies are implemented the estimates may be revised.

The concepts and definitions used in the LFS are agreed by the International Labour Organisation (ILO) – an agency of the United Nations. The definitions are used by European Union member countries and members of the Organisation for Economic Co-operation and Development (OECD). The LFS was carried out every two years from 1973 to 1983. The ILO definition was first used in 1984. This was also the first year in which the survey was conducted on an annual basis with results available for every spring quarter (representing an average of the period from March to May). The survey moved to a continuous basis in spring 1992 in Great Britain and in winter 1994/95 in Northern Ireland, with average quarterly results published four times a year for seasonal quarters: spring (March to May), summer (June to August), autumn (September to November) and winter (December to February). From April 1998, results are published 12 times a year for the average of three consecutive months.

Strengths and limitations of the LFS

The LFS produces coherent labour market information on the basis of internationally standard concepts and definitions. It is a rich source of data on a wide variety of labour market and personal characteristics. It is the most suitable source for making comparisons between countries. The LFS is designed so that households interviewed in each three month period constitute a representative sample of UK households. The survey covers those living in private households and nurses in

National Health Service accommodation. Students living in halls of residence have been included since 1992, as information about them is collected at their parents' address.

However the LFS has its limitations. It is a sample survey and is therefore subject to sampling variability. The survey does not include people living in institutions such as hostels, hotels, boarding houses, mobile home sites or residential homes. 'Proxy' reporting (when members of the household are not present at the interview, another member of the household answers the questions on their behalf) can affect the quality of information on topics such as earnings, hours worked, benefit receipt and qualifications. Around a third of interviews are conducted 'by proxy', usually by a spouse or partner but sometimes by a parent or other near relation. LFS estimates are also potentially affected by non-response.

Sampling Variability

Survey estimates are prone to sampling variability. The easiest way to explain this concept is by example. In the September to November 1997 period, ILO unemployment in Great Britain (seasonally adjusted) stood at 1,847,000. If we drew another sample for the same period we could get a different result, perhaps 1,900,000 or 1,820,000.

In theory, we could draw many samples, and each would give a different result. This is because each sample would be made up of different people who would give different answers to the questions. The spread of these results is the sampling variability. Sampling variability is determined by a number of factors including the sample size, the variability of the population from which the sample is drawn and the sample design. Once we know the sampling variability we can calculate a range of values about the sample estimate that represents the expected variation with a given level of assurance. This is called a confidence interval. For a 95 per cent confidence interval we expect that in 95 per cent of the samples (19 times out of 20) the confidence interval will contain the true value that would be obtained by surveying the entire population. For the example given above, we can be 95 per cent confident that the true value was in the range 1,791,000 to 1,903,000.

Unreliable estimates

Estimates of small numbers have relatively wide confidence intervals making them unreliable. For this reason, the Office for National Statistics (ONS) does not currently publish LFS estimates below 10,000.

Non-response

All surveys are subject to non-response – that is respondents in the sample who either refuse to take part in the survey or who cannot be contacted. Non-response can introduce bias to a survey, particularly if the people not responding have characteristics that are different from those who do respond.

The LFS has a response rate of around 65 per cent to the first interview, and over 90 per cent of those who are interviewed once go on to complete all five interviews. These are relatively high levels for a household survey.

Any bias from non-response is minimised by weighting the results. Weighting (or grossing) converts sample data to represent the full population. In the LFS, the data are weighted separately by age, sex and area of residence to population estimates based on the census. Weighting also adjusts for people not in the survey and thus minimises non-response bias.

LFS concepts and definitions

Discouraged worker - A sub-group of the economically inactive population who said although they would like a job their main reason for not seeking work was because they believed there were no jobs available.

Economically active – People aged 16 and over who are either in employment or unemployed.

Economic activity rate – The number of people who are in employment or unemployed expressed as a percentage of the relevant population.

Economically inactive – People who are neither in employment nor unemployed. These include those who want a job but have not been seeking work in the last four weeks, those who want a job and are seeking work but not available to start, and those who do not want a job.

Employment – People aged 16 and over who did at least one hour of paid work in the reference week (as an employee or self-employed), those who had a job that they were temporarily away from, those on government-supported training and employment programmes, and those doing unpaid family work.

Employees – The division between employees and self employed is based on survey respondents' own assessment of their employment status.

Full Time – The classification of employees, self-employed and unpaid family workers in their main job as full-time or part-time is on the basis of self-assessment. However, people on government supported employment and training programmes that are at college in the reference week are classified, by convention, as part-time.

Government -supported training and employment programmes – Comprise all people aged 16 and over participating in one of the government's employment and training programmes (Youth Training, Training for Work and Community Action), together with those on similar programmes administered by Training and Enterprise Councils in England and Wales, or Local Enterprise Companies in Scotland.

Hours worked – Respondents to the LFS are asked a series of questions enabling the identification of both their usual hours and their actual hours. Total hours include overtime (paid and unpaid) and exclude lunch breaks.

Actual Hours Worked – Actual hours worked statistics measure how many hours were actually worked. These statistics are directly affected by changes in the number of people in employment and in the number of hours that individual works.

Usual Hours Worked – Usual hours worked statistics measure how many hours people usually work per week. Compared with actual hours worked, they are not affected by absences and so can provide a better measure of normal working patterns.

Unemployment – The number of unemployed people in the UK is measured through the LFS following the internationally agreed definition recommended by the International Labour Organisation (ILO), an agency of the United Nations.

Unemployed people are:

Without a job, have actively sought work in the last four weeks and are available to start work in the next two weeks, or

Out of work, have found a job and are waiting to start in the next two weeks

Unemployment (rate) – The number of unemployed people expressed as a percentage of the relevant economically active population.

Unemployment (duration) – The duration of respondents unemployment is defined as the shorter of the following two periods:

Duration of active search for work

Length of time since employment

Part-time – see full-time.

Second jobs – Jobs which LFS respondents hold in addition to a main full-time or part-time job.

Self-employment – See Employees.

Temporary employees – In the LFS these are defined as those employees who say that their main job is non permanent in one of the following ways: fixed period contract, agency temping, casual work, seasonal work or other temporary work.

Unpaid family workers – Persons doing unpaid work for a business they own or for a business that a relative owns.

International Employment Comparisons

All employment rates for European Union (EU) countries published by Eurostat (including the rate for the UK) are based on the population aged 15–64. The rates for Canada and Japan are also based on the population aged 15–64, but the rate for the US is for those aged 16–64. The employment rate for the UK published by ONS is based on the working age population aged 16–64 and therefore takes into account both the current school leaving age and state pension ages.

The unemployment rate published by Eurostat for most EU countries (but not for the UK), are calculated by extrapolating from the most recent LFS data using monthly registered unemployment data. A standard population basis (15–74) is used by Eurostat except for Spain and the UK (16–74). The unemployment rate for the US is based on those aged 16 and over, but the rates for Canada and Japan are for those aged 15 and over. All unemployment rates are seasonally adjusted.

The unemployment rate for the UK published by Eurostat is based on the population aged 16–74 while the unemployment rate for the UK published by ONS is based on those aged 16 and over. There are other minor definitional differences.

Jobseekers allowance claimant count

Under Universal Credit a broader span of claimants are required to look for work than under Jobseeker's Allowance. As Universal Credit Full Service is rolled out in particular areas, the number of people recorded as being on the Claimant Count is therefore likely to rise.

Annual Survey of Hours and Earnings

The Annual Survey of Hours and Earnings (ASHE) is based on a one per cent sample of employee jobs taken from HM Revenue & Customs (HMRC) PAYE records. Information on earnings and paid hours worked is obtained from employers and treated confidentially. ASHE does not cover the self-employed nor does it cover employees not paid during the reference period.

The headline statistics for ASHE are based on the median rather than the mean. The median is the value below which 50 per cent of employees fall. It is ONS's preferred measure of average earnings as it is less affected by a relatively small number of very high earners and the skewed distribution of earnings. It therefore gives a better indication of typical pay than the mean.

The earnings information presented relates to gross pay before tax, National Insurance or other deductions, and excludes payments in kind. With the exception of annual earnings, the results are restricted to earnings relating to the survey pay period and so exclude payments of arrears from another period made during the survey period; any payments due as a result of a pay settlement but not yet paid at the time of the survey will also be excluded.

Average Weekly Earnings

The Average Weekly Earnings (AWE) indicator measures changes in the level of earnings in Great Britain. Average earnings are calculated as the total wages and salaries paid by firms, divided by the number of employees paid. It is given as a level, in pounds per employee per week. Annual growth rates are derived from the level of average weekly earnings.

The AWE data are now published on a SIC 2007 basis, and the historic time series have been re-estimated as a result.

AWE is based on the Monthly wages and Salaries Survey (MWSS). As such, it is a timely indicator of changes in the level of earnings. The survey does not cover businesses with fewer than 20 employees; an adjustment is made to AWE to reflect these businesses. Note that the survey does not include Northern Ireland.

Unlike the previous measure of average earnings (the Average Earnings Index), changes in the composition of the workforce have an impact on AWE. If a high-paying sector of the economy employs more people, other things staying the same, average earnings will increase.

Average Weekly Earnings, like AEI before it, is a measure based on earnings per employee. If the number of paid hours worked per employee change, average earnings will also change.

Trade unions

The statistics relate to all organisations of workers known to the Certification Officer with head offices in Great Britain that fall within the appropriate definition of a trade union in the Trade Union and Labour Relations (Consolidation) Act 1992. Included in the data are home and overseas membership figures of contributory and non-contributory members. Employment status of members is not provided and the figures may therefore include some people who are self-employed, unemployed or retired.

4.1 Labour Force Survey Summary:
economic activity for those aged 16 and over and those aged from 16 to 64

United Kingdom (thousands) seasonally adjusted

All aged 16 & over								
All aged 16 & over level	Total economically active level	Total in employ-ment level	Unemployed level	Economically inactive level	Economic activity rate (%)	Employment rate (%)	Unemploy-ment rate (%)	Economic inactivity rate (%)
People								
MGSL	MGSF	MGRZ	MGSC	MGSI	MGWG	MGSR	MGSX	YBTC
Aug-Oct 2014	51,769	32,802	30,848	1,953	18,967	63.4	59.6	6.0	36.6
Aug-Oct 2015	52,170	33,152	31,436	1,715	19,019	63.5	60.3	5.2	36.5
Aug-Oct 2016	52,535	33,407	31,794	1,613	19,128	63.6	60.5	4.8	36.4
Nov-Jan 2017	52,603	33,421	31,847	1,573	19,183	63.5	60.5	4.7	36.5
Feb-Apr 2017	52,671	33,476	31,956	1,520	19,196	63.6	60.7	4.5	36.4
May-Jul 2017	52,739	33,590	32,133	1,457	19,149	63.7	60.9	4.3	36.3
Aug-Oct 2017	**52,811**	**33,517**	**32,094**	**1,422**	**19,294**	**63.5**	**60.8**	**4.2**	**36.5**
Men								
MGSM	MGSG	MGSA	MGSD	MGSJ	MGWH	MGSS	MGSY	YBTD
Aug-Oct 2014	25,241	17,490	16,404	1,086	7,751	69.3	65.0	6.2	30.7
Aug-Oct 2015	25,465	17,705	16,769	937	7,760	69.5	65.8	5.3	30.5
Aug-Oct 2016	25,672	17,774	16,889	885	7,898	69.2	65.8	5.0	30.8
Nov-Jan 2017	25,707	17,790	16,926	864	7,917	69.2	65.8	4.9	30.8
Feb-Apr 2017	25,742	17,797	16,963	834	7,945	69.1	65.9	4.7	30.9
May-Jul 2017	25,777	17,801	17,010	791	7,976	69.1	66.0	4.4	30.9
Aug-Oct 2017	**25,819**	**17,745**	**16,967**	**777**	**8,074**	**68.7**	**65.7**	**4.4**	**31.3**
Women								
MGSN	MGSH	MGSB	MGSE	MGSK	MGWI	MGST	MGSZ	YBTE
Aug-Oct 2014	26,528	15,311	14,444	867	11,217	57.7	54.4	5.7	42.3
Aug-Oct 2015	26,705	15,446	14,668	779	11,259	57.8	54.9	5.0	42.2
Aug-Oct 2016	26,863	15,634	14,905	728	11,229	58.2	55.5	4.7	41.8
Nov-Jan 2017	26,896	15,630	14,921	709	11,265	58.1	55.5	4.5	41.9
Feb-Apr 2017	26,929	15,679	14,992	686	11,250	58.2	55.7	4.4	41.8
May-Jul 2017	26,962	15,789	15,123	666	11,172	58.6	56.1	4.2	41.4
Aug-Oct 2017	**26,992**	**15,772**	**15,127**	**645**	**11,220**	**58.4**	**56.0**	**4.1**	**41.6**

United Kingdom (thousands) seasonally adjusted

All aged 16 to 64								
All aged 16 to 64 level	Total economically active level	Total in employ-ment level	Unemployed level	Economically inactive level	Economic activity rate (%)	Employment rate (%)	Unemploy-ment rate (%)	Economic inactivity rate (%)
People								
LF2O	LF2K	LF2G	LF2I	LF2M	LF22	LF24	LF2Q	LF2S
Aug-Oct 2014	40,702	31,649	29,715	1,934	9,053	77.8	73.0	6.1	22.2
Aug-Oct 2015	40,904	31,956	30,261	1,695	8,949	78.1	74.0	5.3	21.9
Aug-Oct 2016	41,080	32,174	30,577	1,596	8,907	78.3	74.4	5.0	21.7
Nov-Jan 2017	41,106	32,196	30,639	1,557	8,910	78.3	74.5	4.8	21.7
Feb-Apr 2017	41,132	32,260	30,753	1,507	8,872	78.4	74.8	4.7	21.6
May-Jul 2017	41,157	32,408	30,977	1,432	8,749	78.7	75.3	4.4	21.3
Aug-Oct 2017	**41,181**	**32,317**	**30,917**	**1,400**	**8,864**	**78.5**	**75.1**	**4.3**	**21.5**
Men								
YBTG	YBSL	YBSF	YBSI	YBSO	MGSP	MGSV	YBTJ	YBTM
Aug-Oct 2014	20,191	16,801	15,729	1,072	3,390	83.2	77.9	6.4	16.8
Aug-Oct 2015	20,307	16,978	16,054	924	3,329	83.6	79.1	5.4	16.4
Aug-Oct 2016	20,413	17,025	16,148	877	3,388	83.4	79.1	5.2	16.6
Nov-Jan 2017	20,425	17,053	16,198	855	3,372	83.5	79.3	5.0	16.5
Feb-Apr 2017	20,437	17,051	16,228	823	3,386	83.4	79.4	4.8	16.6
May-Jul 2017	20,450	17,080	16,305	775	3,369	83.5	79.7	4.5	16.5
Aug-Oct 2017	**20,466**	**17,014**	**16,252**	**762**	**3,452**	**83.1**	**79.4**	**4.5**	**16.9**
Women								
LF2P	LF2L	LF2H	LF2J	LF2N	LF23	LF25	LF2R	LF2T
Aug-Oct 2014	20,511	14,848	13,986	862	5,663	72.4	68.2	5.8	27.6
Aug-Oct 2015	20,598	14,978	14,207	771	5,620	72.7	69.0	5.1	27.3
Aug-Oct 2016	20,668	15,149	14,429	719	5,519	73.3	69.8	4.7	26.7
Nov-Jan 2017	20,681	15,143	14,441	702	5,538	73.2	69.8	4.6	26.8
Feb-Apr 2017	20,694	15,209	14,525	685	5,485	73.5	70.2	4.5	26.5
May-Jul 2017	20,707	15,328	14,672	656	5,379	74.0	70.9	4.3	26.0
Aug-Oct 2017	**20,715**	**15,303**	**14,665**	**637**	**5,412**	**73.9**	**70.8**	**4.2**	**26.1**

Source: Labour Force Survey

Labour market statistics enquiries: labour.market@ons.gsi.gov.uk

4.1 Labour Force Survey Summary:
economic activity for those aged 16 and over and those aged from 16 to 64

Note: When comparing quarterly changes ONS recommends comparing with the previous non-overlapping 3-month average time period (eg, compare Apr-Jun with Jan-Mar, not with Mar-May).

The headline employment rate is the number of people aged 16 to 64 in employment divided by the population aged 16 to 64.

The headline unemployment rate is the number of unemployed people (aged 16+) divided by the economically active population (aged 16+).

The economically active population is defined as those in employment plus those who are unemployed.

The headline inactivity rate is the number of economically inactive people aged 16 to 64 divided by the population aged 16 to 64.

Note on headline employment, unemployment and inactivity rates

The headline employment and inactivity rates are based on the population aged 16 to 64 but the headline unemployment rate is based on the economically active population aged 16 and over. The employment and inactivity rates for those aged 16 and over are affected by the inclusion of the retired population in the denominators and are therefore less meaningful than the rates for those aged from 16 to 64. However, for the unemployment rate for those aged 16 and over, no such effect occurs as the denominator for the unemployment rate is the economically active population which only includes people in work or actively seeking and able to work.

Note on headline employment, unemployment and inactivity levels

The headline employment and unemployment levels are for those aged 16 and over; they measure all people in work or actively seeking and able to work. However, the headline inactivity level is for those aged 16 to 64. The inactivity level for those aged 16 and over is less meaningful as it includes elderly people who have retired from the labour force.

4.2 Full-time, part-time and temporary workers

United Kingdom (thousands) seasonally adjusted

People (16+)	All in employment					Full-time and part-time workers[1]							Temporary employees (reasons for temporary working)							Part-time workers (reasons for working part-time)[3]					
	Total	Employees	Self employed	Unpaid family workers	Government supported training & employment programmes[2]	Total people working full-time	Total people working part-time	Employees working full-time	Employees working part-time	Self-employed people working full-time	Self-employed people working part-time	Total workers with second jobs	Total	Total as % of all employees	Could not find permanent job	% that could not find permanent job	Did not want permanent job	Had a contract with period of training	Some other reason	Total[4]	Could not find full-time job	% that could not find full-time job	Did not want full-time job	Ill or disabled	Student or at school
	MGSA	MGRO	MGRR	MGRU	MGRX	YCBF	YCBI	YCBL	YCBO	YCBR	YCBU	YCBX	YCCA	YCCD	YCCG	YCCJ	YCCM	YCCP	YCCS	YCCV	YCCY	YCDB	YCDE	YCDH	YCDK
Nov-Jan 2007	29,202	25,186	3,809	100	107	21,753	7,449	18,762	6,424	2,924	885	1,072	1,529	6.1	399	26.1	442	105	583	7,309	668	9.1	5,235	198	1,166
Dec-Feb 2007	29,175	25,155	3,816	99	105	21,748	7,427	18,760	6,394	2,922	894	1,076	1,522	6.0	397	26.1	440	105	579	7,288	646	8.9	5,255	192	1,158
Jan-Mar 2007	29,194	25,169	3,821	103	100	21,756	7,438	18,782	6,387	2,908	914	1,074	1,534	6.1	405	26.4	435	112	583	7,300	660	9.0	5,257	189	1,162
Feb-Apr 2007	29,232	25,211	3,821	101	100	21,797	7,435	18,829	6,382	2,905	915	1,095	1,514	6.0	403	26.6	426	104	581	7,297	671	9.2	5,264	178	1,152
Mar-May 2007	29,314	25,310	3,801	99	104	21,896	7,419	18,926	6,384	2,907	894	1,100	1,513	6.0	407	26.9	422	101	584	7,278	685	9.4	5,246	176	1,140
Apr-Jun 2007	29,322	25,310	3,803	98	111	21,902	7,420	18,935	6,375	2,907	896	1,107	1,507	6.0	417	27.7	409	97	583	7,271	682	9.4	5,251	174	1,131
May-Jul 2007	29,352	25,319	3,811	104	118	21,921	7,431	18,965	6,355	2,897	914	1,118	1,501	5.9	411	27.4	412	92	586	7,269	697	9.6	5,228	181	1,130
Jun-Aug 2007	29,376	25,330	3,830	102	113	21,947	7,428	18,983	6,347	2,904	926	1,117	1,491	5.9	406	27.2	417	87	582	7,274	697	9.6	5,247	165	1,128
Jul-Sep 2007	29,420	25,380	3,832	97	111	21,989	7,431	19,026	6,354	2,902	930	1,103	1,485	5.9	393	26.4	421	84	587	7,284	694	9.5	5,258	169	1,127
Aug-Oct 2007	29,470	25,422	3,841	97	111	22,038	7,433	19,067	6,355	2,911	930	1,115	1,462	5.8	385	26.3	422	84	571	7,285	701	9.6	5,256	168	1,123
Sep-Nov 2007	29,527	25,471	3,843	103	110	22,054	7,472	19,086	6,385	2,901	942	1,116	1,478	5.8	380	25.7	432	83	583	7,326	698	9.5	5,287	173	1,136
Oct-Dec 2007	29,576	25,521	3,834	106	114	22,075	7,501	19,104	6,417	2,899	936	1,112	1,495	5.9	380	25.4	449	81	585	7,353	729	9.9	5,261	170	1,152
Nov-Jan 2008	29,614	25,546	3,847	109	112	22,083	7,531	19,108	6,438	2,901	947	1,125	1,473	5.8	367	24.9	440	78	588	7,385	739	10.0	5,273	174	1,160
Dec-Feb 2008	29,676	25,557	3,892	110	117	22,130	7,546	19,100	6,457	2,955	937	1,103	1,445	5.7	363	25.1	428	83	572	7,394	728	9.8	5,276	177	1,170
Jan-Mar 2008	29,684	25,582	3,878	109	115	22,134	7,549	19,123	6,459	2,946	932	1,121	1,430	5.6	363	25.4	426	83	558	7,391	705	9.5	5,290	184	1,168
Feb-Apr 2008	29,706	25,623	3,851	115	117	22,160	7,546	19,150	6,473	2,939	911	1,121	1,439	5.6	358	24.9	431	86	564	7,385	696	9.4	5,275	187	1,183
Mar-May 2008	29,749	25,657	3,856	116	119	22,227	7,522	19,216	6,441	2,941	915	1,116	1,420	5.5	357	25.1	404	85	574	7,356	671	9.1	5,281	193	1,165
Apr-Jun 2008	29,722	25,641	3,859	111	110	22,189	7,533	19,176	6,465	2,947	912	1,130	1,396	5.4	348	24.9	404	85	560	7,377	679	9.2	5,283	203	1,168
May-Jul 2008	29,696	25,629	3,854	101	112	22,170	7,526	19,166	6,463	2,942	912	1,131	1,385	5.4	351	25.3	391	88	555	7,375	689	9.3	5,267	215	1,162
Jun-Aug 2008	29,612	25,591	3,826	90	105	22,081	7,531	19,117	6,473	2,913	913	1,120	1,383	5.4	353	25.5	404	83	543	7,387	702	9.5	5,292	216	1,138
Jul-Sep 2008	29,580	25,566	3,818	90	107	22,061	7,519	19,097	6,469	2,915	903	1,126	1,390	5.4	360	25.9	402	88	540	7,372	715	9.7	5,280	207	1,130
Aug-Oct 2008	29,535	25,545	3,794	91	105	22,010	7,525	19,075	6,469	2,886	908	1,127	1,370	5.4	351	25.6	407	83	530	7,377	732	9.9	5,279	197	1,127
Sep-Nov 2008	29,556	25,532	3,830	91	103	21,959	7,597	19,016	6,516	2,900	930	1,145	1,403	5.5	364	26.0	399	84	555	7,447	753	10.1	5,312	199	1,138
Oct-Dec 2008	29,528	25,507	3,828	94	100	21,951	7,578	19,007	6,499	2,898	930	1,142	1,411	5.5	380	26.9	392	86	554	7,429	770	10.4	5,282	198	1,133
Nov-Jan 2009	29,539	25,508	3,845	88	99	21,952	7,587	18,994	6,514	2,918	927	1,152	1,425	5.6	397	27.8	390	91	546	7,441	815	11.0	5,256	191	1,134
Dec-Feb 2009	29,429	25,416	3,826	85	102	21,843	7,586	18,867	6,549	2,919	907	1,155	1,427	5.6	420	29.5	386	84	537	7,457	850	11.4	5,258	183	1,123
Jan-Mar 2009	29,366	25,335	3,844	87	101	21,768	7,598	18,794	6,540	2,918	926	1,161	1,428	5.6	426	29.8	386	81	536	7,466	875	11.7	5,257	184	1,111
Feb-Apr 2009	29,272	25,220	3,860	88	103	21,715	7,557	18,747	6,474	2,911	949	1,156	1,417	5.6	419	29.6	381	87	530	7,423	902	12.1	5,199	190	1,090
Mar-May 2009	29,155	25,096	3,856	101	102	21,585	7,570	18,641	6,455	2,892	964	1,143	1,404	5.6	417	29.7	384	92	512	7,419	936	12.6	5,163	191	1,087
Apr-Jun 2009	29,087	25,043	3,843	98	104	21,479	7,609	18,553	6,490	2,869	974	1,126	1,430	5.7	431	30.2	386	89	524	7,464	963	12.9	5,166	189	1,094
May-Jul 2009	29,018	24,962	3,858	91	107	21,412	7,606	18,462	6,500	2,895	963	1,121	1,429	5.7	443	31.0	371	84	531	7,463	972	13.0	5,175	181	1,086
Jun-Aug 2009	29,076	24,987	3,894	84	110	21,446	7,630	18,466	6,521	2,923	971	1,139	1,435	5.7	446	31.1	376	89	524	7,492	981	13.1	5,176	184	1,099
Jul-Sep 2009	29,069	25,002	3,881	78	109	21,388	7,681	18,434	6,568	2,903	978	1,143	1,450	5.8	464	32.0	377	85	525	7,546	1,001	13.3	5,195	189	1,115
Aug-Oct 2009	29,084	25,019	3,881	83	102	21,349	7,734	18,392	6,627	2,909	971	1,146	1,438	5.7	468	32.5	369	86	516	7,598	1,017	13.4	5,200	190	1,148
Sep-Nov 2009	29,092	25,013	3,896	76	107	21,344	7,748	18,401	6,611	2,898	998	1,129	1,437	5.7	490	34.1	365	76	506	7,609	1,038	13.6	5,199	183	1,144
Oct-Dec 2009	29,102	24,988	3,912	87	115	21,349	7,753	18,382	6,606	2,917	995	1,129	1,445	5.8	497	34.4	362	78	508	7,601	1,038	13.7	5,196	187	1,138
Nov-Jan 2010	29,057	24,947	3,906	86	119	21,291	7,766	18,335	6,612	2,906	999	1,094	1,451	5.8	499	34.4	370	77	505	7,612	1,045	13.7	5,197	185	1,141

4.2 Full-time, part-time and temporary workers

United Kingdom (thousands) seasonally adjusted

People (16+)	All in employment					Full-time and part-time workers[1]							Temporary employees (reasons for temporary working)							Part-time workers (reasons for working part-time)[3]					
	Total	Employees	Self employed	Unpaid family workers	Government supported training & employment programmes[2]	Total people working full-time	Total people working part-time	Employees working full-time	Employees working part-time	Self-employed people working full-time	Self-employed people working part-time	Total workers with second jobs	Total	Total as % of all employees	Could not find permanent job	% that could not find permanent job	Did not want permanent job	Had a contract with period of training	Some other reason	Total[4]	Could not find full-time job	% that could not find full-time job	Did not want full-time job	Ill or disabled	Student or at school
	MGSA	MGRO	MGRR	MGRU	MGRX	YCBF	YCBI	YCBL	YCBO	YCBR	YCBU	YCBX	YCCA	YCCD	YCCG	YCCJ	YCCM	YCCP	YCCS	YCCV	YCCY	YCDB	YCDE	YCDH	YCDK
Dec-Feb 2010	29,024	24,902	3,909	92	122	21,281	7,743	18,320	6,583	2,913	996	1,081	1,481	5.9	513	34.6	366	83	519	7,578	1,051	13.9	5,160	179	1,148
Jan-Mar 2010	29,013	24,844	3,954	90	124	21,234	7,778	18,247	6,598	2,940	1,014	1,067	1,482	6.0	513	34.6	366	79	524	7,612	1,071	14.1	5,187	168	1,140
Feb-Apr 2010	29,048	24,857	3,973	91	126	21,221	7,827	18,221	6,635	2,949	1,024	1,097	1,501	6.0	539	35.9	361	82	519	7,659	1,090	14.2	5,220	172	1,134
Mar-May 2010	29,144	24,960	3,959	93	131	21,265	7,878	18,278	6,683	2,932	1,027	1,134	1,541	6.2	554	36.0	375	80	532	7,710	1,073	13.9	5,280	167	1,148
Apr-Jun 2010	29,192	25,019	3,949	90	133	21,301	7,891	18,315	6,704	2,928	1,021	1,136	1,578	6.3	573	36.3	380	82	543	7,725	1,076	13.9	5,303	172	1,134
May-Jul 2010	29,325	25,094	3,989	103	139	21,335	7,991	18,323	6,771	2,949	1,040	1,134	1,577	6.3	571	36.2	383	80	543	7,811	1,115	14.3	5,311	171	1,174
Jun-Aug 2010	29,339	25,105	3,993	109	132	21,314	8,025	18,291	6,814	2,962	1,031	1,121	1,578	6.3	591	37.5	369	86	532	7,845	1,138	14.5	5,302	170	1,188
Jul-Sep 2010	29,385	25,106	4,046	105	129	21,341	8,044	18,294	6,812	2,990	1,055	1,121	1,588	6.3	600	37.8	370	87	531	7,867	1,153	14.7	5,310	165	1,190
Aug-Oct 2010	29,308	25,101	3,980	99	127	21,289	8,019	18,306	6,795	2,932	1,048	1,114	1,598	6.4	592	37.0	374	92	540	7,843	1,174	15.0	5,311	164	1,142
Sep-Nov 2010	29,284	25,056	4,003	94	131	21,282	8,001	18,285	6,771	2,944	1,059	1,114	1,575	6.3	589	37.4	354	94	538	7,830	1,178	15.0	5,342	172	1,086
Oct-Dec 2010	29,324	25,098	4,009	93	125	21,326	7,999	18,337	6,761	2,938	1,071	1,121	1,546	6.2	577	37.3	340	95	534	7,832	1,192	15.2	5,321	164	1,102
Nov-Jan 2011	29,391	25,138	4,025	100	129	21,387	8,004	18,406	6,732	2,929	1,080	1,151	1,567	6.2	579	37.0	341	93	553	7,828	1,180	15.1	5,349	164	1,080
Dec-Feb 2011	29,442	25,192	4,026	98	126	21,433	8,009	18,472	6,720	2,916	1,096	1,174	1,577	6.3	573	36.3	352	95	557	7,830	1,180	15.1	5,354	172	1,065
Jan-Mar 2011	29,441	25,235	3,983	98	125	21,437	8,005	18,507	6,728	2,883	1,100	1,166	1,595	6.3	579	36.3	357	100	558	7,828	1,188	15.2	5,338	182	1,063
Feb-Apr 2011	29,436	25,217	4,009	93	117	21,414	8,021	18,474	6,743	2,893	1,116	1,153	1,605	6.3	583	36.4	356	94	568	7,859	1,230	15.6	5,321	185	1,069
Mar-May 2011	29,466	25,249	4,022	91	104	21,476	8,016	18,519	6,731	2,891	1,131	1,151	1,605	6.4	582	36.3	363	92	568	7,862	1,259	16.0	5,312	180	1,056
Apr-Jun 2011	29,447	25,247	4,008	96	96	21,452	7,971	18,527	6,720	2,906	1,102	1,136	1,616	6.4	609	37.7	361	88	558	7,822	1,270	16.2	5,268	180	1,047
May-Jul 2011	29,345	25,162	4,001	94	88	21,452	7,893	18,502	6,660	2,915	1,086	1,142	1,568	6.2	578	36.9	354	85	551	7,746	1,287	16.6	5,186	182	1,032
Jun-Aug 2011	29,329	25,069	4,049	100	81	21,441	7,858	18,453	6,616	2,951	1,099	1,123	1,532	6.1	583	38.1	317	82	550	7,715	1,278	16.6	5,179	181	1,026
Jul-Sep 2011	29,281	24,972	4,121	105	83	21,408	7,873	18,387	6,586	2,978	1,142	1,143	1,524	6.1	581	38.1	326	86	532	7,728	1,275	16.5	5,177	190	1,034
Aug-Oct 2011	29,300	24,948	4,154	110	88	21,401	7,900	18,326	6,622	3,021	1,133	1,142	1,539	6.2	607	39.4	326	86	521	7,754	1,286	16.6	5,194	184	1,041
Sep-Nov 2011	29,325	24,965	4,150	114	97	21,381	7,945	18,309	6,656	3,009	1,141	1,146	1,554	6.2	593	38.2	350	90	521	7,797	1,333	17.1	5,187	184	1,048
Oct-Dec 2011	29,342	25,016	4,118	112	97	21,365	7,977	18,317	6,699	2,992	1,125	1,126	1,546	6.2	602	39.0	345	92	507	7,824	1,363	17.4	5,188	188	1,042
Nov-Jan 2012	29,346	25,003	4,127	111	106	21,364	7,982	18,316	6,686	2,995	1,132	1,116	1,560	6.2	611	39.1	343	97	509	7,819	1,395	17.8	5,168	188	1,025
Dec-Feb 2012	29,381	25,007	4,156	99	120	21,345	8,036	18,298	6,708	2,990	1,166	1,147	1,589	6.4	630	39.6	351	97	512	7,874	1,406	17.9	5,191	181	1,048
Jan-Mar 2012	29,454	25,042	4,186	96	129	21,372	8,082	18,319	6,724	2,992	1,194	1,152	1,569	6.3	618	39.4	341	90	521	7,918	1,414	17.9	5,215	191	1,048
Feb-Apr 2012	29,492	25,064	4,195	98	134	21,410	8,081	18,340	6,725	3,010	1,185	1,149	1,557	6.2	610	39.2	333	95	518	7,910	1,413	17.9	5,214	192	1,041
Mar-May 2012	29,564	25,120	4,185	111	148	21,475	8,089	18,381	6,739	3,019	1,166	1,141	1,575	6.3	638	40.5	322	94	521	7,905	1,398	17.7	5,222	191	1,045
Apr-Jun 2012	29,667	25,187	4,222	112	146	21,513	8,154	18,402	6,785	3,032	1,190	1,124	1,615	6.4	644	39.9	331	101	539	7,975	1,427	17.9	5,244	184	1,071
May-Jul 2012	29,744	25,222	4,252	117	153	21,542	8,202	18,414	6,808	3,036	1,216	1,125	1,651	6.5	659	39.9	346	101	545	8,023	1,429	17.8	5,298	181	1,067
Jun-Aug 2012	29,779	25,280	4,232	112	155	21,566	8,213	18,464	6,816	3,013	1,220	1,117	1,637	6.5	648	39.6	343	100	547	8,036	1,418	17.6	5,315	182	1,072
Jul-Sep 2012	29,759	25,255	4,232	110	161	21,568	8,191	18,456	6,800	3,016	1,216	1,109	1,625	6.4	656	40.3	328	93	548	8,015	1,412	17.6	5,316	182	1,054
Aug-Oct 2012	29,744	25,244	4,224	109	167	21,560	8,185	18,440	6,804	3,024	1,200	1,124	1,633	6.5	657	40.2	325	93	559	8,004	1,405	17.6	5,318	185	1,039
Sep-Nov 2012	29,849	25,336	4,236	113	164	21,665	8,184	18,536	6,800	3,035	1,201	1,122	1,651	6.5	655	39.6	339	90	568	8,001	1,406	17.6	5,329	182	1,022
Oct-Dec 2012	29,908	25,369	4,258	115	166	21,736	8,172	18,590	6,779	3,049	1,209	1,146	1,656	6.5	660	39.8	339	97	560	7,988	1,392	17.4	5,337	184	1,010
Nov-Jan 2013	29,891	25,414	4,206	107	164	21,753	8,138	18,643	6,771	3,008	1,198	1,138	1,640	6.5	656	40.0	335	95	554	7,969	1,405	17.6	5,331	185	989
Dec-Feb 2013	29,815	25,339	4,222	102	153	21,699	8,116	18,590	6,749	3,012	1,210	1,121	1,602	6.3	643	40.1	334	101	523	7,959	1,422	17.9	5,302	183	995

4.2 Full-time, part-time and temporary workers

United Kingdom (thousands) seasonally adjusted

People (16+)	All in employment: Total (MGSA)	Employees (MGRO)	Self employed (MGRR)	Unpaid family workers (MGRU)	Government supported training & employment programmes[2] (MGRX)	Total people working full-time (YCBF)	Total people working part-time (YCBI)	Employees working full-time (YCBL)	Employees working part-time (YCBO)	Self-employed people working full-time (YCBR)	Self-employed people working part-time (YCBU)	Total workers with second jobs (YCBX)	Temporary: Total (YCCA)	Total as % of all employees (YCCD)	Could not find permanent job (YCCG)	% that could not find permanent job (YCCJ)	Did not want permanent job (YCCM)	Had a contract with period of training (YCCP)	Some other reason (YCCS)	Part-time: Total[4] (YCCV)	Could not find full-time job (YCCY)	% that could not find full-time job (YCDB)	Did not want full-time job (YCDE)	Ill or disabled (YCDH)	Student or at school (YCDK)
Jan-Mar 2013	29,839	25,398	4,185	102	154	21,734	8,105	18,623	6,774	3,018	1,167	1,111	1,612	6.3	649	40.3	337	104	522	7,942	1,418	17.9	5,278	178	1,014
Feb-Apr 2013	29,884	25,400	4,219	105	160	21,723	8,160	18,612	6,788	3,028	1,191	1,100	1,592	6.3	631	39.7	343	100	517	7,978	1,443	18.1	5,269	182	1,029
Mar-May 2013	29,861	25,411	4,185	103	162	21,725	8,136	18,638	6,773	3,012	1,173	1,122	1,584	6.2	626	39.5	331	101	525	7,946	1,466	18.4	5,223	181	1,016
Apr-Jun 2013	29,938	25,462	4,196	115	165	21,769	8,169	18,672	6,790	3,013	1,184	1,124	1,569	6.2	605	38.5	329	104	531	7,974	1,455	18.2	5,259	189	1,010
May-Jul 2013	29,994	25,519	4,204	108	163	21,867	8,127	18,746	6,773	3,033	1,171	1,132	1,564	6.1	604	38.6	312	112	536	7,944	1,464	18.4	5,223	186	1,009
Jun-Aug 2013	29,994	25,501	4,243	117	164	21,893	8,131	18,731	6,769	3,066	1,176	1,157	1,573	6.2	604	38.4	300	123	547	7,945	1,460	18.4	5,263	193	968
Jul-Sep 2013	30,024	25,501	4,244	113	171	21,944	8,162	18,786	6,794	3,066	1,178	1,160	1,602	6.2	603	37.6	306	126	566	7,972	1,462	18.3	5,258	194	993
Aug-Oct 2013	30,106	25,579	4,297	119	169	21,997	8,226	18,813	6,824	3,095	1,203	1,170	1,598	6.2	596	37.3	318	124	560	8,027	1,462	18.2	5,305	195	1,003
Sep-Nov 2013	30,223	25,638	4,409	115	147	22,111	8,191	18,854	6,777	3,177	1,232	1,184	1,606	6.3	583	36.3	322	129	571	8,009	1,445	18.0	5,296	191	1,022
Oct-Dec 2013	30,302	25,631	4,424	111	141	22,140	8,151	18,862	6,753	3,192	1,232	1,176	1,623	6.3	602	37.1	334	118	568	7,984	1,430	17.9	5,289	190	1,025
Nov-Jan 2014	30,290	25,615	4,501	110	132	22,165	8,164	18,836	6,749	3,247	1,254	1,174	1,620	6.3	594	36.7	349	115	562	8,004	1,436	17.9	5,298	192	1,027
Dec-Feb 2014	30,329	25,586	4,538	115	130	22,243	8,249	18,887	6,822	3,270	1,268	1,157	1,629	6.3	586	36.0	361	116	567	8,090	1,421	17.6	5,397	193	1,032
Jan-Mar 2014	30,492	25,709	4,579	123	123	22,277	8,255	18,919	6,787	3,272	1,307	1,180	1,653	6.4	591	35.7	365	117	580	8,094	1,419	17.5	5,421	192	1,020
Feb-Apr 2014	30,532	25,706	4,574	124	121	22,376	8,264	19,022	6,799	3,270	1,305	1,184	1,681	6.5	600	35.7	379	119	582	8,104	1,405	17.3	5,440	188	1,032
Mar-May 2014	30,640	25,821	4,599	118	120	22,451	8,291	19,102	6,804	3,268	1,331	1,192	1,680	6.5	606	36.1	377	119	579	8,134	1,367	16.8	5,506	194	1,032
Apr-Jun 2014	30,743	25,906	4,606	114	127	22,400	8,305	19,066	6,799	3,262	1,344	1,211	1,653	6.4	603	36.5	378	113	559	8,143	1,347	16.5	5,544	195	1,016
May-Jul 2014	30,705	25,858	4,568	128	131	22,395	8,308	19,171	6,820	3,251	1,317	1,201	1,666	6.4	609	36.5	371	120	567	8,138	1,342	16.5	5,559	181	1,013
Jun-Aug 2014	30,786	25,886	4,527	118	128	22,507	8,279	19,238	6,833	3,258	1,269	1,222	1,685	6.5	610	36.2	386	120	570	8,102	1,349	16.7	5,518	174	1,017
Jul-Sep 2014	30,833	26,004	4,530	122	123	22,566	8,267	19,258	6,820	3,258	1,272	1,203	1,702	6.5	599	35.2	410	123	570	8,092	1,336	16.5	5,524	172	1,016
Aug-Oct 2014	30,848	26,058	4,543	118	121	22,588	8,261	19,263	6,808	3,260	1,283	1,209	1,694	6.5	572	33.8	421	122	578	8,091	1,309	16.2	5,564	175	994
Sep-Nov 2014	30,850	26,066	4,539	103	114	22,565	8,285	19,263	6,832	3,238	1,302	1,200	1,683	6.5	569	33.8	414	121	579	8,133	1,314	16.2	5,577	187	1,008
Oct-Dec 2014	30,944	26,094	4,516	103	110	22,632	8,312	19,359	6,857	3,209	1,307	1,200	1,716	6.5	581	33.8	412	115	609	8,164	1,310	16.1	5,587	189	1,029
Nov-Jan 2015	31,003	26,216	4,542	113	104	22,691	8,312	19,400	6,844	3,223	1,318	1,194	1,681	6.4	573	34.1	394	121	593	8,162	1,324	16.2	5,594	195	1,006
Dec-Feb 2015	31,122	26,244	4,550	113	105	22,745	8,378	19,466	6,888	3,212	1,337	1,219	1,685	6.4	591	35.1	379	120	595	8,226	1,354	16.5	5,616	196	1,013
Jan-Mar 2015	31,156	26,355	4,525	119	102	22,769	8,387	19,491	6,918	3,207	1,318	1,222	1,681	6.4	585	34.8	374	123	599	8,236	1,325	16.1	5,641	213	1,011
Feb-Apr 2015	31,129	26,409	4,495	110	104	22,783	8,346	19,509	6,912	3,194	1,301	1,221	1,681	6.4	583	34.7	374	125	599	8,212	1,308	15.9	5,626	213	1,015
Mar-May 2015	31,057	26,420	4,473	103	97	22,756	8,302	19,514	6,870	3,157	1,316	1,199	1,671	6.3	559	33.5	398	120	594	8,185	1,280	15.6	5,620	217	1,015
Apr-Jun 2015	31,110	26,384	4,518	101	106	22,793	8,317	19,514	6,872	3,195	1,324	1,195	1,642	6.2	566	34.5	382	129	566	8,195	1,291	15.7	5,615	224	1,018
May-Jul 2015	31,188	26,385	4,530	106	111	22,801	8,387	19,545	6,901	3,183	1,347	1,187	1,666	6.3	561	33.7	400	126	578	8,248	1,277	15.5	5,653	229	1,039
Jun-Aug 2015	31,220	26,447	4,513	111	111	22,855	8,366	19,614	6,889	3,170	1,343	1,164	1,644	6.2	579	35.2	399	128	538	8,232	1,268	15.4	5,643	221	1,051
Jul-Sep 2015	31,333	26,503	4,576	111	108	22,894	8,439	19,593	6,961	3,227	1,349	1,162	1,668	6.3	571	34.2	411	127	559	8,310	1,255	15.1	5,715	215	1,077
Aug-Oct 2015	31,436	26,554	4,631	108	101	22,990	8,446	19,661	6,956	3,268	1,363	1,163	1,656	6.2	579	35.0	405	119	552	8,319	1,277	15.3	5,676	218	1,100
Sep-Nov 2015	31,513	26,616	4,653	94	92	23,053	8,460	19,722	6,951	3,273	1,381	1,165	1,647	6.2	570	34.6	392	137	549	8,332	1,240	14.9	5,726	225	1,092
Oct-Dec 2015	31,540	26,673	4,680	88	98	23,081	8,459	19,719	6,947	3,302	1,378	1,142	1,634	6.1	554	33.9	400	137	544	8,325	1,236	14.8	5,706	233	1,089
Nov-Jan 2016	31,541	26,666	4,662	95	93	23,035	8,506	19,676	7,013	3,299	1,363	1,125	1,655	6.2	554	33.5	409	138	554	8,377	1,213	14.5	5,765	237	1,103
Dec-Feb 2016	31,555	26,689	4,688	95	97	23,076	8,479	19,685	6,983	3,323	1,364	1,124	1,648	6.2	542	32.9	413	123	571	8,347	1,190	14.3	5,761	248	1,088
Jan-Mar 2016	31,572	26,665	4,714	100	93	23,094	8,478	19,685	6,980	3,345	1,369	1,120	1,648	6.2	561	34.0	410	116	560	8,349	1,198	14.3	5,745	251	1,096

4.2 Full-time, part-time and temporary workers

United Kingdom (thousands) seasonally adjusted

People (16+)	All in employment					Full-time and part-time workers[1]							Temporary employees (reasons for temporary working)							Part-time workers (reasons for working part-time)[3]					
	Total	Employees	Self employed	Unpaid family workers	Government supported training & employment programmes[2]	Total people working full-time	Total people working part-time	Employees working full-time	Employees working part-time	Self-employed people working full-time	Self-employed people working part-time	Total workers with second jobs	Total	Total as % of all employees	Could not find permanent job	% that could not find permanent job	Did not want permanent job	Had a contract with period of training	Some other reason	Total[4]	Could not find full-time job	% that could not find full-time job	Did not want full-time job	Ill or disabled	Student or at school
	MGSA	MGRO	MGRR	MGRU	MGRX	YCBF	YCBI	YCBL	YCBO	YCBR	YCBU	YCBX	YCCA	YCCD	YCCG	YCCJ	YCCM	YCCP	YCCS	YCCV	YCCY	YCDB	YCDE	YCDH	YCDK
Feb-Apr 2016	31,603	26,693	4,706	110	94	23,091	8,513	19,669	7,024	3,360	1,346	1,127	1,633	6.1	536	32.8	429	128	540	8,370	1,194	14.3	5,779	243	1,094
Mar-May 2016	31,712	26,726	4,778	118	91	23,173	8,539	19,714	7,012	3,400	1,377	1,138	1,615	6.0	511	31.7	428	132	544	8,389	1,187	14.2	5,823	238	1,081
Apr-Jun 2016	31,747	26,757	4,785	120	85	23,183	8,564	19,729	7,028	3,396	1,389	1,131	1,666	6.2	529	31.8	433	135	569	8,417	1,154	13.7	5,870	239	1,100
May-Jul 2016	31,779	26,834	4,754	110	82	23,255	8,524	19,801	7,033	3,396	1,358	1,135	1,651	6.2	524	31.7	430	133	564	8,391	1,149	13.7	5,874	246	1,075
Jun-Aug 2016	31,811	26,829	4,786	124	72	23,234	8,576	19,753	7,077	3,427	1,359	1,167	1,643	6.1	506	30.8	431	129	576	8,436	1,139	13.5	5,915	249	1,094
Jul-Sep 2016	31,810	26,826	4,787	128	68	23,254	8,557	19,793	7,033	3,402	1,386	1,154	1,621	6.0	483	29.8	429	132	578	8,419	1,144	13.6	5,899	252	1,085
Aug-Oct 2016	31,794	26,833	4,762	128	70	23,235	8,559	19,816	7,017	3,353	1,409	1,131	1,614	6.0	489	30.3	426	123	576	8,427	1,151	13.7	5,916	249	1,067
Sep-Nov 2016	31,808	26,822	4,782	128	75	23,262	8,546	19,799	7,023	3,391	1,392	1,120	1,622	6.0	493	30.4	421	111	597	8,415	1,143	13.6	5,887	256	1,078
Oct-Dec 2016	31,845	26,836	4,803	123	83	23,300	8,545	19,807	7,029	3,421	1,382	1,121	1,621	6.0	473	29.1	411	112	626	8,411	1,116	13.3	5,885	257	1,097
Nov-Jan 2017	16,926	13,586	3,229	58	53	14,669	2,258	11,950	1,636	2,670	559	483	729	5.4	226	31.1	168	48	287	2,195	471	21.4	1,129	101	473
Dec-Feb 2017	16,922	13,601	3,216	52	53	14,683	2,239	11,971	1,630	2,666	549	478	735	5.4	219	29.8	179	49	288	2,179	443	20.3	1,131	102	479
Jan-Mar 2017	16,960	13,636	3,214	53	57	14,722	2,237	12,008	1,628	2,665	549	475	726	5.3	222	30.5	182	46	277	2,177	440	20.2	1,153	99	462
Feb-Apr 2017	16,963	13,648	3,220	47	49	14,716	2,247	12,020	1,627	2,658	562	474	731	5.4	217	29.7	182	47	284	2,189	451	20.6	1,176	96	445
Mar-May 2017	16,984	13,679	3,223	42	40	14,727	2,257	12,036	1,643	2,656	567	483	733	5.4	216	29.5	190	52	276	2,210	440	19.9	1,202	90	458
Apr-Jun 2017	17,010	13,697	3,228	46	39	14,772	2,238	12,076	1,621	2,660	568	488	738	5.4	218	29.6	201	52	267	2,189	439	20.1	1,177	91	458
May-Jul 2017	17,010	13,688	3,235	47	40	14,766	2,244	12,075	1,613	2,651	584	485	744	5.4	219	29.5	203	53	268	2,198	454	20.7	1,159	93	466
Jun-Aug 2017	17,002	13,681	3,236	48	37	14,757	2,245	12,073	1,608	2,645	591	478	747	5.5	223	29.8	203	59	263	2,200	446	20.3	1,151	104	472
Jul-Sep 2017	16,978	13,670	3,224	47	37	14,696	2,282	12,031	1,639	2,628	596	496	746	5.5	228	30.5	194	57	267	2,235	453	20.3	1,189	98	469
Aug-Oct 2017	16,967	13,666	3,211	51	40	14,722	2,246	12,064	1,602	2,623	588	496	751	5.5	229	30.5	197	68	257	2,190	437	19.9	1,165	94	467
Sep-Nov 2017	17,081	13,803	3,180	60	37	14,814	2,267	12,172	1,632	2,605	575	497	743	5.4	228	30.7	190	76	249	2,207	454	20.5	1,168	92	464
Oct-Dec 2017	17,035	13,765	3,169	65	36	14,788	2,246	12,141	1,624	2,611	558	494	740	5.4	230	31.0	192	79	240	2,182	443	20.3	1,160	86	465

Source: Labour Force Survey

Note: When comparing quarterly changes ONS recommends comparing with the previous non-overlapping 3-month average time period (eg, compare Apr-Jun with Jan-Mar, not with Mar-May).

1. The split between full-time and part-time employment is based on respondents' self-classification.
2. This series does not include all people on these programmes; it only includes those engaging in any form of work, work experience or work-related training.
3. These series cover Employees and Self-employed only. These series include some temporary employees.
4. The total includes those who did not give a reason for working part-time and it therefore does not equal the sum of the other columns in this section of the table.

4.2 Full-time, part-time and temporary workers

United Kingdom (thousands) seasonally adjusted

Men (16+)	All in employment					Full-time and part-time workers[1]							Temporary employees (reasons for temporary working)							Part-time workers (reasons for working part-time)[3]					
	Total	Employees	Self employed	Unpaid family workers	Government supported training & employment programmes[2]	Total people working full-time	Total people working part-time	Employees working full-time	Employees working part-time	Self-employed people working full-time	Self-employed people working part-time	Total workers with second jobs	Total	Total as % of all employees	Could not find permanent job	% that could not find permanent job	Did not want permanent job	Had a contract with period of training	Some other reason	Total[4]	Could not find full-time job	% that could not find full-time job	Did not want full-time job	Ill or disabled	Student or at school
	MGSA	MGRO	MGRR	MGRU	MGRX	YCBF	YCBI	YCBL	YCBO	YCBR	YCBU	YCBX	YCCA	YCCD	YCCG	YCCJ	YCCM	YCCP	YCCS	YCCV	YCCY	YCDB	YCDE	YCDH	YCDK
Nov-Jan 2007	15,760	12,886	2,773	39	63	14,064	1,696	11,606	1,280	2,416	356	447	712	5.5	209	29.3	187	56	260	1,636	281	17.2	770	83	491
Dec-Feb 2007	15,752	12,885	2,768	40	58	14,052	1,700	11,609	1,276	2,403	366	461	703	5.5	212	30.2	185	51	255	1,642	267	16.2	787	82	495
Jan-Mar 2007	15,758	12,885	2,775	43	56	14,034	1,724	11,597	1,288	2,399	376	463	713	5.5	215	30.1	183	53	261	1,664	274	16.5	798	81	500
Feb-Apr 2007	15,782	12,913	2,772	40	57	14,064	1,718	11,635	1,278	2,392	380	461	696	5.4	211	30.3	177	48	260	1,658	273	16.5	799	80	496
Mar-May 2007	15,846	12,979	2,771	38	59	14,134	1,712	11,703	1,275	2,393	377	457	696	5.4	219	31.4	174	43	261	1,653	277	16.8	804	76	487
Apr-Jun 2007	15,839	12,965	2,775	36	62	14,126	1,713	11,696	1,269	2,393	382	456	693	5.3	223	32.2	172	40	258	1,652	275	16.7	821	75	469
May-Jul 2007	15,846	12,969	2,771	37	68	14,120	1,726	11,705	1,264	2,378	393	460	689	5.3	219	31.8	177	39	255	1,658	281	17.0	832	79	453
Jun-Aug 2007	15,861	12,991	2,771	38	61	14,137	1,724	11,714	1,278	2,388	383	456	694	5.3	218	31.4	185	38	252	1,660	290	17.5	828	70	460
Jul-Sep 2007	15,884	13,027	2,760	37	60	14,141	1,742	11,733	1,294	2,374	386	450	682	5.2	206	30.2	189	37	249	1,680	293	17.4	831	72	471
Aug-Oct 2007	15,903	13,036	2,774	37	56	14,176	1,726	11,754	1,282	2,390	384	458	668	5.1	202	30.3	187	38	241	1,666	292	17.5	817	71	474
Sep-Nov 2007	15,942	13,061	2,783	37	61	14,181	1,762	11,755	1,306	2,385	398	454	684	5.2	203	29.6	192	42	247	1,704	288	16.9	834	72	498
Oct-Dec 2007	15,947	13,057	2,787	39	63	14,182	1,766	11,741	1,317	2,395	392	450	696	5.3	200	28.8	193	45	258	1,709	291	17.0	836	71	498
Nov-Jan 2008	15,968	13,074	2,789	39	66	14,180	1,788	11,733	1,341	2,398	391	453	685	5.2	188	27.4	189	43	265	1,732	292	16.9	858	69	501
Dec-Feb 2008	15,993	13,054	2,825	40	74	14,197	1,796	11,714	1,340	2,435	390	461	663	5.1	181	27.4	181	43	258	1,730	292	16.9	857	69	498
Jan-Mar 2008	16,005	13,070	2,822	40	73	14,209	1,796	11,739	1,331	2,427	395	451	657	5.0	185	28.1	182	39	251	1,726	289	16.7	860	69	494
Feb-Apr 2008	16,024	13,106	2,805	42	70	14,229	1,795	11,769	1,337	2,415	391	458	656	5.0	179	27.3	183	42	252	1,727	287	16.6	860	73	492
Mar-May 2008	16,019	13,101	2,809	40	68	14,254	1,764	11,792	1,309	2,419	390	458	648	4.9	178	27.5	165	45	259	1,699	268	15.8	859	76	479
Apr-Jun 2008	16,011	13,097	2,810	40	64	14,221	1,790	11,758	1,340	2,423	387	461	628	4.8	176	28.1	159	46	247	1,727	271	15.7	864	78	499
May-Jul 2008	15,987	13,089	2,798	36	64	14,178	1,809	11,726	1,363	2,415	383	451	611	4.7	182	29.7	143	47	239	1,746	290	16.6	858	82	503
Jun-Aug 2008	15,927	13,059	2,775	30	64	14,105	1,822	11,685	1,374	2,389	386	451	615	4.7	180	29.2	154	44	238	1,760	303	17.2	858	83	504
Jul-Sep 2008	15,910	13,049	2,769	28	64	14,081	1,829	11,664	1,385	2,387	382	456	633	4.8	184	29.1	158	46	244	1,766	315	17.8	860	81	497
Aug-Oct 2008	15,864	13,029	2,736	33	67	14,024	1,840	11,639	1,390	2,355	381	452	627	4.8	177	28.2	170	42	239	1,772	322	18.2	857	80	498
Sep-Nov 2008	15,885	13,017	2,766	37	65	14,045	1,840	11,634	1,383	2,381	385	471	635	4.9	186	29.4	155	42	251	1,768	333	18.8	847	79	492
Oct-Dec 2008	15,870	12,991	2,774	41	64	14,012	1,858	11,598	1,393	2,386	388	473	639	4.9	195	30.5	152	42	250	1,780	353	19.9	840	80	490
Nov-Jan 2009	15,859	12,977	2,784	38	61	13,995	1,865	11,572	1,405	2,399	385	479	659	5.1	215	32.6	153	47	243	1,790	379	21.2	832	77	485
Dec-Feb 2009	15,786	12,921	2,771	33	61	13,935	1,851	11,504	1,417	2,399	372	483	662	5.1	225	34.1	158	38	240	1,789	385	21.5	830	78	481
Jan-Mar 2009	15,739	12,876	2,773	32	58	13,900	1,839	11,475	1,402	2,394	379	492	664	5.2	232	35.0	155	37	239	1,781	394	22.1	821	77	474
Feb-Apr 2009	15,678	12,807	2,779	33	59	13,841	1,836	11,417	1,391	2,391	388	500	662	5.2	228	34.4	155	40	238	1,779	405	22.8	812	79	464
Mar-May 2009	15,577	12,711	2,774	36	56	13,718	1,859	11,318	1,393	2,369	405	489	649	5.1	219	33.7	153	48	229	1,798	425	23.6	806	80	467
Apr-Jun 2009	15,513	12,658	2,756	40	59	13,646	1,866	11,268	1,390	2,341	415	480	676	5.3	231	34.1	156	47	243	1,805	438	24.3	793	79	472
May-Jul 2009	15,465	12,607	2,761	36	61	13,618	1,847	11,229	1,379	2,356	405	480	680	5.4	231	34.0	151	45	253	1,783	438	24.6	789	72	466
Jun-Aug 2009	15,471	12,608	2,770	33	61	13,616	1,856	11,206	1,402	2,375	395	491	679	5.4	235	34.7	151	47	246	1,797	432	24.0	802	70	472
Jul-Sep 2009	15,447	12,608	2,749	29	61	13,606	1,841	11,215	1,392	2,359	390	489	671	5.3	240	35.8	146	44	241	1,782	433	24.3	785	79	468
Aug-Oct 2009	15,448	12,588	2,769	33	58	13,591	1,857	11,189	1,400	2,373	396	494	672	5.3	245	36.5	140	49	237	1,796	435	24.2	778	79	486
Sep-Nov 2009	15,438	12,593	2,757	32	56	13,568	1,870	11,188	1,405	2,354	403	483	685	5.4	262	38.2	145	43	235	1,808	449	24.8	788	76	479
Oct-Dec 2009	15,436	12,570	2,770	37	60	13,558	1,879	11,168	1,402	2,363	407	485	690	5.5	268	38.8	146	42	236	1,809	447	24.7	799	77	471
Nov-Jan 2010	15,407	12,538	2,773	36	60	13,514	1,893	11,127	1,411	2,360	413	467	683	5.4	259	37.9	147	41	236	1,823	449	24.6	812	75	471

4.2 Full-time, part-time and temporary workers

United Kingdom (thousands)

	All in employment					Full-time and part-time workers[1]							Temporary employees (reasons for temporary working)							Part-time workers (reasons for working part-time)[3]					
Men (16+)	Total	Employ-ees	Self employed	Unpaid family workers	Government supported training & employment pro-grammes[2]	Total people working full-time	Total people working part-time	Employ-ees working full-time	Employ-ees working part-time	Self-employed people working full-time	Self-employed people working part-time	Total workers with second jobs	Total	Total as % of all employ-ees	Could not find perman-ent job	% that could not find perman-ent job	Did not want perman-ent job	Had a contract with period of training reason	Some other reason	Total[4]	Could not find full-time job	% that could not find full-time job	Did not want full-time job	Ill or dis-abled	Student or at school
	MGSA	MGRO	MGRR	MGRU	MGRX	YCBF	YCBI	YCBL	YCBO	YCBR	YCBU	YCBX	YCCA	YCCD	YCCG	YCCJ	YCCM	YCCP	YCCS	YCCV	YCCY	YCDB	YCDE	YCDH	YCDK
Dec-Feb 2010	15,404	12,539	2,763	41	63	13,511	1,893	11,129	1,409	2,354	408	457	709	5.7	273	38.5	149	46	241	1,817	457	25.2	800	72	474
Jan-Mar 2010	15,397	12,516	2,775	41	66	13,486	1,911	11,096	1,420	2,363	411	434	707	5.6	268	37.9	151	43	245	1,832	463	25.3	808	69	479
Feb-Apr 2010	15,417	12,527	2,780	42	69	13,501	1,917	11,103	1,425	2,368	412	452	715	5.7	283	39.6	151	42	239	1,836	463	25.2	807	68	487
Mar-May 2010	15,499	12,593	2,787	42	77	13,551	1,948	11,152	1,441	2,365	422	476	733	5.8	291	39.7	158	41	244	1,863	459	24.6	834	63	497
Apr-Jun 2010	15,539	12,633	2,788	41	77	13,567	1,972	11,165	1,468	2,368	420	472	760	6.0	304	40.0	163	40	253	1,888	468	24.8	847	70	492
May-Jul 2010	15,634	12,684	2,820	43	87	13,622	2,012	11,206	1,478	2,379	441	467	753	5.9	300	39.8	160	40	253	1,919	480	25.0	847	70	508
Jun-Aug 2010	15,654	12,690	2,839	42	83	13,617	2,036	11,182	1,508	2,396	442	458	760	6.0	314	41.3	157	43	246	1,950	492	25.2	851	77	515
Jul-Sep 2010	15,696	12,706	2,864	43	83	13,635	2,061	11,183	1,523	2,414	450	465	756	5.9	319	42.2	150	44	243	1,973	501	25.4	859	72	525
Aug-Oct 2010	15,658	12,726	2,808	43	81	13,590	2,068	11,202	1,524	2,357	451	459	761	6.0	314	41.3	156	48	242	1,975	522	26.4	867	72	500
Sep-Nov 2010	15,644	12,704	2,825	33	82	13,587	2,057	11,189	1,514	2,367	458	462	744	5.9	303	40.7	151	52	238	1,973	520	26.4	878	78	483
Oct-Dec 2010	15,679	12,736	2,834	34	75	13,623	2,057	11,217	1,519	2,375	459	469	718	5.6	289	40.3	147	53	228	1,977	519	26.3	882	73	488
Nov-Jan 2011	15,711	12,774	2,823	35	80	13,672	2,039	11,281	1,493	2,363	460	488	738	5.8	299	40.5	145	51	244	1,953	511	26.2	880	73	471
Dec-Feb 2011	15,717	12,786	2,820	33	77	13,695	2,022	11,322	1,464	2,350	470	509	741	5.8	296	40.0	145	51	249	1,935	513	26.5	871	76	454
Jan-Mar 2011	15,692	12,791	2,795	31	76	13,670	2,022	11,326	1,465	2,323	472	496	752	5.9	300	40.0	146	56	250	1,937	523	27.0	869	81	441
Feb-Apr 2011	15,742	12,817	2,827	28	70	13,689	2,053	11,323	1,494	2,345	481	488	769	6.0	314	40.8	145	51	259	1,975	540	27.3	880	83	452
Mar-May 2011	15,739	12,816	2,832	31	60	13,690	2,049	11,335	1,480	2,336	495	490	771	6.0	311	40.4	149	49	262	1,976	554	28.0	889	74	438
Apr-Jun 2011	15,728	12,795	2,838	38	56	13,687	2,041	11,316	1,479	2,350	488	488	792	6.2	329	41.6	151	46	265	1,967	564	28.7	877	71	437
May-Jul 2011	15,655	12,731	2,836	38	49	13,628	2,027	11,257	1,474	2,353	483	498	760	6.0	304	40.0	154	45	257	1,957	578	29.5	865	71	423
Jun-Aug 2011	15,621	12,688	2,847	40	46	13,647	1,974	11,256	1,432	2,372	475	476	746	5.9	308	41.3	135	42	260	1,907	568	29.8	841	67	415
Jul-Sep 2011	15,591	12,611	2,896	42	42	13,618	1,972	11,199	1,413	2,399	497	487	735	5.8	303	41.3	139	43	250	1,909	555	29.1	851	67	420
Aug-Oct 2011	15,610	12,607	2,918	41	44	13,620	1,990	11,159	1,448	2,433	485	494	737	5.8	313	42.5	134	41	249	1,933	563	29.1	850	68	433
Sep-Nov 2011	15,624	12,610	2,919	44	51	13,599	2,025	11,141	1,470	2,426	493	493	734	5.8	302	41.1	149	41	242	1,963	579	29.5	856	68	440
Oct-Dec 2011	15,632	12,648	2,889	42	53	13,599	2,033	11,161	1,487	2,406	483	480	728	5.8	308	42.4	142	42	236	1,970	595	30.2	852	69	437
Nov-Jan 2012	15,634	12,641	2,892	42	59	13,601	2,034	11,162	1,479	2,408	484	463	732	5.8	313	42.7	140	46	234	1,963	602	30.7	848	73	428
Dec-Feb 2012	15,686	12,677	2,905	37	66	13,589	2,096	11,147	1,530	2,407	498	488	737	5.8	326	44.3	135	46	231	2,029	626	30.9	860	73	456
Jan-Mar 2012	15,720	12,688	2,916	43	74	13,598	2,123	11,154	1,534	2,404	512	487	740	5.8	319	43.1	140	44	237	2,045	635	31.0	867	73	457
Feb-Apr 2012	15,754	12,701	2,930	44	79	13,621	2,132	11,159	1,542	2,423	507	488	739	5.8	311	42.0	138	47	243	2,049	641	31.3	867	75	450
Mar-May 2012	15,801	12,713	2,945	47	97	13,662	2,139	11,164	1,549	2,447	497	485	737	5.8	323	43.8	136	46	233	2,046	633	30.9	869	76	452
Apr-Jun 2012	15,866	12,761	2,970	43	92	13,702	2,165	11,186	1,575	2,465	506	485	753	5.9	325	43.1	137	50	241	2,081	651	31.3	875	73	462
May-Jul 2012	15,874	12,754	2,978	43	99	13,722	2,152	11,185	1,568	2,472	506	476	761	6.0	333	43.7	137	50	242	2,074	643	31.0	881	72	459
Jun-Aug 2012	15,903	12,785	2,970	46	102	13,743	2,159	11,228	1,557	2,453	518	470	755	5.9	325	43.1	135	49	245	2,075	631	30.4	888	75	458
Jul-Sep 2012	15,897	12,767	2,972	50	108	13,754	2,143	11,229	1,537	2,453	519	463	760	6.0	337	44.3	125	45	253	2,056	630	30.7	885	70	447
Aug-Oct 2012	15,906	12,772	2,971	47	116	13,779	2,127	11,249	1,522	2,456	515	464	767	6.0	342	44.5	122	45	258	2,037	631	31.0	889	69	426
Sep-Nov 2012	15,937	12,833	2,947	46	110	13,818	2,119	11,303	1,530	2,447	501	454	780	6.1	346	44.4	131	40	263	2,030	636	31.3	893	62	411
Oct-Dec 2012	15,945	12,836	2,954	45	110	13,833	2,113	11,314	1,522	2,451	503	454	786	6.1	358	45.5	135	43	251	2,025	641	31.7	887	65	402
Nov-Jan 2013	15,925	12,847	2,922	44	113	13,811	2,114	11,325	1,521	2,418	504	440	775	6.0	349	45.1	138	42	246	2,025	637	31.5	902	62	396
Dec-Feb 2013	15,868	12,786	2,938	37	107	13,773	2,095	11,292	1,494	2,415	523	437	765	6.0	342	44.7	140	46	237	2,017	642	31.8	892	62	399

4.2 Full-time, part-time and temporary workers

United Kingdom (thousands) seasonally adjusted

Men (16+)	All in employment					Full-time and part-time workers[1]							Temporary employees (reasons for temporary working)							Part-time workers (reasons for working part-time)[3]					
	Total	Employees	Self employed	Unpaid family workers	Government supported training & employment programmes[2]	Total people working full-time	Total people working part-time	Employees working full-time	Employees working part-time	Self-employed people working full-time	Self-employed people working part-time	Total workers with second jobs	Total	Total as % of all employees	Could not find permanent job	% that could not find permanent job	Did not want permanent job	Had a contract with period of training	Some other reason	Total[4]	Could not find full-time job	% that could not find full-time job	Did not want full-time job	Ill or disabled	Student or at school
	MGSA	MGRO	MGRR	MGRU	MGRX	YCBF	YCBI	YCBL	YCBO	YCBR	YCBU	YCBX	YCCA	YCCD	YCCG	YCCJ	YCCM	YCCP	YCCS	YCCV	YCCY	YCDB	YCDE	YCDH	YCDK
Jan-Mar 2013	15,871	12,806	2,920	36	109	13,788	2,082	11,308	1,498	2,415	505	434	751	5.9	337	45.0	141	41	231	2,003	637	31.8	878	61	404
Feb-Apr 2013	15,881	12,811	2,936	35	109	13,750	2,141	11,283	1,528	2,414	522	447	746	5.8	329	44.1	147	36	234	2,050	665	32.4	879	64	416
Mar-May 2013	15,885	12,831	2,916	33	105	13,761	2,124	11,314	1,518	2,405	511	452	751	5.9	324	43.1	142	42	243	2,029	665	32.7	868	65	406
Apr-Jun 2013	15,933	12,862	2,921	44	107	13,785	2,148	11,341	1,521	2,398	523	457	738	5.7	313	42.4	143	40	242	2,044	659	32.3	883	68	410
May-Jul 2013	15,964	12,895	2,929	40	101	13,831	2,134	11,376	1,519	2,414	515	468	742	5.8	312	42.1	134	49	247	2,033	659	32.4	880	65	408
Jun-Aug 2013	15,995	12,900	2,948	45	101	13,852	2,142	11,377	1,523	2,433	515	490	736	5.7	306	41.6	132	60	237	2,038	661	32.5	891	68	397
Jul-Sep 2013	16,053	12,964	2,946	40	102	13,887	2,166	11,414	1,551	2,431	515	492	754	5.8	309	41.0	132	67	249	2,065	665	32.2	901	72	403
Aug-Oct 2013	16,098	12,969	2,978	44	107	13,922	2,176	11,422	1,547	2,453	525	496	743	5.7	302	40.6	129	61	247	2,072	659	31.8	904	74	412
Sep-Nov 2013	16,157	12,975	3,043	44	94	13,982	2,174	11,434	1,541	2,501	542	502	760	5.9	298	39.2	131	62	269	2,083	657	31.6	905	75	426
Oct-Dec 2013	16,133	12,950	3,049	46	88	13,965	2,168	11,406	1,544	2,508	541	509	773	6.0	301	38.9	142	63	268	2,085	641	30.7	915	76	435
Nov-Jan 2014	16,168	12,934	3,109	46	78	14,001	2,167	11,389	1,545	2,561	548	511	768	5.9	299	39.0	147	60	262	2,093	648	31.0	915	76	433
Dec-Feb 2014	16,239	12,982	3,133	44	81	14,064	2,175	11,421	1,561	2,587	546	515	764	5.9	295	38.7	153	57	258	2,107	640	30.4	932	78	436
Jan-Mar 2014	16,264	12,989	3,144	56	76	14,077	2,187	11,421	1,567	2,598	546	517	777	6.0	301	38.8	147	58	270	2,113	642	30.4	950	80	421
Feb-Apr 2014	16,314	13,046	3,134	61	73	14,155	2,159	11,498	1,548	2,597	537	508	778	6.0	304	39.1	149	61	264	2,085	612	29.3	954	82	421
Mar-May 2014	16,358	13,095	3,136	61	67	14,197	2,161	11,547	1,547	2,593	542	504	776	5.9	310	39.9	147	61	259	2,090	600	28.7	965	90	417
Apr-Jun 2014	16,336	13,068	3,137	57	74	14,194	2,143	11,553	1,515	2,582	555	509	769	5.9	313	40.8	144	57	254	2,070	590	28.5	964	85	413
May-Jul 2014	16,325	13,064	3,124	56	82	14,168	2,156	11,536	1,528	2,573	551	492	777	5.9	319	41.0	141	63	254	2,079	589	28.3	977	86	408
Jun-Aug 2014	16,378	13,133	3,115	55	75	14,224	2,154	11,586	1,547	2,582	533	520	797	6.1	323	40.6	155	58	261	2,080	591	28.4	972	80	416
Jul-Sep 2014	16,409	13,167	3,119	54	70	14,264	2,146	11,624	1,542	2,593	526	510	800	6.1	310	38.7	171	63	256	2,068	583	28.2	974	79	415
Aug-Oct 2014	16,404	13,181	3,102	53	68	14,270	2,134	11,648	1,532	2,580	522	520	807	6.1	291	36.0	185	66	265	2,054	566	27.5	983	75	411
Sep-Nov 2014	16,400	13,201	3,084	48	67	14,241	2,158	11,647	1,554	2,555	530	495	798	6.0	297	37.2	177	65	260	2,084	573	27.5	990	84	420
Oct-Dec 2014	16,441	13,269	3,060	47	64	14,294	2,146	11,725	1,544	2,534	526	489	826	6.2	311	37.7	172	61	281	2,070	565	27.3	994	84	411
Nov-Jan 2015	16,474	13,286	3,079	49	61	14,326	2,149	11,736	1,550	2,553	525	483	807	6.1	310	38.5	161	63	272	2,076	573	27.6	997	85	405
Dec-Feb 2015	16,557	13,341	3,105	51	60	14,374	2,183	11,776	1,565	2,565	540	518	812	6.1	319	39.3	155	66	272	2,105	589	28.0	1,019	84	402
Jan-Mar 2015	16,585	13,390	3,082	53	60	14,389	2,196	11,806	1,584	2,545	537	524	810	6.1	312	38.6	159	67	272	2,121	579	27.3	1,030	88	412
Feb-Apr 2015	16,544	13,364	3,068	51	61	14,367	2,177	11,790	1,574	2,532	537	536	804	6.0	313	38.9	159	64	268	2,110	563	26.7	1,029	88	416
Mar-May 2015	16,491	13,345	3,039	45	62	14,359	2,132	11,797	1,548	2,508	531	530	791	5.9	294	37.2	174	55	267	2,080	548	26.4	1,009	90	418
Apr-Jun 2015	16,534	13,342	3,086	45	62	14,374	2,160	11,795	1,547	2,525	560	542	768	5.8	298	38.8	164	60	246	2,107	558	26.5	1,024	94	418
May-Jul 2015	16,588	13,390	3,091	42	65	14,393	2,195	11,836	1,554	2,515	576	525	775	5.8	285	36.8	171	57	262	2,130	555	26.0	1,036	96	428
Jun-Aug 2015	16,636	13,461	3,071	38	66	14,426	2,210	11,891	1,570	2,492	580	498	758	5.6	295	38.9	170	59	234	2,149	557	25.9	1,052	88	433
Jul-Sep 2015	16,676	13,452	3,120	41	63	14,438	2,238	11,851	1,601	2,543	577	491	792	5.9	298	37.7	182	58	253	2,178	556	25.5	1,073	81	451
Aug-Oct 2015	16,769	13,525	3,145	41	58	14,527	2,241	11,913	1,612	2,571	574	491	787	5.8	308	39.2	182	47	250	2,185	560	25.6	1,055	86	469
Sep-Nov 2015	16,803	13,553	3,156	48	46	14,562	2,241	11,950	1,602	2,577	580	501	774	5.7	297	38.3	175	60	242	2,182	543	24.9	1,073	90	461
Oct-Dec 2015	16,834	13,564	3,169	48	53	14,583	2,251	11,944	1,620	2,602	567	494	769	5.7	287	37.3	182	61	238	2,187	547	25.0	1,062	95	462
Nov-Jan 2016	16,829	13,563	3,170	45	51	14,555	2,274	11,917	1,646	2,601	568	495	773	5.7	289	37.4	185	63	235	2,214	539	24.3	1,086	99	472
Dec-Feb 2016	16,848	13,543	3,200	51	54	14,594	2,255	11,922	1,621	2,633	567	483	772	5.7	285	36.8	184	55	249	2,188	530	24.2	1,064	103	471
Jan-Mar 2016	16,848	13,538	3,210	50	50	14,609	2,240	11,923	1,615	2,649	561	479	772	5.7	297	38.5	173	51	251	2,176	528	24.3	1,047	110	472

4.2 Full-time, part-time and temporary workers

United Kingdom (thousands) seasonally adjusted

	All in employment					Full-time and part-time workers[1]							Temporary employees (reasons for temporary working)							Part-time workers (reasons for working part-time)[3]					
Men (16+)	Total	Employees	Self employed	Unpaid family workers	Government supported training & employment programmes[2]	Total people working full-time	Total people working part-time	Employees working full-time	Employees working part-time	Self-employed people working full-time	Self-employed people working part-time	Total workers with second jobs	Total	Total as % of all employees	Could not find permanent job	% that could not find permanent job	Did not want permanent job	Had a contract with period of training	Some other reason	Total[4]	Could not find full-time job	% that could not find full-time job	Did not want full-time job	Ill or disabled	Student or at school
	MGSA	MGRO	MGRR	MGRU	MGRX	YCBF	YCBI	YCBL	YCBO	YCBR	YCBU	YCBX	YCCA	YCCD	YCCG	YCCJ	YCCM	YCCP	YCCS	YCCV	YCCY	YCDB	YCDE	YCDH	YCDK
Feb-Apr 2016	16,860	13,533	3,222	54	52	14,623	2,237	11,917	1,616	2,671	551	483	768	5.7	273	35.6	184	66	245	2,167	531	24.5	1,037	100	473
Mar-May 2016	16,879	13,524	3,247	57	52	14,624	2,256	11,901	1,623	2,687	560	487	748	5.5	256	34.2	180	70	243	2,182	533	24.4	1,079	92	456
Apr-Jun 2016	16,913	13,551	3,253	58	51	14,616	2,297	11,903	1,648	2,677	576	471	781	5.8	272	34.9	178	76	254	2,224	534	24.0	1,110	95	463
May-Jul 2016	16,921	13,609	3,214	49	49	14,675	2,246	11,976	1,633	2,664	549	480	776	5.7	271	35.0	177	76	252	2,182	514	23.6	1,113	97	440
Jun-Aug 2016	16,961	13,614	3,243	59	45	14,680	2,281	11,963	1,651	2,682	561	491	768	5.6	265	34.5	174	70	259	2,212	516	23.3	1,127	101	454
Jul-Sep 2016	16,926	13,611	3,216	60	39	14,659	2,267	11,969	1,642	2,651	565	498	753	5.5	250	33.1	175	68	261	2,206	516	23.4	1,123	103	450
Aug-Oct 2016	16,889	13,607	3,181	59	42	14,597	2,292	11,958	1,649	2,599	581	494	741	5.4	249	33.7	171	59	261	2,231	515	23.1	1,161	101	441
Sep-Nov 2016	16,886	13,576	3,209	56	46	14,605	2,282	11,930	1,646	2,631	577	489	743	5.5	255	34.4	174	50	264	2,223	511	23.0	1,140	104	453
Oct-Dec 2016	16,901	13,569	3,227	54	51	14,620	2,280	11,921	1,648	2,654	572	485	728	5.4	235	32.3	165	49	279	2,220	498	22.4	1,142	103	461
Nov-Jan 2017	16,926	13,586	3,229	58	53	14,669	2,258	11,950	1,636	2,670	559	483	729	5.4	226	31.1	168	48	287	2,195	471	21.4	1,129	101	473
Dec-Feb 2017	16,922	13,601	3,216	52	53	14,683	2,239	11,971	1,630	2,666	549	478	735	5.4	219	29.8	179	49	288	2,179	443	20.3	1,131	102	479
Jan-Mar 2017	16,960	13,636	3,214	53	57	14,722	2,237	12,008	1,628	2,665	549	475	726	5.3	222	30.5	182	46	277	2,177	440	20.2	1,153	99	462
Feb-Apr 2017	16,963	13,648	3,220	47	49	14,716	2,247	12,020	1,627	2,658	562	474	731	5.4	217	29.7	182	47	284	2,189	451	20.6	1,176	96	445
Mar-May 2017	16,984	13,679	3,223	42	40	14,727	2,257	12,036	1,643	2,656	567	483	733	5.4	216	29.5	190	52	276	2,210	440	19.9	1,202	90	458
Apr-Jun 2017	17,010	13,697	3,228	46	39	14,772	2,238	12,076	1,621	2,660	568	488	738	5.4	218	29.6	201	52	267	2,189	439	20.1	1,177	91	458
May-Jul 2017	17,010	13,688	3,235	47	40	14,766	2,244	12,075	1,613	2,651	584	485	744	5.4	219	29.5	203	53	268	2,198	454	20.7	1,159	93	466
Jun-Aug 2017	17,002	13,681	3,236	48	37	14,757	2,245	12,073	1,608	2,645	591	478	747	5.5	223	29.8	203	59	263	2,200	446	20.3	1,151	104	472
Jul-Sep 2017	16,978	13,670	3,224	47	37	14,696	2,282	12,031	1,639	2,628	596	496	746	5.5	228	30.5	194	57	267	2,235	453	20.3	1,189	98	469
Aug-Oct 2017	16,967	13,666	3,211	51	40	14,722	2,246	12,064	1,602	2,623	588	496	751	5.5	229	30.5	197	68	257	2,190	437	19.9	1,165	94	467
Sep-Nov 2017	17,081	13,803	3,180	60	37	14,814	2,267	12,172	1,632	2,605	575	497	743	5.4	228	30.7	190	76	249	2,207	454	20.5	1,168	92	464
Oct-Dec 2017	17,035	13,765	3,169	65	36	14,788	2,246	12,141	1,624	2,611	558	494	740	5.4	230	31.0	192	79	240	2,182	443	20.3	1,160	86	465

Source: Labour Force Survey

Note: When comparing quarterly changes ONS recommends comparing with the previous non-overlapping 3-month average time period (eg, compare Apr-Jun with Jan-Mar, not with Mar-May).

1. The split between full-time and part-time employment is based on respondents' self-classification.
2. This series does not include all people on these programmes; it only includes those engaging in any form of work, work experience or work-related training.
3. These series cover Employees and Self-employed only. These series include some temporary employees.
4. The total includes those who did not give a reason for working part-time and it therefore does not equal the sum of the other columns in this section of the table.

4.2 Full-time, part-time and temporary workers

United Kingdom (thousands) seasonally adjusted

Women (16+)	All in employment					Full-time and part-time workers[1]							Temporary employees (reasons for temporary working)							Part-time workers (reasons for working part-time)[3]					
	Total	Employees	Self employed	Unpaid family workers	Government supported training & employment programmes[2]	Total people working full-time	Total people working part-time	Employees working full-time	Employees working part-time	Self-employed people working full-time	Self-employed people working part-time	Total workers with second jobs	Total	Total as % of all employees	Could not find permanent job	% that could not find permanent job	Did not want permanent job	Had a contract with period of training	Some other reason	Total[4]	Could not find full-time job	% that could not find full-time job	Did not want full-time job	Ill or disabled	Student or at school
	MGSA	MGRO	MGRR	MGRU	MGRX	YCBF	YCBI	YCBL	YCBO	YCBR	YCBU	YCBX	YCCA	YCCD	YCCG	YCCJ	YCCM	YCCP	YCCS	YCCV	YCCY	YCDB	YCDE	YCDH	YCDK
Nov-Jan 2007	13,441	12,299	1,036	62	44	7,689	5,752	7,155	5,144	507	529	625	817	6.6	190	23.2	256	49	322	5,673	388	6.8	4,465	115	675
Dec-Feb 2007	13,423	12,269	1,048	59	47	7,696	5,727	7,151	5,118	520	528	616	818	6.7	185	22.6	254	55	324	5,646	380	6.7	4,468	110	663
Jan-Mar 2007	13,436	12,284	1,047	61	44	7,721	5,715	7,186	5,099	509	538	611	821	6.7	190	23.1	251	58	321	5,636	386	6.9	4,459	107	662
Feb-Apr 2007	13,450	12,298	1,048	61	43	7,734	5,717	7,194	5,104	513	535	634	817	6.6	192	23.5	249	56	320	5,639	399	7.1	4,465	98	655
Mar-May 2007	13,468	12,331	1,030	61	45	7,762	5,706	7,223	5,109	513	517	643	817	6.6	189	23.1	248	58	323	5,626	408	7.3	4,443	101	653
Apr-Jun 2007	13,483	12,344	1,027	62	49	7,776	5,706	7,239	5,106	514	514	651	814	6.6	194	23.9	237	57	326	5,619	406	7.2	4,430	99	662
May-Jul 2007	13,506	12,350	1,040	66	50	7,801	5,705	7,260	5,090	519	521	658	812	6.6	192	23.6	235	54	331	5,611	415	7.4	4,396	102	678
Jun-Aug 2007	13,514	12,339	1,060	64	52	7,810	5,704	7,269	5,070	516	544	661	798	6.5	188	23.5	232	49	329	5,613	407	7.3	4,420	95	668
Jul-Sep 2007	13,536	12,353	1,072	59	52	7,847	5,689	7,293	5,061	529	543	653	803	6.5	186	23.2	232	47	337	5,604	401	7.2	4,427	97	656
Aug-Oct 2007	13,568	12,386	1,066	60	56	7,861	5,706	7,313	5,073	520	546	657	794	6.4	182	23.0	235	46	330	5,619	409	7.3	4,439	97	650
Sep-Nov 2007	13,584	12,410	1,060	66	49	7,874	5,711	7,331	5,079	516	543	661	794	6.4	177	22.3	240	41	336	5,622	409	7.3	4,453	102	638
Oct-Dec 2007	13,628	12,464	1,047	67	51	7,893	5,735	7,364	5,100	503	544	676	799	6.4	179	22.4	256	36	327	5,644	438	7.8	4,426	99	654
Nov-Jan 2008	13,646	12,472	1,058	70	45	7,903	5,743	7,375	5,097	503	556	659	788	6.3	179	22.7	250	35	324	5,653	447	7.9	4,416	105	660
Dec-Feb 2008	13,683	12,503	1,067	70	43	7,932	5,751	7,386	5,117	520	547	665	782	6.3	181	23.2	247	40	314	5,664	437	7.7	4,419	108	671
Jan-Mar 2008	13,678	12,511	1,056	69	42	7,925	5,753	7,384	5,128	519	537	652	773	6.2	178	23.1	244	44	307	5,665	416	7.3	4,431	115	674
Feb-Apr 2008	13,682	12,517	1,045	73	47	7,931	5,751	7,381	5,136	524	521	663	783	6.3	179	22.8	248	44	312	5,657	409	7.2	4,415	114	690
Mar-May 2008	13,730	12,556	1,047	76	51	7,972	5,758	7,424	5,132	523	524	658	772	6.1	179	23.1	239	39	316	5,657	402	7.1	4,422	117	686
Apr-Jun 2008	13,710	12,544	1,048	72	47	7,968	5,742	7,418	5,125	524	524	669	767	6.1	171	22.3	245	38	313	5,650	408	7.2	4,419	126	670
May-Jul 2008	13,709	12,540	1,056	65	48	7,993	5,716	7,441	5,100	527	529	680	774	6.2	169	21.9	248	40	316	5,629	399	7.1	4,410	134	659
Jun-Aug 2008	13,684	12,532	1,052	60	41	7,976	5,709	7,433	5,099	525	527	670	768	6.1	173	22.6	251	39	305	5,626	399	7.1	4,435	133	634
Jul-Sep 2008	13,670	12,518	1,049	61	43	7,980	5,690	7,434	5,084	528	521	670	757	6.0	176	23.2	244	42	295	5,606	399	7.1	4,419	126	633
Aug-Oct 2008	13,671	12,515	1,058	58	38	7,986	5,685	7,437	5,079	532	527	675	743	5.9	174	23.4	237	41	291	5,606	409	7.3	4,423	117	629
Sep-Nov 2008	13,672	12,516	1,064	54	37	7,914	5,758	7,382	5,134	519	545	673	768	6.1	178	23.1	244	42	304	5,679	420	7.4	4,465	120	646
Oct-Dec 2008	13,659	12,515	1,054	53	36	7,939	5,720	7,409	5,106	511	542	670	772	6.2	185	24.0	240	44	303	5,649	417	7.4	4,442	118	643
Nov-Jan 2009	13,680	12,530	1,061	50	39	7,957	5,722	7,421	5,109	519	542	673	766	6.1	182	23.7	237	44	303	5,651	437	7.7	4,424	114	648
Dec-Feb 2009	13,643	12,495	1,055	52	41	7,908	5,735	7,363	5,133	520	535	671	765	6.1	195	25.5	227	46	297	5,667	465	8.2	4,428	105	643
Jan-Mar 2009	13,627	12,458	1,071	55	43	7,868	5,759	7,320	5,139	524	547	669	764	6.1	193	25.3	230	44	297	5,686	481	8.5	4,436	107	638
Feb-Apr 2009	13,594	12,413	1,081	56	45	7,874	5,721	7,330	5,083	520	561	656	755	6.1	191	25.3	226	47	292	5,644	497	8.8	4,387	112	626
Mar-May 2009	13,578	12,386	1,082	64	46	7,867	5,711	7,324	5,062	523	559	655	755	6.1	198	26.2	231	44	282	5,621	511	9.1	4,357	111	620
Apr-Jun 2009	13,575	12,385	1,087	58	46	7,832	5,742	7,285	5,100	527	559	646	754	6.1	200	26.6	230	42	282	5,660	525	9.3	4,373	110	622
May-Jul 2009	13,553	12,355	1,097	55	46	7,793	5,759	7,234	5,121	539	559	641	750	6.1	212	28.3	220	39	279	5,680	534	9.4	4,386	108	620
Jun-Aug 2009	13,604	12,379	1,124	51	50	7,830	5,774	7,260	5,119	548	576	648	756	6.1	211	27.9	225	42	279	5,695	549	9.6	4,374	114	627
Jul-Sep 2009	13,622	12,394	1,132	48	48	7,782	5,840	7,218	5,176	544	588	654	779	6.3	223	28.7	230	41	284	5,764	569	9.9	4,411	110	648
Aug-Oct 2009	13,636	12,430	1,111	50	44	7,759	5,877	7,203	5,227	536	575	652	767	6.2	223	29.0	228	36	279	5,802	583	10.0	4,422	112	662
Sep-Nov 2009	13,654	12,420	1,139	43	52	7,777	5,877	7,213	5,207	544	595	647	752	6.1	228	30.3	220	34	271	5,801	589	10.2	4,412	107	666
Oct-Dec 2009	13,665	12,418	1,142	50	55	7,791	5,874	7,214	5,204	555	588	644	755	6.1	229	30.4	216	36	273	5,792	591	10.2	4,398	110	667
Nov-Jan 2010	13,650	12,409	1,133	50	59	7,777	5,873	7,207	5,201	546	587	627	768	6.2	240	31.2	223	36	269	5,788	596	10.3	4,385	109	671

4.2 Full-time, part-time and temporary workers

United Kingdom (thousands) seasonally adjusted

Women (16+)	All in employment					Full-time and part-time workers[1]							Temporary employees (reasons for temporary working)							Part-time workers (reasons for working part-time)[3]					
	Total	Employees	Self employed	Unpaid family workers	Government supported training & employment programmes[2]	Total people working full-time	Total people working part-time	Employees working full-time	Employees working part-time	Self-employed people working full-time	Self-employed people working part-time	Total workers with second jobs	Total	Total as % of all employees	Could not find permanent job	% that could not find permanent job	Did not want permanent job	Had a contract with period of training	Some other reason	Total	Could not find full-time job	% that could not find full-time job	Did not want full-time job	Ill or disabled	Student or at school
	MGSA	MGRO	MGRR	MGRU	MGRX	YCBF	YCBI	YCBL	YCBO	YCBR	YCBU	YCBX	YCCA	YCCD	YCCG	YCCJ	YCCM	YCCP	YCCS	YCCV	YCCY	YCDB	YCDE	YCDH	YCDK
Dec-Feb 2010	13,620	12,364	1,146	51	59	7,770	5,850	7,190	5,174	559	587	624	772	6.2	240	31.1	217	37	278	5,761	594	10.3	4,360	107	674
Jan-Mar 2010	13,616	12,329	1,180	50	58	7,749	5,867	7,151	5,177	577	603	633	775	6.3	245	31.6	215	37	278	5,780	608	10.5	4,379	99	661
Feb-Apr 2010	13,630	12,329	1,194	50	58	7,720	5,910	7,119	5,211	581	613	645	786	6.4	256	32.6	210	40	280	5,823	627	10.8	4,413	104	647
Mar-May 2010	13,645	12,367	1,172	51	54	7,715	5,930	7,125	5,242	567	605	658	808	6.5	263	32.6	217	39	289	5,847	613	10.5	4,446	104	651
Apr-Jun 2010	13,653	12,387	1,161	49	56	7,734	5,919	7,151	5,236	560	601	664	818	6.6	269	32.9	217	42	290	5,837	608	10.4	4,456	102	642
May-Jul 2010	13,691	12,410	1,169	60	52	7,712	5,979	7,117	5,293	570	599	667	824	6.6	272	33.0	222	40	289	5,892	635	10.8	4,464	101	666
Jun-Aug 2010	13,686	12,415	1,154	67	49	7,697	5,989	7,109	5,306	566	589	662	818	6.6	277	33.9	211	43	286	5,895	646	11.0	4,451	93	673
Jul-Sep 2010	13,689	12,400	1,181	62	46	7,706	5,983	7,111	5,289	576	606	656	832	6.7	281	33.8	220	42	288	5,894	652	11.1	4,451	93	665
Aug-Oct 2010	13,650	12,375	1,172	57	46	7,699	5,950	7,104	5,271	575	597	655	836	6.8	277	33.2	217	44	297	5,868	652	11.1	4,444	92	642
Sep-Nov 2010	13,640	12,352	1,177	61	49	7,695	5,945	7,095	5,257	577	600	652	831	6.7	286	34.4	203	42	300	5,857	658	11.2	4,464	94	603
Oct-Dec 2010	13,645	12,362	1,174	59	50	7,703	5,942	7,120	5,242	562	612	653	828	6.7	287	34.7	193	42	306	5,854	673	11.5	4,439	91	614
Nov-Jan 2011	13,680	12,364	1,202	65	49	7,715	5,965	7,126	5,238	565	636	663	829	6.7	281	33.9	196	43	309	5,875	669	11.4	4,469	91	609
Dec-Feb 2011	13,725	12,406	1,205	65	49	7,739	5,987	7,150	5,256	566	639	666	836	6.7	277	33.1	208	44	308	5,895	668	11.3	4,483	96	611
Jan-Mar 2011	13,749	12,444	1,189	66	50	7,767	5,982	7,181	5,263	561	628	670	843	6.8	279	33.1	211	45	308	5,891	665	11.3	4,469	100	622
Feb-Apr 2011	13,694	12,400	1,182	65	46	7,726	5,968	7,151	5,249	548	634	665	831	6.7	269	32.4	210	43	309	5,884	689	11.7	4,441	102	617
Mar-May 2011	13,728	12,434	1,191	60	43	7,761	5,967	7,183	5,251	555	636	661	834	6.7	271	32.5	214	43	306	5,886	705	12.0	4,423	106	617
Apr-Jun 2011	13,719	12,452	1,171	58	39	7,789	5,930	7,211	5,241	556	615	648	824	6.6	280	34.0	210	41	293	5,855	707	12.1	4,391	109	611
May-Jul 2011	13,690	12,431	1,164	56	39	7,824	5,866	7,245	5,186	562	603	644	807	6.5	274	34.0	200	40	293	5,789	709	12.2	4,321	111	610
Jun-Aug 2011	13,678	12,381	1,202	60	35	7,794	5,884	7,197	5,184	579	623	647	785	6.3	274	34.9	182	40	289	5,808	711	12.2	4,338	114	611
Jul-Sep 2011	13,690	12,361	1,225	63	42	7,789	5,900	7,188	5,173	579	646	655	789	6.4	278	35.2	187	43	282	5,819	720	12.4	4,326	123	614
Aug-Oct 2011	13,690	12,341	1,236	69	44	7,781	5,909	7,167	5,174	588	648	649	802	6.5	293	36.6	192	44	272	5,822	723	12.4	4,344	116	608
Sep-Nov 2011	13,702	12,354	1,232	69	46	7,781	5,920	7,168	5,186	583	649	653	820	6.6	292	35.6	201	49	279	5,835	754	12.9	4,331	116	608
Oct-Dec 2011	13,710	12,368	1,229	70	44	7,766	5,944	7,156	5,202	586	642	646	818	6.6	294	35.9	204	50	271	5,854	768	13.1	4,336	119	605
Nov-Jan 2012	13,712	12,361	1,235	69	47	7,763	5,948	7,154	5,207	586	648	653	828	6.7	298	36.0	203	52	275	5,855	794	13.6	4,319	116	598
Dec-Feb 2012	13,695	12,329	1,250	62	54	7,756	5,939	7,151	5,178	583	668	659	852	6.9	303	35.6	216	51	281	5,845	780	13.3	4,331	108	592
Jan-Mar 2012	13,733	12,355	1,271	53	55	7,774	5,959	7,165	5,190	588	682	665	829	6.7	299	36.1	201	46	284	5,872	779	13.3	4,348	117	591
Feb-Apr 2012	13,738	12,364	1,265	55	55	7,789	5,949	7,181	5,183	587	678	661	818	6.6	299	36.6	195	48	275	5,861	772	13.2	4,346	116	591
Mar-May 2012	13,763	12,407	1,241	64	52	7,813	5,950	7,217	5,190	572	669	656	838	6.8	316	37.7	186	49	288	5,859	765	13.1	4,353	116	592
Apr-Jun 2012	13,801	12,426	1,252	69	54	7,812	5,989	7,216	5,210	567	685	639	862	6.9	320	37.1	194	51	297	5,895	776	13.2	4,369	110	610
May-Jul 2012	13,870	12,468	1,274	74	54	7,821	6,049	7,229	5,239	564	710	650	890	7.1	326	36.6	210	51	303	5,949	786	13.2	4,417	110	608
Jun-Aug 2012	13,876	12,495	1,262	66	53	7,822	6,054	7,236	5,259	560	702	646	882	7.1	323	36.6	208	50	301	5,961	786	13.2	4,427	107	614
Jul-Sep 2012	13,862	12,489	1,260	61	53	7,814	6,048	7,226	5,262	563	697	646	866	6.9	319	36.8	204	48	295	5,959	782	13.1	4,431	112	607
Aug-Oct 2012	13,838	12,472	1,253	62	51	7,781	6,057	7,191	5,281	568	685	660	866	6.9	315	36.4	202	47	301	5,966	774	13.0	4,429	117	612
Sep-Nov 2012	13,912	12,503	1,289	67	54	7,847	6,065	7,233	5,270	588	701	668	871	7.0	308	35.4	208	50	305	5,971	771	12.9	4,436	120	610
Oct-Dec 2012	13,962	12,533	1,304	70	56	7,903	6,059	7,276	5,257	598	706	691	869	6.9	302	34.7	204	54	310	5,963	751	12.6	4,451	119	608
Nov-Jan 2013	13,966	12,567	1,285	63	51	7,942	6,024	7,318	5,249	590	695	698	865	6.9	307	35.5	197	53	308	5,944	768	12.9	4,428	122	594
Dec-Feb 2013	13,947	12,553	1,285	64	46	7,926	6,021	7,299	5,254	597	687	684	837	6.7	301	35.9	194	56	286	5,942	781	13.1	4,411	121	595

4.2 Full-time, part-time and temporary workers

United Kingdom (thousands) seasonally adjusted

Women (16+)	All in employment					Full-time and part-time workers[1]							Temporary employees (reasons for temporary working)							Part-time workers (reasons for working part-time)[3]					
	Total	Employees	Self employed	Unpaid family workers	Government supported training & employment programmes[2]	Total people working full-time	Total people working part-time	Employees working full-time	Employees working part-time	Self-employed people working full-time	Self-employed people working part-time	Total workers with second jobs	Total	Total as % of all employees	Could not find permanent job	% that could not find permanent job	Did not want permanent job	Had a contract with period of other training reason	Some other reason	Total[4]	Could not find full-time job	% that could not find full-time job	Did not want full-time job	Ill or disabled	Student or at school
	MGSA	MGRO	MGRR	MGRU	MGRX	YCBF	YCBI	YCBL	YCBO	YCBR	YCBU	YCBX	YCCA	YCCD	YCCG	YCCJ	YCCM	YCCP	YCCS	YCCV	YCCY	YCDB	YCDE	YCDH	YCDK
Jan-Mar 2013	13,969	12,592	1,265	66	45	7,946	6,023	7,316	5,276	603	662	677	862	6.8	312	36.2	196	62	291	5,938	780	13.1	4,400	117	610
Feb-Apr 2013	13,993	12,589	1,283	69	52	7,973	6,020	7,329	5,260	614	669	653	846	6.7	303	35.8	196	65	283	5,929	778	13.1	4,390	119	612
Mar-May 2013	13,975	12,580	1,268	70	57	7,964	6,011	7,324	5,255	607	661	670	833	6.6	302	36.3	189	59	283	5,917	801	13.5	4,356	116	610
Apr-Jun 2013	14,005	12,600	1,276	71	58	7,985	6,020	7,332	5,268	615	661	667	831	6.6	291	35.1	186	65	289	5,929	796	13.4	4,376	121	600
May-Jul 2013	14,030	12,625	1,275	68	62	8,037	5,994	7,370	5,254	619	656	663	822	6.5	291	35.4	178	63	289	5,911	804	13.6	4,343	121	601
Jun-Aug 2013	14,030	12,600	1,294	72	63	8,041	5,989	7,354	5,246	633	661	667	837	6.6	297	35.5	167	63	310	5,907	799	13.5	4,371	125	571
Jul-Sep 2013	14,053	12,615	1,298	72	68	8,057	5,996	7,372	5,243	634	664	667	848	6.7	294	34.7	177	60	317	5,907	797	13.5	4,357	121	590
Aug-Oct 2013	14,125	12,668	1,320	76	62	8,076	6,049	7,391	5,277	642	678	674	855	6.7	294	34.4	186	63	313	5,955	803	13.5	4,401	121	591
Sep-Nov 2013	14,145	12,656	1,366	70	53	8,129	6,016	7,420	5,237	675	690	682	846	6.7	286	33.8	191	66	303	5,927	788	13.3	4,391	116	596
Oct-Dec 2013	14,157	12,664	1,375	66	53	8,174	5,983	7,456	5,208	685	690	667	850	6.7	302	35.5	192	55	300	5,899	789	13.4	4,374	114	590
Nov-Jan 2014	14,161	12,651	1,391	64	54	8,164	5,997	7,447	5,205	685	706	663	852	6.7	295	34.6	203	55	300	5,911	788	13.3	4,383	116	593
Dec-Feb 2014	14,253	12,727	1,405	71	50	8,179	6,074	7,466	5,261	684	722	642	865	6.8	290	33.6	208	58	308	5,983	781	13.1	4,465	115	596
Jan-Mar 2014	14,268	12,718	1,436	67	47	8,200	6,068	7,498	5,220	674	761	663	876	6.9	289	33.0	218	59	310	5,981	777	13.0	4,472	112	599
Feb-Apr 2014	14,326	12,775	1,440	64	48	8,221	6,105	7,524	5,251	673	767	676	903	7.1	296	32.8	230	59	318	6,018	793	13.2	4,487	106	611
Mar-May 2014	14,385	12,811	1,463	57	53	8,254	6,130	7,555	5,257	675	788	687	904	7.1	296	32.8	230	58	320	6,045	767	12.7	4,541	104	615
Apr-Jun 2014	14,369	12,789	1,469	58	53	8,206	6,162	7,505	5,285	680	789	703	885	6.9	289	32.7	234	56	305	6,074	757	12.5	4,580	110	603
May-Jul 2014	14,378	12,822	1,444	62	49	8,226	6,151	7,530	5,292	678	766	709	889	6.9	290	32.6	229	58	312	6,058	753	12.4	4,582	94	605
Jun-Aug 2014	14,408	12,871	1,412	72	53	8,283	6,125	7,584	5,286	676	736	702	888	6.9	286	32.3	231	62	309	6,022	758	12.6	4,546	94	601
Jul-Sep 2014	14,424	12,892	1,411	69	52	8,302	6,121	7,614	5,278	665	746	693	902	7.0	289	32.1	238	60	314	6,024	753	12.5	4,551	94	601
Aug-Oct 2014	14,444	12,886	1,441	65	52	8,318	6,127	7,610	5,276	680	761	689	887	6.9	282	31.7	236	56	313	6,037	744	12.3	4,582	99	584
Sep-Nov 2014	14,451	12,894	1,455	56	47	8,324	6,126	7,616	5,278	683	772	705	885	6.9	273	30.8	236	57	320	6,050	741	12.3	4,587	103	588
Oct-Dec 2014	14,503	12,947	1,456	55	45	8,337	6,166	7,634	5,313	675	781	712	891	6.9	270	30.3	239	54	328	6,093	745	12.2	4,593	105	618
Nov-Jan 2015	14,529	12,958	1,463	64	44	8,366	6,164	7,665	5,294	670	793	710	874	6.7	263	30.1	233	57	321	6,087	751	12.3	4,597	110	600
Dec-Feb 2015	14,565	13,014	1,445	62	45	8,370	6,195	7,690	5,323	647	797	700	873	6.7	272	31.1	224	54	323	6,121	765	12.5	4,597	112	611
Jan-Mar 2015	14,571	13,020	1,443	65	42	8,379	6,191	7,686	5,334	663	781	698	871	6.7	273	31.3	215	57	327	6,115	746	12.2	4,611	125	599
Feb-Apr 2015	14,585	13,056	1,426	59	43	8,416	6,169	7,718	5,338	662	764	685	877	6.7	271	30.8	215	61	331	6,102	744	12.2	4,598	125	599
Mar-May 2015	14,566	13,039	1,434	58	35	8,397	6,169	7,718	5,321	650	785	669	881	6.8	265	30.1	224	65	327	6,106	732	12.0	4,611	128	597
Apr-Jun 2015	14,576	13,044	1,433	56	44	8,419	6,157	7,719	5,325	669	764	653	874	6.7	268	30.7	218	69	319	6,089	732	12.0	4,591	129	600
May-Jul 2015	14,600	13,056	1,439	59	46	8,408	6,192	7,709	5,347	668	771	663	891	6.8	276	31.0	229	70	317	6,118	722	11.8	4,617	133	611
Jun-Aug 2015	14,584	13,042	1,441	55	46	8,429	6,155	7,723	5,319	678	763	667	886	6.8	284	32.1	228	69	304	6,082	711	11.7	4,590	133	618
Jul-Sep 2015	14,657	13,102	1,456	54	46	8,456	6,202	7,742	5,360	684	772	672	877	6.7	273	31.1	229	69	306	6,132	699	11.4	4,642	134	626
Aug-Oct 2015	14,668	13,092	1,486	47	43	8,463	6,205	7,748	5,344	696	790	672	869	6.6	271	31.1	223	72	303	6,134	717	11.7	4,622	132	631
Sep-Nov 2015	14,710	13,120	1,497	48	46	8,492	6,219	7,771	5,349	696	801	664	873	6.7	273	31.3	216	76	307	6,150	697	11.3	4,653	135	631
Oct-Dec 2015	14,706	13,102	1,511	48	45	8,498	6,208	7,775	5,327	700	811	648	866	6.6	267	30.8	217	75	306	6,139	689	11.2	4,644	137	626
Nov-Jan 2016	14,712	13,126	1,492	51	42	8,480	6,232	7,759	5,367	697	795	630	882	6.7	266	30.1	223	75	319	6,162	674	10.9	4,679	139	631
Dec-Feb 2016	14,707	13,125	1,487	52	43	8,482	6,225	7,763	5,362	690	798	641	876	6.7	257	29.4	229	68	322	6,159	660	10.7	4,697	145	618
Jan-Mar 2016	14,724	13,126	1,504	50	43	8,485	6,238	7,762	5,365	696	808	641	877	6.7	264	30.1	238	66	309	6,173	670	10.9	4,698	141	623

4.2 Full-time, part-time and temporary workers

United Kingdom (thousands) seasonally adjusted

Women (16+)	All in employment: Total	Employees	Self employed	Unpaid family workers	Government supported training & employment programmes[2]	Total people working full-time	Total people working part-time	Employees working full-time	Employees working part-time	Self-employed people working full-time	Self-employed people working part-time	Total workers with second jobs	Temporary: Total	Total as % of all employees	Could not find permanent job	% that could not find permanent job	Did not want permanent job	Had a contract with period of training	Some other reason	Part-time: Total	Could not find full-time job	% that could not find full-time job	Did not want full-time job	Ill or disabled	Student or at school
	MGSA	MGRO	MGRR	MGRU	MGRX	YCBF	YCBI	YCBL	YCBO	YCBR	YCBU	YCBX	YCCA	YCCD	YCCG	YCCJ	YCCM	YCCP	YCCS	YCCV	YCCY	YCDB	YCDE	YCDH	YCDK
Feb-Apr 2016	14,743	13,161	1,485	56	42	8,468	6,276	7,752	5,409	689	795	644	865	6.6	262	30.3	245	62	295	6,204	662	10.7	4,742	143	620
Mar-May 2016	14,833	13,202	1,531	61	39	8,549	6,283	7,812	5,389	713	818	651	867	6.6	255	29.4	249	62	301	6,207	655	10.5	4,744	146	626
Apr-Jun 2016	14,834	13,206	1,532	63	33	8,568	6,267	7,826	5,380	719	814	660	885	6.7	257	29.0	255	59	314	6,193	620	10.0	4,759	145	637
May-Jul 2016	14,857	13,225	1,540	61	32	8,579	6,278	7,825	5,400	731	809	656	875	6.6	253	28.9	254	57	312	6,208	635	10.2	4,762	149	634
Jun-Aug 2016	14,850	13,216	1,543	64	27	8,554	6,295	7,789	5,426	745	798	676	875	6.6	242	27.6	257	59	317	6,224	623	10.0	4,789	147	640
Jul-Sep 2016	14,885	13,215	1,572	69	30	8,594	6,290	7,823	5,392	751	821	656	868	6.6	234	26.9	254	64	317	6,213	628	10.1	4,776	150	635
Aug-Oct 2016	14,905	13,226	1,582	70	29	8,638	6,267	7,858	5,368	754	828	638	873	6.6	240	27.4	255	64	315	6,196	636	10.3	4,755	147	625
Sep-Nov 2016	14,921	13,246	1,573	72	30	8,657	6,264	7,868	5,377	759	814	631	879	6.6	238	27.1	247	61	333	6,191	632	10.2	4,747	152	625
Oct-Dec 2016	14,944	13,267	1,577	69	32	8,680	6,264	7,886	5,381	767	810	636	893	6.7	237	26.6	247	62	347	6,191	618	10.0	4,743	154	637
Nov-Jan 2017	16,926	13,586	3,229	58	53	14,669	2,258	11,950	1,636	2,670	559	483	729	5.4	226	31.1	168	48	287	2,195	471	21.4	1,129	101	473
Dec-Feb 2017	16,922	13,601	3,216	52	53	14,683	2,239	11,971	1,630	2,666	549	478	735	5.4	219	29.8	179	49	288	2,179	443	20.3	1,131	102	479
Jan-Mar 2017	16,960	13,636	3,214	53	57	14,722	2,237	12,008	1,628	2,665	549	475	726	5.3	222	30.5	182	46	277	2,177	440	20.2	1,153	99	462
Feb-Apr 2017	16,963	13,648	3,220	47	49	14,716	2,247	12,020	1,627	2,658	562	474	731	5.4	217	29.7	182	47	284	2,189	451	20.6	1,176	96	445
Mar-May 2017	16,984	13,679	3,223	42	40	14,727	2,257	12,036	1,643	2,656	567	483	733	5.4	216	29.5	190	52	276	2,210	440	19.9	1,202	90	458
Apr-Jun 2017	17,010	13,697	3,228	46	39	14,772	2,238	12,076	1,621	2,660	568	488	738	5.4	218	29.6	201	52	267	2,189	439	20.1	1,177	91	458
May-Jul 2017	17,010	13,688	3,235	47	40	14,766	2,244	12,075	1,613	2,651	584	485	744	5.4	219	29.5	203	53	268	2,198	454	20.7	1,159	93	466
Jun-Aug 2017	17,002	13,681	3,236	48	37	14,757	2,245	12,073	1,608	2,645	591	478	747	5.5	223	29.8	203	59	263	2,200	446	20.3	1,151	104	472
Jul-Sep 2017	16,978	13,670	3,224	47	37	14,696	2,282	12,031	1,639	2,628	596	496	746	5.5	228	30.5	194	57	267	2,235	453	20.3	1,189	98	469
Aug-Oct 2017	16,967	13,666	3,211	51	40	14,722	2,246	12,064	1,602	2,623	588	496	751	5.5	229	30.5	197	68	257	2,190	437	19.9	1,165	94	467
Sep-Nov 2017	17,081	13,803	3,180	60	37	14,814	2,267	12,172	1,632	2,605	575	497	743	5.4	228	30.7	190	76	249	2,207	454	20.5	1,168	92	464
Oct-Dec 2017	17,035	13,765	3,169	65	36	14,788	2,246	12,141	1,624	2,611	558	494	740	5.4	230	31.0	192	79	240	2,182	443	20.3	1,160	86	465

Source: Labour Force Survey

Note: When comparing quarterly changes ONS recommends comparing with the previous non-overlapping 3-month average time period (eg, compare Apr-Jun with Jan-Mar, not with Mar-May).

1. The split between full-time and part-time employment is based on respondents' self-classification.
2. This series does not include all people on these programmes; it only includes those engaging in any form of work, work experience or work-related training.
3. These series cover Employees and Self-employed only. These series include some temporary employees.
4. The total includes those who did not give a reason for working part-time and it therefore does not equal the sum of the other columns in this section of the table.

4.3 Employment by sex and age group: United Kingdom

United Kingdom (thousands) seasonally adjusted

| | Aged 16 and over | | | | | | | | Aged 16-64 | | | | | | | |
| | Employment | | Unemployment | | Activity | | Inactivity | | Employment | | Unemployment | | Activity | | Inactivity | |
	level	rate (%)	level	rate (%)	level	rate (%)	level	rate (%)	level	rate (%)	level	rate (%)	level	rate (%)	level	rate (%)
People	MGRZ	MGSR	MGSC	MGSX	MGSF	MGWG	MGSI	YBTC	LF2G	LF24	LF2I	LF2Q	LF2K	LF22	LF2M	LF2S
Oct-Dec 2014	30,944	59.7	1,870	5.7	32,814	63.3	19,023	36.7	29,820	73.2	1,851	5.8	31,671	77.7	9,064	22.3
Oct-Dec 2015	31,540	60.4	1,688	5.1	33,227	63.6	19,009	36.4	30,340	74.1	1,669	5.2	32,008	78.2	8,929	21.8
Oct-Dec 2016	31,845	60.6	1,585	4.7	33,430	63.6	19,151	36.4	30,645	74.6	1,568	4.9	32,213	78.4	8,885	21.6
Jan-Mar 2017	31,946	60.7	1,527	4.6	33,474	63.6	19,175	36.4	30,747	74.8	1,514	4.7	32,261	78.4	8,862	21.6
Apr-Jun 2017	32,065	60.8	1,485	4.4	33,550	63.6	19,166	36.4	30,904	75.1	1,466	4.5	32,370	78.7	8,778	21.3
Jul-Sep 2017	32,063	60.7	1,429	4.3	33,492	63.4	19,295	36.6	30,883	75.0	1,407	4.4	32,290	78.4	8,883	21.6
Oct-Dec 2017	32,154	60.8	1,463	4.4	33,617	63.6	19,242	36.4	30,970	75.2	1,442	4.4	32,411	78.7	8,785	21.3
Men	MGSA	MGSS	MGSD	MGSY	MGSG	MGWH	MGSJ	YBTD	YBSF	MGSV	YBSI	YBTJ	YBSL	MGSP	YBSO	YBTM
Oct-Dec 2014	16,441	65.0	1,041	6.0	17,482	69.2	7,796	30.8	15,758	78.0	1,027	6.1	16,785	83.1	3,425	16.9
Oct-Dec 2015	16,834	66.0	921	5.2	17,755	69.6	7,748	30.4	16,096	79.2	911	5.4	17,007	83.7	3,320	16.3
Oct-Dec 2016	16,901	65.8	873	4.9	17,773	69.2	7,923	30.8	16,181	79.2	864	5.1	17,045	83.5	3,376	16.5
Jan-Mar 2017	16,960	65.9	844	4.7	17,804	69.2	7,927	30.8	16,232	79.4	833	4.9	17,065	83.5	3,368	16.5
Apr-Jun 2017	17,010	66.0	815	4.6	17,825	69.2	7,941	30.8	16,306	79.8	802	4.7	17,107	83.7	3,338	16.3
Jul-Sep 2017	16,978	65.8	770	4.3	17,748	68.8	8,057	31.2	16,255	79.4	756	4.4	17,011	83.1	3,449	16.9
Oct-Dec 2017	17,035	65.9	778	4.4	17,813	68.9	8,034	31.1	16,306	79.6	762	4.5	17,068	83.4	3,408	16.6
Women	MGSB	MGST	MGSE	MGSZ	MGSH	MGWI	MGSK	YBTE	LF2H	LF25	LF2J	LF2R	LF2L	LF23	LF2N	LF2T
Oct-Dec 2014	14,503	54.6	829	5.4	15,332	57.7	11,226	42.3	14,062	68.5	825	5.5	14,886	72.5	5,640	27.5
Oct-Dec 2015	14,706	55.0	766	5.0	15,472	57.9	11,261	42.1	14,244	69.1	758	5.1	15,001	72.8	5,609	27.2
Oct-Dec 2016	14,944	55.6	713	4.6	15,657	58.2	11,228	41.8	14,464	70.0	704	4.6	15,168	73.4	5,509	26.6
Jan-Mar 2017	14,986	55.7	683	4.4	15,670	58.2	11,248	41.8	14,515	70.2	680	4.5	15,196	73.4	5,494	26.6
Apr-Jun 2017	15,055	55.9	670	4.3	15,725	58.3	11,226	41.7	14,599	70.5	664	4.4	15,263	73.7	5,440	26.3
Jul-Sep 2017	15,085	55.9	659	4.2	15,744	58.4	11,238	41.6	14,628	70.6	651	4.3	15,279	73.8	5,434	26.2
Oct-Dec 2017	15,119	56.0	685	4.3	15,804	58.5	11,208	41.5	14,663	70.8	680	4.4	15,343	74.0	5,377	26.0

| | Aged 16-17 | | | | | | | | Aged 18-24 | | | | | | | |
| | Employment | | Unemployment | | Activity | | Inactivity | | Employment | | Unemployment | | Activity | | Inactivity | |
	level	rate (%)	level	rate (%)	level	rate (%)	level	rate (%)	level	rate (%)	level	rate (%)	level	rate (%)	level	rate (%)
People	YBTO	YBUA	YBVH	YBVK	YBZL	YCAG	YCAS	LWEX	YBTR	YBUD	YBVN	YBVQ	YBZO	YCAJ	YCAV	LWFA
Oct-Dec 2014	346	23.2	160	31.6	506	33.9	988	66.1	3,479	60.2	591	14.5	4,070	70.4	1,711	29.6
Oct-Dec 2015	364	24.7	134	26.8	498	33.8	975	66.2	3,598	62.5	492	12.0	4,090	71.0	1,668	29.0
Oct-Dec 2016	372	25.8	122	24.6	494	34.2	952	65.8	3,556	62.4	440	11.0	3,996	70.1	1,704	29.9
Jan-Mar 2017	366	25.4	133	26.7	499	34.7	939	65.3	3,554	62.5	423	10.6	3,978	70.0	1,705	30.0
Apr-Jun 2017	340	23.8	111	24.7	451	31.6	978	68.4	3,552	62.7	434	10.9	3,987	70.4	1,679	29.6
Jul-Sep 2017	343	24.1	105	23.5	448	31.5	973	68.5	3,471	61.4	417	10.7	3,888	68.8	1,761	31.2
Oct-Dec 2017	332	23.5	120	26.6	453	32.0	960	68.0	3,481	61.8	420	10.8	3,902	69.2	1,733	30.8
Men	YBTP	YBUB	YBVI	YBVL	YBZM	YCAH	YCAT	LWEY	YBTS	YBUE	YBVO	YBVR	YBZP	YCAK	YCAW	LWFB
Oct-Dec 2014	148	19.3	86	36.7	233	30.5	531	69.5	1,802	61.5	343	16.0	2,145	73.2	787	26.8
Oct-Dec 2015	174	23.1	72	29.3	246	32.7	508	67.3	1,866	63.8	281	13.1	2,147	73.4	779	26.6
Oct-Dec 2016	176	23.7	66	27.3	242	32.6	500	67.4	1,808	62.3	277	13.3	2,085	71.8	818	28.2
Jan-Mar 2017	178	24.1	63	26.1	240	32.6	496	67.4	1,820	62.8	261	12.5	2,081	71.9	815	28.1
Apr-Jun 2017	167	22.8	63	27.3	229	31.3	503	68.7	1,827	63.3	254	12.2	2,081	72.0	808	28.0
Jul-Sep 2017	165	22.7	50	23.3	216	29.6	512	70.4	1,761	61.1	250	12.4	2,011	69.8	871	30.2
Oct-Dec 2017	156	21.5	60	28.0	216	29.9	507	70.1	1,792	62.3	241	11.9	2,034	70.7	842	29.3
Women	YBTQ	YBUC	YBVJ	YBVM	YBZN	YCAI	YCAU	LWEZ	YBTT	YBUF	YBVP	YBVS	YBZQ	YCAL	YCAX	LWFC
Oct-Dec 2014	199	27.3	74	27.2	273	37.4	456	62.6	1,677	58.9	248	12.9	1,925	67.6	924	32.4
Oct-Dec 2015	190	26.5	61	24.4	251	35.0	467	65.0	1,733	61.2	210	10.8	1,943	68.6	889	31.4
Oct-Dec 2016	197	27.9	56	22.1	253	35.8	452	64.2	1,748	62.5	163	8.5	1,911	68.3	886	31.7
Jan-Mar 2017	188	26.8	70	27.2	259	36.9	442	63.1	1,735	62.2	162	8.6	1,897	68.1	890	31.9
Apr-Jun 2017	173	24.8	49	22.0	221	31.8	475	68.2	1,725	62.1	180	9.5	1,906	68.6	871	31.4
Jul-Sep 2017	177	25.6	55	23.6	232	33.5	461	66.5	1,710	61.8	167	8.9	1,877	67.8	890	32.2
Oct-Dec 2017	177	25.6	60	25.3	237	34.3	453	65.7	1,689	61.2	179	9.6	1,868	67.7	891	32.3

4.3 Employment by sex and age group: United Kingdom

United Kingdom (thousands) seasonally adjusted

Aged 25-34 / Aged 35-49

	Aged 25-34 Employment level	rate (%)	Unemployment level	rate (%)	Activity level	rate (%)	Inactivity level	rate (%)	Aged 35-49 Employment level	rate (%)	Unemployment level	rate (%)	Activity level	rate (%)	Inactivity level	rate (%)
People	YBTU	YBUG	YCGM	YCGP	YBZR	YCAM	YCAY	LWFD	YBTX	YBUJ	YCGS	YCGV	YBZU	YCAP	YCBB	LWFG
Oct-Dec 2014	6,998	80.5	381	5.2	7,379	84.9	1,315	15.1	10,836	83.7	421	3.7	11,256	86.9	1,691	13.1
Oct-Dec 2015	7,169	81.7	382	5.1	7,551	86.0	1,225	14.0	10,773	83.5	364	3.3	11,137	86.3	1,767	13.7
Oct-Dec 2016	7,264	82.1	346	4.5	7,610	86.0	1,240	14.0	10,780	83.8	361	3.2	11,142	86.6	1,719	13.4
Jan-Mar 2017	7,307	82.5	340	4.4	7,647	86.3	1,214	13.7	10,780	83.9	342	3.1	11,121	86.6	1,725	13.4
Apr-Jun 2017	7,390	83.3	308	4.0	7,698	86.8	1,174	13.2	10,834	84.4	332	3.0	11,166	87.0	1,666	13.0
Jul-Sep 2017	7,422	83.5	296	3.8	7,718	86.9	1,166	13.1	10,818	84.4	316	2.8	11,134	86.8	1,686	13.2
Oct-Dec 2017	7,423	83.4	298	3.9	7,720	86.8	1,177	13.2	10,854	84.8	325	2.9	11,179	87.3	1,628	12.7
Men	YBTV	YBUH	YCGN	YCGQ	YBZS	YCAN	YCAZ	LWFE	YBTY	YBUK	YCGT	YCGW	YBZV	YCAQ	YCBC	LWFH
Oct-Dec 2014	3,768	87.3	203	5.1	3,972	92.1	343	7.9	5,708	89.3	217	3.7	5,925	92.7	464	7.3
Oct-Dec 2015	3,882	88.9	198	4.8	4,080	93.4	289	6.6	5,700	89.5	177	3.0	5,877	92.3	493	7.7
Oct-Dec 2016	3,926	88.9	177	4.3	4,103	92.9	313	7.1	5,699	89.8	172	2.9	5,871	92.5	478	7.5
Jan-Mar 2017	3,927	88.8	182	4.4	4,109	92.9	313	7.1	5,712	90.1	159	2.7	5,871	92.6	470	7.4
Apr-Jun 2017	3,974	89.8	156	3.8	4,130	93.3	297	6.7	5,730	90.5	160	2.7	5,890	93.0	443	7.0
Jul-Sep 2017	3,990	89.9	152	3.7	4,143	93.4	294	6.6	5,726	90.5	149	2.5	5,876	92.9	451	7.1
Oct-Dec 2017	4,005	90.1	143	3.4	4,148	93.3	299	6.7	5,721	90.5	157	2.7	5,878	93.0	443	7.0
Women	YBTW	YBUI	YCGO	YCGR	YBZT	YCAO	YCBA	LWFF	YBTZ	YBUL	YCGU	YCGX	YBZW	YCAR	YCBD	LWFI
Oct-Dec 2014	3,230	73.7	178	5.2	3,408	77.8	972	22.2	5,128	78.2	204	3.8	5,331	81.3	1,227	18.7
Oct-Dec 2015	3,287	74.6	184	5.3	3,471	78.8	936	21.2	5,073	77.6	187	3.6	5,260	80.5	1,274	19.5
Oct-Dec 2016	3,338	75.3	170	4.8	3,507	79.1	927	20.9	5,081	78.0	190	3.6	5,271	80.9	1,241	19.1
Jan-Mar 2017	3,380	76.1	158	4.5	3,538	79.7	902	20.3	5,068	77.9	183	3.5	5,251	80.7	1,255	19.3
Apr-Jun 2017	3,416	76.9	151	4.2	3,568	80.3	877	19.7	5,104	78.5	172	3.3	5,276	81.2	1,223	18.8
Jul-Sep 2017	3,431	77.1	144	4.0	3,576	80.4	873	19.6	5,091	78.4	167	3.2	5,258	81.0	1,235	19.0
Oct-Dec 2017	3,418	76.8	155	4.3	3,572	80.3	878	19.7	5,133	79.1	167	3.2	5,300	81.7	1,186	18.3

Aged 50-64 / Age 65+

	Aged 50-64 Employment level	rate (%)	Unemployment level	rate (%)	Activity level	rate (%)	Inactivity level	rate (%)	Age 65+ Employment level	rate (%)	Unemployment level	rate (%)	Activity level	rate (%)	Inactivity level	rate (%)
People	LF26	LF2U	LF28	LF2E	LF3A	LF2C	LF2A	LF2W	LFK4	LFK6	K5HU	K5HW	LFK8	LFL2	LFL4	LFL6
Oct-Dec 2014	8,161	69.1	299	3.5	8,460	71.6	3,359	28.4	1,124	10.1	19	1.6	1,143	10.3	9,958	89.7
Oct-Dec 2015	8,435	70.1	297	3.4	8,733	72.6	3,294	27.4	1,200	10.6	19	1.6	1,219	10.8	10,080	89.2
Oct-Dec 2016	8,673	70.8	298	3.3	8,971	73.3	3,271	26.7	1,200	10.5	17	1.4	1,217	10.6	10,266	89.4
Jan-Mar 2017	8,740	71.1	276	3.1	9,016	73.3	3,280	26.7	1,199	10.4	14	1.1	1,213	10.5	10,313	89.5
Apr-Jun 2017	8,789	71.2	280	3.1	9,069	73.4	3,281	26.6	1,161	10.0	19	1.6	1,180	10.2	10,388	89.8
Jul-Sep 2017	8,830	71.2	272	3.0	9,102	73.4	3,297	26.6	1,180	10.2	22	1.8	1,202	10.4	10,411	89.6
Oct-Dec 2017	8,879	71.3	279	3.0	9,158	73.6	3,287	26.4	1,184	10.2	21	1.8	1,205	10.3	10,456	89.7
Men	MGUX	YBUN	MGVM	MGXF	YBZY	MGWQ	MGWB	LWFK	MGVA	YBUQ	MGVP	MGXI	YCAE	MGWT	MGWE	LWFN
Oct-Dec 2014	4,332	74.6	178	4.0	4,510	77.6	1,299	22.4	683	13.5	14	2.1	697	13.8	4,372	86.2
Oct-Dec 2015	4,475	75.7	182	3.9	4,657	78.8	1,252	21.2	738	14.3	11	1.4	748	14.5	4,428	85.5
Oct-Dec 2016	4,572	76.0	172	3.6	4,744	78.9	1,268	21.1	720	13.6	9	1.2	728	13.8	4,546	86.2
Jan-Mar 2017	4,595	76.1	169	3.5	4,764	78.9	1,274	21.1	728	13.7	11	1.4	739	13.9	4,559	86.1
Apr-Jun 2017	4,608	76.0	168	3.5	4,777	78.8	1,288	21.2	704	13.2	13	1.8	718	13.5	4,602	86.5
Jul-Sep 2017	4,611	75.7	155	3.2	4,766	78.3	1,322	21.7	724	13.5	13	1.8	737	13.8	4,607	86.2
Oct-Dec 2017	4,632	75.8	160	3.3	4,792	78.4	1,318	21.6	728	13.6	16	2.2	744	13.9	4,626	86.1
Women	LF27	LF2V	LF29	LF2F	LF3B	LF2D	LF2B	LF2X	LFK5	LFK7	K5HV	K5HX	LFK9	LFL3	LFL5	LFL7
Oct-Dec 2014	3,829	63.7	121	3.1	3,949	65.7	2,060	34.3	441	7.3	*	*	446	7.4	5,586	92.6
Oct-Dec 2015	3,961	64.7	115	2.8	4,076	66.6	2,042	33.4	462	7.5	8	1.8	470	7.7	5,652	92.3
Oct-Dec 2016	4,101	65.8	126	3.0	4,226	67.8	2,003	32.2	480	7.7	9	1.8	489	7.9	5,719	92.1
Jan-Mar 2017	4,145	66.2	107	2.5	4,251	67.9	2,006	32.1	471	7.6	*	*	474	7.6	5,754	92.4
Apr-Jun 2017	4,180	66.5	112	2.6	4,292	68.3	1,993	31.7	456	7.3	6	1.3	462	7.4	5,785	92.6
Jul-Sep 2017	4,218	66.8	118	2.7	4,336	68.7	1,975	31.3	457	7.3	9	1.9	465	7.4	5,804	92.6
Oct-Dec 2017	4,247	67.0	119	2.7	4,366	68.9	1,970	31.1	456	7.2	5	1.1	461	7.3	5,831	92.7

Source: Labour Force Survey

Labour market statistics enquiries: labour.market@ons.gsi.gov.uk

Note: When comparing quarterly changes ONS recommends comparing with the previous non-overlapping 3-month average time period (eg, compare Apr-Jun with Jan-Mar, not with Mar-May).

4.4 All in employment by industry

Standard Industrial Classification (SIC) 2007

United Kingdom (thousands) not seasonally adjusted

	All in employment1	Public sector	Private sector	Agriculture, forestry & fishing A	Mining, energy and water supply B, D, E	Manufacturing C	Construction F	Wholesale, retail & repair of motor vehicles G	Transport & storage H	Accommodation & food services I	Information & communication J	Financial & insurance activities K	Real estate activities L	Professional, scientific & technical activities M	Administrative & support services N	Public admin & defence; social security O	Education P	Human health & social work activities Q	Other services R, S, T
People																			
Jan-Mar 2008	29,596	7,047	22,404	334	452	3,148	2,535	4,184	1,605	1,382	1,024	1,288	263	1,939	1,398	2,011	2,847	3,566	1,509
Apr-Jun 2008	29,637	7,101	22,383	321	446	3,087	2,512	4,195	1,590	1,355	1,033	1,287	253	1,969	1,417	2,035	2,861	3,642	1,519
Jul-Sep 2008	29,685	7,039	22,508	313	463	3,068	2,583	4,248	1,606	1,355	1,063	1,272	259	1,919	1,404	2,033	2,805	3,648	1,533
Oct-Dec 2008	29,593	7,167	22,289	306	507	2,981	2,564	4,221	1,626	1,335	1,036	1,244	252	1,849	1,347	2,037	2,848	3,748	1,584
Jan-Mar 2009	29,277	7,224	21,913	322	489	2,857	2,482	4,107	1,571	1,324	1,035	1,227	273	1,842	1,309	2,009	2,919	3,772	1,579
Apr-Jun 2009	29,003	7,238	21,624	311	483	2,777	2,385	3,989	1,510	1,360	1,015	1,235	268	1,882	1,301	1,967	2,929	3,739	1,583
Jul-Sep 2009	29,170	7,278	21,745	329	491	2,813	2,311	4,031	1,463	1,428	1,020	1,246	265	1,890	1,301	1,977	2,960	3,778	1,599
Oct-Dec 2009	29,167	7,324	21,691	323	482	2,799	2,311	4,017	1,455	1,406	1,014	1,223	254	1,886	1,328	1,944	3,045	3,846	1,579
Jan-Mar 2010	28,924	7,255	21,488	336	472	2,815	2,208	3,943	1,462	1,388	1,011	1,210	263	1,882	1,319	1,930	3,062	3,824	1,549
Apr-Jun 2010	29,110	7,325	21,586	341	463	2,875	2,207	4,004	1,443	1,461	979	1,175	262	1,893	1,326	1,925	3,129	3,848	1,532
Jul-Sep 2010	29,484	7,298	21,973	369	474	2,888	2,234	4,112	1,483	1,503	1,045	1,165	301	1,852	1,341	1,925	3,120	3,821	1,594
Oct-Dec 2010	29,390	7,278	21,929	366	489	2,902	2,238	4,082	1,436	1,427	1,030	1,171	300	1,882	1,379	1,867	3,120	3,884	1,555
Jan-Mar 2011	29,350	7,315	21,842	357	497	2,860	2,212	4,034	1,451	1,417	1,059	1,179	299	1,857	1,299	1,875	3,121	3,967	1,544
Apr-Jun 2011	29,367	7,183	22,021	342	521	2,868	2,198	4,051	1,445	1,458	1,073	1,173	308	1,851	1,321	1,844	3,129	3,931	1,587
Jul-Sep 2011	29,376	7,009	22,216	370	519	2,841	2,237	4,019	1,428	1,498	1,061	1,192	300	1,888	1,339	1,852	2,986	3,981	1,639
Oct-Dec 2011	29,405	6,970	22,276	357	527	2,884	2,172	4,017	1,432	1,499	1,083	1,215	293	1,898	1,353	1,864	3,021	3,927	1,583
Jan-Mar 2012	29,365	6,952	22,209	374	534	2,851	2,148	4,017	1,430	1,438	1,124	1,199	333	1,903	1,369	1,847	3,088	3,865	1,554
Apr-Jun 2012	29,587	6,846	22,517	353	532	2,921	2,171	4,103	1,416	1,496	1,098	1,229	328	1,945	1,357	1,803	3,078	3,871	1,574
Jul-Sep 2012	29,851	6,883	22,722	356	547	2,934	2,175	4,129	1,457	1,527	1,113	1,216	329	1,983	1,372	1,827	3,021	3,948	1,607
Oct-Dec 2012	29,974	6,984	22,752	316	518	2,911	2,150	4,105	1,482	1,523	1,146	1,169	346	1,968	1,376	1,824	3,140	3,992	1,592
Jan-Mar 2013	29,750	6,989	22,504	304	537	2,921	2,102	4,031	1,487	1,504	1,161	1,150	344	2,000	1,354	1,852	3,114	3,985	1,587
Apr-Jun 2013	29,861	6,965	22,638	308	526	2,868	2,155	4,021	1,456	1,504	1,162	1,144	344	2,024	1,423	1,864	3,103	4,029	1,591
Jul-Sep 2013	30,194	7,012	22,933	324	510	2,954	2,188	3,999	1,520	1,566	1,177	1,178	326	2,073	1,450	1,880	3,081	4,036	1,642
Oct-Dec 2013	30,362	6,904	23,226	322	511	2,957	2,200	4,097	1,543	1,555	1,198	1,172	341	2,096	1,437	1,817	3,094	4,068	1,639
Jan-Mar 2014	30,446	6,934	23,311	379	508	2,945	2,229	4,063	1,490	1,579	1,150	1,171	369	2,169	1,388	1,851	3,107	4,076	1,693
Apr-Jun 2014	30,628	6,936	23,497	395	511	2,997	2,217	3,994	1,467	1,633	1,174	1,158	359	2,141	1,441	1,806	3,218	4,105	1,693
Jul-Sep 2014	30,918	6,914	23,808	386	543	3,019	2,242	3,976	1,470	1,665	1,237	1,207	340	2,139	1,487	1,853	3,197	4,062	1,781
Oct-Dec 2014	31,018	6,901	23,914	375	537	3,077	2,268	4,124	1,479	1,594	1,250	1,182	339	2,136	1,452	1,798	3,233	4,142	1,753
Jan-Mar 2015	31,074	6,980	23,915	348	567	3,007	2,266	4,066	1,512	1,618	1,276	1,185	334	2,189	1,414	1,845	3,285	4,170	1,765
Apr-Jun 2015	31,035	6,929	23,913	340	552	3,016	2,196	3,998	1,537	1,663	1,243	1,222	339	2,146	1,499	1,831	3,265	4,155	1,756
Jul-Sep 2015	31,416	6,851	24,377	349	553	2,968	2,223	4,095	1,576	1,698	1,275	1,254	348	2,173	1,539	1,841	3,266	4,162	1,812
Oct-Dec 2015	31,615	6,900	24,528	383	546	2,992	2,264	4,207	1,610	1,638	1,261	1,279	347	2,204	1,563	1,878	3,329	4,130	1,756
Jan-Mar 2016	31,499	6,924	24,414	351	528	3,008	2,271	4,131	1,608	1,663	1,225	1,262	358	2,243	1,525	1,856	3,357	4,143	1,744
Apr-Jun 2016	31,673	6,939	24,564	337	519	3,034	2,315	4,167	1,609	1,707	1,207	1,269	356	2,294	1,496	1,898	3,300	4,158	1,773
Jul-Sep 2016	31,890	6,908	24,830	357	516	3,013	2,290	4,253	1,638	1,766	1,302	1,260	354	2,250	1,511	1,936	3,336	4,100	1,809
Oct-Dec 2016	31,920	6,853	24,898	375	528	2,948	2,316	4,226	1,603	1,768	1,321	1,196	352	2,350	1,530	1,964	3,284	4,148	1,789
Jan-Mar 2017	31,878	6,840	24,876	381	539	2,908	2,336	4,136	1,560	1,747	1,323	1,195	370	2,341	1,551	1,952	3,305	4,234	1,768
Apr-Jun 2017	31,996	6,874	24,998	374	538	2,915	2,327	4,220	1,554	1,807	1,340	1,194	345	2,330	1,573	1,968	3,263	4,187	1,840
Jul-Sep 2017	32,138	6,841	25,176	352	527	2,950	2,367	4,195	1,605	1,784	1,273	1,234	343	2,331	1,589	2,001	3,266	4,184	1,907
Oct-Dec 2017	32,230	6,833	25,271	375	561	2,951	2,330	4,323	1,580	1,727	1,242	1,241	355	2,366	1,592	1,999	3,247	4,224	1,882

4.4 All in employment by industry

United Kingdom (thousands) not seasonally adjusted

	All in employment1	Public sector	Private sector	Agriculture, forestry & fishing A	Mining, energy and water supply B, D, E	Manufacturing C	Construction F	Wholesale, retail & repair of motor vehicles G	Transport & storage H	Accommodation & food services I	Information & communication J	Financial & insurance activities K	Real estate activities L	Professional, scientific & technical activities M	Administrative & support services N	Public admin & defence; social security O	Education P	Human health & social work activities Q	Other services R, S, T
Men																			
Jan-Mar 2008	15,938	2,483	13,374	223	352	2,379	2,204	2,132	1,293	631	722	692	130	1,060	804	1,039	786	756	664
Apr-Jun 2008	15,961	2,509	13,371	222	349	2,354	2,181	2,144	1,274	633	730	675	131	1,082	823	1,053	780	800	659
Jul-Sep 2008	15,990	2,488	13,422	218	370	2,316	2,249	2,177	1,275	620	752	648	143	1,066	802	1,037	741	805	704
Oct-Dec 2008	15,909	2,538	13,293	219	409	2,252	2,233	2,161	1,292	616	744	629	137	1,022	774	1,028	777	830	717
Jan-Mar 2009	15,670	2,572	13,020	247	390	2,144	2,213	2,103	1,255	604	722	624	132	1,069	715	1,003	803	828	722
Apr-Jun 2009	15,464	2,558	12,828	240	383	2,102	2,127	2,056	1,208	620	717	635	124	1,105	710	985	791	812	710
Jul-Sep 2009	15,526	2,572	12,867	252	389	2,128	2,065	2,089	1,171	645	725	642	122	1,094	695	996	831	805	726
Oct-Dec 2009	15,477	2,581	12,814	246	384	2,122	2,070	2,069	1,181	633	722	637	115	1,064	706	995	868	820	719
Jan-Mar 2010	15,326	2,533	12,694	252	380	2,154	1,971	2,037	1,181	643	710	628	118	1,089	707	984	857	792	693
Apr-Jun 2010	15,493	2,576	12,807	264	373	2,207	1,961	2,094	1,160	681	692	622	113	1,092	725	1,003	864	810	697
Jul-Sep 2010	15,773	2,601	13,047	284	381	2,213	1,992	2,159	1,173	684	748	624	132	1,070	760	994	872	804	735
Oct-Dec 2010	15,720	2,582	13,035	283	394	2,214	1,991	2,147	1,133	668	748	640	130	1,097	788	949	843	852	702
Jan-Mar 2011	15,620	2,582	12,930	270	413	2,179	1,955	2,115	1,167	643	756	647	133	1,091	724	963	848	855	697
Apr-Jun 2011	15,685	2,547	13,050	262	428	2,170	1,940	2,112	1,172	651	766	644	144	1,100	752	913	887	863	725
Jul-Sep 2011	15,666	2,453	13,136	271	422	2,167	2,005	2,082	1,149	678	752	650	154	1,105	741	904	842	851	758
Oct-Dec 2011	15,672	2,436	13,158	262	424	2,196	1,939	2,124	1,154	710	788	654	151	1,104	732	922	838	843	714
Jan-Mar 2012	15,649	2,426	13,113	270	434	2,153	1,910	2,093	1,158	678	812	635	177	1,116	741	912	867	823	696
Apr-Jun 2012	15,822	2,358	13,329	257	436	2,203	1,928	2,155	1,140	674	802	660	162	1,146	745	889	866	836	735
Jul-Sep 2012	15,969	2,385	13,424	256	440	2,220	1,942	2,159	1,173	681	803	671	156	1,151	754	922	849	841	746
Oct-Dec 2012	15,985	2,403	13,431	236	404	2,199	1,916	2,160	1,190	677	814	642	164	1,138	765	952	897	853	724
Jan-Mar 2013	15,800	2,378	13,262	220	418	2,226	1,858	2,082	1,202	696	827	638	169	1,178	743	955	865	828	706
Apr-Jun 2013	15,888	2,359	13,373	234	419	2,198	1,894	2,092	1,175	701	848	635	163	1,183	782	953	836	849	715
Jul-Sep 2013	16,121	2,402	13,573	236	400	2,241	1,912	2,109	1,227	750	849	650	149	1,211	804	956	857	860	740
Oct-Dec 2013	16,174	2,321	13,724	233	409	2,225	1,936	2,125	1,251	733	842	653	144	1,242	805	918	847	859	763
Jan-Mar 2014	16,198	2,344	13,743	271	414	2,199	1,948	2,130	1,207	756	819	666	161	1,270	769	927	842	872	767
Apr-Jun 2014	16,290	2,356	13,823	286	414	2,240	1,947	2,064	1,171	764	850	657	152	1,263	816	922	884	889	780
Jul-Sep 2014	16,473	2,373	13,981	291	437	2,255	1,968	2,067	1,152	788	891	677	162	1,256	818	947	901	873	823
Oct-Dec 2014	16,481	2,383	13,979	279	422	2,308	2,003	2,139	1,141	728	892	657	168	1,254	791	924	928	889	794
Jan-Mar 2015	16,525	2,408	14,019	267	444	2,264	1,998	2,139	1,159	755	901	638	173	1,261	776	944	948	918	813
Apr-Jun 2015	16,486	2,359	14,017	259	439	2,274	1,931	2,104	1,195	778	894	670	169	1,232	813	933	940	905	801
Jul-Sep 2015	16,737	2,351	14,287	252	442	2,223	1,968	2,163	1,231	810	930	686	173	1,227	819	915	971	931	846
Oct-Dec 2015	16,874	2,365	14,406	267	432	2,269	2,004	2,224	1,277	789	925	707	174	1,241	834	925	976	912	789
Jan-Mar 2016	16,795	2,334	14,384	252	426	2,291	2,002	2,164	1,281	787	907	710	180	1,278	814	912	938	933	794
Apr-Jun 2016	16,864	2,356	14,421	250	418	2,304	2,034	2,176	1,291	820	873	706	173	1,321	812	927	925	915	780
Jul-Sep 2016	16,982	2,391	14,528	260	413	2,255	2,005	2,239	1,305	814	941	718	174	1,298	810	943	958	915	819
Oct-Dec 2016	16,942	2,363	14,494	287	426	2,203	2,016	2,212	1,276	799	950	683	172	1,351	820	949	900	950	815
Jan-Mar 2017	16,910	2,335	14,487	288	433	2,185	2,046	2,193	1,245	805	943	681	188	1,313	838	940	902	972	794
Apr-Jun 2017	16,964	2,364	14,539	276	429	2,195	2,033	2,246	1,243	823	932	665	174	1,320	843	940	911	949	854
Jul-Sep 2017	17,031	2,316	14,651	264	417	2,183	2,046	2,243	1,274	816	897	716	172	1,341	849	943	920	924	890
Oct-Dec 2017	17,077	2,309	14,690	263	437	2,216	2,024	2,314	1,247	805	877	711	175	1,340	890	948	910	917	870

Standard Industrial Classification (SIC) 2007

4.4 All in employment by industry

United Kingdom (thousands) not seasonally adjusted

Standard Industrial Classification (SIC) 2007

	All in employment1	Public sector	Private sector	Agriculture, forestry & fishing A	Mining, energy and water supply B, D, E	Manufacturing C	Construction F	Wholesale, retail & repair of motor vehicles G	Transport & storage H	Accommodation & food services I	Information & communication J	Financial & insurance activities K	Real estate activities L	Professional, scientific & technical activities M	Administrative & support services N	Public admin & defence; social security O	Education P	Human health & social work activities Q	Other services R, S, T
Women																			
Jan-Mar 2008	13,658	4,563	9,030	112	100	769	331	2,052	311	751	302	595	133	879	594	972	2,061	2,810	845
Apr-Jun 2008	13,677	4,593	9,011	99	97	733	331	2,051	316	722	302	612	122	887	594	982	2,081	2,842	860
Jul-Sep 2008	13,695	4,551	9,086	95	93	753	334	2,071	332	734	310	624	117	853	602	996	2,065	2,843	828
Oct-Dec 2008	13,683	4,629	8,996	88	98	728	331	2,060	334	719	292	614	115	827	574	1,009	2,071	2,917	866
Jan-Mar 2009	13,606	4,652	8,894	75	99	713	270	2,004	316	719	313	603	142	773	594	1,006	2,116	2,944	857
Apr-Jun 2009	13,539	4,680	8,796	71	100	674	258	1,933	302	740	298	600	145	778	591	982	2,138	2,927	872
Jul-Sep 2009	13,644	4,705	8,878	77	102	686	246	1,942	291	783	295	605	143	796	606	981	2,130	2,973	873
Oct-Dec 2009	13,690	4,743	8,877	77	98	677	241	1,948	282	773	292	586	139	822	622	950	2,177	3,026	860
Jan-Mar 2010	13,597	4,721	8,794	84	92	661	237	1,906	281	745	301	583	144	793	613	946	2,206	3,032	856
Apr-Jun 2010	13,616	4,748	8,780	78	89	668	246	1,910	283	780	287	553	149	801	601	923	2,266	3,039	835
Jul-Sep 2010	13,710	4,697	8,926	86	93	676	241	1,953	309	819	297	542	169	782	581	931	2,247	3,017	859
Oct-Dec 2010	13,670	4,696	8,894	84	95	688	247	1,935	303	759	281	531	171	785	592	918	2,277	3,032	853
Jan-Mar 2011	13,730	4,732	8,912	87	85	681	257	1,919	285	774	303	532	167	766	575	913	2,273	3,112	847
Apr-Jun 2011	13,682	4,636	8,972	80	93	698	258	1,939	273	807	307	529	164	751	569	932	2,242	3,068	862
Jul-Sep 2011	13,710	4,556	9,080	98	96	675	232	1,937	279	820	308	542	147	783	598	948	2,144	3,130	881
Oct-Dec 2011	13,733	4,534	9,118	95	103	688	234	1,954	278	789	295	561	143	794	621	942	2,183	3,084	869
Jan-Mar 2012	13,716	4,526	9,095	104	100	698	237	1,924	272	760	312	564	156	787	628	935	2,221	3,042	858
Apr-Jun 2012	13,765	4,488	9,188	96	97	718	243	1,948	275	823	297	569	166	799	611	914	2,211	3,035	839
Jul-Sep 2012	13,882	4,499	9,298	100	107	715	233	1,970	284	846	310	545	173	832	619	905	2,172	3,107	861
Oct-Dec 2012	13,989	4,581	9,321	80	114	712	233	1,945	292	846	332	527	182	830	611	872	2,244	3,139	867
Jan-Mar 2013	13,949	4,611	9,242	84	119	695	244	1,949	285	808	333	512	175	823	611	897	2,250	3,158	880
Apr-Jun 2013	13,972	4,606	9,266	74	107	670	262	1,929	281	803	314	510	180	840	641	911	2,268	3,181	877
Jul-Sep 2013	14,073	4,610	9,359	88	110	713	276	1,891	292	816	328	527	177	862	646	924	2,224	3,176	903
Oct-Dec 2013	14,189	4,583	9,501	88	102	732	264	1,972	292	821	356	518	197	854	632	899	2,248	3,209	876
Jan-Mar 2014	14,248	4,591	9,567	108	94	746	281	1,933	283	823	332	506	208	898	620	924	2,265	3,204	896
Apr-Jun 2014	14,338	4,581	9,674	109	97	757	270	1,930	296	869	324	500	207	878	626	884	2,333	3,216	913
Jul-Sep 2014	14,445	4,541	9,827	95	106	764	274	1,909	318	877	346	530	178	884	669	906	2,296	3,190	958
Oct-Dec 2014	14,537	4,519	9,934	96	115	770	265	1,985	339	866	358	525	171	882	660	875	2,305	3,253	959
Jan-Mar 2015	14,550	4,571	9,897	81	123	743	267	1,927	353	863	375	547	161	928	639	900	2,336	3,252	952
Apr-Jun 2015	14,549	4,571	9,896	81	113	742	265	1,894	342	885	349	552	170	914	687	898	2,325	3,249	955
Jul-Sep 2015	14,679	4,500	10,090	98	111	745	256	1,932	345	888	346	568	175	945	720	926	2,295	3,231	966
Oct-Dec 2015	14,740	4,535	10,122	116	114	722	260	1,983	333	849	336	572	173	964	730	953	2,352	3,218	966
Jan-Mar 2016	14,704	4,590	10,031	99	103	717	270	1,966	327	876	319	552	178	965	711	943	2,419	3,210	950
Apr-Jun 2016	14,808	4,582	10,143	87	101	730	281	1,991	318	886	333	564	183	973	683	971	2,375	3,243	993
Jul-Sep 2016	14,908	4,517	10,302	97	103	758	285	2,014	334	953	361	543	180	952	700	994	2,378	3,185	990
Oct-Dec 2016	14,978	4,491	10,404	88	101	745	300	2,014	327	970	371	513	180	999	710	1,015	2,384	3,198	974
Jan-Mar 2017	14,967	4,505	10,389	93	106	722	290	1,943	315	942	379	514	182	1,028	713	1,012	2,403	3,262	973
Apr-Jun 2017	15,032	4,509	10,458	98	109	721	294	1,974	311	984	408	529	171	1,010	730	1,028	2,351	3,239	986
Jul-Sep 2017	15,107	4,525	10,525	88	110	767	321	1,951	330	968	376	517	171	990	740	1,059	2,346	3,259	1,017
Oct-Dec 2017	15,152	4,525	10,581	112	124	734	306	2,009	334	922	365	530	179	1,026	702	1,052	2,337	3,306	1,011

4.4 All in employment by industry

United Kingdom (thousands) not seasonally adjusted

Source: Labour Force Survey

1 The breakdown by industry sector for Q1 2009 onwards is not entirely consistent with those of previous quarters. This is because:
(a) LFS data on industrial activity were coded directly to SIC 1992 for all quarters up to and including Q4 2008 and then mapped to the new industrial classification, SIC 2007, according to the assumed relationship between the two classifications;
(b) data for Q1 2009 onwards have been coded directly to SIC 2007; and
(c) a new, automatic coding tool was introduced in January 2009.

The effect of these changes on the time series was significant for some of the industry sectors shown. Consequently some adjustments have been made to the pre-2009 estimates to account for the estimated combined effects of the new classification and the new coding tool. This also means that the pre-2009 estimates in this table are not the same as those obtained from LFS microdata.
More information and analysis of these effects are available in the Labour Force Survey User Guide (Volume 1) and from Labour Force Assessment Branch (tel 01633 455839 or email labour.market.assessment@ons.gov.uk).
Includes people with workplace outside UK and those who did not state their industry.
In the LFS the distinction between public and private sector is based on respondents' views about the organisation for which they work. The public sector estimates provided here do not correspond to the official Public Sector Employment estimates which are based on National Accounts definitions.

4.5a International Comparisons of Employment and Unemployment

		Latest period	Employment rate (%)[1][2]	Change on year %			Latest Period	Un-employment rate (%)[3]	Change on month %	Change on year %
Employment rates as published by EUROSTAT: (not seasonally adjusted)					**Unemployment rates as published by EUROSTAT on 31 January 2018** (seasonally adjusted)					
European Union (EU)					**European Union (EU)**					
Austria	YXSN	Jul-Sep 17	72.9	0.3	Austria	ZXDS	Dec 17	5.3	-0.1	-0.4
Belgium	YXSO	Jul-Sep 17	63.4	1.2	Belgium [6]	ZXDI	Dec 17	6.3	-0.3	-1.0
Bulgaria	A495	Jul-Sep 17	68.5	4.3	Bulgaria	A492	Dec 17	6.1	-0.1	-0.6
Croatia	GUMI	Jul-Sep 17	61.0	2.6	Croatia	GUMJ	Dec 17	10.0	-0.3	-2.5
Cyprus	A4AC	Jul-Sep 17	66.6	2.2	Cyprus	A4AN	Dec 17	11.3	0.2	-1.5
Czech Republic	A4AD	Jul-Sep 17	74.1	1.9	Czech Republic	A4AO	Dec 17	2.3	-0.1	-1.2
Denmark	YXSP	Jul-Sep 17	74.9	-0.3	Denmark	ZXDJ	Dec 17	5.6	0.1	-0.5
Estonia	A4AE	Jul-Sep 17	74.8	1.6	Estonia	A4AP	Nov 17	5.4	0.2	-1.3
Finland	YXSQ	Jul-Sep 17	71.0	0.5	Finland	ZXDU	Dec 17	8.7	0.0	0.0
France	YXSR	Jul-Sep 17	65.0	0.4	France	ZXDN	Dec 17	9.2	-0.1	-0.7
Germany	YXSS	Jul-Sep 17	75.6	0.5	Germany	ZXDK	Dec 17	3.6	-0.1	-0.3
Greece	YXST	Jul-Sep 17	54.6	1.6	Greece	ZXDL	Oct 17	20.7	-0.1	-2.6
Hungary	A4AF	Jul-Sep 17	68.7	1.6	Hungary	A4AQ	Nov 17	3.9	-0.1	-0.7
Ireland	YXSU	Jul-Sep 17	68.0	2.6	Ireland	ZXDO	Dec 17	6.2	-0.2	-1.3
Italy	YXSV	Jul-Sep 17	58.4	0.8	Italy	ZXDP	Dec 17	10.8	-0.1	-1.0
Latvia	A4AG	Jul-Sep 17	70.9	2.0	Latvia	A4AR	Dec 17	8.1	0.0	-1.3
Lithuania	A4AH	Jul-Sep 17	70.9	0.9	Lithuania	A4AS	Dec 17	7.1	0.1	-0.5
Luxembourg	YXSW	Jul-Sep 17	67.0	1.8	Luxembourg	ZXDQ	Dec 17	5.6	0.1	-0.6
Malta	A4AI	Jul-Sep 17	68.2	1.6	Malta	A4AT	Dec 17	3.6	0.0	-0.6
Netherlands	YXSX	Jul-Sep 17	76.3	1.0	Netherlands	ZXDR	Dec 17	4.4	0.0	-1.0
Poland	A4AJ	Jul-Sep 17	66.5	1.6	Poland	A4AU	Dec 17	4.4	-0.1	-1.1
Portugal	YXSY	Jul-Sep 17	68.5	2.5	Portugal	ZXDT	Dec 17	7.8	-0.3	-2.4
Romania	A494	Jul-Sep 17	65.3	2.2	Romania	A48Z	Dec 17	4.6	-0.1	-0.9
Slovak Republic	A4AK	Jul-Sep 17	66.4	1.3	Slovak Republic	A4AV	Dec 17	7.4	-0.1	-1.5
Slovenia	A4AL	Jul-Sep 17	70.4	4.0	Slovenia	A4AW	Dec 17	6.2	-0.2	-1.7
Spain	YXSZ	Jul-Sep 17	61.8	1.6	Spain	ZXDM	Dec 17	16.4	-0.2	-2.1
Sweden	YXTA	Jul-Sep 17	78.0	0.7	Sweden [5]	ZXDV	Dec 17	6.5	0.1	-0.4
United Kingdom (*)	ANZ6	Jul-Sep 17	74.2	0.5	United Kingdom (*)	ZXDW	Oct 17	4.3	0.1	-0.4
Total EU [4]	**A496**	**Jul-Sep 17**	**68.2**	**1.1**	**Total EU [4]**	**A493**	**Dec 17**	**7.3**	**0.0**	**-0.9**
Eurozone [4]	YXTC	Jul-Sep 17	66.8	0.9	Eurozone [4]	ZXDH	Dec 17	8.7	0.0	-1.0
Employment rates published by the OECD (seasonally adjusted)					**Unemployment rates as published by national statistical offices** (seasonally adjusted)					
Canada	A48O	Oct-Dec 17	73.8	0.9	Canada	ZXDZ	Jan 18	5.9	0.1	-0.8
Japan	A48P	Oct-Dec 17	75.7	0.9	Japan	ZXDY	Dec 17	2.8	0.1	-0.3
United States	A48Q	Oct-Dec 17	70.3	0.8	United States	ZXDX	Jan 18	4.1	0.0	-0.7

Sources: EUROSTAT, OECD, national statistical offices.
Labour market statistics enquiries: labour.market@ons.gsi.gov.uk

(*) Note: The UK rates shown in this table are as published by EUROSTAT (the EUs statistical office). See Table 1 for the latest rates

1. All employment rates shown in this table are for those aged from 15 to 64 except for the rate for the United States published by OECD which are for those aged from 16 to 64.

2. The employment rates for the EU are published by EUROSTAT and are not seasonally adjusted. EUROSTAT do not publish seasonally adjusted (SA) employment rates but SA rates for some EU countries are published by OECD. These OECD employment rates are available on our website in data table A10 at:
www.ons.gov.uk/employmentandlabourmarket/peopleinwork/employmentandemployeetypes/
datasets/internationalcomparisonsofemploymentandunemploymentratesa10

3. Unemployment rates published by EUROSTAT for most EU countries (but not for the UK), are calculated by extrapolating from the most recent LFS data using monthly registered unemployment data. A standard population basis (15-74) is used by EUROSTAT except for Spain, Italy and the UK (16-74). The unemployment rate for the US is based on those aged 16 and over, but the rates for Canada and Japan are for those aged 15 and over. All unemployment rates shown in this table are seasonally adjusted.

4. The "Total EU" series consist of all 28 EU countries. The Eurozone figures consist of the following EU countries: Austria, Belgium, Cyprus, Estonia, Finland, France, Germany, Greece, Ireland, Italy, Latvia, Lithuania, Luxembourg, Malta, Netherlands, Portugal, Slovak Republic, Slovenia and Spain.

5. The EU unemployment rates are as published on the EUROSTAT database. For Sweden the rates on the database differ from those shown in the EUROSTAT News Release published on 31 January 2018. This is because the figures for Sweden on the database are seasonally adjusted estimates but the figures for Sweden shown in the News Release are the trend component.

6. There is a discontinuity between 2016 Q4 and 2017 Q1 for Belgium due to a methodological break in the Labour Force Survey data.

4.5b Labour Disputes

	Working days lost (thousands)	Working days lost cumulative 12 month totals (thousands)	Number of stoppages	Workers involved (thousands)	Working days lost in the Public Sector (thousands)	Working days lost in the Private Sector (thousands)	Number of stoppages in the Public Sector	Number of stoppages in the Private Sector
	1	2	3	4	5	6	7	8
	BBFW		BLUU	BLUT	F8XZ	F8Y2	F8Y3	F8Y4
2015 Dec	9	170	10	11	1	8	4	6
2016 Dec	19	322	13	4	9	10	7	6
2017 Jan	17	319	17	8	8	9	11	6
2017 Feb	20	322	14	3	1	19	6	8
2017 Mar	18	305	17	5	5	14	8	9
2017 Apr	20	252	14	12	5	15	4	10
2017 May	50	280	11	10	16	34	6	5
2017 Jun	2	267	11	2	1	1	7	4
2017 Jul	50	241	12	3	1	49	3	9
2017 Aug	62	298	10	4	1	61	4	6
2017 Sep	10	300	11	3	1	9	4	7
2017 Oct	8	300	11	3	0	8	1	10
2017 Nov	8	286	11	3	0	8	1	10
2017 Dec	9	276	16	5	4	6	3	13

Source: ONS Labour Disputes Survey
Labour Disputes Statistics Helpline 01633 456724

1. Due to rounding the working days lost for the public and private sector may not add up to the working days lost.
Please note: data for workers is only available from Jan-86

4.6 Civil Service employment; regional distribution by government department[1,2]

All employees
Headcount

31 March 2017

	North East	North West	Yorkshire and The Humber	East Midlands	West Midlands	East	London	South East	South West	Wales	Scotland	Northern Ireland	Overseas	Not reported
Attorney General's Departments														
Attorney General's Office	0	0	0	0	0	0	40	0	0	0	0	0	0	0
Crown Prosecution Service	310	880	800	380	520	410	1,520	560	260	330	0	0	20	0
Crown Prosecution Service Inspectorate	0	0	10	0	0	0	0	0	0	0	0	0	0	0
Government Legal Department	:	10	20	0	0	0	1,750	0	30	0	0	0	0	0
Serious Fraud Office	0	0	0	0	0	0	410	0	0	0	0	0	0	0
Business, Energy and Industrial Strategy														
Department for Business, Energy and Industrial Strategy (excl. agencies)[4]	70	110	:	0	40	80	2,380	0	20	130	90	0	0	0
Advisory Conciliation and Arbitration Service	0	0	60	50	50	40	220	40	40	50	80	0	0	0
Companies House	0	0	0	0	0	0	10	0	0	850	20	20	0	0
Competition and Markets Authority	0	0	0	0	0	0	560	0	0	0	:	0	0	0
Insolvency Service	40	200	90	30	400	90	290	90	90	50	40	0	0	0
Land Registry	520	610	250	690	440	230	340	0	1,190	560	0	10	0	0
Met Office	0	10	20	50	20	50	30	90	1,670	10	100	10	30	0
UK Intellectual Property Office	0	0	0	0	0	0	30	0	0	1,120	0	0	0	0
UK Space Agency	0	0	0	0	0	0	10	0	90	0	0	0	0	0
Cabinet Office														
Cabinet Office (excl. agencies)	:	10	90	10	0	20	2,410	30	10	0	:	0	0	0
Other Cabinet Office agencies														
Crown Commercial Service	0	270	:	0	0	160	180	0	10	100	0	0	0	0
Government in Parliament	0	0	0	0	0	0	100	0	0	0	0	0	0	0
Chancellor's other departments														
Government Actuary's Department	0	0	0	0	0	0	170	0	0	0	0	0	0	0
National Savings and Investments	10	10	0	0	0	0	160	0	0	0	10	0	0	0
Charity Commission														
Charity Commission	0	160	0	0	0	0	70	0	70	10	0	0	0	0
Communities and Local Government														
Department for Communities and Local Government (excl. agencies)	40	80	50	30	70	70	1,050	10	60	0	0	0	0	0
Planning Inspectorate	0	0	0	0	0	0	0	0	650	20	0	0	0	0
Queen Elizabeth II Centre	0	0	0	0	0	0	50	0	0	0	0	0	0	0
Culture, Media and Sport														
Department for Culture Media and Sport	0	0	10	0	0	0	650	0	0	0	0	0	0	0

4.6 Civil Service employment; regional distribution by government department[1,2]

All employees *Headcount*

	North East	North West	Yorkshire and The Humber	East Midlands	West Midlands	East	London	South East	South West	Wales	Scotland	Northern Ireland	Overseas	Not reported
Defence														
Ministry of Defence	260	1,460	2,490	1,410	2,300	3,490	3,210	7,780	6,910	1,120	3,560	1,140	1,140	240
Defence Science and Technology Laboratory	0	0	0	10	0	10	0	1,800	1,950	0	0	0	0	0
Defence Equipment and Support	0	330	20	190	240	230	250	610	8,030	10	570	:	:	330
Royal Fleet Auxiliary	0	0	0	0	0	0	0	0	0	0	0	0	0	1,930
UK Hydrographic Office	0	0	0	0	0	:	0	0	870	0	:	0	0	0
Department for Exiting the European Union														
Department for Exiting the European Union[5]	0	0	0	0	0	0	210	0	0	0	0	0	0	0
Department for International Trade														
Department for International Trade[10]	10	10	10	:	10	0	890	:	:	:	10	0	150	0
Education														
Department for Education	350	230	530	50	140	40	1,860	0	30	:	0	0	0	0
Education Funding Agency	130	110	140	40	170	:	420	0	20	0	0	0	0	0
Standards and Testing Agency	:	:	:	:	50	0	50	0	0	0	0	0	0	0
The National College for Teaching and Leadership	10	100	20	110	50	0	40	0	:	0	0	0	0	0
Skills Funding Agency[6]	40	50	70	10	330	:	90	30	20	0	0	0	0	0
Environment, Food and Rural Affairs														
Department for Environment Food and Rural Affairs (excl. agencies)	30	30	220	10	150	50	1,330	50	60	0	0	0	0	0
Animal and Plant Health Agency	40	190	130	90	280	130	80	620	290	260	130	0	0	0
Centre for Environment Fisheries and Aquaculture Science	0	10	10	0	0	420	0	0	130	0	0	0	0	0
Rural Payments Agency	260	670	180	40	50	40	20	380	300	0	0	:	0	0
Veterinary Medicines Directorate	0	0	0	0	0	0	0	150	0	0	0	0	0	0
ESTYN														
ESTYN	0	0	0	0	0	0	0	0	0	120	0	0	0	0
Food Standards Agency														
Food Standards Agency	0	0	690	0	0	0	290	0	0	30	0	50	0	0
Foreign and Commonwealth Office														
Foreign and Commonwealth Office (excl. agencies)	0	0	0	0	0	0	2,500	290	0	0	0	0	1,700	0
FCO Services	0	0	0	0	0	0	150	530	0	0	0	0	130	0
Wilton Park Executive Agency	0	0	0	0	0	0	0	80	0	0	0	0	0	0

4.6 Civil Service employment; regional distribution by government department[1,2]

All employees   Headcount

	North East	North West	Yorkshire and The Humber	East Midlands	West Midlands	East	London	South East	South West	Wales	Scotland	Northern Ireland	Overseas	Not reported
Health														
Department of Health (excl. agencies)	0	40	440	0	0	0	860	20	0	0	0	0	0	::
Medicines and Healthcare Products Regulatory Agency	0	0	50	0	0	350	920	0	0	0	0	0	0	0
Public Health England	150	420	330	110	370	430	2,030	580	880	10	20	0	10	0
HM Revenue and Customs														
HM Revenue and Customs	11,610	12,620	4,500	3,460	4,200	2,510	9,410	3,450	1,740	4,360	8,900	1,940	0	30
Valuation Office Agency	270	410	460	220	280	240	710	470	370	260	60	0	0	0
HM Treasury														
HM Treasury	0	0	0	0	0	40	1,210	0	0	0	::	0	0	0
Debt Management Office	0	0	0	0	0	0	110	0	0	0	0	0	0	0
Government Internal Audit Agency	30	50	60	10	30	10	180	::	30	30	10	0	0	0
Office for Budget Responsibility	0	0	0	0	0	0	30	0	0	0	0	0	0	0
Home Office														
Home Office[3]	1,230	4,380	2,630	170	870	1,900	11,490	2,990	310	500	780	430	400	10
International Development														
Department for International Development	0	0	0	0	0	0	990	0	0	0	730	0	510	0
Justice														
Ministry of Justice (excl. agencies)	20	100	60	30	110	20	2,190	20	30	70	10	0	0	0
Criminal Injuries Compensation Authority	0	0	0	0	0	0	::	0	0	0	310	0	0	0
Her Majesty's Courts and Tribunals Service	740	2,220	1,310	1,430	1,580	1,130	3,820	1,460	980	910	210	0	0	0
Legal Aid Agency	290	250	50	190	120	20	330	20	90	50	0	0	0	0
National Offender Management Service	2,870	6,170	5,740	4,530	4,300	4,430	5,430	7,090	3,400	1,930	0	0	0	0
Office of the Public Guardian	0	0	0	320	800	0	20	0	0	0	0	0	::	0
The National Archives														
The National Archives[7]	0	0	0	0	0	0	600	0	0	0	0	0	0	0
National Crime Agency														
National Crime Agency	70	880	210	140	360	130	1,520	510	230	50	40	80	160	80
Northern Ireland Office														
Northern Ireland Office	0	0	0	0	0	0	40	0	0	0	0	50	0	0
Office for Standards in Education														
Office for Standards in Education, Children's Services and Skills	0	470	100	150	120	80	370	0	230	0	0	0	0	0

4.6 Civil Service employment; regional distribution by government department[1,2]

All employees   *Headcount*

	North East	North West	Yorkshire and The Humber	East Midlands	West Midlands	East	London	South East	South West	Wales	Scotland	Northern Ireland	Overseas	Not reported
Office of Gas and Electricity Markets														
Office of Gas and Electricity Markets	0	0	0	0	0	0	750	0	0	..	130	0	0	0
Office of Rail and Road														
Office of Rail and Road[8]	..	20	10	..	10	..	210	10	10	..	20	0	0	0
Office of Qualifications and Examinations Regulation														
Ofqual	0	0	0	0	190	0	0	0	0	0	0	0	0	0
Office of Water Services														
Office of Water Services[9]	0	0	0	0	130	0	60	0	0	0	0	0	0	0
Scotland Office														
Scotland Office (incl. Office of the Advocate General for Scotland)	0	0	0	0	0	0	40	0	0	0	80	0	0	0
Scottish Government														
Scottish Government (excl. agencies)	0	0	0	0	0	0	0	0	0	0	5,450	0	10	0
Accountant in Bankruptcy	0	0	0	0	0	0	0	0	0	0	140	0	0	0
Crown Office and Procurator Fiscal Service	0	0	0	0	0	0	0	0	0	0	1,740	0	0	0
Disclosure Scotland	0	0	0	0	0	0	0	0	0	0	330	0	0	0
Education Scotland	0	0	0	0	0	0	0	0	0	0	270	0	0	0
Food Standards Scotland	0	0	0	0	0	0	0	0	0	0	170	0	0	0
National Records of Scotland	0	0	0	0	0	0	0	0	0	0	380	0	0	0
Office of the Scottish Charity Regulator	0	0	0	0	0	0	0	0	0	0	50	0	0	0
Registers of Scotland	0	0	0	0	0	0	0	0	0	0	1,130	0	0	0
Revenue Scotland	0	0	0	0	0	0	0	0	0	0	50	0	0	0
Scottish Courts and Tribunals Service	0	0	0	0	0	0	0	0	0	0	1,690	0	0	0
Scottish Housing Regulator	0	0	0	0	0	0	0	0	0	0	50	0	0	0
Scottish Prison Service	0	0	0	0	0	0	0	0	0	0	4,560	0	0	0
Scottish Public Pensions Agency	0	0	0	0	0	0	0	0	0	0	310	0	0	0
Student Awards Agency for Scotland	0	0	0	0	0	0	0	0	0	0	250	0	0	0
Transport Scotland	0	0	0	0	0	0	0	0	0	0	400	0	0	0
Transport														
Department for Transport (excl. agencies)	20	0	10	20	..	0	2,010	190	10	10	0	0	0	0
Driver and Vehicle Licensing Agency	..	..	..	..	..	..	10	..	..	6,010	10	0	0	0
Driver and Vehicle Standards Agency	340	450	430	500	380	410	420	450	510	450	330	0	0	0
Maritime and Coastguard Agency	30	20	50	50	..	20	20	560	70	120	150	40	0	0
Vehicle Certification Agency	0	0	0	40	..	..	0	..	100	0	0	0	10	0

4.6 Civil Service employment; regional distribution by government department[1,2]

All employees / Headcount

	North East	North West	Yorkshire and The Humber	East Midlands	West Midlands	East	London	South East	South West	Wales	Scotland	Northern Ireland	Overseas	Not reported
United Kingdom Statistics Authority														
United Kingdom Statistics Authority	0	0	0	0	0	0	60	1,960	..	2,010	10	0	0	0
UK Export Finance														
UK Export Finance	..	..	..	..	..	..	250	..	..	..	..	0	0	0
UK Supreme Court														
UK Supreme Court	0	0	0	0	0	0	50	0	0	0	0	0	0	0
Wales Office														
Wales Office	0	0	0	0	0	0	20	0	0	20	0	0	0	0
Welsh Government														
Welsh Government	0	0	0	0	0	0	10	0	0	5,380	0	0	10	0
Work and Pensions														
Department for Work and Pensions	8,990	17,160	8,690	4,290	7,350	4,020	7,950	5,180	5,080	5,450	9,620	..	0	280
The Health and Safety Executive	70	920	350	470	90	130	150	100	80	100	250	0	0	0
All employees	28,850	52,100	31,380	19,260	26,590	21,400	78,070	38,200	36,950	32,440	43,220	3,760	4,280	2,900

Source: Annual Civil Service Employment Survey

1 Numbers are rounded to the nearest ten, and cells containing between one and five employees are represented by ".."

2 Workplace postcode data are used to derive geographical information.

3 Home Office estimates for 31 March 2017 include staff paid via the Foreign and Commonwealth Office employee records system (PRISM).

4 Business, Innovation & Skills became Department for Business, Energy and Industrial Strategy from July 2016. Department of Energy and Climate Change and National Measurement and Regulation Office also became part of Department for Business, Energy and Industrial Strategy.

5 The Department for Exiting the European Union (DExEU) was formed in July 2016. DExEU employment has been reported separately for the first time in this release. The total DExEU headcount was approximately 350 as at 31 March 2017. As all DExEU employees are on loan, some still remain on home department records. To avoid double counting, DExEU has only reported the employees officially transferred as at 31 March 2017.

6 Skills Funding Agency previously reported under Business, Innovation and Skills departmental hierarchy is now reported under Education departmental hierarchy.

7 The National Archives, a non-ministerial department, is now reported as a standalone entry. It was previously reported under the Culture, Media and Sport departmental hierarchy.

8 Office of Rail and Road previously reported under the Transport departmental hierarchy is now reported as a standalone department, which is in line with other regulators.

9 Office of Water Services previously reported under the Environment, Food and Rural Affairs departmental hierarchy is now reported as a standalone non-ministerial department.

10 Department for International Trade is a new department formed July 2016 in the wake of the referendum vote to leave the European Union and replaces UK Trade and Investments.

4.7 Unemployment by age and duration

United Kingdom (thousands) seasonally adjusted

	All aged 16 & over							All aged 16 - 64						
	All level	rate (%)[1]	Up to 6 months level	Over 6 and up to 12 months level	All over 12 months level	% over 12 months	All over 24 months level	All level	rate (%)[1]	Up to 6 months level	Over 6 and up to 12 months level	All over 12 months level	% over 12 months	All over 24 months level
People	MGSC	MGSX	YBWF	YBWG	YBWH	YBWI	YBWL	LF2I	LF2Q	LF2Y	LF32	LF34	LF36	LF38
Jul-Sep 2014	1,961	6.0	954	317	690	35.2	377	1,940	6.1	948	317	675	34.8	371
Aug-Oct 2014	1,953	6.0	937	332	685	35.1	377	1,934	6.1	931	330	672	34.8	371
Sep-Nov 2014	1,924	5.9	950	317	657	34.1	355	1,904	6.0	938	312	653	34.3	350
Oct-Dec 2014	1,870	5.7	930	302	638	34.1	340	1,851	5.8	920	298	633	34.2	333
Nov-Jan 2015	1,858	5.7	945	285	628	33.8	339	1,837	5.8	933	281	623	33.9	333
Dec-Feb 2015	1,840	5.6	934	283	623	33.9	352	1,820	5.7	928	280	612	33.6	343
Jan-Mar 2015	1,826	5.5	954	283	589	32.2	342	1,803	5.7	947	278	578	32.0	333
Feb-Apr 2015	1,813	5.5	949	290	574	31.6	327	1,793	5.6	944	286	563	31.4	318
Mar-May 2015	1,848	5.6	974	304	570	30.9	318	1,830	5.8	970	301	559	30.6	311
Apr-Jun 2015	1,849	5.6	967	304	577	31.2	325	1,826	5.7	958	301	567	31.1	318
May-Jul 2015	1,822	5.5	981	295	546	30.0	309	1,799	5.7	974	291	534	29.7	303
Jun-Aug 2015	1,779	5.4	961	290	529	29.7	302	1,755	5.5	953	286	516	29.4	298
Jul-Sep 2015	1,760	5.3	965	279	516	29.3	290	1,740	5.5	959	275	505	29.0	287
Aug-Oct 2015	1,715	5.2	938	265	512	29.8	288	1,695	5.3	931	262	502	29.6	283
Sep-Nov 2015	1,687	5.1	938	258	491	29.1	269	1,668	5.2	925	254	489	29.3	263
Oct-Dec 2015	1,688	5.1	946	252	490	29.0	261	1,669	5.2	934	248	487	29.2	256
Nov-Jan 2016	1,684	5.1	946	257	481	28.6	258	1,665	5.2	934	254	477	28.6	253
Dec-Feb 2016	1,700	5.1	966	264	471	27.7	264	1,679	5.2	956	259	464	27.6	258
Jan-Mar 2016	1,687	5.1	949	270	468	27.7	264	1,665	5.2	937	268	461	27.7	259
Feb-Apr 2016	1,666	5.0	951	254	461	27.7	262	1,643	5.1	939	252	452	27.5	257
Mar-May 2016	1,648	4.9	950	235	462	28.0	258	1,626	5.1	940	234	452	27.8	253
Apr-Jun 2016	1,643	4.9	959	237	447	27.2	253	1,625	5.1	949	235	441	27.2	249
May-Jul 2016	1,635	4.9	957	230	448	27.4	257	1,618	5.0	952	229	437	27.0	252
Jun-Aug 2016	1,668	5.0	968	256	445	26.6	247	1,653	5.1	965	256	433	26.2	243
Jul-Sep 2016	1,618	4.8	945	239	434	26.8	241	1,601	5.0	941	239	421	26.3	237
Aug-Oct 2016	1,613	4.8	951	245	418	25.9	227	1,596	5.0	944	244	408	25.6	223
Sep-Nov 2016	1,602	4.8	958	237	407	25.4	215	1,585	4.9	944	235	406	25.6	213
Oct-Dec 2016	1,585	4.7	937	244	404	25.5	210	1,568	4.9	923	242	403	25.7	208
Nov-Jan 2017	1,573	4.7	935	244	395	25.1	211	1,557	4.8	921	242	394	25.3	209
Dec-Feb 2017	1,550	4.6	912	247	391	25.2	211	1,537	4.8	902	246	389	25.3	208
Jan-Mar 2017	1,527	4.6	896	247	385	25.2	200	1,514	4.7	887	245	382	25.3	198
Feb-Apr 2017	1,520	4.5	890	247	383	25.2	198	1,507	4.7	884	246	378	25.0	195
Mar-May 2017	1,489	4.4	869	239	381	25.6	199	1,475	4.6	865	236	374	25.3	197
Apr-Jun 2017	1,485	4.4	876	236	373	25.1	197	1,466	4.5	868	230	368	25.1	195
May-Jul 2017	1,457	4.3	837	239	382	26.2	207	1,432	4.4	831	231	369	25.8	204
Jun-Aug 2017	1,452	4.3	841	229	382	26.3	211	1,428	4.4	834	223	370	25.9	208
Jul-Sep 2017	1,429	4.3	822	216	391	27.3	213	1,407	4.4	818	212	377	26.8	209
Aug-Oct 2017	1,422	4.2	831	215	376	26.4	205	1,400	4.3	820	211	368	26.3	201
Sep-Nov 2017	1,437	4.3	831	223	383	26.7	202	1,415	4.4	817	216	382	27.0	198
Oct-Dec 2017	1,463	4.4	855	240	368	25.2	195	1,442	4.4	840	233	369	25.6	192
Men	MGSD	MGSY	MGYK	MGYM	MGYO	YBWJ	YBWM	YBSI	YBTJ	YBWP	YBWS	YBWV	YBWY	YBXB
Jul-Sep 2014	1,088	6.2	491	171	426	39.2	233	1,072	6.4	486	170	417	38.9	228
Aug-Oct 2014	1,086	6.2	486	178	422	38.9	233	1,072	6.4	480	175	417	38.9	228
Sep-Nov 2014	1,069	6.1	489	173	408	38.1	225	1,054	6.3	479	168	406	38.5	221
Oct-Dec 2014	1,041	6.0	477	163	401	38.5	221	1,027	6.1	471	160	396	38.6	216
Nov-Jan 2015	1,025	5.9	481	151	393	38.3	223	1,009	6.0	475	148	386	38.3	218
Dec-Feb 2015	1,014	5.8	474	152	388	38.3	234	1,000	5.9	471	150	380	38.0	227
Jan-Mar 2015	1,004	5.7	479	154	371	36.9	232	988	5.9	476	151	361	36.6	226
Feb-Apr 2015	1,007	5.7	486	159	362	36.0	217	992	5.9	483	157	352	35.5	211
Mar-May 2015	1,007	5.8	493	161	354	35.1	208	992	5.9	490	158	345	34.7	203
Apr-Jun 2015	1,014	5.8	498	160	356	35.1	211	1,001	5.9	492	158	350	35.0	207
May-Jul 2015	988	5.6	504	151	332	33.7	194	974	5.8	501	150	323	33.1	190
Jun-Aug 2015	976	5.5	500	152	323	33.2	190	960	5.7	496	150	315	32.8	188
Jul-Sep 2015	962	5.5	503	144	314	32.7	184	948	5.6	499	142	308	32.4	181
Aug-Oct 2015	937	5.3	485	144	307	32.8	186	924	5.4	479	141	303	32.8	182
Sep-Nov 2015	924	5.2	493	141	290	31.4	173	914	5.4	484	138	292	31.9	169
Oct-Dec 2015	921	5.2	490	141	291	31.5	172	911	5.4	482	138	290	31.9	169
Nov-Jan 2016	922	5.2	488	145	290	31.4	170	912	5.4	481	142	288	31.6	168
Dec-Feb 2016	933	5.2	495	149	289	31.0	180	922	5.4	490	146	286	31.0	178
Jan-Mar 2016	915	5.1	473	156	286	31.3	175	905	5.3	468	155	282	31.1	173
Feb-Apr 2016	896	5.0	469	142	284	31.7	174	885	5.2	464	142	280	31.6	172
Mar-May 2016	898	5.1	485	133	280	31.2	173	888	5.2	480	133	276	31.0	170
Apr-Jun 2016	891	5.0	489	132	269	30.2	171	883	5.2	484	131	268	30.3	169
May-Jul 2016	905	5.1	503	130	272	30.0	178	896	5.2	501	131	264	29.5	175
Jun-Aug 2016	901	5.0	490	133	277	30.8	170	894	5.2	488	134	272	30.4	168
Jul-Sep 2016	884	5.0	487	126	270	30.6	168	876	5.1	485	127	264	30.2	165

4.7 Unemployment by age and duration

United Kingdom (thousands) seasonally adjusted

	All aged 16 & over							All aged 16 - 64						
	All level	rate (%)[1]	Up to 6 months level	Over 6 and up to 12 months level	All over 12 months level	% over 12 months	All over 24 months level	All level	rate (%)[1]	Up to 6 months level	Over 6 and up to 12 months level	All over 12 months level	% over 12 months	All over 24 months level
Men *(contd.)*	MGSD	MGSY	MGYK	MGYM	MGYO	YBWJ	YBWM	YBSI	YBTJ	YBWP	YBWS	YBWV	YBWY	YBXB
Aug-Oct 2016	885	5.0	495	129	261	29.5	156	877	5.2	488	129	259	29.5	154
Sep-Nov 2016	883	5.0	500	131	252	28.5	143	873	5.1	489	129	255	29.1	142
Oct-Dec 2016	873	4.9	495	132	246	28.2	138	864	5.1	486	131	247	28.6	138
Nov-Jan 2017	864	4.9	493	128	243	28.1	144	855	5.0	485	126	244	28.5	144
Dec-Feb 2017	851	4.8	483	128	239	28.1	146	842	4.9	477	127	237	28.2	146
Jan-Mar 2017	844	4.7	476	133	236	27.9	138	833	4.9	469	132	232	27.9	137
Feb-Apr 2017	834	4.7	457	144	233	27.9	136	823	4.8	451	143	228	27.7	134
Mar-May 2017	823	4.6	452	142	229	27.8	133	813	4.8	448	141	224	27.6	131
Apr-Jun 2017	815	4.6	451	140	224	27.5	132	802	4.7	444	135	222	27.7	130
May-Jul 2017	791	4.4	422	139	231	29.2	135	775	4.5	418	135	223	28.8	134
Jun-Aug 2017	780	4.4	417	140	223	28.6	127	764	4.5	412	136	217	28.3	126
Jul-Sep 2017	770	4.3	401	136	232	30.1	129	756	4.4	399	133	224	29.7	128
Aug-Oct 2017	777	4.4	424	128	225	29.0	122	762	4.5	416	124	222	29.1	120
Sep-Nov 2017	769	4.3	413	127	229	29.8	127	755	4.4	402	121	231	30.7	125
Oct-Dec 2017	778	4.4	420	137	221	28.4	123	762	4.5	407	132	223	29.3	122
Women	MGSE	MGSZ	MGYL	MGYN	MGYP	YBWK	YBWN	LF2J	LF2R	LF2Z	LF33	LF35	LF37	LF39
Jul-Sep 2014	874	5.7	463	146	264	30.2	144	868	5.8	463	147	258	29.7	143
Aug-Oct 2014	867	5.7	451	153	263	30.3	144	862	5.8	451	155	256	29.7	143
Sep-Nov 2014	855	5.6	461	144	249	29.2	130	850	5.7	459	144	247	29.1	129
Oct-Dec 2014	829	5.4	453	139	237	28.6	119	825	5.5	450	138	237	28.7	118
Nov-Jan 2015	833	5.4	464	134	236	28.3	116	828	5.6	459	133	237	28.6	115
Dec-Feb 2015	826	5.4	460	131	235	28.4	118	820	5.5	457	130	232	28.3	115
Jan-Mar 2015	822	5.3	475	129	218	26.5	110	815	5.5	471	128	217	26.6	107
Feb-Apr 2015	806	5.2	464	130	212	26.3	110	801	5.4	461	129	211	26.3	107
Mar-May 2015	841	5.5	481	143	217	25.8	110	838	5.6	480	143	214	25.6	108
Apr-Jun 2015	835	5.4	470	145	221	26.4	114	826	5.5	466	142	217	26.3	111
May-Jul 2015	834	5.4	477	144	213	25.6	115	825	5.5	473	141	211	25.6	112
Jun-Aug 2015	804	5.2	460	138	206	25.6	112	795	5.3	457	136	201	25.3	110
Jul-Sep 2015	798	5.2	462	135	201	25.2	106	791	5.3	460	133	197	25.0	106
Aug-Oct 2015	779	5.0	453	121	205	26.3	102	771	5.1	452	120	199	25.8	101
Sep-Nov 2015	763	4.9	445	118	201	26.3	96	754	5.0	441	116	197	26.2	94
Oct-Dec 2015	766	5.0	456	111	199	26.0	90	758	5.1	452	110	197	25.9	88
Nov-Jan 2016	762	4.9	458	112	192	25.1	88	753	5.0	453	112	189	25.1	85
Dec-Feb 2016	767	5.0	471	115	181	23.6	83	757	5.0	466	114	178	23.5	80
Jan-Mar 2016	772	5.0	476	114	182	23.6	89	760	5.1	468	113	179	23.6	86
Feb-Apr 2016	770	5.0	482	111	177	23.0	88	758	5.0	475	110	173	22.8	85
Mar-May 2016	749	4.8	465	103	181	24.2	85	738	4.9	460	102	176	23.9	83
Apr-Jun 2016	752	4.8	469	105	178	23.7	82	742	4.9	465	103	173	23.4	80
May-Jul 2016	731	4.7	454	100	177	24.2	79	722	4.8	450	98	173	24.0	77
Jun-Aug 2016	768	4.9	478	122	167	21.8	77	759	5.0	477	122	160	21.1	75
Jul-Sep 2016	734	4.7	457	112	164	22.3	73	725	4.8	456	112	157	21.6	71
Aug-Oct 2016	728	4.7	456	115	157	21.5	71	719	4.7	455	115	149	20.7	69
Sep-Nov 2016	719	4.6	457	106	156	21.7	72	711	4.7	454	105	152	21.3	71
Oct-Dec 2016	713	4.6	442	112	158	22.2	72	704	4.6	437	111	155	22.1	71
Nov-Jan 2017	709	4.5	442	116	152	21.4	67	702	4.6	436	116	151	21.5	65
Dec-Feb 2017	700	4.5	429	119	152	21.7	64	695	4.6	424	118	152	21.9	63
Jan-Mar 2017	683	4.4	420	114	149	21.8	62	680	4.5	418	113	150	22.0	61
Feb-Apr 2017	686	4.4	432	103	151	22.0	62	685	4.5	433	103	149	21.8	61
Mar-May 2017	666	4.2	417	96	152	22.9	66	662	4.3	417	95	150	22.6	66
Apr-Jun 2017	670	4.3	425	96	149	22.2	64	664	4.4	424	95	146	22.0	64
May-Jul 2017	666	4.2	415	100	151	22.6	72	656	4.3	414	97	146	22.3	69
Jun-Aug 2017	672	4.3	424	89	159	23.6	84	664	4.3	423	88	154	23.2	82
Jul-Sep 2017	659	4.2	421	80	158	24.0	84	651	4.3	420	79	152	23.4	81
Aug-Oct 2017	645	4.1	407	88	150	23.3	83	637	4.2	404	87	146	23.0	81
Sep-Nov 2017	668	4.2	418	96	154	23.1	76	660	4.3	415	95	150	22.8	73
Oct-Dec 2017	685	4.3	435	102	147	21.5	72	680	4.4	432	101	146	21.5	70

4.7 Unemployment by age and duration

	16-17							18-24						
People	All level	rate (%)[1]	Up to 6 months level	Over 6 and up to 12 months level	All over 12 months level	% over 12 months level	All over 24 months level	All level	rate (%)[1]	Up to 6 months level	Over 6 and up to 12 months level	All over 12 months level	% over 12 months level	All over 24 months level
	YBVH	YBVK	YBXD	YBXG	YBXJ	YBXM	YBXP	YBVN	YBVQ	YBXS	YBXV	YBXY	YBYB	YBYE
Jul-Sep 2014	162	33.1	115	30	16	10.0	*	587	14.4	305	89	192	32.8	87
Aug-Oct 2014	160	32.8	111	32	17	10.6	*	598	14.7	308	87	202	33.8	96
Sep-Nov 2014	158	32.3	110	29	18	11.7	*	614	15.2	332	95	187	30.4	84
Oct-Dec 2014	160	31.6	113	28	18	11.6	*	591	14.5	325	83	183	31.0	84
Nov-Jan 2015	154	30.8	110	24	20	12.7	*	591	14.5	332	78	180	30.5	83
Dec-Feb 2015	155	30.9	114	22	19	12.1	*	586	14.3	327	80	179	30.6	92
Jan-Mar 2015	146	29.5	108	21	17	11.6	*	584	14.2	343	86	155	26.6	81
Feb-Apr 2015	149	29.9	109	22	18	12.2	*	582	14.2	349	87	145	25.0	71
Mar-May 2015	145	28.6	109	20	15	10.7	*	574	14.1	343	92	140	24.3	66
Apr-Jun 2015	143	28.6	109	18	16	10.9	*	581	14.1	343	93	145	24.9	65
May-Jul 2015	145	28.1	113	20	12	8.4	*	569	13.9	344	89	136	24.0	62
Jun-Aug 2015	144	28.1	111	21	12	8.5	*	536	13.1	327	83	126	23.5	54
Jul-Sep 2015	141	27.0	106	21	14	9.6	*	516	12.6	321	76	120	23.2	51
Aug-Oct 2015	127	24.9	98	18	11	8.5	*	498	12.2	312	71	116	23.2	52
Sep-Nov 2015	129	25.7	97	17	15	11.6	*	503	12.3	316	74	114	22.6	47
Oct-Dec 2015	134	26.8	102	17	14	10.7	*	492	12.0	307	74	111	22.7	44
Nov-Jan 2016	133	26.9	98	21	14	10.6	*	499	12.2	311	78	110	22.1	43
Dec-Feb 2016	129	26.0	101	16	12	9.5	*	502	12.3	323	80	99	19.7	46
Jan-Mar 2016	127	25.6	105	14	8	6.5	*	500	12.2	318	87	95	19.0	46
Feb-Apr 2016	133	27.2	109	14	9	6.9	*	482	11.9	312	79	92	19.0	44
Mar-May 2016	139	28.3	112	17	9	6.7	*	472	11.6	308	72	92	19.5	42
Apr-Jun 2016	143	28.8	116	17	10	7.1	*	475	11.7	313	72	90	19.0	40
May-Jul 2016	143	29.1	113	16	14	9.8	*	480	11.9	321	67	92	19.1	42
Jun-Aug 2016	140	28.7	112	15	13	9.2	*	488	12.0	313	81	94	19.2	41
Jul-Sep 2016	124	26.7	96	15	13	10.2	*	473	11.7	305	78	90	19.0	43
Aug-Oct 2016	126	27.4	100	14	13	10.0	*	458	11.4	289	86	83	18.2	40
Sep-Nov 2016	124	25.7	94	17	13	10.2	*	446	11.1	289	74	83	18.6	33
Oct-Dec 2016	122	24.6	92	17	12	9.9	*	440	11.0	281	78	81	18.5	32
Nov-Jan 2017	113	23.0	87	18	8	6.7	*	438	11.0	281	76	81	18.6	35
Dec-Feb 2017	125	25.1	99	19	7	5.3	*	432	10.9	284	74	75	17.3	34
Jan-Mar 2017	133	26.7	104	20	9	6.5	*	423	10.6	275	71	77	18.3	31
Feb-Apr 2017	132	26.5	104	21	8	5.7	*	428	10.7	277	72	78	18.3	32
Mar-May 2017	118	25.4	92	18	8	6.6	*	438	11.0	277	80	80	18.3	35
Apr-Jun 2017	111	24.7	90	16	5	4.8	*	434	10.9	281	81	73	16.7	33
May-Jul 2017	107	24.1	83	19	6	5.7	*	425	10.7	275	77	73	17.1	30
Jun-Aug 2017	106	23.5	83	18	*	*	*	427	10.9	271	76	80	18.6	29
Jul-Sep 2017	105	23.5	80	19	6	6.2	*	417	10.7	261	71	86	20.5	31
Aug-Oct 2017	110	24.1	86	18	5	4.7	*	410	10.6	264	67	79	19.2	27
Sep-Nov 2017	118	25.7	92	19	7	5.6	*	417	10.6	274	64	79	18.9	27
Oct-Dec 2017	120	26.6	94	20	7	5.5	*	420	10.8	285	64	72	17.1	26
Men	YBVI	YBVL	YBXE	YBXH	YBXK	YBXN	YBXQ	YBVO	YBVR	YBXT	YBXW	YBXZ	YBYC	YBYF
Jul-Sep 2014	79	36.1	56	11	11	13.7	*	357	16.6	168	60	130	36.3	59
Aug-Oct 2014	78	35.5	53	14	11	13.5	*	356	16.6	165	57	134	37.6	63
Sep-Nov 2014	81	36.2	55	16	9	11.7	*	353	16.5	172	57	124	35.0	55
Oct-Dec 2014	86	36.7	62	15	9	10.5	*	343	16.0	174	44	125	36.4	57
Nov-Jan 2015	85	37.0	62	12	10	12.3	*	345	16.0	176	42	127	36.8	64
Dec-Feb 2015	92	39.0	69	12	12	12.6	*	339	15.6	170	42	127	37.4	69
Jan-Mar 2015	86	37.3	66	11	9	11.0	*	340	15.7	182	48	111	32.5	61
Feb-Apr 2015	87	37.3	66	11	10	11.1	*	343	15.9	187	53	103	30.1	53
Mar-May 2015	81	34.8	63	10	8	9.3	*	338	15.8	187	52	98	29.1	49
Apr-Jun 2015	80	34.3	61	11	8	10.1	*	338	15.7	188	51	100	29.5	47
May-Jul 2015	81	33.4	62	11	8	9.6	*	333	15.5	193	46	94	28.3	42
Jun-Aug 2015	77	32.6	57	13	8	10.3	*	317	14.6	188	44	85	27.0	36
Jul-Sep 2015	79	32.4	56	14	9	11.9	*	293	13.6	180	36	77	26.4	34
Aug-Oct 2015	74	29.9	53	14	7	10.1	*	278	12.9	170	33	74	26.7	35
Sep-Nov 2015	75	30.3	51	12	12	16.4	*	289	13.4	180	39	70	24.3	32
Oct-Dec 2015	72	29.3	51	11	11	14.6	*	281	13.1	172	40	68	24.3	27
Nov-Jan 2016	73	30.5	47	15	10	14.2	*	288	13.4	176	42	70	24.4	27
Dec-Feb 2016	71	29.2	50	12	9	12.7	*	285	13.3	174	45	66	23.2	30
Jan-Mar 2016	70	28.7	51	11	8	10.8	*	282	13.2	171	47	64	22.8	30
Feb-Apr 2016	69	29.4	53	10	7	9.7	*	275	13.0	170	45	60	21.8	28
Mar-May 2016	73	31.3	58	10	*	*	*	273	12.9	168	42	63	23.1	28
Apr-Jun 2016	77	32.4	63	11	*	*	*	282	13.4	176	44	62	21.9	30
May-Jul 2016	81	34.5	65	11	6	6.8	*	297	14.1	190	44	63	21.2	33
Jun-Aug 2016	79	33.4	62	11	6	7.1	*	291	13.8	180	46	65	22.2	32
Jul-Sep 2016	67	30.7	51	11	5	7.4	*	288	13.7	178	45	65	22.6	33

4.7 Unemployment by age and duration

United Kingdom (thousands) seasonally adjusted

	16-17							18-24						
	All level	rate (%)[1]	Up to 6 months level	Over 6 and up to 12 months level	All over 12 months level	% over 12 months level	All over 24 months level	All level	rate (%)[1]	Up to 6 months level	Over 6 and up to 12 months level	All over 12 months level	% over 12 months level	All over 24 months level
Men *(contd.)*	YBVI	YBVL	YBXE	YBXH	YBXK	YBXN	YBXQ	YBVO	YBVR	YBXT	YBXW	YBXZ	YBYC	YBYF
Aug-Oct 2016	64	30.0	52	9	*	*	*	285	13.6	169	55	62	21.6	31
Sep-Nov 2016	65	27.5	52	10	*	*	*	280	13.4	168	52	60	21.4	23
Oct-Dec 2016	66	27.3	52	11	*	*	*	277	13.3	168	54	55	20.0	21
Nov-Jan 2017	65	27.0	52	11	*	*	*	270	12.9	163	51	56	20.6	22
Dec-Feb 2017	63	26.3	50	9	*	*	*	269	12.8	170	47	52	19.4	25
Jan-Mar 2017	63	26.1	50	9	5	7.3	*	261	12.5	161	47	53	20.4	22
Feb-Apr 2017	64	26.7	51	10	*	*	*	257	12.4	153	47	57	22.1	26
Mar-May 2017	62	26.8	49	9	*	*	*	267	12.8	155	57	54	20.4	28
Apr-Jun 2017	63	27.3	51	9	*	*	*	254	12.2	151	54	50	19.6	25
May-Jul 2017	58	26.8	47	8	*	*	*	249	12.1	145	55	49	19.7	21
Jun-Aug 2017	57	25.8	46	9	*	*	*	247	12.1	138	54	55	22.1	20
Jul-Sep 2017	50	23.3	41	8	*	*	*	250	12.4	134	52	63	25.4	23
Aug-Oct 2017	54	24.9	44	8	*	*	*	249	12.4	142	48	60	23.9	20
Sep-Nov 2017	58	26.5	47	8	*	*	*	242	11.8	143	42	57	23.7	20
Oct-Dec 2017	60	28.0	47	11	*	*	*	241	11.9	152	40	50	20.6	20
Women	YBVJ	YBVM	YBXF	YBXI	YBXL	YBXO	YBXR	YBVP	YBVS	YBXU	YBXX	YBYA	YBYD	YBYG
Jul-Sep 2014	83	30.7	59	19	6	6.6	*	230	11.9	137	29	63	27.3	28
Aug-Oct 2014	82	30.7	58	18	6	7.9	*	242	12.6	143	31	68	28.2	32
Sep-Nov 2014	77	29.0	55	13	9	11.7	*	262	13.7	160	38	63	24.3	29
Oct-Dec 2014	74	27.2	52	13	10	12.9	*	248	12.9	150	39	59	23.7	26
Nov-Jan 2015	69	25.5	48	12	9	13.1	*	246	12.7	156	37	53	21.7	19
Dec-Feb 2015	63	23.7	45	11	7	11.3	*	248	12.8	157	38	53	21.3	23
Jan-Mar 2015	60	22.6	42	10	7	12.5	*	244	12.6	161	38	45	18.4	20
Feb-Apr 2015	62	23.3	42	11	9	13.9	*	239	12.3	162	34	42	17.6	18
Mar-May 2015	63	23.2	46	10	8	12.4	*	236	12.2	156	39	42	17.6	16
Apr-Jun 2015	63	23.6	48	8	8	11.9	*	242	12.4	155	42	45	18.6	18
May-Jul 2015	64	23.4	51	8	*	*	*	236	12.1	152	42	42	17.8	20
Jun-Aug 2015	66	24.2	54	8	*	*	*	219	11.3	139	39	41	18.5	18
Jul-Sep 2015	61	22.2	50	7	*	*	*	223	11.5	141	39	42	19.0	17
Aug-Oct 2015	53	20.1	45	5	*	*	*	220	11.4	142	37	42	18.8	17
Sep-Nov 2015	54	21.2	46	5	*	*	*	215	11.0	136	35	44	20.3	16
Oct-Dec 2015	61	24.4	51	6	*	*	*	210	10.8	134	33	43	20.5	17
Nov-Jan 2016	60	23.6	51	6	*	*	*	211	10.8	135	37	40	18.9	16
Dec-Feb 2016	58	22.9	51	*	*	*	*	217	11.1	149	35	33	15.2	16
Jan-Mar 2016	57	22.5	53	*	*	*	*	218	11.2	147	40	31	14.2	17
Feb-Apr 2016	63	25.2	56	5	*	*	*	207	10.7	142	33	32	15.3	16
Mar-May 2016	66	25.6	54	7	5	7.6	*	199	10.2	140	30	29	14.7	14
Apr-Jun 2016	66	25.6	53	7	7	10.3	*	193	9.9	137	27	29	14.8	10
May-Jul 2016	61	24.0	48	*	8	13.7	*	183	9.4	131	24	29	15.7	9
Jun-Aug 2016	61	24.3	50	*	7	11.9	*	197	10.1	133	35	29	14.7	9
Jul-Sep 2016	57	23.2	46	*	8	13.5	*	185	9.5	127	33	25	13.4	10
Aug-Oct 2016	62	25.0	48	5	9	14.6	*	173	8.9	120	31	22	12.5	9
Sep-Nov 2016	59	23.9	42	6	11	17.9	*	166	8.6	121	22	23	14.0	11
Oct-Dec 2016	56	22.1	40	6	9	16.7	*	163	8.5	114	24	26	15.9	12
Nov-Jan 2017	48	19.2	35	7	6	11.7	*	168	8.9	117	25	26	15.2	13
Dec-Feb 2017	62	24.0	49	9	*	*	*	163	8.7	114	27	23	13.8	9
Jan-Mar 2017	70	27.2	55	12	*	*	*	162	8.6	114	24	24	14.9	8
Feb-Apr 2017	68	26.3	53	11	*	*	*	171	9.0	124	25	22	12.7	7
Mar-May 2017	56	24.0	43	9	*	*	*	171	8.9	122	23	26	15.1	8
Apr-Jun 2017	49	22.0	39	7	*	*	*	180	9.5	130	27	23	12.8	8
May-Jul 2017	49	21.5	36	10	*	*	*	176	9.3	130	22	24	13.5	9
Jun-Aug 2017	49	21.3	36	9	*	*	*	181	9.6	133	22	25	13.9	9
Jul-Sep 2017	55	23.6	39	11	5	8.8	*	167	8.9	127	18	22	13.2	8
Aug-Oct 2017	56	23.5	43	10	*	*	*	161	8.6	122	20	19	11.9	7
Sep-Nov 2017	60	25.1	45	11	*	*	*	175	9.3	131	22	21	12.3	7
Oct-Dec 2017	60	25.3	47	9	*	*	*	179	9.6	133	24	22	12.4	6

4.7 Unemployment by age and duration

United Kingdom (thousands) seasonally adjusted

	25-49							50 and over						
	All level	rate (%)[1]	Up to 6 months level	Over 6 and up to 12 months level	All over 12 months level	% over 12 months level	All over 24 months level	All level	rate (%)[1]	Up to 6 months level	Over 6 and up to 12 months level	All over 12 months level	% over 12 months level	All over 24 months level
People	MGVI	MGXB	YBYH	YBYK	YBYN	YBYQ	YBYT	YBVT	YBVW	YBYW	YBYZ	YBZC	YBZF	YBZI
Jul-Sep 2014	872	4.7	385	156	331	38.0	183	340	3.6	148	42	150	44.2	103
Aug-Oct 2014	860	4.6	376	166	317	36.9	174	336	3.5	141	46	149	44.3	104
Sep-Nov 2014	828	4.4	375	147	305	36.9	167	324	3.4	133	46	146	45.0	100
Oct-Dec 2014	802	4.3	370	144	287	35.8	155	317	3.3	121	47	149	46.9	101
Nov-Jan 2015	808	4.3	378	136	294	36.4	160	306	3.2	125	47	134	43.8	94
Dec-Feb 2015	793	4.2	364	133	295	37.2	164	307	3.2	129	48	130	42.4	94
Jan-Mar 2015	785	4.2	376	129	280	35.6	164	310	3.2	127	47	137	44.1	95
Feb-Apr 2015	781	4.2	365	134	282	36.2	167	302	3.1	127	47	128	42.4	89
Mar-May 2015	813	4.4	387	137	289	35.5	169	317	3.3	135	55	127	40.0	84
Apr-Jun 2015	803	4.3	375	135	293	36.5	176	322	3.3	140	59	123	38.3	83
May-Jul 2015	780	4.2	379	124	277	35.5	162	327	3.3	144	63	120	36.7	83
Jun-Aug 2015	768	4.1	383	125	260	33.9	156	331	3.4	141	60	131	39.4	90
Jul-Sep 2015	775	4.2	399	124	252	32.5	146	328	3.3	140	58	130	39.7	92
Aug-Oct 2015	757	4.1	392	120	245	32.4	141	333	3.4	137	56	140	42.1	93
Sep-Nov 2015	726	3.9	383	117	226	31.1	132	329	3.3	142	50	136	41.5	88
Oct-Dec 2015	746	4.0	399	116	231	31.0	135	316	3.2	139	45	133	41.9	81
Nov-Jan 2016	737	3.9	399	118	219	29.8	125	316	3.2	138	40	138	43.6	87
Dec-Feb 2016	733	3.9	397	118	218	29.8	124	336	3.4	145	50	141	42.0	91
Jan-Mar 2016	712	3.8	379	116	216	30.4	122	348	3.5	147	53	148	42.6	93
Feb-Apr 2016	705	3.8	382	111	212	30.1	123	345	3.4	148	50	148	42.7	94
Mar-May 2016	707	3.8	391	101	215	30.4	124	329	3.3	139	46	145	44.0	93
Apr-Jun 2016	708	3.8	394	108	206	29.0	125	317	3.1	136	40	141	44.5	88
May-Jul 2016	693	3.7	382	107	204	29.4	120	320	3.2	140	41	139	43.4	92
Jun-Aug 2016	720	3.8	400	119	200	27.8	114	321	3.2	142	41	138	43.0	89
Jul-Sep 2016	708	3.8	400	108	199	28.1	109	313	3.1	143	37	132	42.3	87
Aug-Oct 2016	710	3.8	411	108	191	26.9	107	319	3.1	151	37	131	40.9	78
Sep-Nov 2016	723	3.9	421	106	196	27.1	106	309	3.0	153	40	116	37.5	73
Oct-Dec 2016	708	3.8	407	107	193	27.3	101	316	3.1	157	42	117	37.2	75
Nov-Jan 2017	717	3.8	417	104	196	27.3	104	306	3.0	150	45	110	36.1	70
Dec-Feb 2017	689	3.7	384	109	197	28.6	106	305	3.0	145	46	113	37.2	71
Jan-Mar 2017	681	3.6	384	106	191	28.1	101	290	2.8	133	49	107	37.1	69
Feb-Apr 2017	663	3.5	369	102	191	28.8	99	297	2.9	139	52	107	35.9	67
Mar-May 2017	642	3.4	363	94	184	28.7	98	292	2.8	137	46	109	37.3	67
Apr-Jun 2017	640	3.4	366	87	187	29.2	100	299	2.9	139	52	108	36.1	64
May-Jul 2017	611	3.2	336	88	186	30.5	105	313	3.1	142	54	117	37.2	71
Jun-Aug 2017	616	3.3	352	82	181	29.5	105	303	3.0	135	52	117	38.4	74
Jul-Sep 2017	612	3.2	355	80	177	28.9	105	294	2.9	126	47	121	41.2	76
Aug-Oct 2017	605	3.2	349	82	173	28.7	104	297	2.9	131	47	119	39.9	74
Sep-Nov 2017	605	3.2	336	93	176	29.1	98	298	2.9	129	47	122	41.0	77
Oct-Dec 2017	622	3.3	346	103	174	27.9	94	300	2.9	131	53	116	38.8	75
Men	MGVJ	MGXC	YBYI	YBYL	YBYO	YBYR	YBYU	YBVU	YBVX	YBYX	YBZA	YBZD	YBZG	YBZJ
Jul-Sep 2014	444	4.5	177	77	190	42.8	107	208	4.0	89	23	96	46.0	65
Aug-Oct 2014	444	4.5	182	83	179	40.4	98	209	4.0	86	25	98	47.1	69
Sep-Nov 2014	430	4.3	180	74	176	40.9	98	206	4.0	82	25	98	47.9	69
Oct-Dec 2014	420	4.2	173	77	170	40.5	96	193	3.7	68	28	97	50.4	66
Nov-Jan 2015	414	4.2	173	71	170	41.0	98	181	3.5	70	26	85	47.1	60
Dec-Feb 2015	406	4.1	166	71	168	41.5	103	178	3.4	69	27	82	46.0	61
Jan-Mar 2015	396	4.0	163	68	165	41.6	106	183	3.5	69	28	86	47.2	63
Feb-Apr 2015	400	4.0	163	68	169	42.3	107	177	3.4	70	28	80	45.2	55
Mar-May 2015	404	4.1	166	68	171	42.2	106	184	3.5	77	30	77	42.0	51
Apr-Jun 2015	411	4.1	171	66	174	42.3	112	185	3.5	78	33	75	40.5	51
May-Jul 2015	391	3.9	170	60	162	41.3	102	183	3.5	80	34	69	37.6	50
Jun-Aug 2015	389	3.9	178	60	151	38.8	98	192	3.6	77	36	79	41.1	56
Jul-Sep 2015	392	4.0	185	60	147	37.6	93	197	3.7	82	34	80	40.7	57
Aug-Oct 2015	383	3.9	183	60	139	36.3	89	202	3.8	78	37	86	42.8	60
Sep-Nov 2015	364	3.7	180	58	126	34.6	84	196	3.6	83	32	81	41.4	56
Oct-Dec 2015	375	3.8	188	59	129	34.3	86	193	3.6	79	31	83	43.0	56
Nov-Jan 2016	367	3.7	184	60	123	33.5	83	194	3.6	80	29	86	44.1	58
Dec-Feb 2016	370	3.7	188	59	124	33.4	86	206	3.8	83	33	91	43.9	62
Jan-Mar 2016	356	3.6	174	60	122	34.2	80	207	3.8	77	38	92	44.6	62
Feb-Apr 2016	348	3.5	172	54	122	35.1	81	203	3.7	74	34	95	46.8	63
Mar-May 2016	354	3.5	185	51	119	33.6	79	198	3.6	74	30	94	47.5	65
Apr-Jun 2016	344	3.4	178	53	113	32.9	77	187	3.4	73	24	91	48.3	63
May-Jul 2016	335	3.3	175	51	109	32.4	73	191	3.5	73	24	94	49.5	70
Jun-Aug 2016	345	3.4	177	53	115	33.3	74	185	3.4	71	22	92	49.6	63
Jul-Sep 2016	347	3.5	187	47	113	32.6	73	182	3.3	72	23	87	47.9	61

4.7 Unemployment by age and duration

	25-49							50 and over						
Men *(contd.)*	All level	rate (%)[1]	Up to 6 months level	Over 6 and up to 12 months level	All over 12 months level	% over 12 months	All over 24 months level	All level	rate (%)[1]	Up to 6 months level	Over 6 and up to 12 months level	All over 12 months level	% over 12 months	All over 24 months level
	MGVJ	MGXC	YBYI	YBYL	YBYO	YBYR	YBYU	YBVU	YBVX	YBYX	YBZA	YBZD	YBZG	YBZJ
Aug-Oct 2016	347	3.5	195	44	108	31.2	71	188	3.4	79	21	87	46.4	54
Sep-Nov 2016	359	3.6	202	45	111	31.0	69	179	3.3	77	23	79	43.9	50
Oct-Dec 2016	348	3.5	195	44	109	31.3	68	181	3.3	80	22	79	43.7	49
Nov-Jan 2017	354	3.6	200	41	113	32.0	76	174	3.2	77	25	72	41.3	46
Dec-Feb 2017	341	3.4	181	45	115	33.7	78	179	3.3	83	27	69	38.5	44
Jan-Mar 2017	341	3.4	183	46	112	32.8	72	179	3.3	82	31	66	36.8	44
Feb-Apr 2017	327	3.3	169	52	106	32.5	68	185	3.4	84	35	66	35.8	42
Mar-May 2017	317	3.2	168	48	101	31.9	63	178	3.2	80	28	69	39.0	43
Apr-Jun 2017	317	3.2	169	45	102	32.3	64	181	3.3	80	31	70	38.6	43
May-Jul 2017	292	2.9	146	44	101	34.7	65	192	3.5	83	31	77	40.3	49
Jun-Aug 2017	296	3.0	157	43	96	32.4	62	180	3.3	76	33	71	39.5	44
Jul-Sep 2017	302	3.0	160	47	95	31.4	61	168	3.1	67	29	72	43.0	46
Aug-Oct 2017	302	3.0	166	41	95	31.5	60	173	3.1	73	31	69	39.9	42
Sep-Nov 2017	294	2.9	151	47	96	32.7	60	175	3.2	72	30	74	41.9	46
Oct-Dec 2017	300	3.0	151	52	97	32.3	56	176	3.2	70	35	72	40.8	46
Women	MGVK	MGXD	YBYJ	YBYM	YBYP	YBYS	YBYV	YBVV	YBVY	YBYY	YBZB	YBZE	YBZH	YBZK
Jul-Sep 2014	428	4.9	208	79	141	32.9	76	132	3.0	59	19	55	41.3	38
Aug-Oct 2014	416	4.7	194	83	138	33.2	76	128	2.9	55	22	51	39.6	35
Sep-Nov 2014	398	4.5	196	73	129	32.5	70	119	2.7	51	20	47	39.9	31
Oct-Dec 2014	382	4.4	197	67	117	30.6	59	125	2.8	54	19	52	41.5	34
Nov-Jan 2015	393	4.5	205	64	124	31.6	62	125	2.8	56	21	49	39.0	34
Dec-Feb 2015	387	4.4	198	62	127	32.7	62	129	2.9	60	21	48	37.6	34
Jan-Mar 2015	390	4.5	213	61	115	29.5	59	128	2.9	58	19	51	39.7	32
Feb-Apr 2015	381	4.4	202	66	113	29.7	60	125	2.8	58	19	48	38.4	33
Mar-May 2015	409	4.7	222	69	118	28.9	63	133	3.0	58	25	49	37.3	32
Apr-Jun 2015	393	4.5	204	69	120	30.5	64	137	3.1	62	26	48	35.3	31
May-Jul 2015	389	4.5	210	64	116	29.7	60	144	3.2	64	29	51	35.5	34
Jun-Aug 2015	379	4.4	204	65	109	28.9	58	140	3.1	63	25	52	37.0	33
Jul-Sep 2015	383	4.4	214	64	105	27.4	53	131	2.9	57	24	50	38.1	34
Aug-Oct 2015	374	4.3	209	60	106	28.3	52	132	2.9	58	19	54	41.0	33
Sep-Nov 2015	362	4.1	203	59	100	27.5	48	132	2.9	59	18	55	41.6	32
Oct-Dec 2015	371	4.2	211	58	103	27.6	48	123	2.7	60	14	50	40.3	25
Nov-Jan 2016	369	4.2	215	59	96	26.0	42	121	2.7	58	11	52	42.8	29
Dec-Feb 2016	363	4.2	210	59	95	26.1	38	129	2.8	62	17	50	38.9	29
Jan-Mar 2016	356	4.1	206	56	95	26.5	42	141	3.1	70	15	56	39.7	31
Feb-Apr 2016	358	4.1	210	57	90	25.2	43	142	3.1	74	16	52	36.9	30
Mar-May 2016	353	4.0	207	50	96	27.3	45	132	2.8	65	16	51	38.7	27
Apr-Jun 2016	364	4.2	216	55	93	25.4	48	129	2.8	63	16	50	38.9	25
May-Jul 2016	358	4.1	207	55	95	26.7	47	129	2.8	68	17	44	34.3	22
Jun-Aug 2016	374	4.3	223	66	85	22.8	40	135	2.9	71	18	46	33.9	25
Jul-Sep 2016	361	4.1	213	62	86	23.8	36	131	2.8	71	14	45	34.6	26
Aug-Oct 2016	363	4.1	216	64	83	22.9	37	131	2.8	72	16	43	33.1	25
Sep-Nov 2016	364	4.1	219	61	84	23.2	37	130	2.8	76	17	37	28.7	23
Oct-Dec 2016	359	4.1	212	63	84	23.5	33	134	2.9	77	19	38	28.4	26
Nov-Jan 2017	363	4.1	217	63	82	22.6	28	131	2.8	72	21	38	29.2	24
Dec-Feb 2017	348	4.0	202	64	82	23.5	28	126	2.7	63	19	44	35.3	27
Jan-Mar 2017	341	3.9	201	60	79	23.3	28	110	2.3	51	18	41	37.5	25
Feb-Apr 2017	335	3.8	200	51	85	25.2	31	112	2.4	55	17	40	36.0	25
Mar-May 2017	325	3.7	196	46	83	25.6	35	114	2.4	57	18	40	34.7	24
Apr-Jun 2017	323	3.7	197	41	85	26.2	36	118	2.5	59	21	38	32.2	21
May-Jul 2017	319	3.6	190	44	85	26.6	40	121	2.5	59	23	39	32.3	22
Jun-Aug 2017	320	3.6	195	39	85	26.7	43	123	2.6	59	19	45	36.9	30
Jul-Sep 2017	311	3.5	195	33	82	26.5	44	126	2.6	60	18	49	38.9	30
Aug-Oct 2017	303	3.4	184	41	78	25.8	43	124	2.6	58	16	50	39.9	32
Sep-Nov 2017	311	3.5	184	47	80	25.7	38	122	2.5	57	17	49	39.7	31
Oct-Dec 2017	322	3.6	195	51	76	23.7	38	124	2.6	61	19	45	36.0	29

Source: Labour Force Survey

Note: When comparing quarterly changes ONS recommends comparing with the previous non-overlapping 3-month average time period
(eg, compare Apr-Jun with Jan-Mar, not with Mar-May).

[1] Denominator = economically active for that age group.

* Sample size too small for reliable estimate.

4.8 Regional Labour Force Survey Summary (employment, unemployment, economic activity, inactivity)

Headline estimates for November 2017 to January 2018

Thousands, seasonally adjusted

Area Codes	Area Names	Economically active [1] Aged 16+ Level	Aged 16-64 Rate (%)[2]	Employment Aged 16+ Level	Aged 16-64 Rate (%)[2]	Unemployment Aged 16+ Level	Aged 16+ Rate (%)[3]	Economically inactive Aged 16-64 Level	Aged 16-64 Rate (%)[2]
K02000001	United Kingdom	33,701	78.8	32,248	75.3	1,453	4.3	8,723	21.2
K03000001	Great Britain	32,825	79.0	31,399	75.5	1,426	4.3	8,397	21.0
E92000001	England	28,550	79.2	27,316	75.7	1,234	4.3	7,204	20.8
E12000001	North East	1,283	76.0	1,216	71.9	67	5.2	395	24.0
E12000002	North West	3,565	77.2	3,418	74.0	147	4.1	1,020	22.8
E12000003	Yorkshire and The Humber	2,713	78.0	2,576	74.0	136	5.0	742	22.0
E12000004	East Midlands	2,375	78.3	2,276	75.0	98	4.1	633	21.7
E12000005	West Midlands	2,845	76.6	2,704	72.7	141	5.0	837	23.4
E12000006	East	3,180	81.4	3,048	77.9	132	4.1	700	18.6
E12000007	London	4,944	78.7	4,697	74.7	247	5.0	1,296	21.3
E12000008	South East	4,762	82.0	4,603	79.2	159	3.3	1,000	18.0
E12000009	South West	2,884	82.5	2,778	79.3	106	3.7	581	17.5
W92000004	Wales	1,513	76.4	1,439	72.6	73	4.8	447	23.6
S92000003	Scotland	2,763	78.1	2,644	74.8	118	4.3	745	21.9
N92000002	Northern Ireland	876	72.2	849	69.8	28	3.2	326	27.8

Change on quarter (change since August to October 2017)

Note: Changes on quarter at regional level are particularly subject to sampling variability and should be interpreted in the context of changes over several quarters rather than in isolation.

Thousands, seasonally adjusted

Area Codes	Area Names	Economically active [1] Aged 16+ Level	Aged 16-64 Rate (%)[2]	Employment Aged 16+ Level	Aged 16-64 Rate (%)[2]	Unemployment Aged 16+ Level	Aged 16+ Rate (%)[3]	Economically inactive Aged 16-64 Level	Aged 16-64 Rate (%)[2]
K02000001	United Kingdom	192	0.3	168	0.3	24	0.0	-136	-0.3
K03000001	Great Britain	173	0.3	143	0.2	30	0.1	-122	-0.3
E92000001	England	168	0.4	145	0.3	23	0.1	-123	-0.4
E12000001	North East	-23	-1.3	-13	-0.7	-10	-0.7	21	1.3
E12000002	North West	-1	0.1	1	0.1	-2	-0.1	-5	-0.1
E12000003	Yorkshire and The Humber	61	1.6	56	1.4	5	0.1	-54	-1.6
E12000004	East Midlands	25	0.7	25	0.7	0	-0.1	-20	-0.7
E12000005	West Midlands	8	-0.2	17	0.0	-9	-0.3	9	0.2
E12000006	East	19	0.5	2	0.0	17	0.5	-17	-0.5
E12000007	London	35	0.3	28	0.2	7	0.1	-16	-0.3
E12000008	South East	-4	-0.3	-19	-0.5	14	0.3	15	0.3
E12000009	South West	49	1.7	48	1.7	1	0.0	-57	-1.7
W92000004	Wales	9	0.0	7	-0.3	3	0.1	0	0.0
S92000003	Scotland	-4	0.0	-8	-0.2	5	0.2	1	0.0
N92000002	Northern Ireland	19	1.2	25	1.7	-6	-0.8	-14	-1.2

Change on year (change since November 2016 to January 2017)

Thousands, seasonally adjusted

Area Codes	Area Names	Economically active [1] Aged 16+ Level	Aged 16-64 Rate (%)[2]	Employment Aged 16+ Level	Aged 16-64 Rate (%)[2]	Unemployment Aged 16+ Level	Aged 16+ Rate (%)[3]	Economically inactive Aged 16-64 Level	Aged 16-64 Rate (%)[2]
K02000001	United Kingdom	276	0.4	402	0.8	-127	-0.4	-158	-0.4
K03000001	Great Britain	287	0.5	391	0.8	-104	-0.4	-176	-0.5
E92000001	England	263	0.5	364	0.8	-101	-0.4	-156	-0.5
E12000001	North East	-6	0.1	15	1.3	-21	-1.6	-2	-0.1
E12000002	North West	5	0.2	28	0.7	-23	-0.6	-11	-0.2
E12000003	Yorkshire and The Humber	16	1.0	22	1.1	-6	-0.2	-32	-1.0
E12000004	East Midlands	-6	-0.7	-3	-0.6	-4	-0.1	23	0.7
E12000005	West Midlands	-3	-0.1	15	0.4	-19	-0.6	6	0.1
E12000006	East	28	1.1	33	1.3	-5	-0.2	-39	-1.1
E12000007	London	104	0.8	133	1.3	-29	-0.7	-32	-0.8
E12000008	South East	72	0.6	75	0.7	-3	-0.1	-30	-0.6
E12000009	South West	52	1.2	46	1.0	6	0.1	-38	-1.2
W92000004	Wales	-2	-0.3	-8	-0.7	6	0.4	5	0.3
S92000003	Scotland	26	0.7	35	1.0	-9	-0.4	-25	-0.7
N92000002	Northern Ireland	-12	-1.5	12	0.5	-23	-2.6	18	1.5

The Labour Force Survey is tabulated by region of residence.
1. Economically active = Employment plus Unemployment.
2. Denominator = all persons aged 16 to 64.
3. Denominator = total economically active

Source: Labour Force Survey

4.9 Claimant count rates: by region

ONS Crown Copyright Reserved [from Nomis on 20 January 2020]

Sex: Total
Rate: Workplace-based estimates
Seasonally adjusted

Area		Jan - Dec 2006 (inclusive)		Jan - Dec 2007 (inclusive)		Jan - Dec 2008 (inclusive)		Jan - Dec 2009 (inclusive)	
		number	rate	number	rate	number	rate	number	rate
England	E92000001	785,550	2.8	723,100	2.6	754,108	2.7	1,275,583	4.5
United Kingdom	K02000001	944,967	2.9	864,467	2.6	906,083	2.7	1,527,683	4.6
North East	E12000001	50,233	4.0	49,275	3.9	53,717	4.3	83,775	6.6
North West	E12000002	115,550	3.3	110,092	3.1	119,708	3.4	191,708	5.3
Yorkshire and The Humber	E12000003	87,358	3.3	81,158	3.0	88,033	3.3	149,817	5.6
East Midlands	E12000004	61,975	2.8	58,483	2.6	61,408	2.7	108,300	4.8
West Midlands	E12000005	108,325	3.8	102,525	3.6	106,167	3.7	173,775	6.2
East	E12000006	65,583	2.2	61,225	2.1	63,625	2.2	115,992	3.9
London	E12000007	166,833	3.4	145,450	2.9	137,658	2.7	211,150	4.1
South East	E12000008	81,675	1.8	71,950	1.6	76,942	1.7	149,225	3.3
South West	E12000009	48,017	1.8	42,942	1.6	46,850	1.7	91,842	3.3
Wales	W92000004	44,217	3.0	40,692	2.8	45,517	3.1	77,133	5.2
Scotland	S92000003	87,258	3.1	76,300	2.7	78,633	2.7	125,958	4.5
Northern Ireland	N92000002	27,875	3.2	24,375	2.8	27,825	3.1	49,008	5.5

Area		Jan - Dec 2010 (inclusive)		Jan - Dec 2011 (inclusive)		Jan - Dec 2012 (inclusive)		Jan - Dec 2013 (inclusive)	
		number	rate	number	rate	number	rate	number	rate
England	E92000001	1,230,708	4.4	1,258,225	4.5	1,301,583	4.6	1,157,878	4.1
United Kingdom	K02000001	1,496,358	4.5	1,534,408	4.6	1,585,575	4.7	1,421,326	4.2
North East	E12000001	81,417	6.5	85,367	6.9	93,425	7.6	84,494	6.8
North West	E12000002	184,017	5.1	190,200	5.3	200,275	5.6	179,102	5.0
Yorkshire and The Humber	E12000003	148,008	5.5	153,267	5.7	164,075	6.1	149,861	5.6
East Midlands	E12000004	101,492	4.5	103,142	4.5	108,842	4.7	95,520	4.2
West Midlands	E12000005	163,667	5.9	164,025	5.9	164,983	5.8	149,178	5.3
East	E12000006	111,067	3.8	111,800	3.8	115,075	3.8	101,690	3.4
London	E12000007	217,283	4.3	228,000	4.5	226,708	4.2	202,011	3.7
South East	E12000008	139,858	3.1	137,242	3.0	139,333	3.0	119,459	2.6
South West	E12000009	83,900	3.0	85,183	3.1	88,867	3.2	76,563	2.7
Wales	W92000004	73,192	5.1	74,850	5.2	79,600	5.5	72,896	5.0
Scotland	S92000003	135,733	5.0	141,508	5.1	141,433	5.1	127,998	4.6
Northern Ireland	N92000002	56,725	6.2	59,825	6.6	62,958	7.1	62,555	7.0

Area		Jan - Dec 2014 (inclusive)		Jan - Dec 2015 (inclusive)		Jan - Dec 2016 (inclusive)		Jan - Dec 2017 (inclusive)	
		number	rate	number	rate	number	rate	number	rate
England	E92000001	828,677	2.8	631,171	2.1	615,959	2.0	645,382	2.1
United Kingdom	K02000001	1,036,114	3.0	797,916	2.3	772,818	2.2	795,606	2.2
North East	E12000001	61,856	5.0	47,595	3.8	50,055	4.1	52,799	4.3
North West	E12000002	128,476	3.5	106,321	2.9	104,886	2.8	108,953	2.9
Yorkshire and The Humber	E12000003	110,499	4.1	82,997	3.0	75,162	2.7	75,647	2.7
East Midlands	E12000004	67,760	2.9	49,353	2.1	45,197	1.9	46,092	1.9
West Midlands	E12000005	109,445	3.8	81,154	2.8	82,132	2.8	85,375	2.8
East	E12000006	68,618	2.2	49,607	1.6	47,056	1.5	49,258	1.5
London	E12000007	148,091	2.6	114,913	2.0	111,375	1.9	119,424	2.0
South East	E12000008	81,401	1.7	60,382	1.2	59,959	1.2	64,213	1.3
South West	E12000009	52,532	1.8	38,848	1.3	40,137	1.4	43,621	1.5
Wales	W92000004	56,857	3.8	45,380	3.0	43,025	2.8	39,601	2.6
Scotland	S92000003	96,569	3.4	78,407	2.8	77,755	2.7	80,175	2.8
Northern Ireland	N92000002	54,012	6.0	42,959	4.8	36,079	3.9	30,449	3.3

Source: Office for National Statistics

From May 2013 onwards these figures are not designated as National Statistics.
From May 2013 onwards these figures are considered Experimental Statistics.

Under Universal Credit a broader span of claimants are required to look for work than under Jobseeker's Allowance. As Universal Credit Full Service is rolled out in particular areas, the number of people recorded as being on the Claimant Count is therefore likely to rise.
Rate figures for dates from 2015 onwards are calculated using mid-2015 workforce estimates.

4.10 Claimant count by Sex and Unitary and Local Authority

ONS Crown Copyright Reserved [from Nomis on 20 January 2020]
Age 16+

Area	Claimant Count December 2017						Change on year					
	Levels			Claimants as a proportion of residents aged 16-64			Levels			Claimants as a proportion of residents aged 16-64		
	Total	Male	Female	Total	Male	Female	Total	Male	Female	Total	Male	Female
United Kingdom	793,430	488,545	304,885	1.9	2.4	1.5	46,010	15,325	30,680	0.1	0.1	0.2
Great Britain	764,875	469,645	295,230	1.9	2.3	1.5	49,450	18,650	30,800	0.1	0.1	0.2
England and Wales	685,840	416,775	269,065	1.9	2.3	1.5	45,325	16,660	28,670	0.2	0.1	0.2
England	648,130	392,870	255,260	1.9	2.3	1.5	47,215	18,005	29,210	0.2	0.1	0.2
North East	53,915	34,605	19,310	3.3	4.2	2.3	3,300	1,135	2,165	0.3	0.1	0.3
Darlington	2,245	1,460	780	3.5	4.6	2.4	70	15	55	0.2	0.1	0.2
County Durham	8,410	5,320	3,090	2.6	3.3	1.9	575	320	255	0.2	0.2	0.2
Hartlepool	3,815	2,405	1,415	6.6	8.5	4.8	1,125	600	530	1.9	2.1	1.8
Middlesbrough	3,985	2,730	1,255	4.5	6.2	2.8	-245	-235	-10	-0.2	-0.5	0.0
Northumberland	4,745	3,005	1,740	2.5	3.2	1.8	105	55	55	0.1	0.1	0.1
Redcar and Cleveland	2,875	1,955	920	3.5	4.9	2.2	-360	-295	-65	-0.5	-0.8	-0.1
Stockton-on-Tees	3,775	2,490	1,285	3.1	4.1	2.1	-145	-180	35	-0.1	-0.3	0.1
Tyne and Wear (Met County)	24,065	15,240	8,830	3.3	4.2	2.4	2,165	860	1,310	0.3	0.2	0.3
Gateshead	3,745	2,355	1,390	2.9	3.7	2.2	255	50	205	0.2	0.1	0.4
Newcastle upon Tyne	7,475	4,630	2,840	3.7	4.5	2.9	1,980	1,045	930	1.0	1.0	0.9
North Tyneside	3,245	2,160	1,085	2.5	3.5	1.7	30	-45	75	0.0	0.0	0.2
South Tyneside	4,150	2,645	1,510	4.4	5.8	3.2	35	-60	100	0.0	-0.1	0.2
Sunderland	5,450	3,450	2,000	3.1	4.0	2.2	-135	-130	-5	-0.1	-0.1	0.0
North West	108,860	65,880	42,980	2.4	2.9	1.9	8,370	1,985	6,385	0.2	0.1	0.3
Blackburn with Darwen	2,750	1,720	1,030	3.0	3.7	2.3	370	185	185	0.4	0.4	0.5
Blackpool	3,780	2,430	1,350	4.4	5.6	3.2	280	45	235	0.3	0.1	0.6
Cheshire East	2,910	1,725	1,190	1.3	1.5	1.0	520	205	320	0.2	0.1	0.2
Cheshire West and Chester	3,240	1,930	1,310	1.6	1.9	1.2	455	210	245	0.2	0.2	0.2
Halton	2,765	1,565	1,200	3.5	4.0	3.0	590	260	330	0.8	0.7	0.9
Warrington	2,720	1,555	1,165	2.1	2.4	1.8	770	325	445	0.6	0.5	0.7
Cumbria	5,215	3,250	1,965	1.8	2.2	1.3	825	365	465	0.3	0.3	0.3
Allerdale	1,480	880	595	2.6	3.1	2.0	435	205	230	0.8	0.8	0.7
Barrow-in-Furness	1,030	690	335	2.5	3.4	1.6	-90	-70	-25	-0.2	-0.3	-0.1
Carlisle	950	570	380	1.4	1.8	1.1	160	60	100	0.2	0.2	0.3
Copeland	1,165	725	435	2.8	3.5	2.1	255	115	135	0.7	0.7	0.7
Eden	245	150	95	0.8	1.0	0.6	40	20	20	0.1	0.1	0.1
South Lakeland	350	225	125	0.6	0.8	0.4	30	25	5	0.1	0.1	0.0
Greater Manchester (Met County)	45,040	27,125	17,920	2.5	3.0	2.0	1,680	-130	1,815	0.1	-0.1	0.2
Bolton	5,475	3,270	2,210	3.1	3.7	2.5	350	60	300	0.2	0.0	0.3
Bury	2,750	1,655	1,095	2.4	2.9	1.9	100	30	65	0.1	0.1	0.2
Manchester	9,335	5,635	3,700	2.4	2.9	2.0	-330	-455	120	-0.1	-0.2	0.1
Oldham	5,175	2,930	2,245	3.6	4.1	3.1	1,340	600	740	0.9	0.8	1.0
Rochdale	3,670	2,160	1,505	2.7	3.2	2.2	-40	-110	65	0.0	-0.2	0.1
Salford	4,190	2,565	1,620	2.5	3.0	2.0	-55	-160	100	-0.1	-0.3	0.1
Stockport	3,265	2,085	1,180	1.8	2.4	1.3	35	-75	110	0.0	-0.1	0.1
Tameside	3,635	2,180	1,455	2.6	3.1	2.1	45	-30	75	0.0	-0.1	0.1
Trafford	2,470	1,470	1,000	1.7	2.1	1.4	250	90	160	0.2	0.2	0.3
Wigan	5,080	3,175	1,905	2.5	3.1	1.9	-10	-85	70	0.0	-0.1	0.1
Lancashire	15,160	9,050	6,110	2.1	2.5	1.7	1,845	615	1,230	0.3	0.2	0.4
Burnley	2,280	1,280	1,000	4.3	4.8	3.7	675	295	380	1.3	1.1	1.4
Chorley	1,085	670	410	1.5	1.8	1.2	25	-20	40	0.0	-0.1	0.1
Fylde	645	410	235	1.4	1.8	1.1	40	20	20	0.1	0.1	0.1
Hyndburn	1,310	785	525	2.7	3.2	2.1	165	75	95	0.4	0.3	0.4
Lancaster	2,395	1,410	985	2.7	3.2	2.2	550	240	310	0.6	0.6	0.7
Pendle	1,120	680	440	2.0	2.5	1.6	60	30	30	0.1	0.1	0.1
Preston	2,115	1,310	805	2.3	2.8	1.8	280	120	160	0.3	0.3	0.4
Ribble Valley	275	155	120	0.8	0.9	0.7	55	25	30	0.2	0.2	0.2
Rossendale	905	535	370	2.1	2.5	1.7	75	30	45	0.2	0.2	0.2
South Ribble	890	515	370	1.3	1.6	1.1	75	0	65	0.1	0.1	0.2
West Lancashire	1,155	685	465	1.7	2.0	1.3	-120	-140	15	-0.1	-0.4	0.0
Wyre	995	615	385	1.6	2.0	1.2	-30	-65	40	0.0	-0.2	0.1

4.10 Claimant count by Sex and Unitary and Local Authority

ONS Crown Copyright Reserved [from Nomis on 20 January 2020]
Age 16+

| Area | Claimant Count December 2017 | | | | | | Change on year | | | | | |
| | Levels | | | Claimants as a proportion of residents aged 16-64 | | | Levels | | | Claimants as a proportion of residents aged 16-64 | | |
	Total	Male	Female	Total	Male	Female	Total	Male	Female	Total	Male	Female
Merseyside (Met County)	**25,275**	**15,530**	**9,745**	**2.8**	**3.5**	**2.1**	**1,030**	**-95**	**1,125**	**0.1**	**0.0**	**0.2**
Knowsley	3,035	1,720	1,320	3.2	3.9	2.7	160	-30	200	0.1	0.0	0.4
Liverpool	10,980	6,865	4,115	3.3	4.1	2.5	390	-5	395	0.1	0.0	0.3
Sefton	4,340	2,685	1,655	2.7	3.4	2.0	335	70	265	0.3	0.2	0.3
St. Helens	2,855	1,680	1,175	2.6	3.1	2.1	-105	-200	95	-0.1	-0.3	0.2
Wirral	4,065	2,585	1,480	2.1	2.7	1.5	250	75	175	0.1	0.0	0.2
Yorkshire and The Humber	**75,285**	**47,135**	**28,150**	**2.2**	**2.8**	**1.7**	**3,655**	**1,215**	**2,440**	**0.1**	**0.1**	**0.2**
East Riding of Yorkshire	**2,855**	**1,845**	**1,010**	**1.5**	**1.9**	**1.0**	**-25**	**-35**	**10**	**0.0**	**0.0**	**0.0**
Kingston upon Hull, City of	**6,140**	**3,965**	**2,180**	**3.6**	**4.6**	**2.6**	**150**	**75**	**80**	**0.1**	**0.1**	**0.1**
North East Lincolnshire	**3,275**	**2,065**	**1,210**	**3.4**	**4.3**	**2.5**	**-40**	**-55**	**15**	**0.0**	**-0.1**	**0.1**
North Lincolnshire	**2,285**	**1,405**	**880**	**2.2**	**2.7**	**1.7**	**10**	**-80**	**90**	**0.0**	**-0.1**	**0.2**
York	**1,350**	**825**	**525**	**1.0**	**1.2**	**0.8**	**440**	**215**	**220**	**0.3**	**0.3**	**0.4**
North Yorkshire	**5,090**	**3,065**	**2,025**	**1.4**	**1.7**	**1.1**	**965**	**495**	**470**	**0.3**	**0.3**	**0.2**
Craven	415	245	170	1.3	1.5	1.0	130	65	65	0.4	0.4	0.4
Hambleton	675	400	280	1.3	1.5	1.1	200	95	110	0.4	0.4	0.5
Harrogate	1,285	725	560	1.4	1.5	1.2	405	195	210	0.5	0.4	0.5
Richmondshire	445	230	215	1.3	1.2	1.5	155	75	80	0.5	0.4	0.6
Ryedale	490	295	195	1.6	1.9	1.2	130	95	35	0.5	0.6	0.2
Scarborough	1,125	765	360	1.8	2.5	1.1	-80	-25	-55	-0.1	-0.1	-0.2
Selby	650	405	245	1.2	1.5	0.9	20	-10	30	0.0	-0.1	0.1
South Yorkshire (Met County)	**21,565**	**13,635**	**7,930**	**2.4**	**3.1**	**1.8**	**710**	**20**	**690**	**0.0**	**0.0**	**0.2**
Barnsley	4,080	2,390	1,690	2.7	3.2	2.2	865	410	455	0.6	0.6	0.6
Doncaster	4,695	2,900	1,795	2.4	3.0	1.9	-35	-85	50	-0.1	-0.1	0.1
Rotherham	4,090	2,605	1,485	2.5	3.3	1.8	-25	-65	35	-0.1	0.0	0.0
Sheffield	8,700	5,735	2,965	2.3	3.0	1.6	-95	-245	150	0.0	-0.1	0.1
West Yorkshire (Met County)	**32,720**	**20,335**	**12,385**	**2.2**	**2.8**	**1.7**	**1,445**	**590**	**860**	**0.1**	**0.1**	**0.1**
Bradford	8,540	5,285	3,255	2.6	3.2	2.0	-245	-180	-65	-0.1	-0.1	0.0
Calderdale	3,615	2,150	1,465	2.8	3.3	2.2	1,065	520	545	0.8	0.8	0.8
Kirklees	5,750	3,535	2,215	2.1	2.6	1.6	365	150	215	0.1	0.1	0.1
Leeds	10,975	7,020	3,955	2.1	2.8	1.5	130	25	105	0.0	0.0	0.0
Wakefield	3,840	2,340	1,495	1.8	2.2	1.4	130	70	60	0.0	0.0	0.0
East Midlands	**46,115**	**28,525**	**17,590**	**1.5**	**1.9**	**1.2**	**4,705**	**2,430**	**2,280**	**0.1**	**0.1**	**0.2**
Derby	**2,595**	**1,570**	**1,030**	**1.6**	**1.9**	**1.3**	**405**	**240**	**170**	**0.2**	**0.3**	**0.2**
Leicester	**3,915**	**2,385**	**1,530**	**1.7**	**2.0**	**1.3**	**310**	**175**	**135**	**0.2**	**0.1**	**0.1**
Nottingham	**6,905**	**4,600**	**2,305**	**3.0**	**3.9**	**2.1**	**370**	**260**	**110**	**0.1**	**0.2**	**0.1**
Rutland	**130**	**80**	**50**	**0.6**	**0.7**	**0.5**	**30**	**20**	**10**	**0.2**	**0.2**	**0.1**
Derbyshire	**6,460**	**3,985**	**2,475**	**1.3**	**1.6**	**1.0**	**780**	**370**	**410**	**0.1**	**0.1**	**0.2**
Amber Valley	895	575	325	1.2	1.5	0.8	5	5	5	0.0	0.0	0.0
Bolsover	660	395	270	1.3	1.6	1.1	-20	-15	0	-0.1	-0.1	0.0
Chesterfield	1,265	810	455	1.9	2.5	1.4	140	45	95	0.2	0.2	0.3
Derbyshire Dales	255	145	115	0.6	0.7	0.5	5	-15	20	0.0	-0.1	0.0
Erewash	1,585	940	645	2.2	2.7	1.8	470	220	250	0.6	0.7	0.7
High Peak	630	390	240	1.1	1.4	0.8	55	40	15	0.1	0.2	0.0
North East Derbyshire	740	470	270	1.2	1.6	0.9	20	25	0	0.0	0.1	0.0
South Derbyshire	425	265	160	0.7	0.8	0.5	100	65	35	0.2	0.2	0.1
Leicestershire	**3,875**	**2,235**	**1,640**	**0.9**	**1.0**	**0.8**	**810**	**410**	**400**	**0.2**	**0.1**	**0.2**
Blaby	430	255	175	0.7	0.9	0.6	65	40	25	0.1	0.2	0.1
Charnwood	885	540	340	0.8	0.9	0.6	35	0	25	0.1	0.0	0.0
Harborough	420	240	180	0.8	0.9	0.6	165	90	75	0.3	0.4	0.2
Hinckley and Bosworth	870	490	380	1.3	1.5	1.1	245	135	110	0.4	0.4	0.3
Melton	510	275	235	1.7	1.8	1.5	230	120	105	0.8	0.8	0.7
North West Leicestershire	500	295	205	0.8	1.0	0.7	85	35	50	0.1	0.2	0.2
Oadby and Wigston	260	140	120	0.7	0.8	0.7	-15	-10	0	-0.1	-0.1	0.0
Lincolnshire	**7,585**	**4,865**	**2,725**	**1.7**	**2.2**	**1.2**	**540**	**370**	**175**	**0.1**	**0.2**	**0.1**
Boston	610	385	225	1.5	1.8	1.1	15	5	10	0.1	0.0	0.1
East Lindsey	1,915	1,215	705	2.5	3.2	1.8	190	150	40	0.3	0.4	0.1
Lincoln	1,575	1,095	480	2.3	3.3	1.4	115	80	40	0.1	0.2	0.1
North Kesteven	610	380	225	0.9	1.1	0.7	-25	-35	5	0.0	-0.1	0.1
South Holland	635	370	265	1.2	1.4	1.0	55	50	5	0.1	0.2	0.1
South Kesteven	1,170	710	455	1.4	1.7	1.1	160	65	90	0.2	0.1	0.3
West Lindsey	1,070	705	365	1.9	2.6	1.3	30	50	-20	0.0	0.2	0.0

4.10 Claimant count by Sex and Unitary and Local Authority

ONS Crown Copyright Reserved [from Nomis on 20 January 2020]
Age 16+

Area	Claimant Count December 2017						Change on year					
	Levels			Claimants as a proportion of residents aged 16-64			Levels			Claimants as a proportion of residents aged 16-64		
	Total	Male	Female	Total	Male	Female	Total	Male	Female	Total	Male	Female
Northamptonshire	7,015	4,095	2,920	1.5	1.8	1.3	980	365	615	0.2	0.2	0.3
Corby	1,035	575	460	2.3	2.6	2.1	410	195	215	0.9	0.8	1.0
Daventry	790	420	370	1.6	1.6	1.5	245	80	165	0.5	0.2	0.7
East Northamptonshire	520	315	205	0.9	1.1	0.7	40	10	25	0.0	0.0	0.1
Kettering	1,040	625	415	1.7	2.1	1.3	125	55	70	0.2	0.2	0.2
Northampton	2,540	1,530	1,010	1.8	2.1	1.4	80	0	85	0.1	0.0	0.1
South Northamptonshire	290	165	125	0.5	0.6	0.4	20	0	20	0.0	0.0	0.0
Wellingborough	800	460	340	1.7	2.0	1.4	65	20	45	0.1	0.1	0.2
Nottinghamshire	7,630	4,720	2,910	1.5	1.9	1.2	475	225	250	0.1	0.1	0.1
Ashfield	1,495	930	560	1.9	2.4	1.4	85	40	40	0.1	0.1	0.1
Bassetlaw	1,130	675	455	1.6	1.9	1.3	5	5	-5	0.0	0.0	0.0
Broxtowe	1,020	635	380	1.5	1.8	1.1	85	60	20	0.2	0.2	0.1
Gedling	1,070	675	400	1.5	1.9	1.1	50	30	30	0.1	0.1	0.1
Mansfield	1,370	865	510	2.0	2.6	1.5	145	65	85	0.2	0.2	0.2
Newark and Sherwood	935	535	400	1.3	1.5	1.1	65	5	60	0.1	0.0	0.2
Rushcliffe	610	400	205	0.9	1.1	0.6	40	15	20	0.1	0.0	0.1
West Midlands	84,415	51,940	32,475	2.3	2.8	1.8	4,360	1,625	2,740	0.1	0.0	0.1
Herefordshire, County of	1,145	695	450	1.0	1.2	0.8	-30	-35	0	0.0	-0.1	0.0
Shropshire	2,040	1,270	770	1.1	1.3	0.8	160	50	110	0.1	0.0	0.1
Stoke-on-Trent	3,665	2,305	1,365	2.3	2.8	1.7	600	355	250	0.4	0.4	0.3
Telford and Wrekin	1,840	1,125	715	1.7	2.0	1.3	165	70	95	0.2	0.1	0.2
Staffordshire	5,425	3,280	2,145	1.0	1.2	0.8	690	295	395	0.1	0.1	0.1
Cannock Chase	735	435	300	1.2	1.4	1.0	35	10	25	0.1	0.0	0.1
East Staffordshire	655	415	240	0.9	1.1	0.7	80	60	20	0.1	0.1	0.1
Lichfield	435	265	170	0.7	0.9	0.6	50	20	35	0.1	0.1	0.1
Newcastle-under-Lyme	1,010	625	385	1.2	1.5	0.9	120	55	65	0.1	0.1	0.1
South Staffordshire	825	485	335	1.2	1.4	1.0	145	60	80	0.2	0.2	0.2
Stafford	725	455	270	0.9	1.1	0.7	110	45	65	0.2	0.1	0.2
Staffordshire Moorlands	535	320	215	0.9	1.1	0.7	95	35	60	0.1	0.1	0.2
Tamworth	510	280	235	1.1	1.2	1.0	60	15	55	0.2	0.1	0.3
Warwickshire	4,315	2,565	1,750	1.2	1.5	1.0	870	525	340	0.2	0.3	0.2
North Warwickshire	400	255	145	1.0	1.3	0.7	10	35	-25	0.0	0.2	-0.2
Nuneaton and Bedworth	1,420	850	570	1.8	2.2	1.4	75	75	0	0.1	0.2	0.0
Rugby	1,250	685	570	1.9	2.1	1.8	385	200	190	0.6	0.6	0.6
Stratford-on-Avon	595	315	280	0.8	0.9	0.8	325	150	175	0.4	0.4	0.5
Warwick	650	465	190	0.7	1.0	0.4	75	70	5	0.1	0.1	0.0
West Midlands (Met County)	61,435	37,840	23,595	3.4	4.1	2.6	1,555	210	1,345	0.1	-0.1	0.1
Birmingham	30,585	19,165	11,420	4.2	5.3	3.1	770	80	690	0.1	0.0	0.2
Coventry	3,860	2,420	1,445	1.6	2.0	1.2	-175	-70	-100	-0.1	-0.1	-0.2
Dudley	6,290	3,830	2,460	3.3	4.0	2.5	715	300	415	0.4	0.3	0.4
Sandwell	6,870	4,140	2,725	3.4	4.1	2.7	-170	-220	45	-0.1	-0.2	0.0
Solihull	2,460	1,405	1,050	1.9	2.2	1.6	260	75	180	0.2	0.1	0.3
Walsall	4,800	2,790	2,010	2.8	3.3	2.3	-5	-100	100	0.0	-0.1	0.1
Wolverhampton	6,575	4,090	2,485	4.0	5.0	3.1	160	145	15	0.0	0.2	0.0
Worcestershire	4,545	2,865	1,680	1.3	1.6	0.9	355	155	200	0.1	0.1	0.1
Bromsgrove	660	425	235	1.1	1.5	0.8	95	65	30	0.1	0.3	0.1
Malvern Hills	495	310	185	1.1	1.4	0.8	65	35	30	0.1	0.1	0.1
Redditch	865	515	345	1.6	2.0	1.3	40	20	20	0.1	0.1	0.1
Worcester	970	645	325	1.5	1.9	1.0	-30	-25	0	0.0	-0.1	0.0
Wychavon	700	440	260	1.0	1.2	0.7	55	25	30	0.1	0.0	0.1
Wyre Forest	860	530	330	1.5	1.8	1.1	125	35	90	0.2	0.1	0.3
East	49,275	29,500	19,775	1.3	1.6	1.0	3,350	1,140	2,210	0.1	0.1	0.1
Bedford	2,670	1,530	1,135	2.5	2.9	2.2	775	365	400	0.7	0.7	0.8
Central Bedfordshire	1,475	880	595	0.8	1.0	0.7	205	100	105	0.1	0.1	0.1
Luton	2,495	1,490	1,005	1.8	2.1	1.5	165	100	65	0.1	0.1	0.1
Peterborough	1,705	995	710	1.4	1.6	1.2	230	130	100	0.2	0.2	0.2
Southend-on-Sea	2,605	1,590	1,015	2.3	2.9	1.8	390	175	210	0.3	0.4	0.4
Thurrock	1,790	940	850	1.7	1.8	1.6	-185	-160	-25	-0.1	-0.3	0.0
Cambridgeshire	2,895	1,805	1,090	0.7	0.9	0.5	240	130	110	0.0	0.1	0.0

4.10 Claimant count by Sex and Unitary and Local Authority

ONS Crown Copyright Reserved [from Nomis on 20 January 2020]
Age 16+

| | Claimant Count December 2017 | | | | | | Change on year | | | | | |
| | Levels | | | Claimants as a proportion of residents aged 16-64 | | | Levels | | | Claimants as a proportion of residents aged 16-64 | | |
Area	Total	Male	Female	Total	Male	Female	Total	Male	Female	Total	Male	Female
Cambridge	705	450	255	0.8	1.0	0.6	-15	-35	20	0.0	0.0	0.0
East Cambridgeshire	320	195	125	0.6	0.7	0.5	30	20	10	0.1	0.0	0.1
Fenland	650	385	265	1.1	1.3	0.9	100	50	50	0.2	0.2	0.2
Huntingdonshire	755	470	285	0.7	0.8	0.5	95	70	20	0.1	0.1	0.0
South Cambridgeshire	465	305	160	0.5	0.6	0.3	35	25	10	0.0	0.0	0.0
Essex	**11,395**	**6,610**	**4,790**	**1.3**	**1.5**	**1.1**	**-125**	**-385**	**265**	**0.0**	**-0.1**	**0.1**
Basildon	1,840	1,050	790	1.6	1.9	1.3	-90	-90	-5	-0.1	-0.1	-0.1
Braintree	975	560	415	1.1	1.2	0.9	-20	-40	20	0.0	-0.1	0.1
Brentwood	370	235	130	0.8	1.0	0.5	-10	5	-20	0.0	0.0	-0.1
Castle Point	495	255	245	0.9	1.0	0.9	-160	-130	-25	-0.3	-0.5	-0.1
Chelmsford	1,135	690	445	1.0	1.3	0.8	-5	-35	30	-0.1	0.0	0.0
Colchester	1,325	785	540	1.1	1.3	0.9	-30	-60	30	0.0	-0.1	0.0
Epping Forest	945	525	420	1.2	1.3	1.0	-5	10	-15	0.0	0.0	-0.1
Harlow	1,145	630	515	2.1	2.4	1.9	195	60	130	0.3	0.3	0.5
Maldon	350	195	155	0.9	1.0	0.8	40	10	30	0.1	0.0	0.1
Rochford	425	230	195	0.8	0.9	0.8	10	-25	35	0.0	-0.1	0.2
Tendring	2,130	1,290	835	2.7	3.4	2.1	-55	-100	40	-0.1	-0.3	0.1
Uttlesford	260	160	100	0.5	0.6	0.4	10	0	10	0.0	0.0	0.1
Hertfordshire	**8,110**	**4,905**	**3,205**	**1.1**	**1.3**	**0.9**	**445**	**210**	**235**	**0.1**	**0.0**	**0.1**
Broxbourne	730	380	350	1.2	1.3	1.1	15	-20	40	0.0	-0.1	0.1
Dacorum	1,090	630	460	1.1	1.3	0.9	70	-10	80	0.0	0.0	0.1
East Hertfordshire	605	375	230	0.7	0.8	0.5	-20	-25	10	0.0	-0.1	0.0
Hertsmere	765	465	300	1.2	1.5	0.9	5	10	-5	0.0	0.0	0.0
North Hertfordshire	910	545	365	1.1	1.3	0.9	15	15	0	0.0	0.0	0.0
St Albans	825	485	340	0.9	1.1	0.7	115	40	75	0.1	0.1	0.1
Stevenage	925	610	320	1.6	2.2	1.1	90	70	25	0.1	0.3	0.1
Three Rivers	535	330	205	0.9	1.2	0.7	40	15	25	0.0	0.1	0.1
Watford	840	540	305	1.3	1.7	1.0	55	70	-10	0.1	0.2	0.0
Welwyn Hatfield	880	545	335	1.1	1.4	0.8	60	50	10	0.1	0.2	0.0
Norfolk	**7,600**	**4,715**	**2,885**	**1.4**	**1.8**	**1.1**	**470**	**145**	**325**	**0.1**	**0.1**	**0.1**
Breckland	780	465	320	1.0	1.1	0.8	-55	-65	15	0.0	-0.2	0.0
Broadland	475	300	180	0.6	0.8	0.5	-15	-20	10	-0.1	-0.1	0.0
Great Yarmouth	2,885	1,705	1,175	5.0	5.9	4.1	535	220	310	1.0	0.8	1.1
King`s Lynn and West Norfolk	845	520	325	1.0	1.2	0.7	20	10	10	0.1	0.0	0.0
North Norfolk	485	320	165	0.9	1.2	0.6	-30	-25	0	0.0	-0.1	0.0
Norwich	1,520	1,030	490	1.6	2.2	1.0	-40	-5	-35	0.0	0.0	-0.1
South Norfolk	615	380	235	0.8	1.0	0.6	60	35	25	0.1	0.1	0.1
Suffolk	**615**	**380**	**235**	**0.8**	**1.0**	**0.6**	**-5,175**	**-3,330**	**-1,845**	**-0.5**	**-0.7**	**-0.3**
Babergh	435	250	185	0.8	1.0	0.7	35	10	25	0.0	0.1	0.1
Forest Heath	325	195	130	0.8	0.9	0.7	50	35	15	0.1	0.1	0.1
Ipswich	1,700	1,120	580	1.9	2.5	1.3	0	-30	30	0.0	-0.1	0.0
Mid Suffolk	450	290	160	0.8	1.0	0.5	15	25	-15	0.1	0.1	-0.1
St Edmundsbury	810	505	305	1.2	1.4	0.9	170	105	70	0.3	0.3	0.2
Suffolk Coastal	485	300	185	0.7	0.8	0.5	20	-5	25	0.1	-0.1	0.1
Waveney	2,330	1,390	940	3.5	4.3	2.8	455	195	255	0.7	0.6	0.8
London	**119,465**	**67,825**	**51,635**	**2.0**	**2.3**	**1.7**	**8,030**	**3,700**	**4,325**	**0.1**	**0.2**	**0.1**
Inner London	**56,150**	**32,320**	**23,830**	**2.2**	**2.5**	**1.9**	**5,185**	**2,545**	**2,640**	**0.2**	**0.2**	**0.2**
Camden	2,690	1,565	1,125	1.5	1.7	1.3	45	40	5	0.0	0.0	0.0
City of London	65	40	25	1.3	1.4	1.1	10	5	5	0.2	0.2	0.1
Hackney	4,820	2,840	1,980	2.4	2.9	2.0	175	105	70	0.0	0.1	0.1
Hammersmith and Fulham	4,075	2,250	1,825	3.1	3.4	2.8	990	490	505	0.7	0.7	0.8
Haringey	4,465	2,710	1,755	2.4	2.8	1.9	-70	5	-75	0.0	0.0	0.0
Islington	3,700	2,195	1,505	2.1	2.5	1.7	85	70	15	0.0	0.1	0.0
Kensington and Chelsea	1,850	1,075	775	1.7	2.0	1.5	235	140	95	0.2	0.3	0.2
Lambeth	5,880	3,540	2,340	2.4	2.9	2.0	330	190	140	0.1	0.2	0.2
Lewisham	5,185	3,015	2,170	2.5	2.9	2.0	285	90	195	0.2	0.1	0.1
Newham	4,535	2,540	1,995	1.9	1.9	1.8	5	-15	20	0.0	0.0	0.0
Southwark	7,160	3,920	3,240	3.1	3.4	2.8	1,670	785	885	0.7	0.7	0.7
Tower Hamlets	6,090	3,405	2,685	2.7	2.8	2.5	1,360	630	725	0.6	0.4	0.6
Wandsworth	3,185	1,840	1,345	1.4	1.6	1.1	60	5	60	0.1	0.0	0.0
Westminster	2,450	1,390	1,065	1.4	1.5	1.3	0	10	-5	0.0	0.0	0.0

4.10 Claimant count by Sex and Unitary and Local Authority

ONS Crown Copyright Reserved [from Nomis on 20 January 2020]
Age 16+

	Claimant Count December 2017						Change on year					
	Levels			Claimants as a proportion of residents aged 16-64			Levels			Claimants as a proportion of residents aged 16-64		
Area	Total	Male	Female	Total	Male	Female	Total	Male	Female	Total	Male	Female
Outer London	**63,310**	**35,505**	**27,805**	**1.8**	**2.1**	**1.6**	**2,840**	**1,155**	**1,685**	**0.0**	**0.1**	**0.1**
Barking and Dagenham	3,110	1,655	1,450	2.3	2.5	2.1	-65	-95	30	-0.1	-0.2	0.0
Barnet	3,425	2,040	1,380	1.4	1.6	1.1	-90	-10	-85	0.0	-0.1	-0.1
Bexley	2,130	1,215	915	1.4	1.6	1.1	105	115	-10	0.1	0.1	-0.1
Brent	4,775	2,970	1,810	2.2	2.6	1.7	-290	-160	-125	-0.1	-0.2	-0.1
Bromley	2,215	1,240	975	1.1	1.3	0.9	-55	-35	-20	0.0	0.0	0.0
Croydon	8,700	4,690	4,010	3.5	3.9	3.1	2,140	1,015	1,125	0.9	0.8	0.8
Ealing	4,900	2,920	1,980	2.2	2.5	1.8	150	105	45	0.1	0.1	0.1
Enfield	4,425	2,385	2,040	2.1	2.3	1.9	-80	-45	-30	0.0	0.0	0.0
Greenwich	4,030	2,315	1,715	2.1	2.4	1.8	160	140	20	0.1	0.1	0.0
Harrow	1,795	975	820	1.1	1.2	1.0	-45	-60	15	-0.1	-0.1	0.0
Havering	2,375	1,350	1,025	1.5	1.7	1.3	-195	-85	-110	-0.1	-0.2	-0.1
Hillingdon	2,520	1,415	1,105	1.3	1.4	1.1	-60	-40	-20	0.0	-0.1	-0.1
Hounslow	4,825	2,525	2,300	2.7	2.7	2.7	945	405	540	0.5	0.4	0.7
Kingston upon Thames	1,270	740	525	1.1	1.3	0.9	80	30	40	0.1	0.1	0.1
Merton	2,330	1,275	1,055	1.7	1.9	1.5	155	45	110	0.1	0.1	0.1
Redbridge	2,400	1,360	1,040	1.2	1.4	1.1	-340	-175	-165	-0.2	-0.2	-0.1
Richmond upon Thames	1,485	820	665	1.2	1.4	1.0	85	0	85	0.1	0.1	0.1
Sutton	2,695	1,375	1,320	2.1	2.2	2.0	470	160	310	0.4	0.3	0.5
Waltham Forest	3,910	2,240	1,670	2.1	2.4	1.8	-225	-160	-65	-0.1	-0.1	-0.1
South East	**66,165**	**40,205**	**25,960**	**1.2**	**1.4**	**0.9**	**7,030**	**3,105**	**3,925**	**0.1**	**0.1**	**0.1**
Bracknell Forest	**520**	**325**	**195**	**0.7**	**0.8**	**0.5**	**-50**	**-10**	**-40**	**0.0**	**-0.1**	**-0.1**
Brighton and Hove	**3,060**	**1,880**	**1,180**	**1.5**	**1.8**	**1.2**	**110**	**35**	**80**	**0.0**	**0.0**	**0.1**
Isle of Wight	**1,565**	**1,020**	**545**	**1.9**	**2.6**	**1.3**	**-45**	**-45**	**0**	**-0.1**	**-0.1**	**0.0**
Medway	**3,245**	**2,015**	**1,230**	**1.8**	**2.3**	**1.4**	**-100**	**-60**	**-35**	**-0.1**	**0.0**	**0.0**
Milton Keynes	**2,395**	**1,430**	**965**	**1.4**	**1.7**	**1.1**	**205**	**70**	**135**	**0.1**	**0.1**	**0.1**
Portsmouth	**2,255**	**1,400**	**855**	**1.6**	**1.9**	**1.2**	**265**	**110**	**160**	**0.2**	**0.2**	**0.2**
Reading	**1,645**	**1,005**	**640**	**1.5**	**1.8**	**1.2**	**60**	**25**	**30**	**0.1**	**0.1**	**0.1**
Slough	**1,200**	**705**	**500**	**1.3**	**1.4**	**1.1**	**-95**	**-50**	**-40**	**-0.1**	**-0.2**	**0.0**
Southampton	**3,745**	**2,270**	**1,475**	**2.2**	**2.5**	**1.8**	**1,525**	**785**	**740**	**0.9**	**0.8**	**0.9**
West Berkshire	**575**	**340**	**235**	**0.6**	**0.7**	**0.5**	**-15**	**-5**	**-10**	**0.0**	**0.0**	**0.0**
Windsor and Maidenhead	**630**	**395**	**240**	**0.7**	**0.9**	**0.5**	**-100**	**-60**	**-40**	**-0.1**	**-0.1**	**-0.1**
Wokingham	**500**	**315**	**185**	**0.5**	**0.6**	**0.4**	**-105**	**-65**	**-40**	**-0.1**	**-0.2**	**0.0**
Buckinghamshire	**2,845**	**1,735**	**1,110**	**0.9**	**1.1**	**0.7**	**225**	**95**	**125**	**0.1**	**0.1**	**0.1**
Aylesbury Vale	985	595	390	0.8	1.0	0.6	105	65	40	0.1	0.1	0.0
Chiltern	405	255	150	0.7	0.9	0.5	45	35	10	0.0	0.1	0.0
South Bucks	265	150	115	0.6	0.7	0.5	-35	-35	0	-0.1	-0.2	0.0
Wycombe	1,190	740	450	1.1	1.4	0.8	110	35	75	0.1	0.1	0.1
East Sussex	**5,725**	**3,360**	**2,360**	**1.8**	**2.2**	**1.5**	**1,460**	**680**	**775**	**0.5**	**0.5**	**0.5**
Eastbourne	1,145	710	440	1.9	2.4	1.4	100	65	45	0.2	0.2	0.1
Hastings	2,340	1,375	965	4.1	4.9	3.3	925	460	465	1.6	1.7	1.6
Lewes	695	435	260	1.2	1.5	0.9	45	15	30	0.1	0.0	0.1
Rother	960	525	430	1.9	2.1	1.7	360	150	205	0.7	0.6	0.8
Wealden	580	320	265	0.6	0.7	0.6	25	0	30	0.0	0.0	0.1
Hampshire	**6,985**	**4,280**	**2,705**	**0.8**	**1.0**	**0.6**	**1,095**	**565**	**530**	**0.1**	**0.1**	**0.1**
Basingstoke and Deane	940	560	380	0.9	1.0	0.7	130	50	80	0.2	0.1	0.2
East Hampshire	415	260	155	0.6	0.8	0.4	15	10	5	0.0	0.1	0.0
Eastleigh	730	430	305	0.9	1.1	0.7	225	105	120	0.3	0.3	0.2
Fareham	510	325	185	0.7	0.9	0.5	115	80	40	0.1	0.2	0.1
Gosport	690	425	265	1.3	1.6	1.0	155	80	75	0.3	0.3	0.3
Hart	295	185	110	0.5	0.6	0.4	55	25	30	0.1	0.0	0.1
Havant	1,120	710	405	1.5	2.0	1.1	85	70	10	0.1	0.2	0.1
New Forest	805	485	320	0.8	1.0	0.6	175	80	95	0.2	0.2	0.2
Rushmoor	520	305	215	0.8	1.0	0.7	-15	-40	25	0.0	-0.1	0.1
Test Valley	500	290	210	0.7	0.8	0.6	75	30	45	0.1	0.1	0.2
Winchester	460	300	160	0.6	0.8	0.4	85	75	10	0.1	0.2	0.0
Kent	**16,400**	**9,900**	**6,500**	**1.7**	**2.1**	**1.4**	**1,615**	**625**	**990**	**0.1**	**0.1**	**0.2**
Ashford	1,295	785	510	1.7	2.1	1.3	170	65	105	0.2	0.2	0.3
Canterbury	1,350	830	520	1.3	1.6	1.0	20	5	20	0.0	0.0	0.0
Dartford	755	410	345	1.1	1.2	1.0	15	-5	20	0.0	0.0	0.0
Dover	1,915	1,135	780	2.8	3.3	2.2	515	210	305	0.8	0.6	0.8

4.10 Claimant count by Sex and Unitary and Local Authority
ONS Crown Copyright Reserved [from Nomis on 20 January 2020]
Age 16+

Area	Claimant Count December 2017						Change on year					
	Levels			Claimants as a proportion of residents aged 16-64			Levels			Claimants as a proportion of residents aged 16-64		
	Total	Male	Female	Total	Male	Female	Total	Male	Female	Total	Male	Female
Gravesham	1,330	805	525	2.0	2.5	1.6	90	25	65	0.1	0.1	0.2
Maidstone	1,195	695	500	1.2	1.4	1.0	10	-10	15	0.0	0.0	0.1
Sevenoaks	525	320	205	0.7	0.9	0.6	15	5	10	0.0	0.0	0.1
Shepway	1,435	945	490	2.2	2.9	1.5	80	70	15	0.1	0.2	0.0
Swale	1,950	1,105	845	2.2	2.5	1.9	35	-40	75	0.0	-0.1	0.2
Thanet	3,275	2,005	1,270	4.0	5.0	3.0	540	225	310	0.7	0.5	0.7
Tonbridge and Malling	745	460	290	1.0	1.2	0.7	15	-5	25	0.1	0.0	0.0
Tunbridge Wells	620	405	215	0.9	1.1	0.6	100	80	20	0.2	0.2	0.1
Oxfordshire	**2,975**	**1,830**	**1,145**	**0.7**	**0.8**	**0.5**	**330**	**160**	**165**	**0.1**	**0.0**	**0.0**
Cherwell	520	325	195	0.6	0.7	0.4	20	10	15	0.1	0.0	0.0
Oxford	1,090	690	400	1.0	1.2	0.8	110	40	70	0.1	0.1	0.2
South Oxfordshire	455	250	205	0.5	0.6	0.5	55	15	40	0.0	0.0	0.1
Vale of White Horse	540	345	195	0.7	0.9	0.5	125	85	35	0.2	0.2	0.1
West Oxfordshire	370	215	155	0.6	0.7	0.5	20	10	10	0.1	0.1	0.1
Surrey	**4,840**	**2,920**	**1,920**	**0.7**	**0.8**	**0.5**	**345**	**145**	**200**	**0.1**	**0.0**	**0.0**
Elmbridge	490	305	185	0.6	0.8	0.4	75	45	25	0.1	0.2	0.0
Epsom and Ewell	340	240	100	0.7	1.0	0.4	20	50	-30	0.0	0.2	-0.1
Guildford	530	320	210	0.5	0.6	0.4	-5	-10	10	-0.1	-0.1	0.0
Mole Valley	280	180	100	0.5	0.7	0.4	20	5	20	0.0	0.0	0.1
Reigate and Banstead	705	425	280	0.8	1.0	0.6	50	20	30	0.1	0.1	0.0
Runnymede	365	200	165	0.6	0.7	0.6	35	-10	45	0.0	-0.1	0.2
Spelthorne	560	320	240	0.9	1.0	0.8	0	0	0	0.0	0.0	0.0
Surrey Heath	335	195	140	0.6	0.7	0.5	25	0	25	0.0	0.0	0.1
Tandridge	460	245	215	0.9	1.0	0.8	95	45	50	0.2	0.2	0.2
Waverley	360	225	130	0.5	0.6	0.4	15	-5	15	0.0	0.0	0.1
Woking	420	260	160	0.7	0.8	0.5	25	5	20	0.1	0.0	0.0
West Sussex	**5,060**	**3,080**	**1,980**	**1.0**	**1.2**	**0.8**	**305**	**95**	**210**	**0.1**	**0.0**	**0.1**
Adur	500	290	210	1.3	1.6	1.1	60	20	40	0.1	0.1	0.2
Arun	1,130	705	420	1.3	1.6	0.9	145	95	45	0.2	0.2	0.1
Chichester	665	420	245	1.0	1.3	0.7	20	10	10	0.1	0.1	0.0
Crawley	890	510	375	1.2	1.4	1.1	-40	-40	0	-0.1	-0.1	0.1
Horsham	610	375	235	0.7	0.9	0.6	75	20	55	0.1	0.0	0.2
Mid Sussex	410	240	170	0.5	0.5	0.4	-20	-30	15	0.0	-0.1	0.1
Worthing	855	540	315	1.3	1.7	0.9	55	20	35	0.1	0.1	0.1
South West	**44,635**	**27,255**	**17,380**	**1.3**	**1.6**	**1.0**	**4,415**	**1,675**	**2,740**	**0.1**	**0.1**	**0.1**
Bath and North East Somerset	**1,980**	**1,130**	**850**	**1.6**	**1.9**	**1.4**	**570**	**255**	**315**	**0.4**	**0.4**	**0.5**
Bournemouth, Christchurch and Poole	**3,235**	**2,055**	**1,180**	**1.3**	**1.7**	**1.0**	**15**	**-25**	**40**	**0.0**	**0.0**	**0.0**
Bristol, City of	**5,030**	**3,195**	**1,830**	**1.6**	**2.0**	**1.2**	**-75**	**-80**	**0**	**0.0**	**-0.1**	**0.0**
Cornwall	**4,255**	**2,700**	**1,555**	**1.3**	**1.7**	**0.9**	**-530**	**-350**	**-175**	**-0.2**	**-0.2**	**-0.1**
Isles of Scilly	**5**	**5**	**0**	**0.4**	**0.6**	**#**	**0**	**0**	**0**	**-0.1**	**-0.2**	**#**
North Somerset	**1,660**	**985**	**675**	**1.3**	**1.6**	**1.1**	**435**	**210**	**225**	**0.3**	**0.3**	**0.4**
Plymouth	**3,390**	**2,210**	**1,180**	**2.0**	**2.6**	**1.4**	**-85**	**-75**	**-10**	**-0.1**	**-0.1**	**0.0**
South Gloucestershire	**1,470**	**900**	**570**	**0.8**	**1.0**	**0.7**	**-15**	**0**	**-10**	**-0.1**	**0.0**	**0.0**
Swindon	**2,745**	**1,540**	**1,205**	**1.9**	**2.2**	**1.7**	**1,105**	**560**	**545**	**0.7**	**0.8**	**0.8**
Torbay	**1,620**	**1,075**	**545**	**2.1**	**2.8**	**1.4**	**120**	**30**	**85**	**0.2**	**0.1**	**0.2**
Wiltshire	**3,305**	**1,910**	**1,395**	**1.1**	**1.3**	**0.9**	**780**	**335**	**445**	**0.3**	**0.2**	**0.3**
Devon	**4,220**	**2,660**	**1,560**	**0.9**	**1.2**	**0.7**	**30**	**20**	**10**	**0.0**	**0.0**	**0.0**
East Devon	640	410	230	0.8	1.1	0.6	5	15	-15	0.0	0.0	0.0
Exeter	815	515	300	0.9	1.2	0.7	-50	-35	-15	-0.1	0.0	0.0
Mid Devon	430	275	155	0.9	1.2	0.6	5	20	-15	0.0	0.1	-0.1
North Devon	575	350	225	1.0	1.3	0.8	55	15	40	0.0	0.1	0.1
South Hams	380	220	155	0.8	0.9	0.6	65	20	40	0.1	0.1	0.1
Teignbridge	670	410	260	0.9	1.1	0.7	0	5	-5	0.0	0.0	0.0
Torridge	470	320	150	1.2	1.7	0.8	-25	-20	-10	-0.1	-0.1	0.0
West Devon	240	160	80	0.8	1.0	0.5	-20	-5	-15	0.0	-0.1	-0.1
Dorset	**2,165**	**1,305**	**860**	**0.9**	**1.1**	**0.7**	**430**	**195**	**230**	**0.1**	**0.0**	**0.1**
Christchurch	255	160	95	1.0	1.2	0.7	5	10	-5	0.1	0.1	0.0
East Dorset	330	205	125	0.7	0.9	0.5	30	5	20	0.1	0.0	0.1
North Dorset	310	160	150	0.8	0.8	0.8	65	25	40	0.2	0.2	0.3
Purbeck	200	115	85	0.8	0.9	0.6	15	0	10	0.1	0.0	0.0
West Dorset	445	260	185	0.8	1.0	0.6	55	5	50	0.1	0.1	0.1
Weymouth and Portland	625	405	215	1.6	2.1	1.1	15	0	10	0.0	0.0	0.0

4.10 Claimant count by Sex and Unitary and Local Authority

ONS Crown Copyright Reserved [from Nomis on 20 January 2020]
Age 16+

| Area | Claimant Count December 2017 | | | | | | Change on year | | | | | |
| | Levels | | | Claimants as a proportion of residents aged 16-64 | | | Levels | | | Claimants as a proportion of residents aged 16-64 | | |
	Total	Male	Female	Total	Male	Female	Total	Male	Female	Total	Male	Female
Gloucestershire	3,640	2,305	1,340	1.0	1.2	0.7	-310	-225	-75	0.0	-0.1	0.0
Cheltenham	645	430	215	0.9	1.2	0.6	-105	-60	-45	-0.1	-0.1	-0.1
Cotswold	310	195	115	0.6	0.8	0.4	-5	5	-10	0.0	0.0	-0.1
Forest of Dean	560	335	220	1.1	1.3	0.9	-25	-45	15	0.0	-0.2	0.1
Gloucester	1,170	770	405	1.4	1.9	1.0	-145	-75	-60	-0.2	-0.2	-0.1
Stroud	535	325	210	0.8	0.9	0.6	45	0	45	0.1	0.0	0.1
Tewkesbury	420	250	170	0.8	1.0	0.6	-75	-50	-25	-0.1	-0.1	-0.1
Somerset	6,175	3,445	2,730	1.9	2.2	1.7	2,210	990	1,220	0.7	0.7	0.8
Mendip	1,315	720	595	2.0	2.2	1.7	510	230	280	0.8	0.7	0.8
Sedgemoor	1,810	1,000	810	2.5	2.8	2.2	430	170	260	0.6	0.5	0.7
South Somerset	1,385	770	620	1.4	1.6	1.3	550	245	315	0.5	0.5	0.7
Taunton Deane	1,325	765	560	1.9	2.2	1.6	540	260	280	0.8	0.7	0.8
West Somerset	340	195	145	1.8	2.2	1.5	180	90	90	0.9	1.0	0.9
Wales	37,710	23,905	13,805	2.0	2.5	1.4	-1,890	-1,345	-545	-0.1	-0.1	-0.1
Anglesey, Isle of	945	600	350	2.4	3.0	1.7	-170	-140	-25	-0.4	-0.7	-0.2
Gwynedd	1,270	830	440	1.7	2.2	1.2	-170	-140	-30	-0.2	-0.4	-0.1
Conwy	1,280	855	430	1.9	2.6	1.3	-155	-90	-55	-0.3	-0.3	-0.1
Denbighshire	1,040	680	365	1.9	2.5	1.3	-145	-95	-45	-0.2	-0.3	-0.2
Flintshire	1,795	1,015	775	1.9	2.2	1.6	315	100	210	0.3	0.2	0.4
Wrexham	1,365	840	525	1.6	2.0	1.3	-100	-90	-10	-0.2	-0.2	0.0
Powys	675	455	220	0.9	1.2	0.6	-120	-70	-50	-0.1	-0.2	-0.1
Ceredigion	530	350	180	1.2	1.6	0.8	10	15	-10	0.0	0.2	-0.1
Pembrokeshire	1,470	985	485	2.0	2.8	1.3	-195	-115	-80	-0.3	-0.3	-0.3
Carmarthenshire	1,820	1,180	640	1.7	2.2	1.1	-175	-75	-100	-0.1	-0.1	-0.2
Swansea	3,195	2,140	1,055	2.0	2.7	1.4	90	75	10	0.0	0.1	0.0
Neath Port Talbot	2,005	1,220	790	2.3	2.8	1.8	20	-20	45	0.0	0.0	0.1
Bridgend	1,740	1,095	650	1.9	2.4	1.5	-100	-35	-60	-0.2	-0.1	-0.1
Vale of Glamorgan	1,275	840	435	1.6	2.1	1.1	-70	-45	-25	-0.1	-0.2	-0.1
Cardiff	4,750	3,145	1,605	1.9	2.6	1.3	-225	-145	-75	-0.1	-0.1	-0.1
Rhondda Cynon Taff	2,930	1,760	1,170	2.0	2.4	1.5	-320	-225	-95	-0.2	-0.3	-0.2
Merthyr Tydfil	960	595	365	2.6	3.2	1.9	-5	20	-20	0.0	0.1	-0.1
Caerphilly	2,845	1,735	1,110	2.5	3.1	2.0	-205	-100	-105	-0.2	-0.2	-0.1
Blaenau Gwent	1,170	695	475	2.7	3.2	2.2	-340	-240	-100	-0.7	-1.1	-0.4
Torfaen	1,600	960	640	2.8	3.5	2.2	400	210	190	0.7	0.8	0.6
Monmouthshire	575	370	200	1.0	1.4	0.7	-35	-20	-20	-0.1	0.0	-0.1
Newport	2,460	1,555	905	2.6	3.3	1.9	-205	-125	-80	-0.3	-0.3	-0.2
Scotland	79,035	52,870	26,165	2.3	3.1	1.5	4,120	1,990	2,130	0.2	0.1	0.1
Aberdeen City	3,110	2,220	890	2.0	2.7	1.1	-310	-205	-105	-0.1	-0.3	-0.2
Aberdeenshire	2,045	1,405	640	1.2	1.7	0.8	-320	-210	-110	-0.2	-0.2	-0.1
Angus	1,520	995	525	2.2	2.9	1.5	115	25	90	0.2	0.1	0.3
Argyll and Bute	900	610	295	1.7	2.3	1.2	45	20	30	0.1	0.1	0.2
Clackmannanshire	1,160	720	440	3.6	4.5	2.7	320	160	155	1.0	1.0	1.0
Dumfries and Galloway	1,580	1,020	555	1.8	2.4	1.2	125	100	20	0.2	0.3	0.0
Dundee City	3,175	2,210	965	3.2	4.6	1.9	-40	-110	70	-0.1	-0.2	0.1
East Ayrshire	2,525	1,690	835	3.3	4.5	2.1	15	-25	40	0.0	-0.1	0.1
East Dunbartonshire	1,070	680	390	1.6	2.1	1.2	345	195	150	0.5	0.6	0.5
East Lothian	1,765	1,000	765	2.7	3.2	2.3	495	225	270	0.7	0.7	0.8
East Renfrewshire	645	440	205	1.1	1.6	0.7	-40	-20	-20	-0.1	-0.1	-0.1
Edinburgh, City of	4,480	3,070	1,410	1.3	1.7	0.8	-450	-210	-240	-0.1	-0.2	-0.1
Na h-Eileanan Siar	270	210	60	1.7	2.6	0.8	-35	-25	-10	-0.2	-0.3	-0.1
Falkirk	2,190	1,455	730	2.1	2.9	1.4	0	-10	0	-0.1	0.0	0.0
Fife	5,635	3,800	1,840	2.4	3.3	1.5	190	85	115	0.1	0.0	0.0
Glasgow City	13,455	9,245	4,215	3.1	4.3	1.9	310	230	85	0.1	0.1	0.0
Highland	3,380	2,150	1,230	2.3	3.0	1.7	915	490	425	0.6	0.7	0.6
Inverclyde	2,360	1,490	870	4.7	6.2	3.4	1,145	650	495	2.3	2.7	1.9
Midlothian	1,405	885	520	2.5	3.3	1.8	545	305	240	0.9	1.1	0.8
Moray	1,010	665	345	1.7	2.2	1.2	50	-10	60	0.1	0.0	0.2
North Ayrshire	3,145	2,025	1,120	3.8	5.1	2.6	-150	-110	-40	-0.1	-0.3	-0.1
North Lanarkshire	5,645	3,800	1,845	2.6	3.5	1.6	35	60	-30	0.0	0.0	-0.1
Orkney Islands	115	80	35	0.9	1.2	0.5	-15	-10	-5	-0.1	-0.1	-0.1
Perth and Kinross	1,020	680	335	1.1	1.5	0.7	25	-10	30	0.0	0.0	0.0
Renfrewshire	3,075	2,120	955	2.7	3.8	1.6	375	235	140	0.3	0.4	0.2

83

4.10 Claimant count by Sex and Unitary and Local Authority

ONS Crown Copyright Reserved [from Nomis on 20 January 2020]
Age 16+

Area	Claimant Count December 2017						Change on year					
	Levels			Claimants as a proportion of residents aged 16-64			Levels			Claimants as a proportion of residents aged 16-64		
	Total	Male	Female	Total	Male	Female	Total	Male	Female	Total	Male	Female
Scottish Borders	1,030	685	340	1.5	2.1	1.0	-50	-40	-15	-0.1	-0.1	0.0
Shetland Islands	95	80	15	0.7	1.1	0.2	-25	-15	-10	-0.1	-0.2	-0.2
South Ayrshire	1,565	1,040	520	2.3	3.2	1.5	-120	-105	-20	-0.2	-0.3	0.0
South Lanarkshire	4,695	3,065	1,630	2.3	3.1	1.6	440	175	260	0.2	0.2	0.3
Stirling	1,165	755	410	1.9	2.6	1.3	220	150	70	0.3	0.5	0.2
West Dunbartonshire	1,970	1,360	610	3.4	4.9	2.0	10	20	-5	0.0	0.1	0.0
West Lothian	1,835	1,230	605	1.6	2.1	1.0	-40	-20	-25	0.0	-0.1	-0.1
Northern Ireland	**28,555**	**18,900**	**9,655**	**2.4**	**3.2**	**1.6**	**-3,440**	**-3,320**	**-120**	**-0.3**	**-0.6**	**0.0**
Antrim and Newtownabbey	1,795	1,225	575	2.0	2.8	1.3	-115	-110	5	-0.2	-0.3	0.0
Armagh City, Banbridge and Craigavon	2,100	1,385	720	2.2	2.9	1.4	-460	-315	-140	0.3	0.3	0.1
Belfast	2,360	1,455	905	1.8	2.2	1.4	-5,745	-4,470	-1,275	-1.8	-3.2	-0.5
Causeway Coast and Glens	7,010	4,910	2,100	3.1	4.5	1.8	4,305	3,080	1,220	0.1	0.4	-0.2
Derry City and Strabane	2,450	1,585	865	2.7	3.5	1.9	-2,620	-1,875	-740	-2.6	-3.9	-1.4
Fermanagh and Omagh	4,510	3,035	1,475	4.7	6.5	3.0	2,815	1,920	900	2.4	3.4	1.4
Lisburn and Castlereagh	1,435	910	520	2.0	2.5	1.5	-55	-125	65	0.3	0.1	0.5
Mid and East Antrim	1,265	830	440	1.4	1.9	1.0	-900	-675	-220	-1.1	-1.6	-0.5
Mid Ulster	2,015	1,345	670	2.3	3.2	1.5	550	425	125	0.7	1.2	0.3
Newry, Mourne and Down	1,350	795	555	1.5	1.7	1.2	-1,170	-950	-220	-0.8	-1.5	-0.2
Ards and North Down	2,265	1,430	835	2.0	2.6	1.5	-50	-220	170	-0.4	-0.9	0.2

Source: ONS Labour Market Statistics; Nomisweb

These figures are suppressed as value is 1 or 2.

Under Universal Credit a broader span of claimants are required to look for work than under Jobseeker's Allowance. As Universal Credit Full Service is rolled out in particular areas, the number of people recorded as being on the Claimant Count is therefore likely to rise.

This dataset only includes claimant records with information to sufficiently classify them within the dataset. Totals from this dataset may not tally with the total number of claim records shown elsewhere as a small number of records do not have this information.

All data are rounded to the nearest 5 and may not precisely add to the sum of the number of people claiming JSA, published on Nomis, and the number of people claiming Universal Credit required to seek work, published by DWP, due to independent rounding.

4.11a Weekly pay - Gross (£) - For all employee jobs[a]: United Kingdom, 2017

Description	Code	Number of jobs[b] (000's)	Median	Annual % change	Mean	Annual % change	Percentiles									
							10	20	25	30	40	60	70	75	80	90
ALL EMPLOYEES		26,241	448.5	2.3	537.9	2.4	144.4	238.0	283.8	316.5	380.0	527.0	624.5	682.9	752.0	972.6
All Industries and Services		26,233	448.5	2.3	537.9	2.4	144.4	238.0	283.8	316.5	380.0	527.0	624.5	683.0	752.0	972.6
All Index of Production Industries		2,798	556.6	2.1	639.8	1.2	304.0	368.3	398.3	427.5	488.9	636.9	724.2	774.3	844.1	1,054.1
All Manufacturing		2,434	544.4	2.2	626.0	1.4	300.0	362.6	390.2	420.0	479.1	622.9	709.3	764.7	827.2	1,018.9
All Service Industries		22,369	430.2	2.4	522.1	2.7	132.1	219.0	261.8	299.0	360.6	505.4	603.3	667.0	733.1	958.2
AGRICULTURE, FORESTRY AND FISHING	A	139	383.1	3.1	422.0	4.4	142.4	227.0	280.6	305.9	345.0	427.0	481.5	522.7	566.9	689.9
Crop and animal production, hunting and related service activities	1	125	374.6	2.7	398.9	3.8	137.3	223.4	272.7	299.5	343.3	420.5	470.8	502.1	548.5	670.8
Forestry and logging	2	9	480.6	1.3	621.7	13.4	x	306.6	345.4	371.5	413.8	527.0	x	x	x	x
Fishing and aquaculture	3	x	x		x		x	x	x	x	x	x	x	x	x	x
MINING AND QUARRYING	B	40	740.0	5.0	895.0	5.9	445.7	542.8	562.5	611.6	684.4	797.8	942.6	1,017.8	1,119.5	x
Mining of coal and lignite	5	x	735.8		725.8	-13.7	x	x	x	x	x	x	x	x	x	x
Extraction of crude petroleum and natural gas	6	6	x		1,519.2	11.0	x	743.7	823.9	914.1	1,105.4	x	x	x	x	x
Mining of metal ores	7	..														
Other mining and quarrying	8	23	649.4	2.0	700.5	2.4	421.7	496.6	538.2	548.1	596.3	713.3	760.0	782.1	813.1	x
Mining support service activities	9	10	854.6	16.7	973.0	11.2	x	569.9	601.0	651.7	762.5	993.1	1,110.7	x	x	x
MANUFACTURING	C	2,434	544.4	2.2	626.0	1.4	300.0	362.6	390.2	420.0	479.1	622.9	709.3	764.7	827.2	1,018.9
Manufacture of food products	10	383	428.5	2.2	510.7	1.3	259.9	307.5	322.5	341.1	383.1	486.0	558.9	607.4	668.4	833.7
Manufacture of beverages	11	47	654.6	6.4	759.3	5.0	338.3	438.5	468.9	495.2	571.2	742.5	848.9	903.0	989.2	x
Manufacture of tobacco products	12	:														
Manufacture of textiles	13	49	388.5	-1.6	469.6	-2.0	218.6	290.1	304.4	320.0	349.3	412.3	479.9	524.0	587.5	x
Manufacture of wearing apparel	14	18	293.0	4.4	381.1	6.4	134.9	172.8	184.5	191.8	255.4	330.9	369.0	x	x	x
Manufacture of leather and related products	15	10	356.3	1.7	419.4	6.6	x	282.0	290.8	296.8	310.7	404.6	447.5	x	x	x
Manufacture of wood and of products of wood and cork, except furniture; manufacture of articles of straw and plaiting materials	16	54	461.9	0.7	508.6	2.3	280.3	345.1	363.1	383.1	417.9	519.7	574.9	622.7	669.8	x
Manufacture of paper and paper products	17	60	547.9	1.9	666.3	8.0	321.4	383.2	409.6	438.8	489.4	596.7	691.8	751.9	819.7	x
Printing and reproduction of recorded media	18	91	480.2	1.7	549.9	-2.2	264.3	337.4	363.8	395.3	439.1	527.1	603.1	655.0	707.5	x
Manufacture of coke and refined petroleum products	19	11	953.9	-0.9	978.4	-2.2	499.0	666.3	726.3	814.7	872.5	1,046.1	1,139.1	1,161.6	x	x
Manufacture of chemicals and chemical products	20	83	609.7	0.7	704.4	3.5	317.0	383.8	417.5	461.6	522.3	685.5	782.6	859.4	931.4	x
Manufacture of basic pharmaceutical products and pharmaceutical preparations	21	55	683.7	-1.0	841.5	-14.7	391.9	461.5	512.8	546.2	605.2	775.5	888.9	953.7	1,088.0	x
Manufacture of rubber and plastic products	22	159	478.4	4.0	564.4	3.1	302.9	345.2	369.2	387.2	428.5	526.6	604.3	647.9	689.1	843.6
Manufacture of other non-metallic mineral products	23	84	548.1	6.1	597.5	6.1	300.0	366.7	404.9	432.8	488.7	613.2	679.4	710.2	764.8	x
Manufacture of basic metals	24	74	689.3	4.5	723.0	5.2	376.0	462.0	501.9	538.4	633.6	751.7	830.3	881.6	910.9	1,049.7
Manufacture of fabricated metal products, except machinery and equipment	25	280	526.0	4.7	588.5	3.5	306.7	370.0	400.0	422.5	473.8	587.6	662.7	700.9	762.2	929.5
Manufacture of computer, electronic and optical products	26	128	618.5	0.8	683.9	-0.5	342.7	406.4	439.7	477.0	536.6	690.4	801.8	862.1	905.2	1,114.6
Manufacture of electrical equipment	27	84	508.9	-1.7	590.8	-4.3	299.5	342.1	372.1	413.7	462.0	572.4	666.8	714.8	766.6	972.1
Manufacture of machinery and equipment n.e.c.	28	197	599.8	2.6	673.8	4.5	338.3	417.8	450.5	480.0	540.9	672.5	760.5	803.0	853.4	1,035.4
Manufacture of motor vehicles, trailers and semi-trailers	29	184	684.1	2.7	756.3	2.0	372.7	458.0	494.0	531.8	615.1	760.5	857.0	923.7	997.7	1,254.6
Manufacture of other transport equipment	30	175	757.1	3.2	799.3	2.3	448.5	548.6	580.4	624.9	696.9	821.7	891.0	928.6	971.5	1,136.3
Manufacture of furniture	31	73	423.1	3.0	492.1	-0.6	292.8	327.1	342.1	357.0	386.2	460.0	513.3	566.6	608.5	x
Other manufacturing	32	65	472.7	6.8	541.1	3.8	230.0	324.7	345.0	362.0	402.8	546.2	632.3	673.8	730.5	x
Repair and installation of machinery and equipment	33	72	598.0	1.2	673.1	-2.9	320.1	404.5	441.1	477.7	519.9	671.1	766.6	838.1	909.1	x
ELECTRICITY, GAS, STEAM AND AIR CONDITIONING SUPPLY	D	164	712.2	-0.1	807.1	0.8	369.1	463.3	512.3	556.5	641.2	813.6	944.3	1,022.6	1,107.3	1,345.0
Electricity, gas, steam and air conditioning supply	35	164	712.2	-0.1	807.1	0.8	369.1	463.3	512.3	556.5	641.2	813.6	944.3	1,022.6	1,107.3	1,345.0

4.11a Weekly pay - Gross (£) - For all employee jobs[a]: United Kingdom, 2017

Description	Code	Number of jobs[b] (000's)	Median	Annual % change	Mean	Annual % change	10	20	25	30	40	60	70	75	80	90
WATER SUPPLY; SEWERAGE, WASTE MANAGEMENT AND REMEDIATION ACTIVITIES	E	160	553.0	-0.5	614.5	-1.7	324.3	380.6	411.2	442.8	496.4	619.3	688.0	724.7	765.3	948.7
Water collection, treatment and supply	36	42	618.9	2.6	680.4	0.8	388.6	459.6	484.4	509.6	561.4	670.5	728.1	751.4	818.5	x
Sewerage	37	16	600.4	-0.9	594.8	-5.8	337.4	381.8	420.6	498.0	550.4	652.6	702.7	720.2	740.4	x
Waste collection, treatment and disposal activities; materials recovery	38	100	516.9	-1.2	592.2	-2.1	304.2	359.4	381.9	411.2	461.5	588.9	656.0	705.9	763.5	949.9
Remediation activities and other waste management services	39	x	499.3		548.7	6.6	x	x	x	x	440.1	x	x	x	x	x
CONSTRUCTION	F	926	560.0	3.2	630.3	2.9	230.4	360.0	400.2	435.4	498.0	630.0	718.7	770.4	833.8	1,054.7
Construction of buildings	41	277	594.8	5.0	692.1	3.1	255.5	382.3	429.0	460.0	520.0	677.9	801.9	869.5	948.7	1,212.2
Civil engineering	42	180	611.5	4.0	691.8	4.2	302.2	410.4	450.0	482.8	540.8	688.1	787.0	833.2	907.4	1,112.6
Specialised construction activities	43	470	520.0	0.5	570.5	2.0	216.9	337.9	373.7	406.5	465.7	582.3	660.8	698.1	753.7	914.1
WHOLESALE AND RETAIL TRADE; REPAIR OF MOTOR VEHICLES AND MOTORCYCLES	G	3,767	353.0	3.8	431.1	3.5	120.0	176.0	206.8	240.6	304.9	409.7	484.4	532.4	591.5	804.9
Wholesale and retail trade and repair of motor vehicles and motorcycles	45	477	452.3	4.3	512.0	6.6	209.3	310.2	334.7	354.7	402.4	503.3	563.2	597.9	646.0	821.9
Wholesale trade, except of motor vehicles and motorcycles	46	1,060	484.2	3.0	594.8	2.4	262.2	332.6	354.9	378.8	430.0	554.6	644.3	701.3	777.9	1,053.4
Retail trade, except of motor vehicles and motorcycles	47	2,230	275.5	3.0	336.0	2.9	96.2	135.9	155.9	177.0	223.1	320.4	373.0	409.0	459.4	631.2
TRANSPORTATION AND STORAGE	H	1,137	528.9	-0.1	603.6	1.1	283.6	371.4	403.9	425.3	472.8	591.5	671.3	718.7	781.4	1,004.0
Land transport and transport via pipelines	49	512	537.3	-0.2	585.4	2.1	249.7	374.8	408.3	436.1	479.5	598.3	668.4	718.7	774.7	972.9
Water transport	50	13	623.4	18.5	724.9	20.6	257.5	396.7	433.5	477.1	540.8	709.0	789.8	889.6	x	x
Air transport	51	73	640.3	5.3	781.5	3.6	326.2	382.3	418.6	454.8	551.3	730.6	862.8	943.8	1,032.2	x
Warehousing and support activities for transportation	52	333	546.1	-3.7	644.0	-0.8	305.2	373.9	401.4	427.4	486.8	620.4	702.6	756.6	839.5	1,110.0
Postal and courier activities	53	206	468.9	0.4	513.3	1.3	277.4	349.5	389.8	413.5	434.4	507.9	557.9	593.6	637.1	766.5
ACCOMMODATION AND FOOD SERVICE ACTIVITIES	I	1,501	243.1	4.5	284.2	6.2	61.0	106.7	122.9	144.7	189.8	297.1	345.0	373.7	412.0	520.1
Accommodation	55	328	303.4	2.5	331.1	3.6	72.1	135.3	170.1	202.3	264.3	334.9	377.2	402.9	443.6	574.9
Food and beverage service activities	56	1,173	225.0	5.1	271.0	6.7	60.0	102.0	119.8	135.0	179.4	279.4	331.7	364.0	402.1	505.9
INFORMATION AND COMMUNICATION	J	1,047	677.5	1.0	791.5	2.5	299.0	418.6	461.6	498.3	577.8	781.5	915.6	991.9	1,095.7	1,389.5
Publishing activities	58	128	565.4	0.8	674.2	2.0	250.3	378.4	413.1	446.2	498.3	661.2	780.5	859.2	934.6	1,158.0
Motion picture, video and television programme production, sound recording and music publishing activities	59	65	510.8	2.8	652.2	4.1	115.0	188.3	267.4	345.6	427.0	589.3	766.5	845.9	957.2	x
Programming and broadcasting activities	60	47	767.7	-2.0	862.4	-2.9	452.0	538.3	575.0	619.0	679.9	858.2	944.3	1,012.3	1,094.8	x
Telecommunications	61	223	689.5	2.8	779.4	2.5	349.3	456.7	498.9	548.4	603.0	769.3	883.0	964.9	1,042.8	1,295.3
Computer programming, consultancy and related activities	62	510	700.3	0.5	815.3	1.4	299.0	421.6	474.1	507.6	597.9	804.9	942.1	1,034.9	1,149.6	1,449.5
Information service activities	63	74	766.6	5.8	941.4	15.1	342.0	448.8	484.5	550.8	670.8	933.9	1,125.8	1,241.8	1,388.1	x
FINANCIAL AND INSURANCE ACTIVITIES	K	982	677.1	7.0	925.6	6.2	313.0	388.2	426.8	471.1	562.9	823.1	1,017.6	1,149.9	1,298.7	1,811.2
Financial service activities, except insurance and pension funding	64	524	719.8	9.6	978.4	7.3	318.4	402.3	442.4	488.0	594.1	891.2	1,102.0	1,223.1	1,394.1	1,896.5
Insurance, reinsurance and pension funding, except compulsory social security	65	100	661.1	5.0	816.1	6.8	335.1	401.5	433.8	479.1	564.9	766.6	905.3	984.2	1,108.3	x
Activities auxiliary to financial services and insurance activities	66	358	633.7	4.6	879.0	4.1	302.8	368.0	402.5	444.6	534.9	761.8	942.2	1,060.3	1,204.0	1,776.9
REAL ESTATE ACTIVITIES	L	370	474.1	2.8	533.6	0.7	172.5	287.5	325.9	356.5	411.5	519.1	584.1	636.5	689.9	918.9
Real estate activities	68	370	474.1	2.8	533.6	0.7	172.5	287.5	325.9	356.5	411.5	519.1	584.1	636.5	689.9	918.9
PROFESSIONAL, SCIENTIFIC AND TECHNICAL ACTIVITIES	M	1,796	594.1	3.3	715.9	3.7	233.4	347.9	390.1	432.5	516.6	689.9	804.9	881.4	976.2	1,327.7

4.11a Weekly pay - Gross (£) - For all employee jobs[a]: United Kingdom, 2017

Description	Code	Number of jobs[b] (000's)	Median	Annual % change	Mean	Annual % change	Percentiles									
							10	20	25	30	40	60	70	75	80	90
Legal and accounting activities	69	551	544.5	5.2	677.8	3.9	220.4	314.3	345.0	383.3	460.0	642.0	766.6	852.8	944.8	1,327.3
Activities of head offices; management consultancy activities	70	389	605.6	9.7	773.1	5.9	230.0	347.0	384.6	430.8	504.8	708.8	851.6	951.9	1,068.1	1,456.5
Architectural and engineering activities; technical testing and analysis	71	414	649.9	2.8	744.0	4.4	316.2	421.6	469.5	505.3	574.9	738.5	843.3	919.9	1,014.1	1,282.9
Scientific research and development	72	144	723.7	0.1	817.1	-0.9	363.9	491.7	533.3	575.9	646.0	799.6	906.4	992.6	1,078.7	1,362.2
Advertising and market research	73	121	594.1	1.6	703.3	0.8	204.8	383.1	418.7	447.0	517.5	675.3	766.6	832.9	921.3	x
Other professional, scientific and technical activities	74	124	527.5	9.2	611.2	6.3	156.4	287.5	341.1	382.7	460.0	586.6	691.1	766.6	853.2	1,149.9
Veterinary activities	75	52	389.3	4.5	467.7	4.7	156.0	217.5	249.4	280.8	328.6	453.7	560.1	602.9	670.8	x
ADMINISTRATIVE AND SUPPORT SERVICE ACTIVITIES	N	1,652	356.6	2.7	431.4	3.3	103.4	178.2	219.8	258.7	309.9	414.0	490.5	536.6	593.5	772.9
Rental and leasing activities	77	147	488.2	-0.2	570.9	1.3	240.0	345.0	364.1	387.1	434.5	547.2	626.7	664.7	710.8	900.4
Employment activities	78	604	333.6	3.8	393.2	3.8	118.0	195.1	229.9	258.4	297.1	374.2	433.1	476.1	524.1	699.2
Travel agency, tour operator and other reservation service and related activities	79	81	423.0	-1.9	530.2	0.2	194.6	274.4	289.8	313.9	357.8	485.8	557.4	610.7	699.5	x
Security and investigation activities	80	109	390.6	0.2	407.9	1.0	100.0	204.3	262.2	304.5	345.1	437.0	496.3	526.9	566.4	667.0
Services to buildings and landscape activities	81	443	295.2	12.8	348.2	5.0	75.0	106.9	123.8	148.5	206.4	356.4	430.5	480.7	536.6	690.3
Office administrative, office support and other business support activities	82	269	415.7	-1.1	557.6	5.8	154.5	250.1	287.5	307.0	357.4	502.7	594.1	651.6	747.4	1,024.4
PUBLIC ADMINISTRATION AND DEFENCE; COMPULSORY SOCIAL SECURITY	O	1,439	561.9	0.4	589.2	0.1	249.7	356.5	388.1	426.1	487.3	629.3	715.7	752.4	796.4	932.4
Public administration and defence; compulsory social security	84	1,439	561.9	0.4	589.2	0.1	249.7	356.5	388.1	426.1	487.3	629.3	715.7	752.4	796.4	932.4
EDUCATION	P	3,685	444.0	2.0	494.1	1.6	117.9	210.9	251.1	290.4	364.0	533.2	635.5	686.1	733.1	899.3
Education	85	3,685	444.0	2.0	494.1	1.6	117.9	210.9	251.1	290.4	364.0	533.2	635.5	686.1	733.1	899.3
HUMAN HEALTH AND SOCIAL WORK ACTIVITIES	Q	3,899	401.4	1.3	484.1	1.2	158.2	231.7	265.3	292.3	344.4	465.8	546.2	601.5	662.9	834.4
Human health activities	86	2,525	461.4	-0.4	555.6	0.6	191.7	280.3	313.4	344.5	402.1	537.3	621.7	675.1	731.6	928.6
Residential care activities	87	647	312.6	4.8	355.0	3.7	140.0	193.6	218.2	239.9	277.5	349.6	403.6	433.8	476.8	614.3
Social work activities without accommodation	88	727	306.5	4.0	350.6	4.2	101.8	160.6	184.6	215.3	265.9	350.7	419.3	457.0	503.5	646.7
ARTS, ENTERTAINMENT AND RECREATION	R	557	310.1	1.1	428.6	0.1	45.3	102.1	135.2	165.1	240.0	368.1	445.0	494.7	560.2	767.2
Creative, arts and entertainment activities	90	52	467.7	10.5	609.6	10.1	x	183.5	220.7	275.7	379.9	553.1	636.1	720.8	780.0	x
Libraries, archives, museums and other cultural activities	91	64	409.4	5.4	455.0	8.4	120.0	204.3	243.5	297.2	347.9	477.6	546.1	604.3	639.0	x
Gambling and betting activities	92	104	340.0	5.2	386.8	4.7	138.3	190.5	221.8	247.3	300.0	376.5	430.3	463.7	521.6	654.7
Sports activities and amusement and recreation activities	93	337	252.5	-4.2	408.7	-4.7	30.9	65.3	84.5	109.9	175.1	325.8	387.1	431.2	497.8	731.4
OTHER SERVICE ACTIVITIES	S	461	354.6	0.3	424.9	1.6	102.8	156.4	187.0	220.0	292.5	427.3	490.1	544.7	598.1	805.7
Activities of membership organisations	94	225	422.0	2.8	465.0	2.6	76.5	160.0	205.8	259.7	351.6	473.1	546.0	594.1	657.7	866.1
Repair of computers and personal and household goods	95	36	472.9	-0.8	597.0	2.7	192.5	311.9	340.1	361.5	422.5	556.3	667.6	732.4	826.4	x
Other personal service activities	96	201	284.3	0.8	349.2	-0.4	112.6	142.8	166.9	186.9	226.5	332.4	397.7	441.3	483.4	632.4
ACTIVITIES OF HOUSEHOLDS AS EMPLOYERS; UNDIFFERENTIATED GOODS-AND SERVICES-PRODUCING ACTIVITIES OF HOUSEHOLDS FOR OWN USE	T	75	146.4	5.8	218.4	7.3	32.4	52.1	65.1	75.0	111.5	199.9	275.0	318.6	365.4	505.1
Activities of households as employers of domestic personnel	97	75	146.4	5.8	218.4	7.3	32.4	52.1	65.1	75.0	111.5	199.9	275.0	318.6	365.4	505.1

4.11a Weekly pay - Gross (£) - For all employee jobs[a]: United Kingdom, 2017

Description	Code	Number of jobs[b] (000's)	Median	Annual % change	Mean	Annual % change	Percentiles									
							10	20	25	30	40	60	70	75	80	90
Undifferentiated goods- and services-producing activities of private households for own use	98	:														
ACTIVITIES OF EXTRATERRITORIAL ORGANISATIONS AND BODIES	U	x	x		556.5	-0.4	x	x	x	x	x	x	x	x	x	x
Activities of extraterritorial organisations and bodies	99	x	x		556.5	-0.4	x	x	x	x	x	x	x	x	x	x
NOT CLASSIFIED		8	345.6		454.1		x	x	x	x	307.5	x	x	x	x	x

Source: Annual Survey of Hours and Earnings, Office for National Statistics.

a Employees on adult rates whose pay for the survey pay-period was not affected by absence.

b Figures for Number of Jobs are for indicative purposes only and should not be considered an accurate estimate of employee job counts.

KEY - The colour coding indicates the quality of each estimate; jobs, median, mean and percentiles but not the annual percentage change.

The quality of an estimate is measured by its coefficient of variation (CV), which is the ratio of the standard error of an estimate to the estimate.

Key	Statistical robustness
CV <= 5%	Estimates are considered precise
CV > 5% and <= 10%	Estimates are considered reasonably precise
CV > 10% and <= 20%	Estimates are considered acceptable
x = CV > 20%	Estimates are considered unreliable for practical purposes
.. = disclosive	
: = not applicable	
- = nil or negligible	

4.11b Hourly pay - Gross (£) - For all employee jobs[a]: United Kingdom, 2017

Description	Code	Number of jobs[b] (000's)	Median	Annual % change	Mean	Annual % change	10	20	25	30	40	60	70	75	80	90
ALL EMPLOYEES		26,241	12.47	2.5	16.16	2.7	7.62	8.46	8.96	9.50	10.76	14.53	17.21	18.83	20.81	27.19
All Industries and Services		26,233	12.47	2.5	16.16	2.7	7.62	8.46	8.96	9.50	10.76	14.53	17.21	18.84	20.81	27.19
All Index of Production Industries		2,798	13.80	2.2	16.40	1.6	8.21	9.50	10.09	10.74	12.17	15.72	18.20	19.78	21.61	27.40
All Manufacturing		2,434	13.53	2.3	16.07	1.9	8.11	9.32	9.95	10.54	11.96	15.34	17.79	19.28	21.12	26.46
All Service Industries		22,369	12.21	2.3	16.16	2.9	7.58	8.34	8.78	9.30	10.52	14.37	17.09	18.75	20.76	27.27
AGRICULTURE, FORESTRY AND FISHING	A	139	9.42	0.5	11.33	4.1	7.50	7.74	8.00	8.27	8.83	10.18	11.34	12.11	13.30	17.23
Crop and animal production, hunting and related service activities	1	125	9.22	1.1	10.66	3.3	7.50	7.67	7.98	8.13	8.67	10.00	10.96	11.77	12.50	16.00
Forestry and logging	2	9	12.63	0.8	17.20	15.6	x	9.09	9.48	10.43	12.02	13.81	15.85	x	x	x
Fishing and aquaculture	3	x	x		x		x	x	x	9.91	10.33	x	x	x	x	x
MINING AND QUARRYING	B	40	18.26	11.9	21.76	7.7	10.24	11.63	12.46	13.18	15.64	20.59	24.31	27.04	29.93	x
Mining of coal and lignite	5	x	14.84		16.23		x	x	x	x	x	x	x	x	x	x
Extraction of crude petroleum and natural gas	6	6	x		42.13	14.1	x	22.87	x	25.07	30.26	x	x	x	x	x
Mining of metal ores	7	..														
Other mining and quarrying	8	23	14.50	7.4	16.17	4.0	9.26	10.44	10.84	11.27	12.58	16.71	18.69	19.61	20.76	x
Mining support service activities	9	10	22.80	20.3	25.03	12.4	x	14.37	15.49	16.44	19.36	26.41	28.84	x	x	x
MANUFACTURING	C	2,434	13.53	2.3	16.07	1.9	8.11	9.32	9.95	10.54	11.96	15.34	17.79	19.28	21.12	26.46
Manufacture of food products	10	383	10.29	2.3	12.98	2.5	7.55	8.00	8.26	8.62	9.38	11.53	13.36	14.48	16.17	20.59
Manufacture of beverages	11	47	17.10	7.6	20.62	4.8	9.35	11.50	12.19	13.04	14.97	19.06	22.68	23.80	25.85	x
Manufacture of tobacco products	12	:														
Manufacture of textiles	13	49	9.99	2.0	12.58	0.0	7.50	8.00	8.20	8.49	9.24	10.58	11.96	12.85	14.56	x
Manufacture of wearing apparel	14	18	8.49	3.5	12.47	10.4	x	x	7.50	7.59	7.99	9.36	x	x	x	x
Manufacture of leather and related products	15	10	9.50	4.9	11.87	7.7	7.50	7.93	8.08	8.32	8.54	10.37	11.23	x	x	x
Manufacture of wood and of products of wood and cork, except furniture; manufacture of articles of straw and plaiting materials	16	54	11.16	1.5	12.57	2.6	7.81	8.85	9.15	9.61	10.19	12.17	13.50	14.38	15.57	x
Manufacture of paper and paper products	17	60	13.63	2.4	17.04	7.8	8.51	9.77	10.26	11.00	12.21	15.10	17.41	18.95	21.20	x
Printing and reproduction of recorded media	18	91	12.14	1.4	14.52	-3.2	7.97	9.07	9.67	10.16	11.34	13.52	15.40	16.60	17.79	x
Manufacture of coke and refined petroleum products	19	11	25.41	-1.2	24.88	-3.1	13.19	17.23	18.60	21.53	23.02	27.93	28.84	29.84	x	x
Manufacture of chemicals and chemical products	20	83	15.65	2.5	18.37	4.7	8.44	9.76	10.73	11.73	13.59	17.56	20.56	22.26	24.94	x
Manufacture of basic pharmaceutical products and pharmaceutical preparations	21	55	18.23	-0.3	22.50	-13.8	10.19	12.51	13.53	14.61	16.30	20.58	24.17	25.92	30.39	x
Manufacture of rubber and plastic products	22	159	11.51	3.1	14.14	3.0	8.00	8.80	9.19	9.50	10.46	13.00	14.72	15.75	16.83	21.19
Manufacture of other non-metallic mineral products	23	84	12.63	4.9	14.78	6.7	8.23	9.39	9.87	10.36	11.55	14.20	15.89	17.72	18.84	x
Manufacture of basic metals	24	74	16.62	1.9	18.05	4.7	9.83	11.71	12.45	13.39	15.11	18.35	20.66	22.71	23.34	27.74
Manufacture of fabricated metal products, except machinery and equipment	25	280	12.79	2.6	14.79	3.2	8.41	9.59	10.10	10.59	11.59	14.10	15.67	16.92	18.32	23.35
Manufacture of computer, electronic and optical products	26	128	15.81	-2.0	18.01	-0.9	9.13	10.76	11.51	12.47	13.99	18.09	21.05	22.28	23.77	29.71
Manufacture of electrical equipment	27	84	13.11	-1.5	15.19	-3.9	8.20	9.23	9.91	10.70	12.00	14.16	15.87	17.52	18.91	x
Manufacture of machinery and equipment n.e.c.	28	197	15.20	5.0	17.35	5.1	9.27	10.94	11.63	12.39	13.81	17.00	19.14	20.60	21.84	27.39
Manufacture of motor vehicles, trailers and semi-trailers	29	184	17.41	4.5	18.97	2.7	9.64	11.35	12.50	13.37	15.26	19.66	21.85	22.99	24.30	31.18
Manufacture of other transport equipment	30	175	19.54	3.5	20.74	2.6	11.73	14.10	14.88	15.97	17.82	21.23	23.19	24.37	25.41	30.21
Manufacture of furniture	31	73	10.54	5.4	12.53	0.8	7.75	8.38	8.65	9.00	9.67	11.43	12.60	13.97	14.98	x
Other manufacturing	32	65	12.41	3.1	15.02	3.5	8.00	9.11	9.35	9.94	11.08	14.02	16.33	17.87	19.43	x
Repair and installation of machinery and equipment	33	72	14.61	2.8	17.14	-2.1	9.20	10.90	11.46	12.00	13.61	16.12	18.82	20.05	22.76	x
ELECTRICITY, GAS, STEAM AND AIR CONDITIONING SUPPLY	D	164	18.83	1.4	21.42	0.3	10.65	12.58	13.69	14.66	16.99	21.17	23.97	26.15	28.51	35.21
Electricity, gas, steam and air conditioning supply	35	164	18.83	1.4	21.42	0.3	10.65	12.58	13.69	14.66	16.99	21.17	23.97	26.15	28.51	35.21

4.11b Hourly pay - Gross (£) - For all employee jobs[a]: United Kingdom, 2017

Description	Code	Number of jobs[b] (000's)	Median	Annual % change	Mean	Annual % change	10	20	25	30	40	60	70	75	80	90
WATER SUPPLY; SEWERAGE, WASTE MANAGEMENT AND REMEDIATION ACTIVITIES	E	160	12.99	-0.1	15.15	-1.2	8.43	9.55	10.02	10.63	11.71	14.73	16.62	17.83	19.38	24.91
Water collection, treatment and supply	36	42	15.57	0.7	17.59	0.3	10.28	12.12	12.64	13.23	14.46	16.92	18.16	19.56	20.39	x
Sewerage	37	16	14.45	0.7	15.54	-4.8	9.07	10.00	11.09	11.82	13.52	15.91	18.49	19.33	x	x
Waste collection, treatment and disposal activities; materials recovery	38	100	11.54	1.2	14.16	-0.8	8.00	9.00	9.45	9.86	10.70	12.71	15.01	16.22	17.67	24.01
Remediation activities and other waste management services	39	x	x		14.79	4.7	x	x	x	x	11.76	x	x	x	x	x
CONSTRUCTION	F	926	13.74	3.3	16.16	3.6	8.39	10.00	10.50	11.16	12.50	15.16	17.14	18.46	20.21	26.03
Construction of buildings	41	277	14.94	6.7	18.11	3.8	8.75	10.27	11.08	11.89	13.22	16.95	19.78	21.45	24.07	31.18
Civil engineering	42	180	14.33	5.2	17.03	6.3	8.75	10.53	11.15	11.89	12.98	16.00	18.21	20.00	21.35	28.22
Specialised construction activities	43	470	13.00	0.5	14.68	2.4	8.10	9.49	10.00	10.63	11.84	14.29	15.77	16.71	17.66	22.03
WHOLESALE AND RETAIL TRADE; REPAIR OF MOTOR VEHICLES AND MOTORCYCLES	G	3,767	9.56	3.4	13.14	3.1	7.50	7.69	7.92	8.15	8.70	10.70	12.43	13.59	15.10	21.15
Wholesale and retail trade and repair of motor vehicles and motorcycles	45	477	10.98	4.5	13.11	6.5	7.50	8.02	8.46	8.89	9.98	12.03	13.45	14.31	15.36	19.78
Wholesale trade, except of motor vehicles and motorcycles	46	1,060	12.35	2.8	15.80	2.3	7.87	8.79	9.33	9.82	10.95	14.13	16.53	18.31	20.46	28.78
Retail trade, except of motor vehicles and motorcycles	47	2,230	8.57	3.3	11.51	2.5	7.50	7.55	7.65	7.80	8.15	9.23	10.19	11.07	12.18	16.73
TRANSPORTATION AND STORAGE	H	1,137	12.43	0.6	15.17	2.2	8.45	9.65	10.10	10.59	11.32	13.77	15.87	17.17	19.08	26.57
Land transport and transport via pipelines	49	512	12.04	2.3	14.22	3.8	8.27	9.33	9.79	10.14	11.05	13.25	15.12	16.46	18.18	25.56
Water transport	50	13	16.29	16.0	19.86	21.1	9.79	10.58	11.57	12.83	14.37	18.17	20.24	22.28	x	x
Air transport	51	73	17.73	-1.0	22.66	3.2	9.65	11.73	12.62	13.85	15.52	20.88	26.11	28.78	31.87	x
Warehousing and support activities for transportation	52	333	13.17	-2.4	16.27	0.1	8.40	9.50	10.04	10.60	11.80	14.86	17.17	18.71	20.99	28.91
Postal and courier activities	53	206	11.52	0.3	13.22	1.5	8.94	10.37	10.63	10.66	10.94	12.47	13.46	14.20	15.01	18.30
ACCOMMODATION AND FOOD SERVICE ACTIVITIES	I	1,501	7.85	4.7	9.84	6.8	6.15	7.23	x	7.50	7.51	8.30	9.00	9.50	10.09	12.87
Accommodation	55	328	8.12	1.8	10.45	5.4	6.75	7.50	x	7.50	7.76	8.73	9.59	10.16	11.19	14.99
Food and beverage service activities	56	1,173	7.75	4.0	9.64	7.1	6.10	7.20	7.50	x	7.50	8.15	8.85	9.30	9.92	12.46
INFORMATION AND COMMUNICATION	J	1,047	18.37	1.6	21.65	1.1	9.47	11.87	12.89	13.98	16.01	20.95	24.53	26.56	29.32	36.98
Publishing activities	58	128	15.94	-1.8	20.09	1.2	9.14	11.24	12.16	12.84	14.30	18.34	21.90	23.54	25.73	32.28
Motion picture, video and television programme production, sound recording and music publishing activities	59	65	14.36	5.0	19.92	6.8	7.35	7.89	8.72	9.50	11.75	16.43	20.44	22.96	25.68	x
Programming and broadcasting activities	60	47	21.42	-0.9	24.35	-1.3	12.98	15.64	16.18	17.47	19.40	23.13	25.98	28.79	30.55	x
Telecommunications	61	223	17.58	2.6	19.04	-5.0	9.75	12.23	13.29	14.43	15.83	19.72	22.35	24.50	26.59	33.16
Computer programming, consultancy and related activities	62	510	19.16	0.8	22.72	1.9	9.83	12.10	13.18	14.37	16.77	21.80	25.55	27.80	30.66	39.24
Information service activities	63	74	20.81	6.8	25.21	13.6	9.80	12.07	13.29	15.09	18.05	24.75	30.12	32.65	35.47	x
FINANCIAL AND INSURANCE ACTIVITIES	K	982	19.43	7.0	26.93	5.6	9.87	11.58	12.57	13.65	16.29	23.51	28.68	32.43	36.70	50.47
Financial service activities, except insurance and pension funding	64	524	20.51	8.1	28.66	6.2	10.24	11.98	12.96	14.00	17.16	25.32	31.25	34.85	39.61	53.75
Insurance, reinsurance and pension funding, except compulsory social security	65	100	19.17	6.6	23.51	7.4	10.00	11.79	12.75	13.88	16.03	21.95	25.93	28.03	31.43	x
Activities auxiliary to financial services and insurance activities	66	358	18.26	5.4	25.40	3.9	9.45	11.00	12.01	12.97	15.33	21.55	26.60	29.86	33.51	47.86
REAL ESTATE ACTIVITIES	L	370	13.02	1.7	15.92	0.8	8.29	9.39	9.97	10.50	11.86	14.31	16.28	17.39	19.17	26.33
Real estate activities	68	370	13.02	1.7	15.92	0.8	8.29	9.39	9.97	10.50	11.86	14.31	16.28	17.39	19.17	26.33

4.11b Hourly pay - Gross (£) - For all employee jobs[a]: United Kingdom, 2017

Description	Code	Number of jobs[b] (000's)	Median	Annual % change	Mean	Annual % change	Percentiles									
							10	20	25	30	40	60	70	75	80	90
PROFESSIONAL, SCIENTIFIC AND TECHNICAL ACTIVITIES	M	1,796	16.65	4.0	20.76	3.6	8.97	10.71	11.72	12.69	14.48	19.17	22.48	24.62	27.15	36.60
Legal and accounting activities	69	551	16.00	4.4	20.75	4.5	8.76	10.38	11.14	12.09	13.93	18.93	22.44	24.62	27.36	38.23
Activities of head offices; management consultancy activities	70	389	17.04	8.6	22.25	5.3	8.90	10.65	11.67	12.68	14.63	19.94	23.82	26.23	29.74	41.04
Architectural and engineering activities; technical testing and analysis	71	414	16.96	1.0	20.13	3.1	9.62	11.68	12.70	13.41	15.19	19.36	22.48	24.53	26.66	33.90
Scientific research and development	72	144	19.76	1.1	22.80	-0.5	11.18	13.92	14.82	15.86	17.94	21.99	25.15	27.08	29.79	38.37
Advertising and market research	73	121	16.47	2.3	20.62	1.5	9.10	10.86	11.75	12.55	14.32	18.91	21.42	23.06	25.41	36.92
Other professional, scientific and technical activities	74	124	14.42	4.7	18.45	6.5	8.41	10.00	10.42	11.39	13.03	16.63	19.45	21.28	23.20	31.45
Veterinary activities	75	52	11.10	0.6	14.32	2.6	7.55	8.13	8.46	8.90	9.96	13.39	15.36	16.81	18.33	x
ADMINISTRATIVE AND SUPPORT SERVICE ACTIVITIES	N	1,652	9.84	3.6	13.13	4.3	7.50	7.70	8.00	8.25	8.94	11.07	12.98	14.29	15.65	21.00
Rental and leasing activities	77	147	11.93	3.5	14.28	2.2	7.79	8.66	9.07	9.60	10.54	12.88	14.84	16.26	17.73	22.57
Employment activities	78	604	9.73	5.2	12.24	6.4	7.50	7.65	7.95	8.24	8.86	10.83	12.66	13.95	15.29	20.58
Travel agency, tour operator and other reservation service and related activities	79	81	11.60	-0.7	15.28	0.4	7.88	8.57	8.86	9.21	10.25	13.14	15.06	16.77	18.24	x
Security and investigation activities	80	109	9.29	4.6	10.82	6.4	7.60	7.98	8.17	8.38	8.82	9.91	10.69	11.45	12.02	14.90
Services to buildings and landscape activities	81	443	8.65	5.5	11.84	1.1	x	x	7.50	7.65	8.07	9.55	10.91	12.04	13.42	17.35
Office administrative, office support and other business support activities	82	269	11.74	2.1	16.49	5.8	7.53	8.24	8.64	9.04	10.20	13.77	16.25	17.79	19.53	28.11
PUBLIC ADMINISTRATION AND DEFENCE; COMPULSORY SOCIAL SECURITY	O	1,439	15.39	2.4	16.81	1.7	9.60	10.81	11.73	12.38	13.80	17.13	18.85	19.61	20.75	24.50
Public administration and defence; compulsory social security	84	1,439	15.39	2.4	16.81	1.7	9.60	10.81	11.73	12.38	13.80	17.13	18.85	19.61	20.75	24.50
EDUCATION	P	3,685	14.81	3.3	17.36	2.0	8.25	9.18	9.71	10.40	12.37	17.52	20.34	21.88	23.71	28.49
Education	85	3,685	14.81	3.3	17.36	2.0	8.25	9.18	9.71	10.40	12.37	17.52	20.34	21.88	23.71	28.49
HUMAN HEALTH AND SOCIAL WORK ACTIVITIES	Q	3,899	12.29	0.6	15.27	1.4	7.85	8.60	9.11	9.55	10.76	14.48	16.57	17.97	19.13	23.88
Human health activities	86	2,525	14.56	0.0	17.40	1.1	8.85	9.96	10.50	11.34	12.87	16.56	18.38	19.72	21.16	26.73
Residential care activities	87	647	8.84	4.7	10.75	2.9	7.50	7.70	7.85	8.00	8.37	9.50	10.76	11.70	13.18	16.48
Social work activities without accommodation	88	727	9.64	3.8	11.82	3.3	7.50	7.80	8.00	8.25	8.83	10.70	12.28	13.39	14.64	18.26
ARTS, ENTERTAINMENT AND RECREATION	R	557	9.75	3.7	15.40	1.1	7.40	7.62	7.98	8.24	8.84	11.09	12.94	14.24	16.04	21.64
Creative, arts and entertainment activities	90	52	13.90	8.8	20.07	13.2	8.07	8.99	9.55	10.19	12.15	16.55	18.67	19.93	21.86	x
Libraries, archives, museums and other cultural activities	91	64	12.16	8.1	14.71	7.7	7.64	8.49	9.05	9.60	10.91	13.67	15.66	16.71	17.96	x
Gambling and betting activities	92	104	8.82	3.6	11.38	4.1	7.50	7.61	7.77	8.00	8.41	9.40	10.73	11.42	12.79	x
Sports activities and amusement and recreation activities	93	337	9.47	4.2	16.37	-3.4	7.03	7.50	7.70	8.03	8.62	10.44	12.34	13.29	14.95	21.15
OTHER SERVICE ACTIVITIES	S	461	10.79	2.4	14.25	3.2	7.50	7.77	8.16	8.53	9.58	12.18	14.38	15.70	17.25	23.64
Activities of membership organisations	94	225	12.53	2.3	16.27	3.3	7.85	8.81	9.40	10.00	11.11	14.32	16.45	17.92	19.99	25.91
Repair of computers and personal and household goods	95	36	12.59	1.6	16.61	7.1	7.72	9.02	9.70	10.43	11.44	15.52	17.16	19.04	21.93	x
Other personal service activities	96	201	8.86	2.7	11.59	1.1	7.18	x	7.50	7.50	8.12	9.79	10.94	11.92	13.25	17.81
ACTIVITIES OF HOUSEHOLDS AS EMPLOYERS; UNDIFFERENTIATED GOODS-AND SERVICES-PRODUCING ACTIVITIES OF HOUSEHOLDS FOR OWN USE	T	75	9.49	4.6	10.91	7.0	7.65	8.01	8.42	8.50	9.00	10.00	10.40	10.97	11.82	x
Activities of households as employers of domestic personnel	97	75	9.49	4.6	10.91	7.0	7.65	8.01	8.42	8.50	9.00	10.00	10.40	10.97	11.82	x

4.11b Hourly pay - Gross (£) - For all employee jobs[a]: United Kingdom, 2017

Description	Code	Number of jobs[b] (000's)	Median	Annual % change	Mean	Annual % change	Percentiles									
							10	20	25	30	40	60	70	75	80	90
Undifferentiated goods- and services-producing activities of private households for own use	98	:														
ACTIVITIES OF EXTRATERRITORIAL ORGANISATIONS AND BODIES	U	x	x		15.64	0.8	x	x	x	x	x	x	x	x	x	x
Activities of extraterritorial organisations and bodies	99	x	x		15.64	0.8	x	x	x	x	x	x	x	x	x	x
NOT CLASSIFIED		8	8.57	1.9	12.83		x	x	x	7.51	8.08	x	x	x	x	x

Source: Annual Survey of Hours and Earnings, Office for National Statistics.

a Employees on adult rates whose pay for the survey pay-period was not affected by absence.

b Figures for Number of Jobs are for indicative purposes only and should not be considered an accurate estimate of employee job counts.

KEY - The colour coding indicates the quality of each estimate; jobs, median, mean and percentiles but not the annual percentage change.

The quality of an estimate is measured by its coefficient of variation (CV), which is the ratio of the standard error of an estimate to the estimate.

Key	Statistical robustness
CV <= 5%	Estimates are considered precise
CV > 5% and <= 10%	Estimates are considered reasonably precise
CV > 10% and <= 20%	Estimates are considered acceptable
x = CV > 20%	Estimates are considered unreliable for practical purposes
.. = disclosive	
: = not applicable	
- = nil or negligible	

4.12a - Weekly pay - Gross (£) - For all employee jobs[a]: United Kingdom, 2017

Description	Number of jobs[b] (thousand)	Median	Annual % change	Mean	Annual % change	Percentiles									
						10	20	25	30	40	60	70	75	80	90
ALL EMPLOYEES	26,241	448.5	2.3	537.9	2.4	144.4	238.0	283.8	316.5	380.0	527.0	624.5	682.9	752.0	972.6
Male	13,257	540.6	1.9	646.9	2.2	225.0	335.0	367.6	402.0	469.2	624.8	728.0	786.1	862.7	1,136.8
Female	12,984	358.3	2.6	426.6	2.7	115.0	178.3	209.6	240.9	301.5	425.0	505.1	556.8	621.8	795.1
Full-Time	18,913	550.0	2.1	661.1	2.5	318.9	372.3	399.8	428.0	484.3	628.0	720.3	774.3	845.3	1,092.0
Part-Time	7,327	182.0	2.8	219.7	2.1	57.5	96.6	115.0	127.6	155.8	210.4	241.5	265.1	295.1	412.9
Male Full-Time	11,413	590.9	2.3	715.9	2.4	338.5	400.0	430.4	460.0	521.0	672.9	768.6	834.3	911.6	1,191.3
Male Part-Time	1,844	171.6	2.9	219.4	1.1	48.5	84.0	100.8	117.6	147.2	197.8	227.9	250.0	281.8	410.4
Female Full-Time	7,500	493.2	2.6	577.7	2.8	300.0	343.8	364.1	385.9	436.3	560.0	643.4	689.9	748.4	926.2
Female Part-Time	5,483	186.4	2.8	219.9	2.4	60.1	101.3	119.6	131.7	158.1	214.4	246.1	269.9	299.0	413.9

a Employees on adult rates whose pay for the survey pay-period was not affected by absence.   Source: Annual Survey of Hours and Earnings, Office for National Statistics.
b Figures for Number of Jobs are for indicative purposes only and should not be considered an accurate estimate of employee job counts.
The quality of an estimate is measured by its coefficient of variation (CV), which is the ratio of the standard error of an estimate to the estimate.

4.12b Hourly pay - Excluding overtime (£) - For all employee jobs[a]: United Kingdom, 2017

Description	Code	Number of jobs[b] (000's)	Median	Annual % change	Mean	Annual % change	10	20	25	30	40	60	70	75	80	90
ALL EMPLOYEES		26,241	12.42	2.8	16.20	2.7	7.61	8.44	8.94	9.46	10.72	14.48	17.16	18.78	20.77	27.18
All Industries and Services		26,233	12.42	2.8	16.20	2.7	7.61	8.44	8.94	9.47	10.73	14.48	17.16	18.78	20.78	27.18
All Index of Production Industries		2,798	13.60	2.2	16.36	1.6	8.09	9.34	9.97	10.58	12.00	15.50	18.02	19.64	21.48	27.29
All Manufacturing		2,434	13.32	2.5	16.03	1.9	8.00	9.17	9.79	10.41	11.77	15.14	17.65	19.17	21.04	26.44
All Service Industries		22,369	12.18	2.4	16.22	2.9	7.57	8.33	8.77	9.28	10.50	14.36	17.07	18.71	20.74	27.27
AGRICULTURE, FORESTRY AND FISHING	A	139	9.22	0.3	11.26	3.9	7.50	7.63	7.91	8.09	8.72	10.00	11.15	12.00	13.17	17.22
Crop and animal production, hunting and related service activities	1	125	9.02	0.2	10.56	3.1	7.50	7.54	7.77	8.00	8.50	9.74	10.65	11.50	12.41	15.93
Forestry and logging	2	9	12.63	0.8	17.26	16.4	x	9.04	9.44	10.25	12.02	13.81	15.85	x	x	x
Fishing and aquaculture	3	x	x		x		x	x	x	9.29	10.06	x	x	x	x	x
MINING AND QUARRYING	B	40	18.13	11.8	21.99	7.9	9.77	11.06	11.99	13.11	15.37	20.59	24.51	26.95	29.70	x
Mining of coal and lignite	5	x	13.92		15.54		x	x	x	x	x	x	x	x	x	x
Extraction of crude petroleum and natural gas	6	6	x		42.13	14.7	x	22.87	x	25.07	30.26	x	x	x	x	x
Mining of metal ores	7	..														
Other mining and quarrying	8	23	14.29	6.9	16.16	4.1	8.95	9.88	10.35	10.77	12.16	16.07	18.34	19.71	x	x
Mining support service activities	9	10	22.80	20.3	24.87	11.7	x	14.32	15.12	16.36	19.00	26.41	28.84	x	x	x
MANUFACTURING	C	2,434	13.32	2.5	16.03	1.9	8.00	9.17	9.79	10.41	11.77	15.14	17.65	19.17	21.04	26.44
Manufacture of food products	10	383	10.17	2.3	12.97	2.5	7.50	7.89	8.15	8.44	9.16	11.45	13.14	14.33	16.09	20.45
Manufacture of beverages	11	47	17.10	7.2	20.73	5.1	9.24	11.36	12.09	12.99	15.01	19.06	22.62	23.98	25.85	x
Manufacture of tobacco products	12	:														
Manufacture of textiles	13	49	9.83	1.3	12.55	0.1	7.50	7.88	8.09	8.42	9.19	10.57	11.86	12.77	14.51	x
Manufacture of wearing apparel	14	18	8.38	2.2	12.49	11.3	x	x	x	7.52	7.99	9.32	x	x	x	x
Manufacture of leather and related products	15	10	9.50	4.9	11.88	8.0	7.50	7.88	8.01	8.31	8.54	10.35	11.23	x	x	x
Manufacture of wood and of products of wood and cork, except furniture; manufacture of articles of straw and plaiting materials	16	54	10.90	-0.3	12.48	2.7	7.75	8.70	9.02	9.48	10.02	11.97	13.15	14.09	15.37	x
Manufacture of paper and paper products	17	60	13.40	0.8	17.11	7.9	8.30	9.51	10.03	10.74	11.96	15.07	17.39	18.79	21.17	x
Printing and reproduction of recorded media	18	91	12.10	1.2	14.70	-2.1	7.97	9.04	9.59	10.00	11.27	13.40	15.38	16.54	17.79	x
Manufacture of coke and refined petroleum products	19	11	24.74	-2.6	24.62	-2.8	13.03	17.23	18.60	21.14	22.71	26.94	28.00	29.29	x	x
Manufacture of chemicals and chemical products	20	83	15.45	3.1	18.38	4.8	8.08	9.60	10.58	11.49	13.53	17.37	20.15	22.25	24.95	x
Manufacture of basic pharmaceutical products and pharmaceutical preparations	21	55	18.15	1.3	22.49	-14.1	9.92	12.35	13.38	14.43	16.30	20.22	23.94	25.53	30.39	x
Manufacture of rubber and plastic products	22	159	11.44	3.1	14.09	2.7	7.91	8.74	9.05	9.36	10.33	12.94	14.50	15.44	16.56	21.03
Manufacture of other non-metallic mineral products	23	84	12.49	4.7	14.73	6.7	8.02	9.21	9.69	10.10	11.35	14.09	15.84	17.55	18.83	x
Manufacture of basic metals	24	74	16.24	0.0	18.14	4.9	9.59	11.52	12.29	13.24	14.95	18.22	20.65	22.65	23.34	27.74
Manufacture of fabricated metal products, except machinery and equipment	25	280	12.45	3.3	14.66	3.0	8.24	9.42	9.96	10.44	11.31	13.83	15.24	16.63	18.14	23.28
Manufacture of computer, electronic and optical products	26	128	15.77	-1.5	17.98	-0.9	8.95	10.64	11.44	12.30	13.80	17.89	20.80	22.21	23.77	29.71
Manufacture of electrical equipment	27	84	12.93	-1.2	15.12	-4.1	8.07	9.04	9.71	10.50	11.74	13.77	15.70	17.27	18.91	x
Manufacture of machinery and equipment n.e.c.	28	197	14.88	4.8	17.33	5.1	9.14	10.76	11.50	12.18	13.53	16.77	19.11	20.56	21.74	27.31
Manufacture of motor vehicles, trailers and semi-trailers	29	184	17.27	4.8	18.84	3.0	9.48	11.27	12.24	13.10	15.00	19.50	21.63	22.86	24.20	31.49
Manufacture of other transport equipment	30	175	19.42	3.6	20.74	2.5	11.64	13.73	14.60	15.66	17.65	21.21	23.17	24.35	25.40	30.19
Manufacture of furniture	31	73	10.39	4.6	12.47	1.2	7.69	8.25	8.50	8.94	9.52	11.29	12.50	13.67	14.85	x
Other manufacturing	32	65	12.37	3.0	14.97	3.3	7.93	9.00	9.32	9.86	11.06	13.74	16.14	17.81	19.23	x
Repair and installation of machinery and equipment	33	72	14.43	2.8	16.97	-2.3	8.99	10.58	11.32	11.97	13.32	15.97	18.41	19.90	22.48	x
ELECTRICITY, GAS, STEAM AND AIR CONDITIONING SUPPLY	D	164	18.50	1.2	21.09	-0.2	10.53	12.46	13.53	14.61	16.73	20.75	23.63	25.71	27.88	34.63
Electricity, gas, steam and air conditioning supply	35	164	18.50	1.2	21.09	-0.2	10.53	12.46	13.53	14.61	16.73	20.75	23.63	25.71	27.88	34.63

4.12b Hourly pay - Excluding overtime (£) - For all employee jobs[a]: United Kingdom, 2017

Description	Code	Number of jobs[b] (000's)	Annual Median	% change	Annual Mean	% change	Percentiles 10	20	25	30	40	60	70	75	80	90
WATER SUPPLY; SEWERAGE, WASTE MANAGEMENT AND REMEDIATION ACTIVITIES	E	160	12.73	-1.0	15.10	-1.5	8.21	9.44	9.92	10.37	11.36	14.44	16.41	17.66	19.09	24.50
Water collection, treatment and supply	36	42	15.42	1.5	17.55	0.3	10.22	11.83	12.42	13.08	14.03	16.60	18.16	18.91	20.34	x
Sewerage	37	16	14.28	-1.9	15.54	-5.8	9.07	10.00	10.87	11.11	13.32	15.79	18.49	19.33	x	x
Waste collection, treatment and disposal activities; materials recovery	38	100	11.27	1.0	14.07	-1.1	7.74	8.80	9.20	9.58	10.46	12.33	14.92	16.17	17.55	24.01
Remediation activities and other waste management services	39	x	x		14.84	4.1	x	x	x	x	11.77	x	x	x	x	x
CONSTRUCTION	F	926	13.55	3.3	16.09	3.5	8.33	9.96	10.46	11.03	12.38	15.00	16.95	18.24	20.14	26.03
Construction of buildings	41	277	14.84	6.3	18.13	3.5	8.70	10.25	11.04	11.83	13.16	16.86	19.78	21.35	23.94	31.13
Civil engineering	42	180	14.22	5.3	17.02	6.2	8.75	10.50	11.00	11.63	12.82	15.67	18.05	19.95	21.25	27.84
Specialised construction activities	43	470	12.88	1.3	14.51	2.3	8.00	9.41	10.00	10.50	11.74	14.06	15.50	16.41	17.50	21.90
WHOLESALE AND RETAIL TRADE; REPAIR OF MOTOR VEHICLES AND MOTORCYCLES	G	3,767	9.55	4.3	13.22	3.3	7.50	7.69	7.91	8.14	8.70	10.69	12.40	13.54	15.03	21.15
Wholesale and retail trade and repair of motor vehicles and motorcycles	45	477	10.93	4.8	13.12	7.0	7.50	8.00	8.42	8.82	9.94	12.00	13.39	14.20	15.27	19.71
Wholesale trade, except of motor vehicles and motorcycles	46	1,060	12.27	2.4	15.85	2.4	7.83	8.74	9.25	9.74	10.87	14.03	16.43	18.26	20.44	28.77
Retail trade, except of motor vehicles and motorcycles	47	2,230	8.60	4.7	11.62	2.8	7.50	7.55	7.65	7.80	8.17	9.25	10.23	11.10	12.21	16.77
TRANSPORTATION AND STORAGE	H	1,137	12.37	0.7	15.26	2.3	8.40	9.53	10.00	10.52	11.29	13.74	15.86	17.14	18.98	26.55
Land transport and transport via pipelines	49	512	11.98	2.7	14.24	4.0	8.19	9.19	9.65	10.00	11.00	13.15	15.14	16.39	18.14	25.29
Water transport	50	13	16.29	16.0	20.14	22.3	9.59	10.49	11.88	12.83	14.37	18.17	20.20	x	x	x
Air transport	51	73	17.58	-0.1	22.64	3.2	9.65	11.72	12.49	13.84	15.40	20.88	25.80	28.70	31.74	x
Warehousing and support activities for transportation	52	333	13.13	-2.6	16.37	0.1	8.36	9.45	9.96	10.49	11.67	14.88	17.14	18.64	20.99	28.85
Postal and courier activities	53	206	11.52	0.3	13.33	1.8	8.90	10.34	10.57	10.65	10.95	12.50	13.55	14.27	15.12	18.36
ACCOMMODATION AND FOOD SERVICE ACTIVITIES	I	1,501	7.84	4.6	9.84	6.8	6.15	7.23	x	7.50	7.51	8.29	9.00	9.49	10.08	12.87
Accommodation	55	328	8.13	2.0	10.46	5.4	6.75	7.50	x	7.50	7.75	8.72	9.58	10.16	11.18	14.99
Food and beverage service activities	56	1,173	7.75	4.2	9.64	7.1	6.10	7.20	x	x	7.50	8.15	8.85	9.30	9.92	12.43
INFORMATION AND COMMUNICATION	J	1,047	18.30	1.5	21.65	1.0	9.47	11.81	12.84	13.93	15.85	20.91	24.52	26.49	29.27	36.97
Publishing activities	58	128	15.94	-1.6	20.10	1.3	9.14	11.27	12.19	12.86	14.30	18.39	21.90	23.54	25.68	32.28
Motion picture, video and television programme production, sound recording and music publishing activities	59	65	14.36	5.0	19.90	6.0	7.35	7.89	8.73	9.50	11.69	16.43	20.44	22.96	25.68	x
Programming and broadcasting activities	60	47	21.28	-1.3	24.33	-1.2	12.98	15.62	16.18	17.29	19.28	23.03	25.98	28.79	30.55	x
Telecommunications	61	223	17.18	2.4	18.95	-5.2	9.75	12.09	12.97	14.35	15.42	19.54	22.16	24.43	26.52	33.11
Computer programming, consultancy and related activities	62	510	19.16	0.7	22.75	1.8	9.79	12.08	13.17	14.37	16.77	21.87	25.55	27.79	30.66	39.29
Information service activities	63	74	20.80	6.8	25.20	13.7	9.77	12.02	13.21	15.08	17.89	24.75	29.80	32.22	35.47	x
FINANCIAL AND INSURANCE ACTIVITIES	K	982	19.42	7.4	27.03	5.7	9.85	11.53	12.52	13.57	16.29	23.51	28.70	32.31	36.63	50.52
Financial service activities, except insurance and pension funding	64	524	20.51	8.3	28.79	6.4	10.25	11.92	12.94	13.95	17.08	25.35	31.18	34.83	39.71	53.75
Insurance, reinsurance and pension funding, except compulsory social security	65	100	19.04	6.9	23.49	7.3	9.91	11.75	12.72	13.81	16.03	21.94	25.95	28.00	31.57	41.66
Activities auxiliary to financial services and insurance activities	66	358	18.24	5.6	25.48	3.9	9.45	10.94	11.83	12.93	15.33	21.52	26.63	29.83	33.46	47.89
REAL ESTATE ACTIVITIES	L	370	13.00	1.5	15.96	0.9	8.29	9.39	9.97	10.50	11.85	14.30	16.19	17.39	19.14	26.19
Real estate activities	68	370	13.00	1.5	15.96	0.9	8.29	9.39	9.97	10.50	11.85	14.30	16.19	17.39	19.14	26.19

4.12b Hourly pay - Excluding overtime (£) - For all employee jobs[a]: United Kingdom, 2017

Description	Code	Number of jobs[b] (000's)	Median	Annual % change	Mean	Annual % change	Percentiles 10	20	25	30	40	60	70	75	80	90
PROFESSIONAL, SCIENTIFIC AND TECHNICAL ACTIVITIES	M	1,796	16.62	4.2	20.81	3.6	8.95	10.68	11.71	12.60	14.43	19.17	22.47	24.58	27.14	36.61
Legal and accounting activities	69	551	16.00	4.4	20.78	4.5	8.75	10.38	11.11	12.08	13.89	18.92	22.43	24.62	27.33	38.23
Activities of head offices; management consultancy activities	70	389	17.05	9.4	22.30	5.1	8.81	10.55	11.63	12.52	14.64	19.83	23.85	26.04	29.74	40.81
Architectural and engineering activities; technical testing and analysis	71	414	16.93	0.9	20.21	3.3	9.58	11.64	12.55	13.33	15.10	19.41	22.41	24.50	26.70	33.90
Scientific research and development	72	144	19.73	0.9	22.82	-0.6	10.92	13.89	14.80	15.79	17.88	21.99	25.15	27.08	29.79	38.37
Advertising and market research	73	121	16.46	2.2	20.64	1.5	9.03	10.88	11.75	12.55	14.32	18.91	21.42	23.06	25.41	36.92
Other professional, scientific and technical activities	74	124	14.37	4.7	18.49	6.6	8.41	10.00	10.41	11.36	13.01	16.65	19.45	21.28	23.20	31.45
Veterinary activities	75	52	11.10	0.4	14.39	2.9	7.54	8.13	8.43	8.84	9.93	13.42	15.32	16.77	18.44	x
ADMINISTRATIVE AND SUPPORT SERVICE ACTIVITIES	N	1,652	9.78	3.0	13.13	4.3	7.50	7.66	7.95	8.23	8.91	11.00	12.92	14.22	15.59	20.99
Rental and leasing activities	77	147	11.72	2.3	14.26	2.4	7.74	8.57	9.02	9.50	10.32	12.69	14.83	16.05	17.50	22.57
Employment activities	78	604	9.63	4.8	12.24	6.5	7.50	7.60	7.86	8.16	8.78	10.75	12.59	13.85	15.29	20.44
Travel agency, tour operator and other reservation service and related activities	79	81	11.60	-0.4	15.31	0.6	7.85	8.56	8.82	9.21	10.25	13.15	15.06	16.77	18.24	x
Security and investigation activities	80	109	9.27	4.3	10.82	6.6	7.59	7.98	8.16	8.38	8.83	9.91	10.64	11.46	12.03	14.93
Services to buildings and landscape activities	81	443	8.62	5.1	11.79	0.8	x	x	7.50	7.63	8.05	9.50	10.89	12.00	13.27	17.31
Office administrative, office support and other business support activities	82	269	11.67	1.5	16.54	5.8	7.51	8.21	8.62	9.00	10.16	13.76	16.18	17.69	19.59	28.11
PUBLIC ADMINISTRATION AND DEFENCE; COMPULSORY SOCIAL SECURITY	O	1,439	15.39	1.9	16.80	1.3	9.60	10.81	11.70	12.34	13.78	17.13	18.81	19.57	20.60	24.54
Public administration and defence; compulsory social security	84	1,439	15.39	1.9	16.80	1.3	9.60	10.81	11.70	12.34	13.78	17.13	18.81	19.57	20.60	24.54
EDUCATION	P	3,685	14.78	3.3	17.40	2.0	8.24	9.16	9.70	10.38	12.34	17.51	20.33	21.84	23.69	28.50
Education	85	3,685	14.78	3.3	17.40	2.0	8.24	9.16	9.70	10.38	12.34	17.51	20.33	21.84	23.69	28.50
HUMAN HEALTH AND SOCIAL WORK ACTIVITIES	Q	3,899	12.31	0.7	15.28	1.4	7.85	8.60	9.11	9.55	10.77	14.50	16.57	17.98	19.11	23.84
Human health activities	86	2,525	14.56	0.0	17.34	1.0	8.86	9.97	10.51	11.34	12.89	16.56	18.38	19.69	21.16	26.70
Residential care activities	87	647	8.84	4.7	10.82	3.1	7.50	7.67	7.83	8.00	8.37	9.50	10.82	11.83	13.22	16.49
Social work activities without accommodation	88	727	9.66	3.9	11.89	3.3	7.50	7.80	8.00	8.25	8.84	10.77	12.32	13.43	14.71	18.32
ARTS, ENTERTAINMENT AND RECREATION	R	557	9.72	3.5	15.55	1.3	7.44	7.60	7.92	8.21	8.80	11.08	12.96	14.26	16.02	21.76
Creative, arts and entertainment activities	90	52	13.61	6.5	20.15	13.1	8.07	8.99	9.53	10.01	12.10	16.40	18.65	19.93	21.97	x
Libraries, archives, museums and other cultural activities	91	64	12.16	9.1	14.72	7.6	7.64	8.48	9.05	9.58	10.90	13.69	15.63	16.71	18.13	x
Gambling and betting activities	92	104	8.79	3.3	11.53	4.3	7.50	7.57	7.73	8.00	8.43	9.38	10.73	11.47	12.79	x
Sports activities and amusement and recreation activities	93	337	9.45	4.4	16.52	-3.1	7.05	7.50	7.67	8.00	8.55	10.44	12.23	13.33	14.99	21.23
OTHER SERVICE ACTIVITIES	S	461	10.75	2.1	14.24	3.1	7.50	7.77	8.15	8.52	9.55	12.13	14.37	15.62	17.17	23.59
Activities of membership organisations	94	225	12.52	2.2	16.28	3.2	7.85	8.79	9.39	10.01	11.11	14.33	16.45	17.86	19.88	25.87
Repair of computers and personal and household goods	95	36	12.63	1.4	16.61	7.0	7.72	8.99	9.68	10.32	11.45	15.56	17.05	18.93	21.75	x
Other personal service activities	96	201	8.84	2.5	11.56	0.9	7.18	x	7.50	7.50	8.12	9.75	10.87	11.88	13.17	17.54
ACTIVITIES OF HOUSEHOLDS AS EMPLOYERS; UNDIFFERENTIATED GOODS-AND SERVICES-PRODUCING ACTIVITIES OF HOUSEHOLDS FOR OWN USE	T	75	9.47	4.3	10.91	7.0	7.65	8.00	8.42	8.50	9.00	10.00	10.39	10.96	11.79	x
Activities of households as employers of domestic personnel	97	75	9.47	4.3	10.91	7.0	7.65	8.00	8.42	8.50	9.00	10.00	10.39	10.96	11.79	x

4.12b Hourly pay - Excluding overtime (£) - For all employee jobs[a]: United Kingdom, 2017

Description	Code	Number of jobs[b] (000's)	Median	Annual % change	Mean	Annual % change	Percentiles									
							10	20	25	30	40	60	70	75	80	90
Undifferentiated goods- and services-producing activities of private households for own use	98	:														
ACTIVITIES OF EXTRATERRITORIAL ORGANISATIONS AND BODIES	U	x	x		15.61	0.8	x	x	x	x	x	x	x	x	x	x
Activities of extraterritorial organisations and bodies	99	x	x		15.61	0.8	x	x	x	x	x	x	x	x	x	x
NOT CLASSIFIED		8	8.57	1.9	12.64		x	x	x	7.51	8.08	x	x	x	x	x

Source: Annual Survey of Hours and Earnings, Office for National Statistics.

a Employees on adult rates whose pay for the survey pay-period was not affected by absence.

b Figures for Number of Jobs are for indicative purposes only and should not be considered an accurate estimate of employee job counts.

KEY - The colour coding indicates the quality of each estimate; jobs, median, mean and percentiles but not the annual percentage change.

The quality of an estimate is measured by its coefficient of variation (CV), which is the ratio of the standard error of an estimate to the estimate.

Key	Statistical robustness
CV <= 5%	Estimates are considered precise
CV > 5% and <= 10%	Estimates are considered reasonably precise
CV > 10% and <= 20%	Estimates are considered acceptable
x = CV > 20%	Estimates are considered unreliable for practical purposes
.. = disclosive	
: = not applicable	
- = nil or negligible	

4.13 Average weekly earnings: main industrial sectors Great Britain
Great Britain
Standard Industrial Classification 2007

	Whole economy		Manufacturing		Construction		Services		Distribution Hotels and Restuarants	
	Actual	Seasonally adjusted	Actual	Seasonally adjusted	Actual	Seasonally adjusted	Actual	Seasonally adjusted	Actual	Seasonally adjusted
	KA46	KAB9	K55I	K5CA	K55L	K5CD	K55O	K5BZ	K55R	K5CG
2000	313	313	363	364	370	370	298	297	212	212
2001	329	330	377	377	399	398	315	314	221	220
2002	340	340	390	391	409	409	326	324	229	229
2003	350	351	405	405	427	427	336	335	234	234
2004	366	366	425	424	439	439	352	351	242	242
2005	382	383	440	440	452	452	369	369	251	252
2006	400	401	457	457	480	480	387	386	260	260
2007	420	420	476	475	512	512	406	405	276	276
2008	436	435	490	490	520	521	422	421	283	283
2009	435	435	495	496	525	525	420	420	286	286
2010	445	444	516	515	525	525	430	430	293	293
2011	455	455	524	522	534	534	442	441	296	296
2012	461	461	533	532	538	538	447	447	304	304
2013	466	466	545	544	536	535	452	452	312	311
2014	472	471	555	554	542	541	457	456	316	315
2015	482	482	564	563	555	554	468	468	328	328
2016	494	493	576	576	583	582	479	478	336	336
2017	506	505	588	587	591	590	490	490	344	343

	Finance and Business Industries		Private Sector		Public Sector		Private Sector Excl Financial Services	
	Actual	Seasonally adjusted	Actual	Seasonally adjusted	Actual	Seasonally adjusted	Actual	Seasonally adjusted
	K55U	K5C4	KA4O	KAC4	KA4R	KAC7	KA4U	KAD8
2000	380	382	312	312	314	314	314	314
2001	408	408	328	328	331	330	331	330
2002	414	414	338	338	344	343	344	342
2003	424	424	347	348	360	360	360	359
2004	447	448	363	363	376	375	375	374
2005	474	475	378	379	396	395	395	394
2006	507	507	397	398	410	409	410	408
2007	534	532	419	419	424	422	424	421
2008	560	557	434	434	439	437	438	436
2009	540	541	429	429	452	452	450	448
2010	566	566	438	437	467	467	459	459
2011	597	596	448	448	479	479	468	467
2012	597	596	454	454	487	487	476	475
2013	598	598	460	460	490	489	480	479
2014	599	599	467	466	493	492	487	486
2015	616	616	479	479	497	497	492	492
2016	631	630	492	491	505	505	501	500
2017	647	646	504	504	513	512	509	508

Source: Office for National Statistics: 01633 456780

4.14a Average Weekly Earnings by Industry (Not Seasonally Adjusted)

All figures are in pounds (£)

CDID	Agriculture, Forestry and Fishing (A) Average Weekly Earnings	of which Bonuses	Arrears	Mining and Quarrying (B) Average Weekly Earnings	Of which Bonuses	Arrears	Manufacturing - Food Products, Beverages and Tobacco (C1) Average Weekly Earnings	Of which Bonuses	Arrears	Manufacturing - Textiles, Leather and Clothing (C2) Average Weekly Earnings	Of which Bonuses	Arrears
	K57A	K57B	K57C	K57D	K57E	K57F	K57G	K57H	K57I	K57J	K57K	K57L
2013 Jan	339	6	0	1207	226	0	467	7	8	387	10	0
2013 Feb	344	6	0	1197	198	3	482	16	0	392	13	0
2013 Mar	342	3	0	1534	538	3	533	46	2	408	27	0
2013 Apr	351	4	0	1266	271	1	491	14	0	397	15	0
2013 May	347	3	0	1183	165	0	483	5	1	393	10	0
2013 Jun	335	3	0	1088	57	2	489	9	1	397	10	0
2013 Jul	336	2	0	1083	39	0	479	7	1	395	8	0
2013 Aug	335	3	0	1098	45	6	475	5	0	395	13	0
2013 Sep	342	3	0	1127	51	1	493	18	0	407	16	0
2013 Oct	346	2	0	1133	52	3	484	10	0	398	10	1
2013 Nov	335	3	0	1112	48	1	485	4	0	404	10	0
2013 Dec	390	53	0	1144	84	0	527	44	0	411	35	0
2014 Jan	340	3	0	1181	114	2	490	6	0	389	8	0
2014 Feb	350	5	0	1200	142	8	503	20	1	388	10	0
2014 Mar	352	2	1	1614	574	3	517	53	0	393	16	0
2014 Apr	345	3	0	1297	233	0	483	9	0	385	11	0
2014 May	334	3	0	1265	176	1	478	5	1	378	6	0
2014 Jun	335	5	0	1137	77	0	486	16	0	384	7	0
2014 Jul	346	13	0	1117	59	1	477	8	0	382	8	0
2014 Aug	330	2	0	1105	44	3	474	4	0	378	6	0
2014 Sep	335	5	0	1118	54	0	491	19	0	385	9	0
2014 Oct	342	3	0	1125	62	0	473	6	0	388	11	0
2014 Nov	339	3	0	1094	45	0	491	22	0	395	8	0
2014 Dec	390	42	1	1134	92	1	514	42	0	411	23	0
2015 Jan	348	13	0	1119	103	0	478	10	0	398	13	0
2015 Feb	347	4	0	1155	126	0	487	18	0	397	11	0
2015 Mar	364	7	0	1507	481	0	514	51	0	414	27	0
2015 Apr	367	4	0	1211	170	1	480	13	0	402	21	1
2015 May	355	5	0	1130	92	0	473	9	0	390	12	0
2015 Jun	350	8	0	1124	72	0	469	9	0	393	7	0
2015 Jul	361	17	0	1077	36	1	467	6	0	399	12	0
2015 Aug	360	9	0	1089	38	0	463	2	1	393	6	0
2015 Sep	358	7	0	1107	59	0	485	18	0	399	10	0
2015 Oct	354	6	0	1097	49	0	478	9	0	400	10	0
2015 Nov	370	3	0	1094	61	0	481	10	1	400	11	0
2015 Dec	377	16	0	1108	71	0	518	38	3	421	33	0
2016 Jan	377	20	0	1170	132	0	481	9	1	399	9	0
2016 Feb	373	4	0	1331	303	2	492	18	1	388	8	0
2016 Mar	387	6	0	1621	592	2	558	79	1	396	23	0
2016 Apr	388	11	0	1220	178	2	498	16	0	401	15	0
2016 May	371	5	0	1086	51	2	496	11	0	392	10	0
2016 Jun	362	3	0	1103	71	1	485	8	0	388	7	0
2016 Jul	374	14	0	1090	53	0	484	7	0	389	7	0
2016 Aug	362	3	0	1073	43	1	482	3	1	384	5	0
2016 Sep	382	2	0	1058	51	3	508	26	0	390	7	0
2016 Oct	389	5	0	1066	40	1	491	6	0	392	9	0
2016 Nov	375	2	0	1086	62	1	490	5	0	401	11	0
2016 Dec	410	37	0	1105	73	1	517	27	0	417	29	0
2017 Jan	376	4	0	1153	125	0	486	7	0	403	7	0
2017 Feb	386	5	0	1426	402	1	499	18	0	396	10	0
2017 Mar	387	1	1	1395	368	1	562	82	0	421	31	0
2017 Apr	396	6	0	1280	240	1	514	21	0	405	13	0
2017 May	389	5	0	1090	49	0	499	10	0	403	8	0
2017 Jun	391	4	0	1090	48	0	499	10	1	407	10	0
2017 Jul	394	15	0	1084	51	0	494	8	1	431	30	0
2017 Aug	402	7	0	1062	25	0	493	3	1	405	6	0
2017 Sep	404	4	0	1093	57	0	516	26	1	409	10	0
2017 Oct	399	4	0	1094	53	0	500	7	2	410	8	1
2017 Nov	393	5	0	1095	52	2	498	6	2	410	7	0
2017 Dec	419	37	0	1109	65	0	518	25	2	440	39	0
2018 Jan	388	3	0	1191	150	0	496	7	1	403	8	1
2018 Feb	384	3	0	1401	359	1	495	7	1	409	14	0
2018 Mar	380	3	0	1441	393	0	542	48	1	431	34	0
2018 Apr	402	12	0	1234	191	3	517	21	0	417	16	0
2018 May	400	4	0	1098	53	0	506	9	1	407	6	0
2018 Jun	395	3	0	1145	94	2	500	7	1	412	8	0
2018 Jul	395	2	0	1089	37	1	511	14	2	421	13	0
2018 Aug	405	5	0	1096	35	0	502	7	0	411	8	0
2018 Sep	394	4	0	1105	40	1	523	23	0	422	12	0
2018 Oct	409	6	0	1097	42	0	512	9	0	419	7	0
2018 Nov	408	3	0	1128	64	0	511	7	0	425	14	0
2018 Dec	414	24	0	1107	52	0	546	39	1	471	56	0
2019 Jan	400	2	0	1181	136	0	508	8	0	421	8	0
2019 Feb	394	3	0	1426	378	4	504	6	1	425	14	0
2019 Mar	415	11	0	1409	357	1	546	49	0	436	26	0

p = Provisional   r = Revised

Source: Monthly Wages and Salaries Survey
Inquiries: Email: earnings@ons.gov.uk   Tel: 01633 456120

4.14a Average Weekly Earnings by Industry (Not Seasonally Adjusted)

All figures are in pounds (£)

CDID	Manufacturing - Chemicals and Man-made Fibres (C3)			Manufacturing - Basic Metals and Metal Products (C4)			Manufacturing - Engineering and Allied Industries (C5)			Other Manufacturing (C6)		
	Average Weekly Earnings	Of which		Average Weekly Earnings	Of which		Average Weekly Earnings	Of which		Average Weekly Earnings	Of which	
		Bonuses	Arrears		Bonuses	Arrears		Bonuses	Arrears		Bonuses	Arrears
	K57M	K57N	K57O	K57P	K57Q	K57R	K57S	K57T	K57U	K57V	K57W	K57X
2013 Jan	696	21	6	523	12	6	593	19	0	472	12	0
2013 Feb	700	31	0	518	17	3	604	28	0	479	16	0
2013 Mar	946	270	0	547	41	1	655	76	1	511	44	0
2013 Apr	811	149	1	534	25	0	621	35	1	492	26	1
2013 May	690	22	2	526	20	0	603	16	1	479	13	0
2013 Jun	693	15	0	524	21	0	632	42	1	477	11	0
2013 Jul	689	12	0	518	16	0	614	22	2	478	12	0
2013 Aug	696	17	0	502	6	0	600	15	0	476	9	1
2013 Sep	685	20	1	517	11	0	589	6	0	478	11	1
2013 Oct	684	16	1	526	12	0	598	11	1	481	10	1
2013 Nov	686	16	0	533	16	0	608	14	1	486	15	2
2013 Dec	704	28	1	551	33	9	614	23	1	499	25	0
2014 Jan	670	20	0	527	14	0	629	30	3	485	13	1
2014 Feb	682	23	0	533	19	0	629	25	2	488	17	1
2014 Mar	1019	351	1	549	33	0	679	75	1	525	52	1
2014 Apr	779	114	1	532	18	0	640	31	1	504	28	1
2014 May	692	14	1	535	17	1	623	20	1	494	15	1
2014 Jun	689	15	1	541	22	0	655	45	3	494	13	0
2014 Jul	704	34	1	539	18	0	626	22	0	487	13	1
2014 Aug	700	28	0	518	6	1	608	8	1	488	12	0
2014 Sep	696	24	1	536	12	4	609	9	0	491	12	1
2014 Oct	687	15	1	542	14	0	614	9	0	489	11	0
2014 Nov	669	13	0	544	15	0	618	10	1	491	13	0
2014 Dec	686	31	0	562	38	1	649	40	2	505	30	0
2015 Jan	678	25	0	539	17	0	629	15	2	489	19	0
2015 Feb	712	38	1	532	12	6	631	18	1	490	16	0
2015 Mar	1030	348	1	569	44	6	700	81	2	540	69	0
2015 Apr	793	117	1	551	21	0	656	30	2	508	31	0
2015 May	705	13	0	538	11	1	638	17	1	492	13	0
2015 Jun	707	17	0	552	18	0	679	56	1	500	19	0
2015 Jul	730	44	2	558	16	3	640	17	1	508	22	1
2015 Aug	706	15	1	540	6	1	625	9	1	504	17	0
2015 Sep	719	19	4	553	11	0	624	7	0	495	10	0
2015 Oct	718	24	1	555	13	1	627	8	0	497	12	1
2015 Nov	702	15	0	558	16	0	631	9	0	505	14	0
2015 Dec	734	42	1	574	37	0	650	25	0	523	34	0
2016 Jan	702	14	2	549	10	0	647	18	1	504	18	0
2016 Feb	710	22	0	557	19	1	644	18	1	504	19	0
2016 Mar	1125	430	1	577	40	0	699	71	1	539	49	1
2016 Apr	796	89	0	567	25	0	673	34	1	521	27	0
2016 May	737	23	0	573	29	1	657	17	0	509	16	0
2016 Jun	739	16	1	572	24	0	692	50	0	512	20	0
2016 Jul	752	26	0	556	10	0	669	22	1	511	17	1
2016 Aug	709	16	0	550	9	0	656	12	1	504	13	0
2016 Sep	715	19	0	561	10	1	647	8	1	507	13	1
2016 Oct	713	28	0	555	8	1	652	11	0	505	11	0
2016 Nov	711	27	0	558	17	0	653	10	1	514	19	0
2016 Dec	739	47	0	573	40	0	677	31	4	526	32	1
2017 Jan	707	30	0	559	14	2	665	16	1	509	17	0
2017 Feb	727	32	6	565	22	0	670	25	1	511	21	1
2017 Mar	969	274	1	588	44	0	744	92	1	537	43	0
2017 Apr	825	114	4	575	26	0	687	33	1	523	29	1
2017 May	742	32	1	572	23	1	666	16	1	512	15	0
2017 Jun	729	24	1	565	21	0	702	45	1	517	19	1
2017 Jul	732	31	1	576	27	0	688	30	2	514	21	1
2017 Aug	733	14	0	556	5	1	669	15	2	508	12	1
2017 Sep	737	13	1	554	7	2	669	11	0	513	13	0
2017 Oct	734	18	1	561	9	0	683	23	1	512	13	0
2017 Nov	760	34	1	576	16	0	678	14	2	522	20	1
2017 Dec	773	45	0	599	38	5	707	39	1	531	30	1
2018 Jan	740	11	0	563	13	0	697	23	1	516	17	0
2018 Feb	758	31	0	568	16	0	697	27	1	517	18	0
2018 Mar	981	249	0	598	46	1	799	127	2	551	47	2
2018 Apr	913	179	1	587	27	1	711	36	1	531	23	1
2018 May	744	19	0	580	23	2	694	24	1	526	17	0
2018 Jun	747	17	5	583	23	1	722	48	1	527	17	0
2018 Jul	735	8	0	576	20	1	707	30	1	534	23	0
2018 Aug	743	14	0	560	8	1	680	9	1	525	12	0
2018 Sep	750	23	0	566	8	1	684	8	1	523	12	0
2018 Oct	729	4	1	571	10	1	695	21	0	527	12	1
2018 Nov	747	21	0	580	14	2	691	11	1	530	17	1
2018 Dec	770	49	0	585	23	1	717	37	3	544	33	0
2019 Jan	772	12	1	581	15	1	693	17	0	544	24	1
2019 Feb	776	21	1	591	19	1	698	21	0	535	19	1
2019 Mar	981	217	0	619	46	2	792	109	1	565	52	0

p = Provisional   r = Revised

Source: Monthly Wages and Salaries Survey
Inquiries: Email: earnings@ons.gov.uk   Tel: 01633 456120

4.14a Average Weekly Earnings by Industry (Not Seasonally Adjusted)

All figures are in pounds (£)

CDID	Electricity, Gas and Water Supply (D, E) Average Weekly Earnings	Of which Bonuses	Of which Arrears	Construction (F) Average Weekly Earnings	Of which Bonuses	Of which Arrears	Wholesale Trade (G46) Average Weekly Earnings	Of which Bonuses	Of which Arrears	Retail Trade and Repairs (G45 & G47) Average Weekly Earnings	Of which Bonuses	Of which Arrears
	K57Y	K57Z	K582	K583	K584	K585	K586	K587	K588	K589	K58A	K58B
2013 Jan	601	20	0	519	12	1	547	57	0	272	13	0
2013 Feb	613	26	4	530	13	1	549	62	0	273	19	0
2013 Mar	656	74	0	550	28	1	575	80	2	294	35	0
2013 Apr	641	48	1	557	28	1	583	81	0	282	21	1
2013 May	612	24	0	537	13	1	541	41	1	285	23	0
2013 Jun	668	71	1	541	20	1	555	52	0	282	16	0
2013 Jul	642	54	0	533	15	1	562	59	1	285	18	0
2013 Aug	616	24	1	528	11	1	550	50	0	287	21	0
2013 Sep	602	11	2	542	23	1	538	38	0	278	11	0
2013 Oct	613	18	2	526	12	1	545	37	0	283	18	0
2013 Nov	617	14	7	534	16	1	536	42	0	277	13	0
2013 Dec	619	20	3	540	24	1	565	64	0	276	13	0
2014 Jan	624	19	1	546	15	1	583	59	1	283	14	0
2014 Feb	625	18	1	533	14	0	543	48	0	280	14	0
2014 Mar	697	83	5	554	39	0	604	108	1	306	38	0
2014 Apr	667	49	12	531	18	0	556	61	1	296	24	0
2014 May	628	23	2	526	12	0	526	31	0	298	24	0
2014 Jun	676	72	1	541	21	0	547	53	0	288	14	0
2014 Jul	663	57	3	551	27	0	539	48	1	288	15	0
2014 Aug	637	27	2	531	14	1	526	41	1	288	14	0
2014 Sep	621	18	2	551	28	1	533	37	0	288	11	0
2014 Oct	623	14	2	537	14	0	533	39	0	294	17	0
2014 Nov	637	18	8	547	22	0	529	38	0	290	16	0
2014 Dec	624	13	1	556	33	1	557	65	1	287	13	0
2015 Jan	627	21	1	534	15	0	556	59	0	303	17	0
2015 Feb	646	37	6	538	12	0	555	65	0	298	15	0
2015 Mar	697	89	2	583	44	0	668	169	1	326	39	0
2015 Apr	670	56	4	546	20	0	551	57	0	317	25	0
2015 May	642	24	2	555	28	0	542	39	0	316	26	0
2015 Jun	694	71	2	551	19	0	552	51	0	307	18	0
2015 Jul	665	51	1	568	29	0	560	57	0	309	20	0
2015 Aug	646	29	1	546	13	0	545	39	0	304	16	0
2015 Sep	626	15	3	559	26	0	529	35	0	303	15	0
2015 Oct	625	11	1	553	16	1	530	36	0	306	19	0
2015 Nov	631	17	2	561	20	1	536	40	1	302	16	0
2015 Dec	633	13	4	565	33	0	575	71	1	302	16	0
2016 Jan	627	16	1	566	21	1	556	51	0	307	18	0
2016 Feb	660	33	4	565	19	0	586	71	0	306	17	0
2016 Mar	713	90	1	598	48	1	642	120	1	326	35	0
2016 Apr	708	77	1	577	27	0	577	49	0	328	30	1
2016 May	661	18	5	584	37	0	560	31	1	326	29	1
2016 Jun	721	66	12	571	22	0	570	44	1	316	15	1
2016 Jul	706	63	2	590	34	1	568	43	0	322	24	1
2016 Aug	670	26	2	567	14	0	559	36	0	316	16	1
2016 Sep	658	14	2	589	32	0	559	35	1	315	14	1
2016 Oct	659	13	2	603	43	0	570	45	0	320	21	1
2016 Nov	672	20	2	592	27	1	568	44	0	319	18	0
2016 Dec	667	19	5	595	39	1	582	60	0	316	18	0
2017 Jan	663	20	1	574	15	1	588	45	1	319	20	0
2017 Feb	675	28	2	579	22	0	610	79	1	312	16	0
2017 Mar	734	82	1	605	50	1	675	137	1	333	34	1
2017 Apr	744	90	1	576	22	0	581	49	0	339	31	0
2017 May	676	21	5	577	21	0	561	32	1	339	31	0
2017 Jun	736	81	2	581	21	1	566	38	0	326	18	0
2017 Jul	705	50	2	593	26	1	571	40	1	321	21	0
2017 Aug	668	16	1	581	15	0	569	38	0	319	16	0
2017 Sep	667	17	2	611	38	0	559	32	0	326	17	0
2017 Oct	661	10	2	605	31	0	569	41	0	329	21	1
2017 Nov	665	12	1	602	26	0	589	50	1	326	21	0
2017 Dec	672	12	6	611	30	1	614	72	0	319	16	0
2018 Jan	671	15	2	600	19	0	597	53	0	324	16	0
2018 Feb	683	30	1	596	18	0	603	68	0	327	21	0
2018 Mar	736	82	2	641	57	0	690	155	1	342	30	0
2018 Apr	759	93	3	609	21	0	610	72	0	342	27	0
2018 May	690	18	6	606	18	0	584	43	0	360	42	0
2018 Jun	733	69	1	609	18	0	576	34	0	333	16	0
2018 Jul	719	48	4	612	21	0	584	40	1	337	19	0
2018 Aug	688	15	4	607	18	0	581	33	0	334	15	0
2018 Sep	680	8	3	629	38	0	580	31	1	330	14	0
2018 Oct	683	9	1	625	34	0	608	59	0	337	20	0
2018 Nov	700	15	8	628	24	0	588	40	1	336	20	0
2018 Dec	691	17	2	644	41	0	618	63	0	329	15	0
2019 Jan	686	13	1	623	19	0	612	47	0	331	18	0
2019 Feb	704	28	6	627	22	0	628	76	0	333	21	0
2019 Mar	738	69	1	661	55	0	665	115	2	345	29	0

p = Provisional   r = Revised

Source: Monthly Wages and Salaries Survey
Inquiries: Email: earnings@ons.gov.uk   Tel: 01633 456120

4.14a Average Weekly Earnings by Industry (Not Seasonally Adjusted)

All figures are in pounds (£)

	Transport and Storage			Accommodation and Food Service Activities			Information and Communication			Financial & Insurance Activities		
	(H)			(I)			(J)			(K)		
	Average Weekly Earnings	Of which		Average Weekly Earnings	Of which		Average Weekly Earnings	Of which		Average Weekly Earnings	Of which	
		Bonuses	Arrears		Bonuses	Arrears		Bonuses	Arrears		Bonuses	Arrears
CDID	K58F	K58G	K58H	K58C	K58D	K58E	K5E9	K5EA	K5EB	K58I	K58J	K58K
2013 Jan	519	7	0	212	5	1	744	74	2	1120	351	1
2013 Feb	523	14	0	216	8	0	773	99	1	1511	736	1
2013 Mar	539	26	1	217	6	0	799	131	1	1528	750	0
2013 Apr	543	21	5	219	8	0	775	95	2	1045	259	1
2013 May	534	9	1	224	9	0	742	59	2	956	168	0
2013 Jun	558	34	1	219	4	0	795	108	5	1037	257	0
2013 Jul	531	10	1	221	5	0	744	61	1	853	73	0
2013 Aug	531	12	1	221	3	0	755	68	1	840	64	0
2013 Sep	526	4	1	220	4	0	745	63	1	863	86	0
2013 Oct	531	6	2	220	3	0	736	51	2	828	50	1
2013 Nov	535	7	5	222	6	0	729	47	1	829	52	1
2013 Dec	566	34	1	227	8	1	740	51	1	927	140	0
2014 Jan	527	7	0	222	6	1	757	72	2	1125	350	0
2014 Feb	627	38	68	224	7	0	780	93	2	1632	841	3
2014 Mar	550	29	1	223	7	0	815	135	2	1619	813	3
2014 Apr	550	20	1	223	6	0	788	102	2	928	129	1
2014 May	545	13	1	229	10	0	750	67	1	948	154	1
2014 Jun	562	33	2	223	8	0	804	111	2	979	186	1
2014 Jul	552	23	0	223	6	0	741	46	3	853	59	2
2014 Aug	534	5	0	222	4	0	767	72	2	890	65	2
2014 Sep	534	5	1	219	4	1	730	37	1	920	90	1
2014 Oct	543	13	1	222	4	0	753	55	1	884	52	3
2014 Nov	543	7	1	224	8	1	753	56	2	885	53	0
2014 Dec	573	31	0	232	11	1	780	61	2	1041	198	2
2015 Jan	541	9	3	227	5	1	790	80	4	1174	337	1
2015 Feb	547	18	0	231	10	1	819	102	3	1568	721	5
2015 Mar	548	19	1	233	9	1	883	163	1	1785	940	1
2015 Apr	556	20	4	229	8	1	795	76	1	1045	185	2
2015 May	550	12	0	237	11	1	794	77	2	1011	154	1
2015 Jun	571	37	1	230	5	1	821	103	1	990	125	7
2015 Jul	543	10	1	233	5	1	797	78	0	954	93	3
2015 Aug	550	10	2	231	4	0	807	95	1	949	90	1
2015 Sep	546	8	1	228	5	0	758	49	2	949	89	1
2015 Oct	544	6	0	233	5	0	784	75	1	923	62	1
2015 Nov	547	10	1	234	8	0	767	57	2	915	48	1
2015 Dec	572	32	0	239	12	0	785	72	3	1083	205	2
2016 Jan	549	8	2	231	5	0	805	93	1	1252	379	0
2016 Feb	555	20	1	234	8	0	792	89	1	1441	559	5
2016 Mar	570	34	0	237	11	0	848	143	1	1927	1050	5
2016 Apr	560	22	0	233	7	0	789	75	2	1058	168	4
2016 May	546	10	1	238	10	0	791	78	1	1070	181	1
2016 Jun	574	29	6	233	6	0	804	87	2	1072	184	1
2016 Jul	551	11	2	236	5	0	798	80	1	1018	124	2
2016 Aug	558	16	1	237	4	0	798	82	1	954	71	2
2016 Sep	552	9	1	237	5	0	770	50	1	960	76	2
2016 Oct	547	7	1	236	4	0	783	63	1	956	71	1
2016 Nov	558	12	2	237	8	0	784	61	1	945	57	0
2016 Dec	581	35	1	243	10	0	796	68	1	1105	217	1
2017 Jan	549	10	1	234	4	1	824	94	1	1255	375	0
2017 Feb	563	21	4	238	10	0	817	94	1	1520	629	2
2017 Mar	570	29	1	243	11	0	878	149	2	2140	1247	5
2017 Apr	568	19	0	238	6	0	810	83	1	1092	190	1
2017 May	570	19	1	242	12	0	795	64	2	1096	188	7
2017 Jun	578	28	0	239	6	0	850	117	1	1172	263	2
2017 Jul	565	12	1	241	6	0	806	70	1	1011	100	1
2017 Aug	572	18	1	241	5	0	822	83	2	1007	87	4
2017 Sep	567	13	2	241	6	0	781	47	1	1053	134	2
2017 Oct	560	6	1	241	5	1	810	69	4	997	79	1
2017 Nov	567	10	1	244	8	0	809	64	1	975	53	1
2017 Dec	590	31	1	251	10	0	832	82	1	1194	263	3
2018 Jan	566	9	0	239	4	0	845	86	1	1334	400	1
2018 Feb	573	18	3	246	10	0	842	98	2	1608	664	3
2018 Mar	584	31	1	251	12	0	907	160	1	2170	1223	3
2018 Apr	578	16	2	247	5	0	825	69	1	1143	185	1
2018 May	581	14	1	254	11	0	816	60	2	1107	150	3
2018 Jun	591	23	1	251	5	0	891	125	4	1188	241	1
2018 Jul	587	15	2	253	5	0	843	75	1	1053	96	2
2018 Aug	593	22	1	256	5	0	859	86	5	1046	87	2
2018 Sep	585	10	1	253	6	0	831	63	1	1061	102	1
2018 Oct	580	6	1	254	4	0	888	116	1	1062	101	2
2018 Nov	589	7	3	254	6	0	836	62	1	1019	55	1
2018 Dec	617	35	2	264	14	1	868	94	1	1223	248	4
2019 Jan	608	8	1	252	4	0	873	98	1	1389	427	1
2019 Feb	601	20	1	253	8	0	888	117	2	1609	642	3
2019 Mar	619	38	1	262	13	0	963	191	1	2183	1213	2

p = Provisional   r = Revised

Source: Monthly Wages and Salaries Survey
Inquiries: Email: earnings@ons.gov.uk   Tel: 01633 456120

4.14a Average Weekly Earnings by Industry (Not Seasonally Adjusted)

All figures are in pounds (£)

CDID	Real Estate Activities (L) Average Weekly Earnings	Of which Bonuses	Of which Arrears	Professional, Scientific & Technical Activities (M) Average Weekly Earnings	Of which Bonuses	Of which Arrears	Administrative and Support Service Activities (N) Average Weekly Earnings	Of which Bonuses	Of which Arrears	Public Administration (O) Average Weekly Earnings	Of which Bonuses	Of which Arrears
	K58L	K58M	K58N	K5EC	K5ED	K5EE	K5EF	K5EG	K5EH	K58O	K58P	K58Q
2013 Jan	477	47	2	640	22	0	342	11	0	546	1	1
2013 Feb	457	32	1	663	35	1	344	12	0	535	0	0
2013 Mar	512	81	1	719	87	2	366	37	0	536	3	0
2013 Apr	511	79	0	686	54	2	351	20	0	537	1	0
2013 May	481	45	0	653	32	1	351	15	0	537	0	0
2013 Jun	484	46	0	657	38	1	339	8	0	540	3	0
2013 Jul	501	65	0	671	46	2	349	15	0	543	6	0
2013 Aug	481	44	1	642	23	0	345	9	0	540	1	4
2013 Sep	473	44	0	650	29	1	337	9	0	547	7	0
2013 Oct	480	48	0	650	24	1	338	10	0	537	1	0
2013 Nov	466	41	0	647	22	1	337	8	0	537	1	0
2013 Dec	491	50	0	692	65	0	357	25	0	540	2	1
2014 Jan	467	39	0	647	21	1	352	18	1	545	1	1
2014 Feb	459	32	0	655	33	0	342	16	0	541	1	1
2014 Mar	537	101	1	727	100	1	363	34	0	548	2	4
2014 Apr	502	65	2	673	44	1	345	17	0	553	7	3
2014 May	487	51	0	664	41	1	348	12	0	547	1	1
2014 Jun	494	50	2	678	55	1	348	13	0	549	3	0
2014 Jul	515	76	1	669	46	1	354	13	0	549	4	3
2014 Aug	475	38	0	650	26	1	358	10	0	547	3	1
2014 Sep	475	30	1	651	26	1	360	10	0	555	7	0
2014 Oct	487	40	1	660	26	1	367	12	0	552	1	1
2014 Nov	478	34	0	655	24	1	357	10	0	549	1	1
2014 Dec	518	67	1	716	83	1	370	26	0	554	1	1
2015 Jan	484	35	0	653	21	1	362	14	0	551	0	1
2015 Feb	482	38	1	690	53	0	373	23	1	551	2	1
2015 Mar	600	155	1	763	122	1	390	34	0	548	2	0
2015 Apr	484	39	1	679	45	1	379	18	0	552	1	1
2015 May	502	61	0	666	36	1	383	13	1	552	1	0
2015 Jun	509	61	3	677	45	1	381	13	0	551	2	0
2015 Jul	536	91	1	683	49	1	378	13	0	554	4	0
2015 Aug	476	33	1	665	33	1	378	10	0	552	2	1
2015 Sep	477	30	0	653	20	0	373	11	0	561	8	1
2015 Oct	480	35	1	660	29	1	379	12	0	556	2	1
2015 Nov	470	27	0	658	22	1	377	11	0	553	1	1
2015 Dec	489	42	0	722	78	1	389	22	0	550	1	0
2016 Jan	479	31	0	670	25	1	389	17	1	556	0	1
2016 Feb	486	36	0	696	57	1	399	22	0	555	1	0
2016 Mar	652	198	1	773	128	3	413	37	0	553	2	1
2016 Apr	518	63	0	717	67	1	404	18	0	566	6	0
2016 May	514	64	0	667	32	0	400	21	0	557	1	0
2016 Jun	510	54	1	678	43	1	401	18	0	561	1	1
2016 Jul	538	88	0	690	62	1	400	17	0	556	1	0
2016 Aug	477	27	2	663	33	2	394	10	0	570	8	1
2016 Sep	478	34	0	653	22	1	396	11	0	562	2	0
2016 Oct	478	32	0	652	29	1	400	15	0	566	2	1
2016 Nov	472	25	0	656	27	1	398	10	0	564	1	1
2016 Dec	490	40	1	679	50	0	415	24	1	562	0	0
2017 Jan	486	34	0	670	28	1	402	15	0	563	0	0
2017 Feb	485	35	1	677	41	1	416	27	0	562	0	0
2017 Mar	620	169	1	740	101	0	451	60	0	561	1	0
2017 Apr	496	39	0	686	41	1	415	20	0	567	7	0
2017 May	518	62	2	681	33	1	412	14	0	561	1	0
2017 Jun	509	54	1	695	49	1	420	24	0	563	1	0
2017 Jul	538	85	1	708	60	2	410	12	1	564	1	2
2017 Aug	478	28	1	675	30	1	413	16	0	569	6	1
2017 Sep	481	31	0	676	32	0	413	14	0	564	3	1
2017 Oct	485	30	0	674	30	0	412	15	0	572	3	1
2017 Nov	476	27	1	673	30	1	411	11	0	568	2	2
2017 Dec	497	36	0	707	65	1	427	28	0	564	1	1
2018 Jan	505	36	0	678	36	1	412	15	0	572	0	0
2018 Feb	507	36	0	692	37	1	422	26	0	568	0	0
2018 Mar	654	183	1	766	109	1	446	51	0	569	1	1
2018 Apr	520	45	1	703	48	1	419	22	0	578	7	0
2018 May	537	63	1	685	30	1	415	16	0	572	1	1
2018 Jun	541	64	1	697	43	1	415	13	1	577	1	3
2018 Jul	566	86	0	717	63	1	432	24	2	574	2	1
2018 Aug	508	29	1	684	33	0	419	12	3	575	2	4
2018 Sep	511	31	1	695	34	1	415	11	0	575	1	2
2018 Oct	520	33	0	695	32	1	431	24	0	580	3	1
2018 Nov	519	32	0	702	40	1	421	14	0	586	2	3
2018 Dec	539	48	1	741	74	1	429	19	0	579	1	1
2019 Jan	515	32	0	709	37	0	443	25	0	581	2	2
2019 Feb	515	32	1	724	54	1	438	23	0	589	1	1
2019 Mar	677	188	1	791	111	2	466	48	1	593	4	6

p = Provisional   r = Revised

Source: Monthly Wages and Salaries Survey
Inquiries: Email: earnings@ons.gov.uk   Tel: 01633 456120

4.14a Average Weekly Earnings by Industry (Not Seasonally Adjusted)

All figures are in pounds (£)

CDID	Education (P) Average Weekly Earnings	Of which Bonuses	Of which Arrears	Health and Social Work (Q) Average Weekly Earnings	Of which Bonuses	Of which Arrears	Arts, Entertainment and Recreation (R) Average Weekly Earnings	Of which Bonuses	Of which Arrears	Other Service Activities (S) Average Weekly Earnings	Of which Bonuses	Of which Arrears
	K58R	K58S	K58T	K58U	K58V	K58W	K5EI	K5EJ	K5EK	K58X	K58Y	K58Z
2013 Jan	397	1	1	406	0	0	366	16	0	332	11	0
2013 Feb	394	0	0	398	1	0	356	5	0	328	11	0
2013 Mar	395	2	0	400	1	0	358	19	0	333	20	0
2013 Apr	398	0	0	407	1	0	364	18	0	336	13	0
2013 May	399	1	0	404	1	0	367	19	0	335	9	0
2013 Jun	399	1	0	403	1	0	375	18	0	337	6	0
2013 Jul	401	1	0	402	1	0	363	10	0	335	9	0
2013 Aug	405	1	1	398	0	0	366	13	0	336	8	0
2013 Sep	408	0	0	400	1	0	365	19	0	329	6	0
2013 Oct	408	1	1	400	1	0	349	4	0	340	5	5
2013 Nov	406	1	1	400	1	0	350	13	0	342	9	1
2013 Dec	410	1	1	402	1	0	346	14	0	346	9	0
2014 Jan	402	0	0	404	1	0	363	12	1	347	12	1
2014 Feb	404	0	0	401	1	0	334	6	0	420	76	0
2014 Mar	405	3	0	399	1	0	329	13	0	373	28	2
2014 Apr	406	1	1	404	1	0	351	14	0	353	14	0
2014 May	407	1	2	404	1	0	344	13	0	352	11	1
2014 Jun	406	1	0	404	1	0	354	15	0	348	7	0
2014 Jul	408	1	0	401	0	0	350	7	0	353	12	0
2014 Aug	415	1	0	399	1	0	338	6	1	357	11	0
2014 Sep	418	1	0	404	1	0	344	16	0	357	7	0
2014 Oct	415	1	0	403	0	0	333	8	1	356	8	0
2014 Nov	413	1	0	404	1	0	336	8	1	341	9	0
2014 Dec	420	2	2	405	1	0	341	18	1	348	14	0
2015 Jan	413	1	0	406	1	0	351	14	1	344	15	0
2015 Feb	411	0	0	404	0	0	330	11	0	346	17	0
2015 Mar	413	3	0	404	1	0	341	22	0	365	33	0
2015 Apr	413	1	0	408	1	0	347	25	0	353	18	0
2015 May	412	0	0	410	1	0	336	11	0	350	13	0
2015 Jun	415	2	0	410	0	0	366	18	0	353	11	0
2015 Jul	416	1	0	408	0	0	372	17	0	358	15	0
2015 Aug	421	1	0	404	0	0	365	20	0	355	15	0
2015 Sep	424	1	0	408	1	0	381	29	0	347	10	0
2015 Oct	420	1	0	404	0	0	347	11	0	348	9	0
2015 Nov	421	1	2	407	1	0	346	7	0	349	10	0
2015 Dec	422	1	1	408	1	0	358	15	3	356	17	0
2016 Jan	418	2	1	408	1	0	365	10	0	375	23	1
2016 Feb	418	1	0	409	0	1	356	10	0	372	20	0
2016 Mar	418	1	0	408	1	1	361	14	0	386	31	0
2016 Apr	419	1	0	413	0	0	382	37	0	396	27	1
2016 May	417	1	0	413	1	0	363	17	0	384	14	0
2016 Jun	421	1	1	415	0	1	376	20	0	389	19	0
2016 Jul	419	1	1	412	0	0	399	15	0	388	14	0
2016 Aug	426	1	0	412	0	0	398	35	0	389	17	0
2016 Sep	430	1	0	413	1	0	394	25	0	385	13	0
2016 Oct	427	1	1	413	0	0	377	13	0	390	12	0
2016 Nov	425	1	1	416	2	1	383	10	1	394	18	1
2016 Dec	427	1	1	415	1	0	380	17	0	403	20	0
2017 Jan	423	1	0	415	1	0	386	20	0	382	17	0
2017 Feb	423	0	0	415	1	0	449	91	0	385	17	0
2017 Mar	424	1	0	414	1	0	376	23	0	400	27	0
2017 Apr	427	1	0	420	1	0	376	20	0	405	28	0
2017 May	427	1	0	424	1	1	378	16	0	390	14	0
2017 Jun	428	1	0	422	1	0	407	28	1	387	10	1
2017 Jul	430	2	0	421	1	0	420	28	0	399	20	2
2017 Aug	439	1	1	420	0	0	403	21	0	390	16	0
2017 Sep	441	1	1	423	1	0	407	21	1	375	8	0
2017 Oct	440	1	1	425	1	0	391	12	0	370	15	0
2017 Nov	438	2	1	427	2	0	396	12	0	375	13	0
2017 Dec	440	2	1	426	1	0	396	17	0	384	16	0
2018 Jan	436	1	1	426	1	0	374	13	1	383	15	0
2018 Feb	437	1	0	425	0	0	380	11	0	381	19	0
2018 Mar	438	2	0	426	1	0	392	28	0	399	25	0
2018 Apr	439	1	0	433	1	0	375	13	0	392	21	0
2018 May	441	1	1	433	1	0	372	10	0	382	12	0
2018 Jun	443	1	1	432	1	0	388	18	0	383	12	1
2018 Jul	445	2	0	434	1	1	394	28	0	402	27	0
2018 Aug	451	1	0	442	0	9	404	21	1	385	12	0
2018 Sep	453	1	0	435	1	0	404	33	0	384	11	0
2018 Oct	453	1	1	439	1	0	392	16	1	385	12	1
2018 Nov	453	2	2	441	1	1	391	12	1	387	15	0
2018 Dec	454	2	2	442	1	0	393	15	0	386	14	0
2019 Jan	444	1	0	443	1	0	414	44	0	394	19	0
2019 Feb	447	1	0	439	1	0	388	15	0	379	16	0
2019 Mar	450	1	4	442	2	0	398	26	1	388	26	0

p = Provisional   r = Revised

Source: Monthly Wages and Salaries Survey
Inquiries: Email: earnings@ons.gov.uk   Tel: 01633 456120

4.14b Average Weekly Earnings (nominal) - Regular Pay
(Great Britain, seasonally adjusted)

Standard Industrial Classification (2007)

	Whole Economy			Private sector [3 4 5]			Public sector [3 4 5]			Services, SIC 2007 sections G-S		
	Weekly Earnings (£)	% changes year on year Single month	3 month average [2]	Weekly Earnings (£)	% changes year on year Single month	3 month average [2]	Weekly Earnings (£)	% changes year on year Single month	3 month average [2]	Weekly Earnings (£)	% changes year on year Single month	3 month average [2]
	KAI7	KAI8	KAI9	KAJ2	KAJ3	KAJ4	KAJ5	KAJ6	KAJ7	K5DL	K5DM	K5DN
Jan 05	354	4.2	4.0	346	3.9	3.8	382	5.0	4.5	339	4.4	4.2
Feb 05	354	4.2	4.1	347	4.2	4.1	382	4.3	4.5	341	4.7	4.5
Mar 05	357	4.6	4.3	349	4.4	4.2	386	5.3	4.9	343	5.1	4.7
Apr 05	357	4.1	4.3	349	3.8	4.1	389	5.4	5.0	344	4.8	4.8
May 05	358	4.2	4.3	349	3.7	4.0	391	5.7	5.4	345	4.7	4.8
Jun 05	359	4.2	4.2	351	3.9	3.8	391	5.2	5.4	346	4.7	4.7
Jul 05	361	4.4	4.3	353	4.2	3.9	392	5.2	5.4	348	5.0	4.8
Aug 05	363	4.4	4.3	354	4.1	4.1	394	5.0	5.1	349	4.8	4.8
Sep 05	364	4.6	4.5	355	4.4	4.2	396	5.2	5.1	350	5.0	4.9
Oct 05	365	4.2	4.4	356	4.0	4.2	397	5.1	5.1	351	4.4	4.7
Nov 05	366	4.2	4.3	357	4.0	4.1	399	5.2	5.2	351	4.5	4.6
Dec 05	368	3.7	4.0	358	3.3	3.8	400	5.5	5.3	352	3.8	4.2
Jan 06	369	4.1	4.0	360	4.0	3.8	400	4.7	5.1	354	4.4	4.2
Feb 06	369	4.3	4.0	361	4.0	3.8	402	5.3	5.2	355	4.1	4.1
Mar 06	370	3.6	4.0	362	3.7	3.9	400	3.6	4.5	356	3.8	4.1
Apr 06	371	3.9	3.9	363	4.2	4.0	400	3.1	4.0	356	3.6	3.9
May 06	373	4.0	3.8	365	4.5	4.1	401	2.6	3.1	357	3.7	3.7
Jun 06	375	4.2	4.0	366	4.5	4.4	404	3.4	3.0	360	4.1	3.8
Jul 06	374	3.6	3.9	366	3.7	4.2	405	3.5	3.2	360	3.4	3.7
Aug 06	375	3.4	3.7	367	3.6	3.9	405	2.9	3.3	360	3.3	3.6
Sep 06	377	3.5	3.5	369	3.9	3.7	406	2.7	3.0	362	3.4	3.3
Oct 06	379	4.0	3.7	371	4.4	4.0	408	2.8	2.8	364	4.0	3.5
Nov 06	380	4.0	3.8	372	4.4	4.2	410	2.8	2.8	365	4.0	3.8
Dec 06	381	4.1	4.1	374	4.5	4.4	411	3.0	2.8	367	4.2	4.1
Jan 07	382	3.7	4.0	373	3.8	4.2	414	3.7	3.1	366	3.5	3.9
Feb 07	384	3.9	3.9	376	4.2	4.1	413	2.7	3.1	369	4.1	3.9
Mar 07	385	4.1	3.9	377	4.3	4.1	415	3.7	3.3	370	4.1	3.9
Apr 07	386	4.0	4.0	378	4.2	4.2	414	3.3	3.2	371	4.2	4.1
May 07	388	4.2	4.1	381	4.4	4.3	416	3.7	3.6	374	4.6	4.3
Jun 07	390	4.2	4.1	383	4.5	4.3	417	3.2	3.4	375	4.3	4.4
Jul 07	392	4.6	4.3	385	5.1	4.7	417	2.8	3.2	376	4.5	4.5
Aug 07	393	4.9	4.6	386	5.2	4.9	419	3.5	3.1	378	4.8	4.6
Sep 07	394	4.5	4.7	387	4.8	5.1	420	3.3	3.2	378	4.6	4.6
Oct 07	394	3.9	4.4	387	4.1	4.7	421	3.1	3.3	379	3.9	4.4
Nov 07	396	4.2	4.2	389	4.5	4.5	422	3.1	3.2	381	4.3	4.3
Dec 07	397	4.0	4.0	389	4.1	4.3	425	3.4	3.2	381	4.0	4.1
Jan 08	398	4.1	4.1	390	4.5	4.4	426	2.9	3.1	382	4.3	4.2
Feb 08	400	4.2	4.1	392	4.4	4.3	429	3.7	3.3	384	4.1	4.1
Mar 08	402	4.3	4.2	394	4.4	4.4	431	3.8	3.5	386	4.3	4.2
Apr 08	404	4.6	4.4	396	4.6	4.4	432	4.4	4.0	389	4.9	4.4
May 08	403	3.8	4.2	395	3.7	4.2	431	3.7	4.0	388	3.7	4.3
Jun 08	404	3.5	4.0	397	3.7	4.0	429	2.9	3.7	388	3.5	4.0
Jul 08	405	3.4	3.6	397	3.2	3.5	434	4.0	3.6	390	3.7	3.6
Aug 08	406	3.3	3.4	398	3.1	3.3	434	3.6	3.5	392	3.6	3.6
Sep 08	407	3.3	3.3	399	3.1	3.2	436	3.8	3.8	392	3.6	3.7
Oct 08	408	3.6	3.4	401	3.6	3.3	436	3.5	3.6	393	3.9	3.7
Nov 08	409	3.2	3.4	401	3.0	3.3	437	3.6	3.6	394	3.3	3.6

4.14b Average Weekly Earnings (nominal) - Regular Pay
(Great Britain, seasonally adjusted)

Standard Industrial Classification (2007)

	Whole Economy			Private sector [3][4][5]			Public sector [3][4][5]			Services, SIC 2007 sections G-S		
	Weekly Earnings (£)	% changes year on year		Weekly Earnings (£)	% changes year on year		Weekly Earnings (£)	% changes year on year		Weekly Earnings (£)	% changes year on year	
		Single month	3 month average [2]		Single month	3 month average [2]		Single month	3 month average [2]		Single month	3 month average [2]
	KAI7	KAI8	KAI9	KAJ2	KAJ3	KAJ4	KAJ5	KAJ6	KAJ7	K5DL	K5DM	K5DN
Dec 08	409	3.2	3.3	401	3.1	3.2	438	3.0	3.4	394	3.3	3.5
Jan 09	409	2.8	3.1	401	2.8	3.0	439	3.0	3.2	394	3.0	3.2
Feb 09	410	2.6	2.9	402	2.4	2.8	442	3.1	3.0	396	3.1	3.1
Mar 09	410	2.0	2.5	402	2.0	2.4	439	2.0	2.7	395	2.3	2.8
Apr 09	411	1.9	2.2	402	1.7	2.0	444	2.8	2.6	397	2.0	2.5
May 09	412	2.3	2.0	403	1.9	1.9	446	3.4	2.7	397	2.6	2.3
Jun 09	412	2.0	2.1	402	1.4	1.7	448	4.3	3.5	398	2.4	2.3
Jul 09	411	1.4	1.9	400	0.6	1.3	450	3.8	3.8	397	1.8	2.3
Aug 09	411	1.3	1.6	401	0.4	0.8	452	4.0	4.0	397	1.4	1.9
Sep 09	413	1.4	1.4	401	0.5	0.5	452	3.7	3.8	398	1.5	1.6
Oct 09	413	1.0	1.2	401	0.1	0.3	453	3.7	3.8	398	1.1	1.3
Nov 09	413	1.0	1.1	401	0.0	0.2	455	3.9	3.8	398	1.1	1.2
Dec 09	415	1.3	1.1	403	0.4	0.2	455	3.9	3.8	399	1.2	1.1
Jan 10	417	1.8	1.4	405	1.0	0.5	457	4.0	3.9	401	1.8	1.4
Feb 10	416	1.5	1.6	404	0.6	0.7	459	3.8	3.9	401	1.2	1.4
Mar 10	418	2.1	1.8	407	1.3	1.0	457	4.1	4.0	403	1.8	1.6
Apr 10	417	1.4	1.7	405	0.6	0.9	459	3.4	3.8	402	1.2	1.4
May 10	417	1.2	1.6	404	0.4	0.8	461	3.5	3.6	402	1.1	1.4
Jun 10	418	1.4	1.3	405	0.8	0.6	459	2.6	3.1	403	1.2	1.2
Jul 10	420	2.2	1.6	407	1.9	1.0	462	2.7	2.9	405	2.1	1.5
Aug 10	421	2.3	2.0	409	2.2	1.6	462	2.3	2.5	406	2.3	1.9
Sep 10	422	2.2	2.2	409	1.9	2.0	465	2.9	2.6	407	2.3	2.2
Oct 10	422	2.2	2.2	409	2.0	2.0	466	2.9	2.7	407	2.3	2.3
Nov 10	423	2.4	2.3	410	2.2	2.0	467	2.8	2.8	408	2.6	2.4
Dec 10	423	1.9	2.2	409	1.6	1.9	468	2.9	2.8	408	2.4	2.5
Jan 11	426	2.2	2.2	413	2.0	1.9	470	2.9	2.9	412	2.6	2.5
Feb 11	425	2.1	2.1	412	1.9	1.8	470	2.6	2.8	411	2.5	2.5
Mar 11	425	1.7	2.0	411	1.2	1.7	472	3.2	2.9	411	2.1	2.4
Apr 11	426	2.1	2.0	412	1.9	1.6	471	2.6	2.8	412	2.5	2.4
May 11	427	2.3	2.0	413	2.3	1.8	472	2.2	2.7	413	2.7	2.4
Jun 11	426	2.0	2.1	413	1.9	2.0	472	2.7	2.5	412	2.3	2.5
Jul 11	427	1.6	2.0	414	1.7	2.0	471	1.9	2.3	412	1.8	2.3
Aug 11	427	1.5	1.7	414	1.4	1.7	473	2.3	2.3	413	1.6	1.9
Sep 11	429	1.7	1.6	416	1.8	1.6	474	1.9	2.0	414	1.7	1.7
Oct 11	430	1.9	1.7	417	1.9	1.7	475	2.1	2.1	415	2.1	1.8
Nov 11	430	1.8	1.8	418	2.0	1.9	476	1.8	1.9	416	1.9	1.9
Dec 11	431	1.9	1.9	418	2.3	2.1	475	1.4	1.7	416	2.0	2.0
Jan 12	430	1.0	1.6	418	1.2	1.8	473	0.7	1.3	416	1.0	1.6
Feb 12	432	1.7	1.5	420	1.9	1.8	477	1.3	1.1	418	1.7	1.5
Mar 12	433	1.9	1.5	421	2.2	1.8	478	1.3	1.1	418	1.8	1.5
Apr 12	433	1.7	1.7	420	2.0	2.0	478	1.4	1.3	418	1.5	1.7
May 12	434	1.7	1.7	421	1.9	2.0	479	1.6	1.4	419	1.5	1.6
Jun 12	435	1.9	1.8	422	2.2	2.0	482	2.2	1.7	420	1.9	1.6
Jul 12	435	1.8	1.8	422	1.9	2.0	483	2.7	2.2	420	1.8	1.7
Aug 12	436	2.1	2.0	423	2.1	2.0	487	3.1	2.7	422	2.2	2.0
Sep 12	435	1.5	1.8	422	1.6	1.8	484	2.2	2.6	421	1.6	1.9
Oct 12	435	1.2	1.6	422	1.3	1.7	484	1.8	2.4	421	1.2	1.7

4.14b Average Weekly Earnings (nominal) - Regular Pay
(Great Britain, seasonally adjusted)

Standard Industrial Classification (2007)

	Whole Economy			Private sector [3 4 5]			Public sector [3 4 5]			Services, SIC 2007 sections G-S		
	Weekly Earnings (£)	% changes year on year		Weekly Earnings (£)	% changes year on year		Weekly Earnings (£)	% changes year on year		Weekly Earnings (£)	% changes year on year	
		Single month	3 month average [2]		Single month	3 month average [2]		Single month	3 month average [2]		Single month	3 month average [2]
	KAI7	KAI8	KAI9	KAJ2	KAJ3	KAJ4	KAJ5	KAJ6	KAJ7	K5DL	K5DM	K5DN
Nov 12	436	1.3	1.4	424	1.5	1.5	484	1.8	2.0	422	1.4	1.4
Dec 12	436	1.2	1.3	423	1.2	1.3	484	1.9	1.8	422	1.2	1.3
Jan 13	434	1.0	1.2	422	1.0	1.2	483	2.0	1.9	420	1.1	1.3
Feb 13	435	0.7	1.0	423	0.8	1.0	481	1.0	1.6	420	0.5	1.0
Mar 13	436	0.7	0.8	424	0.7	0.8	483	1.1	1.4	420	0.4	0.7
Apr 13	438	1.1	0.9	425	1.2	0.9	485	1.4	1.2	423	1.1	0.7
May 13	438	1.0	0.9	426	1.1	1.0	486	1.5	1.3	423	0.9	0.8
Jun 13	438	0.8	1.0	426	1.0	1.1	484	0.5	1.1	423	0.8	0.9
Jul 13	439	0.9	0.9	427	1.2	1.1	485	0.4	0.8	424	0.9	0.9
Aug 13	438	0.5	0.9	427	0.9	1.1	484	-0.7	0.0	423	0.9	0.7
Sep 13	438	0.7	0.7	426	0.9	1.0	485	0.1	-0.1	423	0.3	0.6
Oct 13	439	0.9	0.7	428	1.2	1.0	487	0.6	0.0	424	0.6	0.6
Nov 13	439	0.5	0.7	428	0.8	1.0	486	0.3	0.3	423	0.4	0.6
Dec 13	441	1.1	0.9	429	1.5	1.2	487	0.8	0.6	426	1.0	0.7
Jan 14	441	1.6	1.1	431	2.0	1.4	487	0.9	0.7	425	1.2	0.8
Feb 14	440	1.1	1.3	428	1.3	1.6	487	1.1	1.0	424	0.8	1.0
Mar 14	439	0.7	1.1	427	0.7	1.3	490	1.4	1.2	424	0.8	1.0
Apr 14	440	0.4	0.7	429	0.7	0.9	487	0.4	1.0	424	0.3	0.7
May 14	440	0.5	0.5	430	0.9	0.8	488	0.3	0.7	425	0.5	0.5
Jun 14	441	0.7	0.5	430	0.9	0.9	488	0.3	0.5	426	0.6	0.5
Jul 14	442	0.6	0.6	431	0.9	0.9	489	0.7	0.6	426	0.5	0.5
Aug 14	443	1.0	0.8	432	1.3	1.0	489	1.0	0.8	427	1.0	0.7
Sep 14	446	1.7	1.1	435	2.2	1.5	490	1.0	0.9	430	1.7	1.1
Oct 14	447	1.8	1.5	437	2.3	1.9	490	0.6	0.9	432	1.8	1.5
Nov 14	445	1.6	1.7	436	2.0	2.1	489	0.7	0.8	430	1.7	1.8
Dec 14	447	1.5	1.6	437	1.8	2.0	492	0.9	0.7	433	1.6	1.7
Jan 15	448	1.5	1.5	439	1.9	1.9	491	0.8	0.8	434	2.0	1.8
Feb 15	450	2.3	1.8	440	2.7	2.1	493	1.1	0.9	435	2.7	2.1
Mar 15	451	2.8	2.2	442	3.5	2.7	492	0.5	0.8	436	2.9	2.5
Apr 15	452	2.7	2.6	442	3.1	3.1	493	1.4	1.0	436	2.9	2.8
May 15	452	2.7	2.7	443	3.2	3.3	493	1.2	1.0	437	2.9	2.9
Jun 15	453	2.9	2.7	444	3.2	3.2	494	1.3	1.3	438	3.0	2.9
Jul 15	454	2.9	2.8	445	3.4	3.3	495	1.2	1.2	439	3.0	3.0
Aug 15	454	2.5	2.7	445	2.9	3.2	494	1.2	1.2	438	2.6	2.9
Sep 15	454	1.8	2.4	444	2.0	2.8	497	1.5	1.3	438	1.8	2.5
Oct 15	454	1.6	2.0	445	1.8	2.2	496	1.3	1.3	439	1.6	2.0
Nov 15	455	2.2	1.9	446	2.4	2.0	497	1.6	1.5	440	2.1	1.8
Dec 15	457	2.0	1.9	448	2.4	2.2	497	1.1	1.4	441	2.0	1.9
Jan 16	459	2.4	2.2	450	2.5	2.4	499	1.5	1.4	443	2.2	2.1
Feb 16	460	2.3	2.2	452	2.6	2.5	500	1.4	1.4	445	2.3	2.1
Mar 16	461	2.1	2.3	452	2.3	2.5	500	1.7	1.6	445	2.0	2.2
Apr 16	463	2.5	2.3	454	2.6	2.5	503	1.9	1.7	448	2.6	2.3
May 16	462	2.1	2.2	453	2.3	2.4	500	1.4	1.6	446	1.9	2.2
Jun 16	464	2.3	2.3	455	2.5	2.5	503	1.8	1.7	448	2.2	2.2
Jul 16	464	2.1	2.2	456	2.3	2.4	502	1.4	1.5	448	2.0	2.0
Aug 16	465	2.3	2.3	456	2.5	2.4	503	1.8	1.7	448	2.3	2.2
Sep 16	466	2.7	2.4	458	3.0	2.6	503	1.3	1.5	450	2.6	2.3

4.14b Average Weekly Earnings (nominal) - Regular Pay
(Great Britain, seasonally adjusted)

Standard Industrial Classification (2007)

	Whole Economy			Private sector [3][4][5]			Public sector [3][4][5]			Services, SIC 2007 sections G-S		
	Weekly Earnings (£)	% changes year on year		Weekly Earnings (£)	% changes year on year		Weekly Earnings (£)	% changes year on year		Weekly Earnings (£)	% changes year on year	
		Single month	3 month average [2]		Single month	3 month average [2]		Single month	3 month average [2]		Single month	3 month average [2]
	KAI7	KAI8	KAI9	KAJ2	KAJ3	KAJ4	KAJ5	KAJ6	KAJ7	K5DL	K5DM	K5DN
Oct 16	466	2.6	2.5	458	2.9	2.8	503	1.4	1.5	450	2.5	2.5
Nov 16	467	2.7	2.7	460	3.0	3.0	505	1.5	1.4	452	2.8	2.6
Dec 16	467	2.3	2.5	459	2.5	2.8	504	1.5	1.5	451	2.3	2.5
Jan 17	468	1.9	2.3	460	2.2	2.6	505	1.3	1.4	452	1.9	2.3
Feb 17	468	1.7	2.0	460	1.8	2.2	506	1.2	1.3	452	1.7	1.9
Mar 17	469	1.7	1.8	461	1.9	2.0	507	1.2	1.2	453	1.9	1.8
Apr 17	471	1.8	1.7	463	2.1	2.0	507	0.8	1.1	455	1.7	1.7
May 17	472	2.3	2.0	464	2.4	2.2	511	2.2	1.4	457	2.5	2.0
Jun 17	473	2.0	2.1	465	2.3	2.3	509	1.2	1.4	457	2.1	2.1
Jul 17	474	2.1	2.1	466	2.2	2.3	510	1.7	1.7	458	2.2	2.3
Aug 17	475	2.3	2.1	467	2.5	2.3	512	1.7	1.5	459	2.4	2.2
Sep 17	476	2.3	2.2	469	2.5	2.4	512	1.7	1.7	460	2.3	2.3
Oct 17	477	2.4	2.3	469	2.5	2.5	515	2.3	1.9	461	2.5	2.4
Nov 17	478	2.3	2.3	471	2.4	2.5	514	1.7	1.9	462	2.3	2.3
Dec 17	480	2.7	2.5	472	2.9	2.6	515	2.0	2.0	463	2.5	2.4
Jan 18	480	2.7	2.6	472	2.8	2.7	517	2.3	2.0	463	2.6	2.5

Source: Monthly Wages & Salaries Survey

Inquiries: E-mail: labour.market@ons.gov.uk

Earnings enquiries: 01633 456773

1. Estimates of regular pay exclude bonuses and arrears of pay.

2. The three month average figures are the changes in the average seasonally adjusted values for the three months ending with the relevant month compared with the same period a year earlier.

3. From July 2009 Royal Bank of Scotland Group plc is classified to the public sector; for earlier time periods it is classified to the private sector. Between July 2009 and March 2014 Lloyds Banking Group plc is classified to the public sector; it is classified to the private sector for earlier and later time periods.

4. Between June 2010 and May 2012 English Further Education Corporations and Sixth Form College Corporations are classified to the public sector. Before June 2010 and after May 2012 they are classified to the private sector.

5. From October 2013 Royal Mail plc is classified to the private sector; previously it is in the public sector.

4.14b Average Weekly Earnings (nominal) - Regular Pay
(Great Britain, seasonally adjusted)

	Finance and business services, SIC 2007 sections K-N			Public sector excluding financial services [4][5]			Manufacturing, SIC 2007 section C			Construction, SIC 2007 section F			Wholesaling, retailing, hotels & restaurants, SIC 2007 sections G & I		
	Weekly Earnings (£)	% changes year on year		Weekly Earnings (£)	% changes year on year		Weekly Earnings (£)	% changes year on year		Weekly Earnings (£)	% changes year on year		Weekly Earnings (£)	% changes year on year	
		Single month	3 month average [2]		Single month	3 month average [2]		Single month	3 month average [2]		Single month	3 month average [2]		Single month	3 month average [2]
	K5DO	K5DP	K5DQ	KAK6	KAK7	KAK8	K5DU	K5DV	K5DW	K5DX	K5DY	K5DZ	K5E2	K5E3	K5E4
Jan 05	407	4.2	3.8	382	5.0	4.5	415	4.0	4.2	430	3.0	1.5	234	4.5	4.4
Feb 05	411	5.5	4.6	382	4.3	4.5	416	3.9	3.9	416	-0.4	1.0	235	4.5	4.5
Mar 05	412	5.4	5.0	386	5.3	4.9	419	3.9	3.9	421	-0.6	0.7	235	4.6	4.5
Apr 05	413	5.2	5.4	388	5.4	5.0	420	4.5	4.1	420	-0.6	-0.5	236	4.1	4.4
May 05	414	4.5	5.0	391	5.7	5.4	420	3.7	4.0	423	0.3	-0.3	234	3.0	3.9
Jun 05	413	4.7	4.8	390	5.2	5.4	422	3.6	3.9	424	1.6	0.4	236	3.3	3.5
Jul 05	419	5.9	5.0	392	5.2	5.4	423	3.4	3.5	426	1.7	1.2	237	3.6	3.3
Aug 05	419	5.3	5.3	394	5.1	5.1	426	3.7	3.6	434	2.3	1.9	240	4.6	3.8
Sep 05	420	5.3	5.5	396	5.2	5.1	427	3.7	3.6	434	4.0	2.7	239	4.0	4.1
Oct 05	421	4.9	5.1	397	5.1	5.1	427	3.2	3.5	438	5.4	3.9	241	4.3	4.3
Nov 05	421	4.7	5.0	398	5.2	5.2	428	3.1	3.3	440	5.5	5.0	242	4.6	4.3
Dec 05	420	3.5	4.4	399	5.5	5.3	430	3.7	3.3	437	4.1	5.0	242	3.4	4.1
Jan 06	425	4.5	4.3	400	4.7	5.1	429	3.3	3.4	443	3.0	4.2	243	3.7	3.9
Feb 06	423	3.1	3.7	402	5.4	5.2	435	4.5	3.8	447	7.3	4.8	243	3.5	3.5
Mar 06	429	4.0	3.8	400	3.6	4.5	434	3.5	3.8	445	5.9	5.3	243	3.1	3.4
Apr 06	429	3.9	3.7	400	3.1	4.0	436	3.7	3.9	451	7.2	6.8	242	2.6	3.1
May 06	430	4.0	4.0	401	2.6	3.1	439	4.3	3.8	457	7.9	7.0	245	4.6	3.4
Jun 06	433	4.9	4.3	404	3.4	3.0	437	3.6	3.9	458	8.0	7.7	245	3.8	3.7
Jul 06	433	3.3	4.0	405	3.5	3.2	435	2.7	3.5	463	8.6	8.2	245	3.5	4.0
Aug 06	435	3.8	4.0	405	2.9	3.3	439	3.2	3.2	458	5.6	7.4	245	2.3	3.2
Sep 06	436	3.6	3.6	406	2.7	3.0	441	3.4	3.1	462	6.5	6.9	248	3.7	3.1
Oct 06	437	4.0	3.8	408	2.8	2.8	443	3.7	3.4	465	6.2	6.1	250	3.6	3.2
Nov 06	437	3.9	3.8	409	2.8	2.8	444	3.7	3.6	464	5.4	6.0	252	4.3	3.9
Dec 06	441	5.0	4.3	411	3.0	2.9	443	3.0	3.5	466	6.8	6.1	253	4.8	4.2
Jan 07	434	2.2	3.7	414	3.7	3.1	449	4.6	3.8	466	5.3	5.8	251	3.5	4.2
Feb 07	440	4.0	3.7	413	2.7	3.1	450	3.4	3.7	462	3.4	5.1	254	4.8	4.3
Mar 07	442	3.1	3.1	414	3.7	3.4	450	4.5	4.2	470	5.6	4.7	255	5.1	4.5
Apr 07	445	3.7	3.6	413	3.3	3.2	453	3.3	3.7	469	4.1	4.4	257	5.9	5.3
May 07	448	4.2	3.6	415	3.6	3.5	451	2.9	3.6	471	3.1	4.3	256	4.8	5.3
Jun 07	450	4.0	3.9	416	3.1	3.3	453	3.7	3.3	475	3.8	3.7	258	5.3	5.4
Jul 07	452	4.5	4.2	417	2.8	3.1	455	4.7	3.8	495	7.1	4.7	259	5.6	5.2
Aug 07	453	4.1	4.2	419	3.5	3.1	455	3.6	4.0	497	8.5	6.5	260	6.2	5.7
Sep 07	456	4.8	4.5	420	3.3	3.2	455	3.2	3.8	499	7.8	7.8	260	5.0	5.6
Oct 07	457	4.6	4.5	421	3.1	3.3	457	3.2	3.3	491	5.7	7.3	261	4.4	5.2
Nov 07	461	5.3	4.9	422	3.1	3.2	457	3.0	3.1	494	6.5	6.6	261	3.7	4.3
Dec 07	462	4.7	4.8	425	3.4	3.2	458	3.3	3.2	494	5.8	6.0	260	2.8	3.6
Jan 08	459	5.8	5.2	426	2.8	3.1	462	2.8	3.0	483	3.7	5.3	262	4.4	3.6
Feb 08	465	5.5	5.3	428	3.6	3.3	464	3.3	3.1	492	6.5	5.3	262	3.0	3.4
Mar 08	465	5.2	5.5	430	3.8	3.4	465	2.5	2.8	489	4.1	4.8	265	3.7	3.7
Apr 08	468	5.2	5.3	431	4.4	3.9	466	3.6	3.1	494	5.3	5.3	268	4.2	3.6
May 08	466	4.0	4.8	431	3.7	4.0	466	3.3	3.2	496	5.4	4.9	266	3.6	3.8
Jun 08	472	4.7	4.6	428	2.9	3.7	468	3.1	3.3	499	4.9	5.2	265	2.9	3.6
Jul 08	475	5.1	4.6	433	4.0	3.6	469	3.0	3.2	499	0.8	3.7	264	2.0	2.8
Aug 08	478	5.4	5.1	434	3.6	3.5	468	2.9	3.0	493	-0.9	1.6	266	2.1	2.3
Sep 08	478	4.7	5.1	435	3.8	3.8	468	2.9	2.9	500	0.3	0.1	267	2.9	2.3
Oct 08	479	4.7	4.9	436	3.5	3.6	470	2.8	2.9	504	2.6	0.7	266	2.2	2.4
Nov 08	479	4.0	4.5	437	3.6	3.6	473	3.5	3.1	503	1.8	1.6	265	1.5	2.2

4.14b Average Weekly Earnings (nominal) - Regular Pay
(Great Britain, seasonally adjusted)

Standard Industrial Classification (2007)

	Finance and business services, SIC 2007 sections K-N			Public sector excluding financial services [4][5]			Manufacturing, SIC 2007 section C			Construction, SIC 2007 section F			Wholesaling, retailing, hotels & restaurants, SIC 2007 sections G & I		
	Weekly Earnings (£)	% changes year on year		Weekly Earnings (£)	% changes year on year		Weekly Earnings (£)	% changes year on year		Weekly Earnings (£)	% changes year on year		Weekly Earnings (£)	% changes year on year	
		Single month	3 month average [2]		Single month	3 month average [2]		Single month	3 month average [2]		Single month	3 month average [2]		Single month	3 month average [2]
	K5DO	K5DP	K5DQ	KAK6	KAK7	KAK8	K5DU	K5DV	K5DW	K5DX	K5DY	K5DZ	K5E2	K5E3	K5E4
Dec 08	477	3.3	4.0	438	3.0	3.4	472	3.2	3.2	503	1.9	2.1	268	2.9	2.2
Jan 09	479	4.4	3.9	439	3.0	3.2	471	2.1	2.9	505	4.6	2.8	268	2.2	2.2
Feb 09	480	3.4	3.7	441	3.1	3.1	470	1.2	2.2	503	2.3	2.9	271	3.5	2.8
Mar 09	482	3.6	3.8	439	2.0	2.7	470	1.3	1.5	504	3.0	3.3	269	1.7	2.4
Apr 09	482	3.1	3.4	443	2.8	2.7	472	1.3	1.2	506	2.3	2.6	268	0.1	1.7
May 09	483	3.6	3.4	445	3.3	2.7	474	1.6	1.4	504	1.6	2.3	269	1.1	1.0
Jun 09	483	2.5	3.1	447	4.3	3.5	477	2.0	1.6	496	-0.7	1.1	269	1.3	0.8
Jul 09	483	1.6	2.5	446	2.8	3.5	473	0.8	1.4	503	0.7	0.6	269	2.0	1.4
Aug 09	483	1.2	1.7	448	3.1	3.4	476	1.7	1.5	507	2.9	1.0	269	1.3	1.5
Sep 09	483	1.2	1.3	447	2.8	2.9	478	2.1	1.5	506	1.1	1.6	271	1.2	1.5
Oct 09	484	1.0	1.1	448	2.9	2.9	480	2.2	2.0	510	1.2	1.7	271	1.9	1.4
Nov 09	487	1.6	1.3	450	2.8	2.8	481	1.7	2.0	510	1.3	1.2	271	2.2	1.8
Dec 09	487	2.0	1.5	451	2.8	2.8	491	4.0	2.6	513	2.0	1.5	273	1.9	2.0
Jan 10	490	2.3	1.9	452	2.9	2.9	491	4.2	3.3	512	1.3	1.5	272	1.7	1.9
Feb 10	486	1.3	1.8	453	2.7	2.8	493	4.8	4.3	512	1.8	1.7	274	1.1	1.6
Mar 10	495	2.7	2.1	452	3.1	2.9	496	5.4	4.8	510	1.1	1.4	276	2.4	1.8
Apr 10	492	1.9	2.0	455	2.5	2.8	494	4.5	4.9	512	1.2	1.3	273	2.1	1.9
May 10	493	2.0	2.2	456	2.4	2.6	493	4.1	4.7	506	0.4	0.9	272	1.4	2.0
Jun 10	492	1.9	1.9	453	1.5	2.1	494	3.7	4.1	504	1.6	1.1	274	2.2	1.9
Jul 10	500	3.5	2.4	457	2.5	2.1	493	4.4	4.1	502	-0.2	0.6	275	2.0	1.9
Aug 10	502	3.8	3.1	457	2.0	2.0	496	4.2	4.1	502	-1.1	0.1	276	2.6	2.3
Sep 10	504	4.3	3.9	459	2.7	2.4	496	3.7	4.1	504	-0.3	-0.5	276	2.0	2.2
Oct 10	506	4.7	4.2	460	2.6	2.4	496	3.2	3.7	508	-0.5	-0.6	274	1.2	1.9
Nov 10	508	4.3	4.4	462	2.7	2.7	496	3.2	3.4	506	-0.7	-0.5	275	1.4	1.5
Dec 10	511	5.0	4.7	463	2.8	2.7	497	1.2	2.5	502	-2.2	-1.1	274	0.3	1.0
Jan 11	518	5.6	5.0	464	2.8	2.8	498	1.5	1.9	514	0.4	-0.8	276	1.3	1.0
Feb 11	513	5.5	5.4	465	2.5	2.7	497	0.9	1.2	516	0.8	-0.3	275	0.2	0.6
Mar 11	513	3.7	4.9	466	3.1	2.8	499	0.6	1.0	514	0.9	0.7	275	-0.2	0.4
Apr 11	515	4.7	4.7	465	2.3	2.6	498	0.9	0.8	509	-0.5	0.4	277	1.4	0.5
May 11	519	5.4	4.6	465	2.0	2.5	499	1.1	0.8	512	1.3	0.6	278	2.1	1.1
Jun 11	519	5.3	5.1	465	2.5	2.3	500	1.1	1.0	515	2.3	1.0	277	0.8	1.4
Jul 11	519	3.8	4.8	464	1.6	2.1	501	1.6	1.3	516	2.8	2.1	277	0.7	1.2
Aug 11	517	3.0	4.0	466	2.1	2.1	502	1.2	1.3	518	3.3	2.8	277	0.5	0.7
Sep 11	520	3.3	3.3	467	1.6	1.8	504	1.7	1.5	518	2.7	2.9	277	0.5	0.6
Oct 11	525	3.7	3.3	467	1.5	1.8	503	1.5	1.5	516	1.6	2.5	279	1.6	0.9
Nov 11	524	3.3	3.4	467	1.1	1.4	506	1.9	1.7	518	2.2	2.2	281	2.2	1.4
Dec 11	528	3.3	3.4	466	0.4	1.1	505	1.6	1.6	516	2.8	2.2	280	2.5	2.1
Jan 12	525	1.3	2.7	466	0.9	0.8	504	1.2	1.5	518	0.7	1.9	282	2.2	2.3
Feb 12	527	2.7	2.5	469	0.9	0.7	505	1.7	1.5	519	0.5	1.4	283	2.9	2.5
Mar 12	530	3.3	2.4	470	0.8	0.7	508	1.9	1.6	524	1.8	1.0	283	2.8	2.6
Apr 12	528	2.5	2.8	470	1.0	0.9	509	2.2	1.9	523	2.7	1.7	283	1.9	2.5
May 12	528	1.7	2.5	470	1.2	1.0	511	2.4	2.1	529	3.3	2.6	284	2.0	2.5
Jun 12	530	2.1	2.1	474	2.1	1.4	511	2.2	2.3	523	1.6	2.6	286	3.3	2.4
Jul 12	528	1.7	1.8	476	2.6	2.0	512	2.1	2.2	522	1.2	2.0	285	3.1	2.8
Aug 12	529	2.4	2.1	479	2.7	2.5	513	2.2	2.2	516	-0.5	0.8	285	2.7	3.0
Sep 12	528	1.4	1.9	477	2.1	2.5	513	1.7	2.0	515	-0.6	0.1	285	2.8	2.8
Oct 12	526	0.2	1.3	477	2.1	2.3	514	2.1	2.0	515	-0.2	-0.4	285	2.3	2.6

4.14b Average Weekly Earnings (nominal) - Regular Pay
(Great Britain, seasonally adjusted)

	Finance and business services, SIC 2007 sections K-N			Public sector excluding financial services [4,5]			Manufacturing, SIC 2007 section C			Construction, SIC 2007 section F			Wholesaling, retailing, hotels & restaurants, SIC 2007 sections G & I		
	Weekly Earnings (£)	% changes year on year		Weekly Earnings (£)	% changes year on year		Weekly Earnings (£)	% changes year on year		Weekly Earnings (£)	% changes year on year		Weekly Earnings (£)	% changes year on year	
		Single month	3 month average [2]		Single month	3 month average [2]		Single month	3 month average [2]		Single month	3 month average [2]		Single month	3 month average [2]
	K5DO	K5DP	K5DQ	KAK6	KAK7	KAK8	K5DU	K5DV	K5DW	K5DX	K5DY	K5DZ	K5E2	K5E3	K5E4
Nov 12	527	0.5	0.7	477	2.0	2.1	513	1.4	1.7	516	-0.3	-0.4	287	2.4	2.5
Dec 12	526	-0.5	0.1	476	1.9	2.0	516	2.1	1.9	512	-0.8	-0.4	286	2.0	2.2
Jan 13	525	0.0	0.0	476	2.0	2.0	514	2.0	1.8	504	-2.6	-1.2	284	0.7	1.7
Feb 13	526	-0.2	-0.2	473	1.0	1.7	518	2.5	2.2	516	-0.6	-1.4	284	0.3	1.0
Mar 13	526	-0.8	-0.3	475	1.1	1.4	521	2.6	2.3	518	-1.1	-1.4	286	1.3	0.8
Apr 13	527	-0.1	-0.4	477	1.5	1.2	521	2.4	2.5	527	0.6	-0.4	288	1.8	1.1
May 13	526	-0.3	-0.4	478	1.6	1.4	521	2.1	2.3	524	-1.0	-0.5	288	1.5	1.5
Jun 13	523	-1.2	-0.5	477	0.6	1.2	523	2.3	2.2	518	-1.0	-0.5	291	1.8	1.7
Jul 13	525	-0.4	-0.6	478	0.4	0.9	522	2.0	2.1	515	-1.3	-1.1	293	2.7	2.0
Aug 13	523	-1.1	-0.9	477	-0.3	0.2	522	1.6	2.0	520	0.9	-0.5	293	2.8	2.4
Sep 13	522	-1.2	-0.9	478	0.4	0.2	522	1.7	1.8	520	1.0	0.2	293	2.8	2.7
Oct 13	522	-0.7	-1.0	481	0.9	0.3	524	2.1	1.8	514	-0.2	0.5	293	3.0	2.9
Nov 13	522	-1.1	-1.0	480	0.7	0.7	527	2.8	2.2	515	-0.2	0.2	293	1.9	2.6
Dec 13	526	0.1	-0.6	482	1.2	1.0	528	2.3	2.4	519	1.5	0.4	294	2.8	2.6
Jan 14	522	-0.5	-0.5	482	1.4	1.1	530	3.0	2.7	528	4.8	2.0	297	4.7	3.1
Feb 14	519	-1.4	-0.6	482	1.8	1.5	533	2.9	2.8	518	0.4	2.2	293	3.3	3.6
Mar 14	524	-0.3	-0.7	482	1.4	1.5	529	1.6	2.5	508	-1.9	1.1	293	2.2	3.4
Apr 14	521	-1.2	-0.9	482	1.2	1.4	530	1.7	2.1	515	-2.3	-1.3	292	1.6	2.4
May 14	522	-0.9	-0.8	483	1.2	1.2	530	1.8	1.7	514	-1.8	-2.0	294	2.2	2.0
Jun 14	523	-0.1	-0.7	484	1.4	1.3	532	1.8	1.8	518	0.0	-1.4	293	0.7	1.5
Jul 14	525	-0.1	-0.3	484	1.2	1.3	529	1.3	1.6	522	1.2	-0.2	292	-0.1	0.9
Aug 14	530	1.3	0.4	484	1.4	1.3	530	1.6	1.6	520	-0.1	0.4	294	0.3	0.3
Sep 14	534	2.5	1.2	486	1.5	1.4	532	2.0	1.6	523	0.6	0.6	297	1.2	0.5
Oct 14	539	3.3	2.3	486	1.0	1.3	533	1.7	1.7	524	1.9	0.8	297	0.9	0.8
Nov 14	535	2.5	2.8	485	1.1	1.2	532	0.9	1.5	523	1.5	1.3	297	1.3	1.2
Dec 14	535	1.6	2.5	488	1.4	1.2	532	0.9	1.1	526	1.4	1.6	298	1.4	1.2
Jan 15	536	2.7	2.3	487	1.1	1.2	532	0.5	0.8	517	-2.2	0.2	302	1.7	1.5
Feb 15	538	3.7	2.7	488	1.4	1.3	536	0.4	0.6	525	1.4	0.2	303	3.3	2.1
Mar 15	542	3.3	3.2	489	1.4	1.3	535	1.0	0.7	533	4.9	1.3	304	4.0	3.0
Apr 15	543	4.2	3.8	489	1.4	1.4	537	1.4	1.0	528	2.7	3.0	303	3.7	3.7
May 15	544	4.3	4.0	489	1.2	1.4	537	1.2	1.2	528	2.7	3.4	305	3.6	3.8
Jun 15	545	4.3	4.3	491	1.4	1.4	538	1.2	1.3	531	2.5	2.6	304	3.8	3.7
Jul 15	546	4.0	4.2	491	1.3	1.3	541	2.2	1.5	536	2.8	2.7	306	4.5	4.0
Aug 15	546	3.1	3.8	491	1.5	1.4	542	2.2	1.9	535	3.0	2.8	306	4.1	4.1
Sep 15	546	2.2	3.1	493	1.6	1.5	542	2.0	2.1	533	1.9	2.6	303	2.3	3.6
Oct 15	549	1.8	2.3	493	1.4	1.5	543	1.8	2.0	537	2.5	2.5	305	2.7	3.0
Nov 15	551	3.1	2.3	494	1.7	1.6	545	2.3	2.0	538	2.8	2.4	306	3.1	2.7
Dec 15	555	3.8	2.9	494	1.1	1.4	547	2.7	2.3	536	1.8	2.4	308	3.3	3.0
Jan 16	557	3.8	3.5	495	1.6	1.5	547	2.7	2.6	542	4.9	3.2	308	1.8	2.7
Feb 16	557	3.4	3.7	496	1.6	1.4	547	2.0	2.5	546	4.0	3.6	311	2.6	2.5
Mar 16	557	2.7	3.3	497	1.7	1.6	547	2.4	2.4	547	2.7	3.9	310	2.0	2.1
Apr 16	561	3.3	3.2	499	2.0	1.8	551	2.6	2.3	549	3.8	3.5	313	3.0	2.5
May 16	555	2.0	2.7	496	1.5	1.7	554	3.1	2.7	549	3.9	3.5	313	2.7	2.6
Jun 16	559	2.5	2.6	499	1.8	1.7	553	2.8	2.8	550	3.4	3.7	314	3.4	3.0
Jul 16	557	2.0	2.2	497	1.4	1.5	555	2.7	2.9	552	2.9	3.4	314	2.8	3.0
Aug 16	557	1.9	2.1	500	1.8	1.6	554	2.3	2.6	556	3.8	3.4	316	3.4	3.2
Sep 16	558	2.2	2.0	500	1.3	1.5	554	2.2	2.4	557	4.5	3.7	317	4.4	3.5

Standard Industrial Classification (2007)

4.14b Average Weekly Earnings (nominal) - Regular Pay
(Great Britain, seasonally adjusted)

Standard Industrial Classification (2007)

	Finance and business services, SIC 2007 sections K-N			Public sector excluding financial services [4][5]			Manufacturing, SIC 2007 section C			Construction, SIC 2007 section F			Wholesaling, retailing, hotels & restaurants, SIC 2007 sections G & I		
	Weekly Earnings (£)	% changes year on year		Weekly Earnings (£)	% changes year on year		Weekly Earnings (£)	% changes year on year		Weekly Earnings (£)	% changes year on year		Weekly Earnings (£)	% changes year on year	
		Single month	3 month average [2]		Single month	3 month average [2]		Single month	3 month average [2]		Single month	3 month average [2]		Single month	3 month average [2]
	K5DO	K5DP	K5DQ	KAK6	KAK7	KAK8	K5DU	K5DV	K5DW	K5DX	K5DY	K5DZ	K5E2	K5E3	K5E4
Oct 16	557	1.5	1.9	500	1.5	1.5	554	2.0	2.2	559	4.2	4.2	317	4.0	3.9
Nov 16	560	1.7	1.8	501	1.5	1.4	553	1.6	1.9	561	4.3	4.3	319	4.2	4.2
Dec 16	560	0.9	1.4	501	1.4	1.5	554	1.3	1.6	559	4.3	4.3	317	3.2	3.8
Jan 17	563	1.1	1.2	502	1.3	1.4	556	1.7	1.5	557	2.7	3.8	318	3.3	3.6
Feb 17	562	1.0	1.0	502	1.2	1.3	556	1.8	1.6	557	1.9	3.0	317	2.1	2.9
Mar 17	563	1.2	1.1	503	1.2	1.2	558	1.9	1.8	550	0.6	1.7	319	2.8	2.7
Apr 17	566	0.8	1.0	503	0.8	1.1	558	1.3	1.7	556	1.4	1.3	320	2.4	2.4
May 17	569	2.4	1.5	507	2.2	1.4	559	1.0	1.4	558	1.7	1.2	319	2.0	2.4
Jun 17	569	1.8	1.7	505	1.2	1.4	561	1.3	1.2	560	1.9	1.7	320	1.9	2.1
Jul 17	571	2.5	2.3	506	1.8	1.7	560	0.8	1.0	564	2.3	2.0	318	1.1	1.7
Aug 17	573	2.8	2.4	508	1.7	1.6	564	1.9	1.3	568	2.2	2.1	320	1.2	1.4
Sep 17	573	2.7	2.7	508	1.7	1.7	565	2.0	1.6	573	2.9	2.5	323	1.9	1.4
Oct 17	573	2.8	2.8	511	2.3	1.9	567	2.4	2.1	574	2.6	2.6	322	1.7	1.6
Nov 17	573	2.3	2.6	510	1.8	1.9	569	2.8	2.4	571	1.8	2.4	324	1.7	1.8
Dec 17	574	2.5	2.6	511	2.1	2.0	571	3.1	2.8	581	4.0	2.8	325	2.5	2.0
Jan 18	574	2.0	2.3	513	2.4	2.1	571	2.7	2.9	580	4.1	3.3	324	2.1	2.1

Source: Monthly Wages & Salaries Survey

Inquiries: Email: labour.market@ons.gov.uk

Earnings enquiries: 01633 456773

1. Estimates of regular pay exclude bonuses and arrears of pay.

2. The three month average figures are the changes in the average seasonally adjusted values for the three months ending with the relevant month compared with the same period a year earlier.

3. From July 2009 Royal Bank of Scotland Group plc is classified to the public sector; for earlier time periods it is classified to the private sector. Between July 2009 and March 2014 Lloyds Banking Group plc is classified to the public sector; it is classified to the private sector for earlier and later time periods.

4. Between June 2010 and May 2012 English Further Education Corporations and Sixth Form College Corporations are classified to the public sector. Before June 2010 and after May 2012 they are classified to the private sector.

5. From October 2013 Royal Mail plc is classified to the private sector; previously it is in the public sector.

4.14c Average Weekly Earnings (nominal) - Bonus Pay
(Great Britain, seasonally adjusted)

Standard Industrial Classification (2007)

	Whole Economy			Private sector [2][3][4]			Public sector [2][3][4]			Services, SIC 2007 sections G-S		
	Weekly Earnings (£)	% changes year on year		Weekly Earnings (£)	% changes year on year		Weekly Earnings (£)	% changes year on year		Weekly Earnings (£)	% changes year on year	
		Single month	3 month average [1]		Single month	3 month average [1]		Single month	3 month average [1]		Single month	3 month average [1]
	KAF4	KAF5	KAF6	KAF7	KAF8	KAF9	KAG2	KAG3	KAG4	K5CS	K5CT	K5CU
Jan 05	23	13.6	17.7	28	5.7	10.6	3	-3.4	12.4	21	10.4	21.1
Feb 05	22	18.7	17.0	25	19.8	11.8	3	3.4	3.9	19	25.1	20.6
Mar 05	22	2.4	11.1	25	2.6	8.7	3	-11.0	-4.1	20	3.0	11.9
Apr 05	22	6.4	8.7	26	8.0	9.6	3	19.4	3.3	22	-0.9	7.5
May 05	23	11.2	6.6	25	0.8	3.7	3	17.8	7.8	21	12.8	4.7
Jun 05	22	7.4	8.3	25	8.2	5.6	4	16.3	17.8	21	8.6	6.5
Jul 05	23	11.7	10.1	26	13.8	7.4	4	35.0	22.8	21	12.5	11.3
Aug 05	26	26.3	15.2	29	26.6	16.2	4	33.0	27.9	25	28.4	16.6
Sep 05	24	8.5	15.3	27	8.8	16.2	4	55.7	40.9	23	9.8	16.7
Oct 05	25	-2.2	9.9	27	1.8	11.8	4	42.9	43.3	23	-2.3	11.0
Nov 05	26	14.3	6.5	29	15.8	8.7	4	23.6	40.4	24	14.2	6.9
Dec 05	25	14.7	8.4	28	17.0	11.2	5	36.0	34.2	24	15.2	8.7
Jan 06	24	4.2	11.0	28	-1.7	9.8	5	69.9	43.1	21	0.9	10.1
Feb 06	25	18.3	12.2	29	17.8	10.4	4	58.8	54.3	23	22.3	12.5
Mar 06	27	24.7	15.5	31	24.5	12.9	4	45.8	58.3	26	29.3	17.1
Apr 06	26	17.1	20.0	30	18.0	20.1	5	35.6	46.0	25	15.1	22.0
May 06	27	18.3	20.0	30	16.7	19.7	5	42.4	41.0	25	19.3	21.0
Jun 06	29	32.7	22.6	33	32.2	22.2	5	28.2	35.2	28	35.4	23.1
Jul 06	29	27.0	25.9	33	29.1	26.0	5	22.3	30.5	28	32.5	29.0
Aug 06	28	8.3	21.9	32	9.7	23.0	5	5.8	18.0	27	8.5	24.5
Sep 06	25	5.3	13.1	29	5.1	14.2	3	-22.5	0.7	24	4.8	14.7
Oct 06	27	9.5	7.7	31	14.6	9.8	5	17.7	0.1	26	12.3	8.5
Nov 06	28	9.4	8.1	33	11.8	10.5	3	-19.4	-7.7	27	11.4	9.5
Dec 06	31	26.3	15.0	37	30.9	19.1	5	7.0	2.8	34	40.6	21.7
Jan 07	33	41.0	25.2	37	35.4	25.8	4	-27.8	-13.8	29	38.5	29.8
Feb 07	32	26.8	31.2	37	28.5	31.5	3	-29.3	-17.2	30	27.1	35.4
Mar 07	30	9.8	25.2	33	7.9	23.4	5	13.7	-15.6	28	8.4	23.8
Apr 07	28	8.0	14.7	34	11.9	15.9	5	4.0	-4.0	28	11.9	15.5
May 07	30	11.6	9.8	34	14.0	11.2	5	8.9	8.8	28	10.6	10.3
Jun 07	30	4.0	7.8	35	4.6	9.9	6	23.2	12.0	29	2.7	8.2
Jul 07	31	6.3	7.2	36	7.9	8.6	5	4.2	12.1	30	5.0	5.9
Aug 07	31	10.1	6.8	36	12.8	8.4	4	-1.5	8.7	30	11.8	6.4
Sep 07	34	33.1	15.8	39	35.4	18.0	4	26.8	8.2	34	40.3	18.0
Oct 07	30	11.4	17.8	36	16.0	21.0	2	-50.1	-12.8	31	21.1	23.9
Nov 07	32	16.0	19.8	39	18.0	22.7	5	57.9	1.1	31	15.9	25.3
Dec 07	28	-9.5	5.3	35	-6.6	8.3	4	-7.5	-9.0	29	-14.5	5.4
Jan 08	36	7.4	4.3	39	3.8	4.5	4	-3.0	10.8	30	1.2	-0.4
Feb 08	33	3.0	0.5	39	4.6	0.6	5	49.1	9.2	31	2.7	-4.1
Mar 08	33	9.9	6.7	35	6.0	4.7	3	-28.4	0.5	30	9.5	4.3
Apr 08	29	3.7	5.5	37	10.3	6.9	4	-22.2	-7.0	30	9.3	7.0
May 08	34	13.8	9.2	40	17.3	11.2	5	-5.5	-18.5	32	13.6	10.8
Jun 08	30	-0.4	5.7	36	3.4	10.3	5	-12.0	-13.0	30	3.7	8.8
Jul 08	31	-0.8	4.1	36	0.4	6.9	6	24.0	1.1	30	0.7	5.9

4.14c Average Weekly Earnings (nominal) - Bonus Pay
(Great Britain, seasonally adjusted)

Standard Industrial Classification (2007)

	Whole Economy			Private sector [2 3 4]			Public sector [2 3 4]			Services, SIC 2007 sections G-S		
	Weekly Earnings (£)	% changes year on year		Weekly Earnings (£)	% changes year on year		Weekly Earnings (£)	% changes year on year		Weekly Earnings (£)	% changes year on year	
		Single month	3 month average [1]		Single month	3 month average [1]		Single month	3 month average [1]		Single month	3 month average [1]
	KAF4	KAF5	KAF6	KAF7	KAF8	KAF9	KAG2	KAG3	KAG4	K5CS	K5CT	K5CU
Aug 08	30	-4.6	-2.0	35	-1.8	0.6	5	17.9	8.5	30	-0.8	1.2
Sep 08	29	-15.6	-7.3	34	-13.3	-5.2	5	16.6	19.6	29	-13.3	-4.8
Oct 08	30	-2.1	-7.7	36	-0.6	-5.4	5	86.5	32.5	32	4.8	-3.4
Nov 08	26	-20.3	-12.9	32	-18.2	-10.9	6	25.4	35.3	27	-15.5	-8.2
Dec 08	26	-7.3	-10.2	33	-4.4	-8.0	5	19.9	36.4	29	-0.7	-3.9
Jan 09	23	-36.1	-22.3	24	-37.2	-20.5	5	37.7	27.0	29	-3.2	-6.6
Feb 09	17	-50.0	-32.4	21	-47.4	-30.6	6	3.4	18.9	19	-37.2	-13.9
Mar 09	26	-21.5	-36.0	26	-27.3	-37.6	6	81.0	36.8	18	-39.2	-26.7
Apr 09	27	-6.8	-26.9	37	-1.9	-25.9	6	48.8	40.3	24	-20.5	-32.3
May 09	26	-24.8	-18.2	29	-26.3	-18.5	4	-11.7	33.8	24	-24.9	-28.1
Jun 09	26	-15.2	-16.1	32	-11.6	-13.5	4	-23.8	0.5	22	-26.8	-24.0
Jul 09	24	-21.4	-20.7	30	-17.6	-18.7	4	-38.0	-25.4	24	-20.3	-24.0
Aug 09	24	-19.7	-18.8	30	-14.2	-14.4	3	-41.1	-34.6	24	-18.4	-21.8
Sep 09	25	-13.9	-18.4	30	-11.0	-14.3	6	14.8	-22.7	25	-14.9	-17.9
Oct 09	24	-17.8	-17.2	30	-15.7	-13.7	5	0.2	-9.5	25	-22.3	-18.7
Nov 09	25	-2.4	-11.7	32	2.5	-8.4	5	-20.7	-3.0	27	1.5	-12.7
Dec 09	25	-6.6	-9.3	33	-0.3	-4.9	4	-33.3	-18.8	29	-1.3	-8.2
Jan 10	21	-6.5	-5.1	23	-7.7	-1.3	6	16.6	-13.3	24	-17.5	-6.0
Feb 10	24	45.9	6.7	32	56.9	12.4	5	10.0	-3.0	29	49.2	5.2
Mar 10	25	-2.9	8.3	35	34.8	26.6	4	-30.6	-3.7	26	41.2	18.0
Apr 10	24	-12.2	5.1	29	-20.2	16.0	5	-6.7	-10.8	20	-18.3	20.4
May 10	26	-0.1	-5.2	33	13.7	6.1	6	29.9	-6.3	25	2.5	5.6
Jun 10	24	-5.0	-5.9	28	-13.2	-7.8	5	36.5	16.8	25	15.3	-0.8
Jul 10	23	-4.5	-3.2	28	-7.0	-2.5	5	27.0	31.1	23	-5.2	3.8
Aug 10	23	-1.8	-3.8	28	-8.2	-9.5	4	41.7	34.7	23	-5.8	1.0
Sep 10	25	1.8	-1.5	31	3.5	-3.9	4	-35.6	1.9	25	2.3	-2.8
Oct 10	25	2.9	1.0	31	1.2	-1.2	6	26.6	3.6	25	0.6	-0.9
Nov 10	26	1.7	2.1	31	-4.7	-0.1	6	31.7	4.4	26	-3.0	-0.1
Dec 10	26	6.7	3.7	31	-6.3	-3.4	5	37.0	31.3	26	-11.1	-4.7
Jan 11	27	24.8	10.4	34	50.9	8.9	5	-20.4	11.4	33	38.3	6.5
Feb 11	28	14.4	14.9	33	1.2	11.1	6	7.1	3.4	28	-3.3	6.2
Mar 11	27	8.0	15.3	35	1.2	13.7	6	47.7	8.2	28	8.8	13.3
Apr 11	25	4.0	8.8	31	6.5	2.8	5	-3.7	15.2	22	10.7	4.7
May 11	26	-0.3	3.9	33	-0.7	2.2	6	-8.1	9.5	25	2.7	7.2
Jun 11	30	22.9	8.8	35	26.4	10.0	6	21.8	3.1	31	22.8	12.2
Jul 11	31	37.1	19.2	35	26.1	16.1	17	272.7	86.8	32	42.6	22.1
Aug 11	27	15.5	25.1	33	18.0	23.5	5	5.9	98.6	28	20.9	28.5
Sep 11	26	1.7	17.6	31	0.2	14.2	5	26.5	109.0	26	2.2	21.1
Oct 11	27	6.3	7.7	33	6.2	7.8	4	-32.9	-4.9	28	9.1	10.4
Nov 11	27	4.4	4.2	32	4.9	3.8	5	-11.3	-10.5	27	4.1	5.1
Dec 11	26	-0.6	3.3	31	1.0	4.0	5	-1.6	-16.0	26	3.0	5.4
Jan 12	25	-6.7	-1.0	31	-10.1	-1.7	7	46.5	8.9	27	-17.0	-4.4
Feb 12	26	-7.4	-4.9	31	-5.4	-5.0	4	-31.3	1.9	26	-6.9	-7.8

4.14c Average Weekly Earnings (nominal) - Bonus Pay
(Great Britain, seasonally adjusted)

Standard Industrial Classification (2007)

	Whole Economy			Private sector [2 3 4]			Public sector [2 3 4]			Services, SIC 2007 sections G-S		
	Weekly Earnings (£)	% changes year on year		Weekly Earnings (£)	% changes year on year		Weekly Earnings (£)	% changes year on year		Weekly Earnings (£)	% changes year on year	
		Single month	3 month average [1]		Single month	3 month average [1]		Single month	3 month average [1]		Single month	3 month average [1]
	KAF4	KAF5	KAF6	KAF7	KAF8	KAF9	KAG2	KAG3	KAG4	K5CS	K5CT	K5CU
Mar 12	26	-4.0	-6.1	33	-6.4	-7.3	5	-21.2	-5.9	28	-2.7	-9.3
Apr 12	27	7.3	-1.6	33	6.3	-2.1	4	-13.8	-22.4	26	17.5	1.4
May 12	26	0.4	1.1	32	-3.4	-1.4	6	9.2	-9.6	25	-0.5	3.9
Jun 12	27	-8.2	-0.7	33	-6.4	-1.4	6	-9.3	-5.0	28	-10.4	0.6
Jul 12	29	-8.0	-5.6	34	-0.9	-3.6	4	-74.2	-45.1	30	-7.5	-6.5
Aug 12	28	4.8	-4.2	33	-0.1	-2.5	8	72.7	-35.4	29	5.9	-4.4
Sep 12	27	6.3	0.5	32	3.9	0.9	6	17.9	-32.3	28	9.8	2.0
Oct 12	27	-0.4	3.5	31	-3.7	0.0	9	120.7	67.6	27	-2.1	4.4
Nov 12	27	0.4	2.0	32	-1.0	-0.3	6	10.4	43.8	27	1.0	2.8
Dec 12	26	-1.2	-0.4	30	-2.9	-2.5	8	57.1	56.9	26	0.3	-0.3
Jan 13	27	6.3	1.8	32	4.3	0.1	5	-21.0	11.0	28	1.3	0.9
Feb 13	26	3.2	2.7	32	3.6	1.6	3	-29.3	1.0	27	4.0	1.9
Mar 13	24	-7.2	0.7	30	-9.0	-0.6	5	-2.5	-17.1	26	-6.3	-0.4
Apr 13	41	51.5	16.4	50	50.3	15.3	5	3.9	-8.3	41	61.5	19.1
May 13	30	18.9	21.5	37	17.3	19.6	5	-16.4	-5.8	30	20.8	24.6
Jun 13	29	4.2	24.8	34	5.2	24.4	6	-1.7	-5.4	29	4.3	28.4
Jul 13	28	-4.9	5.6	33	-3.7	6.0	3	-30.7	-15.1	29	-4.0	6.3
Aug 13	28	-0.9	-0.6	34	3.4	1.5	3	-57.6	-33.4	29	-2.3	-0.8
Sep 13	28	4.6	-0.5	34	5.5	1.6	5	-13.0	-37.3	30	4.0	-0.9
Oct 13	27	0.8	1.5	32	2.0	3.6	3	-65.4	-49.6	27	1.5	1.0
Nov 13	27	-0.4	1.7	32	-0.2	2.5	3	-44.8	-45.0	28	1.0	2.2
Dec 13	27	6.1	2.1	33	6.9	2.8	2	-69.7	-61.3	28	4.6	2.3
Jan 14	28	4.4	3.3	33	3.4	3.3	5	-14.7	-46.2	28	2.0	2.5
Feb 14	30	11.7	7.4	35	10.0	6.7	4	58.7	-28.1	30	12.7	6.4
Mar 14	26	7.8	8.0	32	6.1	6.5	5	10.3	10.3	28	6.8	7.1
Apr 14	29	-27.9	-7.0	35	-30.1	-8.9	6	32.2	29.4	30	-27.8	-6.8
May 14	29	-5.1	-11.6	35	-7.1	-13.5	2	-47.6	-1.8	29	-3.8	-11.2
Jun 14	29	-0.2	-13.0	38	9.6	-11.8	2	-56.0	-26.6	28	-3.1	-13.5
Jul 14	26	-4.9	-3.4	31	-7.3	-1.7	5	46.8	-29.3	26	-9.2	-5.3
Aug 14	26	-7.9	-4.3	31	-8.1	-1.9	1	-61.8	-31.4	26	-8.9	-7.1
Sep 14	27	-5.1	-6.0	32	-5.8	-7.1	3	-30.5	-18.8	27	-8.9	-9.0
Oct 14	28	3.2	-3.4	33	2.7	-3.9	3	-7.8	-34.0	29	4.3	-4.7
Nov 14	29	7.1	1.6	34	6.8	1.0	3	-24.4	-22.6	29	5.5	0.1
Dec 14	32	15.6	8.7	37	14.3	7.9	3	49.2	0.9	32	14.8	8.2
Jan 15	28	-0.4	7.4	33	-0.3	6.9	2	-61.6	-24.7	28	-0.1	6.7
Feb 15	28	-4.9	3.2	33	-4.7	2.9	3	-37.0	-29.6	29	-4.3	3.2
Mar 15	29	12.7	2.1	36	12.6	2.3	2	-56.3	-52.1	31	11.8	2.2
Apr 15	29	-0.6	1.9	35	1.0	2.7	2	-60.7	-52.6	30	0.0	2.2
May 15	29	1.5	4.2	35	1.4	4.8	2	-3.4	-48.9	30	1.8	4.4
Jun 15	27	-5.9	-1.7	35	-7.7	-1.9	2	-16.3	-37.7	25	-9.5	-2.4
Jul 15	31	17.6	4.0	37	20.4	3.7	3	-41.3	-25.0	32	22.6	4.5
Aug 15	31	20.9	10.3	37	20.4	9.8	1	8.3	-26.1	32	23.3	11.6
Sep 15	30	9.4	15.9	35	7.8	16.1	6	72.7	7.2	31	13.7	19.8

4.14c Average Weekly Earnings (nominal) - Bonus Pay
(Great Britain, seasonally adjusted)

Standard Industrial Classification (2007)

	Whole Economy			Private sector [2 3 4]			Public sector [2 3 4]			Services, SIC 2007 sections G-S		
	Weekly Earnings (£)	% changes year on year		Weekly Earnings (£)	% changes year on year		Weekly Earnings (£)	% changes year on year		Weekly Earnings (£)	% changes year on year	
		Single month	3 month average [1]		Single month	3 month average [1]		Single month	3 month average [1]		Single month	3 month average [1]
	KAF4	KAF5	KAF6	KAF7	KAF8	KAF9	KAG2	KAG3	KAG4	K5CS	K5CT	K5CU
Oct 15	29	6.7	12.2	35	6.6	11.5	2	-18.1	27.5	31	7.8	14.7
Nov 15	28	-1.7	4.7	34	-1.7	4.1	2	-2.7	21.6	29	-0.1	7.0
Dec 15	31	-3.2	0.4	36	-2.3	0.7	1	-60.0	-30.1	31	-0.6	2.2
Jan 16	31	10.7	1.7	37	11.3	2.2	5	164.7	10.4	31	11.1	3.3
Feb 16	27	-5.8	0.4	32	-5.5	1.0	3	-0.2	10.8	27	-7.1	1.0
Mar 16	30	2.2	2.4	36	1.0	2.2	2	-15.9	36.4	31	0.0	1.2
Apr 16	32	8.2	1.7	38	8.4	1.4	2	4.4	-3.8	33	8.8	0.6
May 16	32	8.2	6.2	38	7.7	5.7	3	26.3	4.9	32	7.6	5.4
Jun 16	28	5.0	7.2	36	2.5	6.2	3	20.0	16.8	27	6.5	7.7
Jul 16	34	8.8	7.4	41	9.9	6.8	2	-24.1	5.5	35	9.9	8.1
Aug 16	30	-3.9	3.2	36	-5.1	2.4	3	82.7	15.3	31	-4.6	3.7
Sep 16	29	-1.1	1.3	35	0.3	1.7	3	-54.5	-26.6	30	-3.7	0.6
Oct 16	31	5.6	0.1	37	5.9	0.3	2	-15.2	-24.6	31	1.0	-2.5
Nov 16	31	11.0	5.1	37	11.2	5.8	3	3.5	-32.5	32	9.6	2.2
Dec 16	29	-4.9	3.7	35	-4.7	4.0	3	95.4	17.1	29	-6.8	1.1
Jan 17	30	-1.7	1.2	36	-1.7	1.4	2	-54.6	-13.7	31	-0.9	0.4
Feb 17	31	15.1	2.2	37	15.6	2.5	2	-28.2	-23.0	31	15.2	1.8
Mar 17	31	4.8	5.6	37	3.5	5.3	2	16.8	-31.7	32	5.5	6.2
Apr 17	31	-3.0	5.1	37	-3.3	4.7	2	-3.8	-7.4	31	-3.9	5.0
May 17	31	-2.4	-0.3	37	-2.2	-0.7	3	-14.9	-2.7	32	0.3	0.6
Jun 17	34	19.5	4.1	43	19.3	4.3	2	-12.7	-10.8	34	25.3	6.2
Jul 17	31	-7.3	2.5	37	-7.9	2.5	2	6.3	-8.4	32	-9.3	3.9
Aug 17	32	5.6	5.2	38	6.3	5.3	2	-9.0	-5.9	33	6.3	5.9
Sep 17	35	21.3	5.9	42	20.9	5.7	2	-13.0	-6.2	36	23.1	5.8
Oct 17	31	-0.8	8.5	37	-1.1	8.5	2	23.5	-1.7	32	2.2	10.3
Nov 17	33	4.4	8.0	39	3.6	7.5	4	62.4	23.8	33	4.4	9.6
Dec 17	32	10.6	4.6	38	11.1	4.4	2	-17.5	22.0	34	14.3	6.8
Jan 18	31	2.2	5.6	37	2.5	5.7	2	-14.6	10.9	32	1.4	6.5

Source: Monthly Wages & Salaries Survey
Inquiries: Email: labour.market@ons.gov.uk
Earnings enquiries: 01633 456773

1. The three month average figures are the changes in the average seasonally adjusted values for the three months ending with the relevant month compared with the same period a year earlier.

2. From July 2009 Royal Bank of Scotland Group plc is classified to the public sector; for earlier time periods it is classified to the private sector. Between July 2009 and March 2014 Lloyds Banking Group plc is classified to the public sector; it is classified to the private sector for earlier and later time periods.

3. Between June 2010 and May 2012 English Further Education Corporations and Sixth Form College Corporations are classified to the public sector. Before June 2010 and after May 2012 they are classified to the private sector.

4. From October 2013 Royal Mail plc is classified to the private sector; previously it is in the public sector.

4.14c Average Weekly Earnings (nominal) - Bonus Pay
(Great Britain, seasonally adjusted)

Standard Industrial Classification (2007)

	Finance and business services, SIC 2007 sections K-N			Public sector excluding financial services [4][5]			Manufacturing, SIC 2007 section C			Construction, SIC 2007 section F			Wholesaling, retailing, hotels & restaurants, SIC 2007 sections G & I		
	Weekly Earnings (£)	% changes year on year		Weekly Earnings (£)	% changes year on year		Weekly Earnings (£)	% changes year on year		Weekly Earnings (£)	% changes year on year		Weekly Earnings (£)	% changes year on year	
		Single month	3 month average [1]		Single month	3 month average [1]		Single month	3 month average [1]		Single month	3 month average [1]		Single month	3 month average [1]
	K5CV	K5CW	K5CX	KAH3	KAH4	KAH5	K5D3	K5D4	K5D5	K5D6	K5D7	K5D8	K5D9	K5DA	K5DB
Jan 05	53	11.0	30.3	2	-5.5	9.4	18	7.8	11.1	23	14.3	11.3	15	13.2	1.6
Feb 05	50	32.9	33.7	2	2.4	1.3	17	19.9	23.6	22	14.0	8.8	15	4.1	-0.2
Mar 05	51	4.4	14.7	2	-6.5	-3.4	17	5.7	10.7	25	10.7	12.9	14	3.4	6.8
Apr 05	53	0.1	10.4	2	10.1	1.8	18	16.1	13.7	23	14.5	13.0	14	-8.6	-0.7
May 05	50	-6.6	-0.9	2	12.5	5.0	16	-11.7	2.5	26	4.6	9.6	15	7.8	0.4
Jun 05	58	27.5	6.0	2	3.1	8.4	16	-0.5	0.5	22	16.8	11.4	14	-2.4	-1.5
Jul 05	53	7.8	8.7	2	30.6	14.8	18	6.8	-2.2	23	10.0	10.0	14	-4.4	0.2
Aug 05	65	77.4	34.1	3	25.8	19.3	18	1.5	2.6	19	11.4	12.7	14	-5.5	-4.1
Sep 05	62	37.0	37.4	3	48.5	34.6	16	-16.8	-3.4	20	41.9	19.1	15	-9.6	-6.6
Oct 05	58	-8.2	27.6	3	51.0	41.3	17	-4.6	-6.9	23	33.6	28.2	15	2.1	-4.6
Nov 05	66	31.9	17.4	2	18.6	39.0	19	13.7	-3.3	25	-1.2	20.3	14	-0.1	-2.8
Dec 05	60	16.8	11.8	3	29.7	32.7	19	-11.3	-1.6	24	20.3	15.4	16	4.7	2.3
Jan 06	56	4.6	17.5	3	71.5	39.3	19	6.9	2.0	25	7.9	8.1	14	-6.1	-0.5
Feb 06	64	28.0	16.2	3	73.6	57.3	21	22.7	4.7	26	17.3	14.9	13	-15.7	-5.7
Mar 06	69	35.8	22.5	3	51.3	65.5	19	13.0	14.1	23	-8.2	5.2	15	7.0	-5.3
Apr 06	68	28.8	30.9	3	35.6	53.0	20	10.8	15.4	23	0.8	2.9	15	1.8	-2.6
May 06	69	39.0	34.4	3	34.9	40.3	19	14.5	12.7	24	-5.0	-4.3	14	-7.3	0.4
Jun 06	94	60.5	43.5	3	25.5	31.9	19	20.1	15.0	26	16.4	3.6	15	5.1	-0.3
Jul 06	80	51.7	51.0	3	17.0	25.5	18	2.8	12.1	27	16.6	8.7	16	18.8	5.2
Aug 06	75	15.5	41.3	3	6.4	15.8	18	-1.9	6.4	27	37.7	22.8	14	3.8	9.2
Sep 06	59	-4.8	19.1	2	-21.6	-0.2	20	25.0	8.0	20	-1.2	17.3	14	-1.4	6.9
Oct 06	72	23.3	11.2	3	17.7	1.1	24	36.7	19.5	24	2.0	11.9	15	0.4	0.9
Nov 06	79	20.6	12.9	2	-22.0	-7.5	19	1.0	20.2	22	-12.0	-4.0	15	1.5	0.2
Dec 06	105	73.7	38.8	3	4.8	1.6	19	2.7	12.8	24	1.4	-3.0	15	-4.3	-0.9
Jan 07	80	43.5	45.2	3	-22.1	-13.3	21	9.7	4.5	24	-2.0	-4.3	16	9.2	1.9
Feb 07	81	27.9	48.1	2	-24.7	-15.1	22	4.5	5.6	44	68.0	23.7	16	27.0	9.5
Mar 07	73	4.8	24.0	3	-0.1	-16.3	20	7.9	7.3	25	9.8	26.7	19	28.7	21.5
Apr 07	76	10.9	14.2	3	-3.7	-10.3	20	1.4	4.6	28	20.0	34.3	18	23.5	26.4
May 07	72	3.9	6.5	3	6.3	0.9	22	17.6	8.8	30	25.7	18.7	19	37.0	29.6
Jun 07	73	-22.5	-4.7	3	20.4	7.6	23	21.3	13.2	35	34.3	27.0	20	37.5	32.6
Jul 07	76	-4.9	-9.2	3	1.5	9.3	23	24.7	21.2	28	5.3	21.5	20	20.6	31.2
Aug 07	79	4.4	-8.7	3	2.2	8.0	22	23.4	23.1	27	0.8	13.3	17	20.6	26.0
Sep 07	89	49.9	13.5	3	25.3	8.4	23	10.9	19.3	31	56.5	17.6	20	41.0	27.1
Oct 07	73	1.5	16.4	2	-50.5	-13.4	22	-5.6	8.2	27	13.0	20.8	21	42.4	34.7
Nov 07	79	-0.9	14.2	3	54.2	-2.3	24	24.2	8.8	31	44.4	36.6	19	30.5	38.0
Dec 07	74	-29.4	-11.9	3	-9.1	-12.1	24	23.3	12.6	33	38.0	31.4	19	23.1	31.9
Jan 08	79	-1.0	-12.3	2	-3.8	8.7	21	2.9	16.4	31	28.6	36.8	17	7.4	20.1
Feb 08	85	4.4	-10.6	2	-13.4	-8.7	22	3.8	9.6	30	-31.8	2.1	18	9.2	13.1
Mar 08	81	11.6	4.8	2	-40.7	-20.2	19	-6.1	0.3	27	8.6	-5.4	22	13.8	10.4
Apr 08	77	2.2	5.9	3	-0.3	-18.8	23	14.0	3.9	25	-10.8	-15.4	18	0.4	7.9
May 08	87	21.1	11.5	3	-8.1	-16.6	34	54.2	21.6	27	-11.1	-5.1	20	4.1	6.2
Jun 08	89	22.1	14.9	3	-13.5	-7.8	24	1.9	23.2	23	-34.7	-19.8	18	-10.2	-2.1
Jul 08	78	2.1	14.9	3	21.3	-0.9	27	16.3	23.6	23	-18.1	-22.0	18	-8.5	-5.1

4.14c Average Weekly Earnings (nominal) - Bonus Pay
(Great Britain, seasonally adjusted)

Standard Industrial Classification (2007)

	Finance and business services, SIC 2007 sections K-N			Public sector excluding financial services [4][5]			Manufacturing, SIC 2007 section C			Construction, SIC 2007 section F			Wholesaling, retailing, hotels & restaurants, SIC 2007 sections G & I		
	Weekly Earnings (£)	% changes year on year		Weekly Earnings (£)	% changes year on year		Weekly Earnings (£)	% changes year on year		Weekly Earnings (£)	% changes year on year		Weekly Earnings (£)	% changes year on year	
		Single month [1]	3 month average [1]		Single month [1]	3 month average [1]		Single month [1]	3 month average [1]		Single month [1]	3 month average [1]		Single month [1]	3 month average [1]
	K5CV	K5CW	K5CX	KAH3	KAH4	KAH5	K5D3	K5D4	K5D5	K5D6	K5D7	K5D8	K5D9	K5DA	K5DB
Aug 08	72	-8.4	4.9	3	19.9	8.1	21	-3.6	5.0	25	-6.3	-21.0	17	-2.6	-7.3
Sep 08	80	-9.3	-5.4	3	16.4	19.2	21	-5.2	2.7	27	-14.5	-13.1	16	-21.3	-11.2
Oct 08	87	19.1	-0.4	3	80.5	32.7	24	5.3	-1.1	21	-24.1	-15.0	15	-26.4	-17.5
Nov 08	63	-19.4	-4.0	4	26.9	35.5	19	-20.4	-7.1	21	-34.2	-24.3	18	-6.5	-18.4
Dec 08	80	7.8	2.0	3	15.7	35.4	21	-11.9	-9.3	20	-40.0	-33.3	16	-12.3	-15.4
Jan 09	76	-3.5	-5.3	3	-1.9	14.2	21	-3.8	-12.3	18	-43.1	-39.1	18	5.5	-4.8
Feb 09	43	-49.5	-16.4	3	60.5	23.1	18	-18.5	-11.5	20	-34.3	-39.2	17	-2.5	-3.4
Mar 09	46	-43.1	-32.5	3	91.9	44.7	21	10.4	-4.7	23	-13.9	-31.1	16	-25.0	-8.8
Apr 09	85	9.1	-28.7	4	49.9	64.3	24	5.1	-1.5	24	-4.5	-18.5	17	-6.7	-12.3
May 09	49	-43.2	-26.7	2	-11.7	36.0	20	-41.2	-14.1	18	-33.8	-17.8	17	-14.2	-15.9
Jun 09	52	-41.0	-26.4	2	-24.4	3.9	18	-22.2	-22.3	19	-14.8	-18.3	16	-9.4	-10.2
Jul 09	61	-20.7	-35.5	2	-45.7	-28.6	18	-33.9	-33.5	19	-18.3	-23.0	16	-11.9	-11.9
Aug 09	67	-7.7	-24.3	2	-54.9	-42.7	18	-13.7	-24.1	19	-24.2	-19.3	18	4.1	-5.9
Sep 09	59	-26.8	-18.7	3	-11.1	-37.9	19	-9.7	-20.3	17	-36.1	-26.6	17	5.1	-1.2
Oct 09	63	-27.6	-21.3	2	-39.9	-35.8	18	-25.8	-16.7	18	-11.2	-24.9	17	9.8	6.3
Nov 09	68	6.8	-17.8	2	-53.5	-35.7	20	7.1	-10.7	21	-0.8	-17.8	17	-5.3	2.8
Dec 09	83	4.3	-7.0	2	-43.5	-46.1	20	-3.7	-8.7	20	-1.4	-4.5	17	0.7	1.4
Jan 10	61	-19.8	-3.3	2	-10.6	-38.7	20	-3.0	-0.1	21	18.7	4.9	18	-1.1	-2.0
Feb 10	73	70.4	9.3	2	-54.1	-38.7	22	18.2	3.2	20	-0.3	5.2	19	7.1	2.2
Mar 10	64	38.8	20.0	4	15.7	-17.7	22	4.2	6.0	24	5.7	7.5	23	41.8	15.3
Apr 10	62	-27.0	14.6	1	-70.9	-39.1	19	-22.9	-2.2	15	-34.9	-10.5	19	9.4	19.1
May 10	72	46.7	10.1	2	-25.9	-30.5	20	-1.0	-7.5	17	-6.4	-12.5	16	-4.1	15.5
Jun 10	66	25.1	7.2	2	-13.2	-43.8	15	-19.4	-14.9	15	-24.1	-23.1	16	-1.3	1.4
Jul 10	56	-8.4	19.0	1	-22.3	-20.6	22	23.0	0.5	16	-14.9	-15.4	15	2.8	-0.9
Aug 10	65	-2.3	3.6	2	1.7	-12.2	22	18.9	7.2	16	-13.2	-17.5	15	-12.3	-3.9
Sep 10	72	22.3	3.4	1	-49.3	-28.5	22	12.7	18.0	16	-5.4	-11.4	16	-3.1	-4.5
Oct 10	71	12.8	10.4	2	-2.7	-22.7	25	43.3	24.4	12	-33.3	-17.5	16	-3.3	-6.3
Nov 10	72	5.6	13.2	2	-4.4	-23.8	20	-2.3	16.7	16	-22.2	-20.7	18	6.6	0.1
Dec 10	68	-18.5	-1.7	1	-10.7	-5.9	22	10.2	15.8	17	-10.6	-21.8	19	13.0	5.4
Jan 11	98	59.4	11.7	1	-39.8	-20.2	23	17.1	8.2	19	-10.5	-14.4	17	-4.7	4.8
Feb 11	74	1.0	10.0	2	20.4	-13.1	21	-2.8	7.9	25	26.4	1.6	18	-3.8	1.2
Mar 11	76	18.5	24.7	2	-57.5	-36.0	22	0.7	4.7	21	-14.2	-0.7	19	-19.7	-10.3
Apr 11	68	9.5	9.3	2	51.2	-19.1	22	15.1	3.8	17	11.7	5.9	17	-9.8	-11.8
May 11	68	-6.0	6.8	1	-20.4	-28.8	20	2.3	5.7	16	-4.9	-4.4	17	6.8	-9.2
Jun 11	86	30.4	10.8	1	-42.0	-11.3	19	31.0	14.8	20	35.8	13.3	23	40.6	11.4
Jul 11	85	50.8	22.8	3	129.1	14.1	22	0.4	9.1	14	-13.1	5.0	20	22.4	23.3
Aug 11	69	6.7	28.3	2	9.9	25.7	19	-12.7	3.3	16	-2.1	6.0	18	13.4	25.6
Sep 11	70	-2.9	15.9	2	12.9	50.1	24	11.4	-0.3	17	4.7	-3.5	18	10.3	15.4
Oct 11	71	0.8	1.3	1	-28.5	-3.6	20	-18.8	-7.3	18	43.9	12.9	18	11.6	11.7
Nov 11	75	4.4	0.7	2	4.2	-5.4	21	6.3	-1.5	22	38.8	27.8	17	-4.3	5.5
Dec 11	69	2.2	2.5	1	-6.4	-10.9	21	-5.7	-7.1	19	6.3	27.7	18	-1.5	1.6
Jan 12	71	-27.5	-9.4	2	41.7	11.6	21	-8.7	-3.1	17	-6.9	11.5	19	13.9	2.4
Feb 12	65	-11.9	-14.2	2	-19.1	1.6	24	14.5	-0.4	18	-27.6	-11.6	19	5.5	5.7

4.14c Average Weekly Earnings (nominal) - Bonus Pay
(Great Britain, seasonally adjusted)

Standard Industrial Classification (2007)

	Finance and business services, SIC 2007 sections K-N			Public sector excluding financial services [4][5]			Manufacturing, SIC 2007 section C			Construction, SIC 2007 section F			Wholesaling, retailing, hotels & restaurants, SIC 2007 sections G & I		
	Weekly Earnings (£)	% changes year on year		Weekly Earnings (£)	% changes year on year		Weekly Earnings (£)	% changes year on year		Weekly Earnings (£)	% changes year on year		Weekly Earnings (£)	% changes year on year	
		Single month	3 month average [1]		Single month	3 month average [1]		Single month	3 month average [1]		Single month	3 month average [1]		Single month	3 month average [1]
	K5CV	K5CW	K5CX	KAH3	KAH4	KAH5	K5D3	K5D4	K5D5	K5D6	K5D7	K5D8	K5D9	K5DA	K5DB
Mar 12	73	-4.6	-15.8	1	-13.0	-0.7	19	-14.9	-3.5	20	-5.1	-14.3	18	-4.0	4.8
Apr 12	75	11.6	-2.1	1	-30.2	-20.9	22	2.6	0.4	20	13.0	-9.0	17	2.7	1.3
May 12	56	-17.8	-3.7	2	9.0	-12.7	25	21.4	2.5	19	16.0	6.9	19	7.8	2.0
Jun 12	66	-22.8	-10.8	2	47.7	2.5	21	8.2	10.6	24	21.8	17.2	20	-10.6	-1.0
Jul 12	75	-11.1	-17.2	2	-48.1	-16.3	21	-4.9	7.9	17	22.9	20.3	18	-11.0	-5.4
Aug 12	74	7.3	-10.0	2	27.2	-10.2	24	28.0	9.7	16	-2.2	14.4	23	28.4	0.7
Sep 12	67	-3.9	-3.2	2	52.1	-4.7	21	-14.5	1.1	15	-9.7	2.5	20	10.8	8.5
Oct 12	66	-7.3	-1.4	3	180.8	78.5	22	6.9	5.1	20	11.7	0.1	20	10.5	16.5
Nov 12	66	-11.4	-7.6	2	19.0	75.9	23	7.8	-0.7	18	-20.4	-7.3	19	11.4	10.9
Dec 12	69	-0.8	-6.6	2	72.7	83.6	20	-5.7	3.0	17	-7.1	-6.5	19	1.0	7.6
Jan 13	72	2.1	-3.5	1	-21.5	19.4	19	-10.5	-2.8	16	-8.0	-12.4	20	2.7	4.9
Feb 13	68	4.3	1.8	1	-25.6	5.0	22	-10.0	-8.8	17	-6.5	-7.2	20	5.8	3.2
Mar 13	66	-8.5	-0.9	2	14.9	-12.2	20	7.4	-5.0	12	-41.2	-19.5	18	0.2	2.9
Apr 13	127	68.5	22.7	2	35.1	5.8	27	22.4	6.1	34	73.4	8.6	25	43.2	15.9
May 13	79	42.2	33.9	2	16.2	21.3	22	-11.2	5.5	19	2.5	11.3	23	21.8	21.5
Jun 13	74	12.5	42.3	2	16.4	21.6	23	7.2	5.5	23	-6.0	21.3	21	1.6	21.1
Jul 13	67	-11.4	11.7	2	4.0	12.0	21	0.7	-1.7	17	-3.2	-2.6	23	29.7	17.0
Aug 13	70	-5.9	-2.2	1	-35.0	-7.8	24	0.3	2.6	20	29.4	4.6	28	24.7	18.4
Sep 13	73	9.1	-3.2	2	-11.1	-15.2	21	1.7	0.9	20	29.9	18.0	22	8.4	20.8
Oct 13	69	4.8	2.4	1	-74.2	-45.0	22	2.6	1.5	16	-20.3	10.3	21	6.4	13.6
Nov 13	69	4.9	6.3	1	-56.0	-50.5	23	-1.4	0.9	17	-3.2	0.1	21	9.6	8.1
Dec 13	70	2.3	4.0	1	-69.6	-68.3	22	11.8	4.0	18	1.7	-7.8	21	11.1	9.0
Jan 14	71	-1.2	1.9	1	-14.3	-51.4	25	32.3	13.3	20	26.5	7.8	21	7.2	9.3
Feb 14	76	11.3	4.0	1	2.3	-37.4	22	3.7	15.4	19	11.3	12.8	16	-20.9	-1.2
Mar 14	68	1.7	3.8	1	-16.9	-10.7	22	10.5	14.9	17	45.4	25.7	20	10.7	-1.5
Apr 14	69	-45.7	-18.9	1	-10.7	-9.6	23	-16.1	-2.1	22	-36.3	-8.1	24	-4.6	-5.4
May 14	72	-8.8	-23.4	1	-32.9	-20.6	22	2.8	-2.4	17	-9.9	-13.7	21	-7.8	-1.5
Jun 14	69	-7.2	-25.1	2	-15.8	-20.1	24	7.5	-2.9	25	8.7	-16.0	22	5.7	-2.5
Jul 14	61	-8.5	-8.2	2	31.7	-6.0	23	10.1	6.8	28	71.4	20.4	20	-14.2	-5.8
Aug 14	68	-2.5	-6.1	1	-36.7	-4.8	21	-12.4	1.1	26	25.8	31.9	21	-25.9	-13.0
Sep 14	70	-4.4	-5.1	1	-32.2	-12.0	24	16.0	3.7	24	20.4	37.2	21	-3.0	-15.4
Oct 14	69	-0.7	-2.6	2	73.4	-12.0	21	-6.3	-1.6	17	6.4	18.5	21	-0.8	-11.4
Nov 14	72	3.5	-0.6	1	63.8	13.6	24	6.3	5.1	23	33.1	20.4	22	4.0	0.0
Dec 14	85	20.1	7.7	2	153.8	93.9	27	23.1	7.6	24	37.4	26.3	21	3.9	2.4
Jan 15	65	-8.4	5.0	1	-42.3	41.4	24	-6.6	6.9	20	-0.5	22.1	23	8.7	5.5
Feb 15	70	-7.2	1.2	4	212.5	99.1	20	-9.5	1.9	16	-12.9	7.1	19	22.4	10.7
Mar 15	71	4.9	-3.8	1	-2.7	51.9	25	10.7	-2.0	20	16.7	0.6	25	28.7	19.5
Apr 15	74	7.2	1.3	1	-76.2	33.4	24	5.7	2.4	23	9.0	4.1	23	-2.5	14.6
May 15	69	-5.3	2.1	0	4.1	-27.7	20	-9.6	2.3	39	125.7	47.2	23	8.2	10.6
Jun 15	54	-21.9	-6.6	1	-27.2	-35.3	27	10.2	2.3	23	-9.0	33.4	23	7.0	4.0
Jul 15	74	20.2	-3.2	1	-41.0	-26.1	23	1.1	0.8	30	6.0	29.8	24	20.1	11.6
Aug 15	84	22.7	6.4	1	4.5	-28.1	23	7.0	6.1	23	-9.4	-3.7	23	8.8	11.8
Sep 15	69	-1.9	13.3	3	79.1	4.7	21	-12.9	-2.0	22	-8.9	-3.6	23	10.7	13.1

4.14c Average Weekly Earnings (nominal) - Bonus Pay
(Great Britain, seasonally adjusted)

Standard Industrial Classification (2007)

	Finance and business services, SIC 2007 sections K-N			Public sector excluding financial services [4][5]			Manufacturing, SIC 2007 section C			Construction, SIC 2007 section F			Wholesaling, retailing, hotels & restaurants, SIC 2007 sections G & I		
	Weekly Earnings (£)	% changes year on year		Weekly Earnings (£)	% changes year on year		Weekly Earnings (£)	% changes year on year		Weekly Earnings (£)	% changes year on year		Weekly Earnings (£)	% changes year on year	
		Single month	3 month average [1]		Single month	3 month average [1]		Single month	3 month average [1]		Single month	3 month average [1]		Single month	3 month average [1]
	K5CV	K5CW	K5CX	KAH3	KAH4	KAH5	K5D3	K5D4	K5D5	K5D6	K5D7	K5D8	K5D9	K5DA	K5DB
Oct 15	72	4.9	8.4	1	-11.5	25.7	24	12.6	1.5	18	10.4	-4.2	21	0.9	6.8
Nov 15	68	-5.6	-1.0	1	-1.3	21.3	21	-11.0	-4.6	21	-7.7	-3.4	22	1.7	4.4
Dec 15	77	-8.7	-3.6	1	-56.7	-26.1	24	-10.9	-4.1	24	0.6	0.2	24	11.1	4.6
Jan 16	75	14.4	-0.9	2	193.9	7.2	22	-5.1	-9.1	28	37.3	8.8	22	-2.7	3.2
Feb 16	62	-10.8	-2.5	2	-55.3	-27.5	22	7.9	-3.6	25	53.7	27.1	21	10.1	5.8
Mar 16	74	5.0	2.6	2	41.4	-2.1	25	0.1	0.6	23	15.0	34.2	22	-14.9	-3.6
Apr 16	80	7.6	0.7	1	315.9	-7.2	23	-4.5	0.8	32	34.2	33.1	22	-6.2	-4.8
May 16	78	13.4	8.6	2	44.7	76.1	26	28.8	6.9	50	29.4	27.3	22	-2.4	-8.1
Jun 16	68	27.0	14.9	2	69.3	89.6	26	-4.2	5.1	26	16.6	27.3	21	-9.3	-6.0
Jul 16	88	20.1	19.7	1	-13.2	30.9	20	-14.8	1.8	35	17.3	22.2	24	0.1	-3.9
Aug 16	72	-14.2	8.3	2	103.6	45.3	25	10.5	-2.9	24	4.1	13.0	21	-5.6	-4.9
Sep 16	67	-2.9	0.4	1	-41.9	-6.3	26	20.0	4.8	26	18.3	13.5	22	-2.7	-2.7
Oct 16	74	2.4	-5.4	1	-8.6	-5.0	21	-9.6	6.5	46	152.6	51.8	24	10.2	0.4
Nov 16	74	9.5	3.0	2	20.5	-17.1	24	14.7	7.7	29	36.3	64.8	24	5.9	4.3
Dec 16	68	-12.4	-0.7	2	106.8	28.9	24	-0.4	1.1	29	20.0	63.6	22	-6.5	2.9
Jan 17	72	-3.4	-2.6	1	-46.4	4.4	23	2.9	5.4	20	-28.7	6.2	22	-0.3	-0.4
Feb 17	70	11.9	-2.2	1	-18.9	-8.9	27	25.7	9.0	30	19.6	2.4	23	6.4	-0.4
Mar 17	79	5.5	4.2	2	-15.2	-28.0	24	-4.3	7.5	24	5.6	-2.3	21	-2.5	1.1
Apr 17	72	-9.7	1.7	2	2.0	-11.4	25	7.3	8.9	26	-18.8	0.3	23	7.3	3.7
May 17	78	0.4	-1.4	2	-12.5	-9.4	24	-7.0	-1.6	27	-45.2	-26.1	23	6.4	3.7
Jun 17	87	28.0	5.2	1	-28.4	-14.2	25	-3.3	-1.3	26	-3.3	-27.3	23	6.9	6.9
Jul 17	75	-14.8	2.7	1	12.8	-12.5	28	42.1	7.9	26	-27.0	-29.5	22	-8.1	1.4
Aug 17	80	11.3	6.2	2	-15.3	-13.5	25	0.3	10.7	27	10.2	-9.2	23	9.4	2.3
Sep 17	98	46.5	11.5	1	-6.6	-5.1	26	0.7	12.1	30	18.1	-2.8	24	5.2	1.8
Oct 17	76	3.1	19.5	1	22.8	-2.1	29	36.1	11.0	32	-30.3	-7.1	23	-3.0	3.7
Nov 17	74	0.1	15.5	3	62.2	27.8	28	16.2	16.6	29	1.7	-8.8	26	10.6	4.2
Dec 17	80	18.8	7.0	1	-16.6	22.7	25	3.6	17.9	21	-27.0	-20.5	23	2.8	3.5
Jan 18	75	4.5	7.5	1	-1.8	16.6	25	9.1	9.7	26	31.0	-1.6	22	-1.5	4.1

Source: Monthly Wages & Salaries Survey
Inquiries: Email: labour.market@ons.gov.uk
Earnings enquiries: 01633 456773

1. The three month average figures are the changes in the average seasonally adjusted values for the three months ending with the relevant month compared with the same period a year earlier.

2. From July 2009 Royal Bank of Scotland Group plc is classified to the public sector; for earlier time periods it is classified to the private sector. Between July 2009 and March 2014 Lloyds Banking Group plc is classified to the public sector; it is classified to the private sector for earlier and later time periods.

3. Between June 2010 and May 2012 English Further Education Corporations and Sixth Form College Corporations are classified to the public sector. Before June 2010 and after May 2012 they are classified to the private sector.

4. From October 2013 Royal Mail plc is classified to the private sector; previously it is in the public sector.

4.14d Average Weekly Earnings in the public sector for selected activities

All figures are in pounds (£)   *Not Seasonally Adjusted*

	Public Administration (Public Sector) (O)			Education (Public Sector) (P)			Health and Social Work (Public Sector) (Q)			Arts, Entertainment and Recreation (Public Sector) (R)		
	Average Weekly Earnings	Of which		Average Weekly Earnings	Of which		Average Weekly Earnings	Of which		Average Weekly Earnings	Of which	
		Bonuses	Arrears		Bonuses	Arrears		Bonuses	Arrears		Bonuses	Arrears
CDID	K5BE	K5BF	K5BG	K5BH	K5BI	K5BJ	K5BK	K5BL	K5BM	K5BN	K5BO	K5BP
2013 Jan	546	1	1	381	0	0	540	0	0	442	0	2
2013 Feb	535	0	0	380	0	0	538	0	0	441	0	0
2013 Mar	536	3	0	379	0	0	538	0	0	451	0	0
2013 Apr	537	1	0	384	0	0	547	0	0	450	3	0
2013 May	537	0	0	383	0	0	546	0	0	447	5	0
2013 Jun	540	3	0	382	0	0	545	0	0	449	4	1
2013 Jul	543	6	0	384	1	0	546	0	0	449	4	3
2013 Aug	540	1	4	386	0	1	536	0	0	442	1	1
2013 Sep	547	7	0	390	0	1	541	0	0	439	1	0
2013 Oct	537	1	0	390	0	1	543	0	0	441	1	1
2013 Nov	537	1	0	387	0	1	543	0	0	450	1	0
2013 Dec	540	2	1	391	0	1	543	0	0	444	0	0
2014 Jan	545	1	1	386	0	0	545	0	0	447	0	3
2014 Feb	541	1	1	388	0	0	544	0	1	450	0	0
2014 Mar	548	2	4	387	0	0	539	0	0	448	0	0
2014 Apr	553	7	3	392	0	2	547	0	0	454	1	0
2014 May	547	1	1	392	0	3	550	0	1	454	2	0
2014 Jun	549	3	0	390	0	0	549	0	0	453	5	0
2014 Jul	549	4	3	391	0	0	547	0	0	451	5	1
2014 Aug	547	3	1	393	1	0	538	0	0	448	6	0
2014 Sep	555	7	0	397	0	0	545	0	0	451	0	0
2014 Oct	552	1	1	395	0	0	540	0	0	448	1	3
2014 Nov	549	1	1	394	0	1	542	0	0	448	2	0
2014 Dec	554	1	1	405	2	3	543	0	0	456	3	1
2015 Jan	551	0	1	394	0	0	545	0	0	452	0	0
2015 Feb	551	2	1	395	0	0	545	0	0	447	0	0
2015 Mar	548	2	0	395	0	0	543	0	0	459	2	5
2015 Apr	552	1	1	398	0	1	550	0	0	455	2	0
2015 May	552	1	0	395	0	0	551	0	0	454	0	0
2015 Jun	552	2	0	394	0	0	555	0	0	451	4	0
2015 Jul	554	4	1	396	0	0	547	0	0	456	7	1
2015 Aug	553	2	1	398	0	0	540	0	0	443	4	0
2015 Sep	561	8	1	403	0	0	548	0	0	447	1	1
2015 Oct	556	2	1	400	0	1	544	0	0	443	1	1
2015 Nov	553	1	1	401	0	2	549	0	0	447	1	0
2015 Dec	550	1	0	403	0	1	550	0	1	455	5	5
2016 Jan	556	0	1	398	0	1	549	0	0	451	0	0
2016 Feb	555	1	0	399	0	0	551	0	1	455	0	3
2016 Mar	553	2	1	401	0	0	546	0	0	455	2	0
2016 Apr	566	6	0	401	0	0	554	0	0	458	2	0
2016 May	557	1	0	398	0	0	552	0	0	461	16	0
2016 Jun	561	1	1	401	0	1	556	0	0	454	4	0
2016 Jul	556	1	0	400	0	1	550	0	0	466	18	2
2016 Aug	570	8	1	404	0	0	547	0	0	455	4	1
2016 Sep	562	2	0	406	0	0	549	0	0	458	0	1
2016 Oct	566	2	1	404	0	1	547	0	0	452	0	0
2016 Nov	564	1	1	406	0	1	551	0	0	460	1	0
2016 Dec	562	0	0	408	0	1	549	0	0	464	0	0
2017 Jan	563	0	0	401	0	0	551	0	0	468	1	4
2017 Feb	562	0	0	402	0	0	551	0	0	461	0	1
2017 Mar	561	1	0	403	0	0	547	0	0	465	1	0
2017 Apr	567	7	0	406	0	0	552	0	0	466	1	0
2017 May	561	1	0	405	0	0	563	0	2	465	0	0
2017 Jun	563	1	0	404	0	0	560	0	0	472	6	1
2017 Jul	564	1	2	405	0	0	557	0	0	472	20	0
2017 Aug	569	6	1	411	0	1	555	0	0	454	0	2
2017 Sep	564	3	1	413	0	1	556	0	0	464	2	2
2017 Oct	572	3	1	413	0	1	556	0	0	470	0	2
2017 Nov	568	2	2	413	0	2	558	0	0	469	1	1
2017 Dec	564	1	1	416	0	2	559	0	0	465	0	0

p = Provisional

r = Revised

Source: Monthly Wages and Salaries Survey
Inquiries: Email: earnings@ons.gov.uk   Tel: 01633 456120

4.15a Weekly pay - Gross (£) - For all employee jobs[a]: United Kingdom, 2017

Description	Code	Number of jobs[b] (000s)	Median	Annual % change	Mean	Annual % change	Percentiles 10	20	25	30	40	60	70	75	80	90
All employees		26,241	448.5	2.3	537.9	2.4	144.4	238.0	283.8	316.5	380.0	527.0	624.5	682.9	752.0	972.6
18-21		1,264	217.7	5.3	233.9	6.5	49.7	83.2	102.0	120.9	164.7	267.9	303.8	324.7	345.0	417.1
22-29		4,553	411.1	2.8	444.0	2.4	167.1	271.1	299.5	322.0	363.7	461.5	517.5	548.4	587.9	714.8
30-39		6,134	508.3	2.0	578.7	2.0	183.6	296.0	333.2	366.4	439.0	586.4	680.3	731.8	790.7	999.9
40-49		6,318	514.2	2.1	622.0	2.5	178.2	280.2	320.9	357.6	432.2	611.6	723.3	783.6	861.4	1,136.8
50-59		5,659	482.0	1.8	593.5	2.0	165.6	259.0	300.0	336.5	405.6	573.7	685.4	747.9	818.2	1,078.4
60+		2,075	358.6	2.8	452.3	3.6	99.3	157.0	188.2	222.1	293.0	430.4	517.5	577.4	649.3	889.0
18-21 ALL OCCUPATIONS		1,264	217.7	5.3	233.9	6.5	49.7	83.2	102.0	120.9	164.7	267.9	303.8	324.7	345.0	417.1
18-21 Managers, directors and senior officials	1	24	372.8	10.1	495.2	39.9	274.4	314.4	334.1	339.2	366.7	402.6	452.3	465.8	484.2	x
18-21 Corporate managers and directors	11	18	383.2	13.7	536.4	48.5	x	334.5	337.7	343.0	369.6	420.1	454.3	x	x	x
18-21 Other managers and proprietors	12	x	343.0	2.4	354.9	12.3	x	x	x	x	x	x	x	x	x	x
18-21 Professional occupations	2	47	369.0	7.0	411.5	10.8	x	x	266.8	287.6	345.0	416.0	457.5	480.5	502.1	x
18-21 Science, research, engineering and technology professionals	21	14	354.6	2.8	402.3	8.1	x	297.4	308.9	337.5	345.0	383.3	447.8	464.2	x	x
18-21 Health professionals	22	9	420.2		x		x	x	x	x	x	441.4	x	x	x	x
18-21 Teaching and educational professionals	23	12	x		x		x	x	x	x	x	x	x	x	x	x
18-21 Business, media and public service professionals	24	13	387.8	7.8	423.9	8.3	x	330.4	x	364.1	370.5	427.4	479.1	x	x	x
18-21 Associate professional and technical occupations	3	95	319.2	2.8	326.6	7.9	54.8	115.4	147.8	237.6	287.6	348.5	383.3	406.4	441.0	528.0
18-21 Science, engineering and technology associate	31	29	330.7	-1.2	340.3	-3.6	x	256.7	284.8	290.7	313.8	361.2	387.9	408.0	440.9	x
18-21 Health and social care associate professionals	32	x	x		x		x	x	x	x	x	x	x	x	x	x
18-21 Protective service occupations	33	x	x		408.9	-11.3	x	x	x	x	x	x	x	x	x	x
18-21 Culture, media and sports occupations	34	16	x		x		x	x	x	x	x	x	x	x	x	x
18-21 Business and public service associate professionals	35	42	334.3	8.0	327.2	9.8	69.2	x	231.5	265.7	298.3	357.4	387.9	403.6	441.6	x
18-21 Administrative and secretarial occupations	4	140	270.0	-1.9	252.1	-1.4	59.1	115.0	144.2	174.9	222.6	299.1	325.0	337.3	356.1	400.6
18-21 Administrative occupations	41	110	287.5	0.0	266.0	0.1	68.5	131.2	157.1	197.2	244.7	306.6	335.4	349.9	368.0	409.8
18-21 Secretarial and related occupations	42	30	205.1	-8.5	200.1	-5.5	35.4	69.8	90.5	109.6	153.7	254.6	283.5	294.0	304.1	x
18-21 Skilled trades	5	96	330.5	8.2	358.4	10.5	200.5	240.0	258.3	274.1	300.2	365.8	413.2	430.8	465.8	540.1
18-21 Skilled agricultural and related trades	51	x	300.0	0.0	329.9	17.8	x	x	x	x	289.9	308.5	x	x	x	x
18-21 Skilled metal, electrical and electronic trades	52	52	348.7	3.3	383.6	5.2	219.6	250.3	268.3	287.5	314.6	387.3	430.8	471.0	498.0	574.9
18-21 Skilled construction and building trades	53	14	331.2	8.3	358.1	8.6	204.7	241.5	255.6	268.2	293.7	390.8	424.4	451.8	x	x
18-21 Textiles, printing and other skilled trades	54	26	307.9	20.0	312.2	24.9	x	202.8	241.1	252.0	284.9	344.9	369.0	394.6	413.4	x
18-21 Caring, leisure and other service occupations	6	167	226.8	7.4	215.0	4.1	44.8	86.9	115.0	137.2	181.4	263.4	286.2	296.6	311.6	360.6
18-21 Caring personal service occupations	61	122	246.6	9.1	229.4	5.7	50.0	113.3	137.3	164.7	209.2	277.7	293.0	303.3	322.5	372.4
18-21 Leisure, travel and related personal service	62	45	155.8	-0.1	176.2	2.2	35.9	57.6	72.7	87.1	126.4	214.0	243.4	268.5	278.4	x
18-21 Sales and customer service occupations	7	259	138.6	1.5	170.4	1.8	49.7	71.0	83.3	94.2	115.1	171.7	212.8	239.9	270.2	330.8
18-21 Sales occupations	71	213	131.3	3.4	161.2	3.1	49.6	69.2	79.9	90.6	109.7	158.5	197.0	222.5	249.2	314.7
18-21 Customer service occupations	72	46	193.5	-19.5	212.6	-7.7	50.8	92.1	107.6	122.5	150.7	260.9	305.3	316.2	332.1	367.7
18-21 Process, plant and machine operatives	8	47	320.5	4.8	320.5	0.7	118.5	210.0	232.1	256.6	293.1	342.5	377.5	393.9	438.0	500.3
18-21 Process, plant and machine operatives	81	36	323.9	4.2	320.7	-2.6	x	219.1	245.9	268.9	295.8	343.7	370.2	394.5	434.5	488.5
18-21 Transport and mobile machine drivers and operatives	82	11	304.3		319.7	14.7	x	x	x	210.4	259.9	334.3	x	x	x	x
18-21 Elementary occupations	9	390	148.8	0.5	176.8	-0.5	37.8	60.8	71.8	86.2	114.6	190.1	240.2	264.3	288.5	341.8
18-21 Elementary trades and related occupations	91	32	300.0	-2.1	296.9	-4.3	96.2	204.9	240.0	253.7	281.0	319.4	350.0	361.8	383.6	x
18-21 Elementary administration and service	92	357	137.6	2.7	165.9	-0.1	36.4	58.0	68.4	81.2	107.7	176.1	220.6	244.9	273.4	325.0

4.15a Weekly pay - Gross (£) - For all employee jobs[a]: United Kingdom, 2017

Description	Code	Number of jobs[b] (000s)	Median	Annual % change	Mean	Annual % change	10	20	25	30	40	60	70	75	80	90
22-29 ALL OCCUPATIONS		4,553	411.1	2.8	444.0	2.4	167.1	271.1	299.5	322.0	363.7	461.5	517.5	548.4	587.9	714.8
22-29 Managers, directors and senior officials	1	228	522.9	7.3	612.2	8.7	306.6	362.0	383.3	411.0	466.5	584.5	685.1	737.4	819.2	1,046.3
22-29 Corporate managers and directors	11	171	574.9	12.6	667.7	12.6	331.7	383.3	414.9	448.5	508.6	653.3	765.9	828.2	919.9	1,111.7
22-29 Other managers and proprietors	12	58	421.9	0.1	448.9	-3.9	243.8	316.9	335.2	353.4	383.3	453.5	502.6	536.6	573.1	x
22-29 Professional occupations	2	934	549.0	-0.8	586.1	0.6	348.5	430.6	460.0	478.1	514.8	594.1	650.5	678.8	728.3	862.4
22-29 Science, research, engineering and technology professionals	21	225	578.4	0.6	625.0	1.9	385.6	460.0	484.6	506.0	544.2	632.4	689.7	734.5	775.4	907.4
22-29 Health professionals	22	243	538.8	-1.6	571.0	1.2	338.8	434.7	456.2	478.8	510.6	579.4	623.3	656.8	697.7	844.1
22-29 Teaching and educational professionals	23	220	502.0	-1.7	502.0	-2.7	238.5	417.2	430.6	433.2	464.6	540.6	583.2	605.1	633.2	706.3
22-29 Business, media and public service professionals	24	246	578.2	-0.9	640.7	1.3	383.2	440.8	464.6	488.4	536.6	640.4	706.3	738.4	781.4	958.2
22-29 Associate professional and technical occupations	3	767	491.8	2.1	541.2	-0.9	316.2	371.9	392.9	417.3	457.2	536.6	581.0	613.3	656.0	770.8
22-29 Science, engineering and technology associate	31	152	472.2	0.1	486.8	1.2	316.2	353.0	371.5	386.5	432.6	506.8	547.1	574.9	596.9	704.7
22-29 Health and social care associate professionals	32	47	406.2	1.7	397.2	0.2	168.9	300.0	328.9	352.0	376.2	432.1	463.3	482.0	514.2	x
22-29 Protective service occupations	33	51	552.4	-3.6	563.8	-1.0	396.3	455.5	473.3	484.0	516.3	603.5	637.5	657.6	673.2	x
22-29 Culture, media and sports occupations	34	66	421.6	1.9	x		109.6	233.2	306.7	346.3	382.8	460.0	483.9	501.7	536.6	x
22-29 Business and public service associate professionals	35	450	517.5	2.7	557.0	1.6	340.9	393.8	418.5	434.9	479.1	560.6	613.3	651.6	689.9	832.1
22-29 Administrative and secretarial occupations	4	530	367.7	2.1	384.1	2.7	161.3	271.1	297.4	315.4	343.0	402.3	441.5	463.8	491.9	574.9
22-29 Administrative occupations	41	449	378.5	2.3	397.1	2.4	184.1	291.4	309.8	327.0	352.2	411.5	451.9	475.8	499.2	577.6
22-29 Secretarial and related occupations	42	81	311.0	2.9	312.8	1.0	104.5	161.4	201.1	229.6	275.2	337.3	371.4	392.9	423.2	498.5
22-29 Skilled trades	5	376	456.9	3.7	485.1	3.5	283.7	337.2	356.9	374.6	412.6	498.2	546.9	580.3	619.8	727.9
22-29 Skilled agricultural and related trades	51	22	356.8	2.9	362.2	0.3	x	300.0	314.6	323.5	339.5	370.1	400.6	413.2	425.9	x
22-29 Skilled metal, electrical and electronic trades	52	193	525.5	4.2	557.4	4.2	343.3	395.7	416.2	444.5	483.9	567.9	621.8	647.0	687.4	808.1
22-29 Skilled construction and building trades	53	63	479.1	0.4	502.1	2.3	300.0	360.0	398.4	410.1	450.0	503.8	548.9	575.0	622.8	x
22-29 Textiles, printing and other skilled trades	54	98	358.9	-0.3	359.0	-0.6	182.7	274.9	292.6	304.0	333.7	386.0	418.1	436.3	456.0	507.2
22-29 Caring, leisure and other service occupations	6	475	293.5	4.8	290.6	4.8	120.8	179.8	201.2	225.8	267.7	319.0	346.6	361.6	382.3	449.5
22-29 Caring personal service occupations	61	387	293.1	5.2	289.0	5.5	126.7	181.7	203.8	227.5	266.8	317.6	345.0	359.1	378.5	443.8
22-29 Leisure, travel and related personal service	62	88	296.0	1.3	297.8	1.6	112.5	162.6	189.4	216.1	273.8	327.9	353.3	372.0	398.0	475.8
22-29 Sales and customer service occupations	7	521	299.0	3.8	297.2	3.7	115.2	156.6	179.6	205.3	259.9	325.8	359.9	378.5	400.0	469.5
22-29 Sales occupations	71	361	270.7	3.8	273.7	4.5	103.8	141.0	158.2	178.0	221.6	305.0	334.9	354.0	376.3	441.2
22-29 Customer service occupations	72	160	345.0	1.9	350.1	1.0	156.8	232.8	273.7	294.2	322.0	370.1	400.1	419.3	440.8	517.5
22-29 Process, plant and machine operatives	8	206	406.2	0.9	436.0	0.3	252.6	300.9	316.0	330.0	364.8	449.6	497.9	523.5	557.4	662.9
22-29 Process, plant and machine operatives	81	138	387.9	-2.3	425.2	-1.3	255.9	300.0	312.1	321.7	352.6	432.5	479.5	506.9	541.3	641.3
22-29 Transport and mobile machine drivers and operatives	82	67	439.3	5.7	458.3	3.6	238.2	308.4	334.8	361.9	401.2	475.7	526.5	548.8	583.0	672.1
22-29 Elementary occupations	9	517	300.0	6.8	291.9	4.3	90.4	148.0	180.0	206.2	263.7	327.3	360.0	378.1	400.4	466.2
22-29 Elementary trades and related occupations	91	72	360.1	1.6	377.4	2.7	261.9	295.0	303.1	316.1	340.4	384.5	422.0	437.9	454.7	537.3
22-29 Elementary administration and service occupations	92	445	284.0	6.0	278.1	4.1	82.5	135.0	161.7	188.5	240.0	315.9	347.1	365.8	386.0	450.1
30-39 ALL OCCUPATIONS		6,134	508.3	2.0	578.7	2.0	183.6	296.0	333.2	366.4	439.0	586.4	680.3	731.8	790.7	999.9
30-39 Managers, directors and senior officials	1	601	745.8	3.3	896.4	3.1	364.9	475.3	515.5	555.8	645.3	862.4	1,029.5	1,146.0	1,255.7	1,619.6
30-39 Corporate managers and directors	11	493	819.4	3.9	964.1	3.8	387.6	512.1	558.3	609.7	703.4	958.2	1,140.1	1,245.7	1,351.3	1,686.4
30-39 Other managers and proprietors	12	108	531.6	-0.1	586.6	3.1	305.1	373.0	416.6	433.9	479.1	585.8	653.6	689.9	745.2	878.1
30-39 Professional occupations	2	1,602	690.9	0.1	742.5	1.3	364.8	485.5	532.9	568.7	633.8	757.6	840.7	885.4	948.7	1,149.9

4.15a Weekly pay - Gross (£) - For all employee jobs[a]: United Kingdom, 2017

Description	Code	Number of jobs[b] (000s)	Annual Median	% change	Annual Mean	% change	10	20	25	30	40	60	70	75	80	90
30-39 Science, research, engineering and technology professionals	21	391	774.2	0.9	832.8	2.1	500.5	592.8	626.9	659.3	718.7	849.2	929.4	981.2	1,034.9	1,228.4
30-39 Health professionals	22	392	594.6	-0.4	676.1	2.3	288.6	410.0	446.9	480.8	541.0	656.5	738.4	792.8	875.8	1,147.6
30-39 Teaching and educational professionals	23	395	681.7	-0.3	655.4	-0.4	293.2	435.8	483.8	539.3	627.2	730.5	783.7	817.4	855.9	957.8
30-39 Business, media and public service professionals	24	425	718.7	0.9	801.7	1.6	426.4	530.3	573.8	601.2	663.2	785.8	878.9	933.2	999.2	1,263.5
30-39 Associate professional and technical occupations	3	1,083	600.4	2.5	669.3	2.6	328.4	425.0	454.2	481.6	538.2	670.6	740.4	779.0	844.3	1,025.5
30-39 Science, engineering and technology associate	31	178	542.0	1.1	579.4	1.6	329.3	400.1	433.2	463.0	505.9	593.4	649.3	697.4	736.6	862.4
30-39 Health and social care associate professionals	32	89	444.3	4.1	440.8	4.2	216.7	302.3	339.4	361.3	406.5	479.1	525.1	544.8	573.9	648.7
30-39 Protective service occupations	33	135	728.3	2.8	702.7	-1.9	459.1	558.6	595.5	622.4	677.4	752.6	788.2	817.4	852.2	908.0
30-39 Culture, media and sports occupations	34	67	488.7	-1.1	598.3	-5.1	123.9	257.7	303.1	352.2	439.6	530.6	586.4	630.6	662.5	x
30-39 Business and public service associate professionals	35	614	639.3	4.2	728.9	3.9	362.1	446.7	479.1	503.2	574.9	705.1	796.0	862.4	936.5	1,185.4
30-39 Administrative and secretarial occupations	4	615	383.7	2.8	411.2	3.5	158.3	221.0	255.0	293.8	345.0	430.3	479.1	514.3	550.5	670.8
30-39 Administrative occupations	41	517	395.1	3.1	423.5	3.5	166.9	236.0	278.9	309.5	355.6	439.8	489.9	521.8	561.9	679.6
30-39 Secretarial and related occupations	42	98	316.7	2.1	346.5	2.0	124.4	174.3	194.4	217.3	258.5	361.2	420.3	450.5	498.6	613.2
30-39 Skilled trades	5	447	506.0	4.8	531.3	4.4	287.4	361.0	390.0	417.8	461.3	556.0	613.3	648.0	682.3	801.2
30-39 Skilled agricultural and related trades	51	21	385.9	1.3	396.4	2.1	281.3	322.0	341.8	351.8	368.4	400.1	422.0	441.0	470.1	x
30-39 Skilled metal, electrical and electronic trades	52	221	585.7	3.7	619.8	3.8	385.5	460.0	481.8	503.8	546.3	638.9	697.3	732.1	766.6	883.9
30-39 Skilled construction and building trades	53	73	520.0	4.0	539.9	4.7	336.1	409.9	427.4	445.9	479.1	558.6	609.1	625.8	650.2	x
30-39 Textiles, printing and other skilled trades	54	131	392.6	7.6	399.6	6.7	179.4	260.4	295.0	313.3	355.3	434.0	466.2	485.5	513.3	604.3
30-39 Caring, leisure and other service occupations	6	496	279.6	2.8	291.9	1.1	115.0	160.3	180.2	203.0	239.9	313.0	350.8	376.0	406.2	480.1
30-39 Caring personal service occupations	61	409	274.1	3.2	285.1	1.7	113.7	161.5	180.4	202.8	239.2	308.2	344.4	367.8	392.9	462.1
30-39 Leisure, travel and related personal service	62	87	302.7	-2.0	324.1	-2.2	119.3	154.0	180.0	205.9	251.1	341.1	396.9	435.9	461.2	553.1
30-39 Sales and customer service occupations	7	421	303.4	3.0	322.6	2.9	119.6	152.0	173.5	198.0	249.7	344.3	387.4	415.2	451.7	557.4
30-39 Sales occupations	71	276	260.2	0.2	287.3	2.0	109.6	135.0	150.1	169.0	212.9	304.0	346.0	370.2	397.6	498.3
30-39 Customer service occupations	72	145	369.3	5.5	389.6	4.2	154.4	219.9	251.2	294.1	334.7	411.5	460.7	492.9	527.8	629.7
30-39 Process, plant and machine operatives	8	317	452.9	1.1	484.8	1.4	270.0	323.7	346.1	365.6	410.7	495.9	549.7	591.5	632.0	750.0
30-39 Process, plant and machine operatives	81	175	419.3	-0.9	453.3	-0.2	259.2	312.5	326.9	344.6	377.2	464.6	515.7	549.8	592.9	702.7
30-39 Transport and mobile machine drivers and operatives	82	143	484.5	1.8	523.5	3.8	287.1	355.7	384.0	410.0	449.5	533.6	597.3	630.6	679.5	800.2
30-39 Elementary occupations	9	551	309.1	5.8	310.9	5.2	84.5	135.5	165.6	191.6	265.3	344.8	384.0	411.8	439.0	531.6
30-39 Elementary trades and related occupations	91	76	378.9	5.0	402.9	6.0	260.0	300.4	314.8	322.5	350.6	409.5	450.3	475.5	503.2	590.0
30-39 Elementary administration and service occupations	92	475	295.6	6.0	296.2	4.6	76.0	123.0	147.9	174.4	232.5	330.0	370.3	396.5	425.2	517.5
40-49 ALL OCCUPATIONS		**6,318**	**514.2**	**2.1**	**622.0**	**2.5**	**178.2**	**280.2**	**320.9**	**357.6**	**432.2**	**611.6**	**723.3**	**783.6**	**861.4**	**1,136.8**
40-49 Managers, directors and senior officials	1	782	866.1	0.7	1,076.9	3.0	383.3	526.2	576.9	640.8	751.3	1,027.4	1,213.2	1,329.0	1,465.0	1,954.8
40-49 Corporate managers and directors	11	669	942.8	3.5	1,146.2	4.6	410.0	572.5	635.7	691.9	804.9	1,113.3	1,289.5	1,400.8	1,571.5	2,052.3
40-49 Other managers and proprietors	12	112	589.8	-0.6	663.9	-3.2	267.1	391.5	423.7	460.0	527.7	674.7	747.4	797.7	862.4	x
40-49 Professional occupations	2	1,490	733.1	0.2	804.6	1.8	334.6	478.7	536.1	583.2	670.8	817.3	912.9	973.9	1,054.1	1,296.5
40-49 Science, research, engineering and technology professionals	21	340	862.4	3.0	907.8	2.5	513.1	636.1	679.0	715.9	792.6	939.5	1,025.0	1,092.3	1,149.9	1,336.4
40-49 Health professionals	22	404	634.3	0.9	741.8	1.5	264.9	399.8	441.2	486.4	554.8	691.5	784.2	824.8	899.7	1,381.5
40-49 Teaching and educational professionals	23	395	725.8	0.1	713.7	0.8	269.5	430.5	473.6	540.6	639.0	782.3	856.7	910.2	958.1	1,116.3
40-49 Business, media and public service professionals	24	351	775.5	1.1	879.2	2.2	404.3	536.9	590.3	627.9	697.2	864.0	981.2	1,054.1	1,136.8	1,423.4
40-49 Associate professional and technical occupations	3	995	631.6	2.6	707.7	3.6	332.7	433.1	469.2	498.3	568.9	711.8	785.8	834.0	896.2	1,136.3

124

4.15a Weekly pay - Gross (£) - For all employee jobs[a]: United Kingdom, 2017

Description	Code	Number of jobs[b] (000s)	Median	Annual % change	Mean	Annual % change	Percentiles 10	20	25	30	40	60	70	75	80	90
40-49 Science, engineering and technology associate	31	158	580.6	0.9	609.1	1.5	335.8	413.9	449.3	478.0	524.4	628.3	689.0	722.8	768.8	892.7
40-49 Health and social care associate professionals	32	89	463.7	4.0	453.6	1.9	196.3	281.3	321.7	356.2	410.5	494.0	532.6	553.0	582.6	698.3
40-49 Protective service occupations	33	149	762.9	1.6	776.3	1.3	549.6	611.2	656.2	697.1	732.5	805.8	845.8	870.8	909.4	1,019.9
40-49 Culture, media and sports occupations	34	44	469.8	-9.2	486.2	-7.2	x	x	x	x	404.1	548.5	616.6	641.1	727.6	x
40-49 Business and public service associate professionals	35	555	652.6	3.2	775.7	5.2	364.0	446.9	479.3	517.5	578.4	738.0	843.3	910.3	1,015.7	1,331.1
40-49 Administrative and secretarial occupations	4	707	368.5	1.2	405.7	2.4	154.8	211.8	239.2	270.8	325.0	414.0	471.2	505.5	549.6	689.9
40-49 Administrative occupations	41	562	383.3	1.4	422.7	3.2	158.5	226.7	258.7	290.9	342.1	431.5	484.8	519.5	570.9	711.2
40-49 Secretarial and related occupations	42	145	310.1	-1.1	340.1	-2.0	138.3	177.2	196.9	217.3	258.5	352.0	402.0	430.4	472.0	582.6
40-49 Skilled trades	5	457	505.0	0.5	543.6	1.3	278.8	353.6	380.0	409.6	460.0	564.2	627.9	663.1	708.7	834.0
40-49 Skilled agricultural and related trades	51	23	377.9	-2.6	420.6	3.0	x	306.6	332.0	340.6	361.0	406.6	445.0	456.3	x	x
40-49 Skilled metal, electrical and electronic trades	52	233	604.6	3.1	642.5	0.9	395.9	454.5	479.2	499.2	554.5	650.8	709.1	751.8	795.4	931.2
40-49 Skilled construction and building trades	53	75	527.8	1.4	559.1	1.0	334.9	425.4	445.3	461.0	495.9	563.4	608.2	648.0	686.6	x
40-49 Textiles, printing and other skilled trades	54	127	353.7	4.4	375.6	4.4	171.2	224.9	252.9	282.2	325.0	389.3	435.0	468.6	490.7	583.9
40-49 Caring, leisure and other service occupations	6	588	282.8	2.6	302.4	3.3	119.6	175.2	195.0	214.8	247.8	321.9	361.3	383.1	409.7	490.4
40-49 Caring personal service occupations	61	509	279.2	3.1	293.1	2.9	118.9	176.8	197.3	215.5	246.8	315.1	352.4	374.4	400.0	472.7
40-49 Leisure, travel and related personal service	62	79	323.9	2.0	362.4	5.4	119.7	166.5	183.0	201.1	266.1	365.9	437.1	471.2	513.0	695.4
40-49 Sales and customer service occupations	7	349	287.4	3.0	319.2	3.1	120.0	158.7	179.1	195.4	240.1	325.8	374.0	399.2	431.2	551.9
40-49 Sales occupations	71	247	247.7	1.4	280.9	3.6	116.2	144.4	159.5	177.3	207.7	292.3	330.0	357.5	385.1	478.6
40-49 Customer service occupations	72	102	371.6	1.9	411.7	2.5	165.0	238.5	265.6	298.2	336.3	413.1	463.9	499.6	545.0	677.2
40-49 Process, plant and machine operatives	8	380	473.9	-0.6	506.9	-0.3	281.3	333.5	355.8	380.3	425.0	528.0	587.0	624.8	660.8	766.7
40-49 Process, plant and machine operatives	81	171	437.8	-0.8	474.5	-1.5	273.8	312.4	332.8	347.7	393.1	489.0	547.2	581.0	623.6	731.9
40-49 Transport and mobile machine drivers and operatives	82	210	502.6	-0.5	533.4	0.5	295.7	360.4	385.9	409.4	455.2	556.7	616.2	650.1	683.3	823.9
40-49 Elementary occupations	9	569	287.1	5.0	293.3	3.3	70.9	116.9	136.1	159.4	221.1	331.1	376.6	405.6	432.2	527.6
40-49 Elementary trades and related occupations	91	72	371.8	3.4	390.9	1.4	213.1	290.0	300.0	312.1	342.1	402.7	442.7	463.6	497.3	586.7
40-49 Elementary administration and service occupations	92	497	256.4	3.7	279.2	3.2	65.8	106.2	121.9	145.1	196.2	314.6	362.3	393.5	420.6	515.5
50-59 ALL OCCUPATIONS		5,659	482.0	1.8	593.5	2.0	165.6	259.0	300.0	336.5	405.6	573.7	685.4	747.9	818.2	1,078.4
50-59 Managers, directors and senior officials	1	720	841.3	4.3	1,053.3	4.9	345.0	493.9	555.3	607.7	721.6	972.5	1,172.9	1,306.7	1,455.1	1,916.5
50-59 Corporate managers and directors	11	612	899.8	4.3	1,116.9	5.0	360.5	531.7	586.7	651.9	766.6	1,050.3	1,264.9	1,391.4	1,539.5	2,011.6
50-59 Other managers and proprietors	12	108	602.7	3.2	693.2	4.7	316.0	391.0	428.9	475.0	536.3	685.9	780.6	840.9	896.7	x
50-59 Professional occupations	2	1,224	723.7	1.7	786.4	1.4	305.6	451.3	513.1	563.4	653.2	793.4	889.6	935.9	1,017.0	1,257.7
50-59 Science, research, engineering and technology professionals	21	252	840.0	1.8	898.7	1.7	509.9	625.7	669.9	697.9	774.0	923.7	1,017.3	1,073.3	1,134.8	1,365.0
50-59 Health professionals	22	371	626.5	0.6	728.6	0.6	264.4	372.7	417.2	459.4	547.8	689.2	783.2	805.1	869.3	1,176.0
50-59 Teaching and educational professionals	23	329	730.5	0.7	727.8	1.3	242.7	424.1	479.0	541.0	652.3	783.7	868.0	904.7	960.2	1,152.4
50-59 Business, media and public service professionals	24	272	749.4	3.4	832.1	1.5	355.6	491.5	555.1	599.7	678.3	824.8	926.6	996.6	1,073.2	1,312.7
50-59 Associate professional and technical occupations	3	701	603.9	2.3	682.1	2.1	303.7	408.6	446.6	478.8	540.7	682.9	766.6	816.0	881.6	1,120.7
50-59 Science, engineering and technology associate	31	135	589.6	0.1	623.2	0.6	352.6	414.4	441.7	474.9	529.6	653.0	712.7	760.1	805.8	962.0
50-59 Health and social care associate professionals	32	84	448.4	1.8	447.0	1.3	187.4	269.7	304.3	345.5	403.0	489.1	531.4	562.0	596.4	675.1
50-59 Protective service occupations	33	61	730.7	-0.5	725.6	-1.2	394.6	520.1	560.0	582.0	664.7	775.8	837.3	870.3	896.9	x
50-59 Culture, media and sports occupations	34	27	442.6	0.4	482.7	-2.9	x	x	x	254.2	367.6	496.8	568.4	620.7	x	x
50-59 Business and public service associate professionals	35	393	657.9	2.9	759.5	2.8	355.8	446.7	483.9	516.0	576.1	732.3	826.8	895.1	980.0	1,295.4

4.15a Weekly pay - Gross (£) - For all employee jobs[a]: United Kingdom, 2017

Description	Code	Number of jobs[b] (000s)	Annual Median	% change	Annual Mean	% change	Percentiles 10	20	25	30	40	60	70	75	80	90
50-59 Administrative and secretarial occupations	4	763	362.0	1.0	392.3	2.2	154.9	208.3	233.2	265.4	317.2	404.2	456.7	489.5	532.7	668.3
50-59 Administrative occupations	41	579	377.1	-0.2	409.3	1.3	156.5	220.4	251.8	287.5	337.5	421.9	476.4	508.2	549.9	688.5
50-59 Secretarial and related occupations	42	184	306.6	3.6	338.9	5.3	138.0	179.9	201.4	220.1	262.4	345.0	392.3	430.4	460.0	587.2
50-59 Skilled trades	5	400	529.6	3.2	554.3	2.2	266.7	362.2	396.0	425.5	476.9	580.1	647.1	682.8	725.1	854.5
50-59 Skilled agricultural and related trades	51	26	369.0	-0.5	381.9	-0.5	230.9	299.9	309.7	322.4	345.7	396.0	424.2	439.5	457.2	x
50-59 Skilled metal, electrical and electronic trades	52	228	604.9	-1.1	644.6	0.8	401.9	460.2	491.6	517.5	556.2	662.0	719.7	756.1	795.7	920.6
50-59 Skilled construction and building trades	53	61	548.1	2.7	580.7	1.2	360.6	439.0	465.7	485.7	513.8	582.7	636.9	675.0	710.3	x
50-59 Textiles, printing and other skilled trades	54	86	338.0	3.6	347.7	3.4	150.0	204.4	229.9	252.4	301.4	376.0	411.7	442.5	467.4	533.8
50-59 Caring, leisure and other service occupations	6	546	295.1	3.7	310.9	2.7	122.8	179.6	201.9	224.9	260.1	333.5	372.3	391.8	420.6	498.3
50-59 Caring personal service occupations	61	461	290.6	4.8	300.7	3.0	123.8	180.6	202.3	224.8	257.6	326.1	364.7	384.8	407.1	477.9
50-59 Leisure, travel and related personal service	62	84	327.4	1.0	367.0	1.6	118.2	171.7	201.4	225.0	284.2	372.1	431.3	467.8	511.9	656.9
50-59 Sales and customer service occupations	7	350	255.0	2.3	297.5	4.7	120.0	155.8	171.2	186.9	221.7	297.4	339.7	369.0	398.2	517.4
50-59 Sales occupations	71	263	231.4	5.3	261.4	5.9	112.5	144.4	158.2	172.5	201.7	265.1	302.3	320.0	345.9	437.4
50-59 Customer service occupations	72	87	367.6	4.4	406.6	3.3	164.7	212.0	240.9	277.0	325.7	405.3	462.8	501.9	552.8	694.7
50-59 Process, plant and machine operatives	8	403	471.1	2.1	506.7	2.7	273.6	333.8	359.9	385.6	427.2	523.9	579.2	618.2	661.9	792.4
50-59 Process, plant and machine operatives	81	163	438.2	1.4	482.1	2.0	270.8	313.9	335.6	354.1	397.4	490.4	544.9	584.4	631.1	764.3
50-59 Transport and mobile machine drivers and operatives	82	240	495.1	3.3	523.5	2.9	276.5	349.0	383.2	405.9	450.0	542.1	600.0	638.7	677.3	801.0
50-59 Elementary occupations	9	553	281.1	2.7	290.8	2.2	77.9	119.6	138.2	163.7	221.9	321.7	371.0	403.0	433.2	525.4
50-59 Elementary trades and related occupations	91	67	374.5	0.6	397.0	-0.8	188.0	286.1	300.0	312.4	341.3	411.8	457.8	485.0	516.8	609.4
50-59 Elementary administration and service occupations	92	487	256.2	2.7	276.3	2.5	74.8	112.5	127.2	149.5	198.6	305.6	350.8	385.7	420.7	510.3
60+ ALL OCCUPATIONS		2,075	358.6	2.8	452.3	3.6	99.3	157.0	188.2	222.1	293.0	430.4	517.5	577.4	649.3	889.0
60+ Managers, directors and senior officials	1	244	642.0	10.6	821.9	7.5	191.7	300.0	359.4	422.3	532.8	768.9	958.2	1,065.3	1,190.9	1,629.0
60+ Corporate managers and directors	11	198	689.9	10.9	874.9	8.6	191.7	312.8	382.3	450.4	574.9	845.5	1,047.8	1,149.9	1,264.9	1,745.7
60+ Other managers and proprietors	12	45	496.9	4.2	590.7	5.1	191.8	276.4	305.6	342.3	439.5	587.2	671.5	745.3	786.2	x
60+ Professional occupations	2	361	574.9	5.9	647.7	2.6	153.7	279.0	328.9	373.3	470.2	676.6	785.1	847.1	937.8	1,158.8
60+ Science, research, engineering and technology professionals	21	63	731.0	-1.7	773.3	-0.4	337.9	498.6	555.8	586.4	670.8	843.2	935.9	1,004.6	1,059.3	x
60+ Health professionals	22	97	478.8	7.0	616.0	5.9	198.2	282.9	317.2	349.1	411.3	572.2	675.1	720.4	792.9	x
60+ Teaching and educational professionals	23	110	441.8	-2.9	564.5	-1.0	93.9	153.9	195.7	251.6	342.7	588.5	731.8	792.3	899.3	1,152.7
60+ Business, media and public service professionals	24	91	633.8	10.0	696.4	5.5	219.2	353.7	410.6	453.5	536.3	721.0	834.0	912.1	980.3	x
60+ Associate professional and technical occupations	3	196	490.8	0.3	549.7	1.8	175.6	290.6	327.5	357.4	422.5	555.8	643.8	698.1	759.8	949.9
60+ Science, engineering and technology associate	31	38	517.2	4.5	573.3	5.3	301.6	354.6	391.6	408.7	460.3	577.9	671.1	720.0	775.1	x
60+ Health and social care associate professionals	32	24	326.8	5.2	345.1	-2.2	x	161.0	203.8	228.4	273.2	393.7	445.3	471.5	488.3	x
60+ Protective service occupations	33	7	540.8	2.0	613.2	12.1	x	378.0	392.9	434.0	487.9	670.3	x	x	x	x
60+ Culture, media and sports occupations	34	13	x		335.0	-13.3	x	x	x	x	x	x	x	x	x	x
60+ Business and public service associate professionals	35	115	527.8	0.7	604.3	1.5	218.3	325.1	350.1	384.8	460.6	608.5	694.5	748.9	814.3	1,025.5
60+ Administrative and secretarial occupations	4	313	275.9	2.2	306.7	1.9	84.1	143.3	157.2	183.7	224.2	333.8	383.3	414.1	448.5	561.4
60+ Administrative	41	228	288.7	0.7	315.2	1.8	77.3	141.7	157.3	188.9	234.2	346.9	396.1	428.6	464.9	571.3
60+ Secretarial and related occupations	42	86	236.8	0.8	284.1	2.2	93.8	143.6	156.1	173.6	203.2	288.8	340.6	370.2	404.7	514.7
60+ Skilled trades occupations	5	149	465.0	2.3	479.0	1.8	183.6	291.9	320.7	350.7	413.9	515.6	576.3	612.8	658.3	767.1
60+ Skilled agricultural and related trades	51	12	259.5	-18.4	262.0	-10.3	x	116.3	135.1	157.2	x	311.2	337.3	352.6	x	x
60+ Skilled metal, electrical and electronic trades	52	79	549.2	-1.0	569.2	1.5	327.6	400.3	428.4	459.3	504.4	596.3	657.4	686.2	717.0	817.8

4.15a Weekly pay - Gross (£) - For all employee jobs[a]: United Kingdom, 2017

Description	Code	Number of jobs[b] (000s)	Annual % Median change		Annual % Mean change		Percentiles									
							10	20	25	30	40	60	70	75	80	90
60+ Skilled construction and building trades	53	23	478.2	1.3	504.3	3.7	264.9	350.6	394.6	424.7	451.5	513.4	562.4	612.4	643.7	x
60+ Textiles, printing and other skilled trades	54	35	310.7	-0.9	333.7	2.5	127.2	183.0	199.2	224.3	272.6	345.9	393.2	415.2	438.8	x
60+ Caring, leisure and other service occupations	6	199	245.3	-0.1	262.9	1.6	78.7	130.4	148.7	165.1	206.7	290.6	331.9	357.5	387.4	456.9
60+ Caring personal service occupations	61	155	247.1	-0.4	261.7	0.7	80.9	136.2	151.8	170.2	207.9	288.4	331.1	355.2	384.9	451.2
60+ Leisure, travel and related personal service	62	43	239.5	5.7	267.0	5.0	73.0	115.6	129.6	143.1	193.2	297.6	338.7	369.0	392.7	x
60+ Sales and customer service occupations	7	155	184.0	3.0	220.2	2.3	86.4	114.2	123.7	134.4	157.5	218.0	253.5	282.6	312.0	401.0
60+ Sales occupations	71	128	171.6	3.1	201.5	3.1	81.5	108.8	120.0	127.7	149.5	199.2	232.7	252.1	282.6	361.1
60+ Customer service occupations	72	28	277.8	1.3	306.6	0.6	120.6	147.9	167.7	191.3	225.7	316.2	374.7	402.8	430.3	x
60+ Process, plant and machine operatives	8	192	390.3	6.2	408.0	5.5	132.5	201.2	246.4	283.7	339.6	442.9	499.8	535.7	576.6	687.7
60+ Process, plant and machine operatives	81	60	388.3	6.1	426.7	6.8	161.4	267.5	295.9	311.2	353.6	444.5	501.0	538.5	583.8	677.2
60+ Transport and mobile machine drivers and operatives	82	132	393.1	6.7	399.5	5.0	119.8	177.9	222.3	258.6	330.6	442.7	499.2	533.2	575.5	692.2
60+ Elementary occupations	9	266	197.8	1.2	242.1	2.5	52.8	84.2	99.0	115.3	149.5	271.0	326.8	355.9	391.1	480.6
60+ Elementary trades and related occupations	91	31	348.8	2.0	355.6	0.3	138.2	218.8	272.4	294.1	320.8	381.4	425.4	439.3	464.4	x
60+ Elementary administration and service	92	235	176.8	0.7	227.3	2.4	50.0	78.5	92.1	105.6	135.0	234.8	302.7	334.0	373.1	467.3
Not Classified		:														

a Employees on adult rates whose pay for the survey pay-period was not affected by absence.

Source: Annual Survey of Hours and Earnings, Office for National Statistics.

b Figures for Number of Jobs are for indicative purposes only and should not be considered an accurate estimate of employee job counts.

KEY - The colour coding indicates the quality of each estimate; jobs, median, mean and percentiles but not the annual percentage change.

The quality of an estimate is measured by its coefficient of variation (CV), which is the ratio of the standard error of an estimate to the estimate.

Key	Statistical robustness
CV <= 5%	Estimates are considered precise
CV > 5% and <= 10%	Estimates are considered reasonably precise
CV > 10% and <= 20%	Estimates are considered acceptable
x = CV > 20%	Estimates are considered unreliable for practical purposes
.. = disclosive	
: = not applicable	
- = nil or negligible	

4.15b Hourly pay - Gross (£) - For all employee jobs[a]: United Kingdom, 2017

Description	Code	Number of jobs[b] (000s)	Median	Annual % change	Mean	Annual % change	Percentiles 10	20	25	30	40	60	70	75	80	90
All employees		26,241	12.47	2.5	16.16	2.7	7.62	8.46	8.96	9.50	10.76	14.53	17.21	18.83	20.81	27.19
18-21		1,264	7.81	4.4	8.74	5.7	5.87	6.70	7.05	7.15	7.50	8.20	8.71	9.10	9.57	11.27
22-29		4,553	10.87	3.1	12.72	2.8	7.54	8.16	8.50	8.89	9.79	12.27	13.84	14.73	15.86	19.50
30-39		6,134	13.93	2.3	16.74	2.4	7.98	9.08	9.73	10.41	12.05	16.06	18.58	20.00	21.78	27.86
40-49		6,318	14.32	2.7	18.27	2.9	8.00	9.14	9.78	10.44	12.12	16.86	19.74	21.55	23.77	31.22
50-59		5,659	13.47	1.9	17.68	2.3	7.93	8.95	9.51	10.09	11.55	15.85	18.84	20.64	22.66	30.00
60+		2,075	11.56	2.3	15.60	3.8	7.65	8.33	8.71	9.19	10.15	13.28	15.80	17.54	19.58	26.68
18-21 ALL OCCUPATIONS		1,264	7.81	4.4	8.74	5.7	5.87	6.70	7.05	7.15	7.50	8.20	8.71	9.10	9.57	11.27
18-21 Managers, directors and senior officials	1	24	9.36	6.7	12.76	40.3	x	8.18	8.55	8.63	9.11	9.81	10.54	11.54	12.19	x
18-21 Corporate managers and directors	11	18	9.63	9.3	13.55	46.3	x	8.62	8.64	8.80	9.25	10.11	10.85	x	x	x
18-21 Other managers and proprietors	12	x	8.18	3.2	9.83	19.5	x	x	x	x	x	x	x	x	x	x
18-21 Professional occupations	2	47	10.61	7.7	13.21	21.4	7.89	9.00	9.18	9.20	10.14	11.79	12.78	13.20	14.70	x
18-21 Science, research, engineering and technology professionals	21	14	10.13	10.3	10.74	8.6	x	8.06	8.79	9.20	9.23	10.42	11.37	12.48	x	x
18-21 Health professionals	22	9	11.98	3.3	14.84	29.1	x	x	9.93	10.26	11.59	12.57	x	x	x	x
18-21 Teaching and educational professionals	23	12	11.76	-26.1	x		x	9.00	x	9.01	9.14	x	x	x	x	x
18-21 Business, media and public service professionals	24	13	10.54	11.0	11.34	8.9	x	9.22	9.37	9.53	10.31	11.48	x	x	x	x
18-21 Associate professional and technical occupations	3	95	9.27	0.9	10.76	6.6	6.95	7.67	7.91	8.16	8.69	10.00	11.02	11.50	12.14	14.31
18-21 Science, engineering and technology associate professionals	31	29	9.10	-1.8	9.56	-1.3	6.69	7.67	7.84	8.13	8.37	9.93	10.41	11.01	11.35	x
18-21 Health and social care associate professionals	32	x	9.07	7.4	9.89	7.4	x	x	x	x	8.59	9.26	x	x	x	x
18-21 Protective service occupations	33	x	11.59	7.0	11.88	4.7	x	x	x	x	x	x	x	x	x	x
18-21 Culture, media and sports occupations	34	16	10.00	3.5	17.21	38.0	6.83	8.48	8.72	8.86	9.29	10.84	12.19	x	x	x
18-21 Business and public service associate professionals	35	42	9.03	0.6	10.19	2.6	7.00	7.51	7.67	7.93	8.46	9.79	10.58	11.16	11.96	x
18-21 Administrative and secretarial occupations	4	140	8.32	2.7	8.81	1.1	6.06	7.05	7.34	7.50	7.90	8.80	9.43	9.75	10.14	11.68
18-21 Administrative occupations	41	110	8.53	3.7	8.97	0.8	6.23	7.19	7.47	7.60	8.04	9.01	9.67	10.01	10.39	12.27
18-21 Secretarial and related occupations	42	30	7.84	3.1	8.09	3.3	5.61	6.71	7.05	7.20	7.50	8.03	8.44	8.74	9.01	x
18-21 Skilled trades occupations	5	96	8.09	6.2	8.82	5.6	5.60	6.50	6.99	7.06	7.57	8.77	9.58	10.03	10.66	12.52
18-21 Skilled agricultural and related trades	51	x	7.62	-1.7	8.12	5.5	x	x	x	x	7.43	8.06	x	x	x	x
18-21 Skilled metal, electrical and electronic trades	52	52	8.44	5.2	9.21	4.6	5.60	6.34	6.96	7.05	7.69	9.49	10.30	10.71	11.44	13.46
18-21 Skilled construction and building trades	53	14	8.19	2.5	9.09	5.6	5.60	6.35	6.93	7.05	7.67	10.00	10.62	10.82	11.39	x
18-21 Textiles, printing and other skilled trades	54	26	7.70	7.1	7.97	9.9	5.72	6.72	7.01	7.21	7.50	8.13	8.50	8.71	8.96	x
18-21 Caring, leisure and other service occupations	6	167	7.90	6.1	8.03	5.0	6.03	7.04	7.10	7.31	7.50	8.23	8.56	8.88	9.30	10.30
18-21 Caring personal service occupations	61	122	7.91	5.7	8.13	5.5	6.44	7.06	7.20	7.42	7.58	8.25	8.61	9.00	9.35	10.45
18-21 Leisure, travel and related personal service occupations	62	45	7.71	8.2	7.71	4.0	5.60	6.42	6.75	7.05	7.50	8.12	8.43	8.69	8.94	x
18-21 Sales and customer service occupations	7	259	7.72	4.6	8.04	2.8	5.80	6.56	7.05	7.10	7.50	8.00	8.37	8.50	8.75	9.72
18-21 Sales occupations	71	213	7.67	4.9	7.95	2.7	5.76	6.50	7.00	7.05	7.50	7.97	8.25	8.48	8.64	9.62
18-21 Customer service occupations	72	46	7.93	3.5	8.37	2.7	5.92	7.05	7.14	7.38	7.65	8.28	8.70	8.99	9.30	9.98
18-21 Process, plant and machine operatives	8	47	8.14	3.8	8.79	2.0	6.59	7.19	7.47	7.50	7.79	8.75	9.27	9.63	9.97	11.48
18-21 Process, plant and machine operatives	81	36	8.10	2.3	8.67	0.3	6.87	7.19	x	7.50	7.79	8.66	9.25	9.62	9.93	10.93
18-21 Transport and mobile machine drivers and operatives	82	11	8.35	10.8	9.22	8.4	x	7.17	7.23	7.46	7.77	8.85	9.29	9.67	x	x
18-21 Elementary occupations	9	390	7.41	5.9	7.64	1.9	5.60	6.15	6.42	6.69	7.05	7.50	7.85	8.01	8.36	9.20
18-21 Elementary trades and related occupations	91	32	7.80	2.0	8.20	0.6	6.17	7.05	7.22	7.45	7.50	8.08	8.50	8.75	9.12	x
18-21 Elementary administration and service occupations	92	357	7.28	4.0	7.56	2.1	5.60	6.11	6.39	6.61	7.05	7.50	7.76	8.00	8.25	9.07

4.15b Hourly pay - Gross (£) - For all employee jobs[a]: United Kingdom, 2017

Description	Code	Number of jobs[b] (000s)	Median	Annual % change	Mean	Annual % change	Percentiles									
							10	20	25	30	40	60	70	75	80	90
22-29 ALL OCCUPATIONS		4,553	10.87	3.1	12.72	2.8	7.54	8.16	8.50	8.89	9.79	12.27	13.84	14.73	15.86	19.50
22-29 Managers, directors and senior officials	1	228	13.10	6.0	15.84	8.1	8.33	9.24	9.77	10.22	11.62	14.81	17.38	19.37	21.47	28.11
22-29 Corporate managers and directors	11	171	14.39	9.4	17.26	11.0	8.68	9.84	10.47	11.26	12.81	16.87	19.97	21.87	24.71	30.36
22-29 Other managers and proprietors	12	58	10.31	-0.4	11.64	-1.9	7.69	8.47	8.66	8.84	9.58	11.36	12.48	13.22	14.15	x
22-29 Professional occupations	2	934	15.69	0.9	16.79	1.7	10.96	12.56	13.22	13.59	14.52	16.92	18.54	19.52	20.60	24.21
22-29 Science, research, engineering and technology professionals	21	225	15.37	0.6	16.54	2.5	10.50	12.13	12.83	13.36	14.29	16.81	18.27	19.29	20.49	23.50
22-29 Health professionals	22	243	14.79	-0.5	15.97	2.2	11.47	12.51	13.05	13.44	13.99	15.92	16.96	17.95	19.26	23.11
22-29 Teaching and educational professionals	23	220	16.89	0.7	17.27	1.1	11.93	13.30	14.28	14.44	15.65	18.02	19.50	20.23	21.14	24.45
22-29 Business, media and public service professionals	24	246	15.87	-0.3	17.48	1.4	10.39	12.03	12.77	13.24	14.65	17.29	19.14	20.23	21.36	25.57
22-29 Associate professional and technical occupations	3	767	13.08	1.5	14.71	-0.8	9.02	10.14	10.65	11.17	12.12	14.19	15.34	16.19	17.24	20.44
22-29 Science, engineering and technology associate professionals	31	152	12.20	0.1	12.72	0.2	8.47	9.48	9.83	10.25	11.24	13.17	14.17	14.72	15.36	17.62
22-29 Health and social care associate professionals	32	47	11.36	1.1	11.82	-1.2	8.72	9.28	9.69	10.13	10.57	11.96	12.67	13.44	13.87	x
22-29 Protective service occupations	33	51	13.85	0.8	14.16	1.4	10.97	11.69	12.14	12.55	13.25	14.77	15.49	15.87	16.66	x
22-29 Culture, media and sports occupations	34	66	12.03	4.9	x		8.37	9.58	10.00	10.29	11.31	12.80	13.47	14.23	14.98	x
22-29 Business and public service associate professionals	35	450	13.81	2.1	15.07	1.8	9.33	10.54	11.11	11.64	12.68	15.00	16.48	17.50	18.50	22.16
22-29 Administrative and secretarial occupations	4	530	10.20	2.0	11.45	4.3	7.70	8.39	8.68	9.01	9.58	11.06	12.23	12.82	13.68	16.36
22-29 Administrative occupations	41	449	10.44	2.4	11.67	4.0	7.87	8.57	8.93	9.19	9.78	11.44	12.46	13.14	13.97	16.82
22-29 Secretarial and related occupations	42	81	9.01	3.0	10.12	3.8	7.50	7.68	7.92	8.08	8.50	9.63	10.22	10.86	11.56	13.68
22-29 Skilled trades occupations	5	376	10.85	4.8	11.94	5.1	7.69	8.42	8.75	9.08	9.99	12.00	13.05	13.72	14.48	16.94
22-29 Skilled agricultural and related trades	51	22	9.09	3.2	9.52	1.5	7.50	7.98	8.15	8.30	8.54	9.50	10.00	10.24	10.62	x
22-29 Skilled metal, electrical and electronic trades	52	193	12.66	5.3	13.34	5.0	8.59	9.65	10.06	10.61	11.74	13.51	14.53	15.46	16.31	18.71
22-29 Skilled construction and building trades	53	63	11.72	1.5	12.27	2.7	8.15	9.26	9.78	10.01	10.98	12.27	13.00	13.73	14.40	x
22-29 Textiles, printing and other skilled trades	54	98	8.52	1.3	9.25	3.7	7.50	7.56	7.78	8.00	8.25	9.00	9.55	9.90	10.07	11.81
22-29 Caring, leisure and other service occupations	6	475	8.66	3.2	9.28	3.3	7.50	7.63	7.78	7.97	8.28	9.14	9.74	10.10	10.59	12.24
22-29 Caring personal service occupations	61	387	8.67	3.2	9.19	3.0	7.50	7.65	7.81	7.99	8.29	9.15	9.72	10.08	10.53	12.05
22-29 Leisure, travel and related personal service occupations	62	88	8.62	4.5	9.69	4.6	7.47	7.50	7.64	7.80	8.20	9.12	9.83	10.20	10.97	13.00
22-29 Sales and customer service occupations	7	521	8.55	3.2	9.54	3.1	7.49	7.58	7.68	7.81	8.16	9.10	9.77	10.19	10.73	12.64
22-29 Sales occupations	71	361	8.28	3.9	9.16	2.8	7.45	7.50	7.61	7.70	8.00	8.67	9.33	9.75	10.18	12.00
22-29 Customer service occupations	72	160	9.34	1.5	10.27	3.3	7.50	7.91	8.16	8.40	8.85	9.95	10.65	11.12	11.73	13.73
22-29 Process, plant and machine operatives	8	206	9.61	2.1	10.70	1.3	7.50	7.90	8.13	8.34	8.91	10.41	11.36	12.01	12.74	14.86
22-29 Process, plant and machine operatives	81	138	9.51	0.2	10.73	0.6	7.50	7.78	8.00	8.27	8.78	10.29	11.46	12.12	12.96	15.13
22-29 Transport and mobile machine drivers and operatives	82	67	9.85	5.3	10.64	2.5	7.50	8.07	8.30	8.50	9.19	10.65	11.29	11.74	12.30	x
22-29 Elementary occupations	9	517	8.16	5.3	9.00	4.5	7.40	x	7.50	7.58	7.83	8.53	9.11	9.49	9.94	11.22
22-29 Elementary trades and related occupations	91	72	8.63	3.8	9.38	3.8	7.50	7.54	7.69	7.85	8.23	9.21	10.00	10.28	10.75	12.21
22-29 Elementary administration and service occupations	92	445	8.08	5.1	8.92	4.6	7.32	x	7.50	7.55	7.80	8.47	8.99	9.35	9.72	11.00
30-39 ALL OCCUPATIONS		6,134	13.93	2.3	16.74	2.4	7.98	9.08	9.73	10.41	12.05	16.06	18.58	20.00	21.78	27.86
30-39 Managers, directors and senior officials	1	601	19.55	3.0	23.67	2.2	9.86	12.05	13.15	14.24	16.75	22.99	27.61	30.59	33.90	44.00
30-39 Corporate managers and directors	11	493	21.79	4.8	25.56	2.9	10.42	13.17	14.39	15.79	18.48	25.50	30.53	33.34	36.79	46.51
30-39 Other managers and proprietors	12	108	13.35	-1.3	15.20	2.1	8.74	9.93	10.39	10.79	11.98	14.78	16.67	17.80	19.33	22.83
30-39 Professional occupations	2	1,602	20.23	0.9	22.01	1.9	13.41	15.53	16.37	17.14	18.65	22.01	24.29	25.64	27.26	32.34
30-39 Science, research, engineering and technology professionals	21	391	20.67	2.0	22.22	2.8	13.69	15.84	16.77	17.59	19.16	22.55	24.75	25.93	27.38	32.32

4.15b Hourly pay - Gross (£) - For all employee jobs[a]: United Kingdom, 2017

Description	Code	Number of jobs[b] (000s)	Median	Annual % change	Mean	Annual % change	10	20	25	30	40	60	70	75	80	90
30-39 Health professionals	22	392	18.01	0.0	20.68	2.7	12.90	14.56	15.00	15.72	16.87	19.47	21.41	23.06	25.00	31.67
30-39 Teaching and educational professionals	23	395	22.30	0.1	22.56	0.9	14.69	17.61	18.60	19.53	20.98	23.81	25.61	26.65	27.92	31.35
30-39 Business, media and public service professionals	24	425	19.77	0.8	22.51	1.5	12.92	15.33	16.15	16.85	18.21	21.52	24.09	25.63	27.54	34.71
30-39 Associate professional and technical occupations	3	1,083	16.10	3.3	18.45	3.8	10.32	12.08	12.70	13.34	14.68	17.86	19.39	20.44	21.72	27.38
30-39 Science, engineering and technology associate professionals	31	178	14.41	2.8	15.51	2.8	9.58	11.00	11.76	12.27	13.28	15.58	16.96	18.03	19.38	22.15
30-39 Health and social care associate professionals	32	89	12.96	2.2	13.42	1.1	9.27	10.37	10.78	11.18	11.95	13.77	14.78	15.30	15.81	17.46
30-39 Protective service occupations	33	135	18.21	6.3	17.61	3.0	12.86	14.37	15.16	15.96	17.24	18.83	19.46	19.78	20.28	21.41
30-39 Culture, media and sports occupations	34	67	14.00	-0.4	18.90	-0.8	9.26	10.87	11.56	12.35	13.02	15.33	16.86	17.50	18.45	x
30-39 Business and public service associate professionals	35	614	17.23	4.1	20.15	4.5	10.92	12.45	13.22	14.05	15.39	19.06	21.41	23.02	24.97	31.73
30-39 Administrative and secretarial occupations	4	615	11.23	2.2	12.89	3.1	8.06	8.99	9.30	9.67	10.30	12.28	13.68	14.46	15.50	19.26
30-39 Administrative occupations	41	517	11.44	1.8	13.09	3.0	8.25	9.17	9.50	9.89	10.51	12.50	13.90	14.76	15.76	19.69
30-39 Secretarial and related occupations	42	98	10.00	0.8	11.73	2.8	7.60	8.10	8.39	8.72	9.35	10.91	11.96	12.86	13.89	17.08
30-39 Skilled trades occupations	5	447	12.14	4.0	13.08	4.2	8.00	9.02	9.54	10.00	11.12	13.23	14.49	15.20	16.01	18.44
30-39 Skilled agricultural and related trades	51	21	9.72	1.6	10.26	3.9	7.94	8.60	8.75	8.87	9.18	10.10	10.73	11.60	11.86	x
30-39 Skilled metal, electrical and electronic trades	52	221	14.18	4.9	14.78	3.2	9.67	11.11	11.69	12.15	13.14	15.04	16.33	17.02	17.87	20.71
30-39 Skilled construction and building trades	53	73	12.75	6.3	13.08	6.2	9.06	10.00	10.48	11.00	11.76	13.48	14.35	14.95	15.51	x
30-39 Textiles, printing and other skilled trades	54	131	9.34	5.2	10.42	5.9	7.50	7.80	8.00	8.25	8.73	9.99	10.93	11.46	11.89	13.51
30-39 Caring, leisure and other service occupations	6	496	9.21	2.4	10.09	2.6	7.52	7.92	8.08	8.29	8.71	9.83	10.49	11.02	11.62	13.37
30-39 Caring personal service occupations	61	409	9.20	2.7	9.91	2.6	7.57	7.93	8.10	8.29	8.72	9.78	10.42	10.89	11.49	13.07
30-39 Leisure, travel and related personal service occupations	62	87	9.37	3.3	10.94	2.5	7.50	7.76	8.00	8.30	8.68	10.10	11.13	11.78	12.40	15.51
30-39 Sales and customer service occupations	7	421	8.95	2.2	10.56	3.1	7.50	7.65	7.79	7.99	8.44	9.68	10.68	11.34	12.28	14.98
30-39 Sales occupations	71	276	8.44	3.0	9.80	2.5	7.50	7.53	7.63	7.73	8.05	8.85	9.59	10.00	10.72	13.06
30-39 Customer service occupations	72	145	10.53	3.2	11.85	3.9	7.69	8.35	8.67	9.04	9.77	11.50	12.61	13.32	14.33	16.98
30-39 Process, plant and machine operatives	8	317	10.46	1.7	11.64	2.8	7.64	8.32	8.69	9.00	9.72	11.40	12.38	12.97	13.76	16.41
30-39 Process, plant and machine operatives	81	175	10.19	1.4	11.35	1.9	7.51	8.17	8.46	8.71	9.31	11.09	12.23	12.85	13.76	16.36
30-39 Transport and mobile machine drivers and operatives	82	143	10.95	3.9	11.98	3.8	7.80	8.66	9.04	9.49	10.15	11.66	12.47	13.01	13.75	16.45
30-39 Elementary occupations	9	551	8.50	4.1	9.69	4.3	x	7.50	7.62	7.78	8.13	9.03	9.75	10.23	10.72	12.57
30-39 Elementary trades and related occupations	91	76	8.96	3.8	9.93	5.7	7.50	7.77	8.00	8.18	8.54	9.52	10.40	10.91	11.57	13.27
30-39 Elementary administration and service occupations	92	475	8.45	4.6	9.64	4.0	x	7.50	7.57	7.73	8.04	8.98	9.67	10.12	10.63	12.41
40-49 ALL OCCUPATIONS		6,318	14.32	2.7	18.27	2.9	8.00	9.14	9.78	10.44	12.12	16.86	19.74	21.55	23.77	31.22
40-49 Managers, directors and senior officials	1	782	23.02	1.8	28.87	2.1	10.64	14.10	15.43	16.99	19.92	27.04	32.22	35.49	39.22	52.62
40-49 Corporate managers and directors	11	669	24.77	3.0	30.67	3.7	11.50	15.19	16.88	18.45	21.47	29.25	34.32	37.49	41.81	55.52
40-49 Other managers and proprietors	12	112	15.69	0.1	18.01	-4.4	9.16	10.55	11.28	12.09	14.08	17.44	19.54	20.95	22.22	x
40-49 Professional occupations	2	1,490	21.85	1.3	24.42	2.1	13.88	16.43	17.50	18.28	20.00	23.97	26.44	28.16	30.28	37.03
40-49 Science, research, engineering and technology professionals	21	340	23.07	1.6	24.59	2.7	14.72	17.61	18.54	19.33	21.17	25.10	27.32	28.89	30.64	35.48
40-49 Health professionals	22	404	18.92	0.3	23.18	1.8	13.44	15.00	15.72	16.56	17.96	20.88	22.49	24.01	26.70	40.13
40-49 Teaching and educational professionals	23	395	24.02	0.7	25.04	1.6	15.33	18.61	19.58	20.83	22.50	25.71	27.91	29.17	30.70	35.96
40-49 Business, media and public service professionals	24	351	21.65	2.0	24.97	2.1	12.92	16.10	17.03	17.89	19.63	23.95	26.96	28.80	30.90	38.95
40-49 Associate professional and technical occupations	3	995	17.01	2.8	19.47	3.8	10.50	12.46	13.21	13.89	15.37	18.60	20.44	21.47	23.09	29.74
40-49 Science, engineering and technology associate professionals	31	158	15.30	2.1	16.31	1.6	9.65	11.20	12.01	12.75	13.89	16.61	17.94	18.84	19.94	23.28

4.15b Hourly pay - Gross (£) - For all employee jobs[a]: United Kingdom, 2017

Description	Code	Number of jobs[b] (000s)	Median	Annual % change	Mean	Annual % change	10	20	25	30	40	60	70	75	80	90
40-49 Health and social care associate professionals	32	89	13.30	2.1	14.13	0.4	9.49	10.57	11.03	11.49	12.44	14.21	15.04	15.63	16.77	18.93
40-49 Protective service occupations	33	149	18.86	1.8	18.92	3.7	13.88	15.66	16.60	17.46	18.31	19.53	20.47	21.05	21.43	23.01
40-49 Culture, media and sports occupations	34	44	15.00	-0.3	16.92	1.5	9.65	10.88	11.73	12.33	13.51	16.32	18.21	19.45	20.44	x
40-49 Business and public service associate professionals	35	555	17.76	2.8	21.49	4.8	10.93	12.78	13.53	14.37	15.97	19.79	22.61	24.51	27.23	36.00
40-49 Administrative and secretarial occupations	4	707	11.28	2.0	13.24	3.9	8.23	9.09	9.39	9.78	10.41	12.29	13.75	14.54	15.69	19.81
40-49 Administrative occupations	41	562	11.52	2.0	13.54	4.7	8.45	9.23	9.62	9.99	10.65	12.67	14.06	15.00	16.13	20.60
40-49 Secretarial and related occupations	42	145	10.24	1.7	11.95	-0.1	7.93	8.49	8.74	9.07	9.66	11.19	12.07	12.88	13.84	17.38
40-49 Skilled trades occupations	5	457	12.29	1.6	13.62	1.5	8.06	9.12	9.59	10.09	11.20	13.53	14.96	15.67	16.58	19.75
40-49 Skilled agricultural and related trades	51	23	10.12	1.3	11.56	5.5	7.66	8.57	9.03	9.19	9.53	11.00	12.00	12.36	x	x
40-49 Skilled metal, electrical and electronic trades	52	233	14.25	1.5	15.14	0.5	9.61	10.98	11.50	12.00	13.12	15.30	16.58	17.48	18.33	21.81
40-49 Skilled construction and building trades	53	75	13.00	3.2	13.85	3.4	9.00	10.08	10.78	11.36	12.17	14.09	15.00	15.55	16.10	x
40-49 Textiles, printing and other skilled trades	54	127	9.19	5.1	10.52	3.6	7.50	7.80	8.00	8.20	8.74	9.80	10.59	11.23	11.96	14.37
40-49 Caring, leisure and other service occupations	6	588	9.34	3.3	10.31	3.2	7.65	8.05	8.24	8.43	8.85	9.99	10.65	11.17	11.75	13.63
40-49 Caring personal service occupations	61	509	9.29	2.8	10.01	2.4	7.68	8.06	8.25	8.42	8.84	9.90	10.52	10.98	11.56	13.22
40-49 Leisure, travel and related personal service occupations	62	79	9.84	7.1	12.20	7.6	7.50	7.97	8.19	8.50	9.00	10.58	11.79	12.81	14.33	x
40-49 Sales and customer service occupations	7	349	8.75	3.0	10.66	2.9	7.50	7.62	7.71	7.84	8.28	9.50	10.42	11.13	11.88	14.97
40-49 Sales occupations	71	247	8.38	4.0	9.83	3.8	7.50	7.51	7.62	7.70	7.98	8.79	9.48	9.86	10.51	12.66
40-49 Customer service occupations	72	102	10.66	1.5	12.38	1.4	7.65	8.22	8.63	9.15	9.91	11.56	12.81	13.55	14.64	17.95
40-49 Process, plant and machine operatives	8	380	10.84	1.5	12.00	1.1	7.70	8.46	8.83	9.17	10.00	11.72	12.78	13.56	14.41	17.27
40-49 Process, plant and machine operatives	81	171	10.48	0.1	11.81	0.1	7.59	8.09	8.50	8.79	9.62	11.58	12.94	13.75	14.63	17.42
40-49 Transport and mobile machine drivers and operatives	82	210	11.02	1.8	12.14	1.9	7.93	8.76	9.10	9.52	10.25	11.83	12.70	13.35	14.20	16.97
40-49 Elementary occupations	9	569	8.47	3.5	9.82	3.7	x	7.50	7.61	7.79	8.08	9.06	9.81	10.30	10.82	12.86
40-49 Elementary trades and related occupations	91	72	8.91	3.3	9.93	3.7	7.50	7.61	7.80	8.00	8.40	9.50	10.47	10.98	11.47	13.50
40-49 Elementary administration and service occupations	92	497	8.44	3.9	9.79	3.6	x	7.50	7.60	7.77	8.04	9.00	9.75	10.16	10.71	12.67
50-59 ALL OCCUPATIONS		5,659	13.47	1.9	17.68	2.3	7.93	8.95	9.51	10.09	11.55	15.85	18.84	20.64	22.66	30.00
50-59 Managers, directors and senior officials	1	720	22.29	3.5	28.57	3.7	10.21	13.56	14.96	16.16	19.12	25.95	30.77	34.73	39.09	52.39
50-59 Corporate managers and directors	11	612	23.84	3.0	30.27	3.6	10.77	14.42	15.67	17.31	20.44	28.00	33.48	37.07	41.56	54.68
50-59 Other managers and proprietors	12	108	16.30	2.6	18.86	4.5	9.04	10.62	11.75	12.70	14.25	18.25	20.92	22.57	24.08	x
50-59 Professional occupations	2	1,224	21.36	1.0	24.19	1.4	13.92	16.28	17.24	18.02	19.64	23.39	25.70	27.30	29.48	36.00
50-59 Science, research, engineering and technology professionals	21	252	22.72	3.0	24.34	1.8	14.48	17.22	18.17	19.07	20.86	24.94	27.26	28.76	30.35	35.57
50-59 Health professionals	22	371	19.00	0.1	23.30	0.8	13.66	15.25	16.05	16.73	18.01	20.80	21.98	23.47	25.06	36.00
50-59 Teaching and educational professionals	23	329	24.01	0.2	25.28	0.6	14.99	18.40	19.58	20.24	22.38	25.47	27.68	29.40	30.77	36.09
50-59 Business, media and public service professionals	24	272	20.96	2.3	24.03	2.4	12.66	15.49	16.46	17.44	19.16	23.01	25.50	27.02	29.46	36.09
50-59 Associate professional and technical occupations	3	701	16.45	2.7	19.08	2.3	10.41	12.07	12.77	13.35	14.79	18.23	20.29	21.37	23.01	29.58
50-59 Science, engineering and technology associate professionals	31	135	15.43	0.8	16.54	0.5	9.80	11.05	11.87	12.53	13.91	17.00	18.50	19.56	20.72	24.69
50-59 Health and social care associate professionals	32	84	13.19	1.5	14.27	1.9	9.57	10.75	11.17	11.74	12.33	14.27	15.24	16.14	16.82	18.82
50-59 Protective service occupations	33	61	18.34	1.7	18.62	5.1	11.97	14.27	14.96	15.64	17.19	19.51	20.93	21.33	21.65	x
50-59 Culture, media and sports occupations	34	27	15.65	4.4	18.15	-1.1	10.01	11.62	11.98	12.95	13.95	16.94	19.43	20.91	22.96	x
50-59 Business and public service associate professionals	35	393	17.52	1.3	21.01	2.1	10.95	12.73	13.29	14.19	15.67	19.62	22.04	23.98	26.41	34.73
50-59 Administrative and secretarial occupations	4	763	11.03	2.3	12.85	3.2	8.13	8.97	9.25	9.63	10.22	12.00	13.26	14.18	15.18	19.11

4.15b Hourly pay - Gross (£) - For all employee jobs[a]: United Kingdom, 2017

Description	Code	Number of jobs[b] (000s)	Median	Annual % change	Mean	Annual % change	10	20	25	30	40	60	70	75	80	90
50-59 Administrative occupations	41	579	11.37	1.8	13.19	2.9	8.30	9.19	9.57	9.90	10.49	12.46	13.76	14.66	15.63	19.78
50-59 Secretarial and related occupations	42	184	10.02	1.9	11.70	4.4	7.88	8.35	8.63	8.96	9.46	10.87	11.71	12.29	13.21	16.56
50-59 Skilled trades occupations	5	400	13.07	3.5	14.15	2.3	8.32	9.50	10.07	10.68	11.76	14.20	15.46	16.33	17.12	20.40
50-59 Skilled agricultural and related trades	51	26	9.79	-0.7	10.50	0.2	7.57	8.11	8.29	8.75	9.32	10.27	10.78	11.34	11.94	x
50-59 Skilled metal, electrical and electronic trades	52	228	14.67	-0.1	15.67	1.4	10.30	11.50	12.12	12.78	13.73	15.81	16.90	17.72	18.71	21.97
50-59 Skilled construction and building trades	53	61	13.50	5.1	14.35	2.4	9.84	11.00	11.57	11.80	12.67	14.36	15.07	15.79	16.98	x
50-59 Textiles, printing and other skilled trades	54	86	9.19	2.5	10.27	3.0	7.50	7.91	8.15	8.37	8.69	9.69	10.30	10.91	11.47	13.66
50-59 Caring, leisure and other service occupations	6	546	9.48	2.4	10.46	1.9	7.65	8.07	8.30	8.51	9.00	10.06	10.82	11.32	11.91	13.80
50-59 Caring personal service occupations	61	461	9.44	2.3	10.15	2.0	7.66	8.06	8.28	8.50	8.98	10.04	10.72	11.14	11.71	13.35
50-59 Leisure, travel and related personal service occupations	62	84	9.71	1.8	12.11	1.8	7.60	8.13	8.40	8.57	9.11	10.61	11.79	12.44	13.57	x
50-59 Sales and customer service occupations	7	350	8.50	3.7	10.25	3.9	7.50	7.60	7.65	7.76	8.09	9.01	9.81	10.42	11.30	14.29
50-59 Sales occupations	71	263	8.22	4.8	9.42	5.4	7.50	7.50	7.62	7.66	7.90	8.56	9.06	9.47	9.92	12.00
50-59 Customer service occupations	72	87	10.56	1.8	12.41	1.2	7.63	8.04	8.46	8.79	9.55	11.58	12.94	14.05	15.21	18.53
50-59 Process, plant and machine operatives	8	403	10.88	2.5	12.07	3.5	7.75	8.50	8.92	9.34	10.02	11.68	12.69	13.43	14.42	17.41
50-59 Process, plant and machine operatives	81	163	10.87	1.8	12.22	2.8	7.61	8.36	8.69	9.10	9.84	11.85	13.10	14.00	15.08	18.24
50-59 Transport and mobile machine drivers and operatives	82	240	10.88	2.8	11.98	4.0	7.89	8.57	9.06	9.47	10.13	11.60	12.49	13.14	13.89	16.68
50-59 Elementary occupations	9	553	8.47	2.8	9.84	4.0	7.50	7.50	7.63	7.80	8.11	9.06	9.95	10.48	10.96	12.90
50-59 Elementary trades and related occupations	91	67	9.32	2.6	10.29	2.6	7.50	7.62	7.83	8.05	8.57	10.00	10.70	11.28	11.97	14.20
50-59 Elementary administration and service occupations	92	487	8.42	3.4	9.76	4.3	x	7.50	7.62	7.78	8.07	8.94	9.77	10.35	10.81	12.73
60+ ALL OCCUPATIONS		2,075	11.56	2.3	15.60	3.8	7.65	8.33	8.71	9.19	10.15	13.28	15.80	17.54	19.58	26.68
60+ Managers, directors and senior officials	1	244	18.69	4.2	24.84	5.5	8.89	11.33	12.45	13.44	16.00	22.40	27.47	30.33	33.34	45.72
60+ Corporate managers and directors	11	198	20.00	6.1	26.37	7.1	9.00	11.74	12.75	14.10	16.97	24.15	29.46	32.08	35.50	50.47
60+ Other managers and proprietors	12	45	15.23	-0.5	18.10	0.4	8.60	9.96	10.84	12.07	13.02	16.87	19.15	20.61	22.12	x
60+ Professional occupations	2	361	20.86	2.0	23.21	2.0	12.36	14.87	16.07	17.16	18.77	22.60	25.19	27.10	29.02	35.51
60+ Science, research, engineering and technology professionals	21	63	21.15	0.0	22.70	1.2	13.49	15.56	16.64	17.83	19.47	23.61	26.33	27.47	29.08	x
60+ Health professionals	22	97	18.30	1.6	22.23	0.7	12.94	14.55	14.87	15.49	17.16	19.60	21.17	22.21	24.29	x
60+ Teaching and educational professionals	23	110	24.00	1.2	25.94	1.7	13.36	17.69	18.96	19.83	22.25	25.95	28.27	30.00	31.22	38.13
60+ Business, media and public service professionals	24	91	19.43	6.1	22.22	4.8	11.11	13.06	14.00	15.80	17.84	21.48	24.37	25.69	28.16	35.34
60+ Associate professional and technical occupations	3	196	14.81	2.8	17.52	3.6	9.45	10.78	11.61	12.10	13.39	16.32	18.20	19.72	21.19	27.13
60+ Science, engineering and technology associate professionals	31	38	14.46	3.3	15.97	4.9	9.51	10.53	11.26	11.72	13.03	15.94	17.31	18.12	20.39	x
60+ Health and social care associate professionals	32	24	12.52	-0.9	13.93	-0.7	9.05	9.86	10.43	11.11	11.87	13.44	14.70	15.57	16.07	x
60+ Protective service occupations	33	7	16.22	13.0	17.08	13.8	x	11.16	12.67	13.20	14.58	18.23	x	x	x	x
60+ Culture, media and sports occupations	34	13	13.08	-9.5	16.34	4.1	9.44	10.01	10.48	11.14	12.52	14.69	16.18	x	x	x
60+ Business and public service associate professionals	35	115	15.81	3.2	18.77	2.6	9.61	11.56	12.07	12.85	14.15	17.32	19.72	21.20	23.03	30.81
60+ Administrative and secretarial occupations	4	313	10.53	2.2	12.12	3.3	7.98	8.67	9.05	9.24	10.00	11.49	12.63	13.36	14.34	17.80
60+ Administrative occupations	41	228	10.85	3.5	12.38	3.6	8.08	8.88	9.19	9.51	10.07	11.80	12.93	13.72	14.61	18.26
60+ Secretarial and related occupations	42	86	10.00	0.3	11.43	2.3	7.75	8.28	8.58	8.87	9.29	10.61	11.73	12.52	13.34	16.23
60+ Skilled trades occupations	5	149	12.06	2.5	13.30	2.9	8.00	9.03	9.45	10.00	11.13	13.16	14.33	15.04	15.84	18.82
60+ Skilled agricultural and related trades	51	12	8.91	-1.2	9.61	-0.8	7.50	7.65	7.75	7.91	8.34	9.38	10.04	10.29	x	x
60+ Skilled metal, electrical and electronic trades	52	79	13.79	1.7	14.55	1.8	9.43	10.96	11.42	11.89	12.96	14.70	15.69	16.34	17.49	20.39
60+ Skilled construction and building trades	53	23	12.61	5.3	13.56	5.8	9.20	10.80	11.03	11.26	11.96	13.42	14.29	14.91	15.44	x

4.15b Hourly pay - Gross (£) - For all employee jobs[a]: United Kingdom, 2017

Description	Code	Number of jobs[b] (000s)	Median	Annual % change	Mean	Annual % change	Percentiles									
							10	20	25	30	40	60	70	75	80	90
60+ Textiles, printing and other skilled trades	54	35	9.25	5.7	10.69	2.8	7.50	7.93	8.00	8.34	8.80	9.73	10.52	11.39	11.95	x
60+ Caring, leisure and other service occupations	6	199	9.30	0.9	9.98	-0.2	7.60	8.05	8.27	8.46	8.89	9.95	10.60	11.08	11.67	13.41
60+ Caring personal service occupations	61	155	9.33	0.6	9.87	-0.9	7.60	8.08	8.30	8.50	8.96	9.96	10.65	11.08	11.61	13.48
60+ Leisure, travel and related personal service occupations	62	43	9.15	2.7	10.42	2.5	7.51	8.00	8.18	8.40	8.62	9.77	10.45	11.06	11.72	x
60+ Sales and customer service occupations	7	155	8.27	4.1	9.47	2.6	7.50	7.50	7.60	7.66	7.89	8.55	9.21	9.58	10.09	12.01
60+ Sales occupations	71	128	8.07	4.6	9.04	4.0	x	7.50	7.53	7.62	7.79	8.50	8.84	9.20	9.59	11.21
60+ Customer service occupations	72	28	9.41	-6.1	11.10	-1.4	7.59	7.80	7.91	8.21	8.81	10.25	11.38	12.14	12.89	x
60+ Process, plant and machine operatives	8	192	9.98	2.4	11.25	3.9	7.50	8.00	8.23	8.54	9.20	10.83	11.75	12.25	12.95	15.54
60+ Process, plant and machine operatives	81	60	10.25	2.7	11.71	4.4	7.60	8.09	8.27	8.61	9.35	11.42	12.50	13.10	13.74	16.44
60+ Transport and mobile machine drivers and operatives	82	132	9.85	2.5	11.04	3.9	7.50	7.93	8.21	8.50	9.11	10.55	11.50	12.00	12.50	14.73
60+ Elementary occupations	9	266	8.30	3.7	9.52	3.0	x	7.50	7.56	7.70	7.95	8.69	9.51	9.94	10.50	12.15
60+ Elementary trades and related occupations	91	31	8.99	0.0	9.96	0.9	7.50	7.70	7.86	8.00	8.50	9.74	10.66	11.16	11.50	x
60+ Elementary administration and service occupations	92	235	8.24	3.7	9.43	3.3	x	7.50	7.53	7.65	7.93	8.59	9.32	9.77	10.27	11.92
Not Classified		:														

Source: Annual Survey of Hours and Earnings, Office for National Statistics.

a Employees on adult rates whose pay for the survey pay-period was not affected by absence.

b Figures for Number of Jobs are for indicative purposes only and should not be considered an accurate estimate of employee job counts.

KEY - The colour coding indicates the quality of each estimate; jobs, median, mean and percentiles but not the annual percentage change.

The quality of an estimate is measured by its coefficient of variation (CV), which is the ratio of the standard error of an estimate to the estimate.

Key	Statistical robustness
CV <= 5%	Estimates are considered precise
CV > 5% and <= 10%	Estimates are considered reasonably precise
CV > 10% and <= 20%	Estimates are considered acceptable
x = CV > 20%	Estimates are considered unreliable for practical purposes
.. = disclosive	
: = not applicable	
- = nil or negligible	

4.16 Median weekly and hourly earnings for all employees[a] by age group: United Kingdom April 2007 to 2017

	18-21	22-29	30-39	40-49	50-59	60+
Median gross weekly earnings						
All	JRG9	JRH2	JRH3	JRH4	JEH5	JRH6
2007	265.5	387.8	509.0	517.3	479.1	418.7
2008	271.6	400.0	532.7	539.9	504.1	437.5
2009	277.5	407.1	541.8	550.5	514.0	446.3
2010	277.4	411.2	547.8	559.6	528.2	457.3
2011	277.8	406.6	553.7	564.7	531.8	466.1
2012	279.7	412.0	557.7	573.1	535.9	476.1
2013	287.5	420.6	562.9	579.9	551.0	490.4
2014	290.0	424.8	566.0	588.2	556.1	491.3
2015	199.3	383.2	489.1	491.9	458.0	338.9
2016	206.8	400.0	498.3	503.7	473.4	348.8
2017	217.7	411.1	508.3	514.2	482.0	358.6
Men	JRH8	JRH9	JRI2	JRI3	JRI4	JRI5
2007	275.9	402.5	539.0	574.9	534.4	440.9
2008	280.0	416.7	566.3	599.1	563.6	462.6
2009	285.7	421.6	571.1	605.2	569.7	469.0
2010	285.9	421.2	573.7	613.7	582.7	483.0
2011	288.2	413.7	574.9	618.9	586.7	496.3
2012	295.4	420.6	574.9	622.9	598.2	508.3
2013	299.0	430.6	579.6	637.8	613.3	527.5
2014	301.7	435.6	578.6	641.3	621.1	528.0
2015	233.0	411.6	556.5	618.8	593.9	443.1
2016	252.8	430.4	574.8	632.4	609.5	458.6
2017	260.6	442.4	582.6	645.9	618.7	465.9
Women	JRI7	JRI8	JRI9	JRJ2	JRJ3	JRJ4
2007	254.3	374.1	460.6	420.3	395.6	356.1
2008	258.8	384.7	480.9	437.3	419.7	376.4
2009	268.3	392.9	497.9	457.7	432.8	382.1
2010	268.3	401.3	507.9	472.2	440.9	389.0
2011	264.7	398.5	519.4	477.1	449.1	399.6
2012	266.3	402.5	527.0	484.4	446.0	407.4
2013	276.2	405.7	533.8	494.1	463.5	416.1
2014	279.2	414.0	537.3	507.3	465.4	412.0
2015	166.6	347.5	401.0	365.2	342.8	235.7
2016	167.4	364.1	408.2	373.6	364.1	251.9
2017	177.8	373.5	416.2	384.2	367.3	258.6

4.16 Median weekly and hourly earnings for all employee jobs[a] by age group: United Kingdom April 2007 to 2017

	18-21	22-29	30-39	40-49	50-59	60+
Median hourly earnings (excluding overtime)						
All	JRJ6	JRJ7	JRJ8	JRJ9	JRK2	JRK3
2007	6.60	9.80	12.77	12.77	11.87	10.09
2008	6.75	10.12	13.34	13.31	12.53	10.57
2009	7.00	10.43	13.82	13.83	12.95	10.97
2010	7.00	10.34	13.91	14.01	13.19	11.18
2011	6.97	10.23	14.07	14.16	13.34	11.46
2012	7.00	10.34	14.23	14.37	13.55	11.83
2013	7.20	10.60	14.37	14.63	13.89	12.13
2014	7.25	10.76	14.37	14.82	14.00	12.22
2015	7.08	10.03	13.31	13.49	12.68	10.88
2016	7.45	10.51	13.55	13.90	13.17	11.25
2017	7.80	10.84	13.89	14.27	13.43	11.51
Men	JRK5	JRK6	JRK7	JRK8	JRK9	JRL2
2007	6.65	9.80	13.10	13.77	12.79	10.31
2008	6.85	10.13	13.70	14.37	13.52	10.89
2009	7.09	10.45	14.15	14.95	13.91	11.25
2010	7.05	10.26	14.07	14.95	14.25	11.50
2011	7.07	10.08	14.14	15.14	14.31	11.84
2012	7.24	10.22	14.28	15.27	14.68	12.15
2013	7.27	10.57	14.43	15.62	15.00	12.61
2014	7.42	10.73	14.35	15.69	15.13	12.78
2015	7.25	10.26	14.05	15.41	14.97	12.26
2016	7.50	10.87	14.34	15.86	15.39	12.56
2017	7.90	11.25	14.68	16.30	15.65	12.84
Women	JRL4	JRL5	JRL6	JRL7	JRL8	JRL9
2007	6.55	9.79	12.28	11.14	10.54	9.48
2008	6.64	10.12	12.78	11.57	11.10	9.82
2009	6.93	10.40	13.29	12.21	11.52	10.23
2010	6.90	10.48	13.66	12.54	11.84	10.40
2011	6.76	10.45	13.99	12.77	12.02	10.64
2012	6.82	10.53	14.18	12.93	11.99	10.99
2013	7.15	10.64	14.28	13.18	12.38	11.10
2014	7.06	10.81	14.37	13.53	12.41	10.99
2015	7.00	9.83	12.41	11.55	10.84	9.60
2016	7.39	10.15	12.62	11.88	11.45	10.00
2017	7.73	10.40	12.82	12.19	11.50	10.19

Source: Annual Survey of Hours and Earnings, Office for National Statistics

a. Employees on adult rates whose pay for the survey pay-period was not affected by absence.

4.17 Trade Unions 2017-18

The trade unions from which returns were received, recorded a total membership of 6,875,231. The 13 unions with a membership of over 100,000, accounted for 5,824,111 members or 84.7% of the total. Returns received in the period show the distribution of trade union membership by size is as follows:

Trade unions: distribution by size

			Number of Unions		Membership of all Unions	
Number of Members	*Number of Returns*	*Membership*	*Per cent*	*Cumulative Per cent*	*Per cent*	*Cumulative Per cent*
Under 100	23	458	15.0	15.0	0.0	0.0
100-499	27	6,640	17.7	32.7	0.1	0.1
500-999	18	12,582	11.8	44.4	0.2	0.3
1,000-2,499	25	39,145	16.3	60.8	0.6	0.9
2,500-4,999	8	31,284	5.2	66.0	0.5	1.3
5,000-9,999	8	61,638	5.2	71.2	0.9	2.2
10,000-14,999	5	61,360	3.3	74.5	0.9	3.1
15,000-24,999	10	201,055	6.5	81.1	2.9	6.0
25,000-49,999	12	395,250	7.9	88.9	5.7	11.8
50,000-99,999	4	241,708	2.6	91.5	3.5	15.3
100,000-249,999	6	947,328	3.9	95.4	13.8	29.1
250,000 and over	7	4,876,783	4.6	100.0	70.9	100.0
Total	153	6,875,231	100		100	

The trade union membership of 6,875,231 recorded in this annual report compares to 6,865,056 reported in the previous annual report. This indicates an increase of 10,175 members or 0.15%. However, this included 147,330 retired members of UNISON that had not been included in previous totals. Without this figure the total figure would have been 6,727,901 or a decrease of 2%. The total recorded membership of around 7.0 million compares with a peak of 13.2 million in 1979, a fall of about 48%.

4.17 Trade Unions 2017-18

The following table shows the trade unions whose membership has increased or decreased by more than 5,000 since the previous reporting period.

Trade Unions: Changes in Membership over 5,000 members

	Total Membership		
	2016-2017	*2015-2016*	*% changes*
Increases			
Royal College of Nursing	452,669	444,685	+1.79
UNISON: The Public Service Union	1,397,803[1]	1,255,653	+11.32
Decreases			
British Medical Association	161,708	169,908	-4.83
Association of Teachers and Lecturers	192,646	200,631	-3.98
Public and Commercial Services Union	185,785	195,091	-4.77
Unite the Union	1,282,671	1,382,126	-7.19

[1] This figure includes 147,330 retired members that had not been included in the previous year's totals

The annual returns submitted by unions to the Certification Officer require each union to provide figures for both total membership and members who pay contributions. There can be significant differences between these figures. This is usually the result of total membership figures including retired and unemployed members, members on long term sick and maternity/child care leave and those on career breaks. The returns submitted by unions during this reporting period show that the total number of contributing members was around 94.6% of the total number of members. This compared to 94.2% in the preceding year.

Source: Annual Report of the Certification Officer 2017/18

this page is intentionally blank

Social protection

Chapter 5

Social Protection

(Tables 5.2 to 5.11, 5.13 and 5.15 to 5.19)

Tables 5.2 to 5.6, 5.9 to 5.11 and 5.13 to 5.19 give details of contributors and beneficiaries under the National Insurance and Industrial Injury Acts, supplementary benefits and war pensions.

There are five classes of National Insurance Contributions (NICs):

Class 1 Earnings-related contributions paid on earnings from employment. Employees pay primary Class 1 contributions and employers pay secondary Class 1 contributions. Payment of Class 1 contributions builds up entitlement to contributory benefits which include Basic State Pension; Additional State Pension (State Earnings Related Pension Scheme SERPS and from April 2002, State Second Pension, S2P); Contribution Based Jobseeker's Allowance; Bereavement Benefits and Employment and Support Allowance.

Class 1A or 1B Employers pay these directly on their employee's expenses or benefits

Primary class 1 contributions stop at State Pension age, but not Class 1 secondary contributions paid by employers. There are reduced contribution rates where the employee contracts out of S2P (previously SERPS). They still receive a Basic State Pension but an Occupational or Personal Pension instead of the Additional State Second Pension.

Class 2 Flat rate contributions paid by the self-employed whose profits are above the small earnings exception. Payment of Class 2 contributions builds up entitlement to the contributory benefits which include Basic State Pension; Bereavement Benefits; Maternity Allowance and the Employment and Support Allowance, but not Additional State Second Pension or Contribution Based Jobseeker's Allowance (JSA).

Class 2 contributions stop at State Pension age.

Class 3 Flat rate voluntary contributions, which can be paid by someone whose contribution record is insufficient. Payment of Class 3 contributions builds up entitlement to contributory benefits which include Basic State Pension and Bereavement Benefits. (Tables 5.2 to 5.11, 5.13 and 5.15 to 5.19) Tables 5.2 to 5.6, 5.9 to 5.11 and 5.13 to 5.19 give details of contributors and beneficiaries under the National Insurance and Industrial Injury Acts, supplementary benefits and war pensions.

Class 4 Profit-related contributions paid by the self employed in addition to Class 2 contributions. Class 4 contributions paid by self-employed people with a profit over £8,164 don't usually count towards state benefits.

National Insurance Credits

In addition to paying, or being treated as having paid contributions, a person can be credited with National Insurance contributions (NIC) credits. Contribution credits help to protect people's rights to State Retirement Pension and other Social Security Benefits.

A person is likely to be entitled to contributions credits if they are: a student in full time education or training, in receipt of Jobseeker's Allowance, unable to work due to sickness or disability, entitled to Statutory Maternity Pay or Statutory Adoption Pay, or they have received Carer's Allowance.

National Insurance Credits eligibility can be checked at: https://www.gov.uk/national-insurance-credits/eligibility

Jobseeker's Allowance (Table 5.6)

Jobseeker's Allowance (JSA) is a benefit payable to unemployed people. In general, to be entitled to Jobseeker's Allowance, a person must be available for work for at least 40 hours a week, be actively seeking work, and have entered into a Jobseeker's Agreement with Jobcentre Plus. JSA has two routes of entry: Contrbution-based which depends mainly upon national insurance contributions and income-based which depends mainly on a means test. Some claimants can qualify by either route. In practice they receive income-based JSA but have an under lying entitlement to the contribution-based element.

Employment and Support Allowance, and Incapacity Benefit (Table 5.7)

Incapacity Benefit replaced Sickness Benefit and Invalidity Benefit from 13 April 1995. The first condition for entitlement to these contributory benefits is that the claimants are incapable of work because of illness or disablement. The second is that they satisfy the contribution conditions, which depend on contributions paid as an employed (Class 1) or self-employed person (Class 2). Under Sickness and Invalidity Benefits the contribution conditions were automatically treated as satisfied if a person was incapable of work because of an industrial accident or prescribed disease. Under Incapacity Benefit those who do not satisfy the contribution conditions do not have them treated as satisfied. Class 1A contributions paid by employers are in respect of the benefit of cars provided for the private use of employees, and the free fuel provided for private use. These contributions do not provide any type of benefit cover.

Since 6 April 1983, most people working for an employer and paying National Insurance contributions as employed persons receive Statutory Sick Pay (SSP) from their employer when they are off work sick. Until 5 April 1986 SSP was payable for a maximum of eight weeks, since this date SSP has been payable for 28 weeks. People who do not work for an employer, and employees who are excluded from the SSP scheme, or those who have run out of SSP before reaching the maximum of 28 weeks and are still sick, can claim benefit. Any period of SSP is excluded from the tables.

Spells of incapacity of three days or less do not count as periods of interruption of employment and are excluded from the tables. Exceptions are where people are receiving regular weekly treatment by dialysis or treatment by radiotherapy, chemotherapy or plasmapheresis where two days in any six consecutive days make up a period of interruption of employment, and those whose incapacity for work ends within three days of the end of SSP entitlement.

At the beginning of a period of incapacity, benefit is subject to three waiting days, except where there was an earlier spell of incapacity of more than three days in the previous eight weeks. Employees entitled to SSP for less than 28 weeks and who are still sick can get Sickness Benefit or Incapacity Benefit Short Term (Low) until they reach a total of 28 weeks provided they satisfy the conditions.

After 28 weeks of SSP and/or Sickness Benefit (SB), Invalidity Benefit (IVB) was payable up to pension age for as long as the incapacity lasted. From pension age, IVB was paid at the person's State Pension rate, until entitlement ceased when SP was paid, or until deemed pension age (70 for a man, 65 for a woman). People who were on Sickness or Invalidity Benefit on 12 April 1995 were automatically transferred to Incapacity Benefit, payable on the same basis as before.

For people on Incapacity Benefit under State Pension age there are two short-term rates: the lower rate is paid for the first 28 weeks of sickness and the higher rate for weeks 29 to 52. From week 53 the Long Term rate Incapacity Benefit is payable. The Short Term rate Incapacity Benefit is based on State Pension entitlement for people over State Pension age and is paid for up to a year if incapacity began before pension age.

The long-term rate of Incapacity Benefit applies to people under State Pension age who have been sick for more than a year. People with a terminal illness, or who are receiving the higher rate care component of Disability Living Allowance, will get the Long Term rate. The Long Term rate is not paid for people over pension age.

Under Incapacity Benefit, for the first 28 weeks of incapacity, people previously in work will be assessed on the 'own occupation' test – the claimant's ability to do their own job. Otherwise, incapacity will be based on a personal capability assessment, which will assess ability to carry out a range of work-related activities. The test will apply after 28 weeks of incapacity or from the start of the claim for people who did not previously have a job. Certain people will be exempted from this test.

The tables exclude all men aged over 65 and women aged over 60 who are in receipt of State Pension, and all people over deemed pension age (70 for a man and 65 for a woman), members of the armed forces, mariners while at sea, and married women and certain widows who have chosen not to be insured for sickness benefit. The tables include a number of individuals who were unemployed prior to incapacity.

The Short Term (Higher) and Long Term rates of Incapacity Benefit are treated as taxable income. There were transitional provisions for people who were on Sickness or Invalidity Benefit on 12 April 1995. They were automatically transferred to Incapacity Benefit, payable on the same basis as before. Former IVB recipients continue to get Additional Pension entitlement, but frozen at 1994 levels. Also their IVB is not subject to tax. If they were over State Pension age on 12 April 1995 they may get Incapacity Benefit for up to five years beyond pension age.

Employment and Support Allowance (ESA) replaced Incapacity Benefit and Income Support paid on the grounds of incapacity for new claims from 27 October 2008. ESA consists of two phases. The first, the assessment phase rate, is paid for the first 13 weeks of the claim whilst a decision is made on the claimants capability through the 'Work Capability Asessment'. The second, or main phase begins after 14 weeks, but only if the 'Work Capability Assesment' has deemed the claimants illness or disability as a limitation on their ability to work.

Within the main phase there are two groups, 'The Work Related Activity Group' and 'The Support Group'. If a claimant is placed in the first, they are expected to take part in work focused interviews with a personal advisor. They will be given support to help them prepare for work and on gaining work will receive a work related activity component in addition to their basic rate. If the claimant is placed in the second group due to their illness or disability having a severe effect upon their ability to work, the claimant will not be expected to work at all, but can do so on a voluntary basis. These claimants will receive a support component in addition to their basic rate.

Child Benefits (Table 5.9a and 5.9b)

You get child benefit if you are responsible for a child under 16 (or under 20 if they stay in approved education or training.

Approved education

Education must be full-time (more than an average of 12 hours a week supervised study or course-related work experience) and can include:

A levels or similar - eg Pre-U, International Baccalaureate

Scottish Highers

NVQs and other vocational qualifications up to level 3

home education - if started before your child turned 16

traineeships in England

Courses are not approved if paid for by an employer or 'advanced', eg a university degree or BTEC Higher National Certificate.

Approved training

Approved training should be unpaid and can include:

Access to Apprenticeships in England

Foundation Apprenticeships or Traineeships in Wales

Employability Fund programmes or Get Ready for Work (if started before 1 April 2013) in Scotland

Training for Success, Pathways to Success or Collaboration and Innovation Programme in Northern Ireland

Courses that are part of a job contract are not approved.

Guardian's Allowance is an additional allowance for people bringing up a child because one or both of their parents has died. They must be getting Child Benefit (CB) for the child. The table shows the number of families in the UK in receipt of CB. The numbers shown in the table are estimates based on a random 5 per cent sample of awards current at 31 August, and are therefore subject to sampling error. The figures take no account of new claims, or revisions to claims that were received or processed after 31 August, even if they are backdated to start before 31 August.

Child and Working Tax Credits (New Tax Credits) (Table 5.10 and 5.11)

Child and Working Tax Credits (CTC and WTC) replaced Working Families' Tax Credit (WFTC) from 6th April 2003. CTC and WTC are claimed by individuals, or jointly by couples, whether or not they have children.

CTC provides support to families for the children (up to the 31 August after their 16th birthday) and the 'qualifying' young people (in full-time non-advanced education until their 20th birthday) for which they are responsible. It is paid in addition to CB.

WTC tops up the earnings of families on low or moderate incomes. People working for at least 16 hours a week can claim it if they: (a) are responsible for at least one child or qualifying young person, (b) have a disability which puts them at a disadvantage in getting a job or, (c) in the first year of work, having returned to work aged at least 50 after a period of at least six months receiving out-of-work benefits. Other adults also qualify if they are aged at least 25 and work for at least 30 hours a week.

Bereavement Benefits (Table 5.12 and 5.13)

Widow's Benefit (WB) is payable to women widowed on or after 11th April 1988 and up to and including 8th April 2001. Statistics are based on WB claims in payment in Great Britain on the reference date; this excludes cases in Northern Ireland.

Widowed Mother's Allowance: Weekly benefit payable to a widowed mother (providing she was widowed before 8th April 2001) if her husband paid enough NIC and receiving Child Benefit for one of her children, or her husband was in receipt of Child Benefit, or she was expecting her husbands baby, or if she was widowed before 11th April 1988 and has a young person under 19 living with her for whom she was receiving Child Benefit.

Widow's Pension: Weekly benefit payable to widows without dependant children providing she was widowed before 8th April 2001. A widow may be able to get widows pension if her husband paid enough NIC. She must be 45 or over when her husband died or when her Widowed Mother's Allowance ends, she cannot get Widow's Pension at the same time as Widowed Mother's Allowance.

Universal Credit (UC)

Universal Credit is a benefits payment that was designed to simplify the benefits system and to incentivise paid work. Universal Credit is gradually replacing the following benefits:

- Child Tax Credit
- Housing Benefit
- Income Support
- income-based Jobseeker's Allowance (JSA)
- income-related Employment and Support Allowance (ESA)
- Working Tax Credit

5.1 National Insurance Fund (Great Britain)

For the year ended 31 March

£ 000

	2013	2014	2015	2016	2017
Receipts					
Opening balance[1]	38,593,953	29,082,990	23,195,862	20,935,278	23,173,893
National Insurance Contributions	79,119,934	82,236,514	84,112,562	86,461,626	95,852,994
Treasury Grant[2]	-	-	4,600,000	9,600,000	0
Compensation for statutory pay recoveries	2,559,760	2,319,000	2,465,000	2,634,000	2,285,000
Income from Investment Account	161,550	125,749	89,443	85,747	79,134
State Scheme Premiums[3]	30,861	56,408	32,622	29,318	42,974
Other receipts	36,164	29,586	23,292	786	2,138
Redundancy receipts[4]	38,320	39,995	36,932	30,758	37,469
Total Receipts	**81,946,589**	**84,807,252**	**91,359,851**	**98,842,235**	**98,299,709**
Less					
Payments					
Benefit payments	87,464,810	88,933,118	91,759,523	94,656,169	97,827,895
of which					
State Pension[5]	*80,008,745*	*82,522,101*	*85,893,497*	*88,688,966*	*91,725,346*
Incapacity Benefit	*3,355,345*	*1,213,380*	*233,704*	*61,610*	*8,824*
Employment and Support Alllowance (con	*2,312,374*	*3,554,301*	*4,130,282*	*4,460,807*	*4,701,380*
Jobseeker's Allowance (Contributory)	*669,184*	*533,630*	*375,940*	*307,176*	*265,374*
Bereavement Benefits	*598,431*	*587,821*	*579,061*	*569,630*	*561,317*
Maternity Allowance	*395,522*	*397,608*	*420,518*	*438,501*	*437,088*
Christmas Bonus	*123,308*	*122,356*	*124,521*	*127,458*	*126,452*
Guardian's Allowance	*1,901*	*1,921*	*2,000*	*2,021*	*2,114*
Personal Pensions[6]	2,124,560	15,913	1,448	-	-
Administrative costs	916,875	903,502	806,386	801,220	694,527
Redundancy payments	453,577	356,069	276,708	285,014	286,980
Transfers to Northern Ireland NIF	334,000	315,000	609,000	690,000	533,500
Other payments	163,730	170,778	167,370	171,217	195,949
Total Payments	**91,457,552**	**90,694,380**	**93,620,435**	**96,603,620**	**99,538,851**
Excess of payments over receipts	(9,510,963)	(5,887,128)	(2,260,584)	(2,238,615)	(1,239,142)
Opening balance	38,593,953	29,082,990	23,195,862	20,935,278	23,173,893
Less excess of payments over receipts	(9,510,963)	(5,887,128)	(2,260,584)	(2,238,615)	(1,239,142)
Closing balance	**29,082,990**	**23,195,862**	**20,935,278**	**23,173,893**	**21,934,751**

Source: HM Revenue and Customs, Department for Work and Pensions

1. Opening balance has been restated based on analysis of prior year data as better management information has become available.

2. In 2016-17 the government made provision for a 5% Treasury Grant in the annual re-rating regulations. However no Treasury Grant was required as the balance, which is regularly monitored, remained above the £16.3 billion (being 16.7% of estimated benefit expenditure) minimum working balance advised by GAD.

3. State Scheme Premiums are payable to the Fund in respect of employed persons' who cease to be covered, in certain specified circumstances, by a contracted-out pension scheme.

4. The receipts represent amounts recovered from employers.

5. State pension is for people who have reached state pension age and is based on NICs paid, treated as paid or credited. The state pension scheme of basic and additional state pension was replaced by the new single tier state pension for people reaching state pension age on 6 April 2016.

6. On 5 April 2012 the abolition of contracting-out on a defined contribution basis took place resulting in these contributions no longer being received. As the payments were made a year in arrears, from April 2013, the number of transactions has greatly reduced as only late payments and recoveries are being dealt with. n 2016-17 and 2015-16, recoveries exceeded payments.

5.2 National Insurance Contributions

For the year ended 31 March

	Notes	2013	2014	2015	2016	2017	2018
Contributions							
Class 1 (employed earner)	i	75,873,021	79,067,796	80,814,248	83,138,168	92,145,709	97,171,387
Class 1A and 1B	ii	1,047,965	1,118,263	1,078,274	1,119,605	1,227,184	1,298,615
Class 2 (Self-employed flat rate)	iii	341,361	327,180	353,608	191,127	400,864	321,873
Class 3 (Voluntary contributions)	iv	40,274	31,627	23,129	29,897	12,809	69,102
Class 3A (Voluntary contributions)	v	-	-	-	58,277	98,784	40,298
Class 4 (Self-employed earnings related	vi	1,817,313	1,689,648	1,842,303	1,924,552	1,967,644	2,325,659
		79,119,934	**82,236,514**	**84,112,562**	**86,461,626**	**95,852,994**	**101,226,934**

Source: HMRC National Insurance Fund Account Great Britain

Notes

i. **Class 1** contributions comprise two parts: primary contributions payable by employees which are approximately
40% of the total Class 1 figure, and secondary contributions payable by employers, which are approximately 60%.

ii. **Class 1A** contributions are paid by employers on most benefits provided to employees. Employers pay Class 1A
contributions to HMRC via the PAYE scheme with their Class 1 contributions.
Class 1B contributions are payable by employers where they have entered into a PAYE settlement agreement for tax enabling
them to settle their National Insurance and income tax liability in a lump sum after the end of the tax year. The figures for Class 1A
and Class 1B have been combined.

iii. **Class 2** self-employed persons pay flat rate weekly contributions. On 6 September 2018, the Government
announced that they would no longer proceed with their plan to abolish Class 2 NICs in April 2019.

iv. **Class 3** voluntary flat rate contributions are paid to maintain contributors' National Insurance record for certain benefit and/or pension purposes.

v. **Class 3A** allows pensioners who have reached state pension age before 6 April 2016 to boost their retirement
incomes by making voluntary payments of NICs. The scheme was available for 18 months between October 2015
and April 2017.

vi. **Class 4** self employed persons pay earnings related contributions. From 6 April 2018, Class 4 contributions will be reformed to
include a new threshold.

5.3 Main Features of National Insurance Contributions (NCIS) 1999-2000 to 2018-19

	Rate in 1999-2000	Rate in 2000-2001	Rate in 2001-2002	Rate in 2002-2003	Rate in 2003-2004	Rate in 2004-2005	Rate in 2005-2006	Rate in 2006-2007	Rate in 2007-2008	Rate in 2008-2009
Class 1										
Lower earnings limit (LEL) – a week	£66	£67	£72	£75	£77	£79	£82	£84	£87	£90
Primary threshold (PT) – a week	-	£76	£87	£89	£89	£91	£94	£97	£100	£105
Secondary threshold (ST) – a week	£83	£84	£87	£89	£89	£91	£94	£97	£100	£105
Upper Secondary Threshold (under 21) (UST) – a week (1)	-	-	-	-	-	-	-	-	-	-
Apprentice Upper Secondary Threshold (apprentice under 25) (AUST) – a week (2)	-	-	-	-	-	-	-	-	-	-
Upper accruals Point (UAP) – a week (3)										
Upper earnings limit (UEL) – a week (4)	£500	£535	£575	£585	£595	£610	£630	£645	£670	£770
Primary contributions (employee)										
Main contribution rate (PT to UEL) (5)	10.0%	10.0%	10.0%	10.0%	11.0%	11.0%	11.0%	11.0%	11.0%	11.0%
Additional contribution rate (above UEL)	-	-	-	-	1.0%	1.0%	1.0%	1.0%	1.0%	1.0%
Contracted out rebate (LEL to UAP/UEL) (6) (7)	1.6%	1.6%	1.6%	1.6%	1.6%	1.6%	1.6%	1.6%	1.6%	1.6%
Reduced rate for married women and widow optants (8)	3.85%	3.85%	3.85%	3.85%	4.85%	4.85%	4.85%	4.85%	4.85%	4.85%
Secondary contributions (employer)										
Contribution rate (above ST)	12.2%	12.2%	11.9%	11.8%	12.8%	12.8%	12.8%	12.8%	12.8%	12.8%
Contracted out rebate (LEL to UAP/UEL) (7)										
- COSRS	3.0%	3.0%	3.0%	3.5%	3.5%	3.5%	3.5%	3.5%	3.7%	3.7%
- COMPS (9)	0.6%	0.6%	0.6%	1.0%	1.0%	1.0%	1.0%	1.0%	1.4%	1.4%
Class 1A and 1B										
Contribution rate (10)	12.2%	12.2%	11.9%	11.8%	12.8%	12.8%	12.8%	12.8%	12.8%	12.8%
Class 2										
Flat-rate contribution – a week	£6.55	£2.00	£2.00	£2.00	£2.00	£2.05	£2.10	£2.10	£2.20	£2.30
Small earnings exception / Small Profits Threshold – a year (13)	£3,770	£3,825	£3,955	£4,025	£4,095	£4,215	£4,345	£4,465	£4,635	£4,825
Class 3										
Flat-rate contribution – a week (11)	£6.45	£6.55	£6.75	£6.85	£6.95	£7.15	£7.35	£7.55	£7.80	£8.10
Class 4										
Lower profits limit (LPL) – a year	£7,530	£4,385	£4,535	£4,615	£4,615	£4,745	£4,895	£5,035	£5,225	£5,435
Upper profits limit (UPL) – a year (4)	£26,000	£27,820	£29,900	£30,420	£30,940	£31,720	£32,760	£33,540	£34,840	£40,040
Main contribution rate (LPL to UPL)	6%	7%	7%	7%	8%	8%	8%	8%	8%	8%
Additional contribution rate (above UPL)	-	-	-	-	1%	1%	1%	1%	1%	1%

Source: HM Revenue & Customs

5.3 Main Features of National Insurance Contributions (NCIS) 1999-2000 to 2018-19

	Rate in 2009-2010	Rate in 2010-11	Rate in 2011-12	Rate in 2012-13 (12)	Rate in 2013-14 (12)	Rate in 2014-15 (12)	Rate in 2015-16 (12)	Rate in 2016-17 (12)	Rate in 2017-18 (12)	Rate in 2018-19 (12)
Class 1										
Lower earnings limit (LEL) – a week	£95	£97	£102	£107	£109	£111	£112	£112	£113	£116
Primary threshold (PT) – a week	£110	£110	£139	£146	£149	£153	£155	£155	£157	£162
Secondary threshold (ST) – a week	£110	£110	£136	£144	£148	£153	£156	£156	£157	£162
Upper Secondary Threshold (under 21) (UST) – a week (1)	-	-	-	-	-	-	£815	£827	£866	£892
Apprentice Upper Secondary Threshold (apprentice under 25) (AUST) – a week (2)	-	-	-	-	-	-	-	£827	£866	£892
Upper accruals Point (UAP) – a week (3)	£770	£770	£770	£770	£770	£770	£770	-	-	-
Upper earnings limit (UEL) – a week (4)	£844	£844	£817	£817	£797	£805	£815	£827	£866	£892
Primary contributions (employee)										
Main contribution rate (PT to UEL) (5)	11.0%	11.0%	12.0%	12.0%	12.0%	12.0%	12.0%	12.0%	12.0%	12.0%
Additional contribution rate (above UEL)	1.0%	1.0%	2.0%	2.0%	2.0%	2.0%	2.0%	2.0%	2.0%	2.0%
Contracted out rebate (LEL to UAP/UEL) (6) (7)	1.6%	1.6%	1.6%	1.4%	1.4%	1.4%	1.4%	-	-	-
Reduced rate for married women and widow optants (8)	4.85%	4.85%	5.85%	5.85%	5.85%	5.85%	5.85%	5.85%	5.85%	5.85%
Secondary contributions (employer)										
Contribution rate (above ST)	12.8%	12.8%	13.8%	13.8%	13.8%	13.8%	13.8%	13.8%	13.8%	13.8%
Contracted out rebate (LEL to UAP/UEL) (7)										
- COSRS	3.7%	3.7%	3.7%	3.4%	3.4%	3.4%	3.4%	-	-	-
- COMPS (9)	1.4%	1.4%	1.4%	-	-	-	-	-	-	-
Class 1A and 1B										
Contribution rate (10)	12.8%	12.8%	13.8%	13.8%	13.8%	13.8%	13.8%	13.8%	13.8%	13.8%
Class 2										
Flat-rate contribution – a week	£2.40	£2.40	£2.50	£2.65	£2.70	£2.75	£2.80	£2.80	£2.85	£2.95
Small earnings exception / Small Profits Threshold – a year (13)	£5,075	£5,075	£5,315	£5,595	£5,725	£5,885	£5,965	£5,965	£6,025	£6,205
Class 3										
Flat-rate contribution – a week (11)	£12.05	£12.05	£12.60	£13.25	£13.55	£13.90	£14.10	£14.10	£14.25	£14.65
Class 4										
Lower profits limit (LPL) – a year	£5,715	£5,715	£7,225	£7,605	£7,755	£7,956	£8,060	£8,060	£8,164	£8,424
Upper profits limit (UPL) – a year (4)	£43,875	£43,875	£42,475	£42,475	£41,450	£41,865	£42,385	£43,000	£45,000	£46,350
Main contribution rate (LPL to UPL)	8%	8%	9%	9%	9%	9%	9%	9%	9%	9%
Additional contribution rate (above UPL)	1%	1%	2%	2%	2%	2%	2%	2%	2%	2%

Source: HM Revenue & Customs

5.3 Main Features of National Insurance Contributions (NCIS) 1999-2000 to 2018-19

(1) From 6 April 2015 employers with employees under 21 years old do not pay Class 1 secondary National Insurance contributions (NICs) on earnings up to the Upper Secondary Threshold (UST) for those employees.

An employer must choose the appropriate National Insurance contribution category letter to have the exemption applied

(2) From 6 April 2016 employers of apprentices under 25 years old who are following an approved UK government statutory apprenticeship framework do not pay Class 1 secondary National Insurance contributions(NICs) on earnings up to the Apprentices Upper Secondary Threshold (AUST) for those apprentices. An employer must choose the appropriate National Insurance contribution category letter to have the exemption applied.

(3) The upper accruals point was introduced in April 2009 until April 2015-16. It is no longer needed after the contracting out rebates are abolished from 2016-17 onwards.

(4) From April 2009 the upper earnings limit and upper profits limit were aligned to the income tax higher rate threshold.

(5) Between LEL and UEL for 1999-2000.

(6) For Appropriate Personal Pension Schemes (APPS) both employer and employee pay NICs at the full contracted-out rate and in the following tax year on submission of the end-of-year returns HMRC pay an age related rebate direct to the schemes. The employee's share of this rebate is 1.6%.

(7) Up to and including 2008-09, the rebate applies between the LEL and the UEL. From 2009-10 onwards the rebate applies between the LEL and UAP. The rebates are abolished from 2016-17 onwards.

(8) Married women opting to pay contributions at the reduced rate earn no entitlement to contributory National Insurance benefits as a result of these contributions.

No women have been allowed to exercise this option since 1977.

(9) For employers operating a COMPS, in addition to the reduction shown in secondary Class 1 contributions, in the following tax year on submission of end-of-year returns, HMRC pay an additional "top-up" rebate direct to the scheme. As with APPS, this rebate is age related. COMPs are abolished from 2012-13 onwards.

(10) From April 2000 the Class 1A liability for employers was extended from company cars and fuel to include other taxable benefits not already attracting a Class 1 liability.

Class 1A and Class 1B contributions are paid in the year following accrual.

(11) Class 3 contribution rules changed in 2009-10 to allow those reaching state pension age before April 2015 with 20 qualifying years to purchase up to 6 additional years.

(12) From 2012-13 the default indexation assumption for NICs is CPI (excluding the secondary threshold up until 2016-17).

(13) The Small Profits Threshold replaced the Small Earnings Exception on 6 April 2015.

Notes:

Class 1 National Insurance Contributions (NICs)

Class 1 NICs are earnings related contributions paid by employed earners and their employers. Liability starts at age 16 and ends at Sate Pension age for earners; employers continue to pay beyond State Pension age. Up to April 2016 the contributions were paid at either the contracted-out rate or the not contracted-out rate. The contracted-out rate, abolished in April 2016, was payable payable only where the employee was a member of a contracted-out occupational scheme in place of State Second Pension (formerly SERPS).

Class 1 NICs are collected by HMRC along with income tax under the Pay As You Earn (PAYE) scheme.

Class 1A National Insurance Contributions (NICs)

Class 1A NICs are paid only by employers on the value of most taxable benefits-in-kind provided to employees, such as private use of company cars and fuel, private medical insurance, accommodation and loan benefits. They do not give any benefit rights.

Class 1B National Insurance Contributions (NICs)

Class 1B NICs were introduced on 6 April 1999. Like Class 1A they are also paid only by employers and cover PAYE Settlement Agreements (PSA) under which employers agree to meet the income tax liability arising on a restricted range of benefits. Class 1B is payable on the value of the items included in the PSA that would otherwise attract a Class 1 or Class 1A liability and the value of the income tax met by the employer. They do not give any benefit rights.

Class 2 National Insurance Contributions (NICs)

Class 2 contributions are a flat rate weekly liability payable by all self-employed people over 16 (up to State Pension age) with profits above the Small Profits Threshold. Self-employed people with profits below the Small Profits Threshold may pay Class 2 contributions voluntary. Voluntary payments of Class 2 NICs are typically collected through self-assessment but can usually be paid up to six years after the tax year. Class 4 NICs may also have to be paid by the self-employed if their profits for the year are over the lower profits limit (see below).

Class 3 National Insurance Contributions (NICs)

Class 3 NICs may be paid voluntarily by people aged 16 and over (but below State Pension age) to help them qualify for State Pension if their contribution record would not otherwise be sufficient. Contributions are flat rate and can be paid up to six years after the year in which they are due.

Class 4 National Insurance Contributions (NICs)

Class 4 NICS are paid by the self-employed whose profits are above the lower profits limit. They are profit related and do not count for any benefits themselves.

5.4 Proposed benefit and pension rates 2017-2018

(Weekly rates unless otherwise shown)	RATES 2016	RATES 2017
ATTENDANCE ALLOWANCE		
higher rate	82.30	83.10
lower rate	55.10	55.65
BEREAVEMENT BENEFIT		
Bereavement payment (lump sum)	2000.00	2000.00
Widowed parent's allowance	112.55	113.70
Bereavement Allowance		
standard rate	112.55	113.70
age-related		
age 54	104.67	105.74
53	96.79	97.78
52	88.91	89.82
51	81.04	81.86
50	73.16	73.91
49	65.28	65.95
48	57.40	57.99
47	49.52	50.03
46	41.64	42.07
45	33.77	34.11

BENEFIT CAP - Rates introduced November 2016

Reduction in annual level of Benefit Cap (Greater London)		
Couples (with or without children) or single claimants with a child of qualifying age	23000.00	
Single adult households without children	15410.00	
Reduction in annual level of Benefit Cap (Rest of Great Britain)		
Couples (with or without children) or single claimants with a child of qualifying age	20000.00	
Single adult households without children	13400.00	
Monthly equivalent (Greater London)		
Couples (with or without children) or single claimants with a child of qualifying age	1916.67	
Single adult households without children	1284.17	
Monthly equivalent (Rest of Great Britain)		
Couples (with or without children) or single claimants with a child of qualifying age	1666.67	
Single adult households without children	1116.67	
Weekly equivalent (Greater London)		
Couples (with or without children) or single claimants with a child of qualifying age	442.31	
Single adult households without children	296.35	

5.4 Proposed benefit and pension rates 2017-2018

(Weekly rates unless otherwise shown)	RATES 2016	RATES 2017
Weekly equivalent (Rest of Great Britain)		
Couples (with or without children) or single claimants with a child of qualifying age	384.62	
Single adult households without children	257.69	
CAPITAL LIMITS - rules common to Income Support, income based Jobseeker's Allowance,		
income-related Employment and Support Allowance, Pension Credit, and Housing Benefit, and Universal Credit		
unless stated otherwise		
upper limit	16000.00	16000.00
upper limit - Pension Credit and those getting Housing Benefit and Pension Credit Guarantee Credit	No limit	No limit
Amount disregarded - all benefits except Pension Credit and Housing Benefit for those above the qualifying age for Guarantee Credit	6000.00	6000.00
Amount disregarded - Pension Credit and Housing Benefit for those above the qualifying age for Pension Credit	10000.00	10000.00
Child disregard (not Pension Credit, Employment and Support Allowance nor Housing Benefit)	3000.00	3000.00
Amount disregarded (living in RC/NH)	10000.00	10000.00

Tariff income
$\qquad$ £1 for every £250, or part thereof, between the amount of
$\qquad$ capital disregarded and the capital upper limit

Tariff income - Pension Credit and Housing Benefit where
claimant / partner is over Guarantee Credit qualifying age
$\qquad$ £1 for every £500, or part thereof, above or between the
amount of
$\qquad$ capital disregarded and any capital upper limit applicable

	RATES 2016	RATES 2017
CARER'S ALLOWANCE	62.10	62.70

DEDUCTIONS - rules common to Income Support, Jobseeker's Allowance,
Employment and Support Allowance, Pension Credit and Housing Benefit
unless stated otherwise

Non-dependant deductions from housing benefit and from IS, JSA(IB), ESA(IR) and Pension Credit

	RATES 2016	RATES 2017
aged 25 and over in receipt of IS and JSA(IB), or any age in receipt of main phase ESA(IR), aged 18 or over, not in remunerative work	14.65	14.80
aged 18 or over and in remunerative work		

5.4 Proposed benefit and pension rates 2017-2018

(Weekly rates unless otherwise shown)	RATES 2016	RATES 2017
- gross income: less than £136	14.65	14.80
- gross income: £136 to £199.99	33.65	34.00
- gross income: £200 to £258.99	46.20	46.65
- gross income: £259 to £345.99	75.60	76.35
- gross income: £346 to £429.99	86.10	86.95
- gross income: £430 and above	94.50	95.45
Deductions from housing benefit		
Service charges for fuel		
heating	28.80	28.80
hot water	3.35	3.35
lighting	2.30	2.30
cooking	3.35	3.35
Amount ineligible for meals		
three or more meals a day		
single claimant	26.85	27.10
each person in family aged 16 or over	26.85	27.10
each child under 16	13.60	13.75
less than three meals a day		
single claimant	17.85	18.05
each person in family aged 16 or over	17.85	18.05
each child under 16	9.00	9.10
breakfast only - claimant and each member of the family	3.30	3.35
Amount for personal expenses (not HB)	24.00	24.25
Third party deductions from IS, JSA(IB), ESA(IR) and Pension Credit for;		
arrears of housing, fuel and water costs Council Tax etc. and deductions for ELDS and ILS.	3.70	3.70
child support, contribution towards maintenance (CTM)		
standard deduction	7.40	7.40
lower deduction	3.70	3.70
arrears of Community Charge		
court order against claimant	3.70	3.70
court order against couple	5.75	5.75
fine or compensation order		
standard rate	5.00	5.00
lower rate	3.70	3.70
Maximum deduction rates for recovery of overpayments (not JSA(C)/ESA(C))		
ordinary overpayments	11.10	11.10
Fraud Overpayments		
Housing Benefit/ CTB only	18.50	18.50
Benefits (not HB or Council Tax)	29.60	29.60

5.4 Proposed benefit and pension rates 2017-2018

(Weekly rates unless otherwise shown)	RATES 2016	RATES 2017
Deductions from JSA(C) and ESA (C)		
Arrears of Comm. Charge & overpayment recovery		
Age 16 - 24	19.30	19.30
Age 25 +	24.36	24.36
Arrears of Council Tax & Fines		
Age 16 - 24	23.16	23.16
Age 25 +	29.24	29.24
Maximum deduction for arrears of Child Maintenance		
Age 16 - 24	19.30	19.30
Age 25 +	24.36	24.36

DEPENDENCY INCREASES

Adult dependency increases for spouse or person looking after children - payable with;

	RATES 2016	RATES 2017
State Pension on own insurance (Cat A)	65.70	66.35
State Pension (non-contributory, Cat C)	39.30	39.70
long term Incapacity Benefit	61.20	61.80
Unemployability Supplement.	62.10	62.70
Severe Disablement Allowance	36.75	37.10
Carer's Allowance	36.55	36.90
short-term Incapacity Benefit (over state pension age)	58.90	59.50
short-term Incapacity Benefit (under State Pension age)	47.65	48.15

	RATES 2016	RATES 2017
Child Dependency Increases - payable with;		
State Pension; Widowed Mothers/Parents Allowance;	11.35	11.35
short-term Incapacity benefit - higher rate or over state pension age;		
long-term Incapacity Benefit; Carer's Allowance; Severe Disablement		
Unemployability Supplement.		

	RATES 2016	RATES 2017
NB - The rate of child dependency increase is adjusted where it is payable for the eldest child for whom child benefit is also paid. The weekly rate in such cases is reduced by the difference (less £3.65) between the ChB rates for the eldest and subsequent children.	8.00	8.00

DISABILITY LIVING ALLOWANCE

	RATES 2016	RATES 2017
Care Component		
Highest	82.30	83.10
Middle	55.10	55.65
Lowest	21.80	22.00
Mobility Component		

5.4 Proposed benefit and pension rates 2017-2018

(Weekly rates unless otherwise shown)	RATES 2016	RATES 2017
Higher	57.45	58.00
Lower	21.80	22.00

DISREGARDS

Housing Benefit

Earnings disregards

standard (single claimant)	5.00	5.00
couple	10.00	10.00
higher (special occupations/circumstances)	20.00	20.00
lone parent	25.00	25.00
childcare charges	175.00	175.00
childcare charges (2 or more children)	300.00	300.00
permitted work higher	115.50	120.00
permitted work lower	20.00	20.00

Other Income disregards

adult maintenance disregard	15.00	15.00
war disablement pension and war widows pension	10.00	10.00
widowed mothers/parents allowance	15.00	15.00
Armed Forces Compensation Scheme	10.00	10.00
student loan	10.00	10.00
student's covenanted income	5.00	5.00
income from boarders (plus 50% of the balance)	20.00	20.00
additional earnings disregard	17.10	17.10
income from subtenants (£20 fixed from April 08)	20.00	20.00

Income Support, income-based Jobseeker's Allowance,
Income-related Employment and Support Allowance
(ESA(IR)) and Pension Credit

Earnings disregards

standard (single claimant) (not ESA(IR))	5.00	5.00
couple (not ESA(IR))	10.00	10.00
higher (special occupations/circumstances)	20.00	20.00
partner of claimant (ESA(IR))	20.00 (maximum)	20.00 (maximum)

Other Income disregards

war disablement pension and war widows pension	10.00	10.00
widowed mothers/parents allowance	10.00	10.00
Armed Forces Compensation Scheme	10.00	10.00
student loan (not Pension Credit)	10.00	10.00
student's covenanted income (not Pension Credit)	5.00	5.00
income from boarders (plus 50% of the balance)	20.00	20.00
income from subtenants (£20 fixed from April 08)	20.00	20.00

EARNINGS RULES

Carer's Allowance	110.00	116.00
Limit of earnings from councillor's allowance	115.50	120.00

5.4 Proposed benefit and pension rates 2017-2018

(Weekly rates unless otherwise shown)	RATES 2016	RATES 2017
Permitted work earnings limit - higher	115.50	120.00
- lower	20.00	20.00
Industrial injuries unemployability supplement permitted earnings level (annual amount)	6006.00	6240.00
Earnings level at which adult dependency (ADI) increases are affected with:		
short-term incapacity benefit where claimant is		
(a) under state pension age	47.65	48.15
(b) over state pension age	58.90	59.50
state pension, long term incapacity benefit, severe disablement allowance, unemployability supplement - payable when dependant		
(a) is living with claimant	73.10	73.10
(b) still qualifies for the tapered earnings rule	45.09	45.09
Earnings level at which ADI is affected when dependant is not living with claimant;		
state pension,	65.70	66.35
long-term incapacity benefit,	61.20	61.80
unemployability supplement,	62.10	62.70
severe disablement allowance	36.75	37.10
Carer's allowance	36.55	36.90
Earnings level at which child dependency increases are affected		
for first child	230.00	230.00
additional amount for each subsequent child	30.00	30.00
Pension income threshold for incapacity benefit	85.00	85.00
Pension income threshold for contributory Employment Support Allowance	85.00	85.00

EMPLOYMENT AND SUPPORT ALLOWANCE

Personal Allowances		
Single		
under 25	57.90	57.90
25 or over	73.10	73.10
lone parent		
under 18	57.90	57.90
18 or over	73.10	73.10
couple		
both under 18	57.90	57.90

5.4 Proposed benefit and pension rates 2017-2018

(Weekly rates unless otherwise shown)	RATES 2016	RATES 2017
both under 18 with child	87.50	87.50
both under 18 (main phase)	73.10	73.10
both under 18 with child (main phase)	114.85	114.85
one 18 or over, one under 18 (certain conditions apply)	114.85	114.85
both over 18	114.85	114.85
claimant under 25, partner under 18	57.90	57.90
claimant 25 or over, partner under 18	73.10	73.10
claimant (main phase), partner under 18	73.10	73.10
Premiums		
enhanced disability		
single	15.75	15.90
couple	22.60	22.85
severe disability		
single	61.85	62.45
couple (lower rate)	61.85	62.45
couple (higher rate)	123.70	124.90
carer	34.60	34.95
pensioner		
single with WRAC	53.45	57.20
single with support component	46.30	49.70
single with no component	82.50	86.25
couple with WRAC	93.65	99.35
couple with support component	86.50	91.85
couple with no component	122.70	128.40
Components		
Work-related Activity	29.05	29.05
Support	36.20	36.55
HOUSING BENEFIT		
Personal allowances		
single		
under 25	57.90	57.90
25 or over	73.10	73.10
entitled to main phase ESA	73.10	73.10
lone parent		
under 18	57.90	57.90
18 or over	73.10	73.10
entitled to main phase ESA	73.10	73.10
couple		
both under 18	87.50	87.50

5.4 Proposed benefit and pension rates 2017-2018

(Weekly rates unless otherwise shown)	RATES 2016	RATES 2017
one or both 18 or over	114.85	114.85
claimant entitled to main phase ESA	114.85	114.85
dependent children	66.90	66.90
pensioner		
single/lone parent has attained the qualifying age for Pension Credit but under 65.	155.60	159.35
couple – one or both has attained the qualifying age for Pension Credit but both under 65	237.55	243.25
single / lone parent - 65 and over	168.70	172.55
couple - one or both 65 and over	252.30	258.15
polygamous marriage		
for the claimant and the other party to the marriage where no members of the marriage have attained the age of 65	237.55	243.25
for each additional spouse who is a member of the same household as the claimant and no members of the marriage have attained the age of 65	81.95	83.90
for the claimant and the other party to the marriage where one or more of the members of the marriage are aged 65 or over	252.30	258.15
for each additional spouse who is a member of the same household as the claimant and one or more of the members of the marriage are aged 65 or over	83.60	85.60
Premiums		
family	17.45	17.45
family (lone parent rate)	22.20	22.20
disability		
single	32.25	32.55
couple	45.95	46.40
enhanced disability		
single	15.75	15.90
disabled child	24.43	24.78
couple	22.60	22.85
severe disability		
single	61.85	62.45
couple (lower rate)	61.85	62.45
couple (higher rate)	123.70	124.90
disabled child	60.06	60.90
carer	34.60	34.95
ESA components		
work-related activity	29.05	29.05
support	36.20	36.55

5.4 Proposed benefit and pension rates 2017-2018

(Weekly rates unless otherwise shown)	RATES 2016	RATES 2017
INCAPACITY BENEFIT		
Long-term Incapacity Benefit	105.35	106.40
Short-term Incapacity Benefit (under state pension age)		
lower rate	79.45	80.25
higher rate	94.05	95.00
Short-term Incapacity Benefit (over state pension age)		
lower rate	101.10	102.10
higher rate	105.35	106.40
Increase of Long-term Incapacity Benefit for age		
higher rate	11.15	11.25
lower rate	6.20	6.25
Invalidity Allowance (Transitional)		
higher rate	11.15	11.25
middle rate	6.20	6.25
lower rate	6.20	6.25
INCOME SUPPORT		
Personal Allowances		
Single		
under 25	57.90	57.90
25 or over	73.10	73.10
lone parent		
under 18	57.90	57.90
18 or over	73.10	73.10
Couple		
both under 18	57.90	57.90
both under 18 - higher rate	87.50	87.50
one under 18, one under 25	57.90	57.90
one under 18, one 25 and over	73.10	73.10
both 18 or over	114.85	114.85
dependent children	66.90	66.90
Premiums		
family / lone parent	17.45	17.45
pensioner (applies to couples only)	122.70	128.40
Disability		
Single	32.25	32.55
Couple	45.95	46.40
enhanced disability		

5.4 Proposed benefit and pension rates 2017-2018

(Weekly rates unless otherwise shown)	RATES 2016	RATES 2017
Single	15.75	15.90
disabled child	24.43	24.78
couple	22.60	22.85
severe disability		
Single	61.85	62.45
couple (lower rate)	61.85	62.45
couple (higher rate)	123.70	124.90
disabled child	60.06	60.90
Carer	34.60	34.95
Relevant sum for strikers	40.50	40.50

INDUSTRIAL DEATH BENEFIT

	RATES 2016	RATES 2017
Widow's pension		
higher rate	119.30	122.30
lower rate	35.79	36.69
Widower's pension	119.30	122.30

INDUSTRIAL INJURIES DISABLEMENT BENEFIT

	RATES 2016	RATES 2017
Standard rate		
100%	168.00	169.70
90%	151.20	152.73
80%	134.40	135.76
70%	117.60	118.79
60%	100.80	101.82
50%	84.00	84.85
40%	67.20	67.88
30%	50.40	50.91
20%	33.60	33.94
Maximum life gratuity (lump sum)	11150.00	11260.00
Unemployability Supplement	103.85	104.90
increase for early incapacity		
higher rate	21.50	21.70
middle rate	13.90	14.00
lower rate	6.95	7.00
Maximum reduced earnings allowance	67.20	67.88
Maximum retirement allowance	16.80	16.97
Constant attendance allowance		
exceptional rate	134.40	135.80
intermediate rate	100.80	101.85

5.4 Proposed benefit and pension rates 2017-2018

(Weekly rates unless otherwise shown)	RATES 2016	RATES 2017
normal maximum rate	67.20	67.90
part-time rate	33.60	33.95
Exceptionally severe disablement allowance	67.20	67.90

JOBSEEKER'S ALLOWANCE

	RATES 2016	RATES 2017
Contribution based JSA - Personal rates		
under 25	57.90	57.90
25 or over	73.10	73.10
Income-based JSA - personal allowances		
under 25	57.90	57.90
25 or over	73.10	73.10
lone parent		
under 18	57.90	57.90
18 or over	73.10	73.10
couple		
both under 18	57.90	57.90
both under 18 - higher rate	87.50	87.50
one under 18, one under 25	57.90	57.90
one under 18, one 25 and over	73.10	73.10
both 18 or over	114.85	114.85
dependent children	66.90	66.90
Premiums		
family / lone parent	17.45	17.45
pensioner		
single	82.50	86.25
couple	122.70	128.40
disability		
single	32.25	32.55
couple	45.95	46.40
enhanced disability		
single	15.75	15.90
disabled child	24.43	24.78
couple	22.60	22.85
severe disability		
single	61.85	62.45
couple (lower rate)	61.85	62.45
couple (higher rate)	123.70	124.90
disabled child	60.06	60.90

5.4 Proposed benefit and pension rates 2017-2018

(Weekly rates unless otherwise shown)	RATES 2016	RATES 2017
carer	34.60	34.95
Prescribed sum for strikers	40.50	40.50
MATERNITY ALLOWANCE		
Standard rate	139.58	140.98
MA threshold	30.00	30.00
PENSION CREDIT		
Standard minimum guarantee		
single	155.60	159.35
couple	237.55	243.25
Additional amount for severe disability		
single	61.85	62.45
couple (one qualifies)	61.85	62.45
couple (both qualify)	123.70	124.90
Additional amount for carers	34.60	34.95
Savings credit		
threshold - single	133.82	137.35
threshold - couple	212.97	218.42
maximum - single	13.07	13.20
maximum - couple	14.75	14.90
Amount for claimant and first spouse in polygamous marriage	237.55	243.25
Additional amount for additional spouse	81.95	83.90
Non-State Pensions (for Pension Credit purposes)		
Statutory minimum increase to non-state pensions	0.00%	1.00%
PERSONAL INDEPENDENCE PAYMENT		
Daily living component		
Enhanced	82.30	83.10
Standard	55.10	55.65
Mobility component		
Enhanced	57.45	58.00
Standard	21.80	22.00
SEVERE DISABLEMENT ALLOWANCE		
Basic rate	74.65	75.40
Age-related addition (from Dec 90)		
Higher rate	11.15	11.25
Middle rate	6.20	6.25
Lower rate	6.20	6.25

5.4 Proposed benefit and pension rates 2017-2018

(Weekly rates unless otherwise shown)	RATES 2016	RATES 2017
STATE PENSION		
New State Pension		
Full rate	155.65	159.55
Transitional rate below full rate	0.00%	2.5056%
Protected Payment	0.00%	1.00%
Increments - own (based on deferred new State Pension)	0.00%	1.00%
Increments - inherited (based on deferred old State Pension)	0.00%	1.00%
Old State Pension		
Category A or B basic pension	119.30	122.30
Category B (lower) basic pension - spouse or civil partner's insurance	71.50	73.30
Category C or D - non-contributory	71.50	73.30
Additional pension	0.00%	1.00%
Maximum additional pension (own + inherited)	165.60	167.26
Increments to:-		
Basic pension	0.00%	1.00%
Additional pension	0.00%	1.00%
Graduated Retirement Benefit (GRB)	0.00%	1.00%
Inheritable lump sum	0.00%	1.00%
Contracted-out Deduction from AP in respect of pre-April 1988 contracted-out earnings	Nil	Nil
Contracted-out Deduction from AP in respect of contracted-out earnings from April 1988 to 1997	0.00%	1.00%
Graduated Retirement Benefit (unit)	0.1330	0.1343
Increase of long term incapacity for age	0.00%	1.00%
Addition at age 80	0.25	0.25
Increase of Long-term incapacity for age		
higher rate	21.50	21.70
lower rate	10.80	10.90
Invalidity Allowance (Transitional) for State Pension recipients		
higher rate	21.50	21.70
middle rate	13.90	14.00
lower rate	6.95	7.00
STATUTORY ADOPTION PAY		
Earnings threshold	112.00	113.00
Standard Rate	139.58	140.98

5.4 Proposed benefit and pension rates 2017-2018

(Weekly rates unless otherwise shown)	RATES 2016	RATES 2017
STATUTORY MATERNITY PAY		
Earnings threshold	112.00	113.00
Standard rate	139.58	140.98
STATUTORY PATERNITY PAY		
Earnings threshold	112.00	113.00
Standard Rate	139.58	140.98
STATUTORY SHARED PARENTAL PAY		
Earnings threshold	112.00	113.00
Standard rate	139.58	140.98
STATUTORY SICK PAY		
Earnings threshold	112.00	113.00
Standard rate	88.45	89.35
UNIVERSAL CREDIT (monthly rates)		
Universal Credit Minimum Amount	0.01	0.01
Universal Credit Amounts		
Standard allowance		
Single		
Single under 25	251.77	251.77
Single 25 or over	317.82	317.82
Couple		
Joint claimants both under 25	395.20	395.20
Joint claimants, one or both 25 or over	498.89	498.89
Child amounts		
First child	277.08	277.08
Second/ subsequent child	231.67	231.67
Disabled child additions		
Lower rate addition	126.11	126.11
Higher rate addition	367.92	372.30
Limited Capability for Work amount	126.11	126.11
Limited Capability for Work and Work-Related Activity amount	315.60	318.76
Carer amount	150.39	151.89
Childcare costs amount		
Maximum for one child	646.35	646.35
Maximum for two or more children	1108.04	1108.04
Non-dependants' housing cost contributions	69.37	70.06

5.4 Proposed benefit and pension rates 2017-2018

(Weekly rates unless otherwise shown)	RATES 2016	RATES 2017
Work allowances		
Higher work allowance (no housing amount)		
One or more dependent children or limited capability for work		
Lower work allowance	397.00	397.00
One or more dependent children or limited capability for work	192.00	192.00
Assumed income from capital for every £250 or part thereof, between capital disregard and upper capital limit	4.35	4.35
UC Daily Reduction		
100% reduction - High, medium or low level sanctions apply - claimants aged 18 or over		
Single		
Single under 25	8.20	8.20
Single 25 or over	10.40	10.40
Couple		
Joint claimants both under 25 (per sanctioned claimant)	6.40	6.40
Joint claimants, one or both 25 or over and one is sanctioned (per sanctioned claimant)	8.20	8.20
40% reduction - Lowest level sanction applies		
Single		
Single under 25	3.30	3.30
Single 25 or over	4.10	4.10
Couple		
Joint claimants both under 25 (per sanctioned claimant)	2.50	2.50
Joint claimants, one or both 25 or over (per sanctioned claimant)	3.20	3.20
Third Party Deductions at 5% of UC Standard Allowance (excludes deductions for rent and service charges included in rent)		
Single		
Single under 25	12.59	12.59
Single 25 or over	15.89	15.89
Couple		
Joint claimants both under 25	19.76	19.76
Joint claimants, one or both 25 or over	24.94	24.94
Maximum deductions for Fines	108.35	108.35
Minimum deductions for rent and service charges included in rent at 10% of UC Standard Allowance (10% minimum introduced from Nov 2014)		
Single		
Single under 25	25.18	25.18
Single 25 or over	31.78	31.78
Couple		
Joint claimants both under 25	39.52	39.52

5.4 Proposed benefit and pension rates 2017-2018

(Weekly rates unless otherwise shown)	RATES 2016	RATES 2017
Joint claimants, one or both 25 or over	49.89	49.89
Maximum deductions for rent and service charges included in rent at 20% of UC Standard Allowance (20% maximum introduced from Nov 2014)		
Single		
Single under 25	50.35	50.35
Single 25 or over	63.56	63.56
Couple		
Joint claimants both under 25	79.04	79.04
Joint claimants, one or both 25 or over	99.78	99.78
Overall Maximum deduction Rate at 40% of UC Standard Allowance:		
Single		
Single under 25	100.71	100.71
Single 25 or over	127.13	127.13
Couple		
Joint claimants both under 25	158.08	158.08
Joint claimants, one or both 25 or over	199.56	199.56
Fraud Overpayments, Recoverable Hardship Payments and Administrative Penalties at 40% of UC Standard Allowance		
Single		
Single under 25	100.71	100.71
Single 25 or over	127.13	127.13
Couple		
Joint claimants both under 25	158.08	158.08
Joint claimants, one or both 25 or over	199.56	199.56
Ordinary Overpayments and Civil Penalties at 15% of UC Standard Allowance		
Single		
Single under 25	37.77	37.77
Single 25 or over	47.67	47.67
Couple		
Joint claimants both under 25	59.28	59.28
Joint claimants, one or both 25 or over	74.83	74.83
Ordinary Overpayments and Civil Penalties at 25% of UC Standard Allowance if claimant's and/or partner's earnings are over the Work Allowance		
Single		
Single under 25	62.94	62.94
Single 25 or over	79.46	79.46
Couple		

5.4 Proposed benefit and pension rates 2017-2018

(Weekly rates unless otherwise shown)	RATES 2016	RATES 2017
Joint claimants both under 25	98.80	98.80
Joint claimants, one or both 25 or over	124.72	124.72
WIDOW'S BENEFIT		
Widowed mother's allowance	112.55	113.70
Widow's pension		
standard rate	112.55	113.70
age-related		
age 54 (49)	104.67	105.74
53 (48)	96.79	97.78
52 (47)	88.91	89.82
51 (46)	81.04	81.86
50 (45)	73.16	73.91
49 (44)	65.28	65.95
48 (43)	57.40	57.99
47 (42)	49.52	50.03
46 (41)	41.64	42.07
45 (40)	33.77	34.11

Source: Department for Work and Pensions

Note: For deaths occurring before 11 April 1988
refer to age-points shown in brackets.

Note: The Cat C equivalent in Widow's Pension (code: WPE) is
still linked to the rate of Category C State Pension. Not relevant to
the Order.

5.5 Number of Persons claiming benefits: Caseloads by age groups, thousands

	2006/07 Outturn	2007/08 Outturn	2008/09 Outturn	2009/10 Outturn	2010/11 Outturn	2011/12 Outturn	2012/13 Outturn	2013/14 Outturn	2014/15 Outturn	2015/16 Outturn	2016/17 Outturn
Benefits directed at Children											
Attendance Allowance (in payment)	-	-	-	-	-	-	-	-	-	-	-
Child Benefit & One Parent Benefit	-	-	-	-	-	-	-	-	-	-	-
number of children covered	292	300	310	322	331	339	350	362	381	406	426
Disability Living Allowance	292	300	310	322	331	339	350	362	381	406	426
of which in payment	292	300	310	322	331	339	350	362	381	406	426
of which entitlement without payment	-	-	-	-	-	-	-	-	-	-	-
Mobility Allowance	-	-	-	-	-	-	-	-	-	-	-
Benefits Directed at People of Working Age											
Armed Forces Independence Payment								1	1	1	1
Attendance Allowance (in payment)	-	-	-	-	-	-	-	-	-	-	-
Bereavement related benefits	147	129	117	108	103	100	97	95	92	92	90
Carer's Allowance	485	496	519	550	584	615	650	675	730	793	844
of which in payment	432	442	462	491	523	558	595	631	679	740	780
of which entitlement without payment	52	54	57	59	61	57	55	44	52	52	65
Christmas Bonus - non-contributory	1,920	2,003	3,164	3,246	3,227	3,212	3,240	3,215	3,329	3,565	3,310
Council Tax Benefit	2,570	2,533	2,566	2,940	3,172	3,254	3,368	-	-	-	-
Disability Living Allowance	1,657	1,692	1,733	1,780	1,819	1,847	1,877	1,866	1,761	1,566	1,237
of which in payment	1,641	1,676	1,715	1,761	1,800	1,828	1,858	1,846	1,742	1,547	1,221
of which entitlement without payment	16	16	17	19	19	20	20	20	19	18	17
Disability Working Allowance	-	-	-	-	-	-	-	-	-	-	-
Employment and Support Allowance	-	-	136	391	579	811	1,365	1,912	2,235	2,356	2,381
of which contributory	-	-	59	145	199	262	364	492	507	489	483
of which contributory and income-based	-	-	6	22	38	59	103	180	248	325	379
of which income-based	-	-	56	168	275	424	784	1,116	1,340	1,398	1,381
of which credits only	-	-	16	56	67	65	114	124	141	144	137
Family Credit	-	-	-	-	-	-	-	-	-	-	-
Housing benefits	2,518	2,527	2,625	2,981	3,224	3,356	3,502	3,520	3,457	3,360	3,230
Incapacity Benefit, Invalidity Benefit & Sickness Benefit	2,443	2,415	2,332	2,031	1,827	1,577	965	366	133	74	52
of which in payment	1,456	1,413	1,346	1,177	1,054	909	566	192	38	10	3
of which credits only	987	1,002	986	854	773	668	399	174	95	65	49
Income Support	2,135	2,117	2,087	1,935	1,803	1,619	1,254	939	799	706	632
Industrial Injuries benefits	176	170	164	157	152	146	137	137	131	126	123
Jobseeker's Allowance	927	818	1,025	1,538	1,415	1,515	1,507	1,273	885	643	503
of which contributory	157	142	244	321	234	212	178	145	111	77	66
of which contributory and income-based	15	13	21	30	22	19	18	15	11	9	7
of which income-based	691	609	695	1,072	1,069	1,208	1,242	1,045	710	522	392
of which credits only	64	54	66	114	91	77	69	68	53	35	38
Maternity Allowance	30	44	54	56	54	57	60	58	59	62	61
Mobility Allowance	-	-	-	-	-	-	-	-	-	-	-
Personal Independence Payment	-	-	-	-	-	-	-	12	184	569	967
of which in payment	-	-	-	-	-	-	-	12	182	563	956
of which entitlement without payment	-	-	-	-	-	-	-	-	2	6	11
Severe Disablement Allowance	234	221	210	200	191	184	177	167	134	75	25
Statutory Maternity Pay	154	193	245	250	269	266	275	271	264	267	259
Unemployment Benefit	-	-	-	-	-	-	-	-	-	-	-
Universal Credit (Total)								2	13	130	376
Benefits Directed at Pensioners											
Attendance Allowance	1,666	1,700	1,737	1,776	1,782	1,756	1,710	1,641	1,617	1,602	1,594
of which in payment	1,489	1,528	1,568	1,607	1,619	1,597	1,553	1,490	1,462	1,459	1,445
of which entitlement without payment	177	172	169	169	163	160	158	151	155	143	148
Bereavement related benefits	21	19	14	11	8	6	5	3	2	1	1
Carer's Allowance	319	356	388	411	420	417	405	396	378	378	365
of which in payment	26	28	30	31	30	26	22	22	20	19	18
of which entitlement without payment	292	327	357	381	390	391	383	374	358	359	347
Christmas Bonus - contributory	12,586	12,728	11,754	12,123	12,239	12,335	12,346	12,245	12,459	12,757	12,642
Council Tax Benefit	2,510	2,535	2,592	2,631	2,633	2,620	2,544	-	-	-	-
Disability Living Allowance	903	949	991	1,031	1,055	1,066	1,079	1,079	1,072	1,046	967
of which in payment	897	942	984	1,023	1,046	1,057	1,070	1,069	1,062	1,036	957
of which entitlement without payment	6	6	7	8	9	9	9	10	10	10	10
Housing benefits	1,503	1,509	1,541	1,566	1,574	1,576	1,551	1,505	1,464	1,417	1,364
Incapacity Benefit, Invalidity Benefit & Sickness Benefit	-	-	-	-	-	-	-	-	-	-	-
of which in payment	-	-	-	-	-	-	-	-	-	-	-
of which credits only	-	-	-	-	-	-	-	-	-	-	-
Industrial Injuries benefits	159	171	172	176	182	185	187	189	188	187	185

5.5 Number of Persons claiming benefits: Caseloads by age groups, thousands

	2006/07 Outturn	2007/08 Outturn	2008/09 Outturn	2009/10 Outturn	2010/11 Outturn	2011/12 Outturn	2012/13 Outturn	2013/14 Outturn	2014/15 Outturn	2015/16 Outturn	2016/17 Outturn
Income Support	-	-	-	-	-	-	-	-	-	-	-
Mobility Allowance	-	-	-	-	-	-	-	-	-	-	-
Over-75 TV Licence	3,982	3,993	4,079	4,206	4,236	4,277	4,316	4,414	4,493	4,360	4,516
Pension Credit	2,729	2,732	2,724	2,736	2,718	2,649	2,505	2,380	2,228	2,098	1,903
Personal Independence Payment	-	-	-	-	-	-	-	1	17	25	88
of which in payment	-	-	-	-	-	-	-	1	16	24	87
of which entitlement without payment	-	-	-	-	-	-	-	-	-	-	1
State Pension	11,715	11,938	12,160	12,410	12,566	12,667	12,810	12,888	12,958	12,957	12,859
of which contributory	11,692	11,914	12,134	12,382	12,537	12,634	12,774	12,846	12,912	12,908	12,877
of which basic element	11,617	11,837	12,053	12,285	12,460	12,556	12,737	12,814	12,887	12,890	12,681
of which earnings-related element ("Additional Pension", "SERPS" or "S2P")	7,963	8,324	8,673	9,052	9,320	9,549	9,848	10,021	10,207	10,335	10,250
of which Graduated Retirement Benefit	9,594	9,842	10,087	10,358	10,563	10,702	10,874	10,984	11,096	11,145	10,995
of which lump sums (covering all contributory elements)	8	24	46	58	66	59	56	56	50	47	49
of which new State Pension (excluding protected payments)											196
of which new State Pension Protected Payments (including inherited elements)											59
of which non-contributory only ("Category D")	23	25	26	28	29	33	35	42	46	48	51
Severe Disablement Allowance	42	42	41	40	39	36	34	31	29	38	35
Winter Fuel Payments	11,750	12,123	12,421	12,681	12,783	12,686	12,683	12,585	12,467	12,215	12,025

Source: Department for Work and Pensions (DWP)

Caseloads:

Caseload figures represent an average over the full financial year; caseloads for benefits which commenced or ended during a financial year still reflect an average over the full year.

Revisions:

Some figures for past years may have changed since previous publications and from other sources owing to the incorporation of more up-to-date information improvements in the methodology used to break down expenditure totals between sub-groups, or minor definitional differences.

Accounting basis:

Figures for 1999/00 onwards are on a Resource Accounting and Budgeting basis. There may be small differences between figures quoted in these tables and those quoted in Department for Work and Pensions Accounts.

Treatment of Universal Credit:

Universal Credit was introduced in October 2013. This will gradually replace Income Support, income-based Jobseeker's Allowance, income-based Employment and Support Allowance and Housing Benefit, along with Child Tax Credit and Working Tax Credit (delivered by Her Majesty's Revenue & Customs at present).

Benefit changes over time:

Where one benefit has been renamed or superseded by another, and the earlier benefit is not shown separately, the most recent name for the benefit is shown, but the figures cover all relevant benefits, as follows:

Bereavement benefits include Widows' Benefits and Bereavement Support Payment.
Carer's Allowance includes Invalid Care Allowance.
Child Benefit includes Family Allowance, and also One Parent Benefit where not shown separately.
Council Tax Benefit includes Community Charge Benefit and Rate Rebate.
Family Credit includes Family Income Supplement.
Income Support includes Supplementary Benefit and National Assistance.
Severe Disablement Allowance includes Non-Contributory Invalidity Pension.

Rounding: Caseloads stated in these tables are rounded to the nearest 1,000, as such totals may not sum due to this rounding.

5.6 Jobseeker's Allowance[1,2,3] claimants: by benefit entitlement
Great Britain

As at May of each year   Thousands

		2007	2008	2009	2010	2011	2012	2013 *	2014 *	2015 *	2016	2017
All Persons												
All with benefit - total	KXDX	730.8	718	1316.4	1237.3	1298.3	1377	1432.6	1035.3	719.1	562.4	474.0
Contribution-based JSA only	KXDY	113.6	127.8	341.8	205.3	182.7	159.5	142.4	97.3	72.7	63.7	57.7
Contribution based JSA & income-based JSA	KXDZ	11.9	12.8	34.6	21.1	19.9	16.8	14.9	10.9	7.1	8.3	7.9
Income-based JSA only payment	KXEA	605.3	577.4	940	1010.9	1095.7	1200.7	1156	842.4	583.2	446.0	370.6
No benefit in payment	KXEB	76.4	69.9	126.6	117.3	105.8	107.2	119.4	84.7	56.1	44.3	37.8
Males												
All with benefit - total	KXED	537.8	529.9	978.9	890.9	879.8	930	938.4	665.7	460.9	351.1	286.4
Contribution-based JSA only	KXEE	79.6	90.6	248.7	143.4	118.9	103.5	94.1	62	45.8	39.7	35.1
Contribution based JSA & income-based JSA	KXEF	10.7	11.7	31.2	18.1	16.9	13.7	12.6	8.9	5.9	6.7	6.0
Income-based JSA only payment	KXEG	447.5	427.6	698.9	729.5	744	812.8	753	539.1	374.1	277.2	222.9
No benefit in payment	KXEH	51.7	46.7	88.8	82.2	69.3	71.6	78.7	55.8	35.1	27.5	22.3
Females												
All with benefit - total	KXEJ	193	188.1	337.6	346.4	418.4	447	494.3	369.5	258.2	211.3	187.6
Contribution-based JSA only	KXEK	34	37.2	93.1	61.9	63.7	56	48.3	35.3	26.9	24.1	22.6
Contribution based JSA & income-based JSA	KXEL	1.2	1.2	3.4	3.0	3	3.1	2.3	2	1.3	1.7	1.8
Income-based JSA only payment	KXEM	157.8	149.8	241.1	281.4	351.7	387.9	403	303.3	209.1	168.8	147.7
No benefit in payment	KXEN	24.8	23.2	37.7	35.1	36.5	35.5	40.7	28.9	21	16.8	15.5

Sources: Department for Work and Pensions

1. Jobseeker's Allowance (JSA) is a benefit payable to unemployed people. In general, to be entitled to Jobseeker's Allowance, a person must be available for work for at least 40 hours a week, be actively seeking work, and have entered into a Jobseeker's Agreement with Jobcentre Plus.
JSA has two routes of entry: contrbution-based which depends mainly upon national insurance contributions and income-based which depends mainly on a means test. Some claimants can qualify by either route. In practice they receive income-based JSA but have an under lying entitlement to the contribution-based element.

2 Figures are given at May each year and have been derived by applying 5% proportions to 100% totals taken from the DWP 100% Work and Pensions Longitudinal Study (WPLS).

3 Figures are rounded to the nearest hundred and quoted in thousands. They may not sum due to rounding and are subject to sampling variation.

* 5% sample data - DWP recommends that, where the detail is only available on the 5% sample data, the proportions derived should be applied to the overall 100% total for the benefit.

5.7 Employment and Support Allowance and Incapacity Benefit claimants: by sex, age and duration of claim

Great Britain

As at May   Count

	2010	2011	2012	2013	2014	2015	2016	2017
Male								
All durations: All ages	301455	362360	608526	935744	1150176	1242802	1245241	1221741
All durations:								
Under 18	2479	2259	2271	2019	2050	2032	1991	1839
18-24	41131	49865	66498	81905	97424	106134	102265	95491
25-34	56314	65655	102628	145989	182064	200254	199231	193059
35-44	68818	81733	136130	192386	222224	238832	236166	225826
45-49	35649	43814	81729	128904	154682	163456	162624	156847
50-54	34339	42363	80507	134578	166363	181195	183348	180415
55-59	36254	43905	83117	142023	172524	184783	185164	184766
Over 60	26465	32762	55644	107932	152839	166109	174456	183500
Unknown	6	..	..	6	7	..	..	5
Over six months:								
Under 18	1236	1189	1215	1051	1060	1110	1074	1052
18-24	19451	28819	42317	56118	70828	80459	81034	77740
25-34	26495	36581	57166	96660	138469	162711	166970	164187
35-44	34397	49754	76891	133915	176424	197010	203371	196924
45-49	17933	27191	45594	89026	126235	141176	143902	140384
50-54	17295	26986	44713	91595	136827	158176	164282	163022
55-59	18820	28371	46783	94839	143075	163248	166545	167450
Over 60	15194	22854	34412	72167	128162	150002	159799	168042
Female								
All durations: All ages	225268	299404	524288	807088	989471	1092604	1119183	1131515
All durations:								
Under 18	2184	2003	2056	1733	1748	1897	1804	1582
18-24	31746	38732	51544	62948	74429	81109	79194	73752
25-34	35806	48626	79634	118000	145272	160866	161629	158388
35-44	58042	77175	124690	171653	194602	209405	207171	200167
45-49	35235	47217	86097	128895	151227	159606	158636	154206
50-54	32168	42737	86361	139806	169241	185858	189239	187674
55-59	29399	38294	82127	144170	175266	190798	194372	197110
Over 60	694	4623	11775	39888	77683	103060	127143	158625
Unknown	..	..	..	..	6	6	..	..
Over six months:								
Under 18	956	909	1023	832	828	852	898	778
18-24	14725	22447	32575	42816	52970	60473	62119	59463
25-34	17007	26269	45025	79320	111593	131334	136079	134815
35-44	29034	45882	74290	123122	156426	173807	179403	175394
45-49	18040	29692	50423	92385	125223	138465	140852	138651
50-54	16961	27748	48938	98491	142033	163086	170864	170931
55-59	15893	25270	45892	98324	148542	170254	177561	181020
Over 60	527	3181	8201	27149	66385	92827	117140	146331

Source: Stat-Xplore, Department for Work and Pensions

".." denotes a nil or negligible number of claimants or award amount based on a nil or negligible number of claimants.

Statistical disclosure control has been applied to this table to avoid the release of confidential data. Totals may not sum due to the disclosure control applied.

The move to publishing benefit data via the Stat-Xplore instead of the DWP Tabulation Tool has involved adopting a new disclosure control methodology, in line with other benefits published via this new tool. Although the data still comes from the same source there may be small differences in the outputs displayed using this new tool, when compared to the Tabulation Tool.

5.8 Attendance Allowance - cases in payment: Age and gender of claimant Great Britain
Great Britain

At May each year   Count

	2008	2009	2010	2011	2012	2013	2014	2015	2016	2017
Males: All ages	496,991	516,126	531,070	529,570	525,691	504,351	493,165	498,963	503,439	502,904
Unknown age	23	20	20	24	19	19	21	24	33	34
65-69	23,501	24,354	24,752	23,769	23,252	21,122	20,567	24,254	27,308	26,853
70-74	70,191	73,604	74,890	71,909	68,697	63,516	59,777	60,772	62,890	66,701
75-79	109,051	112,296	114,784	113,594	112,183	107,189	103,958	103,978	101,279	98,789
80-84	135,335	137,612	139,715	137,503	135,277	129,925	125,576	125,627	125,765	124,045
85-89	107,708	116,034	118,027	117,833	116,921	112,561	111,066	110,872	110,900	110,258
90 and over	51,192	52,205	58,883	64,942	69,344	70,022	72,204	73,444	75,261	76,213
Females: All ages	1,049,375	1,068,964	1,082,436	1,068,797	1,046,878	997,006	962,307	950,845	941,899	924,826
Unknown age	20	18	23	26	24	33	30	38	40	63
65-69	29,128	30,021	30,312	28,687	27,147	23,793	22,440	26,566	30,169	29,789
70-74	99,660	103,449	104,453	99,642	93,820	85,257	78,399	78,187	79,738	82,791
75-79	186,646	188,379	188,345	182,140	176,428	166,160	158,411	155,890	151,438	146,091
80-84	277,675	277,658	276,881	269,083	259,667	244,729	231,714	226,030	220,996	215,215
85-89	276,660	290,707	286,518	279,699	271,425	258,583	249,908	244,734	240,270	234,685
90 and over	179,590	178,729	195,903	209,519	218,367	218,457	221,400	219,397	219,245	216,189

Source: Stat-Xplore, Department for Work and Pensions

".." denotes a nil or negligible number of claimants or award amount based on a nil or negligible number of claimants.

Statistical disclosure control has been applied to this table to avoid the release of confidential data. Totals may not sum due to the disclosure control applied.
Description

The move to publishing benefit data via the Stat-Xplore instead of the DWP Tabulation Tool has involved adopting a new disclosure control methodology, in line with other benefits published via this new tool. Although the data still comes from the same source there may be small differences in the outputs displayed using this new tool, when compared to the Tabulation Tool.

5.9a Families and children receiving Child Benefit, in each country and English Region, 2003 to 2017

Time Series	United Kingdom [1]	Great Britain	England and Wales	England	North East	North West	Yorkshire and the Humber	East Midlands	West Midlands	East	London	South East	South West	Wales	Scotland	Northern Ireland	Foreign and not known
Area Codes [2]	KO2000001	KO3000001	KO4000001	E92000001	E12000001	E12000002	E12000003	E12000004	E12000005	E12000006	E12000007	E12000008	E12000009	W92000004	S92000003	N92000002	n/a
Number of families																	
August 2003	7,246,335	7,000,770	6,394,870	6,037,500	318,470	861,775	619,630	517,590	663,400	653,695	876,120	956,080	570,735	357,370	605,900	225,885	19,675
August 2004	7,296,495	7,055,160	6,448,355	6,087,500	317,515	863,070	622,065	520,870	667,175	660,390	894,090	965,480	576,845	360,855	606,805	226,850	14,485
August 2005	7,315,165	7,074,665	6,470,575	6,110,190	315,855	860,660	622,475	522,195	667,565	664,155	909,045	970,225	578,015	360,385	604,085	226,800	13,705
August 2006	7,413,475	7,129,720	6,528,205	6,168,010	316,665	864,650	626,740	527,105	672,220	671,850	926,055	981,015	581,705	360,195	601,515	230,140	53,615
August 2007	7,475,035	7,212,565	6,605,270	6,241,895	318,020	869,475	631,995	535,775	678,300	683,780	937,480	995,990	591,085	363,375	607,290	230,825	31,650
August 2008	7,582,990	7,320,990	6,708,080	6,341,345	319,815	876,795	640,670	543,350	686,910	696,485	964,180	1,013,595	599,550	366,735	612,910	233,830	28,165
August 2009	7,769,880	7,485,730	6,864,935	6,492,290	324,525	892,240	653,645	554,925	701,070	713,455	1,002,815	1,038,010	611,600	372,650	620,795	238,605	45,545
August 2010	7,841,675	7,557,305	6,935,695	6,562,705	324,265	894,940	657,700	559,645	705,640	723,030	1,028,265	1,051,885	617,340	372,985	621,615	240,985	43,385
August 2011	7,884,760	7,600,115	6,979,465	6,606,285	323,155	895,670	659,240	561,885	708,325	730,180	1,044,355	1,061,870	621,605	373,180	620,650	242,310	42,335
August 2012	7,920,495	7,641,575	7,022,780	6,650,070	321,310	895,845	661,370	564,385	711,110	737,485	1,061,620	1,071,795	625,145	372,705	618,795	243,185	35,735
August 2013	7,550,265	7,279,100	6,691,985	6,328,460	311,725	868,775	643,560	545,720	692,110	690,410	996,490	979,075	600,590	363,525	587,115	239,125	32,040
August 2014	7,461,675	7,195,865	6,619,190	6,259,275	307,860	862,015	640,080	542,575	688,340	681,035	982,060	959,600	595,710	359,910	576,675	237,865	27,945
August 2015	7,416,800	7,153,935	6,584,675	6,227,865	305,000	857,415	636,600	541,540	687,010	678,260	977,940	951,060	593,040	356,810	569,260	236,890	25,970
August 2016	7,396,355	7,139,250	6,573,395	6,219,065	302,920	856,100	635,760	542,215	688,655	677,570	978,825	945,350	591,665	354,335	565,850	236,280	20,825
August 2017	7,376,965	7,121,585	6,558,625	6,206,745	301,265	854,385	635,155	542,275	689,350	675,515	979,760	940,185	588,855	351,880	562,960	235,295	20,085
Number of children																	
August 2003	13,138,075	12,670,975	11,625,050	10,983,290	552,970	1,549,900	1,116,630	934,450	1,219,985	1,200,175	1,613,235	1,754,585	1,041,360	641,755	1,045,925	439,870	27,230
August 2004	13,096,760	12,635,505	11,600,380	10,960,280	544,840	1,534,595	1,109,155	930,920	1,214,695	1,200,175	1,632,425	1,752,995	1,040,475	640,100	1,035,125	435,690	25,565
August 2005	13,111,665	12,654,135	11,626,490	10,988,765	540,940	1,528,255	1,109,150	932,310	1,215,315	1,204,750	1,658,755	1,758,520	1,040,780	637,725	1,027,640	431,995	25,535
August 2006	13,233,320	12,706,365	11,685,995	11,050,975	540,980	1,529,585	1,113,190	936,980	1,219,915	1,212,530	1,686,375	1,768,965	1,042,445	635,020	1,020,370	435,485	91,475
August 2007	13,267,355	12,778,460	11,754,415	11,117,770	540,610	1,529,060	1,117,760	946,090	1,225,025	1,225,485	1,699,215	1,782,530	1,052,000	636,645	1,024,045	433,370	55,525
August 2008	13,340,565	12,857,555	11,831,255	11,194,420	539,840	1,528,890	1,124,420	951,000	1,231,190	1,235,400	1,732,120	1,795,225	1,056,340	636,835	1,026,300	434,390	48,625
August 2009	13,604,375	13,088,240	12,054,140	11,409,950	546,125	1,549,625	1,143,245	967,010	1,251,900	1,258,520	1,794,220	1,827,530	1,071,775	644,190	1,034,095	440,570	75,565
August 2010	13,685,250	13,170,155	12,138,365	11,495,395	544,775	1,551,080	1,147,440	971,690	1,257,180	1,269,870	1,831,965	1,843,465	1,077,930	642,965	1,031,795	443,110	71,985
August 2011	13,721,160	13,207,465	12,179,715	11,537,505	542,680	1,549,475	1,148,450	973,310	1,259,770	1,276,525	1,853,670	1,852,950	1,080,680	642,210	1,027,750	444,285	69,410
August 2012	13,771,635	13,267,355	12,243,960	11,602,370	540,060	1,550,880	1,153,480	976,870	1,265,765	1,284,980	1,880,560	1,865,335	1,084,435	641,590	1,023,390	445,220	59,055
August 2013	13,107,460	12,618,675	11,651,810	11,026,465	525,215	1,505,780	1,124,295	943,980	1,233,780	1,198,215	1,763,895	1,693,670	1,037,630	625,345	966,865	437,440	51,345
August 2014	12,962,175	12,482,260	11,532,980	10,913,100	520,170	1,497,345	1,121,595	938,835	1,229,210	1,181,620	1,738,575	1,656,975	1,028,775	619,885	949,280	435,055	44,860
August 2015	12,895,530	12,420,785	11,482,570	10,867,625	517,125	1,494,870	1,118,980	938,125	1,229,805	1,176,055	1,729,510	1,639,590	1,023,565	614,945	938,215	433,940	40,805
August 2016	12,877,170	12,410,910	11,476,565	10,864,980	515,905	1,497,955	1,121,150	940,315	1,236,365	1,174,760	1,728,685	1,629,015	1,020,835	611,585	934,350	433,310	32,950
August 2017	12,847,100	12,383,285	11,452,910	10,845,455	514,160	1,499,395	1,121,460	940,715	1,239,475	1,169,820	1,726,105	1,618,310	1,016,010	607,460	930,375	431,875	31,940

Source: HM Revenue and Customs

Footnotes

[1] Includes Foreign and not known

[2] New area codes implemented from 1 January 2011; in line with the new GSS Coding and Naming policy.

5.9b Families receiving Child Benefit nationally, in each country and English Region, August 2017

Area names	Area Codes [1]	Number of families, by size						Number of children in these families, by age				
		Total	One child	Two children	Three children	Four children	Five or more children	Total	Under 5	5-10	11-15	16 and over
United Kingdom [2]	K02000001	7,376,965	3,520,110	2,695,220	838,865	234,660	88,110	12,847,100	3,391,530	4,412,885	3,318,180	1,724,505
Great Britain	K03000001	7,121,585	3,406,925	2,601,975	802,680	224,790	85,210	12,383,285	3,269,630	4,256,880	3,201,755	1,655,025
England and Wales	K04000001	6,558,625	3,118,855	2,398,325	747,625	212,315	81,505	11,452,910	3,025,365	3,932,535	2,952,365	1,542,645
England	E92000001	6,206,745	2,949,690	2,269,075	708,225	201,955	77,800	10,845,455	2,867,930	3,726,280	2,794,480	1,456,760
North East	E12000001	301,265	149,205	107,940	32,145	8,720	3,255	514,160	135,130	175,270	133,270	70,485
North West	E12000002	854,385	408,405	305,560	99,130	29,500	11,790	1,499,395	398,640	511,900	386,310	202,545
Yorkshire and the Humber	E12000003	635,155	298,900	230,330	74,260	22,845	8,820	1,121,460	300,110	384,585	288,125	148,640
East Midlands	E12000004	542,275	258,535	200,805	60,080	16,595	6,260	940,715	249,825	322,865	241,740	126,285
West Midlands	E12000005	689,350	319,375	245,770	85,245	27,315	11,645	1,239,475	329,760	423,670	319,510	166,540
East	E12000006	675,515	317,925	256,010	75,510	19,590	6,480	1,169,820	309,555	402,245	301,150	156,870
London	E12000007	979,760	477,470	334,255	116,855	35,515	15,660	1,726,105	453,955	599,540	444,850	227,760
South East	E12000008	940,185	443,875	360,355	101,690	25,865	8,400	1,618,310	423,010	559,130	418,335	217,840
South West	E12000009	588,855	276,005	228,050	63,300	16,010	5,495	1,016,010	267,945	347,075	261,190	139,800
Wales	W92000004	351,880	169,165	129,250	39,405	10,355	3,705	607,460	157,435	206,255	157,885	85,885
Scotland	S92000003	562,960	288,075	203,650	55,055	12,475	3,705	930,375	244,265	324,340	249,385	112,380
Northern Ireland	N92000002	235,295	101,855	86,455	34,635	9,545	2,805	431,875	113,685	144,910	108,430	64,855
Foreign and not known	N/A	20,085	11,325	6,785	1,550	330	95	31,940	8,215	11,095	7,995	4,630

Source: HM Revenue and Customs

Footnotes

[1] New area codes implemented from 1 January 2011; in line with the new GSS Coding and Naming policy.

[2] Includes Foreign and not known.

5.10 Child Tax Credit and Working Tax Credit elements and thresholds

Annual rate (£), except where specified

	2008-09	2009-10	2010-11	2011-12	2012-13	2013-14	2014-15	2015-16	2016-17	2017-18
Child Tax Credit										
Family element	545	545	545	545	545	545	545	545	545	545
Family element, baby addition [1]	545	545	545	-	-	-	-	-	-	-
Child element [2]	2,085	2,235	2,300	2,555	2,690	2,720	2,750	2,780	2,780	2,780
Disabled child additional element [3]	2,540	2,670	2,715	2,800	2,950	3,015	3,100	3,140	3,140	3,175
Severely disabled child element [4]	1,020	1,075	1,095	1,130	1,190	1,220	1,255	1,275	1,275	1,290
Working Tax Credit										
Basic element	1,800	1,890	1,920	1,920	1,920	1,920	1,940	1,960	1,960	1,960
Couples and lone parent element	1,770	1,860	1,890	1,950	1,950	1,970	1,990	2,010	2,010	2,010
30 hour element [5]	735	775	790	790	790	790	800	810	810	810
Disabled worker element	2,405	2,530	2,570	2,650	2,790	2,855	2,935	2,970	2,970	3,000
Severely disabled adult element	1,020	1,075	1,095	1,130	1,190	1,220	1,255	1,275	1,275	1,290
50+ return to work payment [6]										
16 but less than 30 hours per week	1,235	1,300	1,320	1,365	-	-	-	-	-	-
at least 30 hours per week	1,840	1,935	1,965	2,030	-	-	-	-	-	-
Childcare element Maximum eligible costs allowed (£ per week)										
Eligible costs incurred for 1 child	175	175	175	175	175	175	175	175	175	175
Eligible costs incurred for 2+ children	300	300	300	300	300	300	300	300	300	300
Percentage of eligible costs covered	*80%*	*80%*	*80%*	*70%*	*70%*	*70%*	*70%*	*70%*	*70%*	*70%*
Common features										
First income threshold [7]	6,420	6,420	6,420	6,420	6,420	6,420	6,420	6,420	6,420	6,420
First withdrawal rate	*39%*	*39%*	*39%*	*41%*	*41%*	*41%*	*41%*	*41%*	*41%*	*41%*
Second income threshold [8]	50,000	50,000	50,000	40,000	-	-	-	-	-	-
Second withdrawal rate	*1 in 15*	*1 in 15*	*1 in 15*	*41%*	-	-	-	-	-	-
First income threshold for those entitled to Child Tax Credit only [9]	15,575	16,040	16,190	15,860	15,860	15,910	16,010	16,105	16,105	16,105
Income increase disregard [10]	25,000	25,000	25,000	10,000	10,000	5,000	5,000	5,000	2,500	2,500
Income fall disregard [10]	-	-	-	-	2,500	2,500	2,500	2,500	2,500	2,500
Minimum award payable	26	26	26	26	26	26	26	26	26	26

Source: HM Revenue and Customs

Footnotes

[1] Payable to families for any period during which they have one or more children aged under 1. Abolished 6 April 2011.

[2] Payable for each child up to 31 August after their 16th birthday, and for each young person for any period in which they are aged under 20 (under 19 to 2005-06) and in full-time non-advanced education, or under 19 and in their first 20 weeks of registration with the Careers service or Connexions.

[3] Payable in addition to the child element for each disabled child.

[4] Payable in addition to the disabled child element for each severely disabled child.

[5] Payable for any period during which normal hours worked (for a couple, summed over the two partners) is at least 30 per week.

[6] Payable for each qualifying adult for the first 12 months following a return to work.Abolished 6 April 2012.

[7] Income is net of pension contributions, and excludes Child Benefit, Housing benefit, Council tax benefit, maintenance and the first £300 of family income other than from work or benefits. The award is reduced by the excess of income over the first threshold, multiplied by the first withdrawal rate.

[8] For those entitled to the Child Tax Credit, the award is reduced only down to the family element, plus the baby addition where relevant, less the excess of income over the second threshold multiplied by the second withdrawal rate. Abolished 6 April 2012.

[9] Those also receiving Income Support, income-based Jobseeker's Allowance or Pension Credit are passported to maximum CTC with no tapering.

[10] Introduced from 6 April 2012, this drop in income is disregarded in the calculation of Tax Credit awards.

5.11 Number of families and children with Child Tax Credits or Working Tax Credits, by level of award, 2017-2018

Thousands

Area Codes[1]		Total out-of-work families	In-work families				Total in receipt (out-of-work and in-work families)	Number of children in recipient families		
			With children			With no children		Total out-of-work families	In-work families	
			receiving WTC and CTC	Receiving CTC only	Of which, lone parents	Receiving WTC only			Receiving WTC and CTC	Receiving CTC only
United Kingdom[2]	K02000001	1,122	1,556	764	1,100	342	3,784	2,179	2,929	1,610
Great Britain	K03000001	1,074	1,502	735	1,059	326	3,637	2,087	2,831	1,546
England and Wales	K04000001	984	1,394	679	969	296	3,353	1,925	2,649	1,433
England	E92000001	920	1,318	638	913	276	3,152	1,803	2,513	1,348
North East	E12000001	57	67	31	50	20	175	109	121	63
North West	E12000002	126	196	89	134	49	459	248	375	185
Yorkshire and the Humber	E12000003	105	151	72	97	35	363	210	290	151
East Midlands	E12000004	77	115	65	81	27	285	153	213	138
West Midlands	E12000005	118	160	79	102	32	389	240	318	169
East	E12000006	88	124	70	92	23	305	171	230	150
London	E12000007	162	238	76	152	33	509	310	478	157
South East	E12000008	116	161	88	124	30	395	225	297	189
South West	E12000009	71	106	68	79	28	273	138	191	147
Wales	W92000004	64	75	41	56	20	201	122	135	85
Scotland	S92000003	90	108	57	90	30	284	162	182	113
Northern Ireland	N92000002	47	54	28	40	16	144	90	96	63
Foreign and not known		1	1	1	1	-	3	2	2	1

Source: HM Revenue and Customs
Child and Working Tax Credits Statistics: Finalised annual awards 2016 to 2017, Geographical Analysis

Footnotes

[1] Area codes implemented from 1 January 2011 in line with the new GSS Coding and Naming policy.

[2] Includes Foreign and not known

[3] All figures are rounded to the nearest integer therefore not all totals may exactly equal the sums of their respective components.

[4] "Foreign and not known" consists of a small proportion of recipient families and children who do not live within England, Scotland, Northern Island or Wales. They may for instance be a Crown servant posted overseas, or living in a British Crown Dependency. Due to the small size of this population we combine them into one group which also consists of those whom, at the time of publication, either have a UK postcode that does not match to a geographical office region code, or do not have a postcode in the available data.

5.12 Widows' Benefit (excluding bereavement payment[1,2,3]): by type of benefit

Great Britain

Number in reciept of widows benefit as at May each year

	2008	2009	2010	2011	2012	2013	2014	2015	2016	2017
All Widows' benefit (excluding breavement allowance)										
All ages	71,447	56,637	46,221	40,178	34,738	30,175	26,209	22,881	20,301	18,851
Unknown age	..	..	..	..	..	..	..	..	..	..
18-24	..	..	..	..	..	..	..	..	..	..
25-29	15	8	..	..	..	..	..	..	..	..
30-34	192	122	71	41	21	12	5	..	..	..
35-39	1,022	701	472	316	201	128	83	50	26	12
40-44	2,855	2,176	1,611	1,132	777	548	352	211	131	55
45-49	5,867	4,797	3,915	3,146	2,519	1,936	1,422	1,012	663	447
50-54	13,583	11,022	9,161	7,602	6,328	5,231	4,325	3,519	2,839	2,268
55-59	34,480	28,419	23,122	18,771	15,329	12,462	10,126	8,462	7,033	5,841
60-64	13,425	9,402	7,871	9,167	9,550	9,859	9,892	9,622	9,607	10,228
Widow's Pension - not age related										
All ages	6,536	4,451	3,389	2,984	2,636	2,370	2,077	1,795	1,643	1,546
Unknown age	..	..	..	..	..	..	..	..	..	..
18-24	..	..	..	..	..	..	..	..	..	..
25-29	..	..	..	..	..	..	..	..	..	..
30-34	..	..	..	..	..	..	..	..	..	..
35-39	..	..	..	..	..	..	..	..	..	..
40-44	..	..	..	..	..	..	..	..	..	..
45-49	..	..	..	..	..	..	..	..	..	..
50-54	104	43	13	9	5	..	..	..	..	..
55-59	2,553	2,116	1,741	1,372	1,122	919	708	559	459	353
60-64	3,878	2,289	1,633	1,602	1,509	1,451	1,372	1,237	1,187	1,191
Widow's Pension - age related										
All ages	52,638	42,466	35,125	31,235	27,609	24,556	21,851	19,563	17,728	16,799
Unknown age	..	..	..	..	..	..	..	..	..	..
18-24	..	..	..	..	..	..	..	..	..	..
25-29	..	..	..	..	..	..	..	..	..	..
30-34	..	..	..	..	..	..	..	..	..	..
35-39	..	..	..	..	..	..	..	..	..	..
40-44	31	12	5	..	..	..	..	..	..	..
45-49	2,099	1,745	1,382	1,155	1,012	851	678	528	400	305
50-54	10,454	8,496	7,117	5,956	5,041	4,262	3,621	3,032	2,515	2,081
55-59	30,634	25,195	20,479	16,653	13,597	11,104	9,087	7,663	6,419	5,398
60-64	9,425	7,015	6,145	7,472	7,961	8,334	8,460	8,338	8,391	9,016
Widowed Mother's Allowance - with dependants										
All ages	12,209	9,678	7,669	5,920	4,461	3,236	2,265	1,508	919	495
Unknown age	..	..	..	..	..	..	..	..	..	..
18-24	..	..	..	..	..	..	..	..	..	..
25-29	15	8	..	..	..	..	..	..	..	..
30-34	188	120	69	38	21	12	5	..	..	..
35-39	1,021	694	476	316	203	131	81	50	26	12
40-44	2,827	2,166	1,601	1,136	777	548	352	209	132	55
45-49	3,751	3,031	2,521	1,976	1,502	1,078	750	488	265	140
50-54	3,014	2,466	2,024	1,635	1,281	958	692	479	315	179
55-59	1,277	1,093	893	731	601	438	328	236	152	87
60-64	119	92	85	92	76	64	56	41	29	20
Widowed Mother's Allowance - without dependants										
All ages	57	46	40	33	26	22	18	13	10	10
Unknown age	..	..	..	..	..	..	..	..	..	..
18-24	..	..	..	..	..	..	..	..	..	..
25-29	..	..	..	..	..	..	..	..	..	..
30-34	..	..	..	..	..	..	..	..	..	..
35-39	..	..	..	..	..	..	..	..	..	..
40-44	6	5	..	..	..	..	..	..	..	..
45-49	17	15	14	13	7	6	..	..	..	..
50-54	13	10	7	5	7	9	8	5	6	..
55-59	20	14	13	11	10	6	6	..	6	5
60-64	..	..	..	..	7	8	8	..	..	..

Source: Stat-Xplore, Department for Work and Pensions

5.12 Widows' Benefit (excluding bereavement payment[1,2,3]): by type of benefit
Great Britain

Notes

".." denotes a nil or negligible number of claimants or award amount based on a nil or negligible number of claimants.

1. Statistical disclosure control has been applied to this table to avoid the release of confidential data. Totals may not sum due to the disclosure control applied.

2 Caseload (Thousands) All Claimants of Widows Benefit are female. No new claims for WA have been accepted since April 2001 when it was replaced by Bereavement Allowance
3 Figures include overseas cases.

Widow's Benefit (WB) is payable to women widowed on or after 11th April 1988 and up to and including 8th April 2001. Statistics are based on WB claims in payment in Great Britain on the reference date; this excludes cases in Northern Ireland.

Widowed Mother's Allowance: Weekly benefit payable to a widowed mother (providing she was widowed before 8th April 2001) if her husband paid enough NIC and receiving Child Benefit for one of her children, or her husband was in receipt of Child Benefit, or she was expecting her husbands baby, or if she was widowed before 11th April 1988 and has a young person under 19 living with her for whom she was receiving Child Benefit.

Widow's Pension: Weekly benefit payable to widows without dependant children providing she was widowed before 8th April 2001. A widow may be able to get widows pension if her husband paid enough NIC. She must be 45 or over when her husband died or when her Widowed Mother's Allowance ends, she cannot get Widow's Pension at the same time as Widowed Mother's Allowance.

5.13 Bereavement Benefit[1] (excluding bereavement payment): by sex, type of benefit and age of widow/er

Great Britain and Overseas

At May of each year

Count

	Male					Female				
	2013	2014	2015	2016	2017	2013	2014	2015	2016	2017
All Bereavement Benefit (exluding bereavement allowance)										
All ages	18789	18570	18486	18427	16744	48804	49625	50832	51619	49611
18-24	8	..	6	6	6	42	34	34	21	20
25-29	42	56	60	54	39	472	401	393	370	311
30-34	258	259	254	229	223	1774	1759	1692	1640	1528
35-39	791	754	745	747	740	3819	3756	3732	3756	3709
40-44	2103	1939	1839	1724	1606	7663	7430	7291	6905	6552
45-49	3822	3751	3586	3469	3231	11182	11313	11016	11042	10546
50-54	4108	4177	4210	4379	4018	10944	11137	11366	11397	10700
55-59	3572	3598	3654	3784	3433	9790	9628	10083	10003	8985
60-64	4085	4038	4142	4043	3447	3117	4168	5222	6489	7269
Widowed Parents' Allowance with dependants										
All ages	11664	11475	11177	11059	10694	34306	34455	34416	34350	33442
18-24	8	..	6	6	6	37	34	34	20	20
25-29	42	56	60	54	39	472	401	389	368	305
30-34	258	259	251	229	223	1764	1761	1683	1635	1522
35-39	791	754	745	747	740	3809	3743	3728	3750	3704
40-44	2098	1939	1835	1724	1606	7640	7409	7271	6891	6545
45-49	3337	3253	3078	3001	2850	9817	9944	9671	9701	9438
50-54	2977	3036	2996	3096	2980	7307	7552	7762	7905	7804
55-59	1497	1558	1554	1562	1616	3091	3166	3364	3426	3368
60-64	652	630	659	645	638	370	456	516	645	735
Widowed Parents' Allowance without dependants										
All ages	25	17	12	9	7	161	134	118	100	77
18-24	..	..	..	..	..	..	..	..	..	..
25-29	..	..	..	..	..	..	..	..	..	7
30-34	..	..	..	..	..	9	..	6	9	..
35-39	..	..	..	..	..	13	11	8	10	5
40-44	..	..	..	..	..	30	25	22	14	5
45-49	6	10	8	5	5	49	38	24	26	22
50-54	5	7	..	..	..	43	43	31	27	21
55-59	6	..	..	..	..	12	11	15	18	19
60-64	..	..	..	..	..	5	5	6	6	6
Bereavement allowance - age related										
All ages	1755	1786	1894	1935	1599	5381	5371	5405	5282	4462
18-24	..	..	..	..	..	..	..	..	..	..
25-29	..	..	..	..	..	..	..	..	..	..
30-34	..	..	..	..	..	..	..	..	..	..
35-39	..	..	..	..	..	..	..	..	..	..
40-44	..	..	..	..	..	..	..	..	..	..
45-49	476	493	500	464	375	1313	1335	1316	1313	1089
50-54	1121	1131	1211	1281	1038	3593	3543	3572	3459	2873
55-59	155	166	182	190	180	476	498	513	509	503
60-64	..	..	..	..	..	..	..	..	..	..
Bereavement allowance - not age related										
All ages	5342	5286	5402	5429	4446	8955	9666	10896	11887	11628
18-24	..	..	..	..	..	..	..	..	..	..
25-29	..	..	..	..	..	..	..	..	..	..
30-34	..	..	..	..	..	..	..	..	..	..
35-39	..	..	..	..	..	..	..	..	..	..
40-44	..	..	..	..	..	..	..	..	..	..
45-49	..	..	..	..	..	..	..	..	..	..
50-54	..	..	..	..	..	..	..	..	..	..
55-59	1910	1875	1916	2028	1637	6206	5954	6192	6049	5097
60-64	3436	3408	3486	3400	2803	2747	3710	4704	5837	6530

Source: Stat-Xplore, Department for Work and Pensions

5.13 Bereavement Benefit[1] (excluding bereavement payment): by sex, type of benefit and age of widow/er
Great Britain and Overseas

Footnotes

1 Figures are for Great Britain and Overseas and do not include figures for Northern Ireland

".." denotes a nil or negligible number of claimants or award amount based on a nil or negligible number of claimants.

Bereavement Benefit is payable to men/women widowed on or after 9th April 2001.

Widowed Parents Allowance Weekly benefit payable to widowed parents. A widow/widower may be entitled to this if his/her late spouse/civil partner had paid enough NIC and the widow/widower is receiving Child Benefit or can be treated as entitled to Child Benefit, or the late spouse/civil partner was receiving child benefit, or expecting her husband's/civil partner's baby (pregnant from fertility treatment).

Bereavement Allowance: Weekly benefit payable to widows/widowers without dependent children and payable between age 45 and pensionable age. The amount payable to a widow/widower aged between 45 and 54 is related to their age at date of entitlement. The weekly rate is reduced by 7% for each year they are aged under 55, so they will get 93% rate at 54, falling to 30% at age 45. Those aged 55 or over at the date of entitlement will get their full rate of Bereavement Allowance.

Recording and clerical errors can occur within the data source - for this reason, no reliance should be placed on very small numbers obtained through Stat-Xplore.

Statistical disclosure control has been applied to this table to avoid the release of confidential data. Totals may not sum due to the disclosure control applied.

The move to publishing benefit data via the Stat-Xplore instead of the DWP Tabulation Tool has involved adopting a new disclosure control methodology, in line with other benefits published via this new tool. Although the data still comes from the same source there may be small differences in the outputs displayed using this new tool, when compared to the Tabulation Tool.

5.14 State Pension Caseload by Category of Pension, Age and Sex[1]

Great Britain

As at May each year

Count (000s)

	2008	2009	2010	2011	2012	2013	2014	2014	2016	2017
MALE										
Pre-2016 State Pension[2]										
Contributory State Pension[3]										
60-64	0.12	0.14	0.17	0.15	0.17	0.17	0.19	0.17	0.20	0.13
65-69	1223.01	1261.61	1307.80	1362.91	1486.80	1592.27	1558.83	1608.48	1572.26	1215.11
70-74	1083.34	1104.68	1118.81	1114.94	1121.46	1192.21	1149.82	1235.52	1290.69	1415.93
75-79	842.96	850.39	864.04	876.96	894.74	942.28	917.88	955.55	952.21	959.36
80-84	551.82	562.65	577.98	594.47	609.01	633.17	621.47	644.32	656.85	672.11
85-89	285.84	301.36	305.99	312.45	320.48	338.96	326.79	348.22	360.80	371.34
90 and over	95.72	95.61	109.25	122.23	132.09	146.32	138.09	150.65	157.20	162.59
Unknown	0.22	0.26	0.27	0.32	0.37	0.52	0.46	0.62	0.76	0.81
Total all ages	**4083.03**	**4176.69**	**4284.30**	**4384.45**	**4565.14**	**4845.90**	**4713.53**	**4943.54**	**4990.95**	**4797.38**
Non-contributory State Pension[4]										
60-64	..	..	..	..	..	..	..	..	..	..
65-69	..	..	..	..	..	..	..	..	..	..
70-74	..	..	..	..	..	0.01	0.01	0.01	0.01	0.01
75-79	0.07	0.09	0.04	0.10	0.17	0.11	0.24	0.23	0.17	0.03
80-84	3.17	3.55	4.08	4.93	6.05	6.86	7.68	8.22	8.56	8.47
85-89	1.85	1.88	1.90	1.95	2.08	2.48	2.77	3.08	3.57	4.08
90 and over	0.96	0.96	1.02	1.10	1.23	1.37	1.47	1.52	1.52	1.55
Unknown	..	0.01	0.01	0.01	0.01	0.02	0.02	0.03	0.04	0.04
Total all ages	**6.06**	**6.49**	**7.05**	**8.09**	**9.55**	**10.84**	**12.20**	**13.07**	**13.86**	**14.19**
New State Pension:										
60-64	..	..	..	..	..	..	..	..	0.01	0.09
65-69	..	..	..	..	..	..	..	..	46.48	338.27
70-74	..	..	..	..	..	..	..	..	..	..
75-79	..	..	..	..	..	..	..	..	..	..
80-84	..	..	..	..	..	..	..	..	..	..
85-89	..	..	..	..	..	..	..	..	..	..
90 and over	..	..	..	..	..	..	..	..	..	..
Unknown	..	..	..	..	..	..	..	..	0.02	0.12
Total all ages	..	..	..	..	..	..	..	..	**46.50**	**338.47**
Total	**4089.08**	**4183.19**	**4291.35**	**4392.54**	**4574.69**	**4724.36**	**4858.10**	**4956.62**	**5004.82**	**4811.57**
FEMALE										
Pre-2016 State Pension[2]										
Contributory State Pension[3]										
60-64	1589.38	1628.72	1639.14	1521.08	1305.32	928.75	1105.41	758.49	567.98	262.25
65-69	1338.95	1378.26	1423.22	1477.81	1608.72	1720.75	1684.33	1743.82	1762.22	1684.61
70-74	1207.57	1227.05	1238.54	1232.20	1238.38	1307.06	1266.45	1350.45	1403.35	1529.79
75-79	1056.13	1051.16	1051.83	1053.73	1063.81	1103.13	1081.85	1113.80	1108.39	1114.01
80-84	822.96	822.26	829.14	837.33	842.70	843.91	843.99	841.77	844.79	854.31
85-89	564.66	582.60	572.79	565.40	561.34	564.72	558.15	568.77	576.29	580.51
90 and over	289.42	285.25	313.74	339.34	355.00	372.61	360.65	371.31	375.84	376.37
Unknown	0.51	0.48	0.53	0.59	0.85	1.41	1.12	1.71	2.10	2.38
Total all ages	**6869.59**	**6975.76**	**7068.91**	**7027.49**	**6976.14**	**6842.33**	**6901.94**	**6750.11**	**6640.96**	**6404.22**
Non-contributory State Pension[4]										
60-64	..	..	..	..	..	..	..	..	..	..
65-69	..	..	..	..	..	..	..	..	..	..
70-74	..	..	0.01	..	..	0.01	..	0.01	0.01	0.01
75-79	0.22	0.27	0.12	0.22	0.34	0.22	0.57	0.49	0.39	0.05
80-84	8.71	9.79	11.09	12.93	15.17	17.07	18.93	20.80	22.17	22.43
85-89	5.54	5.75	5.64	5.75	6.26	7.29	7.94	8.50	9.26	10.07
90 and over	4.03	3.91	4.01	4.13	4.35	4.97	5.18	5.08	5.12	5.13
Unknown	0.02	0.02	0.04	0.05	0.06	0.08	0.12	0.17	0.22	0.25
Total all ages	**18.53**	**19.76**	**20.92**	**23.09**	**26.19**	**29.64**	**32.74**	**35.03**	**37.15**	**37.93**
New State Pension:										
60-64	..	..	..	..	..	..	..	..	..	81.77
65-69	..	..	..	..	..	..	..	..	..	
70-74	..	..	..	..	..	..	..	..	..	
75-79	..	..	..	..	..	..	..	..	..	
80-84	..	..	..	..	..	..	..	..	..	
85-89	..	..	..	..	..	..	..	..	..	
90 and over	..	..	..	..	..	..	..	..	..	
Unknown	..	..	..	..	..	..	..	..	..	0.05
Total all ages	..	..	..	..	..	..	..	..	..	**81.82**
Total	**6888.11**	**6995.51**	**7089.83**	**7050.57**	**7002.33**	**6931.58**	**6875.07**	**6785.15**	**6678.11**	**6442.14**

Source: Stat-Xplore, Department for Work and Pensions

5.14 State Pension Caseload by Category of Pension, Age and Sex[1]
Great Britain

Footnotes:

1. Excludes Abroad cases

2. Pre-2016 State Pension: This is paid to people reaching their State Pension age (SPa) prior to 6th April 2016. There are two main types of SP: contributory and non-contributory. Contributory SP consists of any combination of a Basic Pension, Additional Pension or Graduated Retirement Benefit. Non-contributory SP consists of a Basic Pension plus any Graduated Retirement Benefit that is due.

3. Includes all categories of contributory state pension (Cat A, Cat B, Cat ABL, Cat BL, Cat AB and GRB only)

4. Includes all categories of non-contributory state pension (Cat C and Cat D)

5. New State Pension: This is paid to people reaching SPa on, or after, 6th April 2016. nSP is a contributory SP made up of an amount based upon the persons NI contributions. Where transitional arrangements apply, some individuals will receive a Protected Payment – which is an amount above the full rate of nSP.

".." denotes a nil or negligible number of claimants or award amount based on a nil or negligible number of claimants.

Statistical disclosure control has been applied to this table to avoid the release of confidential data. Totals may not sum due to the disclosure control applied.

The move to publishing benefit data via the Stat-Xplore instead of the DWP Tabulation Tool has involved adopting a new disclosure control methodology, in line with other benefits published via this new tool. Although the data still comes from the same source there may be small differences in the outputs displayed using this new tool, when compared to the Tabulation Tool.

From November 2017 State Pension data excludes persons with a 'suspended' account and should be treated as a break in series. A suspended account, when the department suspends payment, can occur for a number of reasons. Suspended cases may end and pensions put back into payment, at which point they will be shown in the next extract of data, or the suspension may end and the case terminated.

5.15a: War Pensions: estimated number of pensioners[1] Great Britain

At 31 March each year

Thousands

		2006[2]	2007	2008	2009	2010	2011	2012	2013	2014	2015	2016	2017
Disablement	KADH	182.8	173.85	165.17	157.13	148.95	141.72	134.43	127.6	121.9	116.05	111.23	106.28
Widows and dependants	KADI	41.05	38.69	36.1	33.62	31.45	29.2	27.11	25.1	23.11	21.17	19.55	17.88
Total	KADG	223.85	212.54	201.27	190.75	180.4	170.92	161.54	152.7	145.01	137.22	130.78	124.16

Source: War Pensions Computer System

1 See chapter text. From 1914 war, 1939 war and later service.
2 The discontinuity between 2005 and 2006 is due to improvements in data processing.
An additional validation check was introduced in 2014/15 on the 'not known' disablement pensioners to put them in the correct payment category.
The sum of the sub-totals may not sum to the totals due to rounding

5.15b: War Pensions in payment by type of pension, gender and financial year end, 31 March 2013 to 31 March 2017, numbers[1]

	Awards in payment at:				
	31-Mar-13	31-Mar-14	31-Mar-15	31-Mar-16	31-Mar-17
ALL IN PAYMENT	152,695	145,004	137,216	130,774	124,157
Men	122,106	116,611	110,935	106,295	101,472
Women	30,589	28,393	26,281	24,479	22,685
Disablement Pensioners	127,590	121,900	116,049	111,228	106,282
Men	121,531	116,057	110,409	105,788	100,987
Women	6,059	5,843	5,640	5,440	5,295
War Widow(er)s	24,414	22,446	20,533	18,950	17,311
Men	83	82	72	74	72
Women	24,331	22,364	20,461	18,876	17,239
Other Pensioners	691	658	634	596	564
Men	492	472	454	433	413
Women	199	186	180	163	151

Source: War Pensions Computer System

1 The sum of the sub-totals may not sum to the totals due to rounding.
2 Please note that pensioners can move from disablement pensions to other pensions and vice versa.

5.16 Income support by statistical group: number of claimants receiving weekly payment
Great Britain

As at May each year   Count (1000s)

	2009	2010	2011	2012	2013	2014	2015	2016	2017
All income support claimants	**1990.00**	**1852.30**	**1703.20**	**1417.10**	**1021.70**	**840.77**	**734.76**	**665.68**	**599.91**
Incapacity Benefits	1097.00	996.60	924.90	654.13	328.05	167.62	92.13	45.01	9.17
Lone Parent	720.50	679.20	595.40	577.08	499.63	474.73	441.60	416.48	387.43
Carer	93.80	103.90	115.70	129.75	145.89	156.53	165.37	173.22	176.76
Others on income related benefit	78.90	72.70	67.20	56.10	48.14	41.89	35.66	30.98	26.56

Source: Stat-Xplore, Department for Work and Pensions

Statistical disclosure control has been applied to this table to avoid the release of confidential data. Totals may not sum due to the disclosure control applied.

The move to publishing benefit data via the Stat-Xplore instead of NOMIS has involved adopting a new disclosure control methodology, in line with other benefits published via this new tool. Although the data still comes from the same source there may be small differences in the outputs displayed using this new tool, when compared to NOMIS.

Statistical Group refers to the client group which is based on the combination of benefits being claimed. It is based on receipt of certain other benefits:
• Incapacity Benefits: claimant receives Incapacity Benefit, Severe Disablement Allowance (IBSDA) or Employment and Support Allowance (ESA);
• Lone Parent: claimant receives Income Support with a child under 16 and no partner;
• Carer: claimant receives Carer's Allowance (CA);
• Others on Income Related Benefit: claimant receives Income Support (including IS Disability Premium) or Pension Credit (PC).

Benefits are arranged hierarchically and claimants are assigned to the topmost benefit they receive. Thus a person who is a lone parent and receives Incapacity Benefit would be classified in the Incapacity Benefits statistical group.

Others on Income Related Benefit: After August 2003, there was a sharp decline in this statistical group due to the migration of most existing Minimum Income Guarantee (MIG) claimants to Pension Credit. Some residual cases remain after this date.

Lone Parent Obligations: From November 2008, lone parent obligations (LPOs) were introduced and lone parents with a youngest child aged 12 or over were no longer able to make a new or repeat claim for Income Support solely on the basis of their parental status. Existing IS LPs with a youngest child aged 12 or over had their eligibility removed over a period of time commencing 2nd March 2009. From October 2009 this policy was extended to LPs with a youngest child aged 10 or 11 and from October 2010, to LPs with a youngest child aged 7 or over. The Welfare Reform Act 2012 introduced further changes and from 21 May 2012 lone parents will only be eligible to claim Income Support until their youngest child is five years old. Most effected LPs will leave IS and claim Jobseeker's Allowance (JSA). However, there are exceptions to these rules where the youngest child can legitimately be over the ages mentioned above. Similarly, some former LPs remain on IS for other reasons (e.g. they have a long term caring responsibility and claim Carer's Allowance).

5.17 Pension Credit: number of claimants
Great Britain

As at May each year										Count (1000s)
	2008	2009	2010	2011	2012	2013	2014	2015	2016	2017
All Pension Credit	2719.1	2730.5	2734.1	2674.6	2541.4	2413.7	2270.2	2096.6	1948.5	1822.1
Guarantee Credit only	882.1	925.7	954.3	936.9	1015.8	968.3	950.2	931.4	936.8	881.3
Both Guarantee and Savings Credit	1246.2	1205.2	1202.4	1148.0	964.2	906.5	824.2	739.6	659.7	624.8
Savings Credit only	590.8	599.6	577.4	589.6	561.4	538.8	495.8	425.7	352.0	316.1

Source: Stat-Xplore, Department for Work and Pensions

Statistical disclosure control has been applied to this table to avoid the release of confidential data. Totals may not sum due to the disclosure control applied.

The move to publishing benefit data via the Stat-Xplore instead of the DWP Tabulation Tool has involved adopting a new disclosure control methodology, in line with other benefits published via this new tool. Although the data still comes from the same source there may be small differences in the outputs displayed using this new tool, when compared to the Tabulation Tool.

5.18 Income support: average amounts of benefit[1,2]
Great Britain

As at May										£ per week[3]
	2008	2009	2010	2011	2012	2013	2014	2015	2016	2017
Statistical Group[4]:										
Incapacity Benefits	81.85	89.22	91.81	93.66	93.69	82.12	67.47	63.68	65.93	94.35
Lone Parent	87.37	82.79	79.02	75.94	76.54	74.09	72.75	71.98	70.74	69.99
Carer	70.21	71.85	70.90	69.94	72.23	72.29	72.16	72.82	72.97	74.26
Others on income related benefit	63.02	66.35	67.87	68.82	73.19	74.11	74.86	75.21	75.44	77.1
Total	82.55	85.17	85.01	84.88	83.93	76.42	71.69	71.28	71.21	71.94

Source: Stat-Xplore, Department for Work and Pensions

".." denotes a nil or negligible number of claimants or award amount based on a nil or negligible number of claimants.

1 Figures prior to 2013 are taken from the DWP 100% Source: Department for Work and Pensions (DWP); Work and Pensions Longitudinal Study (WPLS) Work and Pensions Longitudinal Study (WPLS).

From 2013 onwards, data is sourced from Stat-Xplore, DWP. Although the data still comes from the same source there may be small differences in the outputs displayed using this new tool, when compared to NOMIS.

2 The amount of Income Support is affected by the introduction in April 2003 of Child Tax Credit. From that date there were no new child dependency increases awarded to IS claimants, although existing CDIs were transitionally protected.

3 Average amounts are shown as pounds per week and rounded to the nearest penny.

4 Statistical Group refers to the client group which is based on the combination of benefits being claimed. It is based on receipt of certain other benefits:
• Incapacity Benefits: claimant receives Incapacity Benefit, Severe Disablement Allowance (IBSDA) or Employment and Support Allowance (ESA)
• Lone Parent: claimant receives Income Support with a child under 16 and no partner
• Carer: claimant receives Carer's Allowance (CA);
• Others on Income Related Benefit: claimant receives Income Support (including IS Disability Premium) or Pension Credit (PC). After August 2003, there was a sharp decline in this statistical group due to the migration of most existing Minimum Income Guarantee (MIG) claimants to Pension Credit. Some residual cases remain after this date.

Income Support

Income Support (IS) is intended to help cover costs for people on low incomes who do not have to be available for employment. It is being replaced by Universal Credit (UC) and IS cannot be claimed at the same time as UC.
The main groups of people who receive it are lone parents, carer's and people on Incapacity Benefits.

The amount of IS that a claimant can receive depends mainly upon their age; whether they have a partner, whether they have special needs such as a disability or caring responsibilities; and whether they have liabilities for certain types of housing costs such as mortgage interest payments. The maximum amount that a claimant can receive is normally reduced by income from other types of benefits or other sources.

Recording and clerical errors can occur within the data source - for this reason, no reliance should be placed on very small numbers obtained through Stat-Xplore.

Statistical disclosure control has been applied with Stat-Xplore, which guards against the identification of an individual claimant.

5.19 Pension Credit: average weekly amounts of benefit[1]

Great Britain

As at May										£ per week[2]
	2008	2009	2010	2011	2012	2013	2014	2015	2016	2017
Type of Pension Credit										
Guarantee Credit only	85.08	88.87	90.75	92.52	90.13	90.07	89.19	87.59	86.25	86.59
Savings Credit only	13.62	13.71	14.01	13.57	11.37	10.87	9.85	8.51	7.29	7.48
Both Guarantee and Savings Credit	48.29	50.82	51.75	52.04	50.26	49.41	47.72	45.49	43.77	44.07
Residual MIG case	112.71	107.44	90.24	100.51	108.44	97.81	122.8	93.79	74.18	80.93
Total	52.69	55.57	57.39	57.74	57.61	57.12	56.75	56.64	57.61	58.29

Source: Stat-Xplore, Department for Work and Pensions

1 Figures are given in each May and are taken from the DWP 100% Work and Pensions Longitudinal Study (WPLS).
2 Average amounts are shown as pounds per week and rounded to the nearest penny.

Pension Credit:

Pension Credit is an income-related benefit made up of 2 parts - Guarantee Credit and Savings Credit.

Pension Credit (Guarantee Credit) tops up any other income to a standard minimum amount (Standard Minimum Guarantee). There are additional amounts for those who are severely disabled, have caring responsibilities or certain housing costs.

To be entitled to Guarantee Credit, the claimant must have reached the qualifying age (women's state pension age) and live in Great Britain. A couple, where only one has reached the qualifying age can currently chose to claim Pension Credit.

Partner in this term refers to; your husband, your wife, your civil partner or the person you live with as if they were your husband, wife or civil partner.

Pension Credit (Savings Credit) is an extra amount for those 65 and over, who reached state pension age before 5th April 2016 and have made some additional provision for their retirement, for example through a private pension or savings.

Pensioners receive 60 pence of Savings Credit for every pound of qualifying income between the savings credit threshold and the Standard Minimum Guarantee. For those who have income above the Standard Minimum Guarantee and any additional amounts they are entitled to (the Appropriate Amount) the Savings Credit Maximum is reduced by 40 pence for every pound of income above that level.

The savings credit is only applicable to people aged 65 or over. In the case of a couple if either of the couple is aged 65 or over the savings credit may be payable. Some people will be entitled to the guarantee credit, the savings credit or both.

Recording and clerical errors can occur within the data source - for this reason, no reliance should be placed on very small numbers obtained through Stat-Xplore.
Statistical disclosure control has been applied with Stat-Xplore, which guards against the identification of an individual claimant.

External trade and investment

Chapter 6

External trade and investment

External trade (Table 6.1 and 6.3 to 6.6)

The statistics in this section are on the basis of Balance of Payments (BoP). They are compiled from information provided to HM Revenue and Customs (HMRC) by importers and exporters on the basis of Overseas Trade Statistics (OTS) which values exports 'f.o.b.' (free on board) and imports 'c.i.f.' (including insurance and freight). In addition to deducting these freight costs and insurance premiums from the OTS figures, coverage adjustments are made to convert the OTS data to a BoP basis. Adjustments are also made to the level of all exports and European Union (EU) imports to take account of estimated under-recording. The adjustments are set out and described in the annual United Kingdom *Balance of Payments Pink Book* (Office for National Statistics (ONS)). These adjustments are made to conform to the definitions in the 6th edition of the IMF Balance of Payments Manual.

Aggregate estimates of trade in goods, seasonally adjusted and on a BoP basis, are published monthly in the ONS statistical bulletin UK Trade. More detailed figures are available from time series data on the ONS website (www.ons.gov.uk) and are also published in the *Monthly Review of External Trade Statistics*. Detailed figures for EU and non-EU trade on an OTS basis are published in *Overseas trade statistics: United Kingdom trade with the European Community and the world*
(HMRC).

Overseas Trade Statistics

HMRC provide accurate and up to date information via the website:
www.uktradeinfo.com
They also produce publications entitled 'Overseas Trade Statistics'.

Sales of products manufactured in the United Kingdom by Industry Division (Table 6.2)

The industries are grouped according to the 2007 Standard Industrial Classification at 2-digit level. Information in this table relates to products corresponding to a division irrespective of which division the business making the product is classified to.

Table 6.3 to 6.6

The series are now available as datasets in the UK Trade release, which is updated monthly

International trade in services (Tables 6.7 and 6.8)

These data relate to overseas trade in services and cover both production and non-production industries (excluding the public sector). In terms of the types of services traded these include royalties, various forms of consultancy, computing and telecommunications services, advertising and market research and other business services. A separate inquiry covers the film and television industries. The surveys cover receipts from the provision of services to residents of other countries (exports) and payments to residents of other countries for services rendered (imports).

Sources of data

The International Trade in Services (ITIS) surveys (which consist of a quarterly component addressed to the largest businesses and an annual component for the remainder) are based on a sample of companies derived from the Inter-departmental Business Register in addition to a reference list and from 2007 onwards a sample of approximately 5000 contributors from the Annual Business Inquiry (ABI). The companies are asked to show the amounts for their imports and exports against the geographical area to which they were paid or from which they were received, irrespective of where they were first earned.

The purpose of the ITIS survey is to record international transactions which impact on the UK's BoP. Exports and imports of goods are generally excluded, as they will have been counted in the estimate for trade in goods. However earnings from third country trade – that is, from arranging the sale of goods between two countries other than the UK and where the goods never physically enter the UK (known as merchanting) – are included. Earnings from commodity trading are also included. Together, these two comprise trade related services.

Royalties are a large part of the total trade in services collected in the ITIS survey. These cover transactions for items such as printed matter, sound recordings, performing rights, patents, licences, trademarks, designs, copyrights, manufacturing rights, the use of technical know-how and technical assistance.

Balance of Payments (Tables 6.9 to 6.12)

Tables 6.9 to 6.12 are derived from *United Kingdom Balance of Payments: The Pink Book*. The following general notes to the tables provide brief definitions and explanations of the figures and terms used. Further notes are included in the Pink Book.

Summary of Balance of Payments

The BoP consists of the current account, the capital account, the financial account and the International Investment Position (IIP). The current account consists of trade in goods and services, income, and current transfers. Income consists of investment income and compensation of employees. The capital account mainly consists of capital transfers and the financial account covers financial transactions. The IIP covers balance sheet levels of UK external assets and liabilities. Every credit entry in the balance of payments accounts should, in theory, be matched by a corresponding debit entry so that total current, capital and financial account credits should be equal to, and therefore offset by, total debits. In practice there is a discrepancy termed net errors and omissions.

Current account

Trade in goods

The goods account covers exports and imports of goods. Imports of motor cars from Japan, for example, are recorded as debits in the trade in goods account, whereas exports of vehicles manufactured in the UK are recorded as credits. Trade in goods forms a component of the expenditure measure of gross domestic product (GDP).

Trade in services

The services account covers exports and imports of services, for example civil aviation. Passenger tickets for travel on UK aircraft sold abroad, for example, are recorded as credits in the services account, whereas the purchases of airline tickets from foreign airlines by UK passengers are recorded as debits. Trade in services, along with trade in goods, forms a component of the expenditure measure of GDP.

Income

The income account consists of compensation of employees and investment income and is dominated by the latter. Compensation of employees covers employment income from cross-border and seasonal workers which is less significant in the UK than in other countries. Investment income covers earnings (for example, profits, dividends and interest payments and receipts) arising from cross-border investment in financial assets and liabilities. For example, earnings on foreign bonds and shares held by financial institutions based in the UK are recorded as credits in the investment income account, whereas earnings on UK company securities held abroad are recorded as investment income debits. Investment income forms a component of gross national income (GNI) but not GDP.

Current transfers

Current transfers are composed of central government transfers (for example, taxes and payments to and receipts from, the EU) and other transfers (for example gifts in cash or kind received by private individuals from abroad or receipts from the EU where the UK government acts as an agent for the ultimate beneficiary of the transfer). Current transfers do not form a component either of GDP or of GNI. For example, payments to the UK farming industry under the EU Agricultural Guarantee Fund are recorded as credits in the current transfers account, while payments of EU agricultural levies by the UK farming industry are recorded as debits in the current transfers account.

Capital account

Capital account transactions involve transfers of ownership of fixed assets, transfers of funds associated with acquisition or disposal of fixed assets and cancellation of liabilities by creditors without any counterparts being received in return. The main components are migrants transfers, EU transfers relating to fixed capital formation (regional development fund and agricultural guidance fund) and debt forgiveness. Funds brought into the UK by new immigrants would, for example, be recorded as credits in the capital account, while funds sent abroad by UK residents emigrating to other countries would be recorded as debits in the capital account. The size of capital account transactions are quite minor compared with the current and financial accounts.

Financial account

While investment income covers earnings arising from cross-border investments in financial assets and liabilities, the financial account of the balance of payments covers the flows of such investments. Earnings on foreign bonds and shares held by financial institutions based in the UK are, for example, recorded as credits in the investment income account, but the acquisition of such foreign securities by UK-based financial institutions are recorded as net debits in the financial account or portfolio investment abroad. Similarly, the acquisitions of UK company securities held by foreign residents are recorded in the financial account as net credits or portfolio investment in the UK.

International Investment Position

While the financial account covers the flows of foreign investments and financial assets and liabilities, the IIP records the levels of external assets and liabilities. While the acquisition of foreign securities by UK-based financial institutions are recorded in the financial account as net debits, the total holdings of foreign securities by UK-based financial institutions are recorded as levels of UK external assets. Similarly, the holdings of UK company securities held by foreign residents are recorded as levels of UK liabilities.

Foreign direct investment (Tables 6.13 to 6.18)

Direct investment refers to investment that adds to, deducts from, or acquires a lasting interest in an enterprise operating in an economy other than that of the investor – the investor's purpose being to have an effective voice in the management of the enterprise. (For the purposes of the statistical inquiry, an effective voice is taken as equivalent to a holding of 10 per cent or more in the foreign enterprise.) Other investments in which the investor does not have an effective voice in the management of the enterprise are mainly portfolio investments and these are not covered here.

Direct investment is a financial concept and is not the same as capital expenditure on fixed assets. It covers only the money invested in a related concern by the parent company and the concern will then decide how to use the money. A related concern may also raise money locally without reference to the parent company.

The investment figures are published on a net basis; that is they consist of investments net of disinvestments by a company into its foreign subsidiaries, associate companies and branches.

ESA2010

The European System of National and Regional Accounts (ESA 2010) is the newest internationally compatible EU accounting framework for a systematic and detailed description of an economy. The ESA 2010 was published in the Official Journal on 26 June 2013. It was implemented in September 2014.

The ESA 2010 differs in scope as well as in concepts from its predecessor ESA 95 reflecting developments in measuring modern economies, advances in methodological research and the needs of users. The structure of the ESA 2010 is consistent with the worldwide guidelines on national accounting set out in the System of National Accounts 2008 (2008 SNA).

What are the main changes?
• Research and development expenditure is counted as investment. Expenditures on research and development (R&D) have the nature of an investment and contribute to future economic growth. This is the major improvement introduced by ESA 2010.
• Expenditure on weapon systems is counted as investment. The new system recognises the productive potential of expenditure on weapon systems for the external security of a country, over several years. This identifies them as investment.
• A more detailed analysis of pension schemes. A compulsory supplementary table transparently shows the liabilities of all pension schemes, including those of government whether unfunded or funded, in order to improve comparability between countries.
• Goods sent abroad for processing. The value of goods sent abroad for processing will no longer impact on gross exports and imports figures because ESA 2010 uses a change in ownership approach and is no more based on physical movements. ESA2010 just records an export processing service. This will reduce the level of exports and imports, but will not affect the overall current account balance. See https://ec.europa.eu/eurostat/documents/737960/738004/ESA2010-FAQ.pdf/fea21e81-a2cb-421a-8b9e-41aae7d02a14

Sources of data

The figures in Tables 6.13 to 6.18 are based on annual inquiries into foreign direct investment for 2017. These were sample surveys which involved sending approximately 1530 forms to UK businesses investing abroad, and 2370 forms to UK businesses in which foreign parents and associates had invested. The tables also contain some revisions as a result of new information coming to light in the course of the latest surveys. Further details from the latest annual surveys, including analyses by industry and by components of direct investment, are available in business monitor MA4.

Country allocation

The analysis of inward investment is based on the country of ownership of the immediate parent company. Thus, inward investment in a UK company may be attributed to the country of the intervening overseas subsidiary, rather than the country of the ultimate parent. Similarly, the country analysis of outward investment is based on the country of ownership of the immediate subsidiary; for example, to the extent that overseas investment in the UK is channeled through holding companies in the Netherlands, the underlying flow of investment from this country is overstated and the inflow from originating countries is understated.

Further information

More detailed statistics on foreign direct investment are available on request from Office for National Statistics, Telephone: +44 (0)1633 455923, email fdi@ons.gsi.gov.uk

6.1 Trade in goods[1]
United Kingdom
Balance of payments basis

£million and Indices 2013=100

		2005	2006	2007	2008	2009	2010	2011	2012	2013	2014	2015	2016	2017
Value (£ million)														
Exports of goods	**BOKG**	212,801	245,031	222,741	252,326	227,262	266,023	301,571	298,430	301,084	292,092	285,960	297,922	337,466
Imports of goods	**BOKH**	282,044	322,668	310,807	342,251	311,753	361,038	397,579	403,446	418,904	413,199	403,657	432,101	473,361
Balance on trade in goods	**BOKI**	-69,243	-77,637	-88,066	-89,925	-84,491	-95,015	-96,008	-105,016	-117,820	-121,107	-117,697	-134,179	-135,895
Price index numbers[2]														
Exports of goods	**BQKR**	83.8	85.2	84.8	96.4	97.3	103.9	112.2	112.0	113.3	108.1	100.0	105.5	112.1
Imports of goods	**BQKS**	81.3	83.8	84.0	96.0	97.7	102.2	111.1	110.9	111.4	106.6	100.0	103.3	109.9
Terms of trade[3]	**BQKT**	103.1	101.7	101.0	100.4	99.6	101.7	101.0	101.0	101.7	101.4	100.0	102.1	102.0
Volume index numbers[4]														
Exports of goods	**BQKU**	83.5	94.7	87.1	88.4	79.1	88.1	94.0	92.4	91.5	94.5	100.0	99.1	106.2
Imports of goods	**BQKV**	80.8	90.3	87.7	86.2	77.1	86.4	87.7	89.7	92.1	96.2	100.0	104.5	108.0

Source: Office for National Statistics: 01633 455582

1 These are provisional estimates subject to fluctuations when new or amended data become available. See chapter text. Statistics of trade in goods on a balance of payments basis are obtained by making certain adjustments in respect of valuation and coverage to the statistics recorded in the Overseas Trade Statistics. These adjustments are described in detail in The Pink Book.

2. Price index numbers - not seasonally adjusted.

3 Export price index as a percentage of the import price index.

4. Volume index numbers - seasonally adjusted.

This table may show revisions to data going back over time. This is mainly to reflect revised data from HMRC and other data suppliers, later survey data and a re-assessment of seasonal factors.

6.2 Sales of products manufactured in the United Kingdom by Industry Division

Division	SIC(07) Division	2013	2014	2015	2016
					£ millions
Other mining and quarrying	8	2,299	2,315	1,738	1,861
Manufacture of food products	10	66,030	67,053	66,176	65,720
Manufacture of beverages and tobacco products	11 & 12	14,109	S	14,599	13,772
Manufacture of textiles	13	S	S	4,193	4,212
Manufacture of wearing apparel	14	1,634	1,830	1,665	1,652
Manufacture of leather and related products	15	727	681	690	685
Manufacture of wood and of products of wood and cork, except furniture; manufacture of articles of straw and plaiting materials	16	6,097	S	6,534	6,406
Manufacture of paper and paper products	17	S	10,364	10,105	9,870
Printing and reproduction of recorded media	18	8,055	S	7,910	7,532
Manufacture of coke and refined petroleum products	19	S	S	0	0
Manufacture of chemicals and chemical products	20	S	21,431	20,453	20,956
Manufacture of basic pharmaceutical products and pharmaceutical preparations	21	12,047	10,914	10,799	12,478
Manufacture of rubber and plastic products	22	S	S	19,226	19,341
Manufacture of other non-metallic mineral products	23	10,359	11,468	12,436	12,856
Manufacture of basic metals	24	6,825	6,583	5,926	5,523
Manufacture of fabricated metal products, except machinery and equipment	25	24,266	25,551	26,439	26,150
Manufacture of computer, electronic and optical products	26	12,559	12,145	12,166	12,044
Manufacture of electrical equipment	27	S	S	10,288	10,352
Manufacture of machinery and equipment n.e.c.	28	S	27,581	24,756	23,851
Manufacture of motor vehicles, trailers and semi-trailers	29	S	47,382	48,424	54,639
Manufacture of other transport equipment	30	26,471	27,192	27,759	28,258
Manufacture of furniture	31	6,306	6,557	6,709	7,221
Other manufacturing	32	S	4,771	4,932	5,093
Repair and installation of machinery and equipment	33	13,987	14,086	14,221	14,261
Total		354,721	362,545	358,143	364,733

Source: Office for National Statistics
Email: prodcompublications@ons.gov.uk
Tel: (01633) 456720

Note: Information in this table relates to products corresponding to a division irrespective of which division the business making the product is classified to.

S A figure has been suppressed as disclosive

6.3 United Kingdom exports: by commodity[1,2]

Seasonally adjusted

		2007	2008	2009	2010	2011	2012	2013	2014	2015	2016	2017
0. Food and live animals	**BOGG**	**7745**	**9104**	**9504**	**10881**	**12478**	**11651**	**12430**	**12531**	**12151**	**13441**	**15016**
01. Meat and meat preparations	BOGS	859	1203	1278	1453	1749	1638	1697	1707	1477	1592	1845
02. Dairy products and eggs	BQMS	804	895	848	1054	1299	1202	1423	1537	1329	1392	1802
04 & 08. Cereals and animal feeding stuffs	BQMT	1825	2322	2385	2653	2831	2838	2868	2926	3131	3495	3596
05. Vegetables and fruit	BQMU	581	678	736	810	912	857	1022	950	1027	1120	1247
1. Beverages and tobacco	**BQMZ**	**4651**	**5136**	**5431**	**6166**	**7489**	**7268**	**7363**	**6934**	**6771**	**7274**	**7786**
11. Beverages	BQNB	4324	4661	5005	5790	7166	6964	7064	6638	6494	7030	7666
12. Tobacco	BQOW	327	475	426	376	323	304	299	296	277	244	120
2. Crude materials	**BQOX**	**5001**	**6042**	**4578**	**6716**	**8524**	**7875**	**6840**	**6287**	**5350**	**5710**	**7425**
of which:												
24. Wood, lumber and cork	BQOY	146	131	87	115	113	100	109	114	100	104	127
25. Pulp and waste paper	BQOZ	401	471	342	530	625	536	486	474	498	558	691
26. Textile fibres	BQPA	439	470	511	626	739	798	695	636	500	524	690
28. Metal ores	BQPB	3003	3758	2607	4213	5279	4660	4049	3722	3155	3433	4464
3. Fuels	**BOPN**	**24727**	**35898**	**27461**	**35907**	**43796**	**44653**	**41979**	**36005**	**23800**	**21100**	**30059**
33. Petroleum and petroleum products	ELBL	22789	32347	24958	31838	38996	40472	38396	32809	20720	18754	26908
32, 34 & 35. Coal, gas and electricity	BOQI	1938	3551	2503	4069	4800	4181	3583	3196	3080	2346	3151
4. Animal and vegetable oils and fats	**BQPI**	**280**	**321**	**336**	**387**	**420**	**440**	**458**	**411**	**383**	**409**	**526**
5. Chemicals	**ENDG**	**39533**	**44378**	**48021**	**51811**	**54385**	**53769**	**49885**	**48899**	**52726**	**53709**	**58319**
of which:												
51. Organic chemicals	BQPJ	7746	8527	9398	9204	9431	9217	7811	6971	8633	7744	8393
52. Inorganic chemicals	BQPK	2854	3000	2846	3513	3081	2937	2559	1877	1888	1733	1641
53. Colouring materials	CSCE	1746	1905	1740	2009	2229	2023	2045	2084	1911	2092	2402
54. Medicinal products	BQPL	14770	17479	20914	22566	23291	23690	21094	21374	24412	25473	27282
55. Toilet preparations	CSCF	3719	3916	4148	4330	4900	4885	5311	5240	4896	5478	6196
57 & 58. Plastics	BQQA	4772	5003	4553	5488	6142	5834	5719	5819	5556	5864	6679
6. Manufactures classified chiefly by material	**BQQB**	**29450**	**32469**	**24660**	**29523**	**35358**	**32477**	**31856**	**29312**	**27679**	**26990**	**31339**
of which:												
63. Wood and cork manufactures	BQQC	263	247	221	230	265	274	257	264	241	275	296
64. Paper and paperboard manufactures	BQQD	2182	2393	2335	2414	2473	2362	2340	2289	2035	2252	2368
65. Textile manufactures	BQQE	2563	2540	2346	2599	2822	2739	2811	2862	2730	2846	3010
67. Iron and steel	BQQF	6048	6896	4588	5070	6120	5812	6104	6091	4761	3688	4456
68. Non-ferrous metals	BQQG	5859	6887	3974	5879	8642	6806	6991	6885	7027	6629	8607
69. Metal manufactures	BQQH	4752	5161	4351	4636	5036	5046	5023	5479	5353	5605	6110
7. Machinery and transport equipment[3]	**BQQI**	**83892**	**89563**	**79942**	**93491**	**105095**	**105209**	**110794**	**109689**	**110017**	**122513**	**137519**
71-716, 72, 73 & 74. Mechanical machinery	BQQK	29647	32457	29537	33145	39295	39807	41864	40446	39111	41410	48691
716, 75, 76 & 77. Electrical machinery	BQQL	24168	25042	24022	25995	26547	24565	24711	24531	24205	25486	28550
78. Road vehicles	BQQM	21651	22909	17400	23660	27861	28957	31554	32603	32849	38080	42291
79. Other transport equipment	BQQN	8426	9155	8983	10691	11392	11880	12665	12109	13852	17537	17987
8. Miscellaneous manufactures[3]	**BQQO**	**25133**	**27108**	**26447**	**30547**	**33309**	**33582**	**35241**	**37957**	**41813**	**42953**	**47205**
of which:												
84. Clothing	CSCN	3017	3188	3322	3570	4149	4342	4881	5434	5786	6243	6803
85. Footwear	CSCP	526	601	697	830	883	978	1154	1247	1444	1579	1708
87 & 88. Scientific and photographic	BQQQ	7243	8206	8474	9836	10931	10763	11445	11461	11747	12428	13397
9. Other commodities and transactions	**BOQL**	**2112**	**2240**	**984**	**1608**	**2950**	**2464**	**3544**	**5091**	**6062**	**4974**	**3545**
Total United Kingdom exports	**BOKG**	**222524**	**252259**	**227364**	**267037**	**303804**	**299388**	**300390**	**293116**	**286752**	**299073**	**338739**

Source: Office for National Statistics
Email: trade@ons.gov.uk
Tel: 01633 651988

1 See chapter text. The numbers on the left hand side of the table refer to the code numbers of the Standard International Trade Classification, Revision 3, which was introduced in January 1988.

2 Balance of payments consistent basis.

3 Sections 7 and 8 are shown by broad economic category in table G2 of the Monthly Review of External Trade Statistics.

6.4 United Kingdom imports: by commodity [1,2]

Seasonally adjusted

		2007	2008	2009	2010	2011	2012	2013	2014	2015	2016	2017
0. Food and live animals	**BQQR**	**21676**	**25556**	**26663**	**27492**	**29610**	**30693**	**32566**	**32432**	**32563**	**35827**	**38927**
of which:												
01. Meat and meat preparations	BQQS	4052	4641	4947	5041	5697	5663	5846	5976	5848	6195	6686
02. Dairy products and eggs	BQQT	1868	2286	2328	2432	2546	2665	2920	2852	2614	2724	3186
04 & 08. Cereals and animal feeding stuffs	BQQU	2951	3850	3979	4019	4210	4744	5522	5226	5034	5277	6007
05. Vegetables and fruit	BQQV	6321	7154	7191	7564	8101	8181	8449	8401	8745	10330	11082
1. Beverages and tobacco	**BQQW**	**5737**	**6210**	**6444**	**6937**	**7322**	**7379**	**7526**	**7626**	**7546**	**7947**	**8039**
11. Beverages	EGAT	4384	4767	5072	5231	5604	5734	5899	5932	5800	6154	6352
12. Tobacco	EMAI	1353	1443	1372	1706	1718	1645	1627	1694	1746	1793	1687
2. Crude materials	**ENVB**	**8777**	**9661**	**6655**	**9299**	**10606**	**9586**	**9875**	**9681**	**8648**	**8889**	**10279**
of which:												
24. Wood, lumber and cork	ENVC	1837	1399	1177	1497	1452	1514	1811	2186	2324	2569	2884
25. Pulp and waste paper	EQAH	539	647	457	601	654	531	517	516	550	498	555
26. Textile fibres	EQAP	246	272	241	300	446	347	355	384	341	332	424
28. Metalores	EHAA	4080	4917	2470	4151	4941	4250	4330	3793	2856	2739	3083
3. Fuels	**BQAT**	**32738**	**49245**	**36033**	**45430**	**63971**	**68684**	**64185**	**54202**	**38114**	**34254**	**45139**
33. Petroleum and petroleum products	ENXO	27360	38439	28346	36156	49901	54482	49493	43201	29196	26506	34445
32, 34 & 35. Coal, gas and electricity	BPBI	5378	10806	7687	9274	14070	14202	14692	11001	8918	7748	10694
4. Animal and vegetable oils and fats	**EHAB**	**916**	**1408**	**1055**	**1126**	**1483**	**1409**	**1363**	**1184**	**1063**	**1158**	**1547**
5. Chemicals	**ENGA**	**36471**	**39893**	**41965**	**47063**	**50955**	**51311**	**49947**	**52150**	**51603**	**54105**	**60100**
of which:												
51. Organic chemicals	EHAC	8804	8562	8335	9452	10186	10232	8198	6993	6268	6390	7485
52. Inorganic chemicals	EHAE	2739	2760	2799	2947	3639	3325	2529	2272	2326	2002	2120
53. Colouring materials	CSCR	1171	1218	1152	1269	1322	1357	1391	1401	1388	1600	1846
54. Medicinal products	EHAF	11536	12983	16105	17484	18489	19403	19938	23278	23911	25563	27746
55. Toilet preparations	CSCS	3472	3916	4147	4403	4639	4734	5048	5203	5540	5867	6505
57 & 58. Plastics	EHAG	5709	6211	5552	6871	7657	7449	7942	8072	7617	8078	9177
6. Manufactures classified chiefly by material	**EHAH**	**39880**	**41703**	**35672**	**43732**	**49342**	**48730**	**47312**	**44507**	**42919**	**45776**	**52205**
of which:												
63. Wood and cork manufactures	EHAI	1782	1776	1549	1703	1728	1707	1857	2037	2186	2331	2535
64. Paper and paperboard manufactures	EHAJ	5351	5581	5630	6028	6133	5846	5949	6117	5768	5937	6064
65. Textile manufactures	EHAK	4093	4016	3794	4247	4568	4498	4697	4878	5007	5279	5702
67. Iron and steel	EHAL	6066	6646	3928	5243	6584	6284	5422	5915	5063	5192	6484
68. Non-ferrous metals	EHAM	6232	6343	6030	8787	10388	11473	10225	8327	7557	8073	10647
69. Metal manufactures	EHAN	6643	7018	6036	6908	7541	7647	8005	8359	8434	9292	10065
7. Machinery and transport equipment[3]	**EHAO**	**118193**	**120467**	**107500**	**128445**	**128621**	**131098**	**140423**	**146683**	**151767**	**169087**	**180884**
71 - 716, 72, 73 & 74. Mechanical machinery	EHAQ	26124	29083	24415	28783	33002	34522	35937	36969	36223	40765	47382
716, 75, 76 & 77. Electrical machinery	EHAR	46084	47001	44521	50793	51266	50796	52329	51679	53182	57431	62200
78. Road vehicles	EHAS	36811	34051	26737	33474	36399	36274	41182	45766	50006	53857	55870
79. Other transport equipment	EHAT	9174	10332	11827	15395	7954	9506	10975	12269	12356	17034	15432
8. Miscellaneous manufactures[3]	**EHAU**	**45324**	**47751**	**47365**	**51031**	**53866**	**55125**	**58077**	**61926**	**66186**	**69847**	**74110**
of which:												
84. Clothing	CSDR	11957	12716	13465	14125	15126	15141	16304	17052	17970	18946	20020
85. Footwear	CSDS	2583	2760	3021	3459	3491	3800	3929	4280	4753	4933	5286
87 & 88. Scientific and photographic	EHAW	7581	8307	8335	8984	9440	10033	10659	10855	11471	12474	13429
9. Other commodities and transactions	**BQAW**	**1396**	**2029**	**3317**	**2096**	**2391**	**2093**	**8106**	**4796**	**4153**	**4835**	**4544**
Total United Kingdom imports	**BOKH**	**311108**	**343923**	**312669**	**362651**	**398167**	**406108**	**419380**	**415187**	**404562**	**431725**	**475774**

1 See chapter text. The numbers on the left hand side of the table refer to the code numbers of the Standard International Trade Classification, Revision 3, which was introduced in January 1988.

2 Balance of payments consistent basis.

3 Sections 7 and 8 are shown by broad economic category in table G2 of the Monthly Review of External Trade Statistics.

Source: Office for National Statistics
Email: trade@ons.gov.uk
Tel: 01633 651988

6.5 Trade in goods - exports

£ million

		2008	2009	2010	2011	2012	2013	2014	2015	2016	2017
Exports (Credits)											
Europe											
European Union (EU)											
Austria	QBRY	1446	1266	1452	1701	1547	1614	1623	1536	1776	1768
Belgium	QDOH	13521	10912	13504	16279	14630	14196	12942	11686	11732	14060
Bulgaria	QAMF	237	165	238	317	294	380	428	313	469	334
Croatia	QAMM	187	195	143	123	112	100	115	109	135	99
Cyprus	QDNZ	517	610	568	723	467	428	423	329	283	268
Czech Republic	QDLF	1524	1415	1822	1949	1744	1920	2056	1945	2106	2021
Denmark	QBSE	2654	2517	2818	3149	2940	3051	2879	2332	2485	2657
Estonia	QAMN	169	97	157	249	257	295	231	175	201	133
Finland	QBSH	1917	1327	1518	1698	1554	1552	1610	1327	1305	1262
France	QDJA	18195	17349	19489	22590	21212	20934	18847	17904	19516	24248
Germany	QDJD	27932	24327	28141	33157	32371	29489	30831	30487	32403	37135
Greece	QDJG	1676	1643	1376	1191	862	926	991	857	896	887
Hungary	QDLI	1009	826	1066	1230	1113	1220	1227	1250	1319	1348
Ireland	QDJJ	19440	16201	17431	18721	17880	19200	18393	16899	16982	20309
Italy	QDJM	9415	8383	8950	10197	8087	8464	8796	8384	9678	10365
Latvia	QAMO	134	71	121	204	218	348	305	173	192	247
Lithuania	QAMP	246	134	204	251	356	297	274	235	301	465
Luxembourg	QDOK	177	158	224	250	200	222	201	171	180	187
Malta	QDOC	446	389	389	461	398	447	373	359	365	403
Netherlands	QDJP	20028	18588	22092	24140	25262	24866	22918	17486	19052	22043
Poland	QDLL	3007	2788	3831	4414	3517	3728	3882	3626	4202	4991
Portugal	QDJT	1662	1558	1876	1800	1413	1375	1352	1246	1400	1430
Romania	QAMQ	727	661	768	953	930	929	938	971	1018	1146
Slovak Republic	QAMR	435	370	464	550	521	457	447	419	488	470
Slovenia	QAMS	188	145	199	218	188	181	213	169	222	157
Spain	QDJW	10273	9349	10044	9955	8546	8823	8980	8837	9421	10367
Sweden	QDJZ	5242	4220	5623	6413	5882	5781	5597	4439	4578	5285
European Central Bank	QARP	-	-	-	-	-	-	-	-	-	-
EU Institutions	EOAY	-	-	-	-	-	-	-	-	-	-
Total EU28	L87R	142404	125664	144508	162883	152501	151223	146872	133664	142705	164085
European Free Trade Association (EFTA)											
Iceland	QDKW	151	91	101	127	153	136	146	191	566	334
Liechtenstein	EPOW	4	-	5	14	7	1	16	5	3	11
Norway	QDKZ	2980	2936	3178	3535	3764	3622	3880	3426	3307	3223
Switzerland	QDLC	4123	3488	4779	4930	5507	5845	6027	9952	8596	6462
Total EFTA	EPOT	7258	6515	8063	8606	9431	9604	10069	13574	12472	10030
Other Europe											
Albania	QAMC	4	7	8	7	8	11	8	1	5	4
Belarus	QAME	68	44	89	107	107	62	59	45	41	110
Russia	QDLO	4251	2449	3410	4615	4993	4615	4084	2695	2765	3116
Turkey	QDLR	2656	2436	3297	4029	3723	4095	3670	3620	4649	6421
Ukraine	QAMT	590	561	454	546	574	577	352	266	392	365
Serbia and Montenegro	QAMW	87	59	62	83	97	121	107	115	115	88
of which: Serbia	KN2P	81	58	60	82	92	107	100	106	104	82
Montenegro	KN2M	6	1	2	1	5	14	7	9	11	6
Other	BOQE	548	486	780	994	1047	1012	1175	1150	1135	1255
Total Europe	EPLM	157866	138221	160671	181870	172481	171320	166396	155130	164279	185474

6.5 Trade in goods - exports

£ million

		2008	2009	2010	2011	2012	2013	2014	2015	2016	2017
Exports (Credits) *(contd.)*											
Americas											
Argentina	QAOM	318	239	345	409	374	351	297	291	309	309
Brazil	QDLU	1732	1828	2257	2525	2693	2706	2429	2289	2016	1941
Canada	QATH	3293	3402	4252	4973	4279	4108	4137	4128	4997	5935
Chile	QAMG	235	488	598	779	655	1102	474	456	455	571
Colombia	QAML	149	161	214	310	306	328	333	372	256	459
Mexico	QDLX	898	753	953	1019	1132	1146	1085	1337	1339	1692
United States of America	J8V9	35405	34117	37914	40244	40661	40622	40169	47942	48543	51461
Uruguay	QAMU	46	40	61	102	93	147	98	205	109	165
Venezuela	QAMV	275	287	271	321	389	303	207	217	104	62
Other Central American Countries	BOQQ	707	412	597	670	897	546	561	752	811	655
Other	BOQT	152	159	261	375	585	540	364	480	242	283
Total Americas	EPLO	43210	41886	47723	51727	52064	51899	50154	58469	59181	63533
Asia											
China	QDMA	6053	5616	8256	10536	11540	13709	15682	13207	13652	18037
Hong Kong	QDMD	3625	3698	4386	5445	5334	5542	5116	6048	6441	7919
India	QDMG	5001	3115	4554	6619	5472	6292	5410	4100	3456	4465
Indonesia	QDMJ	373	358	438	639	634	673	509	454	541	799
Iran	QAON	421	369	292	177	85	68	86	77	126	207
Israel	QDMM	1273	1103	1342	1611	1448	1362	1108	1087	1151	1238
Japan	QAMJ	3903	3577	4334	4761	4942	4746	4563	4632	5040	6364
Malaysia	QDMP	1124	1037	1290	1474	1520	1617	1541	1396	1342	1463
Pakistan	QDMS	450	446	449	519	533	453	515	515	621	669
Philippines	QDMV	220	242	263	262	296	340	343	389	406	504
Saudi Arabia	QDMY	2213	2365	2495	2718	3062	3364	5112	4722	4621	4213
Singapore	QDNB	2849	2976	3463	3892	4484	4305	3855	3966	4682	5268
South Korea	QDNE	2519	2137	2333	2684	4894	5049	4331	4891	4587	6297
Taiwan	QDNH	872	766	1121	1400	1156	1194	1134	1207	1195	1162
Thailand	QDNK	758	913	1160	1445	1998	1957	1490	1306	1154	1298
Residual Gulf Arabian Countries	BOQW	5861	5544	6414	7410	8294	9324	9467	10016	10465	13478
Other Near & Middle Eastern Countries	QARJ	872	967	1208	1260	1185	1108	1136	1353	1061	882
Other	BORB	647	794	1027	1170	1058	1176	1124	1231	1308	1324
Total Asia	EPLP	39034	36023	44825	54022	57935	62279	62522	60597	61849	75587
Australasia & Oceania											
Australia	QDNN	3127	2973	3401	4439	4704	4107	3901	4010	4162	4975
New Zealand	QDNQ	378	317	409	522	588	607	653	568	718	826
Other	EGIZ	64	58	54	31	51	24	29	28	59	46
Total Australasia & Oceania	EPLQ	3569	3348	3864	4992	5343	4738	4583	4606	4939	5847
Africa											
Egypt	QDNT	937	1021	1215	1138	971	969	1119	1085	1293	1278
Morocco	QAOO	524	284	554	554	628	438	581	487	868	769
South Africa	QDNW	2484	2054	2491	3020	2709	2522	2564	2453	2489	2358
Other North Africa	BORU	744	876	836	594	673	843	791	541	731	605
Other	BOQH	3891	3651	4858	5887	6584	5382	4406	3384	3444	3420
Total Africa	EPLN	8580	7886	9954	11193	11565	10154	9461	7950	8825	8430
International Organisations	EPLR	-	-	-	-	-	-	-	-	-	-
World total	LQAD	252259	227364	267037	303804	299388	300390	293116	286752	299073	338871

Source: Office for National Statistics

6.6 Trade in goods - imports

£ million

		2008	2009	2010	2011	2012	2013	2014	2015	2016	2017
Imports (Debits)											
Europe											
European Union (EU)											
Austria	QBRZ	2343	2311	2625	2927	2621	2943	3109	3016	3171	3255
Belgium	QDOI	16638	15458	17635	19432	18790	20726	21605	21181	23780	25478
Bulgaria	QAMZ	179	155	202	254	272	340	328	337	385	389
Croatia	QANC	78	57	74	73	55	151	53	76	74	77
Cyprus	QDOA	111	59	61	97	133	109	110	135	153	113
Czech Republic	QDLG	3539	3327	4075	4372	4634	4696	4706	4846	5386	5602
Denmark	QBSF	4064	3789	4083	6085	5887	5352	4494	3443	3890	4865
Estonia	QAND	136	107	149	205	214	143	144	165	219	197
Finland	QBTG	2728	2057	2116	2444	2198	2587	2420	2090	2222	2466
France	QDJB	23818	21117	22103	23465	23235	24549	24979	24400	25121	27455
Germany	QDJE	44645	40181	46191	50246	52324	55968	60023	60513	63308	68722
Greece	QDJH	580	492	625	605	661	692	677	680	775	816
Hungary	QDLJ	2471	2469	3216	3065	2626	2593	2431	2481	2650	2714
Ireland	QDJK	12061	12248	12777	13031	12906	11803	11842	12424	13058	14488
Italy	QDJN	13917	11898	13790	13937	14247	15072	16534	15603	16880	18562
Latvia	QANE	343	283	380	384	292	502	365	452	662	557
Lithuania	QANF	322	328	536	606	880	927	1008	808	786	675
Luxembourg	QDOL	791	569	901	869	627	345	452	456	404	339
Malta	QDOD	115	84	176	173	116	103	115	161	180	111
Netherlands	QDJQ	26117	22547	26466	29103	32103	34887	32845	31085	34959	40680
Poland	QDLM	4235	4585	6023	7010	7441	7831	7742	8098	9158	10457
Portugal	QDJU	1677	1382	1722	1764	1729	1928	2243	2321	2642	2971
Romania	QANG	775	775	1219	1247	1234	1373	1459	1505	1685	1910
Slovak Republic	QANH	1573	1559	1578	1470	1541	1778	1890	1953	2488	2451
Slovenia	QANI	294	220	328	335	309	296	283	298	369	331
Spain	QDJX	12115	11490	12218	12698	12116	13376	14395	14876	16236	16436
Sweden	QDKA	7128	5681	6813	7888	9627	7684	7951	7119	6426	7024
European Central Bank	QARQ	-	-	-	-	-	-	-	-	-	-
EU Institutions	EOBS	-	-	-	-	-	-	-	-	-	-
Total EU28	L87T	182793	165228	188082	203785	208818	218754	224203	220522	237067	259141
European Free Trade Association (EFTA)											
Iceland	QDKX	444	455	464	417	389	374	412	445	455	522
Liechtenstein	EPOX	34	20	6	4	-	1	4	2	3	1
Norway	QDLA	21778	16136	20940	26481	24893	19865	17592	12969	13534	19562
Switzerland	QDLD	4429	4481	6011	6874	6389	7854	8111	8566	9921	7945
Total EFTA	EPOU	26685	21092	27421	33776	31671	28094	26119	21982	23913	28030
Other Europe											
Albania	QAMX	-	-	-	-	-	8	6	-	-	-
Belarus	QAMY	150	22	40	58	76	92	63	16	26	42
Russia	QDLP	7274	4937	6124	8515	9259	8236	7292	5243	4972	5819
Turkey	QDLS	5553	5208	5784	5969	6219	6749	7280	7716	8303	8830
Ukraine	QANJ	134	128	260	337	262	370	379	221	280	373
Serbia and Montenegro	QANM	66	53	71	69	56	94	79	117	185	220
of which: Serbia	KN2Q	64	53	71	69	56	94	79	117	185	204
Montenegro	KN2N	2	-	-	-	-	-	-	-	-	16
Other	BOQF	217	298	239	313	393	322	631	231	379	326
Total Europe	EPMM	222872	196966	228021	252822	256754	262719	266052	256048	275125	302781

6.6 Trade in goods - imports

£ million

		2008	2009	2010	2011	2012	2013	2014	2015	2016	2017
Imports (Debits) *(contd.)*											
Americas											
Argentina	QAOP	547	641	635	599	616	618	564	587	652	694
Brazil	QDLV	2895	2466	3047	2655	2390	2472	2202	1963	1921	1771
Canada	QATI	5767	4377	5749	6054	5656	5819	5044	6253	5339	5394
Chile	QANA	576	607	569	608	581	616	597	563	650	718
Colombia	QANB	645	505	678	899	904	771	633	608	473	473
Mexico	QDLY	762	711	971	1028	658	712	874	976	1024	1187
United States of America	J8VA	24499	23745	26104	27689	28450	27052	28882	33751	36410	39133
Uruguay	QANK	92	111	93	96	109	84	69	54	69	40
Venezuela	QANL	788	518	520	481	563	263	288	255	171	117
Other Central American Countries	BOQR	1395	1215	1055	1040	745	1046	922	881	1058	925
Other	BOQU	332	334	334	431	481	444	429	405	563	581
Total Americas	EPMO	38298	35230	39755	41580	41153	39897	40504	46296	48330	51033
Asia											
China	QDMB	22393	24747	29618	30390	30655	34632	35481	35819	39486	43553
Hong Kong	QDME	7484	7175	7941	7105	6766	6991	6578	6467	6508	7719
India	QDMH	4951	5438	6346	6490	6519	9729	8341	6173	6341	7683
Indonesia	QDMK	1144	1185	1342	1280	1237	1201	1088	1069	1241	1100
Iran	QAOQ	53	189	182	359	104	19	33	19	27	13
Israel	QDMN	1063	1004	1469	2122	2263	1836	1151	1009	1004	1056
Japan	QAMK	7952	6278	7389	8040	7791	7348	6819	6753	7997	9822
Malaysia	QDMQ	1863	1613	1801	1739	1655	1619	1699	1863	1817	1821
Pakistan	QDMT	587	649	761	842	781	890	1063	1037	1142	1215
Philippines	QDMW	600	397	502	438	423	385	382	403	481	467
Saudi Arabia	QDMZ	884	882	897	1120	1925	3130	2884	2239	1804	2239
Singapore	QDNC	4204	3808	4181	4026	3698	2834	2593	1950	2631	3225
South Korea	QDNF	3671	2923	2426	2607	3176	3389	4119	4622	4821	4568
Taiwan	QDNI	2574	2176	3040	3306	3599	3477	3327	3195	3144	3430
Thailand	QDNL	2270	2235	2381	2432	2396	2602	2582	2550	2803	3059
Residual Gulf Arabian Countries	BOQX	3049	3333	6109	9940	8122	8237	7251	6306	6444	8598
Other Near & Middle Eastern Countrie	QARK	376	320	279	201	782	303	142	200	185	317
Other	BORD	2887	3378	3571	4368	5883	6647	6571	7762	8746	9251
Total Asia	EPMP	68005	67730	80235	86805	87775	95269	92104	89436	96622	109136
Australasia & Oceania											
Australia	QDNO	2329	2168	2219	2374	2367	2086	1727	1912	2031	2241
New Zealand	QDNR	723	809	801	803	908	849	901	903	850	814
Other	HFKF	154	159	92	185	157	173	198	144	140	152
Total Australasia & Oceania	EPMQ	3206	3136	3112	3362	3432	3108	2826	2959	3021	3207
Africa											
Egypt	QDNU	611	660	610	765	613	746	853	638	636	622
Morocco	QAOR	416	324	324	396	450	520	583	557	941	676
South Africa	QDNX	4661	3839	4357	2881	3252	2402	2404	2857	2985	3064
Other North Africa	BORW	2340	1987	2910	2207	4162	5206	4050	1966	1054	2378
Other	BOQJ	3514	2797	3327	7349	8517	9513	5811	3805	3011	3422
Total Africa	EPMN	11542	9607	11528	13598	16994	18387	13701	9823	8627	10162
International Organisations	EPMR	-	-	-	-	-	-	-	-	-	-
World total	LQBL	343923	312669	362651	398167	406108	419380	415187	404562	431725	476319

Source: Office for National Statistics

6.7 Total International Trade in Services all industries (excluding travel, transport and banking) analysed by product 2016-2017

£ million

	Exports		Imports		Balance	
	2016	2017	2016	2017	2016	2017
Agricultural and Mining Services						
Agricultural, forestry and fishing	15	19	12	17	3	2
Mining and oil and gas extraction services	1,157	955	236	204	921	751
Total Agricultural and Mining services	**1,172**	**974**	**248**	**221**	**924**	**753**
Manufacturing, Maintenance and On-site Processing Services						
Waste treatment and depolution services	117	97	165	205	-49	-108
Manufacturing services on goods owned by others	2,735	3,518	601	925	2,134	2,593
Maintenance and repair services	2,041	2,533	725	794	1,317	1,739
Total Manufacturing, Maintenance and On-site Processing services	**4,893**	**6,148**	**1,491**	**1,924**	**3,402**	**4,224**
Business and Professional Services						
Accountancy, auditing, bookkeeping and tax consulting services	2,395	2,466	1,070	1,303	1,326	1,163
Advertising, market research and public opinion polling services	7,085	8,388	3,495	3,808	3,590	4,581
Business management and management consulting services	8,921	11,149	5,254	5,939	3,666	5,210
Public relations services	386	509	168	259	218	250
Recruitment services	1,613	2,241	760	993	854	1,248
Legal services	4,982	5,091	1,132	1,061	3,850	4,030
Operating leasing services	358	476	503	433	-144	43
Procurement services	217	150	266	315	-49	-165
Property management services	423	414	27	153	396	261
Other business and professional services	4,859	5,032	2,767	2,985	2,092	2,047
Services between related enterprises	14,668	17,609	11,520	14,111	3,147	3,498
Total Business and Professional Services	**45,907**	**53,525**	**26,961**	**31,360**	**18,945**	**22,165**
Research and Development Services						
Provision of R&D services	5,749	7,001	4,906	6,245	843	756
Provision of product development and testing activities	1,343	1,501	327	599	1,015	903
Total Research and Development Services	**7,092**	**8,502**	**5,233**	**6,843**	**1,858**	**1,659**
Intellectual Property						
Trade marks, franchises, brands or design rights						
Outright sales and purchases	724	257	1,141	1,008	-417	-750
Charges or payments for the use of	5,103	6,455	3,805	3,799	1,298	2,656
Copyrighted literary works, sound recordings, films, television prgrammes and databases						
Outright sales and purchases	549	625	322	195	227	430
Charges or payments for the use of	6,459	8,060	4,679	5,424	1,780	2,635
Patents and other intellectual property that are the end result of research and development						
Outright sales and purchases	86	183	74	84	13	99
Charges or payments for the use of	1,455	1,985	919	1,310	536	675
Total intellectual property	**14,376**	**17,566**	**10,939**	**11,821**	**3,436**	**5,745**
Telecommunication, Computer and Information Services						
Postal and courier	1,757	1,718	1,082	1,143	675	575
Telecommunications	8,617	6,887	4,947	4,725	3,670	2,162
Computer Services	8,335	9,451	5,190	5,947	3,146	3,503
Publishing Services	1,032	1,164	151	138	880	1,027
News agency Services	674	807	50	37	624	770
Information Services	2,330	2,757	1,041	1,395	1,289	1,362
Total Telecommunication, Computer and Information Services	**22,744**	**22,785**	**12,461**	**13,385**	**10,283**	**9,399**
Construction Services						
Construction in the UK	381	365	798	1,220	-417	-855
Construction outside the UK	1,593	1,271	612	509	980	762
Total Construction Services	**1,974**	**1,636**	**1,410**	**1,729**	**563**	**-93**

6.7 Total International Trade in Services all industries (excluding travel, transport and banking) analysed by product 2016-2017

£ million

	Exports		Imports		Balance	
	2016	2017	2016	2017	2016	2017
Financial Services						
Financial	18,415	19,971	3,430	3,830	14,985	16,142
Insurance and Pension Services						
Insurance and Pension Services Claims	3,008	3,223			3,008	3,223
Insurance and Pension Services Premiums			450	445	-450	-445
Merchanting and Other Trade related Services						
Merchanting	2,691	7,223	345	304	2,346	6,920
Other trade - related services	1,607	2,138	832	873	775	1,265
Total Merchanting and Other Trade related Services	**4,297**	**9,362**	**1,177**	**1,177**	**3,121**	**8,185**
Personal, Cultural and Recreational Services						
Audio- Visual and related services	1,438	1,841	755	890	684	950
Medical Services	41	35	12	11	28	24
Training and educational services	437	542	84	92	353	450
Heritage and recreational services	440	375	106	156	334	218
Social, domestic and other personal services	130	138	34	26	96	112
Total Personal, Cultural and Recreational Services	**2,486**	**2,931**	**990**	**1,176**	**1,495**	**1,756**
Technical and Scientific Services						
Architectural services	456	571	48	61	408	510
Engineering Services	7,314	7,750	1,963	2,985	5,350	4,765
Scientific and other techinical services inc surveying	1,431	1,632	445	518	986	1,114
Total Technical Services	**9,201**	**9,953**	**2,456**	**3,565**	**6,745**	**6,389**
Other Trade in Services						
Other trade in services	7,092	5,566	1,470	1,256	5,622	4,310
Total International Trade in Services	**142,657**	**162,141**	**68,718**	**78,731**	**73,939**	**83,410**

Source: Office for National Statistics

- Denotes nil or less than £500,000

.. Denotes disclosive data

The sum of constituent items may not always agree exactly with the totals shown due to rounding

Data from 2013 has been collected in accordance with BPM6 regulations

Excludes the activities of travel, transport and banking industries

6.8 Total International Trade in Services (excluding travel, transport and banking) analysed by continents and countries 2015 - 2017

£ million

	Exports			Imports			Balance		
	2015	2016	2017	2015	2016	2017	2015	2016	2017
Europe									
European Union (EU)									
Austria	388	482	887	345	355	444	43	127	443
Belgium	2,078	2,516	3,292	1,264	1,331	1,350	814	1,186	1,942
Bulgaria	159	137	132	89	138	176	70	-1	-44
Croatia	53	72	133	21	23	44	33	49	89
Cyprus	287	413	435	181	135	137	106	278	298
Czech Republic	351	444	507	257	305	385	94	139	122
Denmark	1,427	1,976	2,046	553	638	781	874	1,337	1,265
Estonia	30	38	47	31	36	19	-1	3	27
Finland	604	623	726	181	226	231	423	397	495
France	5,346	6,516	7,242	3,568	4,392	4,917	1,778	2,124	2,325
Germany	7,276	9,719	10,741	4,910	6,395	6,597	2,366	3,324	4,144
Greece	342	425	581	119	106	108	223	319	473
Hungary	260	325	362	140	183	372	120	142	-10
Irish Republic	7,188	9,092	11,573	3,791	4,736	5,340	3,398	4,357	6,233
Italy	3,035	2,680	3,043	1,401	1,664	1,460	1,634	1,016	1,583
Latvia	45	77	56	17	30	23	29	47	33
Lithuania	47	75	101	31	41	59	16	34	42
Luxembourg	2,189	2,589	2,986	1,905	2,078	2,170	284	511	816
Malta	240	410	334	73	153	98	168	256	236
Netherlands	6,854	7,672	9,740	2,720	3,751	4,134	4,134	3,921	5,606
Poland	805	1,004	1,034	619	759	986	186	245	48
Portugal	409	393	426	244	242	337	165	152	89
Romania	230	273	289	142	217	258	87	56	31
Slovakia	127	193	206	85	79	83	42	114	123
Slovenia	..	87	..	..	22	..	..	64	94
Spain	2,138	2,532	2,781	1,333	1,362	1,678	805	1,169	1,102
Sweden	2,123	2,496	2,548	1,824	1,482	1,832	299	1,014	716
EU Institutions	..	11	..	..	-	..	..	11	128
Total European Union (EU)	**44,098**	**53,267**	**62,512**	**25,866**	**30,879**	**34,059**	**18,232**	**22,389**	**28,453**
European Free Trade Association (EFTA)									
Iceland	118	155	..	16	47	..	102	109	-142
Liechtenstein	65	84	82	23	39	69	42	45	13
Norway	1,624	1,708	..	400	559	..	1,224	1,149	1,084
Switzerland	6,863	7,737	8,030	2,603	2,480	2,511	4,261	5,257	5,519
Total EFTA	**8,670**	**9,684**	**10,215**	**3,041**	**3,124**	**3,742**	**5,629**	**6,560**	**6,473**
Other Europe									
Russia	1,080	900	1,294	294	304	342	786	596	951
Channel Islands	1,813	2,017	2,013	264	294	292	1,549	1,724	1,721
Isle of Man	197	239	236	9	40	32	188	199	205
Turkey	808	699	747	370	198	217	438	501	530
Rest of Europe	726	841	1,211	146	292	358	580	549	853
Europe Unallocated	2,126	2,437	2,710	1,494	1,580	1,962	632	858	748
Total Europe	**59,518**	**70,085**	**80,938**	**31,483**	**36,710**	**41,004**	**28,035**	**33,375**	**39,934**
America									
Brazil	781	507	535	184	220	381	597	287	154
Canada	1,476	2,089	2,305	563	707	910	913	1,383	1,396
Mexico	383	370	362	45	82	113	338	288	249
USA	27,991	32,733	38,223	13,880	16,734	19,669	14,110	15,999	18,554
Rest of America	3,094	3,533	4,221	683	808	1,568	2,412	2,725	2,654
America Unallocated	328	444	243	57	100	49	271	343	193
Total America	**34,053**	**39,675**	**45,890**	**15,413**	**18,650**	**22,689**	**18,641**	**21,025**	**23,200**

6.8 Total International Trade in Services (excluding travel, transport and banking) analysed by continents and countries 2015 - 2017

£ million

	Exports			Imports			Balance		
	2015	2016	2017	2015	2016	2017	2015	2016	2017
Asia									
China	1,647	1,600	2,183	861	832	825	785	767	1,359
Hong Kong	1,241	1,330	1,686	760	854	808	481	476	878
India	1,189	1,062	908	1,759	2,066	2,574	-570 -	1,004	-1,666
Indonesia	180	153	335	72	63	48	108	89	286
Israel	452	581	581	375	279	357	77	302	224
Japan	2,370	2,743	2,941	1,702	2,443	3,072	668	300	-130
Malaysia	406	477	415	147	335	127	259	143	288
Pakistan	150	243	196	35	35	39	115	208	157
Philippines	103	123	131	319	214	179	-216	-91	-48
Saudi Arabia	4,613	4,874	3,903	84	132	111	4,530	4,742	3,792
Singapore	2,311	1,767	2,031	1,196	1,389	1,213	1,115	378	819
South Korea	768	851	1,233	207	242	278	561	609	955
Taiwan	298	514	476	86	106	140	212	408	336
Thailand	317	212	287	64	73	71	253	140	217
Rest of Asia	4,634	7,388	7,276	1,188	1,373	1,016	3,446	6,015	6,260
Asia Unallocated	771	859	1,416	209	224	277	562	634	1,139
Total Asia	**21,452**	**24,778**	**25,999**	**9,065**	**10,662**	**11,133**	**12,386**	**14,116**	**14,866**
Australasia and Oceania									
Australia	1,787	1,968	2,194	1,096	824	1,163	691	1,143	1,030
New Zealand	175	223	194	70	62	131	106	161	63
Rest of Australia and Oceania	..	69	57	..	14	15	54	55	42
Oceania Unallocated	..	2	-	..	2	-	1	-1	-
Total Australasia and Oceania	**2,023**	**2,261**	**2,446**	**1,172**	**903**	**1,310**	**851**	**1,359**	**1,135**
Africa									
Nigeria	867	397	457	97	104	88	770	293	369
South Africa	1,037	1,111	1,133	408	424	471	629	687	662
Rest of Africa	3,216	3,242	3,144	520	865	872	2,696	2,377	2,272
Africa Unallocated	216	220	219	114	158	239	102	62	-21
Total Africa	**5,335**	**4,970**	**4,953**	**1,138**	**1,551**	**1,670**	**4,196**	**3,419**	**3,283**
Total Unallocated	850	887	1,917	237	242	925	613	645	992
International Organisations	-	-	-	-	-	-	-	-	-
TOTAL INTERNATIONAL TRADE IN SERVICES	**123,231**	**142,657**	**162,141**	**58,508**	**68,718**	**78,731**	**64,722**	**73,939**	**83,410**

Source: Office for National Statistics

- Denotes nil or less than £500,000

.. Denotes disclosive data

The sum of constituent items may not always agree exactly with the totals shown due to rounding

For a breakdown of geographical groupings, see the International Trade In Services Quality and Methodology Information (QMI)

Excludes the activities of travel, transport and banking industries

Estimates for Croatia have been published for the first time in 2013. Previously this was included in the Rest of Europe.

6.9 Summary of balance of payments in 2016

£ million

	Credits	Debits
1. Current account		
A. Goods and services	547,473	590,486
1. Goods	302,067	437,458
2. Services	245,406	153,028
2.1. Manufacturing physical inputs owned by others	2,094	812
2.2. Maintenance and repair	1,871	692
2.3. Transport	25,950	22,076
2.4. Travel	30,756	47,989
2.5. Construction	1,787	1,069
2.6. Insurance and pension services	17,630	199
2.7. Financial	61,383	10,598
2.8. Intellectual property	12,649	8,847
2.9. Telecommunications, computer and information services	19,036	10,731
2.10. Other business	66,051	43,309
2.11. Personal, cultural and recreational services	3,631	3,539
2.12. Government	2,568	3,167
B. Primary income	136,761	187,178
1. Compensation of employees	1,376	1,735
2. Investment income	132,954	182,116
2.1. Direct investment	57,513	60,378
2.2. Portfolio investment	49,134	89,537
2.3. Other investment	25,207	32,201
2.4. Reserve assets	1,100	
3. Other primary income	2,431	3,327
C. Secondary income	18,302	40,327
1. General government	771	20,839
2. Other sectors	17,531	19,488
Total current account	702,536	817,991
2. Capital account		
1. Capital transfers	759	2,263
2. Acquisition/disposal of non-produced, non-financial assets	1,177	1,017
Total capital account	1,936	3,280

6.9 Summary of balance of payments in 2016

contd.   £ million

	Net acquisition of financial assets	Net incurrence of liabilities
3. Financial account		
1. Direct investment	36,123	220,468
Abroad	36,123	
1.1. Equity capital other than reinvestment of earnings	9,403	
1.2. Reinvestment of earnings	-5,307	
1.3. Debt instruments[1]	32,027	
In United Kingdom		220,468
1.1. Equity capital other than reinvestment of earnings		183,015
1.2. Reinvestment of earnings		11,577
1.3. Debt instruments[2]		25,876
2. Portfolio investment	-180,263	-41,069
Assets	-180,263	
2.1. Equity and investment fund shares	-61,665	
2.2. Debt securities	-118,598	
Liabilities		-41,069
2.1. Equity and investment fund shares		-130,297
2.2. Debt securities		89,228
3. Financial derivatives and employee stock options (net)	21,615	
4. Other investment	154,054	-21,797
Assets	154,054	
4.1. Other equity	253	
4.2. Currency and deposits	110,927	
4.3. Loans	43,231	
4.4. Trade credit and advances	-203	
4.5. Other accounts receivable	-154	
Liabilities		-21,797
4.1. Currency and deposits		21,134
4.2. Loans		-43,905
4.3. Insurance, pensions and standardised guarantee schemes		999
4.4. Trade credit and advances		-
4.5. Other accounts payable		-25
4.6. Special drawing rights		-
5. Reserve assets	6,511	
5.1. Monetary gold	-	
5.2. Special drawing rights	-1,397	
5.3. Reserve position in the IMF	1,926	
5.4. Other reserve assets	5,982	
Total financial account	38,040	157,602

Source: Office for National Statistics

1 Debt instrument transactions on direct investment abroad represents claims on affiliated enterprises plus claims on direct investors.
2 Debt instrument transactions on direct investment in the United Kingdom represents liabilities to direct investors plus liabilities to affiliated enterprises

6.10 Summary of balance of payments: Balances (net credits less net debits)

£ million

			Current account										
	Trade in goods	Trade in services	Total goods and services	Compen-sation of employees	Invest-ment income	Other primary income	Total primary income	Second-ary income	Current balance	Current balance as % of GDP[1]	Capital account	Financial account	Net errors & omissions
	LQCT	KTMS	KTMY	KTMP	HMBM	MT5W	HMBP	KTNF	HBOG	AA6H	FKMJ	-HBNT	HHDH
1946	-101	-274	-375	-20	76	-	56	166	-153	..	-21	-181	-7
1947	-358	-197	-555	-19	140	-	121	123	-311	..	-21	-552	-220
1948	-152	-64	-216	-20	223	-	203	96	83	0.7	-17	58	-8
1949	-138	-43	-181	-20	206	-	186	29	34	0.3	-12	103	81
1950	-55	-4	-59	-21	378	-	357	39	337	2.6	-10	447	120
1951	-693	32	-661	-21	322	-	301	29	-331	-2.3	-15	-426	-80
1952	-274	123	-151	-22	231	-	209	169	227	1.5	-15	229	17
1953	-246	122	-124	-25	207	-	182	143	201	1.2	-13	177	-11
1954	-213	114	-99	-27	227	-	200	55	156	0.9	-13	174	31
1955	-318	41	-277	-27	149	-	122	43	-112	-0.6	-15	-34	93
1956	44	25	69	-30	203	-	173	2	244	1.2	-13	250	19
1957	-39	120	81	-32	223	-	191	-5	267	1.2	-13	313	59
1958	22	117	139	-34	261	-	227	4	370	1.6	-10	411	51
1959	-129	116	-13	-37	233	-	196	-	183	0.7	-5	68	-110
1960	-418	36	-382	-35	201	-	166	-6	-222	-0.8	-6	7	235
1961	-160	48	-112	-35	223	-	188	-9	67	0.2	-12	-23	-78
1962	-122	47	-75	-37	301	-	264	-14	175	0.6	-12	195	32
1963	-142	1	-141	-38	364	-	326	-37	148	0.5	-16	30	-102
1964	-573	-37	-610	-33	365	-	332	-74	-352	-1	-17	-392	-23
1965	-288	-70	-358	-34	405	-	371	-75	-62	-0.2	-18	-49	31
1966	-144	39	-105	-39	358	-	319	-91	123	0.3	-19	-22	-126
1967	-646	151	-495	-39	354	-	315	-118	-298	-0.7	-25	-179	144
1968	-770	333	-437	-48	303	-	255	-119	-301	-0.7	-26	-688	-361
1969	-283	382	99	-47	468	-	421	-109	411	0.8	-23	794	406
1970	-94	443	349	-56	527	-	471	-89	731	1.3	-22	818	109
1971	120	602	722	-63	454	-	391	-90	1023	1.6	-23	1330	330
1972	-829	703	-126	-52	350	-	298	-142	30	-	-35	-477	-472
1973	-2676	883	-1793	-68	970	-	902	-336	-1227	-1.5	-39	-1031	235
1974	-5357	1255	-4102	-92	1010	-	918	-302	-3486	-3.8	-34	-3185	335
1975	-3378	1659	-1719	-102	257	-	155	-313	-1877	-1.6	-36	-1569	344
1976	-4079	2811	-1268	-140	760	-	620	-534	-1182	-0.9	-12	-507	687
1977	-2439	3632	1193	-152	-678	-	-830	-889	-526	-0.3	11	3286	3801
1978	-1710	4127	2417	-140	-300	-	-440	-1420	557	0.3	-79	2655	2177
1979	-3514	4461	947	-130	-342	-	-472	-1777	-1302	-0.6	-103	-864	541
1980	1123	4275	5398	-82	-2268	-	-2350	-1653	1395	0.5	-4	2157	766
1981	2986	4604	7590	-66	-1883	-	-1949	-1219	4422	1.5	-79	5312	969
1982	1614	4062	5676	-95	-2336	-	-2431	-1476	1769	0.6	6	1233	-542
1983	-1892	5175	3283	-89	-1050	-	-1139	-1391	753	0.2	75	3287	2459
1984	-5736	5850	114	-94	-326	-	-420	-1566	-1872	-0.5	107	7130	8895
1985	-3762	8221	4459	-120	-2609	-	-2729	-2924	-1194	-0.3	185	1657	2666
1986	-9973	7854	-2119	-156	71	-	-85	-2094	-4298	-1	-261	122	4681
1987	-12113	8619	-3494	-174	-785	-	-959	-3325	-7778	-1.6	159	-12456	-4837
1988	-22066	6976	-15090	-64	-1234	-	-1298	-3171	-19559	-3.5	-39	-16879	2719
1989	-25262	6949	-18313	-138	-2367	-	-2505	-4085	-24903	-4.1	-56	-15468	9491
1990	-19336	8201	-11135	-110	-4644	-	-4754	-4658	-20547	-3.1	222	-29500	-9175

6.10 Summary of balance of payments: Balances (net credits less net debits)

£ million

			Current account							Current balance as % of GDP[1]	Capital account	Financial account	Net errors & omissions
	Trade in goods	Trade in services	Total goods and services	Compen-sation of employees	Invest-ment income	Other primary income	Total primary income	Second-ary income	Current balance				
	LQCT	KTMS	KTMY	KTMP	HMBM	MT5W	HMBP	KTNF	HBOG	AA6H	FKMJ	-HBNT	HHDH
1991	-11001	8393	-2608	-63	-5691	-	-5754	-943	-9305	-1.3	-55	-2248	7112
1992	-13869	8972	-4897	-49	-990	-	-1039	-5182	-11118	-1.5	35	25	11108
1993	-13979	11276	-2703	35	-2488	-	-2453	-4931	-10087	-1.3	32	-16890	-6835
1994	-12094	11854	-240	-170	1625	-	1455	-5038	-3823	-0.5	-252	-13289	-9214
1995	-13119	15436	2317	-296	-435	-	-731	-7240	-5654	-0.7	97	1701	7258
1996	-14895	16420	1525	93	-2351	-	-2258	-4398	-5131	-0.6	806	-5201	-876
1997	-13060	16897	3837	83	130	686	899	-5951	-1215	-0.1	509	4707	5413
1998	-22484	14619	-7865	-10	9233	819	10042	-8697	-6520	-0.7	-261	7864	14645
1999	-29568	13578	-15990	201	-4289	711	-3377	-7686	-27053	-2.6	-258	-21282	6029
2000	-33688	13611	-20077	150	3568	441	4159	-9759	-25677	-2.4	393	-25053	231
2001	-42207	16245	-25962	66	7628	579	8273	-6518	-24207	-2.1	73	-14083	10051
2002	-49208	16226	-32982	67	14384	968	15419	-8940	-26503	-2.2	-675	-17494	9684
2003	-50992	20999	-29993	59	14784	1272	16115	-10286	-24164	-1.9	-65	-39121	-14892
2004	-61587	27185	-34402	-494	13035	1279	13820	-10519	-31101	-2.4	90	-31339	-328
2005	-70004	34391	-35613	-610	18408	1140	18938	-12199	-28874	-2.1	-843	-40111	-10394
2006	-78807	43549	-35258	-958	1725	816	1583	-11949	-45624	-3.1	-1527	-40291	6860
2007	-89862	51809	-38053	-734	-6996	537	-7193	-13101	-58347	-3.8	-169	-64071	-5555
2008	-94402	49349	-45053	-715	-14271	362	-14624	-13218	-72895	-4.6	252	-76082	-3439
2009	-86414	52961	-33453	-259	-12025	756	-11528	-14785	-59766	-3.9	243	-60974	-1451
2010	-97180	56012	-41168	-389	1349	114	1074	-19596	-59690	-3.8	-725	-72683	-12268
2011	-94779	69608	-25171	-173	6491	229	6547	-20271	-38895	-2.4	-764	-30019	9640
2012	-108687	75323	-33364	-148	-17351	-273	-17772	-20449	-71585	-4.2	-589	-58578	13596
2013	-119783	84416	-35367	-326	-35594	-471	-36391	-25280	-97038	-5.5	-1181	-84683	13536
2014	-123122	86291	-36831	-469	-36691	-654	-37814	-23395	-98040	-5.3	-1927	-89905	10062
2015	-118626	86256	-32370	-89	-41722	-1126	-42937	-22838	-98145	-5.2	-1978	-90892	9231
2016	-135391	92378	-43013	-359	-49162	-896	-50417	-22025	-115455	-5.9	-1344	-119562	-2763

Source: Office for National Statistics

1. Using series YBHA: GDP at current market prices.

209

6.11 Balance of payments: current account

£ million

		1995	1996	1997	1998	1999	2000	2001	2002	2003	2004	2005
Credits												
Exports of goods and services												
Exports of goods	LQAD	153,577	167,196	172,110	163,997	166,539	188,130	189,624	186,776	188,546	191,608	212,053
Exports of services	KTMQ	59,202	65,085	67,743	70,804	76,485	82,451	89,039	93,723	105,288	115,513	130,178
Total exports of goods and services	KTMW	212,779	232,281	239,853	234,801	243,024	270,581	278,663	280,499	293,834	307,121	342,231
Primary income												
Compensation of employees	KTMN	887	911	1,007	840	960	1,032	1,087	1,121	1,116	931	974
Investment income	HMBN	85,797	90,122	96,616	105,916	103,364	134,441	139,925	124,299	124,862	140,789	191,781
Other primary income	MT5S	-	-	3,068	2,937	2,781	2,571	2,679	2,912	3,227	3,449	3,408
Total primary income	HMBQ	86,684	91,033	100,691	109,693	107,105	138,044	143,691	128,332	129,205	145,169	196,163
Secondary income												
General government	FJUM	1,730	2,828	660	390	371	381	431	564	408	707	730
Other sectors	FJUN	10,213	11,372	8,649	8,620	6,393	6,574	6,395	8,137	8,684	9,134	11,331
Total secondary income	KTND	11,943	14,200	9,309	9,010	6,764	6,955	6,826	8,701	9,092	9,841	12,061
Total	HBOE	311,406	337,514	349,853	353,504	356,893	415,580	429,180	417,532	432,131	462,131	550,455
Debits												
Imports of goods and services												
Imports of goods	LQBL	166,562	181,932	185,170	186,481	196,107	221,818	231,831	235,984	239,538	253,195	282,057
Imports of services	KTMR	43,823	48,745	50,846	56,185	62,907	68,840	72,794	77,497	84,289	88,328	95,787
Total exports of goods and services	KTMX	210,385	230,677	236,016	242,666	259,014	290,658	304,625	313,481	323,827	341,523	377,844
Primary income												
Compensation of employees	KTMO	1,183	818	924	850	759	882	1,021	1,054	1,057	1,425	1,584
Investment income	HMBO	86,232	92,473	96,486	96,683	107,653	130,873	132,297	109,915	110,078	127,754	173,373
Other primary income	MT5U	-	-	2,382	2,118	2,070	2,130	2,100	1,944	1,955	2,170	2,268
Total primary income	HMBR	87,415	93,291	99,792	99,651	110,482	133,885	135,418	112,913	113,090	131,349	177,225
Secondary income												
General government	FJUO	4,811	5,081	6,637	9,324	7,742	9,735	6,496	8,869	10,088	10,211	11,941
Other sectors	FJUP	14,495	13,658	8,623	8,383	6,708	6,979	6,848	8,772	9,290	10,149	12,319
Total secondary income	KTNE	19,306	18,739	15,260	17,707	14,450	16,714	13,344	17,641	19,378	20,360	24,260
Total	HBOF	317,106	342,707	351,068	360,024	383,946	441,257	453,387	444,035	456,295	493,232	579,329
Balances												
Trade in goods and services												
Trade in goods	LQCT	-12,985	-14,736	-13,060	-22,484	-29,568	-33,688	-42,207	-49,208	-50,992	-61,587	-70,004
Trade in services	KTMS	15,379	16,340	16,897	14,619	13,578	13,611	16,245	16,226	20,999	27,185	34,391
Total exports of goods and services	KTMY	2,394	1,604	3,837	-7,865	-15,990	-20,077	-25,962	-32,982	-29,993	-34,402	-35,613
Primary income												
Compensation of employees	KTMP	-296	93	83	-10	201	150	66	67	59	-494	-610
Investment income	HMBM	-435	-2,351	130	9,233	-4,289	3,568	7,628	14,384	14,784	13,035	18,408
Other primary income	MT5W	-	-	686	819	711	441	579	968	1,272	1,279	1,140
Total primary income	HMBP	-731	-2,258	899	10,042	-3,377	4,159	8,273	15,419	16,115	13,820	18,938
Secondary income												
General government	FJUQ	-3,081	-2,253	-5,977	-8,934	-7,371	-9,354	-6,065	-8,305	-9,680	-9,504	-11,211
Other sectors	FJUR	-4,282	-2,286	26	237	-315	-405	-453	-635	-606	-1,015	-988
Total secondary income	KTNF	-7,363	-4,539	-5,951	-8,697	-7,686	-9,759	-6,518	-8,940	-10,286	-10,519	-12,199
Total (Current balance)	HBOG	-5,700	-5,193	-1,215	-6,520	-27,053	-25,677	-24,207	-26,503	-24,164	-31,101	-28,874

- = nil or less than a million

Source: **Office for National Statistics**
Email: bop@ons.gsi.gov.uk
Tel: 01633 456333

6.11 Balance of payments: current account

(contd.)

£ million

		2006	2007	2008	2009	2010	2011	2012	2013	2014	2015	2016
Credits												
Exports of goods and services												
Exports of goods	LQAD	243,957	222,964	254,577	229,107	270,196	308,171	301,621	302,169	297,306	288,770	302,067
Exports of services	KTMQ	146,899	159,069	166,866	170,542	175,552	190,691	199,434	217,744	221,619	228,391	245,406
Total exports of goods and services	KTMW	390,856	382,033	421,443	399,649	445,748	498,862	501,055	519,913	518,925	517,161	547,473
Primary income												
Compensation of employees	KTMN	938	984	1,046	1,176	1,097	1,121	1,124	1,094	1,082	1,295	1,376
Investment income	HMBN	247,208	305,951	288,221	176,037	174,500	200,529	171,149	157,759	142,463	131,853	132,954
Other primary income	MT5S	3,221	2,952	3,051	3,411	3,059	3,166	2,625	2,455	2,306	1,961	2,431
Total primary income	HMBQ	251,367	309,887	292,318	180,624	178,656	204,816	174,898	161,308	145,851	135,109	136,761
Secondary income												
General government	FJUM	901	885	867	891	809	404	630	651	696	550	771
Other sectors	FJUN	16,559	10,292	15,435	11,641	12,753	12,089	14,180	17,303	16,047	16,338	17,531
Total secondary income	KTND	17,460	11,177	16,302	12,532	13,562	12,493	14,810	17,954	16,743	16,888	18,302
Total	HBOE	659,683	703,097	730,063	592,805	637,966	716,171	690,763	699,175	681,519	669,158	702,536
Debits												
Imports of goods and services												
Imports of goods	LQBL	322,764	312,826	348,979	315,521	367,376	402,950	410,308	421,952	420,428	407,396	437,458
Imports of services	KTMR	103,350	107,260	117,517	117,581	119,540	121,083	124,111	133,328	135,328	142,135	153,028
Total exports of goods and services	KTMX	426,114	420,086	466,496	433,102	486,916	524,033	534,419	555,280	555,756	549,531	590,486
Primary income												
Compensation of employees	KTMO	1,896	1,718	1,761	1,435	1,486	1,294	1,272	1,420	1,551	1,384	1,735
Investment income	HMBO	245,483	312,947	302,492	188,062	173,151	194,038	188,500	193,353	179,154	173,575	182,116
Other primary income	MT5U	2,405	2,415	2,689	2,655	2,945	2,937	2,898	2,926	2,960	3,087	3,327
Total primary income	HMBR	249,784	317,080	306,942	192,152	177,582	198,269	192,670	197,699	183,665	178,046	187,178
Secondary income												
General government	FJUO	12,493	12,859	12,104	13,876	17,854	18,572	18,879	23,328	21,514	21,094	20,839
Other sectors	FJUP	16,916	11,419	17,416	13,441	15,304	14,192	16,380	19,906	18,624	18,632	19,488
Total secondary income	KTNE	29,409	24,278	29,520	27,317	33,158	32,764	35,259	43,234	40,138	39,726	40,327
Total	HBOF	705,307	761,444	802,958	652,571	697,656	755,066	762,348	796,213	779,559	767,303	817,991
Balances												
Trade in goods and services												
Trade in goods	LQCT	-78,807	-89,862	-94,402	-86,414	-97,180	-94,779	-108,687	-119,783	-123,122	-118,626	-135,391
Trade in services	KTMS	43,549	51,809	49,349	52,961	56,012	69,608	75,323	84,416	86,291	86,256	92,378
Total exports of goods and services	KTMY	-35,258	-38,053	-45,053	-33,453	-41,168	-25,171	-33,364	-35,367	-36,831	-32,370	-43,013
Primary income												
Compensation of employees	KTMP	-958	-734	-715	-259	-389	-173	-148	-326	-469	-89	-359
Investment income	HMBM	1,725	-6,996	-14,271	-12,025	1,349	6,491	-17,351	-35,594	-36,691	-41,722	-49,162
Other primary income	MT5W	816	537	362	756	114	229	-273	-471	-654	-1,126	-896
Total primary income	HMBP	1,583	-7,193	-14,624	-11,528	1,074	6,547	-17,772	-36,391	-37,814	-42,937	-50,417
Secondary income												
General government	FJUQ	-11,592	-11,974	-11,237	-12,985	-17,045	-18,168	-18,249	-22,677	-20,818	-20,544	-20,068
Other sectors	FJUR	-357	-1,127	-1,981	-1,800	-2,551	-2,103	-2,200	-2,603	-2,577	-2,294	-1,957
Total secondary income	KTNF	-11,949	-13,101	-13,218	-14,785	-19,596	-20,271	-20,449	-25,280	-23,395	-22,838	-22,025
Total (Current balance)	HBOG	-45,624	-58,347	-72,895	-59,766	-59,690	-38,895	-71,585	-97,038	-98,040	-98,145	-115,455

Source: **Office for National Statistics**
Email: bop@ons.gsi.gov.uk
Tel: 01633 456333

- = nil or less than a million

6.12 Summary of international investment position, financial account and investment income

£ billion

		2006	2007	2008	2009	2010	2011	2012	2013	2014	2015	2016
Investment abroad												
International investment position												
Direct investment	N2V3	992.3	1148.6	1438.6	1261.9	1343	1363.5	1411.5	1437.4	1387.7	1376.2	1502.7
Portfolio investment	HHZZ	1690.7	1898.4	1690.1	1817.5	1932.3	1812.2	2011.4	2058.5	2227	2233.2	2499.6
Financial derivatives	JX96	853.7	1378.2	4040.2	2176.4	2962.9	3638.2	3093.8	2485.2	2910.9	2410.1	2649.9
Other investment	HLXV	3008.2	3794.6	4312.2	3591.7	3903.7	4187.4	3863.1	3597.3	3624.4	3481	4182.6
Reserve assets	LTEB	22.9	26.7	36.3	40.1	49.7	56.8	61.7	61.4	67.7	87.5	110
Total	HBQA	6567.8	8246.4	11517.3	8887.7	10191.4	11058	10441.5	9639.8	10217.7	9588	10944.7
Financial account transactions[1]												
Direct investment	-N2SV	77	184.1	200.1	-32.8	35.8	50	7.7	28.6	-70.5	-37.9	36.1
Portfolio investment	-HHZC	253.8	131.4	-283.3	65.2	-11.7	-93.9	102.3	-69.5	79.5	-31	-180.3
Financial derivatives (net)	-ZPNN	-20.6	27	121.7	-29.1	-44.8	4.6	-41.6	40.6	19	-84.2	21.6
Other investment	-XBMM	374.9	699.9	-598.3	-307.1	259.8	102.4	-272	-249.7	80.5	-106.5	154.1
Reserve assets	-LTCV	-0.4	1.2	-1.3	5.8	6.1	4.9	7.6	5	7.1	21.1	6.5
Total	-HBNR	684.7	1043.5	-561.2	-298	245.2	68.1	-195.9	-245.2	115.5	-238.5	38
Investment income												
Direct investment	N2QP	91.7	101.1	86.6	74.3	92.9	104.6	87.2	84	74.3	61.8	57.5
Portfolio investment	HLYX	54.9	66.6	69	56	49.5	53.3	50.6	46.6	43.9	47.6	49.1
Other investment	AIOP	100	137.6	131.9	44.9	31.3	41.8	32.6	26.6	23.6	21.6	25.2
Reserve assets	HHCB	0.6	0.6	0.8	0.8	0.7	0.8	0.7	0.6	0.6	0.8	1.1
Total	HMBN	247.2	306	288.2	176	174.5	200.5	171.1	157.8	142.5	131.9	133
Investment in the UK												
International investment position												
Direct investment	N2UG	774.4	788.1	944.3	881.2	948.2	994.1	1250.9	1265.3	1323.7	1275.3	1461.9
Portfolio investment	HLXW	1737.2	2013.6	2026.1	2417.1	2547.1	2607.9	2758.4	2741.3	2880.1	2965.2	3143.5
Financial derivatives	JX97	890.5	1392.2	3915.3	2096.8	2895	3554.9	3032.2	2376.7	2806.4	2391.4	2607.4
Other investment	HLYD	3281	4175.2	4478.3	3735.5	3925.4	4092.2	3887.9	3580.7	3616.5	3303.4	3753.2
Total	HBQB	6683.1	8369.1	11364	9130.6	10315.7	11249.2	10929.4	9964	10626.7	9935.3	10966
Financial account transactions												
Direct investment	N2SA	113	104.6	142.4	6.2	42.3	16.6	29.7	35.8	36.4	38.2	220.5
Portfolio investment	HHZF	180.8	213.1	155.5	201.6	118.3	40.6	-71.7	112.2	65.8	108	-41.1
Other investment	XBMN	431.2	789.9	-783	-444.9	157.3	40.9	-95.3	-308.4	103.2	-293.7	-21.8
Total	HBNS	725	1107.6	-485.1	-237	317.9	98.1	-137.4	-160.5	205.4	-147.6	157.6
Investment income												
Direct investment	N2Q4	51.9	61.7	60.5	50.6	45.1	51.1	52.3	56.2	53.8	52.7	60.4
Portfolio investment	HLZC	80.6	98.7	101.8	82	86.2	89.7	91.7	102	94.4	92.2	89.5
Other investment	HLZN	112.9	152.6	140.2	55.4	41.9	53.2	44.5	35.2	31	28.7	32.2
Total	HMBO	245.5	312.9	302.5	188.1	173.2	194	188.5	193.4	179.2	173.6	182.1
Net investment												
International investment position												
Direct investment	MU7O	217.9	360.5	494.2	380.7	394.8	369.4	160.6	172.2	63.9	100.8	40.7
Portfolio investment	CGNH	-46.5	-115.2	-336	-599.6	-614.8	-795.7	-747	-682.8	-653.1	-732.1	-643.9
Financial derivatives	JX98	-36.8	-14.1	124.9	79.6	67.8	83.2	61.6	108.5	104.5	18.8	42.5
Other investment	CGNG	-272.9	-380.6	-166.1	-143.8	-21.8	95.2	-24.8	16.6	7.9	177.6	429.4
Reserve assets	LTEB	22.9	26.7	36.3	40.1	49.7	56.8	61.7	61.4	67.7	87.5	110
Net investment position	HBQC	-115.3	-122.7	153.3	-242.9	-124.3	-191.2	-487.9	-324.2	-409	-347.3	-21.3
Financial account transactions[1]												
Direct investment	-MU7M	-36	79.5	57.7	-39	-6.5	33.3	-22	-7.2	-106.9	-76.1	-184.3
Portfolio investment	-HHZD	73	-81.8	-438.8	-136.4	-129.9	-134.5	174	-181.7	13.7	-139	-139.2
Financial derivatives	-ZPNN	-20.6	27	121.7	-29.1	-44.8	4.6	-41.6	40.6	19	-84.2	21.6
Other investment	-HHYR	-56.3	-89.9	184.7	137.8	102.5	61.5	-176.6	58.7	-22.8	187.2	175.9
Reserve assets	-LTCV	-0.4	1.2	-1.3	5.8	6.1	4.9	7.6	5	7.1	21.1	6.5
Net transactions	-HBNT	-40.3	-64.1	-76.1	-61	-72.7	-30	-58.6	-84.7	-89.9	-90.9	-119.6
Investment income												
Direct investment	MU7E	39.7	39.4	26.1	23.8	47.8	53.5	34.9	27.8	20.5	9.1	-2.9
Portfolio investment	HLZX	-25.8	-32.1	-32.8	-26	-36.7	-36.4	-41	-55.5	-50.5	-44.6	-40.4
Other investment	CGNA	-12.9	-15	-8.4	-10.5	-10.5	-11.4	-11.9	-8.6	-7.4	-7.1	-7
Reserve assets	HHCB	0.6	0.6	0.8	0.8	0.7	0.8	0.7	0.6	0.6	0.8	1.1
Net earnings	HMBM	1.7	-7	-14.3	-12	1.3	6.5	-17.4	-35.6	-36.7	-41.7	-49.2

1. When downloading data from the Pink Book dataset users should reverse the sign of series that have an identifier that is prefixed with a minus sign.

Source: **Office for National Statistics**
Email: bop@ons.gsi.gov.uk
Tel: 01633 456333

6.13 Net foreign direct investment flows abroad analysed by area and main country, 2008 to 2017 (Directional)

£ million

	2008	2009	2010	2011	2012	2013	2014	2015	2016	2017
EUROPE	**64,811**	**16,642**	**24,436**	**37,212**	**-3,964**	**-20,241**	**-107,766**	**-12,078**	**-30,635**	**-3,756**
EU	54,853	-13,416	20,191	25,441	-5,459	-11,656	-83,140	-13,403	-26,149	-3,023
AUSTRIA	-73	-42	472	-49	-287	14	-131	..	60	198
BELGIUM	1,784	-1,833	7,284	16,053	1,472	-4,054	373	1,610	..	..
BULGARIA	48	16	-101	-10	-43	4	13	26	-2	-5
CROATIA	54	-2	-9	-3	-12	11	11	25	-13	
CYPRUS	470	60	-308	-340	296	36	-83	25	..	-165
CZECH REPUBLIC	387	-189	-175	-42	974	268	26	5	-24	142
DENMARK	3,178	-2,177	479	-1,018	536	-124	-5	-800	801	164
ESTONIA	-21	24	6	-10	-29	15	8	..	..	..
FINLAND	80	-88	-19	-615	-198	3	-107	-43	38	79
FRANCE	8,507	845	5,180	-4,215	-6,925	-9,700	-2,631	1,085	189	-2,275
GERMANY	1,000	3,027	-2,793	3,513	2,593	732	4,222	-1,569	2,198	2,593
GREECE	465	-349	106	-982	-46	35	-33	160	..	141
HUNGARY	186	89	116	-7	161	40	-115	-123	-29	-17
IRISH REPUBLIC	223	3,619	-560	6,372	-4,185	-2,223	-192	-81	-4,481	589
ITALY	508	-3,530	527	657	116	-341	1,883	..	216	..
LATVIA	144	-60	-23	-36	20	50	-10	-9	-11	..
LITHUANIA	..	..	4	4	-22	23	-5	3	-33	..
LUXEMBOURG	6,854	-2,760	6,945	11,717	3,267	-6,291	-76,928	-7,964	1,720	2,208
MALTA	..	-422	-509	-266	350	278	505	216	42	-937
NETHERLANDS	12,118	-15,664	949	-4,777	-4,549	8,966	-7,488	-6,807	-17,830	855
POLAND	-19	923	155	213	832	879	237	1,123	-129	133
PORTUGAL	384	721	604	-252	82	173	-67	-338	-446	59
ROMANIA	241	..	96	-28	133	187	15	82	-154	-144
SLOVAKIA	112	-34	..	95	16	-26	-12	-12	-97	..
SLOVENIA	12	9	..	127	-34	6	-3	-6	8	-55
SPAIN	16,191	2,961	2,182	-371	9	-1,100	-1,526	289	1,220	1,235
SWEDEN	735	1,116	-486	-290	11	997	-1,098	477	-303	1,766
EFTA	2,321	4,613	2,052	1,156	-2,594	3,284	2,045	2,344	-82	..
of which										
NORWAY	1,630	1,884	-311	2,536	-833	-704	134	1,023	1,001	..
SWITZERLAND	742	2,885	2,354	-1,379	-1,744	3,554	1,900	1,323	-1,080	-1,274
OTHER EUROPEAN COUNTRIES	7,637	25,445	2,194	10,614	4,089	-11,869	-26,670	-1,020	-4,404	..
of which										
RUSSIA	3,747	-34	-2,497	373	3,382	..	4,144	764	-408	663
UK OFFSHORE ISLANDS	2,961	24,591	3,222	10,775	493	-2,509	-27,050	-1,615	-3,889	988
THE AMERICAS	**20,391**	**-3,872**	**-19,648**	**45,305**	**11,258**	**40,992**	**8,396**	**-21,418**	**..**	**96,193**
of which										
BERMUDA	4,254	-2,014	-65	-879	-334	1,785	1,269	-362	-503	..
BRAZIL	932	390	1,687	2,283	321	-115	1,646	950	429	57
CANADA	-153	-2,698	-11,664	9,042	-3,638	655	333	..	-936	741
CHILE	-290	-	211	805	160	111	200	-532	166	132
COLOMBIA	169	188	..	-694	-486	115	-15	-120	-150	..
MEXICO	498	29	501	965	654	279	601	698	410	1,012
PANAMA	-2	59	29	58	10	59	-33	..	-54	..
USA	11,486	7,805	-13,691	34,813	12,478	37,199	2,474	-11,227	-4,270	88,465
ASIA	**12,738**	**7,835**	**11,466**	**-22,989**	**1,253**	**-6,498**	**8,666**	**-9,730**	**12,164**	**2,893**
NEAR & MIDDLE EAST COUNTRIES	3,309	1,637	-5,736	-1,123	9,675	-9,139	1,788	..	5,313	..
of which										
GULF ARABIAN COUNTRIES	928	1,559	-734	907	2,469	915	-1,450	..	4,815	1,959
OTHER ASIAN COUNTRIES	9,429	6,197	17,202	-21,866	-8,422	2,641	6,878	5,110	6,851	..
of which										
CHINA	1,116	179	569	1,135	1,702	352	-251	-807	1,221	-733
HONG KONG	1,115	-1,201	1,786	3,545	6,832	277	1,454	3,346	2,748	3,959
INDIA	784	547	2,567	8,977	-165	611	314	651	..	-189
INDONESIA	40	662	3,184	3,933	277	426	92	-164	767	119
JAPAN	980	1,031	-559	1,760	-729	516	-102	..	-97	356
MALAYSIA	312	280	629	800	550	164	1,541	65	..	..
SINGAPORE	2,417	1,572	4,882	-43,882	-18,893	-1,533	1,662	2,737	..	-4,608

6.13 Net foreign direct investment flows abroad analysed by area and main country, 2008 to 2017 (Directional)

£ million

	2008	2009	2010	2011	2012	2013	2014	2015	2016	2017
SOUTH KOREA	672	498	937	370	1,725	196	350	..	1,283	..
THAILAND	275	300	141	164	272	137	204	-32	563	108
AUSTRALASIA & OCEANIA	**9,548**	**-3,399**	**8,530**	**6,655**	**2,902**	**8,406**	**-2,570**	**92**	**..**	**-376**
of which										
AUSTRALIA	8,450	-4,517	8,240	6,012	2,585	8,234	-2,840	122	..	-547
NEW ZEALAND	127	1,140	237	555	257	142	126	-128	..	106
AFRICA	**318**	**1,388**	**6,339**	**-6,522**	**1,655**	**3,243**	**2,523**	**225**	**-6,994**	**-3,538**
of which										
KENYA	56	43	93	-51	-69	39	94	84	33	68
NIGERIA	242	873	220	888	1,545	-940	-213	..	530	358
SOUTH AFRICA	1,394	-1,961	2,038	799	1,610	1,536	883	893	800	926
ZIMBABWE	-6	1	10	25	17	3	-10	..	10	-7
WORLD TOTAL	**107,806**	**18,593**	**31,124**	**59,660**	**13,105**	**25,902**	**-90,751**	**-42,910**	**-27,696**	**91,416**
OECD	78,285	-4,300	8,088	81,558	4,832	35,927	-80,685	-36,365	-27,935	88,160
CENTRAL & EASTERN EUROPE	8	36	13	22	-16	66	41	-65	187	..

Source: Office for National Statistics

The sum of constituent items may not always agree exactly with the totals shown due to rounding.

A negative sign before values indicates a net disinvestment abroad.

.. Indicates data are disclosive

- Indicates nil data

6.14 FDI International investment positions abroad analysed by area and main country, 2008 to 2017 (Directional)

£ million

	2008	2009	2010	2011	2012	2013	2014	2015	2016	2017
EUROPE	670,112	586,151	645,252	643,536	586,061	551,832	525,230	534,045	673,729	704,455
EU	538,044	504,202	564,615	557,436	504,848	454,904	445,048	456,163	578,258	576,327
AUSTRIA	3,069	3,932	1,119	1,036	347	862	993	986	4,564	4,744
BELGIUM	12,769	15,807	43,284	40,991	16,699	12,537	11,032	15,818	13,609	13,224
BULGARIA	141	116	..	95	74	137	160	89	68	55
CROATIA	282	282	246	244	185	248	224	230	246	-77
CYPRUS	626	556	623	739	1,224	702	526	520	3,299	2,193
CZECH REPUBLIC	1,036	531	554	472	1,346	1,749	1,733	1,747	1,952	2,118
DENMARK	11,554	7,321	8,324	7,245	24,489	7,465	7,032	6,050	7,143	6,277
ESTONIA	28	41	70	97	63	109	113	164	182	623
FINLAND	610	417	1,265	1,659	193	721	1,054	735	879	855
FRANCE	56,017	49,796	63,767	58,951	70,853	25,477	57,030	60,488	71,579	78,837
GERMANY	25,711	28,029	23,625	22,222	15,526	20,217	20,698	17,738	25,214	23,767
GREECE	1,186	960	2,125	1,510	1,301	1,612	1,442	544	272	332
HUNGARY	2,328	728	673	825	1,038	673	969	702	1,095	803
IRISH REPUBLIC	24,777	35,008	41,200	45,737	44,223	27,660	31,896	42,839	48,641	61,756
ITALY	12,454	12,735	12,903	12,308	10,836	10,200	11,165	10,950	13,523	12,750
LATVIA	106	10	-17	66	71	143	101	73	123	..
LITHUANIA	..	28	..	45	45	49	60	52	47	31
LUXEMBOURG	134,048	133,212	149,180	151,827	121,881	104,572	108,192	93,837	115,841	112,065
MALTA	1,961	1,884	1,152	487	1,765	243	1,858	2,184	..	405
NETHERLANDS	136,218	131,505	132,179	132,464	128,725	111,882	121,884	128,707	175,433	153,294
POLAND	3,232	4,676	3,874	4,075	4,053	4,585	4,974	6,047	6,276	6,523
PORTUGAL	4,741	4,177	4,432	3,397	3,097	1,056	3,202	3,039	3,452	4,017
ROMANIA	721	578	565	567	591	820	847	884	1,013	..
SLOVAKIA	..	262	275	295	203	267	307	239	351	448
SLOVENIA	..	55	..	343	262	101	77	111	..	..
SPAIN	76,835	47,132	48,454	44,751	36,194	4,781	40,532	45,015	64,837	70,450
SWEDEN	27,345	24,428	24,163	24,988	19,562	20,257	16,947	16,374	16,489	20,077
EFTA	21,392	25,463	26,055	23,789	26,254	28,581	17,025	21,380	29,913	57,489
of which										
NORWAY	4,402	5,417	5,837	4,976	4,820	3,754	2,384	3,358	6,681	5,548
SWITZERLAND	16,723	19,564	20,189	18,787	21,415	19,676	14,608	18,020	23,211	51,910
OTHER EUROPEAN COUNTRIES	110,675	56,485	54,582	62,310	54,958	68,347	63,157	56,502	65,558	70,639
of which										
RUSSIA	11,328	9,973	9,698	8,108	5,705	13,668	13,798	8,266	11,834	9,696
UK OFFSHORE ISLANDS	88,937	36,220	34,530	42,255	38,440	42,019	39,889	38,363	43,952	53,589
THE AMERICAS	337,475	296,105	263,684	288,127	293,686	326,729	367,527	342,900	361,654	377,313
of which										
BERMUDA	25,552	16,943	17,333	17,017	15,391	15,923	26,435	14,877	20,305	14,630
BRAZIL	6,144	4,105	6,121	14,331	14,582	9,241	14,639	15,645	14,373	8,937
CANADA	31,764	29,607	36,461	31,760	30,925	28,074	28,517	17,206	26,899	27,025
CHILE	369	275	574	838	718	877	1,327	1,273	1,297	2,218
COLOMBIA	1,785	2,101	2,661	465	-117	673	602	580	783	465
MEXICO	1,439	607	980	2,544	5,093	898	6,581	7,348	7,531	8,250
PANAMA	..	196	196	318	..	..	..	..	..	..
USA	236,081	222,237	181,511	203,918	198,001	169,146	254,153	237,262	237,827	257,829
ASIA	71,948	76,317	100,563	111,614	107,822	125,021	121,398	138,216	153,010	161,063
NEAR & MIDDLE EAST COUNTRIES	13,745	17,524	22,733	25,977	13,428	28,705	13,415	16,437	24,360	35,623
of which										
GULF ARABIAN COUNTRIES	7,881	11,915	14,980	18,412	10,529	23,899	6,943	9,145	14,340	25,415
OTHER ASIAN COUNTRIES	58,203	58,794	77,829	85,637	94,394	96,316	107,983	121,779	128,650	125,439
of which										
CHINA	4,504	4,614	6,001	6,598	6,501	4,588	9,165	9,862	12,026	9,998
HONG KONG	19,029	16,750	19,434	20,572	42,640	15,197	52,142	55,047	61,411	66,080
INDIA	4,827	10,629	12,345	13,632	5,088	3,547	3,148	12,117	13,233	14,479
INDONESIA	1,093	1,897	2,712	5,976	4,089	3,331	4,303	4,575	5,813	4,641
JAPAN	1,552	2,936	2,697	6,003	6,308	3,776	4,106	4,497	4,322	4,847
MALAYSIA	1,380	879	1,433	1,874	1,910	3,984	3,266	3,213	3,422	3,249

6.14 FDI International investment positions abroad analysed by area and main country, 2008 to 2017 (Directional)

£ million

	2008	2009	2010	2011	2012	2013	2014	2015	2016	2017
SINGAPORE	14,639	8,850	9,791	10,658	9,060	10,939	16,409	18,698	10,698	5,857
SOUTH KOREA	3,376	3,175	3,857	4,192	4,468	2,528	4,652	3,450	6,322	7,194
THAILAND	1,653	1,174	1,308	1,395	1,496	1,151	2,545	1,758	2,382	2,920
AUSTRALASIA & OCEANIA	**19,891**	**19,218**	**34,036**	**39,716**	**43,490**	**50,632**	**21,814**	**29,951**	**42,141**	**36,280**
of which										
AUSTRALIA	17,654	16,191	33,267	38,532	42,388	48,495	20,369	28,248	41,054	35,744
NEW ZEALAND	780	1,624	601	758	664	677	834	1,073	867	840
AFRICA	**19,827**	**36,608**	**33,604**	**35,036**	**42,403**	**36,424**	**42,723**	**38,862**	**44,061**	**34,185**
of which										
KENYA	405	393	466	590	483	348	530	512	537	431
NIGERIA	1,233	2,096	1,854	2,148	1,743	1,169	1,483	4,122	..	..
SOUTH AFRICA	8,395	21,124	13,672	13,782	13,013	8,189	12,949	11,565	13,626	9,595
ZIMBABWE	39	20	5	12	32	14	83	..	109	62
WORLD TOTAL	**1,119,252**	**1,014,399**	**1,077,139**	**1,118,030**	**1,073,462**	**1,090,639**	**1,078,692**	**1,083,975**	**1,274,595**	**1,313,296**
OECD	855,854	809,588	854,859	876,229	822,719	815,767	785,585	780,870	936,136	981,473
CENTRAL & EASTERN EUROPE	7	46	7	25	-8	122	204	388	..	15

Source: Office for National Statistics

The sum of constituent items may not always agree exactly with the totals shown due to rounding.

A negative sign before values indicates a net disinvestment abroad.

.. Indicates data are disclosive

- Indicates nil data

6.15 Earnings from foreign direct investment abroad analysed by area and main country, 2008 to 2017 (Directional)

£ million

	2008	2009	2010	2011	2012	2013	2014	2015	2016	2017
EUROPE	**42,693**	**35,740**	**39,666**	**46,306**	**28,508**	**30,844**	**22,360**	**22,096**	**25,844**	**38,444**
EU	36,782	30,652	30,668	36,523	23,003	24,002	17,248	16,662	20,267	28,000
AUSTRIA	174	253	137	-14	-150	2	-8	12	208	245
BELGIUM	1,548	1,454	1,165	1,853	1,355	533	474	356	1,456	1,309
BULGARIA	-4	5	-	8	-	21	24	15	10	2
CROATIA	28	26	..	-13	-3	..	27	31	28	..
CYPRUS	331	56	-11	-86	133	30	-20	7	..	..
CZECH REPUBLIC	-123	-122	-75	2	210	173	191	133	145	16
DENMARK	543	219	627	541	477	404	360	264	608	601
ESTONIA	6	27	17	22	14	6	16	11	10	..
FINLAND	131	120	184	218	62	67	152	103	106	173
FRANCE	1,956	1,406	3,222	3,566	2,859	795	554	2,563	2,275	3,087
GERMANY	2,783	2,743	1,474	1,392	207	1,126	-9	2,424	2,395	3,491
GREECE	114	204	-45	23	31	40	71	2	19	40
HUNGARY	72	34	21	89	114	71	1	88	114	126
IRISH REPUBLIC	1,892	-800	-2,857	730	1,318	2,429	1,211	1,184	4,673	5,256
ITALY	524	214	687	181	355	436	677	729	562	498
LATVIA	-18	-27	-20	-24	33	28	-6	-8	-4	-7
LITHUANIA	-1	-	3	4	4	2	5	13	6	..
LUXEMBOURG	13,309	11,422	10,353	12,092	7,368	3,769	3,553	1,100	..	2,813
MALTA	42	324	-5	-51	69	..	526	..	150	-28
NETHERLANDS	10,363	7,915	11,820	11,796	4,739	8,462	5,985	3,934	2,171	5,695
POLAND	421	708	382	497	542	586	619	358	554	651
PORTUGAL	289	173	324	59	208	..	146	49	18	131
ROMANIA	108	93	12	61	76	126	103	128	187	88
SLOVAKIA	21	23	..	42	32	37	53	..	35	56
SLOVENIA	14	5	..	-32	-37	15	10	15	17	7
SPAIN	1,229	1,000	602	876	-366	518	988	878	1,420	1,997
SWEDEN	1,029	3,177	2,462	2,689	3,355	2,336	1,544	2,034	2,115	1,379
EFTA	4,392	2,739	5,251	5,090	3,410	4,219	1,755	2,459	2,663	4,126
of which										
NORWAY	602	580	538	472	-415	303	578	4	-173	616
SWITZERLAND	3,786	2,174	4,714	4,617	3,832	3,421	1,175	2,454	2,834	3,508
OTHER EUROPEAN COUNTRIES	1,520	2,348	3,746	4,693	2,094	2,623	3,358	2,975	2,914	6,318
of which										
RUSSIA	1,835	1,509	1,652	2,961	313	1,802	1,326	1,189	798	842
UK OFFSHORE ISLANDS	-86	165	1,620	1,120	1,257	477	1,670	1,422	2,039	4,868
THE AMERICAS	**13,573**	**15,510**	**21,212**	**26,439**	**23,625**	**24,233**	**22,785**	**16,783**	**8,524**	**22,910**
of which										
BERMUDA	1,549	1,183	2,339	1,455	2,159	2,612	2,447	..	706	391
BRAZIL	728	1,212	1,045	1,270	1,052	43	1,246	928	101	458
CANADA	-2,248	-485	471	-325	417	359	263	-462	1,914	1,570
CHILE	601	531	125	322	89	91	95	89	140	329
COLOMBIA	321	137	280	657	-441	36	13	-18	-7	7
MEXICO	293	255	544	1,362	760	169	724	798	728	942
PANAMA	..	66	54	84	61	76	72	178	..	..
USA	9,470	10,927	14,424	20,344	17,314	12,976	15,940	11,615	2,499	16,272
ASIA	**10,271**	**10,967**	**14,636**	**18,208**	**16,888**	**14,851**	**16,525**	**14,446**	**12,134**	**16,544**
NEAR & MIDDLE EAST COUNTRIES	3,661	1,830	2,023	2,645	782	3,172	2,531	1,633	1,305	3,620
of which										
GULF ARABIAN COUNTRIES	1,719	961	762	928	653	1,929	1,216	1,300	1,030	2,917
OTHER ASIAN COUNTRIES	6,611	9,137	12,613	15,563	16,107	11,680	13,994	12,813	10,829	12,924
of which										
CHINA	293	539	778	773	1,291	360	1,223	1,191	2,036	1,417
HONG KONG	1,410	797	2,736	4,179	5,023	1,056	4,845	5,101	3,248	5,539
INDIA	790	1,015	1,512	1,696	1,259	895	1,365	1,568	1,107	1,073
INDONESIA	178	400	560	1,065	1,047	661	772	416	658	648
JAPAN	431	364	299	391	242	517	385	308	976	442
MALAYSIA	526	674	734	536	562	435	988	1,011	514	895
SINGAPORE	959	2,423	2,278	3,158	3,039	-635	1,675	1,279	212	626
SOUTH KOREA	608	381	676	564	764	320	572	230	502	675
THAILAND	-122	164	173	201	244	194	287	203	218	280

6.15 Earnings from foreign direct investment abroad analysed by area and main country, 2008 to 2017 (Directional)

£ million

	2008	2009	2010	2011	2012	2013	2014	2015	2016	2017
AUSTRALASIA & OCEANIA	**4,213**	**4,602**	**6,555**	**1,953**	**3,979**	**3,642**	**3,001**	**2,579**	**2,137**	**4,584**
of which										
AUSTRALIA	3,947	4,181	6,133	1,671	3,682	3,378	2,712	2,249	1,803	3,649
NEW ZEALAND	194	343	331	176	236	208	162	202	214	276
AFRICA	**3,792**	**2,885**	**5,581**	**6,060**	**7,506**	**5,113**	**3,801**	**1,289**	**2,640**	**3,398**
of which										
KENYA	93	68	87	114	5	25	83	117	52	95
NIGERIA	212	64	652	1,031	1,261	311	687	..	176	250
SOUTH AFRICA	1,173	1,025	2,358	2,494	3,180	1,803	1,590	1,725	1,696	1,999
ZIMBABWE	-5	-6	3	22	23	..	12	26	11	17
WORLD TOTAL	**74,542**	**69,704**	**87,649**	**98,966**	**80,507**	**78,682**	**68,473**	**57,193**	**51,279**	**85,880**
OECD	54,204	49,921	59,032	66,564	49,880	52,156	39,401	33,864	31,447	56,308
CENTRAL & EASTERN EUROPE	-	2	-1	6	-3	3	-6	..	..	78

Source: Office for National Statistics

The sum of constituent items may not always agree exactly with the totals shown due to rounding.

A negative sign before values indicates a net disinvestment abroad.

.. Indicates data are disclosive

- Indicates nil data

6.16 Net foreign direct investment flows into the United Kingdom analysed by area and main country, 2008 to 2017 (Directional)

£ million

	2008	2009	2010	2011	2012	2013	2014	2015	2016	2017
EUROPE	**22,115**	**22,584**	**8,957**	**-28,258**	**22,830**	**9,456**	**-8**	**-12,305**	**135,463**	**36,203**
EU	21,267	15,181	-922	-23,682	15,381	-466	2,911	-13,807	129,768	24,291
AUSTRIA	75	89	170	876	-74	21	-87	32	32	-40
BELGIUM	-533	86	416	5,416	2,048	724	-1,947	-540	..	..
BULGARIA	..	..	..	..	-2	-	-1	-2	-	6
CROATIA	..	..	..	..	..	..	..	-	-	-
CYPRUS	73	27	..	-1,639	-192	265	61	388	162	178
CZECH REPUBLIC	1	-	1	-4	..	..	1	21	-	..
DENMARK	303	-432	-601	-822	506	-1,291	281	451	47	..
ESTONIA	..	-	..	..	..	..	..	-	-	-1
FINLAND	41	170	62	42	44	139	46	68	-18	..
FRANCE	-4,291	12,121	-9,065	2,127	6,499	-5,115	4,290	..	2,093	7,041
GERMANY	6,421	4,056	11,221	561	1,358	2,790	976	4,978	2,731	3,076
GREECE	5	30	-193	-387	178	-10	3	21	30	-22
HUNGARY	1	-	-34	5	..	-20	..	7	10	-
IRISH REPUBLIC	2,384	-629	3,408	216	634	98	35	681	-599	997
ITALY	-209	166	467	-1,372	1,082	775	226	228	333	..
LATVIA	..	-	-	-	..	..	..	-1	1	2
LITHUANIA	-	-	-	-	..	..	..	-	1	-
LUXEMBOURG	1,785	-6,548	-4,246	3,693	4,870	-1,040	1,107	-850	18,415	3,326
MALTA	20	14	-16	346	-49	19	5	..	..	-78
NETHERLANDS	9,866	-729	-3,274	-35,732	-1,594	1,251	-2,714	-22,289	..	4,192
POLAND	9	-1	1	23	-	11	12	19	6	-3
PORTUGAL	3	149	94	-81	63	-18	21	13	3	1
ROMANIA	1	-	-	-	..	-	-	-	2	2
SLOVAKIA	..	..	..	-1	-	..	..	21	2	..
SLOVENIA	1	1	-1	..	-	..	-1	-1	-	-
SPAIN	4,530	6,389	2,226	2,578	-84	978	630	2,288	1,955	1,467
SWEDEN	785	222	115	469	329	-25	211	601	-13	-162
EFTA	-4,930	3,209	5,194	-2,221	5,765	8,278	1,955	1,313	2,705	2,411
of which										
NORWAY	289	269	147	253	-384	483	477	..	..	254
SWITZERLAND	-5,258	2,985	4,176	-2,476	6,126	7,960	1,507	..	..	2,160
OTHER EUROPEAN COUNTRIES	5,778	4,194	4,686	-2,355	1,685	1,644	-4,875	188	2,989	9,500
of which										
RUSSIA	1,765	6	..	205	-40	-19	..	..	-39	3
UK OFFSHORE ISLANDS	4,115	4,003	4,322	-2,563	-691	1,662	-4,759	237	2,994	9,599
THE AMERICAS	**24,634**	**39,448**	**25,971**	**40,866**	**6,552**	**17,842**	**7,874**	**27,994**	**29,096**	**22,839**
of which										
BRAZIL	1	-	75	8	5	-4	-3	-2	25	-
CANADA	327	3	3,782	83	1,833	1,277	956	3,527	138	3,496
USA	24,073	37,575	17,131	40,118	4,237	16,528	9,159	21,113	26,780	15,026
ASIA	**3,530**	**-2,845**	**4,729**	**9,721**	**4,922**	**4,606**	**6,692**	**7,869**	**28,651**	**31,039**
NEAR & MIDDLE EAST COUNTRIES	-880	333	1,049	5,070	498	542	..	739	..	732
OTHER ASIAN COUNTRIES	4,410	-3,178	3,680	4,651	4,423	4,065	..	7,130	..	30,308
of which										
CHINA	-23	100	12	..	845	..	604	460	433	650
HONG KONG	902	-323	1,030	878	623	1,009	276	927	..	2,294
INDIA	2,428	108	-8	1,470	94	-124	85	..	-580	1,733
JAPAN	1,001	-2,433	1,608	1,106	1,666	2,240	1,552	1,718	2,127	27,063
SINGAPORE	250	47	-164	1,137	103	-28	2,118	3,754	4,125	-1,546
SOUTH KOREA	214	38	..	-92	146	-18	112	..	179	122
AUSTRALASIA & OCEANIA	**-1,137**	**-1,696**	**-2,083**	**3,175**	**810**	**934**	**36**	**..**	**-1,256**	**2,311**
of which										
AUSTRALIA	-1,102	-1,676	-2,047	3,333	804	..	247	..	-913	2,495
NEW ZEALAND	-19	-16	-34	-20	-5	..	26	19	43	-61
AFRICA	**990**	**95**	**92**	**835**	**-15**	**216**	**399**	**..**	**-2**	**-8**
of which										
SOUTH AFRICA	1,034	125	29	734	-60	216	370	353	-78	-50
WORLD TOTAL	**50,131**	**57,586**	**37,666**	**26,339**	**35,099**	**33,054**	**14,993**	**25,309**	**191,952**	**92,385**
OECD	40,745	51,967	28,179	20,016	30,093	28,461	16,904	13,534	160,222	74,950
CENTRAL & EASTERN EUROPE	..	..	..	..	..	..	..	-	-	-

The sum of constituent items may not always agree exactly with the totals shown due to rounding.
A negative sign before values indicates a net disinvestment in the UK.

Source: Office for National Statistics

.. Indicates data are disclosive
- Indicates nil data

6.17 FDI International investment positions in the United Kingdom analysed by area and main country, 2008 to 2017 (Directional)

£ million

	2008	2009	2010	2011	2012	2013	2014	2015	2016	2017
EUROPE	**363,898**	**387,753**	**398,821**	**431,298**	**540,839**	**557,987**	**600,935**	**544,394**	**685,123**	**744,059**
EU	317,375	344,344	347,469	366,400	452,276	462,342	490,245	440,832	551,766	573,149
AUSTRIA	1,105	931	1,076	2,419	5,235	2,565	2,165	1,810	1,536	1,015
BELGIUM	4,278	3,577	7,133	15,971	10,480	24,841	26,162	21,161	25,651	15,370
BULGARIA	..	..	..	..	..	7	7	5	42	46
CROATIA	..	..	..	..	..	..	..	-	-	-
CYPRUS	501	311	1,827	2,577	4,090	3,605	3,942	4,155	4,117	3,629
CZECH REPUBLIC	19	1	6	23	11	26	11	165	6	..
DENMARK	8,626	4,944	3,079	3,538	5,959	8,861	5,910	6,498	7,385	6,019
ESTONIA	..	-	..	..	..	..	..	-	1	8
FINLAND	727	752	741	838	2,178	1,474	1,401	1,132	600	584
FRANCE	45,752	69,101	61,212	54,725	73,521	66,776	75,797	69,600	66,960	57,603
GERMANY	68,928	61,992	46,752	45,312	62,123	49,290	51,033	50,496	61,018	63,579
GREECE	221	414	661	757	325	511	179	710	251	-108
HUNGARY	20	5	..	115	55	..	-8	144	..	..
IRISH REPUBLIC	8,613	9,248	7,565	10,619	9,008	11,911	13,825	12,309	11,399	16,345
ITALY	4,218	5,068	938	2,175	8,932	13,481	4,226	7,651	5,708	4,856
LATVIA	..	-	-	-	..	..	..	47	..	8
LITHUANIA	-	-	-	-	..	..	..	1	3	8
LUXEMBOURG	26,006	47,050	60,817	52,006	56,610	71,455	71,782	69,698	115,278	116,250
MALTA	143	48	606	271	913	969	514	633	644	1,018
NETHERLANDS	132,646	106,999	117,577	129,166	148,494	147,586	180,188	145,763	211,868	228,236
POLAND	78	18	24	88	76	209	105	290	127	177
PORTUGAL	301	558	761	308	479	515	486	544	..	..
ROMANIA	..	..	..	10	14	9	10	9	6	7
SLOVAKIA	..	..	..	-2	..	6	4	160	..	-1
SLOVENIA	14	..	6	-1	..	7	9	6	3	7
SPAIN	10,731	26,825	32,997	39,895	56,116	49,470	43,062	38,347	30,503	50,216
SWEDEN	4,320	6,382	3,489	5,578	7,593	8,479	9,402	9,497	7,935	7,644
EFTA	22,773	20,078	25,167	31,056	35,368	45,986	46,916	46,420	52,906	55,147
of which										
NORWAY	1,425	1,764	1,600	2,229	4,852	3,861	4,954	4,523	5,523	6,825
SWITZERLAND	19,882	17,354	22,665	25,931	29,630	39,289	39,136	39,065	43,928	45,110
OTHER EUROPEAN COUNTRIES	23,750	23,331	26,186	33,841	53,195	49,659	63,774	57,142	80,451	115,764
of which										
RUSSIA	..	..	817	979	881	1,647	..	..	1,123	899
UK OFFSHORE ISLANDS	21,748	22,579	25,037	32,206	44,824	46,562	61,435	55,394	78,741	114,413
THE AMERICAS	**209,659**	**197,481**	**227,155**	**247,993**	**299,735**	**279,612**	**323,596**	**388,570**	**389,374**	**450,613**
of which										
BRAZIL	10	-	249	-5	-10	121	15	143	266	..
CANADA	18,625	18,833	18,201	17,012	14,606	19,656	20,872	16,924	19,570	14,695
USA	170,369	159,900	185,458	205,925	255,169	217,950	242,070	309,632	293,727	350,963
ASIA	**42,448**	**35,958**	**46,551**	**56,722**	**62,689**	**74,581**	**73,987**	**82,070**	**96,349**	**127,766**
NEAR & MIDDLE EAST COUNTRIES	3,397	2,991	3,646	4,445	3,710	4,900	5,983	6,295	6,749	7,958
OTHER ASIAN COUNTRIES	39,051	32,967	42,905	52,277	58,979	69,681	68,004	75,774	89,600	119,809
of which										
CHINA	199	607	367	767	1,118	..	1,100	2,261	1,678	2,349
HONG KONG	6,979	5,884	7,527	16,972	10,253	11,622	12,799	10,888	..	20,437
INDIA	3,284	1,875	2,749	2,774	2,072	1,676	1,942	2,643	1,791	7,536
JAPAN	24,801	19,263	22,501	26,071	36,036	36,194	39,756	41,260	45,452	77,723
SINGAPORE	1,518	3,454	5,739	3,635	4,064	3,655	8,482	15,363	18,216	9,432
SOUTH KOREA	884	696	2,752	889	2,516	2,389	2,076	1,836	1,887	1,850
AUSTRALASIA & OCEANIA	**7,026**	**11,468**	**8,399**	**10,879**	**8,230**	**3,953**	**11,858**	15,064	**14,004**	**12,143**
of which										
AUSTRALIA	6,750	11,229	8,171	10,022	7,633	..	11,192	14,372	12,436	10,805
NEW ZEALAND	275	239	224	271	579	..	338	526	382	280
AFRICA	**1,994**	**1,009**	**1,403**	**1,769**	**1,368**	**2,345**	**2,887**	2,435	**2,453**	**1,921**
of which										
SOUTH AFRICA	1,558	490	553	819	353	1,274	2,060	1,538	1,536	939
WORLD TOTAL	**625,025**	**633,669**	**682,330**	**748,661**	**912,861**	**918,478**	**1,013,263**	**1,032,534**	**1,187,303**	**1,336,502**
OECD	561,816	574,328	607,930	653,463	799,380	781,800	847,247	865,073	970,940	1,079,122
CENTRAL & EASTERN EUROPE	..	..	..	..	..	..	..	-	-2	-2

The sum of constituent items may not always agree exactly with the totals shown due to rounding.
A negative sign before values indicates a net disinvestment in the UK.

Source: Office for National Statistics

.. Indicates data are disclosive
- Indicates nil data

6.18 Earnings from foreign direct investment in the United Kingdom
analysed by area and main country, 2008 to 2017 (Directional)

£ million

	2008	2009	2010	2011	2012	2013	2014	2015	2016	2017
EUROPE	**14,778**	**25,216**	**23,706**	**20,112**	**22,752**	**27,190**	**26,346**	**21,969**	**21,078**	**22,046**
EU	17,640	20,488	18,038	16,747	21,071	24,951	23,931	18,263	15,681	16,509
AUSTRIA	130	98	104	168	207	141	119	32	124	-28
BELGIUM	697	-169	427	911	961	1,964	1,469	1,547	2,119	1,421
BULGARIA	..	..	..	..	-	1	-	-1	2	11
CROATIA	..	..	..	..	..	..	..	-	-	-
CYPRUS	75	33	104	86	197	235	246	284	358	492
CZECH REPUBLIC	1	-	-	30	..	6	1	8	-	-2
DENMARK	-8	-256	89	-71	156	421	326	618	810	539
ESTONIA	..	-	..	..	..	..	..	-	-	-
FINLAND	92	94	93	77	198	146	149	76	2	54
FRANCE	3,812	10,506	4,135	4,437	2,170	2,326	2,960	2,648	174	1,090
GERMANY	5,905	2,045	4,373	3,582	5,601	6,441	5,615	4,281	5,313	3,871
GREECE	304	190	328	71	122	122	68	53	61	60
HUNGARY	-	..	1	2	1	15	2	12	13	1
IRISH REPUBLIC	2,004	264	-1,089	382	486	493	506	975	335	455
ITALY	596	513	264	151	1,145	1,223	575	273	345	94
LATVIA	..	-	-	-	..	-	..	3	2	3
LITHUANIA	-	-	-	-	..	..	..	-	1	-
LUXEMBOURG	496	882	1,528	754	2,077	1,853	1,146	1,177	96	1,810
MALTA	11	1	31	18	91	71	35	2	17	109
NETHERLANDS	-1,334	652	2,872	3,679	4,337	6,062	6,634	3,375	2,892	3,058
POLAND	4	1	2	4	3	6	8	10	7	5
PORTUGAL	101	71	58	-66	67	38	152	122	105	51
ROMANIA	..	..	..	1	-1	1	1	1	-	-
SLOVAKIA	..	..	-	..	-	..	..	7	-	-
SLOVENIA	..	..	-	1	-1	1	-	-	-	-
SPAIN	4,272	5,281	4,245	2,009	2,677	2,689	3,182	2,106	2,280	2,534
SWEDEN	478	284	471	519	573	698	736	654	626	882
EFTA	-4,587	2,823	1,780	1,380	-410	353	983	1,949	2,739	2,744
of which										
NORWAY	265	12	18	194	398	686	255	205	73	-17
SWITZERLAND	-4,901	2,771	1,004	804	-795	-499	422	1,413	2,066	2,645
OTHER EUROPEAN COUNTRIES	1,726	1,905	3,889	1,984	2,091	1,886	1,432	1,757	2,658	2,793
of which										
RUSSIA	31	3	59	125	-2	129	-44	..	-39	-1
UK OFFSHORE ISLANDS	1,769	1,852	3,809	1,855	1,875	1,661	1,438	1,787	2,651	2,802
THE AMERICAS	**32,264**	**21,404**	**16,914**	**20,972**	**19,829**	**20,844**	**16,947**	**21,508**	**23,241**	**26,204**
of which										
BRAZIL	-	-	28	16	12	17	5	7	21	..
CANADA	1,427	1,012	886	420	779	1,965	1,258	310	1,357	..
USA	29,498	19,135	14,648	18,894	17,889	17,608	12,659	18,236	17,035	21,996
ASIA	**471**	**-2,704**	**-742**	**4,051**	**3,198**	**2,486**	**4,781**	**4,855**	**5,792**	**6,959**
NEAR & MIDDLE EAST COUNTRIES	329	149	181	837	440	649	342	863	189	273
OTHER ASIAN COUNTRIES	142	-2,854	-923	3,213	2,758	1,836	4,439	3,992	5,603	6,685
of which										
CHINA	-20	38	21	34	106	202	207	42	32	-246
HONG KONG	-358	-141	488	-4	589	1,179	949	1,590	1,610	..
INDIA	280	111	167	1,251	117	94	167	148	415	1,498
JAPAN	-375	-3,126	-2,141	-147	1,441	640	1,733	1,604	3,290	3,411
SINGAPORE	161	-215	-101	1,508	230	-218	1,406	485	409	470
SOUTH KOREA	117	54	111	60	99	136	-96	30	105	162
AUSTRALASIA & OCEANIA	**683**	**1,913**	**-134**	**145**	**-99**	**272**	**-14**	**-259**	**132**	**-234**
of which										
AUSTRALIA	697	1,922	-151	159	-128	228	266	-284	344	..
NEW ZEALAND	-11	-5	17	5	20	181	33	22	32	..
AFRICA	**226**	**121**	**73**	**199**	**-37**	**87**	**-49**	**85**	**45**	**-188**
of which										
SOUTH AFRICA	171	119	20	110	-73	20	-108	52	-31	-250
WORLD TOTAL	**48,423**	**45,949**	**39,817**	**45,478**	**45,644**	**50,878**	**48,011**	**48,158**	**50,288**	**54,787**
OECD	44,372	42,328	33,089	37,143	40,498	45,693	40,280	39,552	39,695	44,815
CENTRAL & EASTERN EUROPE	..	..	..	..	..	..	..	-	-	-

The sum of constituent items may not always agree exactly with the totals shown due to rounding.
A negative sign before values indicates a net disinvestment in the UK.

Source: Office for National Statistics

.. Indicates data are disclosive
- Indicates nil data

this page is intentionally blank

Research and development

Chapter 7

Research and development

(Tables 7.1 to 7.5)

Research and experimental development (R&D) is defined for statistical purposes as 'creative work undertaken on a systematic basis in order to increase the stock of knowledge, including knowledge of man, culture and society, and the use of this stock of knowledge to devise new applications'.

R&D is financed and carried out mainly by businesses, the Government, and institutions of higher education. A small amount is performed by non-profit-making bodies. Gross Expenditure on R&D (GERD) is an indicator of the total amount of R&D performed within the UK. Detailed figures are reported each year in a statistical bulletin published in July. Table 7.1 shows the main components of GERD.

ONS conducts an annual survey of expenditure and employment on R&D performed by government, and of government funding of R&D. The survey collects data for the reference period along with future estimates. Until 1993 the detailed results were reported in the *Annual Review of Government Funded R&D*. From 1997 to 2012, the results have appeared in the Science, Engineering and Technology (SET) Statistics published by the Department for Business, Innovation and Skills (BIS). From 2013, the results have appeared in the Government Expenditure on Science, Engineering and Technology Statistics published by ONS and the Department for Business, Energy and Industrial Strategy (BEIS). Table 7.2 gives some broad totals for gross expenditure by government (expenditure before deducting funds received by government for R&D). Table 7.3 gives a breakdown of net expenditure (receipts are deducted).

The ONS conducts an annual survey of R&D in business. Tables 7.4 and 7.5 give a summary of the main trends up to 2017.

Statistics on expenditure and employment on R&D in higher education institutions (HEIs) are based on information collected by Higher Education Funding Councils and the Higher Education Statistics Agency (HESA). In 1994 a new methodology was introduced to estimate expenditure on R&D in HEIs. This is based on the allocation of various Funding Council Grants. Full details of the new methodology are contained in science, engineering and technology (SET) Statistics available on the gov.uk website at: https://www.gov.uk/government/collections/science-engineering-and-technology-statistics

The most comprehensive international comparisons of resources devoted to R&D appear in Main Science and Technology Indicators published by the Organisation for Economic Co-operation and Development (OECD). The Statistical Office of the European Union and the United Nations also compile R&D statistics based on figures supplied by member states.

To make international comparisons more reliable the OECD have published a series of manuals giving guidance on how to measure various components of R&D inputs and outputs. The most important of these is the Frascati Manual, which defines R&D and recommends how resources for R&D should be measured. The UK follows the Frascati Manual as far as possible.

For information on available aggregated data on Research and Development please call Office for National Statistics on 01633 456767.

7.1a EXPENDITURE ON R&D IN THE UK BY PERFORMING AND FUNDING SECTORS, 2016

Current prices	Sector performing the R&D						£ million
	Government	Research Councils	Higher Education	Business Enterprise	Private Non-Profit[1]	Total	Overseas
Sector funding the R&D							
Government	1136	137	483	1,730	98	**3,584**	542
Research Councils	47	554	2,107	5	197	**2,909**	292
Higher Education Funding Councils	-	-	2,207	-	-	**2,207**	-
Higher Education	2	17	299	-	131	**449**	-
Business Enterprise	15	25	350	16,742	18	**17,151**	6,658
Private Non-Profit	13	42	1,242	188	170	**1,655**	-
Overseas	122	60	1,346	3,560	85	**5,174**	-
TOTAL	**1,335**	**837**	**8,035**	**22,224**	**699**	**33,130**	-
of which:							
Civil	1,178	837	7,994	20,658	688	**31,354**	-
Defence	156	-	42	1,567	11	**1,776**	-

Source: Office for National Statistics

1. Prior to 2011 PNP data was estimated. From 2011 data has been collected from a biennial survey with intermediate years being estimated using previous years data.

- denotes nil, figures unavailable or too small to display.

Please note:

Differences may occur between totals and the sum of their independently rounded components.

7.1b EXPENDITURE ON R&D IN THE UK BY SECTOR OF PERFORMANCE:
2007 to 2016

£ million

		2007	2008	2009	2010	2011	2012	2013	2014	2015	2016
Sector performing the R&D											
Current prices											
TOTAL[3]	**GLBA**	**24,696**	**25,345**	**25,632**	**26,173**	**27,452**	**27,257**	**29,015**	**30,577** [†]	**31,766**	**33,130**
Government[3]	GLBK	1,320	1,348	1,406	1,372	1,321	1,391	1,503	1,391	1,321 [†]	1,335
Research Councils	DMRS	1,034	1,041	1,097	1,141	1,035	804	814	819	771 [†]	837
Business Enterprise	GLBL	15,676	15,814	15,532	16,045	17,452	17,409	18,617	19,982 [†]	21,038	22,224
Higher Education	GLBM	6,119	6,545	6,931	6,963	7,117	7,133	7,593	7,835	8,003 [†]	8,035
Private Non-Profit[2]	GLBN	546	595	666	652	526	520	489	549	634	699
As % of GDP		1.59	1.63	1.66	1.64	1.67	1.60	1.64	1.65	1.67	1.67
		2007	2008	2009	2010	2011	2012	2013	2014	2015	2016
Sector performing the R&D											
Constant prices (2015)[1]											
TOTAL[3]	**GLBD**	**28,791**	**28,799**	**28,707**	**28,787**	**29765**	**28,952**	**30,302**	**31,477**	**32,483**	**33,130**
Government[3]	GLBW	1,539	1,532	1,575	1,509	1,432	1,478	1,570	1,432	1,351	1,335
Research Councils	DMSU	1,205	1,183	1,229	1,255	1,122	854	850	843	788	837
Business Enterprise	GLBX	18,275	17,969	17,396	17,648	18,923	18,492	19,443	20,570	21,513	22,224
Higher Education	GLBY	7,134	7,437	7,763	7,659	7,717	7,577	7,930	8,066	8,184	8,035
Private Non-Profit[2]	GLBZ	637	676	746	717	570	552	511	565	648	699

Source: Office for National Statistics

	2007	2008	2009	2010	2011	2012	2013	2014	2015	2016
GDP deflator used to convert current prices to constant prices	85.777	88.008	89.287	90.918	92.228	94.144	95.752	97.140	97.794	100

1 Please note that the latest deflators have been applied to the business research and development estimates in this bulletin which has resulted in small differences being observed between the BERD and GERD publications.

2 Prior to 2011 PNP data was estimated. From 2011 data has been collected from a biennial survey with intermediate years being estimated using previous years data.

3 Estimates of launch investment loan repayments received by government from business have been removed following a review of how these payments should be reported. These loan repayments are in relation to loans given out in previous years and therefore should not be included in current totals of R&D expenditure. The total of loan repayments have been removed from the total performed by government and the UK total for 2013, and 2015, there were no repayments in 2014. In current prices the values removed were 2013 (£212 Million) and 2015 (£112 million).

[†] crosses denote earliest data revision.

Please Note:

Differences may occur between totals and the sum of their independently rounded components.

7.2 EXPENDITURE ON R&D IN THE UK BY PERFORMING AND FUNDING SECTORS, 2017

Current prices		Sector performing the R&D					£ million
	Government	Research Councils	Higher Education[2]	Business Enterprise	Private[1] Non-Profit	**Total**	Overseas
Sector funding the R&D							
Government	1119	101	590	1,793	102	**3,705**	606
Research Councils	54	627	2,246	4	174	**3,106**	332
Higher Education Funding Councils	-	-	2,236	-	-	**2,236**	-
Higher Education	2	16	-	210	13	**241**	-
Business Enterprise	12	23	358	18,285	23	**18,700**	7,990
Private Non-Profit	15	41	1,288	93	359	**1,796**	-
Overseas	128	59	1,455	3,299	84	**5,024**	-
TOTAL	**1,330**	**866**	**8,173**	**23,685**	**754**	**34,808**	-
of which:							
Civil	1,178	866	8,130	22,080	745	**32,999**	-
Defence	152	-	43	1,604	10	**1,809**	-

Source: Office for National Statistics

1 Prior to 2011 PNP data were estimated. From 2011 data has been collected from a biennial survey with non-survey years being estimated using data from survey years.
2 Following further quality assurance of the flow of funding within the higher education sector it was decided to remove the element relating to funding between higher education establishments. The earliest point of revision is 1992.

- denotes nil, figures unavailable or too small to display.

Please note:

Differences may occur between totals and the sum of their independently rounded components.

7.3 UK GOVERNMENT EXPENDITURE ON R&D BY SOCIO-ECONOMIC OBJECTIVE[1], PERCENTAGE SHARE: 2008 TO 2017

Current prices
£ million

	2008	2009	2010	2011	2012	2013	2014	2015	2016	2017
TOTAL	9,235	9,590	9,452	9,174	9,202	10,048	10,253	10,155	10,334 [†]	10,975

	2008	2009	2010	2011	2012	2013	2014	2015	2016	2017
Per cent										
TOTAL	100	100	100	100	100	100	100	100	100	100
General advancement of knowledge: R&D financed from General University Funds	18	19	20	21	21	22	22	23	21	22
Health	24	25	24	25	24	23	22	22	21	21
Defence	22	18	18	14	16	15	17	16	16	15
General advancement of knowledge: R&D financed from other sources	18	16	18	13	13	13	12	12	11	11
Industrial production and technology	1	1	1	3	3	3	4	4	4	7
Transport, telecommunication, other infrastructure	2	4	3	4	3	4	4	4	5	5
Exploration and exploitation of the earth	3	3	3	3	3	3	4	4	4	5
Exploration and exploitation of space	3	4	3	4	4	4	3	4	4	4
Agriculture	1	1	1	2	2	2	2	3	3	3
Energy	2	2	2	1	3	3	3	3	3	3
Political and social systems, structures and processes	3	3	3	3	3	3	2	2	2	2
Environment	2	2	2	3	3	4	3	3	4	1
Culture, recreation, religion and mass media	2	2	2	2	1	1	1	1	1	1
Education	1	1	1	-	-	-	-	-	-	-

Source: Office for National Statistics

- denotes nil, figures unavailable or too small to display.

[†] denotes earliest data revision.

A socio-economic objective is the classification relating to the purpose of the R&D programme or project, i.e. its primary objective.

R&D and related concepts follow internationally agreed standards as published in the Frascati Manual (https://www.oecd.org/sti/inno/frascati-manual.htm).

Please Note:

Differences may occur between totals and the sum of their independently rounded components.

7.4 EXPENDITURE ON R&D PERFORMED IN UK BUSINESSES: BROAD PRODUCT GROUPS, 2006 TO 2017

CURRENT PRICES

£ million

		2006	2007	2008	2009	2010	2011	2012	2013	2014	2015	2016	2017
TOTAL	DLBX	14,144	15,676	15,814	15,532	16,045	17,452	17,409	18,617	19,982	21,018 [†]	22,587	23,685
Manufacturing: Total	DLDF	10,555	11,612	11,672	11,230	11,585	12,462	12,383	13,158	13,721	14,607	15,468 [†]	15,640
Chemicals	DLDG	4,205	4,602	4,985	5,034	5,339	5,437	4,873	4,793	4,639	5,003	5,108 [†]	5,206
Mechanical engineering	DLDH	997	1,124	864	..	902	1,091	1,102	1,156	1,121	1,089	1,032 [†]	1,243
Electrical machinery	DLDI	1,273	1,297	1,400	1,275	1,144	1,200	1,339	1,535	1,649	1,799	1,617 [†]	1,712
Transport	DLDJ	913	1,097	1,344	..	1,468	1,800	2,040	2,398	2,785	3,201	3,845 [†]	4,049
Aerospace	DLDK	1,832	2,070	1,732	1,466	1,437	1,438	1,511	1,639	1,666	1,699	1,904 [†]	1,521
Other manufacturing	DLDL	1,336	1,422	1,348	1,251	1,295	1,495	1,518	1,638	1,860	1,816	1,962 [†]	1,909
Services	DLDM	3,404	3,860	3,904	3,952	4,120	4,563	4,539	4,898	5,577	5,714 [†]	6,436	7,241
Other: Total	LABA	185	205	238	350	340	426	487	561	684	697	683 [†]	804
Agriculture, hunting & forestry; Fishing	LADE	88	..	88	..	102	133	132	121	135	139	132	144
Extractive industries	LADM	59	82	90	140	152	195	172	209	230	206	177 [†]	162
Electricity, gas & water supply; Waste management	LAEB	21	35	40	75	72	68	118	140	168	196	160 [†]	179
Construction	LAEM	17	..	21	..	14	31	64	91	151	156	214 [†]	319

CONSTANT PRICES (2017)

£ million

	2006 [†]	2007	2008	2009	2010	2011	2012	2013	2014	2015	2016	2017
TOTAL	17,213	18,616	18,284	17,707	17,958	19,279	18,852	19,797	20,979	21,892	23,016	23,685
Manufacturing: Total	12,845	13,790	13,495	12,803	12,966	13,767	13,409	13,992	14,406	15,214	15,762	15,640
Chemicals	5,118	5,465	5,764	5,739	5,976	6,006	5,277	5,097	4,870	5,211	5,205	5,206
Mechanical engineering	1,213	1,335	999	..	1,010	1,205	1,193	1,229	1,177	1,134	1,052	1,243
Electrical machinery	1,549	1,540	1,619	1,454	1,280	1,326	1,450	1,632	1,731	1,874	1,648	1,712
Transport	1,111	1,303	1,554	..	1,643	1,988	2,209	2,550	2,924	3,334	3,918	4,049
Aerospace	2,230	2,458	2,003	1,671	1,608	1,589	1,636	1,743	1,749	1,770	1,940	1,521
Other manufacturing	1,626	1,689	1,559	1,426	1,449	1,652	1,644	1,742	1,953	1,891	1,999	1,909
Services	4,143	4,584	4,514	4,505	4,611	5,041	4,915	5,208	5,855	5,952	6,558	7,241
Other: Total	225	243	275	399	381	471	527	597	718	726	696	804
Agriculture, hunting & forestry; Fishing	107	..	102	..	114	147	143	129	142	145	135	144
Extractive industries	72	97	104	160	170	215	186	222	241	215	180	162
Electricity, gas & water supply; Waste management	26	42	46	86	81	75	128	149	176	204	163	179
Construction	21	..	24	..	16	34	69	97	159	162	218	319

Source: Office for National Statistics

	2006	2007	2008	2009	2010	2011	2012	2013	2014	2015	2016	2017
GDP deflator used to convert current pri	82.169 [†]	84.205	86.491	87.716	89.348	90.522	92.345	94.039	95.247	96.009	98.137	100

.. denotes disclosive figures.

[†] denotes earliest data revision.

Differences may occur between totals and the sum of their independently rounded components.
From 2016 onwards civil and defence figures are no longer separated due to limited data availability following changes in disclosure.

7.5a SOURCES OF FUNDS FOR R&D PERFORMED IN UK BUSINESSES: 2014 TO 2017

CURRENT PRICES

		2014	2015	2016	2017
£ million					
TOTAL	**DLBX**	**19,982**	**21,018** [†]	**22,587**	**23,685**
UK Government	DLDO	1,886	1,817 [†]	1,700	1,798
Overseas total of which:	DLHK	4,061	3,801 [†]	3,531	3,271
European Commission grants	DLDQ	64	67	93 [†]	46
Other Overseas	DLDS	3,996	3,734 [†]	3,438	3,224
Other UK Business	DLDU	391	384	507 [†]	564
Own funds	DLDW	13,515	14,805 [†]	16,605	17,721
Other [1]	DLDY	129	211 [†]	244	332

	2014	2015	2016	2017
Per cent				
TOTAL	**100**	**100**	**100**	**100**
UK Government	9	9	8	8
Overseas total of which:	20	18	16	14
European Commission grants	-	-	-	-
Other Overseas	20	18	15	14
Other UK Business	2	2	2	2
Own funds	68	70	74 [†]	75
Other [1]	1	1	1	1

Source: Office for National Statistics

- denotes nil, figures unavailable or too small to display.
[†] denotes earliest data revision.
Differences may occur between totals and the sum of their independently rounded components.

1 'Other' includes funds from UK Private Non-Profit organisations and Higher Education establishments, and from 2011, international organisations.

7.5b SOURCES OF FUNDS FOR R&D PERFORMED IN UK BUSINESSES: CIVIL AND DEFENCE, 2014 TO 2017

CURRENT PRICES £ million

		Civil					Defence			
		2014	2015	2016	2017		2014	2015	2016	2017
UK Government	DLFG	838	821 [†]	715	775	DLFN	1,047	997	985 [†]	1,023
Overseas total of which:	DLHS	3,914	3,651 [†]	3,396	3,143	DLIF	147	151	135	128
European Commission grants	DLFH	63	66	91	46	DLFO	1	1	1	-
Other Overseas	DLFI	3,850	3,585 [†]	3,304	3,097	DLFP	146	149	133	127
Other UK Business	DLFJ	339	328	428	453	DLFQ	53	56	80	110
Own	DLFK	13,225	14,524 [†]	16,245	17,384	DLFR	290	280	359 [†]	337
Other [1]	DLFL	118	203 [†]	237	326	DLFS	11	8	7	6
TOTAL	DLBV	18,435	19,527 [†]	21,021	22,080	DLBW	1,548	1,491	1,566 [†]	1,604

Source: Office for National Statistics

- denotes nil, figures unavailable or too small to display.
.. denotes disclosive figures.
[†] denotes earliest data revision.
Differences may occur between totals and the sum of their independently rounded components.

1 'Other' includes funds from UK Private Non-Profit organisations and Higher Education establishments, and from 2011, international organisations.

231

this page is intentionally blank

Income and wealth

Chapter 8

Personal income, expenditure and wealth

Distribution of total incomes (Table 8.1)

The information shown in Table 8.1 comes from the Survey of Personal Income. This is an annual survey covering approximately 600,000 individuals across the whole of the UK. It is based on administrative data held by HM Revenue & Customs (HMRC) on individuals who could be liable for tax.

The table relates only to those individuals who are taxpayers. The distributions only cover incomes as computed for tax purposes, and above a level which for each year corresponds approximately to the single person's allowance. Incomes below these levels are not shown because the   information about them is incomplete.

Some components of investment income (for example interest and dividends), from which tax has been deducted at source, are not always held on HMRC business systems. Estimates of missing bank and building society interest and dividends from UK companies are included in these tables. The missing investment income is distributed to cases, so that the population as a whole has amounts consistent with evidence from other sources. For example, amounts of tax accounted for by deposit takers and the tendency to hold interest-bearing accounts as indicated by household surveys.

Superannuation contributions are estimated and included in total income. They have been distributed among earners in the Survey of Personal Incomes sample, by a method consistent with information about the number of employees who are contracted in or out of the State Earnings Related Pension Scheme (SERPS) and the proportion of their earnings contributed.

When comparing results of these surveys across years, it should be noted that the Survey of Personal Incomes is not a longitudinal survey. However, sample sizes have increased in recent years to increase precision.

Effects of taxes and benefits, by household type (Table 8.2)

Original income is the total income in cash of all the members of the household before receipt of state benefits or the deduction of taxes. It includes income from employment, self-employment, investment income and occupational pensions. Gross income is original income plus cash benefits received from government (retirement pensions, child benefit, and so on). Disposal income is the income available for consumption; it is equal to gross income less direct taxes (which include income tax, national insurance contributions, and council tax). By further allowing for taxes paid on goods and services purchased, such as VAT, an estimate of post-tax income is derived. These income figures are derived from estimates made by the Office for National Statistics (ONS) based largely on information from the Living Costs and Food Survey (LFC) and published each year on the ONS website.

In Table 8.2, a retired household is defined as one where the combined income of retired members amounts to at least half the total gross income of the household; where a retired person is defined as anyone who describes themselves as retired, or anyone over the minimum National Insurance (NI) pension age describing themselves as 'unoccupied' or 'sick or injured but not intending to seek work.

Children are defined as people aged under 16 or aged between 16 and 19, not married nor in a Civil Partnership, nor living with a partner; and living with parent(s)/guardian(s); and and receiving non-advanced further education or in unwaged-government training

Living Costs and Food Survey (Tables 8.3–8.5)

The Living Costs and Food Survey (LCF) is a sample survey of private households in the UK. The LCF sample is representative of all regions of the UK and of different types of households. The survey is continuous, with interviews spread evenly over the year to ensure that estimates are not biased by seasonal variation. The survey results show how households spend their money; how much goes on food, clothing and so on, how spending patterns vary depending upon income, household composition, and regional location of households. In previous releases, Family Spending has reported on a calendar year basis. However, data used in these tables and subsequent Family Spending releases will report on a financial year basis (April to March). This change has been made to reduce disparities in estimates. The methodology in sampling and collecting Living Costs and Food Survey data has remained consistent to ensure that data in this release remain comparable with that of previous Family Spending releases.

One of the main purposes of the LCF is to define the 'basket of goods' for the Retail Prices Index (RPI) and the Consumer Prices Index (CPI). The RPI has a vital role in the up rating of state pensions and welfare benefits, while the CPI is a key instrument of the Government's monetary policy. Information from the survey is also a major source for estimates of household expenditure in the UK National Accounts. In addition, many other government departments use LCF data as a basis for policy making, for example in the areas of housing and transport. The Department for Environment, Food and Rural Affairs (Defra) uses LCF data to report on trends in food consumption and nutrient intake within the UK. Users of the LCF outside government include independent research institutes, academic researchers and business and market researchers. Like all surveys based on a sample of the population, its results are subject to sampling variability and potentially to some bias due to non-response. The results of the survey are published in an annual report called Family spending in the UK. The report includes a list of definitions used in the survey, items on which information is collected, and a brief account of the fieldwork procedure.

8.1 Distribution of total income before and after tax by gender, 2015-16

Taxpayers only

Numbers: thousands; Amounts: £ million

Range of total income (lower limit) £	Total							
	Before tax, by range of total income before tax				After tax, by range of total income after tax			
	No. of taxpayers	Total income before tax	Total tax	Total income after tax	No. of taxpayers	Total income before tax	Total tax	Total income after tax
10,600 [a]	1,960	22,000	248	21,800	2,400	27,400	377	27,000
12,000	4,310	58,100	2,230	55,900	5,230	73,900	3,380	70,500
15,000	6,160	107,000	7,830	99,300	6,980	132,000	11,000	121,000
20,000	8,050	197,000	21,200	176,000	8,170	225,000	26,200	199,000
30,000	6,990	266,000	35,800	230,000	6,080	270,000	41,300	228,000
50,000	1,780	103,000	20,100	83,100	1,180	89,700	21,200	68,500
70,000	919	76,100	18,600	57,500	577	65,600	19,100	46,500
100,000	467	55,700	16,600	39,000	228	40,800	13,600	27,100
150,000	162	27,700	9,240	18,400	69	18,400	6,580	11,800
200,000	111	26,700	9,510	17,100	53	20,400	7,580	12,800
300,000	65	24,500	9,210	15,300	32	19,700	7,560	12,100
500,000	36	24,900	9,630	15,300	18	19,900	7,730	12,200
1,000,000	19	46,600	18,100	28,500	8	33,200	12,700	20,500
All ranges	31,000	1,040,000	178,000	858,000	31,000	1,040,000	178,000	858,000

Range of total income (lower limit) £	Male							
	Before tax, by range of total income before tax				After tax, by range of total income after tax			
	No. of taxpayers	Total income before tax	Total tax	Total income after tax	No. of taxpayers	Total income before tax	Total tax	Total income after tax
10,600 [a]	851	9,560	115	9,450	1,060	12,000	176	11,900
12,000	1,970	26,600	1,040	25,500	2,430	34,400	1,610	32,800
15,000	3,150	54,800	4,020	50,800	3,710	70,500	5,890	64,600
20,000	4,710	116,000	12,500	103,000	4,930	137,000	16,100	121,000
30,000	4,530	173,000	23,500	150,000	4,070	182,000	28,300	153,000
50,000	1,270	73,500	14,500	58,900	862	65,800	15,700	50,100
70,000	680	56,400	13,900	42,400	436	49,800	14,700	35,200
100,000	358	42,700	12,800	29,900	183	32,800	11,000	21,800
150,000	131	22,400	7,490	14,900	56	15,100	5,400	9,670
200,000	90	21,700	7,780	13,900	44	17,000	6,360	10,600
300,000	54	20,500	7,710	12,800	27	16,900	6,520	10,400
500,000	31	21,500	8,330	13,200	16	17,300	6,730	10,500
1,000,000	17	41,500	16,300	25,300	8	29,900	11,600	18,300
All ranges	17,800	680,000	130,000	550,000	17,800	680,000	130,000	550,000

8.1 Distribution of total income before and after tax by gender, 2015-16

Taxpayers only

Numbers: thousands; Amounts: £ million

Range of total income (lower limit) £	Female							
	Before tax, by range of total income before tax				After tax, by range of total income after tax			
	No. of taxpayers	Total income before tax	Total tax	Total income after tax	No. of taxpayers	Total income before tax	Total tax	Total income after tax
10,600 (a)	1,110	12,400	133	12,300	1,350	15,400	200	15,200
12,000	2,340	31,600	1,190	30,400	2,800	39,500	1,770	37,700
15,000	3,010	52,300	3,810	48,500	3,270	61,600	5,090	56,500
20,000	3,340	81,400	8,640	72,800	3,230	88,600	10,200	78,400
30,000	2,460	92,800	12,300	80,600	2,010	87,800	13,000	74,800
50,000	516	29,800	5,640	24,200	319	23,900	5,440	18,400
70,000	239	19,800	4,710	15,000	141	15,800	4,450	11,400
100,000	109	12,900	3,780	9,160	45	7,930	2,590	5,340
150,000	31	5,300	1,750	3,550	13	3,320	1,180	2,140
200,000	21	4,930	1,730	3,190	9	3,370	1,220	2,150
300,000	11	4,050	1,500	2,550	5	2,780	1,040	1,740
500,000	5	3,410	1,300	2,110	2	2,620	999	1,620
1,000,000	2	5,010	1,840	3,180	1	3,320	1,170	2,150
All ranges	13,200	356,000	48,300	307,000	13,200	356,000	48,300	307,000

Source: Survey of Personal Incomes 2015-16

Footnote

(a) Can include some taxpayers who are not entitled to a Personal Allowance whose total income can be less than the Personal Allowance of £10,600 for 2015-16 (see Annex B for details).

Notes on the Table
Distribution of total income before and after tax by gender, 2015-16

1. This table only covers individuals with some liability to tax.

2. It should be noted that individuals may not necessarily fall into the same total income range for before and after tax breakdowns. Total income before tax is used to assign people to an income range for columns 2 to 5, whereas total income after the deduction of tax is used to assign individuals to an income band for columns 6 to 9.

3. For more information about the SPI and symbols used in this table, please refer to
Personal Income Statistics release: https://www.gov.uk/government/collections/personal-incomes-statistics

8.2 Summary of the effects of taxes and benefits, by household type,[1] 2016/17

	All households	Retired households			
		1 adult Men	1 adult Women	All 1 adult	2 or more adults
Average per household (£ per year)					
Original income	37,270	8,389	7,281	7,651	21,397
plus Cash benefits	6,374	9,965	9,345	9,552	13,620
Gross income	43,645	18,354	16,627	17,204	35,017
less Direct taxes and employees' NIC	8,397	1,954	1,692	1,779	4,695
Disposable income	35,247	16,400	14,935	15,425	30,322
Equivalised[3] disposable income	32,676	24,487	22,356	23,067	29,523
less Indirect taxes	6,518	3,187	2,636	2,820	6,902
Post-tax income	28,730	13,213	12,299	12,604	23,421
plus Benefits in kind	7,414	4,764	5,312	5,129	8,163
Final income	36,144	17,977	17,612	17,734	31,583

	Non-Retired households									
	1 adult Men[2]	1 adult Women[2]	All 1 adult[2]	2 adults[2]	3 or more adults[2]	1 adult with children[3]	2 adults with 1 child	2 adults with 2 children	2 adults with 3 or more children	3 or more adults with children
Average per household (£ per year)										
Original income	25,679	21,493	23,854	52,000	65,122	12,891	51,379	54,559	39,367	61,955
plus Cash benefits	2,696	3,046	2,848	2,897	4,208	11,340	3,853	4,343	10,344	7,460
Gross income	28,375	24,539	26,703	54,897	69,329	24,231	55,232	58,902	49,712	69,415
less Direct taxes and employees' NIC	5,784	5,117	5,493	11,758	13,395	2,209	12,444	13,214	8,954	12,857
Disposable income	22,591	19,422	21,210	43,139	55,935	22,022	42,788	45,687	40,757	56,558
Equivalised[4] disposable income	33,886	29,133	31,815	43,139	38,436	20,903	34,807	31,599	23,751	31,246
less Indirect taxes	4,100	3,576	3,872	7,743	10,092	4,210	7,179	8,014	7,561	10,607
Post-tax income	18,490	15,846	17,338	35,397	45,843	17,812	35,609	37,673	33,197	45,951
plus Benefits in kind	1,867	2,082	1,960	3,867	6,505	11,905	8,305	14,632	23,271	15,808
Final income	20,357	17,928	19,298	39,264	52,348	29,717	43,914	52,306	56,468	61,759

Source: Office for National Statistics

Notes:

1 See Methodology and Coherence section for definitions of retired households, adults and children.

2 Without children.

3 Children are defined as people aged under 16 or aged between 16 and 19, not married nor in a Civil Partnership, nor living with a partner; and living with parent(s)/guardian(s); and and receiving non-advanced further education or in unwaged-government training.

4 Using the modified-OECD scale.

8.3 Income and source of income for all UK households, 1977 to 2017/18

(2017/18 prices[1])

Year[2,3,4]	Number of households in the population	Average annual household income		Sources of income						
		Disposable	Gross	Wages and salaries	Imputed income from benefits in kind	Self-employment income	Private pensions, annuities	Investment income	Other income	Total cash benefits
	000s	£	£	Percentage of gross household income						
1977	-	17,163	22,261	73	-	7	3	3	1	13
1978	-	18,560	23,649	-	-	-	3	3	1	13
1979	-	19,248	24,195	74	-	6	3	3	1	14
1980	-	20,090	25,378	73	-	7	3	3	1	13
1981	-	19,753	25,259	70	-	7	3	4	1	14
1982	-	19,016	24,375	69	-	7	3	4	1	16
1983	-	19,212	24,676	67	-	8	4	4	1	17
1984	-	19,474	24,920	67	-	7	4	4	1	17
1985	-	20,513	26,166	66	-	8	5	4	1	16
1986	-	21,396	27,198	65	-	9	5	4	1	16
1987	-	22,756	28,950	65	-	10	5	5	1	15
1988	-	24,482	30,907	64	-	11	5	5	1	13
1989	-	24,704	30,959	65	-	11	5	5	1	13
1990	-	25,949	32,552	64	1	11	5	6	1	12
1991	-	25,789	32,040	63	1	9	5	7	1	13
1992	-	25,076	31,095	62	1	9	6	6	1	15
1993	-	24,878	30,917	62	1	8	6	5	1	16
1994/95	-	24,878	31,278	62	1	10	6	4	1	16
1995/96	-	24,768	31,199	62	1	9	7	5	1	16
1996/97	24,253	26,013	32,434	63	1	10	6	4	1	15
1997/98	24,556	27,021	33,750	65	1	8	7	4	1	14
1998/99	24,664	27,773	35,023	66	1	8	6	4	1	14
1999/00	25,334	28,825	36,233	64	1	10	7	4	1	13
2000/01	25,030	29,939	37,671	65	1	9	7	4	1	13
2001/02	24,888	31,798	40,073	67	1	8	7	4	1	13
2002/03	24,346	32,161	40,030	66	1	8	7	3	1	14
2003/04	24,670	32,579	41,103	65	1	9	7	3	1	14
2004/05	24,431	33,829	42,733	66	1	8	7	3	1	14
2005/06	24,799	33,986	42,960	65	1	8	7	3	1	14
2006/07	24,836	34,836	44,299	66	1	9	7	3	1	14
2007/08	25,289	34,548	43,938	66	1	9	7	3	1	14
2008/09	25,874	34,147	42,952	65	1	9	7	3	1	15
2009/10	26,053	34,498	43,215	65	1	8	7	2	1	16
2010/11	26,265	34,898	43,623	63	1	9	8	2	1	16
2011/12	26,436	33,181	41,550	65	1	7	9	2	1	16
2012/13	26,823	33,203	41,235	62	1	9	9	3	1	16
2013/14	26,675	33,569	41,520	62	1	8	9	3	1	16
2014/15	26,856	34,793	42,833	63	1	8	9	3	1	15
2015/16	27,204	35,116	43,191	63	1	8	10	3	1	15
2016/17	27,210	36,162	44,777	62	1	9	10	3	1	15
2017/18	27,111	36,870	45,773	63	1	9	9	3	1	14

Source: Office for National Statistics

Notes:

1 Income figures have been deflated to 2016/17 prices using the consumer prices index including owner-occupiers' housing costs (CPIH).

2 For 1978 it is not possible to provide a breakdown of income from employment into employee wages and salaries and self-employment income.

3 Prior to 1990 imputed income from benefits in kind was not included in gross household income.

4 Prior to 1996/97 income figures are unweighted. From 1996/97 onwards income figures are based on weighted data.

8.4 Household expenditure at current[1] prices

UK, financial year ending March 2003 to financial year ending March 2017

	2002-03	2003-04	2004-05	2005-06	2006[2]	2006[3]	2007	2008	2009	2010	2011	2012	2013[4]	2014	2014-15	2015-16	2016-17
Weighted number of households (thousands)	24,350	24,670	24,430	24,800	24,790	25,440	25,350	25,690	25,980	26,320	26,110	26,410	26,840	26,600	26,760	27,220	27,210
Total number of households in sample	6,930	7,050	6,800	6,790	6,650	6,650	6,140	5,850	5,830	5,260	5,690	5,600	5,140	5,130	5,170	4,920	5,040
Total number of persons in sample	16,590	16,970	16,260	16,090	15,850	15,850	14,650	13,830	13,740	12,180	13,430	13,180	12,120	12,120	12,160	11,620	11,960
Total number of adults in sample	12,450	12,620	12,260	12,170	12,000	12,000	11,220	10,640	10,650	9,430	10,330	10,200	9,350	9,440	9,510	8,950	9,260
Weighted average number of persons per household	2.4	2.4	2.4	2.4	2.4	2.3	2.4	2.4	2.3	2.3	2.4	2.3	2.4	2.4	2.4	2.4	2.4
Commodity or service									Average weekly household expenditure (£)								
1 Food and non-alcoholic drinks	42.70	43.50	44.70	45.30	46.90	46.30	48.10	50.70	52.20	53.20	54.80	56.80	58.80	58.80	58.30	56.80	58.00
2 Alcoholic drinks, tobacco and narcotics	11.40	11.70	11.30	10.80	11.10	11.10	11.20	10.80	11.20	11.80	12.00	12.60	12.00	12.30	12.00	11.40	11.90
3 Clothing and footwear	22.30	22.70	23.90	22.70	23.20	23.00	22.00	21.60	20.90	23.40	21.70	23.40	22.60	23.70	23.70	23.50	25.10
4 Housing (net)[5], fuel and power	36.90	39.00	40.40	44.20	47.60	47.50	51.80	53.00	57.30	60.40	63.30	68.00	74.40	72.70	72.80	72.50	72.60
5 Household goods and services	30.20	31.30	31.60	30.00	30.30	29.90	30.70	30.10	27.90	31.40	27.30	28.50	33.10	35.40	36.70	35.50	39.30
6 Health	4.80	5.00	4.90	5.50	5.90	5.80	5.70	5.10	5.30	5.00	6.60	6.40	6.20	7.10	7.00	7.20	7.30
7 Transport	59.20	60.70	59.60	61.70	62.00	60.80	61.70	63.40	58.40	64.90	65.70	64.10	70.40	74.80	73.30	72.70	79.70
8 Communication	10.60	11.20	11.70	11.90	11.70	11.60	11.90	12.00	11.70	13.00	13.30	13.80	14.50	15.50	15.50	16.00	17.20
9 Recreation and culture	56.40	57.30	59.00	57.50	58.50	57.60	57.40	60.10	57.90	58.10	63.90	61.50	63.90	68.80	69.30	68.00	73.50
10 Education	5.20	5.20	6.50	6.60	7.20	7.00	6.80	6.20	7.00	10.00	7.00	6.80	8.80	9.80	9.00	7.00	5.70
11 Restaurants and hotels	35.40	34.90	36.10	36.70	37.90	37.60	37.20	37.70	38.40	39.20	39.70	40.50	40.40	42.50	42.50	45.10	50.10
12 Miscellaneous goods and services	33.10	33.60	34.90	34.60	36.00	35.70	35.30	35.60	35.00	35.90	38.60	38.40	39.10	40.00	40.40	39.70	41.80
1-12 All expenditure groups	348.30	356.20	364.70	367.60	378.30	373.80	379.80	386.30	383.10	406.30	413.90	420.70	444.30	461.20	460.50	455.30	482.20
13 Other expenditure items[6]	57.90	61.90	69.70	75.80	77.60	75.10	79.30	84.60	71.80	67.30	69.70	68.30	73.00	70.10	66.90	73.60	72.00
Total expenditure	406.20	418.10	434.40	443.40	455.90	449.00	459.20	471.00	455.00	473.60	483.60	489.00	517.30	531.30	527.30	528.90	554.20
Average weekly expenditure per person (£)																	
Total expenditure	170.50	177.40	182.00	188.00	192.80	192.00	194.80	199.80	194.40	203.10	205.40	208.70	219.40	221.80	221.20	224.50	233.80

Source: Office for National Statistics

Note: The commodity and service categories are not comparable to those in publications before 2001-02.
1 Data in Table 4.3 have not been deflated to 2016-17 prices and therefore show the actual expenditure for the year they were collected. Because inflation is not taken into account, comparisons between the years should be made with caution.
2 From 2001-02 to this version of 2006, figures shown are based on weighted data using non-response weights based on the 1991 Census and population figures from the 1991 and 2001 Censuses.
3 From this version of 2006 until 2012, figures shown are based on weighted data using non-response weights and population figures based on the 2001 Census.
4 From 2013, figures are based on weighted data using non-response weights based on the 2001 Census and population estimates based on the 2011 Census.
5 Excluding mortgage interest payments, council tax and Northern Ireland rates.
6 An improvement to the imputation of mortgage interest payments has been implemented for 2006 data onwards. This means there is a slight discontinuity between 2006 and earlier years.

8.5 Percentage of households with durable goods
UK, 1998-99 to financial year ending 2017

	Car/ van	Central heating[1]	Washing machine	Tumble dryer	Dish- washer	Micro- wave	Tele- phone	Mobile phone	DVD Player	Home computer	Internet connection
1998-99[2]	72	89	92	51	23	79	95	27	--	33	10
1999-2000	71	90	91	52	23	80	95	44	--	38	19
2000-01	72	91	93	53	25	84	93	47	--	44	32
2001-02[3]	74	92	93	54	27	86	94	64	--	49	39
2002-03	74	93	94	56	29	87	94	70	31	55	45
2003-04	75	94	94	57	31	89	92	76	50	58	49
2004-05	75	95	95	58	33	90	93	78	67	62	53
2005-06	74	94	95	58	35	91	92	79	79	65	55
2006[4]	76	95	96	59	38	91	91	80	83	67	59
2006[5]	74	95	96	59	37	91	91	79	83	67	58
2007	75	95	96	57	37	91	89	78	86	70	61
2008	74	95	96	59	37	92	90	79	88	72	66
2009	76	95	96	58	39	93	88	81	90	75	71
2010	75	96	96	57	40	92	87	80	88	77	73
2011	75	96	97	56	41	92	88	87	88	79	77
2012	75	96	97	56	42	93	88	87	87	81	79
2013	76	96	97	56	42	92	89	92	85	83	82
2014	76	96	97	56	44	92	88	94	83	85	84
2014-15	76	96	97	56	44	92	88	94	83	85	84
2015-16	78	95	97	56	45	93	88	95	80	88	88
2016-17	79	95	97	56	46	93	89	95	78	88	89

Source: Office for National Statistics

Please see background notes for symbols and conventions used in this report.
-- Data not available.
1 Full or partial.
2 From this version of 1998-99, figures shown are based on weighted data and including children's expenditure.
3 From 2001-02 onwards, weighting is based on the population figures from the 2001 census.
4 From 1998-99 to this version of 2006, figures shown are based on weighted data using non-response weights based on the 1991 Census and population figures from the 1991 and 2001 Census.
5 From this version of 2006, figures shown are based on weighted data using updated weights, with non-response weights and population figures based on the 2001 Census.

this page is intentionally blank

Lifestyles

Chapter 9

Lifestyles

Expenditure by the Department for Culture, Media and Sport (Table 9.1)
The department was renamed to the Department for Digital, Culture, Media and Sport in July 2017. This was to reflect the department's increased activity in the Digital sector. The figures in this table are taken from the department's Annual Report and are outturn figures for each of the headings shown (later figures are the estimated outturn). The department's planned expenditure for future years is also shown.

International tourism and holidays abroad (Tables 9.8 and 9.9)
The figures in these tables are compiled using data from the International Passenger Survey (IPS).   A holiday abroad is a visit made for holiday purposes. Business trips and visits to friends and relatives are excluded.

Domestic tourism (Table 9.10)
The figures in this table are compiled using data from the Visit England. Data includes total number of trips taken as well as number of bednights and expenditure, as well as average trip totals.

Gambling (Table 9.12a and b)
The figures in these tables are the latest released by The Gambling Commission at the time of going to press. The statistics come from the data provided by licence holders through their regulatory returns.

9.1 Expenditure by the Department for Culture, Media and Sport

(£'000s)

Resource DEL	2013-14 Outturn	2014-15 Outturn	2015-16 Outturn	2016-17 Outturn	2017-18 Outturn	2018-19 Plans	2019-20 Plans
Support for the Museums and Galleries sector	16,267	16,003	20,314	16,198	24,745	30,167	26,887
Museums and Galleries sponsored ALBs (net)	351,059	326,032	339,528	362,966	363,201	403,029	409,240
Libraries sponsored ALBs (net)	101,374	98,369	113,571	115,172	117,386	124,209	126,201
Support for the Arts sector[1]	(67,219)	(58,465)	(79,113)	(76,979)	(77,566)	(78,341)	18,583
Arts and culture ALBs (net)	448,788	433,475	439,548	442,231	432,551	447,908	369,211
Support for the Sports sector	21,147	18,075	11,159	7,585	11,520	7,124	7,124
Sport sponsored ALBs (net)	112,767	111,006	106,112	128,683	146,677	133,976	130,452
Ceremonial and support for the Heritage sector[2]	16,690	29,456	53,141	48,451	67,739	33,067	30,173
Heritage sponsored ALBs (net)[3]	105,685	115,478	84,350	99,814	76,762	83,889	75,691
The Royal Parks[4]	13,637	14,600	12,320	6,022	-	-	-
Support for the Tourism sector	10	(200)	-	-	-	-	-
Tourism sponsored ALBs (net)[5]	48,200	46,502	66,374	57,095	59,835	57,332	33,732
Support for the Broadcasting and Media sector	15,862	42,315	19,498	28,346	41,100	34,835	19,302
Broadcasting and Media sponsored ALBs (net)[6]	101,810	88,099	95,600	82,204	65,330	43,185	42,111
Administration and Research	36,554	41,748	54,081	57,898	64,333	65,243	64,566
Support for Horseracing and the Gambling sector[7]	(1,603)	(843)	(2,858)	(2,539)	(3,714)	(2,880)	-
Gambling Commission (net)[7]	3,097	1,449	365	1,197	3,222	3,223	310
Olympics - legacy programmes[8]	(18,083)	(33,823)	(55,210)	(30,408)	(26)	-	-
London 2012 (net)[8]	(29,477)	55,715	-	-	-	-	-
Office for Civil Society[9]	163,530	219,954	162,582	255,511	224,460	293,999	317,273
Spectrum Management Receipts[10]	(54,535)	(52,594)	(52,139)	(49,645)	(33,299)	-	-
Total Resource DEL	**1,385,560**	**1,512,351**	**1,389,223**	**1,549,802**	**1,584,256**	**1,679,965**	**1,670,856**

Of which:							
Staff costs[11]	633,746	565,023	550,393	563,808	580,795	*	*
Purchase of goods and services[11]	1,151,471	907,001	565,167	599,948	613,806	*	*
Income from sales of goods and services	(491,390)	(106,416)	(55,343)	(57,591)	(58,480)	(19,032)	(19,030)
Current grants to local government (net)	(6,900)	28,150	46,224	69,566	44,881	-	-
Current grants to persons and non-profit (net)	614,324	754,888	630,433	758,716	790,068	827,296	752,575
Current grants abroad (net)	(1,528)	13,269	(1,820)	(3,787)	(785)	-	-
Subsidies to private sector companies	157		-	-	-	-	-
Subsidies to public corporations	9,663	51,410	44,951	30,227	14,443	-	-
Net public service pensions	-	-	6,763	-	-	-	-
Rentals	32,444	24,410	24,188	19,699	23,832	-	-
Depreciation[12]	158,177	105,770	127,561	143,221	149,256	186,293	198,854
Change in pension scheme liabilities[13]	(44)	-	-	158	-	-	-
Unwinding of discount rate on pension scheme liabilities[13]	-	-	1,389	813	843	-	-
Other resource	(714,560)	(831,154)	(550,683)	(574,976)	(574,403)	(342,118)	(271,353)

Resource AME (£'000s)	2013-14 Outturn	2014-15 Outturn	2015-16 Outturn	2016-17 Outturn	2017-18 Outturn	2018-19 Plans	2019-20 Plans
British Broadcasting Corporation (net)[14]	3,046,611	3,363,160	3,238,249	3,117,377	3,028,455	3,258,566	3,604,113
Provisions, Impairments and other AME spend	18,523	11,157	31,006	14,024	43,569	43,644	-
Levy bodies[15]	(2,721)	4,021	8,139	7,490	(20,942)	2	-
London 2012 (net)	102,138	(38,195)		-	-	-	-
Lottery Grants[16]	1,352,673	1,594,409	1,070,465	1,294,717	1,214,041	1,050,740	1,189,222
Total Resource AME	**4,517,224**	**4,934,552**	**4,347,859**	**4,433,608**	**4,265,123**	**4,352,952**	**4,793,335**
Of which:							
Staff costs[17]	929,732	1,041,637	1,083,741	1,170,477	1,055,940	*	*
Purchase of goods and services[17]	2,469,733	2,672,561	2,647,535	2,412,420	2,478,429	*	*
Income from sales of goods and services	-	-	-	-	-	(298,729)	(181,815)
Current grants to local government (net)	34,896	32,218	24,145	21,787	25,345	35,623	35,623
Current grants to persons and non-profit (net)	1,245,777	1,470,874	949,267	1,260,754	1,177,008	936,993	1,000,901
Current grants abroad (net)	-	-	37	-	-	-	-
Subsidies to public corporations	24	4,823	7,245	-	1,795	-	-
Rentals	104,496	106,848	51,535	34,284	38,377	-	-
Depreciation	274,590	212,367	212,951	221,081	273,033	189,074	-
Take up of provisions	29,932	21,917	46,670	44,783	100,436	2,252	-
Release of provision	(10,526)	(19,185)	(6,886)	-	-		

9.1 Expenditure by the Department for Culture, Media and Sport

(£'000s)

Resource AME (£'000s)	2013-14 Outturn	2014-15 Outturn	2015-16 Outturn	2016-17 Outturn	2017-18 Outturn	2018-19 Plans	2019-20 Plans
Change in pension scheme liabilities[18]	178,831	192,115	165,013	195,426	246,549	-	-
Unwinding of discount rate on pension scheme liabilities[18]	82,412	70,393	43,439	47,882	39,540	15,273	-
Release of provisions covering pension benefits[18]	(14,336)	(9,633)	(6,763)	-	-	-	-
Other resource	(808,337)	(862,383)	(870,070)	(975,286)	(1,171,329)	(526,769)	13,323
Total Resource Budget[19]	**5,902,784**	**6,446,903**	**5,737,082**	**5,983,410**	**5,849,379**	**6,032,917**	**6,464,191**
Of which:							
Depreciation[20]	432,767	318,137	340,512	364,302	422,289	375,367	198,854

Capital DEL (£'000s)	2013-14 Outturn	2014-15 Outturn	2015-16 Outturn	2016-17 Outturn	2017-18 Outturn	2018-19 Plans	2019-20 Plans
Support for the Museums and Galleries sector[21]	100	1,981	1,170	3,118	145	91,749	5,500
Museums and Galleries sponsored ALBs (net)[22]	19,887	42,177	30,031	65,867	62,061	28,815	26,273
Libraries sponsored ALBs (net)	7,173	12,561	3,408	8,050	2,126	3,221	3,221
Support for the Arts sector	3,932	-	723	314	323	3,311	-
Arts and culture ALBs (net)[23]	18,679	14,432	21,413	49,316	61,507	33,334	12,634
Support for the Sports sector	250	-	154	-	-	-	-
Sport sponsored ALBs (net)	30,120	29,019	38,916	37,131	38,800	42,765	43,265
Ceremonial and support for the Heritage sector	6,882	2,182	5,491	5,056	1,291	500	-
Heritage sponsored ALBs (net)[24]	24,095	106,864	17,421	25,401	23,675	22,459	20,359
The Royal Parks[25]	2,620	2,570	3,577	5,201	-	-	-
Tourism sponsored ALBs (net)	357	325	253	1,184	1,208	186	186
Support for the Broadcasting and Media sector[26]	55,198	229,066	213,138	51,779	66,326	252,502	342,080
Broadcasting and Media sponsored ALBs (net)[27]	23,525	4,289	4,720	30,522	91,491	106,057	150,357
Administration and Research	2,215	4,401	1,800	1,424	2,409	125	-
Support for Horseracing and the Gambling sector[28]	9,000	49,896	-	-	-	-	-
Gambling Commission (net)	302	335	633	724	947	-	-
Olympics - legacy programmes[29]	-	-	-	(6,435)	-	-	-
London 2012 (net)[29]	(184,059)	(256,703)	-	-	-	-	-
Office for Civil Society[30]	12,553	20,357	6,136	(3,705)	(1,674)	-	-
Total Capital DEL	**32,829**	**263,752**	**348,984**	**274,947**	**350,635**	**585,024**	**603,875**
Of which:							
Staff costs[31]	11,332	11,450	10,660	-	8,004	*	*
Purchase of goods and services[31]	3,900	3,900	3,900	14,521	6,250	*	*
Release of provision	3,815	-	-	-	-	-	-
Income from sales of goods and services	-	-	-	-	(9,982)	-	-
Capital support for local government (net)	50,452	220,067	202,988	82,509	93,785	-	-
Capital grants to persons & non-profit (net)	(103,067)	(100,388)	(81,098)	(76,394)	71,701	444,816	553,936
Capital grants to private sector companies (net)	(1,737)	41,159	43,204	(247)	18,713	-	-
Capital grants abroad (net)	(202)	-	-	-	-	-	-
Capital support for public corporations	-	80,050	-	-	(1,400)	-	-
Current grants to persons & non-profit (net)	-	-	-	-	7,309	-	-
Subsidies to public corporations	-	-	-	-	3,748	-	-
Purchase of assets	531,860	262,825	191,711	276,438	160,704	126,418	36,149
Income from sales of assets	(384,775)	(260,065)	(43,645)	(3,322)	(7,020)	-	-
Net lending to the private sector and abroad	(10,604)	23,628	31,752	8,186	3,715	-	-
Other capital	(68,145)	(18,874)	(10,488)	(26,744)	(4,892)	-	-

Capital AME (£'000s)	2013-14 Outturn	2014-15 Outturn	2015-16 Outturn	2016-17 Outturn	2017-18 Outturn	2018-19 Plans	2019-20 Plans
British Broadcasting Corporation (net)[32]	127,393	139,462	45,226	143,691	158,060	323,708	170,000
Channel Four Television[33]	-	-	-	-	-	20,000	-
London 2012 (net)	(3,815)	-	-	-	-	-	-
Levy bodies[34]	(1,763)	1,991	(2,079)	(2,737)	(1,924)	-	-
Lottery Grants[35]	523,705	601,444	453,717	503,897	448,775	221,139	447,593
Total Capital AME	**645,520**	**742,897**	**496,864**	**644,851**	**604,911**	**564,847**	**617,593**
Of which:							
Staff costs[36]	20,324	23,078	23,078	-	12,565	*	*
Purchase of goods and services[36]	-	-	-	14,000	6,904	*	*

9.1 Expenditure by the Department for Culture, Media and Sport

(£'000s)

Capital AME (£'000s)	2013-14 Outturn	2014-15 Outturn	2015-16 Outturn	2016-17 Outturn	2017-18 Outturn	2018-19 Plans	2019-20 Plans
Income from sales of goods and services	-	-	-	-	(2,114)	-	-
Release of provision	(3,815)	-	-	-	-	-	-
Capital support for local government (net)	132,060	188,770	21,803	149,469	133,006	-	
Capital grants to persons & non-profit (net)	311,789	329,681	379,110	280,753	270,423	221,000	407,454
Capital grants to private sector companies (net)	-	-	-	-	3,505	159,000	
Capital support for public corporations	18	417	-	-	(20,151)	-	-
Purchase of assets	129,856	124,090	153,770	126,206	119,473	169,526	210,139
Income from sales of assets	(14,264)	(6,342)	(107,851)	(8,067)	(6,818)	-	-
Net lending to the private sector and abroad	21,360	17,612	20,554	33,471	65,874	-	-
Other capital	48,192	65,591	6,400	49,019	22,244	-	-
Total Capital Budget[37]	**678,349**	**1,006,649**	**845,848**	**919,798**	**955,546**	**1,149,871**	**1,221,468**
Total Departmental Spending[38]	**6,148,366**	**7,135,415**	**6,242,418**	**6,538,906**	**6,382,636**	**6,807,421**	**7,486,805**
Of which:							
Total DEL[39]	1,260,212	1,670,333	1,610,646	1,681,528	1,785,635	2,078,696	2,075,877
Total AME[40]	4,888,154	5,465,082	4,631,772	4,857,378	4,597,001	4,728,725	5,410,928

Source: Department for Culture, Media and Sport

These tables present actual expenditure by the department for the years 2013-14 to 2017-18 and planned expenditure for the years 2018-19 to 2019-20 (derived from the Spending Review (SR) 2015 and subsequent fiscal events). The data relates to the department's expenditure on an Estimate and budgeting basis.

The format of the tables is determined by HM Treasury, and the disclosure in Tables 1 and 2 follow that of the Supply Estimate functions.

All years have been restated for the effect of Machinery of Government changes and the change in budgetary treatment of research and development (from resource to capital). Please note that the current year Statement of Parliamentary Supply has not been restated for any other prior period adjustments.

Table 1 Public spending – summarises expenditure on functions administered by the department. Consumption of resources includes Departmental Expenditure Limits (DEL) for administration, programme and capital costs, and Annually Managed Expenditure (AME) both Voted and Non-Voted expenditure. The figures are derived from the OSCAR database and the mappings replicate the lines in SoPS Note 1.

Table 2 Administration budgets – provides a more detailed analysis of the administration costs of the department. It retains the high level functional analysis used in Table 1. The figures are derived from the OSCAR database and the mappings replicate the lines in SoPS Note 1.

Resource DEL

1. Support for the Arts Sector. The income relates to contributions from the Department for Education towards the cost of Music Hubs and other programmes managed by the Arts Council England. The funding profile is agreed on a year by year basis, therefore no income is yet shown in 19-20 plans. The contra expenditure budget in Arts Sponsored Bodies is also not shown in 19-20 plans.

2. Ceremonial and Support for the Heritage sector included funding for World War One commemorations in 2013-14 through to 2017-18 including the Battle of Jutland and the Somme.

3. The Heritage Sponsored Bodies line illustrates a reduction in 2015-16 following the Spending Review 2013. This is partially offset by funding for Church Roof repairs, announced in the 2014 Autumn Statement, via the National Heritage Memorial fund which commenced in 2015-16.

4. On 15 March 2017 The Royal Parks Limited took over the role of managing the parks from The Royal Parks Agency. The charity manages the parks on behalf of the government, however it now receives less exchequer funding than it raises in commercial income, consequently it has been reclassified as outside of central government and removed from the DCMS Supply Estimate. As a result The Royal Parks Agency has no planned expenditure after 2016-17.

5. Tourism sponsored ALBs line shows a drop in expenditure in 2019-20 which reflects the end of the current Discover England funding stream.

6. Broadcasting and Media sponsored ALBs includes the clearance and auction of the 800MHz band, with additional funding in 2014-15 for Superfast Broadband. The reduction in expenditure between 2016-17 and 2017-18 is due to Ofcom becoming self-funding from October 2017 and so does not need funding from the Exchequer from this point.

7. Support for the Horse Racing and Gambling Sector, and the Gambling Commission. The National Lottery Commission and the subsequent income it receives is recorded on a year by year basis.

8. Olympics legacy and London 2012 relate to the staging of the Olympic and Paralympic games 2012. This includes income from the sale of the Olympic Village, residual costs and final settlements with the Greater London Authority (GLA) and Olympic Lottery Distribution Fund (OLDF), most of which concluded by 2016-17.

9. On 15 July 2016 the Office for Civil Society moved from the Cabinet Office to the Department for Digital, Culture Media and Sport via a Machinery of Government transfer. The National Citizens Service participation levels are forecast to increase from 2018-19 onwards.

10. Spectrum Management receipts which partially offset Broadcasting Administration expenditure, will, from 2018-19, be treated as income by Ofcom

11. Figures for Plans for staff costs and purchase of goods and services are redacted to avoid publishing any assumptions about future price or pay movements.

12. Depreciation includes impairments.

13. Pension schemes report under IAS 19 Employee Benefits accounting requirements. These figures, therefore, include cash payments made and contributions received, as well as certain non-cash items.

9.1 Expenditure by the Department for Culture, Media and Sport

(£'000s)

Resource AME

14. British Broadcasting Corporation: BBC Commercial Holdings and its direct subsidiary holding companies have been consolidated in these accounts, reflected in the outturn from 2017-18.

15. Levy Bodies: Levy Expenditure is only recorded at year end via the annual accounts, hence no forward plans data.

16. Lottery Grants: The group accounts exclude the Devolved Administrations and records expenditure on an accruals basis since 2014-15.

17. Figures for Plans for staff costs and purchase of goods and services are redacted to avoid publishing any assumptions about future price or pay movements.

18. Pension schemes report under IAS 19 Employee Benefits accounting requirements. These figures, therefore, include cash payments made and contributions received, as well as certain non-cash items.

19. Total Resource Budget is the sum of the Resource DEL budget and the Resource AME budget, including depreciation.

20. Depreciation includes impairments.

Capital DEL

21. Support for the Museums and Galleries. In the Spending Review 2015 it was announced that the government would invest £150m to support the British Museum, Science Museum and Victoria and Albert Museum to replace out of date museum storage at Blythe House; this programme commenced in 2016-17.

22. Museums and Galleries Sponsored ALBs funding from 2013-14 to 2015-16 illustrate the efficiency savings made by the Museums and Galleries following spending reviews of 2010 and 2013. Additional Capital funding was allocated for 2016-17 in the Spending Review 2015, with reserves access granted to them as part of the new Museums Freedoms programme. The Museums Freedoms Reserves can only be accessed at the Supplementary Estimate stage and so are not yet incorporated in funding data for 2018-19 and 2019-20.

23. Arts and culture ALBs includes funding in 2016-17, 2017-18 and 2018-19 for the Factory Manchester as part of the Northern Powerhouse.

24. Heritage Sponsored ALBs saw an additional £80m allocated in 2014-15 to Historic England (formerly English Heritage) on implementation of the New Model whereby the management of historic bodies was transferred to a charity, the English Heritage Trust.

25. On 15 March 2017 The Royal Parks Limited took over the role of managing the parks from The Royal Parks Agency. The charity manages the parks on behalf of the government, however it now receives less exchequer funding than it raises in commercial income, consequently it has been reclassified as outside of central government and removed from the DCMS Supply Estimate. As a result, The Royal Parks Agency has no planned expenditure after 2016-17.

26. Support for Broadcasting and Media sector is related to Broadband Delivery UK (BDUK) programmes. At the Autumn Statement 2016 it was announced that the Government would invest a further £740m targeted at supporting the market to roll out full-fibre connections and future 5G communications

27. Broadcasting and Media sponsored ALBs. In the 2015 Spending Review it was announced the government would invest up to £550m during this Spending Review period to make the 700 MHz spectrum band available for mobile broadband use. The increase from 2016-17 onwards includes increased funding for clearance and auction of the band.

28. Support for Horse Racing and Gambling sector. Following the sale of the Tote in 2011-12 it was agreed that the proceeds would be returned to the racing industry over a period of years. Initially, at £9m a year and then with the industry it was agreed that the balance of £49.9m be repaid in 2014-15.

29. Olympics legacy and London 2012 relate to the staging of the Olympic and Paralympic games 2012. This includes income from the sale of the Olympic Village, residual costs and final settlements with the GLA and OLDF.

30. On 15 July 2016 the Office for Civil Society moved from the Cabinet Office to the Department for Digital, Culture, Media and Sport via a Machinery of Government transfer. Any future year's expenditure will be reflected at the Supplementary Estimate. The change from net income in 2016-17 relates to loan repayments and fewer capital grants issued.

31. Figures for Plans for staff costs and purchase of goods and services are redacted to avoid publishing any assumptions about future price or pay movements.

Capital AME

32. BBC Capital expenditure is net of property disposals including the sale of Television Centre in White City in 2015-16. BBC Commercial Holdings and its direct subsidiary holding companies have been consolidated in these accounts, reflected in the outturn from 2017-18.

33. Channel Four Television includes £20m budget cover secured to cover commercial borrowings. This facility has not previously been required.

34. Levy Bodies: Levy Expenditure is only recorded at year end via the annual accounts, hence no forward plans data.

35. Lottery Grants: The group accounts excludes the Devolved Administrations and records expenditure on an accruals basis since 2014-15. The funding profile is agreed on a year by year basis, and refined at the Supplementary Estimate for 2018-19.

36. Figures for Plans for staff costs and purchase of goods and services are redacted to avoid publishing any assumptions about future price or pay movements.

Totals

37. Total Capital Budget is the sum of the Capital DEL budget and the Capital AME budget.

38. Total Departmental Spending is the sum of the resource budget and the capital budget less depreciation.

39. Total DEL is the sum of the resource budget DEL and capital budget DEL less depreciation in DEL.

40. Total AME is the sum of the resource budget AME and the capital budget AME less depreciation in AME.

9.2 Estimates of Average Issue Readership of National Daily Newspapers

Great Britain

National Dailies (Mon-Sat. A.I.R.)		2014 Mar	2014 Jun	2014 Sep	2014 Dec	2015 Mar	2015 Jun	2015 Sep	2015 Dec	2016 Mar	2016 Jun	2016 Sep	2016 Dec	2017 Mar	2017 Jun	2017 Sep
The Sun	WSDV	5,685	5,508	5,421	5,347	5,178	4878	4664	4490	4,392	4,317	4,188	3,915	3,653	3,486	3,417
Daily Mail	WSEI	4,074	3,866	3,833	3,745	3,704	3657	3605	3546	3,444	3,456	3,354	3,294	3,215	3,052	2,974
Daily Mirror/Record[1]	WSEH	2,975	2,893	2,908	2,881	2,796	2554	2433	2337	2,287	2,325	2,283	2,151	2,064	1,926	1,884
Daily Mirror	WSEM	2,309	2,230	2,251	2,281	2,211	2029	1953	1848	1,813	1,851	1,818	1,773	1,691	1,566	1,541
Daily Record		688	679	679	624	608	550	496	504	487	486	484	406	406	393	369
The Daily Telegraph	WSEN	1,313	1,261	1,237	1,192	1,119	1154	1150	1165	1,190	1,191	1,183	1,139	1,171	1,101	1,101
The Times	WSES	1,155	1,110	1,047	1,062	995	985	1014	1036	1,054	1,092	1,047	1,014	1,049	1,056	1,072
Daily Express	WSEP	1,114	1,097	1,079	1,079	993	948	845	777	758	767	842	828	838	827	808
Daily Star	WSEQ	1,090	1,039	978	1,010	943	878	838	808	791	793	800	759	769	762	738
The Guardian	WSET	793	748	744	748	761	793	793	814	809	835	865	828	898	886	868
i (Newspaper)		563	584	573	579	549	561	533	536	569	518	541	499	475	509	488
The Independent[2]	WSEU	309	261	264	262	272	284	270	270	285	-	-	-	-	-	-
Financial Times[3]	WSEY	-	-	-	-	-	-	-	-	-	-	-	-	-	-	-
Any national morning[4]	WSEZ	19,092	18,384	18,106	17,929	17,331	16,716	16,161	15,794	15,591	15,306	15,124	14,455	14,165	13,638	13,376

The figures in this table are for printed media only.

Source: National Readership Surveys Ltd

1. The combined figures for the DAILY MIRROR/RECORD are net readership figures for the Daily Mirror and the Daily Record.

2. The last published data for The Independent (print) was for the period ending March 2016, after that it became a digital only title.

3. NRS ceased to publish estimates for Financial Times as of 2014 data, so data cannot be provided for March 2014 onwards.

4. Gross national dailies.

- Data not available

Readership is an estimate of how many readers a publication has. As most publications have more than one reader per copy, the NRS readership estimate is very different from the circulation count (i.e. the count of how many copies of a particular publication are distributed).

Great Britain
Adults 15+.

000's

National Dailies (Mon-Sat. A.I.R.)	PAMCo Individual Brand Reach[1] Oct 17-Sep 18				
	Total	Phone	Tablet	Desktop	Print[2]
The Sun - D	6,557	3,085	484	350	2,866
Daily Mail - D	6,228	2,359	498	996	2,590
Daily Mirror - D	3,442	1,773	298	265	1,155
Daily Record - D	647	250	43	59	310
The Daily Telegraph - D	3,256	1,613	332	481	902
The Times - N	1,418	278	111	83	978
Daily Express - N	1,589	442	242	283	635
Daily Star - N	970	196	71	88	624
The Guardian - D	3,979	1,944	396	1,079	699
i (Newspaper) - D	593	138	20	26	413
The Independent - D	1,778	1,164	195	425	-
Financial Times	-	-	-	-	-
Total	34,054	13,958	2,826	4,301	13817

Source: PAMCo
Email: info@pamco.co.uk

1. From Oct 2017 onwards NRS has been replaced by PAMCo data. For more information see: https://pamco.co.uk/

2. The daily print figure is the daily reach of the 6-day paper (i.e. everyone who has read it 'yesterday') which can be different from the modelled 6-day AIR.

- Data not available

D - digital estimates include traffic from one or more third party platforms (i.e. Facebook Instant Articles, Google AMP, Apple News, Flipboard)

N - digital estimates do not include traffic from one or more third party platforms (i.e. Facebook Instant Articles, Google AMP, Apple News, Flipboard)

9.3 Employment in Creative Industries sub-sectors, by gender [1]

Unit: Thousands
Years: 2012-2016
Coverage: UK

000's

	Sub-sector	2012	2013	2014	2015	2016
Total	1. Advertising and marketing	144	155	167	182	198
	2. Architecture	89	94	101	90	98
	3. Crafts	7	8	8	7	7
	4. Design and designer fashion	117	124	136	132	160
	5. Film, TV, video, radio and photography	240	232	228	231	246
	6. IT, software and computer services	558	574	607	640	674
	7. Publishing	223	198	193	200	193
	8. Museums, Galleries and Libraries	86	85	84	97	92
	9. Music, performing and visual arts	227	244	284	286	291
	Creative Industries	**1,691**	**1,713**	**1,808**	**1,866**	**1,958**
	All UK Sectors	**30,334**	**30,760**	**31,410**	**32,037**	**32,422**

	Sub-sector	2012	2013	2014	2015	2016
Male	1. Advertising and marketing	76	86	97	109	107
	2. Architecture	64	67	74	59	63
	3. Crafts	-	-	-	-	-
	4. Design and designer fashion	68	64	77	76	92
	5. Film, TV, video, radio and photography	143	144	148	139	154
	6. IT, software and computer services	454	461	487	503	525
	7. Publishing	124	101	88	102	102
	8. Museums, Galleries and Libraries	-	-	-	-	-
	9. Music, performing and visual arts	121	118	139	149	154
	Creative Industries	**1,080**	**1,072**	**1,145**	**1,172**	**1,232**
	All UK Sectors	**16,079**	**16,261**	**16,603**	**16,972**	**17,160**

	Sub-sector	2012	2013	2014	2015	2016
Female	1. Advertising and marketing	68	69	70	73	90
	2. Architecture	26	26	27	31	35
	3. Crafts	-	-	-	-	-
	4. Design and designer fashion	49	60	59	56	68
	5. Film, TV, video, radio and photography	97	89	80	92	91
	6. IT, software and computer services	103	112	120	137	148
	7. Publishing	99	97	105	98	91
	8. Museums, Galleries and Libraries	-	-	-	-	-
	9. Music, performing and visual arts	106	125	146	138	137
	Creative Industries	**611**	**641**	**663**	**694**	**726**
	All UK Sectors	**14,255**	**14,499**	**14,806**	**15,065**	**15,262**

Source: Department for Culture, Media and Sport
evidence@culture.gov.uk

Notes

1. Estimates rounded to the nearest 1,000.

Notation

"-" Figure has been suppressed due to disclosiveness

"N/A" The value is not applicable. This could be due to a) no jobs associated with this sector; or b) an inappropriate calculation such as calculating the percentage change of a percentage.

9.4 Proportion of adults who participated in specific free time activities in the 12 months prior to interview by age, 2017/18, England

	16 - 24	25 - 44	45 - 64	65 - 74	75+	All Ages 16 +
	%	%	%	%	%	%
Free time activities						
Spend time with friends/family	91.9	91.8	89.4	90.1	84.2	90.1
Read	54.4	65.4	73.4	80.7	73.4	69.1
Listen to music	90.9	85.2	80.6	76.3	65.6	81.5
Watch TV	84.9	89.7	93.7	95.2	94.6	91.5
Days out or visits to places of interest	67.3	76.7	74.4	73.8	54.2	72.1
Eat out at restaurants	73.4	81.6	77.2	79.5	64.7	77.2
Go to pubs/bars/clubs	55.7	59.3	54.5	45.0	31.0	52.7
DIY	22.4	47.4	52.2	47.2	26.8	43.4
Gardening	14.2	43.0	63.7	72.0	58.1	50.6
Shopping	75.1	77.7	76.4	80.1	70.3	76.5
Play video/computer games	54.6	35.0	18.7	14.0	7.5	27.3
Browse the internet	88.1	87.2	79.8	62.9	31.3	76.5

Source: Taking Part Survey, Department for Digital, Culture, Media & Sport
takingpart@culture.gov.uk

9.5 Films
United Kingdom

Numbers and £ million

	Production of UK films[1,4]		Expenditure on feature films			
	Films produced in the UK (numbers)	Production costs (current prices)	UK Box Office	Video[2] rental	Video retail[2,3]	Video on demand
	KWGD	KWGE	KWHU	KWHV	KWHW	
1998	83	389	547	437	453	33
1999	92	507	563	451	451	40
2000	80	578	583	601	601	50
2001	74	379	645	494	821	65
2002	119	551	755	494	1,175	63
2003	214	1,127	742	462	1,392	68
2004	198	879	770	476	1,557	73
2005	223	608	770	399	1,399	74
2006	208	827	762	340	1,302	67
2007	240	849	821	295	1,440	75
2008	297	723	850	284	1,454	89
2009	338	1,256	944	285	1,311	110
2010	374	1,154	988	278	1,267	127
2011	363	1,328	1,040	262	1,165	142
2012	373	1,009	1,099	221	968	208
2013	370	1,167	1,083	146	940	320
2014	364	1,575	1,058	87	861	397
2015	361	1,570	1,236	62	712	487
2016	327	1,840	1,228	48	589	578
2017	315	2,153	1,280	37	503	681

Source: British Film Institute

1 Includes films with a production budget of £500,000 or more.

2 Video includes only rental and retail of physical discs, and does not include downloads.

3 In 2005 the British Video Association changed its methodology for producing market value which has necessitated a change to historical figures quoted.

4 Prior to 2010, the Research and Statistics Unit tracked all features shooting in the UK with a minimum budget of £500,000. However, evidence from avariety of sources (data on British film certification and the 2008 UK Film Council report Low and Micro-Budget Film Production in the UK) revealed a substantial number of films produced below this budget level. In order to broaden the evidence base, production tracking was extended to include feature films with budgets under £500,000 and data was collected from 2008 onwards.

Figures not adjusted for inflation

9.6 Box office results for the top 20 UK qualifying films released in the UK and Republic of Ireland, 2016

	Title	Country of origin	Box office gross (£ million)	Distributor
1	Rogue One: A Star Wars Story*	UK/USA	65.9	Walt Disney
2	Fantastic Beasts and Where to Find Them*	UK/USA	54.6	Warner Bros
3	Bridget Jones's Baby	UK/USA/Fra	48.2	Universal
4	The Jungle Book	UK/USA	46.2	Walt Disney
5	Jason Bourne	UK/USA	23.4	Universal
6	Doctor Strange	UK/USA	23.2	Walt Disney
7	Absolutely Fabulous: The Movie	UK/USA#	16.1	20th Century Fox
8	Passengers*	UK/USA	12.7	Sony Pictures
9	Miss Peregrine's Home for Peculiar Children	UK/USA	12.3	20th Century Fox
10	London Has Fallen	UK/USA	11.0	Lionsgate
11	The Conjuring 2	UK/USA	11.0	Warner Bros
12	Alice Through the Looking Glass*	UK/USA	10.0	Walt Disney
13	Me Before You	UK/USA	9.7	Warner Bros
14	The Legend of Tarzan	UK/USA	9.2	Warner Bros
15	Eddie the Eagle	UK/USA/Ger#	8.7	Lionsgate
16	Dad's Army	UK/USA#	8.7	Universal
17	The Danish Girl	UK/Ger/Den/Bel	7.5	Universal
18	Now You See Me 2	UK/USA	6.3	eOne Films
19	The Huntsman: Winter's War	UK/USA	5.3	Universal
20	Grimsby	UK/USA#	5.3	Sony Pictures

Notes:
Box office gross = cumulative total up to 19 February 2017.

Source: comScore, BFI RSU analysis

* Film still on release on 19 February 2017.

UK qualifying film made with independent (non-studio) US support or with the independent arm of a US studio.

9.7 Full year accommodation usage figures for 2004-2017 (excluding 2013)

Data	Accommodation	2004	2005	2006	2007	2008	2009	2010	2011	2012	2014	2015	2016	2017
Total Visits (000s)	All staying visits	25,677	28,038	30,654	30,871	30,142	28,208	28,300	29,197	29,282	32,613	34,436	35,814	37,651
	Bed & Breakfast	1,177	1,221	1,242	1,174	1,118	1,065	1,083	996	1,003	1,043	1,349	1,570	1,741
	Free guest with relatives or friends	10,174	11,144	12,081	11,944	12,056	11,165	10,570	10,646	10,811	11,120	11,881	12,633	12,988
	Holiday village/Centre	37	60	44	67	59	77	41	49	59	170	60	63	106
	Hostel/university/school	949	1,096	1,167	1,203	1,138	1,218	1,102	1,180	1,174	1,144	1,267	1,217	1,167
	Hotel/guest house	11,451	12,465	13,939	14,156	13,472	13,001	13,841	14,446	14,406	16,383	16,796	16,886	18,321
	Other	1,815	1,855	2,143	2,073	1,930	1,470	1,518	1,360	1,439	1,971	2,317	2,886	2,708
	Own home	413	412	466	446	418	385	358	392	364	382	401	397	359
	Paying guest family or friends house	610	616	595	567	650	461	465	577	537	583	736	611	474
	Camping/caravan	295	381	425	394	298	333	312	302	284	335	300	292	332
	Rented house/flat	740	806	832	836	868	934	882	958	1,002	1,361	1,421	1,373	1,751
Total Nights (000s)	All staying visits	227,406	249,181	273,417	251,522	245,775	229,391	227,960	234,363	230,149	264,366	272,941	278,057	284,601
	Bed & Breakfast	8,576	7,840	7,411	7,078	6,500	5,804	6,341	5,639	5,307	5,809	8,041	9,566	10,285
	Free guest with relatives or friends	113,736	121,487	133,209	121,272	122,898	111,947	107,642	108,703	107,575	111,173	120,308	120,730	126,233
	Holiday village/Centre	222	346	336	576	329	448	321	321	393	1,522	362	329	577
	Hostel/university/school	16,757	17,863	21,103	15,680	15,443	16,775	16,305	18,670	18,236	18,543	18,796	21,888	17,531
	Hotel/guest house	45,291	51,069	57,540	57,174	55,133	54,246	58,741	60,365	60,081	72,640	73,414	73,926	83,057
	Other	5,546	7,078	6,930	6,452	6,597	4,397	4,982	4,990	4,635	10,428	10,007	11,410	10,831
	Own home	4,638	5,507	7,895	5,169	5,110	4,524	5,068	5,625	4,895	5,412	6,413	5,324	4,651
	Paying guest family or friends house	8,807	8,716	9,692	8,240	7,961	5,423	5,148	6,574	7,433	7,060	8,614	7,379	5,079
	Camping/caravan	3,015	4,705	4,350	4,182	2,988	3,296	2,904	3,229	2,757	3,047	2,763	2,272	3,324
	Rented house/flat	20,818	24,570	24,950	25,695	22,817	22,362	20,482	20,237	18,838	28,731	24,223	25,232	23,033
Total Spend (£m)	All staying visits	12,798	14,011	15,759	15,699	16,058	16,354	16,649	17,666	18,245	21,578	21,787	22,257	24,275
	Bed & Breakfast	481	511	512	488	470	459	522	470	457	562	732	829	923
	Free guest with relatives or friends	3,822	3,972	4,498	4,207	4,501	4,425	4,254	4,371	4,515	4,593	4,878	4,789	5,343
	Holiday village/Centre	9	12	17	27	22	27	23	23	27	84	21	16	39
	Hostel/university/school	739	848	937	829	813	1,055	1,004	1,122	1,083	1,350	1,253	1,445	1,136
	Hotel/guest house	5,826	6,547	7,580	7,916	7,914	7,773	8,577	9,138	9,626	11,254	11,594	11,812	13,546
	Other	263	308	245	328	337	328	272	307	265	529	466	516	540
	Own home	354	357	412	402	429	531	354	377	505	515	532	558	616
	Paying guest family or friends house	345	393	439	385	379	364	340	426	475	445	501	521	305
	Camping/caravan	97	112	142	146	111	144	129	128	111	134	143	98	140
	Rented house/flat	861	951	977	971	1,084	1,245	1,172	1,305	1,181	2,111	1,666	1,674	1,689
Sample	All staying visits	37,321	39,977	40,410	36,432	31,854	41,952	41,948	38,144	37,666	37,359	37,058	35,918	34,192
	Bed & Breakfast	1,570	1,617	1,471	1,266	1,137	1,469	1,518	1,252	1,224	1,129	1,417	1,511	1,460
	Free guest with relatives or friends	15,167	16,363	16,256	14,386	12,831	16,976	16,003	14,233	14,207	13,133	13,090	12,815	11,662
	Holiday village/Centre	45	64	53	59	46	99	49	55	82	177	60	55	92
	Hostel/university/school	1,232	1,372	1,348	1,261	1,153	1,616	1,510	1,382	1,381	1,158	1,230	1,171	1,002
	Hotel/guest house	18,716	19,971	20,585	18,560	15,641	20,885	21,998	19,765	19,458	19,842	19,110	18,330	17,688
	Other	965	893	910	864	873	1,008	1,033	1,136	1,174	1,314	1,649	1,818	1,804
	Own home	730	744	740	645	522	656	637	545	594	528	516	484	426
	Paying guest family or friends house	562	599	527	487	507	529	538	637	548	558	666	541	473
	Camping/caravan	250	322	339	320	298	363	378	379	325	360	383	324	329
	Rented house/flat	1,101	1,161	1,164	1,034	973	1,457	1,397	1,291	1,334	1,607	1,525	1,439	1,658

International Passenger Survey, Office for National Statistics Statistics

Please note for **2013** the accommodation categories were reduced therefore we have included the analysis separately on the next page. ONS reverted back to the full accommodation options for 2014 data onwards, therefore 2014 -2017 data is shown in this table.

9.7 Full year accommodation usage figures by visitors from overseas 2013

		Year
Data	Accommodation	**2013**
Total Visits (000s)	All staying visits	31,064
	Own Home	366
	Hotels or similar	16,852
	Camp/caravan sites	480
	Other Short Term Rented	2,100
	Friends or Relatives	10,569
	Other (Non Rented or Long Term Rented)	2,299
Total Nights (000s)	All staying visits	245,412
	Own Home	5,453
	Hotels or similar	77,553
	Camp/caravan sites	3,980
	Other Short Term Rented	34,680
	Friends or Relatives	105,763
	Other (Non Rented or Long Term Rented)	17,983
Total Spend (£m)	All staying visits	20,938
	Own Home	531
	Hotels or similar	11,956
	Camp/caravan sites	225
	Other Short Term Rented	2,514
	Friends or Relatives	4,656
	Other (Non Rented or Long Term Rented)	1,056
Sample	All staying visits	39,416
	Own Home	566
	Hotels or similar	22,514
	Camp/caravan sites	602
	Other Short Term Rented	2,561
	Friends or Relatives	13,609
	Other (Non Rented or Long Term Rented)	1,910

Source: International Passenger Survey, Office for National Statistics

Please note: For 2013 the accommodation categories were reduced therefore the analysis has been shown separately in this table. ONS reverted back to the full accommodation options for 2014. Therefore, 2014-2017 data is shown in the previous table.

9.8 International tourism[1]

Thousands and £ million

	Visits to the UK by overseas residents (thousands)	Spending in the UK by overseas residents		Visits overseas by UK residents (thousands)	Spending overseas by UK residents	
		Current prices	Constant 1995 prices[2]		Current prices	Constant 1995 prices[2]
2000	25,209	12,805	11,102	56,837	24,251	27,281
2001	22,835	11,306	9,528	58,281	25,332	27,710
2002	24,180	11,737	9,641	59,377	26,962	29,311
2003	24,715	11,855	9,451	61,424	28,550	28,677
2004	27,755	13,047	10,146	64,194	30,285	30,444
2005	29,970	14,248	10,714	66,441	32,154	30,954
2006	32,713	16,002	11,641	69,536	34,411	30,904
2007	32,778	15,960	11,389	69,450	35,013	32,477
2008	31,888	16,323	11,276	69,011	36,838	28,657
2009	29,889	16,592	11,032	58,614	31,694	22,673
2010	29,803	16,899	10,644	55,562	31,820	22,116
2011	30,798	17,998	10,870	56,836	31,701	20,569
2012	31,084	18,640	10,842	56,538	32,450	22,024
2013	32,692	21,258	11,740	57,792	34,510	23,613
2014	34,377	21,849	11,874	60,082	35,537	23,800
2015	36,115	22,072	11,730	65,720	39,028	24,381
2016	37,609	22,543	-	70,815	43,771	-
2017	39,214	24,507	-	72,772	44,840	-

1 See chapter text
2. As of 2016 constant 1995 prices are no longer produced by ONS.

Sources: International Passenger Survey
Office for National Statistics;

9.9 Holidays abroad:[1] by destination

Percentages

		2003	2004	2005	2006	2007	2008	2009	2010	2011	2012	2013	2014	2015	2016	2017
Spain	JTKC	29.8	28.4	27.2	27.8	26.5	26.6	26.5	25.9	25.8	27.0	27.5	28.2	27.1	28.7	30.0
France	JTKD	18.1	17.3	16.6	15.9	16.7	16.7	18.6	18.3	17.4	17.2	16.4	15.5	14.6	12.4	13.0
Greece	JTKF	6.6	5.7	5.1	5.0	5.0	4.2	4.4	4.8	4.8	4.5	4.4	4.5	4.9	4.9	4.5
USA	JTKE	5.5	6.1	6.0	5.1	5.2	5.4	5.4	4.9	5.2	5.2	5.0	5.4	5.4	5.1	4.7
Italy	JTKG	5	5	5.4	5.4	5.6	5.2	4.7	4.6	4.5	5.0	5.3	5.3	5.8	6.2	6.2
Irish Republic	JTKI	3.7	3.8	3.8	4.0	3.3	3.2	3.2	2.5	2.9	2.3	2.2	2.6	2.6	2.9	2.6
Portugal	JTKH	4	3.5	3.6	3.7	4.1	4.8	4.1	4.4	4.5	4.5	4.8	4.7	5.0	5.1	5.0
Cyprus	JTKL	2.7	2.6	2.8	2.4	2.4	2.4	2.1	2.1	2.2	2.0	1.5	1.3	1.3	1.5	1.6
Netherlands	JTKK	2.6	2.6	2.5	2.7	2.4	2.1	2.2	2.2	2.6	2.6	2.7	3.1	3.6	3.8	3.4
Turkey	JTKJ	2.3	2.3	2.7	2.7	2.8	3.7	3.5	4.7	3.7	3.3	2.8	2.9	2.8	1.6	1.9
Belgium	JTKM	2.2	1.8	1.9	2.0	2.2	2.0	1.9	1.7	2	2.5	2.3	2.5	2.2	1.7	1.8
Germany	JTKN	1.2	1.6	1.7	1.7	2.0	1.9	1.7	1.7	2.2	2.0	2.3	2.1	2.4	2.5	2.5
Austria	JTKP	1.1	1.4	1.3	1.2	1.2	1.4	1.4	1.5	1	1.1	1.2	1.0	1.0	1.4	1.0
Malta	JTKO	1	1	1.1	1.0	0.9	0.9	0.8	1.0	1	1.0	1.2	1.1	1.0	1.2	1.0
Other countries	JTKQ	14.2	16.8	18.4	20	19.6	19.5	19.5	19.6	20.2	13.3	13.4	13.3	12.5	11.9	11.7

1 See chapter text.

Sources: International Passenger Survey, Office for National Statistics;
01633 456032

9.10 All tourism trips in Great Britain, 2009 - 2017

	2009	2010	2011	2012	2013	2014	2015 (O)[1]	2015 (RP)[1]	2016	2017
TRIPS (millions)	122.537	115.711	126.635	126.019	122.905	114.242	124.426	125.162	119.455	120.68
BEDNIGHTS (millions)	387.448	361.398	387.329	388.24	373.607	349.546	377.101	379.449	359.557	369.46
EXPENDITURE (£ millions)	£20,971	£19,797	£22,666	£23,976	£23,294	£22,692	£24,825	£24,100	£23,079	£23,683
Av. Trip Length	3.16	3.12	3.06	3.08	3.04	3.06	3.03	3.03	3.01	3.1
Av. £/Night	54	55	59	62	62	65	66*	64	64	64
Av. £/trip	171	171	179	190	190	199	200*	194	193	196

Source: Visit England

1. When comparing between 2015 and 2016 it is recommended to use 2015 (RP) (Reprocessed 2015) figures. When comparing 2015 to earlier years use 2015 (O) (Original 2015) figures.

2. All figures for 2009 -2014 are original figures (O).

This table shows a breakdown of all domestic overnight tourism into trips, nights and spend.
In 2017 the volume of tourism trips increased by +1.0% when compared with 2016. The trend since 2011 however, is marginally a negative one with a small decrease of -0.8% each year in this period.

Since 2011 the volume of nights spent has seen an equal decrease to trips, while spend has increased by +1.4% per annum. The year on year proportional change figures for nights and spend comparing 2017 to 2016 were +2.8% and +2.6%.

9.11 All touism in Great Britain: 2016 - 2017

	Trips				Nights				Spend			
	GB	England	Scotland	Wales	GB	England	Scotland	Wales	GB	England	Scotland	Wales
	Millions				*Millions*				*£millions*			
All tourism - 2016	119.455	99.342	11.514	9.307	359.557	287.702	38.876	32.978	£23,079	£18,492	£2,897	£1,689
All tourism - 2017	120.676	100.622	11.664	9.024	369.455	299.410	39.066	30.979	£23,683	£19,049	£3,006	£1,628
Purpose												
Leisure	103.641	86.214	9.750	8.210	327.147	265.100	33.920	28.128	£19,465	£15,559	£2,443	£1,463
Total holiday	88.024	72.244	8.755	7.508	290.346	232.415	31.227	26.704	£17,404	£13,761	£2,263	£1,380
Holiday\pleasure\leisure	59.149	47.245	6.512	5.712	202.318	157.809	23.542	20.967	£14,134	£11,023	£1,907	£1,203
Visiting friends or relatives - mainly holiday	28.875	24.999	2.243	1.796	88.028	74.606	7.685	5.737	£3,270	£2,737	£356	£177
Visiting friends or relatives - mainly other reason	12.931	11.607	0.798	0.571	31.201	27.735	2.293	1.173	£1,381	£1,212	£113	£56
VFR	41.805	36.606	3.041	2.366	119.229	102.341	9.978	6.909	£4,651	£3,950	£468	£233
Total Business	16.531	14.166	1.827	0.622	37.978	31.637	4.814	1.526	£4,245	£3,556	£553	£136
To attend a conference	1.710	1.487	0.138	0.085	3.436	3.023	0.278	0.135	£398	£335	£44	£19
To attend an exhibition\trade show	0.976	0.877	0.059	0.047	2.164	1.927	0.121	0.116	£282	£250	£24	£8
Travel\transport is my work	0.954	0.782	0.093	0.087	4.149	3.189	0.186	0.773	£234	£201	£20	£13
To do paid work\on business	13.845	11.802	1.629	0.490	32.377	26.688	4.415	1.275	£3,565	£2,971	£485	£109
School trip	0.505	0.375	0.046	0.090	1.015	0.755	0.131	0.129	£135	£83	£34	£18
Other reason	1.633	1.351	0.146	0.147	4.493	3.405	0.414	0.674	£274	£225	£24	£25

Source: Visit England

GB headlines by tourism type [2017]
· During 2017, GB residents took a total of 120.7m overnight tourism trips to destinations in England, Scotland or Wales; amounting to 369.4m nights and £23.7bn was spent during these trips.
· Examining tourism type, the category 'holidays' accounted for 59.1m of these trips and £14.1bn of spend.
· Those who took trips to 'visit friends and relatives' accounted for 41.8m trips and £4.7bn of spend.
· Tourism for the purpose of 'business' accounted for 16.5m trips and £4.2bn of spend

GB trends by country [2016 vs 2017]
· At GB level, a percentage increase in tourism trips taken, of +1.0% was observed between 2016 and 2017. In the same period, nights volume increased by +2.7%, as did expenditure, by +2.6%.
· England saw an overall increase across the year compared to 2016, despite 7 out of the 12 months seeing a reduction in domestic trip volume. This equates to an increase in trip volume from 2016-2017 of +1.3%.
· The pattern in Scotland was one of six months increase and six months decline in domestic trips taken relative to 2016, overall up +1.7% between 2016 and 2017.
· Wales generally saw more months of declining trip volumes than increases compared to 2016, with an overall decrease of -3.2% from 2016 to 2017.

9.12a Gambling industry data
Industry Statistics - November 2018

	Gambling Industry GGY by Sector (£m)										
	Apr 2008 Mar 2009	Apr 2009 Mar 2010	Apr 2010 Mar 2011	Apr 2011 Mar 2012	Apr 2012 Mar 2013	Apr 2013 Mar 2014	Apr 2014 Mar 2015	Apr 2015 Mar 2016	Apr 2016R Mar 2017	Apr 2017R Mar 2018	Apr 2018P Mar 2019
Arcades	480.35	455.96	392.00	381.06	358.71	379.11	387.22	418.06	424.93	422.12	429.65
Betting	2,903.24	2,811.36	2,957.32	3,029.59	3,198.60	3,176.77	3,266.91	3,318.19	3,310.84	3,253.64	3,253.39
Bingo	703.11	627.22	625.58	680.64	700.90	673.44	679.02	693.11	684.28	680.11	677.04
Casinos	796.17	751.13	797.43	872.80	961.41	1,111.06	1,159.79	999.38	1,163.54	1,180.58	1,059.23
Remote betting, bingo and casino - 2005 Act Regulated	816.86	632.22	653.06	710.19	932.61	1,134.66	753.53*				
Remote betting, bingo and casino - 2014 Act Regulated GB Customers only							1485.47**	4,230.76	4,775.18	5,355.33	5,321.11
National Lottery (remote and non-remote)	2,521.50	2,679.20	2,840.20	3,123.90	3,279.50	3,099.80	3,232.10	3,416.80	2,978.60	3,007.80	3,079.30
Lotteries (remote and non-remote)	143.69	158.55	170.12	228.61	273.00	293.79	349.74	379.77	442.43	508.11	541.58
Total	8,364.92	8,115.64	8,435.70	9,026.80	9,704.74	9,868.62	11,313.78	13,456.05	13,779.80	14,407.69	14,361.30

Source: Gambling Commission

Notes on this section:

R - Previously published data revised.

P - Provisional new data.

* The figure represented reflects seven months data under the Gambling Act 2005 from 1 April 2014 to 31 October 2014, and represents the whole regulated market (regardless of where the customer was based).

** The figure represented reflects five months data under the Gambling (Licensing and Advertising) Act from 1 November 2014 to 31 March 2015, and relates to GB customers only.

For National Lottery and Lotteries, figures relate to the proceeds from ticket sales minus the amount given out in prizes, which is a GGY equivalent.

9.12b Remote: Gambling (Licensing and Advertising) Act Data
Industry Statistics - November 2019

GB only remote data (£m)						
Turnover	Nov 2014 Mar 2015	Apr 2015 Mar 2016	Apr 2016[R] Mar 2017	Apr 2017[R] Mar 2018	Apr 2018[P] Mar 2019		
	Total	Total	Total	Total	Proprietary	Revenue Share	Total
Betting	5,322.85	22,892.85	25,866.08	26,899.30	17,470.14	10,165.29	27,635.42
Bingo	330.51	1,468.81	1,403.67	1,307.74	656.13	658.89	1,315.02
Casino	6,761.34	62,227.59	71,778.84	82,521.42	25,796.06	63,884.69	89,680.75
Pool Betting	29.54	60.72	89.71	88.40	84.75	20.83	105.58
Total	12,444.23	86,649.96	99,138.30	110,816.86	44,007.08	74,729.69	118,736.77

GGY	Nov 2014 Mar 2015	Apr 2015 Mar 2016	Apr 2016[R] Mar 2017	Apr 2017[R] Mar 2018	Apr 2018[P] Mar 2019		
	Total	Total	Total	Total	Proprietary	Revenue Share	Total
Betting	456.44	1,565.75	1,748.58	2,056.86	1,121.81	710.01	1,831.82
Betting Exchange	52.44	151.90	171.53	168.85	160.56	5.86	166.42
Bingo	66.98	122.41	161.15	163.93	77.92	98.19	176.11
Casino	899.79	2,364.44	2,661.04	2,936.98	930.20	2,182.87	3,113.07
Pool Betting	9.82	26.27	32.89	28.70	29.00	4.69	33.68
Total	1,485.47	4,230.76	4,775.18	5,355.33	2,319.49	3,001.62	5,321.11

Betting Turnover	Nov 2014 Mar 2015	Apr 2015 Mar 2016	Apr 2016[R] Mar 2017	Apr 2017[R] Mar 2018	Apr 2018[P] Mar 2019		
	Total	Total	Total	Total	Proprietary	Revenue Share	Total
Cricket	238.11	492.39	508.24	749.40	432.01	278.20	710.21
Dogs	86.48	284.36	443.51	586.27	413.62	224.96	638.57
Financials	56.50	234.75	275.94	176.75	0.00	-	0.00
Football	2,119.33	5,890.46	8,188.68	10,445.73	7,419.60	3,568.56	10,988.16
Golf	34.93	182.44	194.16	258.51	181.17	90.52	271.69
Horses	1,631.01	5,354.42	6,403.15	8,886.28	6,246.37	3,393.62	9,639.99
Other	741.97	2,136.55	2,246.48	2,770.45	1,707.05	946.78	2,653.83
Tennis	444.05	1,467.74	1,588.90	2,342.86	1,112.57	1,038.08	2,150.65
Virtual	-	-	-	771.45	42.50	645.41	687.90
Unallocated rev share	-	6,910.45	6,106.72	-	-	-	-
Total	5,352.39	22,953.56	25,955.79	26,987.70	17,554.88	10,186.12	27,741.00

Betting GGY	Nov 2014 Mar 2015	Apr 2015 Mar 2016	Apr 2016[R] Mar 2017	Apr 2017[R] Mar 2018	Apr 2018[P] Mar 2019		
	Total	Total	Total	Total	Proprietary	Revenue Share	Total
Cricket	4.22	21.23	20.80	31.65	26.70	10.46	37.16
Dogs	9.02	28.44	44.41	55.06	36.41	20.25	56.66
Financials	3.10	12.70	37.03	22.00	0.00	-	0.00
Football	160.71	577.80	663.25	1,037.41	671.28	319.92	991.20
Golf	2.62	14.39	14.20	14.25	10.51	3.71	14.21
Horses	115.47	346.06	436.20	614.47	338.43	183.63	522.06
Other	41.43	156.24	159.46	256.97	156.29	65.04	221.33
Tennis	22.80	72.21	89.95	144.16	68.02	52.70	120.72
Virtual	-	-	-	78.44	3.74	64.85	68.59
Unallocated rev share	159.35	514.84	487.69	-	-	-	-
Total	518.70	1,743.91	1,952.99	2,254.42	1,311.37	720.56	2,031.93

Casino Turnover	Nov 2014 Mar 2015	Apr 2015 Mar 2016	Apr 2016[R] Mar 2017	Apr 2017[R] Mar 2018	Apr 2018[P] Mar 2019		
	Total	Total	Total	Total	Proprietary	Revenue Share	Total
Blackjack	-	-	-	2,229.77	2,148.38	7,296.02	9,444.40
Card Game	492.81	8,350.19	8,826.13	6,528.33	-	-	-
Other	223.16	3,464.10	5,055.55	6,601.40	2,278.28	4,902.95	7,181.24
Roulette	-	-	-	4,478.91	2,656.92	15,394.30	18,051.23
Slots	5,171.23	34,938.89	41,270.20	50,391.99	18,712.17	36,291.41	55,003.58
Table Game	874.14	15,474.41	16,626.95	12,291.02	0.31	-	0.31
Total	6,761.34	62,227.59	71,778.84	82,521.42	25,796.06	63,884.69	89,680.75

Casino GGY	Nov 2014 Mar 2015	Apr 2015 Mar 2016	Apr 2016[R] Mar 2017	Apr 2017[R] Mar 2018	Apr 2018[P] Mar 2019		
	Total	Total	Total	Total	Proprietary	Revenue Share	Total
Blackjack	-	-	-	50.44	47.71	168.49	216.20
Card Game	66.45	184.23	203.91	153.71	-	-	-
Other	59.15	132.80	175.10	214.96	51.04	183.46	234.50
Peer to Peer	44.00	103.68	98.35	97.13	68.86	27.09	95.95
Roulette	-	-	-	109.93	86.57	361.66	448.23
Slots	594.31	1,559.10	1,772.71	2,002.10	676.00	1,442.18	2,118.18
Table Game	135.89	384.63	410.97	308.70	0.02	-	0.02
Total	899.79	2,364.44	2,661.04	2,936.98	930.20	2,182.87	3,113.07

Accounts (Millions)	Nov 2014 Mar 2015	Apr 2015 Mar 2016	Apr 2016[R] Mar 2017	Apr 2017[R] Mar 2018	Apr 2018[P] Mar 2019
Funds held in customer accounts	498.49	607.38	880.05	803.16	828.44
New account registrations	8.97	23.94	32.28	32.33	33.28
Active number accounts GC Licensed Facilities	21.78	23.05	30.52	34.22	30.99

9.12b Remote: Gambling (Licensing and Advertising) Act Data

Industry Statistics - November 2019

Total remote data (£m)						
Turnover	Nov 2014 Mar 2015	Apr 2015 Mar 2016	Apr 2016[R] Mar 2017	Apr 2017[R] Mar 2018	Apr 2018[P] Mar 2019		
	Total	Total	Total	Total	Proprietary	Revenue Share	Total
Betting	5,460.64	23,345.92	26,274.83	27,427.58	18,188.37	10,224.19	28,412.56
Bingo	331.21	1,472.77	1,408.10	1,311.49	657.00	662.01	1,319.01
Casino	7,470.54	65,494.81	73,653.53	83,875.06	27,351.22	64,563.61	91,914.83
Pool Betting	92.78	214.63	266.74	267.15	132.51	21.04	153.55
Total	13,355.17	90,528.12	101,603.20	112,881.27	46,329.10	75,470.85	121,799.95
GGY	Nov 2014 Mar 2015	Apr 2015 Mar 2016	Apr 2016[R] Mar 2017	Apr 2017[R] Mar 2018	Apr 2018[P] Mar 2019		
	Total	Total	Total	Total	Proprietary	Revenue Share	Total
Betting	478.71	1,625.29	1,806.65	2,110.63	1,258.93	714.37	1,973.31
Betting Exchange	56.88	171.82	191.00	195.60	164.49	26.42	190.91
Bingo	67.59	123.28	161.05	164.60	78.02	98.79	176.81
Casino	1,042.70	2,529.04	2,843.63	3,034.94	994.24	2,242.98	3,237.22
Pool Betting	37.29	73.95	96.09	129.94	50.08	5.63	55.71
Total	1,683.17	4,523.37	5,098.41	5,635.71	2,545.76	3,088.18	5,633.95
Betting Turnover	Nov 2014 Mar 2015	Apr 2015 Mar 2016	Apr 2016[R] Mar 2017	Apr 2017[R] Mar 2018	Apr 2018[P] Mar 2019		
	Total	Total	Total	Total	Proprietary	Revenue Share	Total
Cricket	240.17	498.71	514.67	761.02	438.14	278.49	716.63
Dogs	95.75	294.32	447.88	592.54	421.25	225.44	646.69
Financials	61.72	255.19	290.03	186.08	0.00	-	0.00
Football	2,182.80	6,022.46	8,328.79	10,670.30	7,701.67	3,592.96	11,294.63
Golf	36.24	186.18	198.68	264.82	186.07	90.91	276.97
Horses	1,681.73	5,479.14	6,523.75	9,045.64	6,408.38	3,403.26	9,811.64
Other	804.55	2,359.37	2,492.12	3,018.75	1,959.46	955.84	2,915.30
Tennis	450.46	1,494.07	1,611.41	2,369.87	1,161.32	1,043.29	2,204.61
Virtual	-	-	-	785.71	44.58	655.05	699.63
Unallocated rev share	-	6,971.12	6,134.23		-	-	-
Total	5,553.42	23,560.55	26,541.57	27,694.73	18,320.88	10,245.23	28,566.11
Betting GGY	Nov 2014 Mar 2015	Apr 2015 Mar 2016	Apr 2016[R] Mar 2017	Apr 2017[R] Mar 2018	Apr 2018[P] Mar 2019		
	Total	Total	Total	Total	Proprietary	Revenue Share	Total
Cricket	4.31	22.10	21.28	32.27	27.40	10.57	37.97
Dogs	10.97	30.51	45.16	55.84	37.44	20.28	57.72
Financials	3.42	13.93	37.88	22.58	0.00	-	0.00
Football	172.16	587.93	676.97	1,057.73	703.88	321.63	1,025.51
Golf	2.73	14.65	14.70	14.38	10.97	3.76	14.72
Horses	123.17	362.09	448.35	627.98	349.42	184.88	534.30
Other	70.61	240.05	255.01	398.62	269.10	85.86	354.97
Tennis	23.68	74.90	91.57	145.99	71.33	52.94	124.28
Virtual	-	-	-	80.78	3.95	66.51	70.45
Unallocated rev share	161.83	524.90	502.82	-	-	-	-
Total	572.88	1,871.05	2,093.74	2,436.18	1,473.50	746.42	2,219.92
Casino Turnover	Nov 2014 Mar 2015	Apr 2015 Mar 2016	Apr 2016[R] Mar 2017	Apr 2017[R] Mar 2018	Apr 2018[P] Mar 2019		
	Total	Total	Total	Total	Proprietary	Revenue Share	Total
Blackjack	-	-	-	2,301.16	2,243.15	7,394.20	9,637.35
Card Game	544.62	8,465.38	8,967.90	6,672.11	-	-	-
Other	243.77	3,674.38	5,284.28	7,169.78	3,516.11	4,926.17	8,442.28
Roulette	-	-	-	4,532.72	2,683.09	15,543.69	18,226.78
Slots	5,721.67	37,260.22	42,240.24	50,806.36	18,908.56	36,699.55	55,608.11
Table Game	960.48	16,094.83	17,161.11	12,392.93	0.31	-	0.31
Total	7,470.54	65,494.81	73,653.53	83,875.06	27,351.22	64,563.61	91,914.83
Casino GGY	Nov 2014 Mar 2015	Apr 2015 Mar 2016	Apr 2016[R] Mar 2017	Apr 2017[R] Mar 2018	Apr 2018[P] Mar 2019		
	Total	Total	Total	Total	Proprietary	Revenue Share	Total
Blackjack	-	-	-	52.86	49.84	173.85	223.69
Card Game	71.68	192.31	207.73	157.35	-	-	-
Other	139.19	212.54	249.28	245.23	99.06	192.54	291.60
Peer to Peer	49.44	113.37	102.62	99.14	70.30	27.60	97.90
Roulette	-	-	-	111.18	87.08	371.15	458.23
Slots	641.75	1,616.72	1,858.74	2,055.49	687.94	1,477.83	2,165.78
Table Game	140.64	394.10	425.26	313.68	0.02	-	0.02
Total	1,042.70	2,529.04	2,843.63	3,034.94	994.24	2,242.98	3,237.22

Accounts (Millions)	Nov 2014 Mar 2015	Apr 2015 Mar 2016	Apr 2016[R] Mar 2017	Apr 2017[R] Mar 2018	Apr 2018[P] Mar 2019
Funds held in customer accounts	519.07	637.63	937.14	876.80	910.57
Active number accounts GC Licensed Facilities	24.08	24.55	32.63	36.24	33.45

Source: Gambling Commission

9.12b Remote: Gambling (Licensing and Advertising) Act Data
Industry Statistics - November 2019

Notes on this section:
1. Since 1 November 2014 the Commission has been collecting GGY derived from revenue share agreements between licensed operators. To mitigate against the risk of double counting this revenue, the Commission requires operators to report only their relevant portion of the revenue share. Details of this arrangement can be seen in the remote regulatory returns consultation responses document, relevant sections are contained within Annex A..
(http://www.gamblingcommission.gov.uk/PDF/consultations/Review-of-remote-regulatory-return-responses-August-2014.pdf).
The Commission is working with operators to ensure reported data complies with regulatory return requirements.
2. Revenue share is defined as GGY which is subject to a contractual arrangement to be shared between two or more Commission licensed operators.
3. Customers gambling on betting exchanges tend to maintain a significant balance in their account as they need to have sufficient funds to cover the liabilities of their bets, rather than just have sufficient funds to cover the stake of the bets they intend to make.
4. Active accounts are those that have been used by customers in the last 12 months. New registrants includes new individual customer registrations that occurred during the period, but may not have gambled.
5. Customers may have accounts with more than one operator and therefore the data relates to accounts rather than the individuals.

9.13 Most Popular Boy and Girl Baby Names in England and Wales, 2016

BOYS				GIRLS			
Rank 1-50	Name	Rank 51-100	Name	Rank 1-49	Name	Rank 50-100	Name
1	OLIVER	51	JAXON	1	OLIVIA	51	MAISIE
2	HARRY	52	LUCA	2	AMELIA	52	HOLLY
3	GEORGE	53	MATTHEW	3	EMILY	53	EMMA
4	JACK	54	HARVEY	4	ISLA	54	GEORGIA
5	JACOB	55	HARLEY	5	AVA	55	AMBER
6	NOAH	55	REGGIE	6	ISABELLA	56	MOLLY
7	CHARLIE	57	TOMMY	7	LILY	57	HANNAH
8	MUHAMMAD	58	JENSON	8	JESSICA	58	ABIGAIL
9	THOMAS	59	LUKE	9	ELLA	59	JASMINE
10	OSCAR	59	MICHAEL	10	MIA	60	LILLY
11	WILLIAM	61	JAYDEN	11	SOPHIA	61	ANNABELLE
12	JAMES	62	JUDE	12	CHARLOTTE	62	ROSE
13	LEO	63	FRANKIE	13	POPPY	63	PENELOPE
14	ALFIE	64	ALBERT	14	SOPHIE	64	AMELIE
15	HENRY	65	STANLEY	15	GRACE	65	VIOLET
16	JOSHUA	66	ELLIOT	16	EVIE	66	BELLA
17	FREDDIE	67	GABRIEL	17	ALICE	67	ARIA
18	ARCHIE	68	MOHAMMAD	18	SCARLETT	68	ZARA
19	ETHAN	69	OLLIE	19	FREYA	69	MARIA
20	ISAAC	70	RONNIE	20	FLORENCE	70	NANCY
21	ALEXANDER	71	LOUIS	21	ISABELLE	71	DARCIE
22	JOSEPH	72	CHARLES	22	DAISY	72	LOTTIE
23	EDWARD	73	BLAKE	23	CHLOE	73	ANNA
24	SAMUEL	74	ELLIOTT	24	PHOEBE	74	SUMMER
25	MAX	75	LEWIS	25	MATILDA	75	MARTHA
26	LOGAN	76	FREDERICK	26	RUBY	76	HEIDI
27	LUCAS	77	NATHAN	27	EVELYN	77	GRACIE
28	DANIEL	78	TYLER	28	SIENNA	78	LUNA
29	THEO	79	JACKSON	29	SOFIA	79	MARYAM
30	ARTHUR	80	RORY	30	EVA	80	BEATRICE
31	MOHAMMED	81	RYAN	31	ELSIE	81	MILA
32	HARRISON	82	CARTER	32	WILLOW	82	DARCEY
33	BENJAMIN	83	DEXTER	33	IVY	83	MEGAN
34	MASON	84	ALEX	34	MILLIE	84	IRIS
35	FINLEY	85	AUSTIN	35	ESME	85	LEXI
36	SEBASTIAN	86	CALEB	36	ROSIE	86	ROBYN
37	ADAM	87	KAI	37	IMOGEN	87	AISHA
38	DYLAN	88	ALBIE	38	ELIZABETH	88	CLARA
39	ZACHARY	89	ELLIS	39	MAYA	88	FRANCESCA
40	RILEY	90	BOBBY	40	LAYLA	90	SARA
41	TEDDY	91	EZRA	41	EMILIA	91	VICTORIA
42	THEODORE	92	LEON	42	LOLA	92	ZOE
43	DAVID	92	ROMAN	43	LUCY	93	JULIA
44	ELIJAH	94	JESSE	44	HARPER	94	ARABELLA
45	JAKE	95	AARON	45	ELIZA	95	MADDISON
46	TOBY	96	IBRAHIM	46	ERIN	96	SARAH
47	LOUIE	97	LIAM	47	ELEANOR	97	FELICITY
48	REUBEN	98	JASPER	48	ELLIE	98	DARCY
49	ARLO	99	FELIX	49	HARRIET	99	LEAH
50	HUGO	100	FINN	49	THEA	100	LYDIA

Source: Office for National Statistics (ONS)

Notes:
These rankings have been produced using the exact spelling of the name given at birth registration. Similar names with different spellings have been counted separately.

Births where the name was not stated have been excluded from these figures. Of the 357,046 baby boys in the 2016 dataset, 12 were excluded for this reason and of the 339,225 baby girls in the 2016 dataset, 8 were excluded for this reason.

The sum of the counts for individual names appearing in Table 2 and Table 3 may not equal the count in Table 1. This is because births where the usual residence of mother was not stated at the time of registration have been excluded from the counts in Table 2 and Table 3.

* denotes new entry to top 100

263

9.14a Libraries overview [1,2]

	2005/06	2006/07	2007/08	2008/09	2009/10	2010/11	2011/12	2012/13	2013/14	2014/15	2015/16	2016/17 [3]
	%	%	%	%	%	%	%	%	%	%	%	%
Has use a public library service in the last 12 months (own-time or voluntary)	48.2	**46.1**	**45.0**	**41.1**	**39.4**	**39.7**	**38.8**	**37.0**	**35.4**	**34.5**	**33.4**	**34.0**
Has used a public library service in the last 12 months (any purpose)	:	:	:	:	:	:	:	:	:	:	:	36.2
Frequency of public library use (own-time or voluntary)												
At least once a week	7.9	**7.2**	**6.7**	**5.9**	**5.4**	**5.8**	**5.7**	**5.0**	**4.9**	**4.9**	**4.4**	:
At least once a month	16.4	15.7	**14.9**	**13.3**	**12.8**	**13.4**	**12.4**	**12.4**	**12.2**	**10.8**	**10.6**	:
3-4 times a year	13.4	12.9	13.0	**11.4**	**10.9**	**11.6**	**12.2**	**10.8**	**9.9**	**10.9**	**10.0**	:
1-2 times a year	10.4	10.3	10.5	**8.9**	**7.9**	**8.9**	**8.4**	**8.7**	**8.5**	**7.9**	**8.3**	:
Has not visited	51.8	**53.9**	**55.0**	**60.5**	**63.0**	**60.3**	**61.2**	**63.0**	**64.6**	**65.5**	**66.6**	:
Frequency of public library use (any purpose)												
At least once a week	:	:	:	:	:	:	:	:	:	:	:	5.5
At least once a month	:	:	:	:	:	:	:	:	:	:	:	11.6
3-4 times a year	:	:	:	:	:	:	:	:	:	:	:	10.0
1-2 times a year	:	:	:	:	:	:	:	:	:	:	:	9.1
Has not visited	:	:	:	:	:	:	:	:	:	:	:	63.8

Symbols
: Not available

Notes
[1] Figures in bold indicate a significant change from 2005/06
[2] Figures exclude people who have used a public library service for the purposes of paid work or academic study
[3] In 2016/17, the question on frequency was asked about public library use for any purpose rather than public library use in own-time or as part of voluntary work. Therefore, estimates of frequency of public library use in own-time or as part of voluntary work are not available for 2016/17.

Revision note for tables 9.14a,b and c

The figures on frequency of library use, which were published on 28 September 2017, have been revised slightly. The original 2016/17 figures were described as frequency of use in own time or as part of voluntary work. However, due to a change in questionnaire filters from April 2016, the survey now only collects frequency information for public library use for any purpose. The figures for frequency of public library use have therefore been revised and are now based on respondents who have used a public library service for any purpose. A new row has also been added to the Overview table showing the proportion of adults who used a public library for any purpose to provide context. All other figures remain as previously published

Source: Department for Culture, Media and Sport - Taking Part Survey 2016/17 Q4
https://www.gov.uk/government/statistics/taking-part-201617-quarter-4-statistical-release

9.14b Proportion who have visited a public library in the last year - area-level breakdown [1,2]

	2005/06	2006/07	2007/08	2008/09	2009/10	2010/11	2011/12	2012/13	2013/14	2014/15	2015/16	2016/17
	%	%	%	%	%	%	%	%	%	%	%	%
deprivation												
1- Most deprived	:	:	:	:	37.6	39.8	37.3	40.1	40.4	37.6	35.8	35.6
2	:	:	:	:	32.8	38.2	37.0	35.7	35.3	34.8	35.3	33.2
3	:	:	:	:	38.3	39.2	39.2	33.5	33.4	33.5	35.8	34.5
4	:	:	:	:	36.1	41.2	39.1	35.0	33.3	37.8	32.2	31.0
5	:	:	:	:	42.1	38.4	40.6	35.3	35.7	**33.9**	35.5	34.0
6	:	:	:	:	38.7	36.4	38.3	37.5	36.1	32.2	**30.4**	33.8
7	:	:	:	:	39.9	40.7	40.4	34.8	**33.3**	31.8	35.1	33.1
8	:	:	:	:	34.6	40.1	33.8	37.1	32.7	34.5	29.6	34.0
9	:	:	:	:	45.4	**38.7**	42.2	41.8	**37.1**	**35.7**	32.9	**34.4**
10- Least deprived	:	:	:	:	46.3	43.5	39.8	**38.4**	**36.8**	33.8	31.4	36.8
Region												
North East	44.9	43.9	**40.1**	37.1	42.9	**38.8**	36.3	38.2	34.1	34.6	32.5	27.1
North West	46.9	46.2	45.0	42.1	41.4	**43.0**	37.5	37.8	35.4	34.0	34.9	35.0
Yorkshire and the Humber	42.1	40.5	36.6	33.7	30.0	33.6	33.4	34.9	33.1	31.0	31.2	27.4
East Midlands	44.7	42.9	44.3	38.8	42.7	35.8	37.7	34.9	32.3	30.9	27.9	29.8
West Midlands	47.6	47.1	43.5	39.3	38.0	36.6	40.1	35.5	34.8	35.0	33.1	34.2
East of England	50.5	47.5	45.5	42.9	39.5	42.4	41.2	37.5	37.0	32.2	33.0	35.8
London	52.6	**49.2**	49.5	43.6	38.1	43.1	43.1	41.9	40.9	40.2	37.4	39.9
South East	51.0	48.3	48.5	44.5	43.5	40.5	37.4	34.6	34.0	34.9	31.9	35.6
South West	47.9	45.1	45.5	41.8	38.8	38.8	40.0	36.3	33.6	34.4	35.2	32.6
Urban	48.5	**46.5**	45.6	41.6	40.1	40.0	39.2	37.8	36.4	35.4	34.2	35.2
Rural	47.1	44.7	42.8	38.9	36.5	38.2	36.9	33.7	31.4	30.9	29.9	28.5
ACORN												
Wealthy Achievers	50.9	48.9	**47.4**	42.1	39.9	40.7	40.6	38.4	33.8	32.9	33.5	N/A
Urban Prosperity	57.3	**51.0**	52.0	41.8	**42.6**	43.9	41.9	39.2	38.3	38.8	37.8	N/A
Comfortably Off	48.5	**46.4**	44.2	41.9	41.9	38.2	37.9	35.4	35.0	34.0	32.6	N/A
Moderate Means	45.3	45.0	42.4	41.5	37.7	41.3	38.9	36.9	35.7	35.1	30.7	N/A
Hard-pressed	40.9	39.8	41.0	37.5	33.9	36.9	35.8	36.4	35.9	34.5	33.9	N/A
Unclassified	61.7	50.0	**62.7**	46.9	*	*	*	*	40.3	*	29.9	N/A
ACORN [3]												
Affluent Achievers	:	:	:	:	:	:	:	:	:	:	:	36.2
Rising Prosperity	:	:	:	:	:	:	:	:	:	:	:	36.1
Comfortable Communities	:	:	:	:	:	:	:	:	:	:	:	31.6
Financially Stretched	:	:	:	:	:	:	:	:	:	:	:	32.5
Urban Adversity	:	:	:	:	:	:	:	:	:	:	:	35.7
Not Private Households	:	:	:	:	:	:	:	:	:	:	:	*
All	48.2	**46.1**	45.0	41.1	39.4	39.7	38.8	37.0	35.4	34.5	33.4	34.0

Symbols

: Data not available

* Data suppressed due to small samples sizes

Notes

[1] Figures in bold indicate a significant change from 2005/06, except for the breakdowns by Index of Multiple Deprivation where figures in bold indicate a significant change since 2010/11

[2] Figures exclude people who have used a public library service for the purposes of paid work or academic study

[3] Following revisions to the ACORN methodology, the ACORN categories applied to the 2016/17 data are different from those applied in earlier survey years and 2016/17 estimates are not comparable with those from earlier years

Source: Department for Culture, Media and Sport - Taking Part Survey 2016/17 Q4
https://www.gov.uk/government/statistics/taking-part-201617-quarter-4-statistical-release

9.14c Proportion who have visited a public library in the last year - demographic breakdown[1,2]

	2005/06	2006/07	2007/08	2008/09	2009/10	2010/11	2011/12	2012/13	2013/14	2014/15	2015/16	2016/17
	%	%	%	%	%	%	%	%	%	%	%	%
Age												
16-24	51.0	**47.1**	**45.4**	**42.8**	**40.0**	34.4	34.5	32.3	33.4	28.5	27.4	29.1
25-44	51.2	50.2	49.5	**43.7**	40.9	44.6	44.0	42.2	40.4	41.3	37.4	38.0
45-64	45.7	44.3	42.1	38.8	39.5	36.0	36.1	33.1	30.9	30.6	31.0	30.2
65-74	46.7	44.4	44.7	42.0	39.3	44.3	35.8	38.6	37.1	35.2	34.8	38.9
75+	42.3	**37.1**	**37.6**	35.0	32.9	37.1	38.9	36.5	33.3	31.6	34.3	34.2
Sex												
Male	43.8	**42.1**	**40.2**	**35.3**	**35.5**	34.3	33.6	31.4	29.7	29.8	29.2	29.6
Female	52.3	**49.9**	**49.6**	**46.5**	**43.2**	44.8	43.8	42.3	40.8	39.0	37.4	38.3
NS-SEC												
Upper socio-economic group	52.1	**50.2**	**48.0**	**43.3**	**43.1**	43.9	42.3	39.7	36.7	35.9	35.5	36.9
Lower socio-economic group	40.1	**38.1**	38.7	**35.1**	32.3	33.6	33.5	33.0	32.4	31.2	30.3	29.0
Employment status												
Not working	49.7	**46.7**	**46.6**	**44.0**	42.4	42.9	41.7	41.0	38.3	37.8	37.1	38.9
Working	47.2	**45.7**	**44.0**	**39.0**	37.4	37.5	36.8	34.3	33.4	32.3	31.0	30.6
Tenure												
Owners	48.7	**46.4**	**44.8**	**41.1**	39.8	40.1	39.2	37.0	34.9	33.8	33.2	34.5
Social rented sector	41.9	40.5	42.2	39.0	36.8	37.0	37.2	38.6	38.1	39.1	37.1	33.6
Private rented sector	53.3	51.0	**49.2**	**43.0**	39.8	40.2	38.7	35.8	35.0	33.4	31.3	33.0
Ethnicity												
White	47.2	**44.9**	**43.6**	**40.1**	37.9	38.3	37.8	35.9	33.8	32.9	31.6	32.3
Black or ethnic minority	57.5	56.7	57.9	**50.2**	50.6	50.0	46.5	45.1	47.8	47.5	45.6	45.4
Religion												
No religion	46.8	45.5	45.3	**40.1**	35.2	37.8	36.8	33.2	34.4	33.2	30.3	30.3
Christian	47.3	**44.8**	**43.7**	**40.3**	39.7	39.2	38.8	37.7	34.3	33.4	33.5	34.5
Other religion	58.2	58.9	57.1	52.7	53.0	**48.2**	44.4	46.9	48.3	49.7	46.7	46.2
Long-standing illness or disability												
No	50.0	**48.0**	**46.0**	**42.1**	**40.4**	40.1	39.2	36.9	36.0	35.7	33.2	33.3
Yes	43.8	**41.5**	42.4	**38.4**	36.8	38.5	37.9	37.3	34.2	31.6	33.8	35.6
All	48.2	**46.1**	**45.0**	**41.1**	39.4	39.7	38.8	37.0	35.4	34.5	33.4	34.0

Notes

[1] Figures in bold indicate a significant change from 2005/06.

[2] Figures exclude people who have used a public library service for the purposes of paid work or academic study

Source: Department for Culture, Media and Sport - Taking Part Survey 2016/17 Q4
https://www.gov.uk/government/statistics/taking-part-201617-quarter-4-statistical-release

9.15 Museum and galleries overview [1]

Proportion of adults who have visited a museum or gallery in the last 12 months, 2005/06 to 2016/17, England

	2005/06	2006/07	2007/08	2008/09	2009/10	2010/11	2011/12	2012/13	2013/14	2014/15	2015/16	2016/17
	%	%	%	%	%	%	%	%	%	%	%	%
Has visited a museum or gallery in the last 12 months [2]	42.3	41.5	**43.5**	43.4	**46.0**	**46.3**	**48.9**	**52.8**	**53.1**	**52.0**	**52.5**	**52.3**
Frequency of attendance [2]												
At least once a week	0.3	**0.4**	0.3	0.4	0.4	**0.5**	**0.5**	**0.6**	0.4	**0.5**	0.4	**0.6**
At least once a month	3.2	2.8	3.2	2.9	3.6	**3.6**	3.4	3.5	3.1	3.1	3.6	**4.1**
3-4 times a year	13.2	12.9	13.8	13.9	14.1	**14.7**	**15.3**	**17.2**	**17.2**	**17.7**	**18.0**	**17.2**
1-2 times a year	25.6	25.3	26.3	26.1	**27.9**	**27.4**	**29.6**	**31.3**	**32.4**	**30.7**	**30.6**	**30.3**
Has not visited	57.7	58.5	**56.5**	56.7	**54.0**	**53.7**	**51.1**	**47.2**	**46.9**	**48.0**	**47.5**	**47.7**
Have been to an museum or gallery in last 12 months [3]												
In your own time	:	:	:	96.8	97.4	97.5	**98.1**	97.4	**97.9**	**98.1**	**98.3**	96.9
For paid work	:	:	:	2.5	2.0	2.6	2.5	2.8	2.1	2.4	2.4	3.2
For academic study	:	:	:	3.0	3.1	2.3	**2.2**	2.6	2.5	**2.1**	**2.0**	2.7
As a part of voluntary work	:	:	:	.6	0.4	0.5	0.5	0.6	0.4	0.6	0.5	0.9
For some other reason	:	:	:	.2	0.1	0.2	0.1	0.2	0.2	0.2	0.1	0.2

Symbols
: Data not available

Notes
[1] Figures in bold indicate a significant change from 2005/06, except for purpose of visit where figures in bold indicate a significant change since 2008/09
[2] Figures exclude people who have visited a museum or gallery for the purpose of paid work or academic study
[3] Figures do not sum to 100 as respondents were able to select multiple options

Source: Department for Culture, Media and Sport - Taking Part Survey 2016/17 Q4
https://www.gov.uk/government/statistics/taking-part-201617-quarter-4-statistical-release

9.16 Participation in voluntary activities, England 2001 to 2017-18

Face-to-Face Estimates

England, 2001 to 2015-16

Percentages — At least once a month

	2001[1]	2003[1]	2005[1]	2007-08[1]	2008-09[1]	2009-10[1]	2010-11[1]	2012-13	2013-14	2014-15	2015-16
Informal volunteering[2]	34	37	37	35	35	29	29	36	35	34	34
Formal volunteering[3]	27	28	29	27	26	25	25	29	27	27	27
Any volunteering[4]	46	50	50	48	47	42	41	49	48	47	47

England, 2001 to 2015-16

Percentages — At least once in last year

	2001[1]	2003[1]	2005[1]	2007-08[1]	2008-09[1]	2009-10[1]	2010-11[1]	2012-13	2013-14	2014-15	2015-16
Informal volunteering[2]	67	63	68	64	62	54	55	62	64	59	60
Formal volunteering[3]	39	42	44	43	41	40	39	44	41	42	41
Any volunteering[4]	74	73	76	73	71	66	65	72	74	69	70
Employer volunteering[5]	n/a	n/a	n/a	n/a	n/a	n/a	n/a	6	8	8	8
Respondents	9,430	8,920	9,195	8,804	8,768	8,712	9,664	6,915	5,105	2,022	3,027

Online/Paper Estimates[6]

England, 2013-14 to 2017-18

Percentages — At least once a month

	2013-14	2014-15	2015-16	2016-17	2017-18
Informal volunteering[2]	31	28	29	27	27
Formal volunteering[3]	27	25	21	22	22
Any volunteering[4]	44	41	39	39	38

England, 2013-14 to 2017-18

Percentages — At least once in last year

	2013-14	2014-15	2015-16	2016-17	2017-18
Informal volunteering[2]	58	54	54	52	53
Formal volunteering[3]	45	40	37	37	38
Any volunteering[4]	70	65	65	63	64
Employer volunteering[5]	6	5	5	6	-
Respondents	10,215	2,323	3,256	10,256	10,217

Source: Community Life Survey 2017/18

[1] Data collected through the Citizenship Survey
[2] Informal volunteering refers to giving unpaid help to individuals who are not a relative.
[3] Formal volunteering refers to giving unpaid help through clubs or organisations
[4] Participated in either formal or informal volunteering
[5] Volunteering undertaken by employees that is encouraged by employers / companies
[6] Demographic breakdowns exclude those with missing answers.

The Community Life Survey (CLS) is a national survey, capturing views on a range of issues critical for supporting stronger communities.
Following thorough testing, the CLS has moved to an online and paper mixed method approach from 2016-17 onwards, with an end to the current face-to-face method.
As this change represents a break in the time series, differences observed between the two data collection modes mean that the data are not comparable; we have published online/paper data for the 3 years 2013-14 to 2015-16 to enable trend analysis.

This data was collected during the methodological testing of moving the survey to an online/paper mixed mode.
More information on this change can be found in the Online/Postal Estimates Technical note:
https://www.gov.uk/government/statistics/community-life-survey-2017-18

We would advise users to use the new online/postal estimates for any analysis.
We do not recommend appending the 2016-17 online/postal estimate onto the face-to-face backseries as any differences may be the result of a change in mode, not a true change.

9.17 UK residents' visits to friends and relatives[1] abroad: by destination

United Kingdom

Percentages[2]

	2006	2009	2011	2012	2013	2014	2015	2016	2017
Irish Republic	15	14	13	10	10	11	10	9	8
France	11	11	10	10	10	11	9	9	9
Poland	6	9	9	9	9	9	10	10	10
Spain	9	8	7	8	8	8	7	7	8
Germany	6	5	5	5	5	5	5	4	5
USA	6	5	5	4	4	4	4	4	3
India	3	4	4	4	4	4	4	4	4
Italy	4	4	3	4	4	4	5	5	5
Netherlands	3	3	3	3	3	3	3	3	2
Pakistan	3	3	3	3	3	3	2	3	3
Other countries	35	36	39	39	39	37	35	42	42
All destinations (=100%) (millions)	12.0	11.6	11.6	11.8	12.3	13.3	14.7	16.6	17.9

Source: International Passenger Survey,
Office for National Statistics

1 As a proportion of all visits to friends and relatives taken abroad by residents of the UK.
Excludes business trips and other miscellaneous visits.

2 Percentages may not add up to 100 per cent due to rounding.

9.18 The Internet

Internet activities, by age group, sex, 2017

Within the last 3 months   %

	Age group						Sex		All
	16-24	25-34	35-44	45-54	55-64	65+	Men	Women	
Sending/receiving emails	93	91	93	88	81	56	84	80	**82**
Finding information about goods and services	69	78	80	81	77	48	73	68	**71**
Social networking (eg Facebook or Twitter)	96	88	83	68	51	27	65	67	**66**
Reading online news, newspapers or magazines	74	78	75	74	57	37	68	60	**64**
Internet banking	74	82	82	68	58	30	67	60	**63**
Looking for health-related information (eg injury, disease, nutrition, improving health etc)	52	68	68	58	48	30	49	57	**53**
Using services related to travel or travel related accommodation	46	59	61	62	58	31	53	51	**52**
Uploading content created by you to a website to be shared	72	61	60	50	40	20	48	49	**48**
Telephoning or making video calls over the internet via a webcam	69	64	60	43	35	20	48	45	**46**
Looking for a job or sending a job application	38	35	28	22	14	0	22	21	**22**
Professional networking	28	33	29	25	14	2	25	17	**21**
Selling goods or services over the internet	25	26	28	22	12	8	22	17	**19**
Posting opinions on civic or political issues	21	22	17	13	10	8	16	13	**15**
Taking part in online consultations or voting on civic or political issues	12	14	16	14	10	7	13	11	**12**

Base: Adults (aged 16+) in Great Britain.   Source: Office for National Statistics

Online purchases, by age group, sex, 2017

Within the last 12 months   %

	Age group						Sex		All
	16-24	25-34	35-44	45-54	55-64	65+	Men	Women	
Clothes or sports goods	73	70	72	63	47	24	53	59	**56**
Household goods (eg furniture, toys etc)	40	65	72	55	49	25	50	50	**50**
Holiday accommodation	37	52	55	54	47	23	47	41	**44**
Travel arrangements (eg transport tickets, car hire)	40	48	47	47	41	20	43	36	**40**
Tickets for events	44	42	46	46	36	14	39	34	**37**
Films, music (including downloads)	45	47	48	36	26	10	39	29	**34**
Books, magazines, newspapers (including e-books and downloads)	25	34	39	31	32	19	29	30	**29**
Food or groceries	24	37	41	25	21	10	23	28	**26**
Electronic equipment (including cameras)	39	33	33	27	19	7	33	18	**25**
Video games software, other computer software and upgrades (including downloads)	40	33	34	22	15	5	31	17	**24**
Telecommunication services	15	25	32	22	18	9	25	15	**20**
Computer hardware	23	20	22	17	11	4	21	10	**15**
Medicine	8	13	16	13	14	8	13	11	**12**
E-learning material	7	10	16	9	6	1	8	7	**8**

Base: Adults (aged 16+) in Great Britain.   Source: Office for National Statistics

Households with internet access, 2005 to 2017

	Year	%		Year	%
GB	2005	55		2011	77
	2006	57		2012	80
	2007	61		2013	83
	2008	65		2014	84
	2009	70		2015	86
	2010	73		2016	89
				2017	90

Source: Office for National Statistics

GB estimates from 2005 to 2017.

2015 to 2017 estimates relate to January, February and April. Previous estimates relate to January to March, except 1998 which relates to April to June and 2005 which relates to May. During 2005, estimates were published at irregular intervals as Topic Based Summaries. These are no longer available on the ONS website.

9.19 Radio Listening

United Kingdom (including Channel Islands and Isle of Man)

Adults aged 15 and over: population[1]: 54,752,000

	Weekly Reach[2] (000s)	Weekly Reach[3] %	Average Hours per head[4]	Average Hours per listener[5]	Total Hours[6] (000s)	Share of Listening in TSA %
Quarterly Summary of Radio Listening - Survey period ending June 2018						
All Radio	48826	89%	18.5	20.8	1015622	100.0%
All BBC Radio	34468	63%	9.6	15.2	525464	51.7%
BBC Local Radio	7874	14%	1.2	8.6	67928	6.7%
All Commercial Radio	35507	65%	8.5	13.1	464279	45.7%
All National Commercial[8]	20574	38%	3.4	8.9	183949	18.1%
All Local Commercial (National TSA)	26591	49%	5.1	10.5	280330	27.6%
All BBC Network Radio[8]	31613	58%	8.4	14.5	457536	45.0%
BBC Radio 1	9236	17%	1.1	6.5	59667	5.9%
BBC Radio 2	14935	27%	3.3	12.2	181477	17.9%
BBC Radio 3	1908	3%	0.2	6.0	11448	1.1%
BBC Radio 4	10998	20%	2.4	11.9	130649	12.9%
BBC Radio 4 (including 4 Extra)	10598	19%	2.2	11.3	119258	11.7%
BBC Radio 4 Extra	1965	4%	0.2	5.8	11391	1.1%
BBC Radio 5 live	5086	9%	0.6	6.9	34854	3.4%
BBC Radio 5 live (inc. sports extra)	4733	9%	0.6	6.6	31306	3.1%
BBC Radio 5 live sports extra	1173	2%	0.1	3.0	3548	0.3%
BBC 6 Music	2444	4%	0.4	9.9	24283	2.4%
1Xtra from the BBC	1033	2%	0.1	4.1	4284	0.4%
BBC Asian Network UK	672	1%	0.1	5.2	3509	0.3%
BBC World Service	1514	3%	0.1	4.9	7353	0.7%

Source: RAJAR/Ipsos MORI/RSMB

STATION SUMMARY ANALYSIS - RAJAR REPORTING PERIOD 2018 Q1

(Report 1690) All Radio, Adults Only

Quarterly Weighted Data

	All Persons	Males						
	Total	Total	15-24	25-34	35-44	45-54	55-64	65+
Unw. Sample	22219	11046	1118	1233	1460	2073	2399	2763
Est. Pop'n[1]	54466	26648	4074	4541	4146	4560	3838	5490
ALL RADIO								
Weekly Reach 000s	49153	23864	3216	3917	3875	4258	3630	4967
Weekly Reach %	90.2	89.6	78.9	86.3	93.5	93.4	94.6	90.5
Total Hours	1024585	537030	45217	78181	87764	108742	94283	122843
Average Hours Per Head	18.8	20.2	11.1	17.2	21.2	23.8	24.6	22.4
Average Hours Per Listener	20.8	22.5	14.1	20	22.6	25.5	26	24.7
Market Share	100	100	100	100	100	100	100	100

	Females						
	Total	15-24	25-34	35-44	45-54	55-64	65+
Unw. Sample	11173	1145	1287	1868	2201	2125	2547
Est. Pop'n[1]	27818	3867	4496	4211	4701	3966	6576
ALL RADIO							
Weekly Reach 000s	25289	3377	3955	3853	4418	3767	5919
Weekly Reach %	90.9	87.3	88	91.5	94	95	90
Total Hours	487555	38727	55030	66930	89049	87220	150599
Average Hours Per Head	17.5	10	12.2	15.9	18.9	22	22.9
Average Hours Per Listener	19.3	11.5	13.9	17.4	20.2	23.2	25.4
Market Share	100	100	100	100	100	100	100

Source: RAJAR

9.19 Radio Listening

United Kingdom (including Channel Islands and Isle of Man)
Adults aged 15 and over: population[1]: 54,752,000

1. The number of people aged 15+ who live within the TSA of a given station
2. The number of people aged 15+ who listen to a radio station for at least 5 minutes in the course of an
3. The Weekly Reach expressed as a percentage of the Population within the TSA.
4. The total hours of listening to a station during the course of a week, averaged across the total adult population of the UK.
5. The total hours of listening to a station during the course of a week, averaged across all those listening to
the station for at least 5 minutes.
6. The overall number of hours of adult listening to a station in the UK in an average week.
7. The percentage of total listening time accounted for by a station in the UK in an average week.
8. Audiences in local analogue areas excluded from 'All BBC Network Radio' and 'All National Commercial' totals.
Total Survey Area (TSA): The area within which a station's audience is measured. This is defined by the
station using postcode
Population: The number of people aged 15+ who live within the TSA of a given station.

**RAJAR Ltd (Radio Joint Audience Research) was set up in 1992 to align, design and operate a single
audience measurement system for the UK radio industry serving both the BBC and licensed commercial
stations. The company is jointly owned by the BBC (British Broadcasting Corporation) and by the
RadioCentre (the trade body representing the vast majority of Commercial Radio stations in the UK). For
more information visit: www.rajar.co.uk**

Environment

Chapter 10

Environment

Air emissions (Table 10.2 to 10.8)

Emissions of air pollutants arise from a wide variety of sources. The National Atmospheric Emissions Inventory (NAEI) is prepared annually for the Government and the devolved administrations by AEA Energy and Environment, with the work being co-ordinated by the Department for Business, Energy & Industrial Strategy (BEIS). Information is available for a range of point sources including the most significant polluters. However, a different approach has to be taken for diffuse sources such as transport and domestic emissions, where this type of information is not available. Estimates for these are derived from statistical information and from research on emission factors for stationary and mobile sources. Although for any given year considerable uncertainties surround the emission estimates for each pollutant, trends over time are likely to be more reliable.

UK national emission estimates are updated annually and any developments in methodology are applied retrospectively to earlier years. Adjustments in the methodology are made to accommodate new technical information and to improve international comparability.

Three different classification systems are used in the tables presented here; a National Accounts basis (Table 10.2), the format required by the Inter-governmental Panel on Climate Change (IPCC) (Table 10.3) and the National Communications (NC) categories (Tables 10.5-10.7).

The NC source categories are detailed below together with details of the main sources of these emissions:

Energy supply total: Power stations, refineries, manufacture of solid fuels and other energy industries, solid fuel transformation, exploration, production and transport of oils, offshore oil and gas – venting and flaring, power stations - FGD, coal mining and handling, and exploration, production and transport of gas.

Business total: Iron and steel – combustion, other industrial combustion, miscellaneous industrial and commercial combustion, energy recovery from waste fuels, refrigeration and air conditioning, foams, fire fighting, solvents, one components foams, and electronics, electrical insulation and sporting goods.

Transport total: Civil aviation (domestic, landing and take off, and cruise), passenger cars, light duty vehicles, buses, HGVs, mopeds & motorcycles, LPG emissions (all vehicles), other road vehicle engines, railways, railways – stationary combustion, national navigation, fishing vessels, military aircraft and shipping, and aircraft – support vehicles.

Residential total: Residential combustion, use of non aerosol consumer products, accidental vehicle fires, and aerosols and metered dose inhalers.

Agriculture total: Stationary and mobile combustion, breakdown of pesticides, enteric fermentation (cattle, sheep, goats, horses, pigs, and deer), wastes (cattle, sheep, goats, horses, pigs, poultry, and deer), manure liquid systems, manure solid storage and dry lot, other manure management, direct soil emission, and field burning of agricultural wastes.

Industrial process total: Sinter production, cement production, lime production, limestone and dolomite use, soda ash production and use, fletton bricks, ammonia production, iron and steel, nitric acid production, adipic acid production, other – chemical industry, halocarbon production, and magnesiun cover gas.

Land-use change: Forest land remaining forest land, forest land biomass burning, land converted to forest land, direct N_2O emissions from N fertilisation of forest land, cropland liming, cropland remaining cropland, cropland biomass burning, land converted to cropland, N_2O emissions from disturbance associated with land-use conversion to cropland, grassland biomass burning, grassland liming, grassland remaining grassland, land converted to grassland, wetlands remaining wetland, Non-CO2 emissions from drainage of soils and wetlands, settlements biomass burning, land converted to settlements, and harvested wood.

Waste management total: Landfill, waste-water handling, and waste incineration.

Atmospheric emissions on a National Accounts basis (Table 10.1)

The air and energy accounts are produced for ONS by AEA Technology plc based on data   compiled for the National Atmospheric Emissions Inventory (NAEI)7 and UK Greenhouse Gas Inventory (GHGI)8. Every year a programme of development work is undertaken to optimise the   methodologies employed in compiling the accounts. Assessments in previous years have   indicated that a number of splits used to apportion road transport source data to more than one   industry should be reviewed. The results of this review have been implemented in the 2011 UK   Environmental Accounts for reference period 2009 and years back to 1990.

The industry breakdown used in the accounts has moved to using the Standard Industrial   Classification 2007 (SIC 2007). Historically, the accounts were based on Environmental Accounts codes (EAcodes) based on SIC 2003. This change will allow the accounts which are broken down by industry to be more readily compared with other economic statistics. A methodology article that outlines this change in more detail was published on the ONS website in May 2011: http//www.statistics.gov.uk/cci/article.asp?id=2694. As a result while names given to the   breakdown maybe similar they are not necessarily the same.

The National Accounts figures in Table 10.2 differ from those on an IPCC basis, in that they   include estimated emissions from fuels purchased by UK resident households and companies   either at home or abroad (including emissions from UK international shipping and aircraft   operators), and exclude emissions in the UK resulting from the activities of non-residents. This   allows for a more consistent comparison with key National Accounts indicators such as Gross   Domestic Product (GDP).

Greenhouse gases include carbon dioxide, methane, nitrous oxide, hydro-fluorocarbons, perfluorocarbons and sulphur hexafluoride which are expressed in thousand tonnes of carbon   dioxide equivalent.

Acid rain precursors include sulphur dioxide, nitrogen oxides and ammonia which are expressed as thousand tonnes of sulphur dioxide equivalent.

Road Transport Emissions (Table 10.2)

Various pollutants are emitted from road transport into the atmosphere. Table 10.2 shows emissions by pollutant generated from combustion by road vehicles.

Greenhouse gases are made up of carbon dioxide, methane and nitrous oxide
Acid rain precursors are made of sulphur dioxide, nitrogen oxides and ammonia

Estimated total emissions of greenhouse gases on an IPCC basis (Table 10.3)

The IPCC classification is used to report greenhouse gas emissions under the UN Framework Convention on Climate Change (UNFCCC) and includes Land Use Change and all emissions from Domestic aviation and shipping, but excludes International aviation and shipping bunkers.   Estimates of the relative contribution to global warming of the main greenhouse gases, or classes   of gases, are presented weighted by their global warming potential.

Greenhouse gas emissions bridging table (Table 10.4)
National Accounts measure to UNFCCC measure
The air and energy accounts are produced for ONS by AEA Technology plc based on data   compiled for the National Atmospheric Emissions Inventory (NAEI)7 and UK Greenhouse Gas   Inventory (GHGI)8. Every year a programme of development work is undertaken to optimise the   methodologies employed in compiling the accounts. Assessments in previous years have indicated that a number of splits used to apportion road transport source data to more than one industry should be reviewed. The results of this review have been implemented in the 2011 UK   Environmental Accounts for reference period 2009 and years back to 1990.   There are a number of formats for the reporting and recording of atmospheric emissions data, including those used by the Department for Business, Energy & Industrial Strategy (BEIS) for reporting greenhouse gases under UNFCCC and the Kyoto Protocol, and for reporting air pollutant emissions to the UN Economic Commission for Europe (UNECE), which differ from the National Accounts consistent measure published by the Office for National Statistics (ONS).
Differences between the National Accounts measure and those for reporting under UNFCCC and   the Kyoto Protocol, following the guidance of the IPCC, are shown in Table 10.4.

Emissions of carbon dioxide (Table 10.5)
Carbon dioxide is the main man-made contributor to global warming, with the highest source of emissions being Transport.

Emissions of methane (Table 10.6)
The overall amount of methane emissions has fallen year on year since 2002, except 2016 when it stayed the same as the previous year. The highest source of emissions of methane for the 6th year running is Agriculture.

Emissions of nitrous oxide (Table 10.7)
Agriculture continues to be the highest source of nitrous oxide emissions for 2016.

Material Flow Account (Table 10.8)
Economy-wide material flow accounts estimate the physical flow of materials through our economy. As well as providing an aggregate overview of the annual extraction of raw materials, they also measure the physical amounts of imports and exports. This information is important in attempting to understand resource productivity. For example, they shed light on the depletion of natural resources and seek to promote a sustainable and more resource-efficient economy.

Annual rainfall (Table 10.9)

Regional rainfall is derived by the Met Office's National Climate Information Centre for the National Hydrological Monitoring Programme at the Centre for Ecology and Hydrology. These monthly area rainfalls are based initially on a subset of rain gauges (circa 350) but are updated after four to five   months with figures using the majority of the UK's rain gauge network.

The regions of England shown in this table correspond to the original nine English regions of the   National Rivers Authority (NRA). The NRA became part of the Environment Agency on its creation   in April 1996. The figures in this table relate to the country of Wales, not the Environment Agency   Welsh Region.

UK weather summary (Table 10.10)

Table 10.10 represents an initial assessment of the weather that was experienced across the UK   and how it compares with the 1961 to 1990 average.

Final averages use quality controlled data from the UK climate network of observing stations. They show the Met Office's best assessment of the weather that was experienced across the UK during the years and how it compares with the 1961 to 1990 average. The columns headed
'Anom' (anomaly) show the difference from, or percentage of, the 1961 to 1990 long-term average.

Biological and chemical quality of rivers and canals (Table 10.11)

Table 10.11 shows a Summary of River Basin District Ecological and Chemical Status from 2009-2016. It looks at each river basin district in terms of Percent of surface water bodies that are at good chemical status or better and also Percent of surface water bodies are at good ecological status/potential or better. Improvements are measured in terms of the number of water bodies meeting good status

Status of Rivers and Canals in Scotland (Table 10.12)

Looks at the overall status of rivers & canals in Scotland, including water quality.

Reservoir stocks in England and Wales (Table 10.13)

Data are collected for a network of major reservoirs (or reservoir groups) in England and Wales for the National Hydrological Monitoring Programme at the Centre for Ecology and Hydrology. Figures   of usable capacity are supplied by the Water PLCs and the Environment Agency at the start of   each month and are aggregated to provide an index of the total reservoir stocks for England and   Wales.

Water Industry Regulatory Capital Values (Table 10.14)

Table 10.14 shows the Water Industry Regulatory Capital Values.

Water pollution incidents (Table 10.15)

The Environment Agency responds to complaints and reported incidents of pollution in England. Each incident is then logged and categorised according to its severity. The category describes the impact of each incident on water, land and air. The impact of an incident on each medium is considered and reported separately. If no impact has occurred for a particular medium, the incident is reported as a category 4. Before 1999, the reporting system was used only for water pollution incidents; thus the total number of substantiated incidents was lower, as it did not include incidents not relating to the water environment.

Bathing waters (Table 10.16)

Table 10.16 shows bathing water compliance results for 2016/17. These results summarise the compliance of coastal and inland bathing waters to the Bathing Water Directive (2006/7/EC) for the whole of the United Kingdom in 2016/17.

Surface and groundwater abstractions (Table 10.17)

Significant changes in the way data is collected and/or reported were made in 1991 (due to the Water Resources Act 1991) and 1999 (commission of National Abstraction Licensing Database). Figures are therefore not strictly comparable with those in previous/intervening years. From 1999, data have been stored and retrieved from one system nationally and are therefore more accurate and reliable. Some regions report licensed and actual abstracts for financial rather than calendar years. As figures represent an average for the whole year expressed as daily amounts, differences between amounts reported for financial and calendar years are small.

Under the Water Act 2003, abstraction of less than 20 m3/day became exempt from the requirement to hold a licence as of 1 April 2005. As a result over 22,000 licences were deregulated, mainly for agricultural or private water supply purposes. However, due to the small volumes involved, this has had a minimal affect on the estimated licensed and actual abstraction totals.   The following changes have occurred in the classification of individual sources:

• Spray irrigation: this category includes small amounts of non-agricultural spray irrigation
• Mineral washing: from 1999 this was not reported as a separate category; licences for 'Mineral washing' are now contained in 'Other industry'
• Private water supply: this was shown as separate category from 1992 and includes private abstractions for domestic use and individual households
• Fish farming, cress growing, amenity ponds: includes amenity ponds, but excludes miscellaneous from 1991

Estimates of remaining recoverable oil and gas reserves (Table 10.18)

Only a small proportion of the estimated remaining recoverable reserves of oil and gas are known with any degree of certainty. The latest oil and gas data for 2016 shows that the upper range of total UK oil reserves was estimated to be around 1.6 billion tonnes, while UK gas reserves were around 1658 billion cubic metres. Of these, proven reserves of oil were 0.3 billion tonnes and proven reserves of gas were 176 billion cubic metres.

Local authority collected (Table 10.19)

Local authority collected includes household and non-household waste that is collected and disposed of by local authorities. It includes regular household collections, specific recycling collections, and special collections of bulky items, waste received at civic amenity sites, and waste collected from non-household sources that come under the control of local authorities.

Waste arisings from households (Table 10.20)

The 'waste from households' calculation was first published by Defra in May 2014. It was introduced for statistical purposes to provide a harmonised UK indicator with a comparable calculation in each of the four UK countries and to provide a consistent approach to report recycling rates at UK level on a calendar year basis under the Waste Framework Directive (2008/98/EC). The waste from household measure is a narrower measure than the
'household waste' measure which was previously used and excludes waste not considered to have come directly from households, such as recycling from street bins, parks and grounds.

Chartered Institute of Environmental Health Survey of Local Authority Noise Enforcement Activity (Table 10.21)

Every year the CIEH collects data on noise complaints made to local authorities and on the enforcement actions consequently taken by them. The data helps to inform policy and practice in environmental protection and public health and, in particular, the survey results provide the source of data for the Public Health England Outcome Indicator on noise. This, part of the Public Health Outcomes Framework, provides recognition of noise as one of the wider determinants of health.

Government revenue from environmental taxes (Table 10.22)

Environmental taxes data are based on the definition outlined in Regulation (EU) No 691/2011 on European environmental economic accounts. The European Statistical Office (Eurostat) define an environmental tax as a tax whose base is a physical unit (for example, a litre of petrol or a passenger flight) that has a proven negative impact on the environment. These taxes are designed to promote environmentally positive behaviour, reduce damaging effects on the environment and generate revenue that can potentially be used to promote further environmental protection.
In 2017, revenue from environmentally related taxes stood at £48,889 million. This corresponded to 2.4% of the UK's gross domestic product (GDP). Looking over the time series as a whole, environmental taxes as a share of GDP has remained at a broadly consistent level of between 2% and 3%

10.1 Atmospheric emissions, 2016[1]

	Thousand tonnes CO₂ equivalent							Thousand tonnes of SO₂ equivalent			
	Total greenhouse gas emissions	Carbon Dioxide (CO_2)	Methane (CH_4)	Nitrous Oxide (N_2O)	Hydrofluoro-carbons (HFCs)	Perfluoro-carbons (PFCs)	Sulphur hexafluoride (SF_6)	Total acid rain precursors	Sulphur Dioxide (SO_2)	Nitrogen Oxides (NO_X)	Ammonia (NH_3)
Agriculture, forestry and fishing	49367	8364	26290	14667	46	0	0	526	3	34	489
Mining and quarrying	21640	19467	1719	446	8	0	0	84	11	73	0
Manufacturing	84136	81071	225	519	1821	354	146	190	85	96	9
Electricity, gas, steam and air conditioning supply	106919	102375	3812	488	21	0	223	120	36	84	0
Water supply; sewerage, waste management and remediation activiti	23203	3651	17992	1385	156	0	18	18	1	10	6
Construction	12810	12017	25	466	302	0	0	38	0	37	0
Wholesale and retail trade; repair of motor vehicles and motorcycles	17343	12577	35	161	4569	0	0	25	0	24	0
Transport and storage	86012	83467	84	1043	1419	0	0	479	114	364	1
Accommodation and food services	3540	2918	9	18	595	0	0	4	0	3	0
Information and communication	1558	1309	2	13	235	0	0	2	0	2	0
Financial and insurance activities	132	12	0	0	120	0	0	0	0	0	0
Real estate activities	747	668	1	2	76	0	0	1	0	1	0
Professional, scientific and technical activities	2319	1873	3	17	425	0	1	3	0	3	0
Administrative and support service activities	3297	2872	2	34	388	0	0	8	0	7	1
Public administration and defence; compulsory social security	4931	4591	6	33	184	0	118	17	2	13	2
Education	2917	2599	6	5	308	0	0	5	2	3	0
Human health and social work activities	5764	4667	10	603	484	0	1	5	0	5	0
Arts, entertainment and recreation	1454	1339	3	7	104	0	0	6	0	2	4
Other service activities	1367	1236	2	5	123	0	0	2	0	2	0
Activities of households as employers; undifferentiated goods and services-producing activities of households for own use	98	91	0	1	6	0	0	0	0	0	0
Consumer expenditure	146723	140961	1339	648	3774	0	0	183	30	122	31
Total	576277	488124	51567	20561	15164	354	507	1717	284	888	544
Of which: emissions from road transport[2]	118041	116857	101	1083	..	..	..	219	..	210	8

Source: Ricardo Energy & Environment, ONS

Notes:

1. Components may not sum to totals due to rounding.
2. Includes emissions from fuel sources which are used by road vehicles (eg HGVs, LGVs, cars and motorcycles) across all industries.

10.1 Atmospheric emissions, 2016[1]

Continued

	Thousand tonnes						Tonnes		
	PM$_{10}$[3]	PM$_{2.5}$[3]	CO	NMVOC	Benzene	1,3-Butadiene	Lead	Cadmium	Mercury
Agriculture, forestry and fishing	20.58	7.73	55.92	202.51	0.19	0.04	0.45	0.04	0.03
Mining and quarrying	12.71	3.09	40.96	64.71	0.55	0.01	0.44	0.06	0.04
Manufacturing	33.13	23.98	321.63	259.23	1.81	0.21	43.07	1.55	1.52
Electricity, gas, steam and air conditioning supply	3.14	2.51	42.03	25.95	0.31	0.00	2.74	0.15	0.85
Water supply; sewerage, waste management and remediation activities	1.09	0.88	9.89	4.74	0.88	0.00	0.23	0.02	0.40
Construction	27.80	5.70	221.94	58.21	0.63	0.16	0.42	0.06	0.02
Wholesale and retail trade; repair of motor vehicles and motorcycles	2.29	1.51	11.91	33.79	0.07	0.02	1.51	0.03	0.02
Transport and storage	17.48	15.62	71.04	26.15	2.37	0.18	1.59	0.20	0.18
Accommodation and food services	0.17	0.15	2.35	0.52	0.01	0.00	0.02	0.00	0.00
Information and communication	0.16	0.12	1.60	0.23	0.01	0.00	0.02	0.00	0.00
Financial and insurance activities	0.00	0.00	0.02	0.00	0.00	0.00	0.00	0.00	0.00
Real estate activities	0.04	0.03	0.49	0.05	0.00	0.00	0.02	0.00	0.00
Professional, scientific and technical activities	0.19	0.16	2.25	0.35	0.01	0.00	0.02	0.00	0.00
Administrative and support service activities	0.48	0.37	21.70	0.86	0.04	0.01	0.04	0.01	0.01
Public administration and defence: compulsory social security	0.53	0.46	8.90	2.20	0.08	0.01	0.07	0.01	0.01
Education	0.50	0.46	4.82	0.14	0.01	0.00	0.72	0.01	0.07
Human health and social work activities	0.15	0.13	4.04	0.38	0.02	0.00	0.01	0.00	0.00
Arts, entertainment and recreation	0.16	0.13	3.37	1.03	0.01	0.00	0.07	0.00	0.01
Other service activities	0.13	0.10	1.17	0.85	0.00	0.00	0.02	0.00	0.59
Activities of households as employers; undifferentiated goods and services-producing activities of households for own use	0.01	0.00	13.52	0.26	0.02	0.00	0.00	0.00	0.00
Consumer expenditure	61.07	55.55	717.82	216.44	6.68	1.24	13.81	1.40	0.34
Total (excluding natural world)	181.79	118.68	1557.38	898.58	13.67	1.90	65.25	3.58	4.07
Of which: emissions from road transport[2]	19.85	13.42	273.84	32.03	0.79	0.20	1.72	0.37	0.23

Source: Ricardo Energy & Environment, ONS

Notes:

1. Components may not sum to totals due to rounding.
2. Includes emissions from fuel sources which are used by road vehicles (eg HGVs, LGVs, cars and motorcycles) across all industries.
3. PM$_{10}$ and PM$_{2.5}$ is particulate matter arising from various sources including fuel combustion, quarrying and construction, and formation of 'secondary' particles in the atmosphere from reactions involving other pollutants - sulphur dioxide, nitrogen oxides, ammonia and NMVOCs.

10.2 Road transport[1] emissions by pollutant, 2002 to 2016
UK resident basis

Weights in thousand tonnes

Pollutant	2002	2003	2004	2005	2006	2007	2008	2009	2010	2011	2012	2013	2014	2015	2016
Greenhouse gases	**120,469.4**	**120,042.5**	**120,772.9**	**121,495.4**	**121,804.9**	**123,474.3**	**119,473.1**	**115,730.2**	**114,911.7**	**113,215.1**	**112,471.9**	**112,024.7**	**113,939.5**	**115,429.1**	**118,040.8**
of which:															
Carbon dioxide	118,637.5	118,362.6	119,202.9	120,012.9	120,416.6	122,152.2	118,355.9	114,728.0	113,917.0	112,212.1	111,450.7	110,978.4	112,849.3	114,296.6	116,857.2
Methane	482.7	425.2	381.1	349.7	315.6	281.4	251.1	187.9	165.1	148.8	133.6	118.7	112.3	106.5	100.7
Nitrous oxide	1,349.2	1,254.7	1,188.8	1,132.8	1,072.7	1,040.7	866.0	814.3	829.6	854.3	887.6	927.7	978.0	1,026.1	1,083.0
Acid rain precursors	**503.8**	**474.0**	**451.7**	**428.6**	**411.6**	**392.3**	**365.3**	**302.9**	**284.9**	**267.5**	**254.0**	**243.7**	**235.9**	**227.0**	**218.8**
of which:															
Sulphur dioxide	3.0	3.0	2.7	2.4	2.1	1.7	1.0	0.5	0.5	0.5	0.5	0.4	0.4	0.5	0.5
Nitrogen Oxides as NO_2	462.3	435.9	416.2	396.0	381.5	364.9	341.5	280.7	265.3	250.6	239.4	231.1	224.8	217.2	210.0
Ammonia	38.5	35.1	32.8	30.1	28.1	25.7	22.8	21.8	19.1	16.5	14.2	12.2	10.7	9.4	8.3
PM_{10}	33.1	32.4	31.7	30.8	29.9	29.1	27.6	26.4	25.7	24.0	22.9	21.9	21.1	20.5	19.9
$PM_{2.5}$	26.9	26.1	25.4	24.5	23.5	22.6	21.3	20.2	19.5	17.8	16.9	15.8	14.9	14.2	13.4
Carbon monoxide	2,160.2	1,961.0	1,777.0	1,578.8	1,405.7	1,218.8	1,091.1	838.6	717.0	589.2	519.7	439.9	367.8	324.7	273.8
Non Methane VOC	298.2	253.3	215.7	183.4	158.6	133.8	117.7	81.6	68.9	57.3	49.5	42.5	37.7	34.9	32.0
Benzene	8.8	8.1	7.3	6.5	5.9	5.0	4.6	2.8	2.4	1.9	1.6	1.3	1.0	0.9	0.8
1,3-Butadiene	2.6	2.2	1.9	1.6	1.4	1.2	1.0	0.7	0.6	0.5	0.4	0.3	0.3	0.2	0.2
Cadmium	0.4	0.4	0.4	0.4	0.4	0.4	0.4	0.4	0.4	0.4	0.4	0.3	0.4	0.4	0.4
Lead	2.0	2.0	2.0	2.1	2.1	1.8	1.7	1.6	1.6	1.6	1.6	1.6	1.6	1.7	1.7
Mercury	0.3	0.3	0.3	0.3	0.3	0.3	0.2	0.2	0.2	0.2	0.2	0.2	0.2	0.2	0.2

Source: Ricardo Energy and Environment, Office for National Statistics

Notes

[1] Emissions from fuel sources which are used by road vehicles across industry groups.

Greenhouse gases are made up of carbon dioxide, methane and nitrous oxide. Weight in carbon dioxide equivalent

Acid rain precursors are made of sulphur dioxide, nitrogen oxides and ammonia. Weight in sulphur dioxide equivalent

All figures are reported to 1 decimal place. Total figures are based on raw data and therefore may not sum due to rounding.

Enquiries about these data can be sent by email to: **environment.accounts@ons.gsi.gov.uk**

10.3 UK greenhouse gas emissions headline results, UK 2002-2017

Million tonnes carbon dioxide equivalent (MtCO$_2$e)

Gas	2002	2003	2004	2005	2006	2007	2008	2009	2010	2011	2012	2013	2014	2015	2016	2017
Net CO$_2$ emissions (emissions minus removals)	545.4	556.1	556.7	553.3	550.4	540.8	527.0	476.2	492.7	449.7	469.9	459.0	419.1	402.5	378.9	378.9
Methane (CH$_4$)	101.8	96.6	92.3	87.6	83.2	79.2	73.5	69.1	64.5	61.8	60.1	55.5	53.1	51.6	51.6	51.6
Nitrous oxide (N$_2$O)	26.3	26.2	26.7	25.8	24.6	24.6	23.9	22.2	22.5	21.7	21.7	21.4	21.9	21.5	21.4	21.4
Hydrofluorocarbons (HFC)	11.3	12.8	11.8	13.1	14.0	14.4	14.9	15.5	16.4	14.9	15.4	15.7	16.0	15.9	15.2	15.2
Perfluorocarbons (PFC)	0.4	0.4	0.4	0.4	0.4	0.3	0.3	0.2	0.3	0.4	0.3	0.3	0.3	0.3	0.4	0.4
Sulphur hexafluoride (SF$_6$)	1.5	1.3	1.1	1.1	0.9	0.8	0.7	0.6	0.7	0.6	0.6	0.5	0.5	0.5	0.5	0.5
Nitrogen trifluoride (NF$_3$)	0.0	0.0	0.0	0.0	0.0	0.0	0.0	0.0	0.0	0.0	0.0	0.0	0.0	0.0	0.0	0.0
Total greenhouse gas emissions	**686.7**	**693.4**	**689.1**	**681.3**	**673.5**	**660.3**	**640.3**	**583.8**	**597.1**	**549.1**	**568.0**	**552.4**	**510.8**	**492.4**	**467.9**	**467.9**

Footnotes:

Source: Deprtment of Energy and Climate Change (DECC)

1. The entire time series is revised each year to take account of methodological improvements

2. In accordance with international reporting and carbon trading protocols, each of these gases are weighted by their global warming potential (GWP), so that total greenhouse gas emissions can be reported on a consistent basis (in carbon dioxide equivalent units). The GWP for each gas is defined as its warming influence relative to that of carbon dioxide. The GWPs used are from Working Group 1 of the IPCC Fourth Assessment Report: Climate Change 2007.

Uncertainty in estimates of UK Greenhouse Gas emissions, UK, Crown Dependencies and Overseas Territories 1990/2017[1]

Pollutant		GWP[2]	1990 emissions[3]	2017 emissions[3]	Uncertainty[4] in 2017 emissions	Range of uncertainty in 2017 emissions		Percentage change between 1990 and 2017	Range of likely % change between 1990 and 2017[5]	
			(thousand tonnes CO2 equivalent)			2.5 percentile	97.5 percentile		2.5 percentile	97.5 percentile
Carbon dioxide[6]	CO$_2$	1	599,366	376,722	3%	365,720	387,399	-37%	-39%	-36%
Methane	CH$_4$	25	133,145	51,927	17%	44,414	61,752	-60%	-70%	-50%
Nitrous oxide	N$_2$O	298	48,331	20,680	18%	17,585	24,828	-56%	-70%	-42%
Hydrofluorocarbons	HFC	12 - 14,800	14,394	14,197	9%	12,883	15,490	-1%	-17%	19%
Perfluorocarbons	PFC	7,390 - 17,340	1,652	372	24%	291	469	-77%	-82%	-72%
Sulphur hexafluoride	SF$_6$	22,800	1,305	525	12%	465	586	-60%	-66%	-53%
Nitrogen trifluoride	NF$_3$	17,200	0.4	0.5	47%	0.3	0.8	36%	-34%	153%
All greenhouse gases weighted by GWP[7]			**798,192**	**464,424**	**3%**	**450,386**	**479,248**	**-42%**	**-45%**	**-39%**

Source: Department of Energy and Climate Change (DECC)

Footnotes:

1. Figures include emissions for the UK, Crown Dependencies and the Overseas Territories. Uncertainties are not calculated for different geographical coverages but would be expected to be similar.

2. The GWP (Global Warming Potential) of a greenhouse gas measures its effectiveness in global warming over 100 years relative to carbon dioxide. The GWPs used in these statistics are from Working Group 1 of the IPCC Fourth Assessment Report: Climate Change 2007.

3. 1990 and 2017 estimates, and the percentage change, are presented as the central estimate from the model. These differ from the actual emissions estimates.

4. Expressed as a percentage relative to the mean value 2017 emissions. Calculated as 0.5*R/E where R is the difference between 2.5 and 97.5 percentiles and E is the mean.

5. Equivalent to a 95 per cent probability that the percentage change between 1990 and 2017 is between the two values shown. Values include uncertainties for overseas territories data.

6. CO$_2$ emissions are net emissions. Total emissions minus removals.

7. Totals will be similar but not exactly the same as those presented for Table 11.

10.4 Summary Greenhouse Gas Emissions Bridging Table showing relationship of Environmental Accounts measure to UNFCCC[1] measure

United Kingdom [5][6][10]

Mass of air emissions per annum in thousand tonnes of carbon dioxide equivalent

	1995	2000	2005	2010	2015	2016	2017
Greenhouse gases[2] - CO_2, CH_4, N_2O, HFCs, PFCs SF_6 and NF_3							
Environmental Account	793,896.22	771,742.63	770,842.63	686,874.23	601,898.15	579,972.21	566,383.08
less							
Bunker emissions[3]	28,046.53	36,929.06	41,724.92	40,796.27	41,566.44	42,327.11	42,798.47
CO2 biomass[4]	6,537.85	7,411.16	12,863.43	22,029.64	38,358.77	41,159.65	43,509.61
Cross-boundary[5]	12,045.38	16,044.83	25,464.83	14,052.74	14,288.87	13,625.18	9,959.46
plus							
Land-Use, Land-Use Change and Forestry (LULUCF)[7]	-1,688.40	-3,853.47	-7,100.49	-9,113.77	-9,733.96	-9,802.51	-9,892.77
BEIS reported (Excluding Crown Dependencies & Overseas Territories)[8][9]	745,578.05	707,504.11	683,688.97	600,881.81	497,950.12	473,057.76	460,222.77
plus							
Crown Dependencies[6] (including net emissions from LULUCF)	2,003.44	2,166.44	1,981.00	1,945.10	1,790.43	1,822.35	1,778.06
Overseas Territories (including net emissions from LULUCF)	1,979.54	2,141.92	2,380.18	2,423.95	2,527.78	2,355.99	2,452.61
UNFCCC reported in the UK Greenhouse Gas Inventory[8]	749,561.03	711,812.46	688,050.14	605,250.86	502,268.32	477,236.10	464,453.44

Source: Ricardo Energy and Environment, Department for Business, Energy & Industrial Strategy, Office for National Statistics

Notes

1 United Nations Framework Convention on Climate Change https://unfccc.int/process/transparency-and-reporting/greenhouse-gas-data/ghg-data-unfccc

2 Carbon dioxide, methane, nitrous oxide, hydrofluorocarbons, perfluorocarbons, sulphur hexafluoride and nitrogen trifluoride expressed as thousand tonnes of carbon dioxide equivalent

3 Bunker emissions include IPCC memo items International Aviation and International Shipping

4 Emissions arising from wood, straw, biogases and poultry litter combustion for energy production

5 Emissions generated by UK households and business transport and travel abroad, net of emissions generated by non-residents travel and transport in the UK

6 Emissions of Crown dependencies; Guernsey, Jersey, Isle of Man.

7 Emissions from deforestation, soils and changes in forest and other woody biomass.

8 https://www.gov.uk/government/collections/final-uk-greenhouse-gas-emissions-national-statistics

9. Calculations for moving from the Environmental Accounts measure to others.

10. This is the UK total for the sum of the 7 pollutants and differs slightly from the Kyoto greenhouse gas basket totals which uses a narrower definition of Land Use, Land Use Change and Forestry and includes emissions from UK Overseas Territories (Gibraltar, the Falkland Islands, the Cayman Islands, Montserrat, Bermuda)

All figures are reported to 2 decimal places. Total figures are based on raw data and therefore may not sum due to rounding.

10.5 Estimated emissions of carbon dioxide by source category, type of fuel and end user category, UK 2001-2016

Million tonnes carbon dioxide equivalent (MtCO$_2$e)

Sector	2001	2002	2003	2004	2005	2006	2007	2008	2009	2010	2011	2012	2013	2014	2015	2016
(a) By source																
Energy supply	214.4	212.4	219.9	218.0	218.9	224.3	219.3	213.0	190.0	197.1	182.6	193.0	181.0	156.3	136.7	113.7
Business	106.2	95.5	98.2	97.6	96.9	94.0	92.0	89.6	76.5	78.2	71.5	72.9	73.3	71.0	69.9	66.0
Transport	130.6	133.3	132.5	133.5	134.3	134.3	135.9	129.9	125.0	123.2	121.1	120.1	118.6	120.0	122.1	124.4
Public	12.2	10.3	10.2	11.2	11.1	10.1	9.4	9.7	8.8	9.4	8.0	8.9	9.1	7.7	8.0	8.2
Residential	87.9	84.4	85.2	87.0	82.5	79.9	76.3	78.2	75.0	84.5	67.3	73.7	74.5	62.0	64.5	67.0
Agriculture	5.4	5.5	6.2	6.2	6.1	5.9	5.7	5.5	5.4	5.4	5.6	5.4	5.2	5.6	5.5	5.5
Industrial processes	15.7	14.8	15.7	16.0	16.4	15.5	16.8	15.1	10.0	10.6	10.1	9.9	12.2	12.3	12.1	9.9
Land use, land use change and forestry	-11.0	-11.4	-12.3	-13.2	-13.2	-13.7	-14.9	-14.2	-14.9	-16.1	-16.6	-14.2	-15.1	-15.9	-16.6	-16.0
Waste management	0.5	0.5	0.5	0.5	0.4	0.3	0.3	0.3	0.3	0.3	0.3	0.3	0.3	0.3	0.3	0.3
Grand Total	**561.8**	**545.4**	**556.1**	**556.7**	**553.3**	**550.4**	**540.8**	**527.0**	**476.2**	**492.7**	**449.7**	**469.9**	**459.0**	**419.1**	**402.5**	**378.9**

Source: Department of Energy and Climate Change

Footnotes:
: means data are not available
1. The entire time series is revised each year to take account of methodological improvements.

10.6 Estimated emissions of methane (CH$_4$) by source category, type of fuel and end user category, UK 2001-2016

Million tonnes carbon dioxide equivalent (MtCO$_2$e)

Sector	2001	2002	2003	2004	2005	2006	2007	2008	2009	2010	2011	2012	2013	2014	2015	2016
(a) By source																
Energy supply	15.2	14.7	13.1	12.8	11.2	10.3	9.9	9.5	9.4	9.1	8.8	8.7	7.5	7.2	6.7	5.7
Business	0.1	0.1	0.1	0.1	0.1	0.1	0.1	0.1	0.1	0.1	0.1	0.1	0.1	0.1	0.1	0.1
Transport	0.5	0.5	0.4	0.4	0.4	0.3	0.3	0.3	0.2	0.2	0.2	0.1	0.1	0.1	0.1	0.1
Public	0.0	0.0	0.0	0.0	0.0	0.0	0.0	0.0	0.0	0.0	0.0	0.0	0.0	0.0	0.0	0.0
Residential	0.9	0.8	0.8	0.7	0.6	0.6	0.6	0.7	0.7	0.9	0.8	0.8	0.9	0.8	0.9	0.9
Agriculture	27.9	27.4	27.6	27.8	27.5	27.0	26.7	26.0	25.7	25.9	25.8	25.7	25.8	26.3	26.3	26.3
Industrial processes	0.1	0.1	0.2	0.1	0.1	0.1	0.1	0.1	0.1	0.1	0.1	0.1	0.1	0.1	0.1	0.1
Land use, land use change and forestry	0.0	0.0	0.1	0.0	0.0	0.0	0.0	0.0	0.0	0.0	0.0	0.1	0.0	0.0	0.0	0.0
Waste management	59.3	58.1	54.4	50.2	47.6	44.5	41.4	36.8	32.8	28.2	26.1	24.5	20.8	18.4	17.4	18.3
Grand Total	**104.3**	**101.8**	**96.6**	**92.3**	**87.6**	**83.2**	**79.2**	**73.5**	**69.1**	**64.5**	**61.8**	**60.1**	**55.5**	**53.1**	**51.6**	**51.6**

Source: Department of Energy and Climate Change

Footnotes:
1. The entire time series is revised each year to take account of methodological improvements.

10.7: Estimated emissions of nitrous oxide (N$_2$O) by source category, type of fuel and end user category, UK 2001-2016

Million tonnes carbon dioxide equivalent (MtCO$_2$e)

Sector	2001	2002	2003	2004	2005	2006	2007	2008	2009	2010	2011	2012	2013	2014	2015	2016
(a) By source																
Energy supply	1.2	1.3	1.3	1.2	1.2	1.3	1.2	1.1	1.0	1.0	1.0	1.2	1.1	1.0	1.0	0.8
Business	1.6	1.6	1.6	1.7	1.6	1.7	1.7	1.6	1.4	1.4	1.3	1.5	1.3	1.3	1.5	1.4
Transport	1.6	1.6	1.5	1.4	1.4	1.3	1.3	1.1	1.0	1.0	1.0	1.1	1.1	1.2	1.2	1.3
Public	0.0	0.0	0.0	0.0	0.0	0.0	0.0	0.0	0.0	0.0	0.0	0.0	0.0	0.0	0.0	0.0
Residential	0.2	0.2	0.2	0.2	0.1	0.1	0.1	0.2	0.2	0.2	0.2	0.2	0.2	0.2	0.2	0.2
Agriculture	16.0	16.0	16.0	15.7	15.5	14.9	14.6	14.4	14.5	14.6	14.7	14.6	14.6	15.1	14.6	14.6
Industrial processes	4.7	2.7	2.9	3.8	3.1	2.6	3.0	2.8	1.4	1.5	0.6	0.3	0.3	0.3	0.3	0.3
Land use, land use change and forestry	2.0	2.0	1.9	1.9	1.8	1.8	1.7	1.6	1.6	1.6	1.6	1.5	1.5	1.5	1.4	1.4
Waste management	1.0	1.0	0.9	1.0	1.0	1.1	1.1	1.1	1.2	1.2	1.3	1.3	1.3	1.3	1.3	1.4
Grand Total	**28.3**	**26.3**	**26.2**	**26.7**	**25.8**	**24.6**	**24.6**	**23.9**	**22.2**	**22.5**	**21.7**	**21.7**	**21.4**	**21.9**	**21.5**	**21.4**

Source: Department of Energy and Climate Change

Footnotes:
1. The entire time series is revised each year to take account of methodological improvements.

10.8 Material flows account for the United Kingdom: Domestic extraction, 2004 to 2016

UK

1,000 Metric tonnes

Domestic extraction	2004	2005	2006	2007	2008	2009	2010	2011	2012	2013	2014	2015	2016
Biomass	**139,307**	**138,420**	**134,642**	**129,458**	**138,406**	**134,338**	**131,555**	**136,493**	**128,897**	**132,684**	**143,435**	**138,504**	**131,647**
Crops	42,742	41,653	39,698	37,156	43,713	42,359	40,118	43,079	37,667	40,479	46,421	43,397	39,062
Crop residues (used), fodder crops and grazed biomass	90,995	91,058	89,357	86,371	89,139	86,296	85,095	86,906	84,635	85,202	89,676	88,171	85,541
Wood	4,915	5,039	4,963	5,311	4,958	5,092	5,730	5,907	5,963	6,371	6,583	6,231	6,341
Wild fish catch and aquatic plants/animals	655	670	625	620	596	591	613	601	632	632	756	705	702
Metal ores (gross ores)	**5**	**4**	**4**	**3**	**3**	**4**	**4**	**4**	**1**	**1**	**1**	**1**	**1**
Iron	1	0	0	0	0	0	0	0	0	0	0	0	0
Non-ferrous metal	4	3	3	3	3	4	4	4	1	1	1	1	1
Non-metallic minerals[1]	**300,018**	**290,601**	**291,300**	**295,217**	**261,116**	**210,290**	**204,761**	**207,706**	**192,183**	**196,181**	**211,206**	**221,210**	**224,211**
Limestone and gypsum	83,334	79,296	81,928	85,191	75,345	61,311	58,185	59,300	56,000	58,100	67,500	71,353	75,046
Clays and kaolin	14,504	14,221	13,437	13,135	11,014	7,226	8,084	8,536	7,491	8,419	8,758	9,084	8,840
Sand and gravel	174,227	170,603	169,275	173,861	155,996	126,935	122,202	123,756	112,228	113,938	121,275	126,300	126,700
Other	27,953	26,481	26,660	23,030	18,762	14,818	16,290	16,114	16,464	15,724	13,673	14,473	13,626
Fossil energy materials /carriers[2]	**216,500**	**193,352**	**175,195**	**165,368**	**158,965**	**144,388**	**136,587**	**114,538**	**98,864**	**89,189**	**87,434**	**92,797**	**90,853**
Coal and other solid energy materials/ carriers[2]	26,046	21,632	19,717	17,674	18,626	18,542	19,103	19,173	17,394	13,712	12,246	9,201	4,178
Crude oil, condensate and natural gas liquids	95,374	84,721	76,578	76,575	71,789	68,199	62,962	51,972	44,561	40,646	39,928	45,288	47,445
Natural gas	95,080	87,000	78,900	71,119	68,551	57,648	54,523	43,393	36,909	34,831	35,260	38,308	39,231
Total domestic extraction[3]	**655,830**	**622,377**	**601,140**	**590,047**	**558,490**	**489,020**	**472,907**	**458,741**	**419,944**	**418,055**	**442,076**	**452,511**	**446,712**

Source: Department for Environment, Food and Rural Affairs; Food and Agriculture Organization of the United Nations; Eurostat; Kentish Cobnuts Association; British Geological Survey

Notes

1. 2015 and 2016 data have been estimated due to unavailable data sources. Alternative data sources are being reviewed. In addition some non-metallic mineral data were not available in 2016. This may mean total figures are a slight underestimate. This is likely to be small, as the missing data was around 0.4% of total domestic extraction in 2015.

2. Data for peat, included in 'coal and other solid energy materials/carriers, was not available in 2016. This may mean total figures are a slight underestimate. This is likely to be small, as peat was around 0.1% of total domestic extraction in 2015.

3. Total domestic extraction for 2016 may be a slight underestimate due to small missing data sources as detailed above. This is estimated to be less than 0.5% of the total.

4. : denotes unavailable data.

5. All figures are reported to 1 decimal place. Total figures are based on raw data and may not sum due to rounding.

Enquiries about these data can be sent by email to: **environment.accounts@ons.gsi.gov.uk**

10.8 Material flows account for the United Kingdom: Imports, 2004 to 2016

UK

1,000 Metric tonnes

Imports	2004	2005	2006	2007	2008	2009	2010	2011	2012	2013	2014	2015	2016
Biomass and biomass products	53,671	53,935	53,710	53,867	52,312	49,423	51,425	50,468	52,387	58,477	59,603	58,982	60,251
Metal ores and concentrates, raw and processed	45,964	43,623	45,291	47,862	42,605	26,955	32,936	33,476	33,856	37,972	40,595	37,718	35,089
Non-metallic minerals, raw and processed	16,989	16,752	16,454	17,160	16,337	12,684	15,078	15,904	13,904	14,505	18,308	17,940	20,292
Fossil energy materials/ carriers, raw and processed	151,906	161,792	170,454	170,004	163,769	158,813	158,337	176,737	190,554	184,907	172,143	158,874	148,465
Other products	18,258	16,915	17,167	17,038	16,469	14,858	16,541	15,611	15,102	15,939	16,631	15,873	17,004
Waste imported for final treatment and disposal	1	1	15	10	13	19	22	5	5	30	17	19	13
Total imports	**286,790**	**293,018**	**303,090**	**305,941**	**291,504**	**262,752**	**274,339**	**292,200**	**305,808**	**311,830**	**307,297**	**289,406**	**281,114**

Exports	2004	2005	2006	2007	2008	2009	2010	2011	2012	2013	2014	2015	2016
Biomass and biomass products	19,118	20,005	20,704	21,374	21,543	20,159	22,413	21,950	22,103	21,239	21,978	22,950	23,758
Metal ores and concentrates, raw and processed	26,677	26,355	27,216	27,335	27,171	20,704	23,502	24,502	24,005	25,316	25,712	24,086	22,899
Non-metallic minerals, raw and processed	21,928	22,634	23,591	22,840	21,170	17,094	16,681	17,034	14,444	15,119	14,436	11,809	11,419
Fossil energy materials/ carriers, raw and processed	114,242	104,307	99,946	95,972	96,421	94,463	101,577	98,556	95,222	91,250	89,308	91,199	91,371
Other products	8,685	9,423	8,504	9,539	8,754	7,721	7,716	7,739	7,705	7,693	7,855	7,740	7,605
Waste exported for final treatment and disposal	4	2	3	0	1	3	3	4	5	124	509	467	689
Total exports	**190,653**	**182,726**	**179,964**	**177,061**	**175,060**	**160,145**	**171,892**	**169,784**	**163,484**	**160,740**	**159,797**	**158,249**	**157,740**

Indicators	2004	2005	2006	2007	2008	2009	2010	2011	2012	2013	2014	2015	2016
Domestic material consumption (domestic extraction plus imports minus exports)[1,2]	**751,967**	**732,669**	**724,267**	**718,926**	**674,935**	**591,626**	**575,355**	**581,157**	**562,268**	**569,144**	**589,576**	**583,668**	**570,086**
Biomass	176,101	174,131	169,680	163,660	171,134	165,603	163,075	167,277	161,303	172,392	183,641	176,864	170,753
Metal ores	19,541	17,450	18,298	20,747	15,616	6,331	9,585	9,101	9,984	12,840	15,096	13,815	12,381
Non-metallic minerals	298,884	287,661	287,607	292,593	259,251	208,399	206,329	209,414	194,197	198,409	218,143	230,376	236,706
Fossil fuels	257,441	253,428	248,682	241,927	228,934	211,293	196,365	195,366	196,785	185,504	172,696	162,613	150,246
Physical trade balance (imports minus exports)	**96,137**	**110,292**	**123,126**	**128,880**	**116,444**	**102,607**	**102,447**	**122,416**	**142,324**	**151,090**	**147,500**	**131,157**	**123,374**
Direct material input (domestic extraction plus imports)[2]	**942,620**	**915,395**	**904,231**	**895,987**	**849,995**	**751,771**	**747,247**	**750,941**	**725,752**	**729,885**	**749,373**	**741,917**	**727,826**

Sources: Office for National Statistics

Notes

1. 'Other' and 'waste for final treatment and disposal' imports and exports have been allocated to categories within Domestic Material Consumption using the same method that is used by Eurostat:

http://ec.europa.eu/eurostat/statistics-explained/images/7/75/Raw_material_equivalents_2015.xlsx.

2. Total material input and consumption for 2016 may be slight underestimates due to small missing data sources on extraction for some minor materials. See domestic material consumption sheet for more information.

3. All figures are reported to 1 decimal place. Total figures are based on raw data and therefore may not sum due to rounding.

Enquiries about these data can be sent by email to: **environment.accounts@ons.gsi.gov.uk**

10.8 Material flows account for the United Kingdom: Per capita, 2004 to 2016

UK

Metric tonnes

Per capita (person)	2004	2005	2006	2007	2008	2009	2010
Mid-year population estimates (persons)	59,950,400	60,413,300	60,827,100	61,319,100	61,823,800	62,260,500	62,759,500
Domestic extraction	10.9	10.3	9.9	9.6	9.0	7.9	7.5
Total imports	4.8	4.9	5.0	5.0	4.7	4.2	4.4
Total exports	3.2	3.0	3.0	2.9	2.8	2.6	2.7
Domestic material consumption (DMC)	12.5	12.1	11.9	11.7	10.9	9.5	9.2
Direct material input (DMI)	15.7	15.2	14.9	14.6	13.7	12.1	11.9

Per capita (person)	2011	2012	2013	2014	2015	2016
Mid-year population estimates (persons)	63,285,100	63,705,000	64,105,700	64,596,800	65,110,000	65,648,100
Domestic extraction	7.2	6.6	6.5	6.8	6.9	6.8
Total imports	4.6	4.8	4.9	4.8	4.4	4.3
Total exports	2.7	2.6	2.5	2.5	2.4	2.4
Domestic material consumption (DMC)	9.2	8.8	8.9	9.1	9.0	8.7
Direct material input (DMI)	11.9	11.4	11.4	11.6	11.4	11.1

Source: Office for National Statistics

Notes

1. Mid-year population estimates available here:

www.ons.gov.uk/peoplepopulationandcommunity/populationandmigration/populationestimates/bulletins/annualmidyearp opulationestimates/mid2017

2. : denotes unavailable data.

3. All figures are reported to 1 decimal place. Total figures are based on raw data and therefore may not sum due to rounding.

Enquiries about these data can be sent by email to: **environment.accounts@ons.gsi.gov.uk**

10.9 Annual rainfall: by region

United Kingdom

Millimetres and percentages

Region		1971 - 2000[2] rainfall average (= 100%) millimetres	2003	2004	2005	2006	2007	2008	2009	2010	2011	2012	2013	2014	2015	2016	2017
United Kingdom	**JSJB**	1084	83	112	100	109	111	120	112	88	108	123	101	120	117	101	104
North West	**JSJC**	1177	85	116	96	114	110	126	113	84	116	136	97	111	128	106	117
Northumbria	**JSJD**	831	80	120	111	101	105	135	116	105	104	144	106	110	124	105	101
Severn Trent[1]	**JSJE**	759	81	110	92	103	123	121	103	84	74	139	100	121	99	103	97
Yorkshire	**JSJF**	814	82	114	96	110	115	130	105	91	89	143	91	114	115	106	101
Anglian	**JSJG**	603	86	115	89	102	118	116	99	97	73	138	93	122	95	101	101
Thames	**JSLK**	700	81	103	79	106	118	115	104	87	79	134	101	130	93	97	95
Southern	**JSLL**	782	85	97	79	101	106	108	109	94	81	130	107	137	100	92	94
Wessex	**JSLM**	866	83	98	89	100	113	116	107	80	84	140	101	128	95	93	95
South West	**JSLN**	1208	78	99	90	92	110	112	110	83	86	133	103	115	104	80	84
England	**JSLO**	819	83	109	91	103	114	120	107	89	87	137	99	120	106	100	101
Wales	**JSLP**	1373	83	108	95	107	108	121	109	82	94	124	98	113	114	100	103
Scotland	**JSLQ**	1440	84	117	110	114	109	120	117	87	131	112	102	122	129	104	107
Northern Ireland	**JSLR**	1111	84	98	96	104	99	114	114	94	115	107	103	117	119	95	103

Sources: Met Office; National Hydrological Monitoring Programme, Centre for Ecology and Hydrology

[1] The regions of England shown in this table correspond to the original nine English regions of the National Rivers Authority (NRA); the NRA became part of the Environment Agency upon its creation in April 1996. The exception to this is the Severn Trent region, part of which (the upper Severn) lies in Wales.

[2] 1971-2000 averages have been derived using arithmetic averages of Met Office areal rainfall.

10.10 UK Annual Weather Summary

	Max Temp		Min Temp		Mean Temp		Sunshine		Rainfall	
	Actual (degrees celsius)	Anomaly (degrees celsius)	Actual (degrees celsius)	Anomaly (degrees celsius)	Actual (degrees celsius)	Anomaly (degrees celsius)	Actual (hours/ day)	Anomaly (%)	Actual (mm)	Anomaly (%)
	WLRL	WLRM	WLRO	WLRP	WLRR	WLRS	WLRX	WLRY	WLSH	WLSI
1990	13.1	1.2	5.8	0.9	9.4	1.1	1490.7	111.4	1172.8	106.7
1991	12.1	0.3	5.1	0.2	8.6	0.3	1302.0	97.3	998.2	90.8
1992	12.3	0.4	5.2	0.4	8.7	0.4	1290.8	96.5	1186.8	107.9
1993	11.8	-0.1	5.0	0.1	8.4	0.0	1218.6	91.1	1121.1	102.0
1994	12.4	0.5	5.5	0.6	8.9	0.6	1366.9	102.2	1184.7	107.7
1995	13.0	1.1	5.4	0.6	9.2	0.9	1588.5	118.7	1023.7	93.1
1996	11.7	-0.1	4.7	-0.1	8.2	-0.2	1403.5	104.9	916.6	83.4
1997	13.1	1.3	5.8	1.0	9.4	1.1	1430.3	106.9	1024.0	93.1
1998	12.6	0.8	5.8	1.0	9.1	0.8	1268.4	94.8	1265.1	115.1
1999	13.0	1.1	5.9	1.0	9.4	1.1	1419.4	106.1	1239.1	112.5
2000	12.7	0.8	5.6	0.8	9.1	0.8	1367.5	102.2	1337.3	121.5
2001	12.4	0.6	5.3	0.5	8.8	0.5	1411.9	105.5	1052.8	95.5
2002	13.0	1.1	6.0	1.2	9.5	1.2	1304.0	97.5	1283.7	116.5
2003	13.5	1.6	5.6	0.7	9.5	1.2	1587.4	118.7	904.2	82.0
2004	13.0	1.2	6.0	1.2	9.5	1.2	1361.4	101.8	1210.1	110.1
2005	13.1	1.2	5.9	1.1	9.5	1.1	1399.2	104.6	1083.0	98.4
2006	13.4	1.5	6.1	1.3	9.7	1.4	1495.9	111.8	1175.9	106.8
2007	13.3	1.4	6.0	1.1	9.6	1.3	1450.7	108.4	1197.1	108.8
2008	12.7	0.8	5.5	0.6	9.1	0.7	1388.8	103.8	1295.0	117.7
2009	12.8	1.0	5.6	0.7	9.2	0.9	1467.4	109.7	1213.3	110.2
2010	11.7	-0.1	4.2	-0.6	8.0	-0.4	1456.0	108.8	950.5	86.4
2011	13.5	1.5	6.0	1.2	9.6	1.3	1406.2	105.0	1172.5	107
2012	12.4	0.5	5.2	0.4	8.8	0.4	1340.5	100	1334.8	121
2013	12.4	0.5	5.2	0.4	8.8	0.5	1421.1	106	1091	99
2014	13.5	1.7	6.3	1.5	9.9	1.6	1426.6	107	1300.5	118
2015	12.9	1.1	5.5	0.7	9.2	0.9	1456.7	109	1272.4	116
2016	13.0	1.1	5.7	0.9	9.3	1.0	1424.9	107	1099.8	100
2017	13.1	1.3	6.0	1.2	9.6	1.2	1374	103	1124.4	102

Source: Met Office

10.11 Summary of River Basin District Ecological and Chemical Status 2009- 2016

Chemical Classification of surface waters at good or better status

Percentage

River Basin District	2009	2010	2011	2012	2013	2014	2015	2015*	2016*
Solway Tweed	50	89	89	88	91	91	95	99	99
Northumbria	50	68	72	68	70	75	66	92	91
Humber	77	79	83	77	76	77	76	97	96
Anglian	85	89	92	87	84	83	84	99	99
Thames	75	76	79	74	72	70	70	99	98
South East	88	91	93	88	88	87	80	98	99
South West	77	80	83	84	79	79	83	96	96
Severn	78	83	88	82	82	83	83	95	97
Dee	75	92	92	87	82	67	100	90	90
North West	70	75	72	71	74	76	79	98	97

Ecological classification of surface water bodies at good or better status

Percentage

River Basin District	2009	2010	2011	2012	2013	2014	2015	2015*	2016*
Solway Tweed	44	43	45	48	41	41	42	41	43
Northumbria	43	41	43	40	42	38	41	26	25
Humber	18	16	17	18	17	16	16	14	14
Anglian	18	19	18	18	17	13	13	10	9
Thames	23	22	22	18	19	14	14	7	6
South East	19	15	16	15	15	13	14	15	17
South West	33	31	34	32	31	27	25	23	23
Severn	29	30	30	30	29	26	27	19	9
Dee	28	26	30	30	25	15	15	0	10
North West	30	31	30	29	30	28	28	22	20

Source: Environment Agency

% of English water bodies only
Improvements measured in terms of the number of water bodies meeting good status
* Cycle 2 figures of each year. Data sourced from Catchment Data Explorer, Environment Agency
(https://environment.data.gov.uk/catchment-planning/). These figures are not directly comparable to those of previous years.
Therefore, 2015 data from this source is also provided to allow comparison with 2016 figures.

Please note: This table has not been updated since the last edition of AOS.
2016 (cycle 2) classifications are the most recent publically available assessments. The environment agency does not have a set of results for 2017 . The next results to be published will be the 2019 classifications due to be published in May 2020.
For more information see: https://environment.data.gov.uk/catchment-planning/.
Or contact the national customer contact centre by emailing: national.requests@environment-agency.gov.uk

10.12 Overall status of rivers & canals in Scotland, 2014 - 2016

Status of surface waters assessed for:	No. of river water bodies[1,2]														
	2014					2015					2016				
	High	Good	Moderate	Poor	bad	High	Good	Moderate	Poor	bad	High	Good	Moderate	Poor	bad
Overall status	270	1251	639	447	132	272	1255	660	424	128	279	1243	687	403	131
Overall ecology	296	1054	727	471	191	298	1072	732	449	188	305	1056	760	432	190
Overall hydrology	1890	268	448	31	101	1870	286	448	34	100	1876	273	456	35	102

	Pass		Fail		Pass		Fail		Pass		Fail	
Overall chemistry	2731		8		2734		5		2735		8	

Length of river water bodies (km)	8332	11782	4370.7	505.7	122.8	8332	11782	4370.7	505.7	122.8	8332	11782	4370.7	505.7	122.8
% of length of river water bodies	33.2%	46.9%	17.4%	2.0%	0.5%	33.2%	46.9%	17.4%	2.0%	0.5%	33.2%	46.9%	17.4%	2.0%	0.5%

Source: Water Classification Hub, Scottish Environment Protection Agency
https://www.sepa.org.uk/data-visualisation/water-classification-hub/

Notes:
1. Data includes Solway Tweed river basin district and all categories of surface water bodies (coastal, lake, river and transitional).
2. Data excludes scottish marine regions.

10.13 Monthly reservoir stocks for England & Wales[1]

Percentages

		2005	2006	2007	2008	2009	2010	2011	2012	2013	2014	2015	2016	2017
January	JTAS	92.3	88.7	93.7	95.7	95.3	92.4	89.8	90.4	96.7	98.2	94.8	97	85.2
February	JTAT	92.1	91.2	96.7	95.6	93.4	91.8	93.7	92.0	93.8	98.1	95.5	95.6	93
March	JTAU	93.6	96.2	95.2	97.3	94.5	94.1	92.2	89.2	92.0	96.3	95.6	94.3	96.5
April	JTAV	95.0	93.4	91.9	95.1	92.0	91.9	88.8	94.0	94.5	95.4	93.1	94.5	91.1
May	JTAW	93.0	94.4	91.1	92.6	93.3	86.2	87.4	93.6	94.5	96.1	93.9	91.6	87.1
June	JTAX	85.6	88.4	94.4	90.6	88.6	79.0	86.7	97.8	90.8	91.3	88.7	89.6	85.6
July	JTAY	77.9	77.2	93.5	92.0	91.0	77.3	84.7	96.8	83.7	82.9	84.1	88	81.6
August	JTAZ	71.5	70.7	88.3	92.5	89.7	75.5	79.9	96.0	82.1	81.4	82.7	87	81.3
September	JTBA	67.4	67.8	86.1	90.9	84.0	81.1	81.0	95.3	77.7	73.5	79.7	83.6	82.1
October	JTBB	77.2	80.0	81.2	93.8	82.3	82.0	80.0	94.3	87.6	79.8	74.6	78.4	82.9
November	JTBC	83.8	89.8	82.4	93.1	93.0	86.9	79.7	95.8	88.3	83.5	89.4	81.6	82.7
December	JTBD	85.9	92.2	89.8	92.4	90.6	84.3	88.6	97.6	95.4	92.0	95.3	83.7	88.7

1 Reservoir stocks are the percentage of useable capacity based on a representative
selection of reservoirs; the percentages relate to the end of each month.

Sources: Water PLCs;
Environment Agency;
National Hydrological Monitoring Programme, Centre for Ecology and Hydrology: 01491 838800

10.14 Water Industry Regulatory Capital Values

RCV roll forward for indexation (£ million)	
RCV at 31 March 2019 as published in April 2018[1]	73,445
Indexation	1,750
RCV at 31 March 2019 in March 2019 prices	75,239

Wholesale water RCV (£ million)	2015-16	2016-17	2017-18	2018-19	2019-20
Opening RCV	29,965	30,449	31,101	31,660	32,013
RCV additions (from totex)	1,752	1,934	1,834	1,653	1,419
Less RCV run-off and depreciation	-1,268	-1,281	-1,272	-1,298	-1,324
Other adjustments[2]	0	-2	-2	-2	-2
Closing RCV	30,449	31,101	31,660	32,013	32,107
Average RCV (year average)	30,017	30,581	31,183	31,636	31,858

Wholesale water RCV breakdown (£ million)	2015-16	2016-17	2017-18	2018-19	2019-20
2015 RCV	28,733	27,563	26,479	25,443	24,446
Totex RCV	1,716	3,538	5,181	6,570	7,661
Total	30,449	31,101	31,660	32,013	32,107

Wholesale wastewater RCV (£ million)	2015-16	2016-17	2017-18	2018-19	2019-20
Opening RCV	40,694	41,284	41,971	42,667	43,226
RCV additions (from totex)	2,387	2,490	2,502	2,377	1,970
Less RCV run-off and depreciation	-1,797	-1,803	-1,806	-1,818	-1,810
Other adjustments[2]	0	0	0	0	0
Closing RCV	41,284	41,971	42,667	43,226	43,387
Average RCV (year average)	40,731	41,366	42,053	42,677	43,034

Wholesale wastewater RCV breakdown (£ million)	2015-16	2016-17	2017-18	2018-19	2019-20
2015 RCV	38,939	37,266	35,682	34,178	32,763
Totex RCV	2,345	4,705	6,985	9,049	10,624
Total	41,284	41,971	42,667	43,226	43,387

Source: OFWAT

Notes

1 Presented in March 2018 prices

2 Impact of the reduction in allowed totex and the associated RCV run-off as included in the CMA's final determination for Bristol Water
Link to CMA final determination: https://www.gov.uk/cma-cases/bristol-water-plc-price-determination#final-determination

10.15 Summary of pollution incidents by region and incident category, England 2017

Region	Water						Land						Air					
	Cat 1	Cat 2	Cat 3	Cat 4	Total	Cat 1 and 2 Total	Cat 1	Cat 2	Cat 3	Cat 4	Total	Cat 1 and 2 Total	Cat 1	Cat 2	Cat 3	Cat 4	Total	Cat 1 and 2 Total
Anglian	3	41	5	27	76	44	0	13	9	54	76	13	0	20	8	48	76	20
Midlands	7	30	8	15	60	37	0	13	9	38	60	13	0	12	7	41	60	12
North East	11	36	1	14	62	47	0	6	5	51	62	6	1	9	2	50	62	10
North West	6	29	7	15	57	35	1	14	2	40	57	15	1	8	6	42	57	9
South East	9	40	8	20	77	49	0	14	9	54	77	14	0	18	9	50	77	18
South West	6	45	5	33	89	51	0	12	10	67	89	12	0	28	5	56	89	28
Total	**42**	**221**	**34**	**124**	**421**	**263**	**1**	**72**	**44**	**304**	**421**	**73**	**2**	**95**	**37**	**287**	**421**	**97**

Source: Environment Agency

Details of all pollution incidents reported to the Environment Agency are held on the National Incident Reporting System. This dataset only includes substantiated and closed environmental protection incidents, where the environment impact level is either category 1 (major) ,category 2 (significant), category 3 (minor) or category 4 (no impact) to at least 1 media (i.e. water, land or air).

10.16 Bathing Water Complaince Results 2016/2017

Year: 2016[1]

EA Area Name	Overall Water Body Class					Total Surface Water Bodies	Percentage of Total (%)				
	High	Good	Moderate	Poor	Bad		High	Good	Moderate	Poor	Bad
Cambridgeshire and Bedfordshire	0	23	144	16	4	187	0.0	12.3	77.0	8.6	2.1
Cumbria and Lancashire	1	159	251	45	11	467	0.2	34.0	53.7	9.6	2.4
Derbyshire Nottinghamshire and Leicestershire	0	26	152	49	5	232	0.0	11.2	65.5	21.1	2.2
Devon and Cornwall	2	116	276	55	4	453	0.4	25.6	60.9	12.1	0.9
Essex Norfolk and Suffolk	0	16	163	45	3	227	0.0	7.0	71.8	19.8	1.3
Greater Manchester Merseyside and Cheshire	1	7	203	49	10	270	0.4	2.6	75.2	18.1	3.7
Hertfordshire and North London	0	2	81	22	6	111	0.0	1.8	73.0	19.8	5.4
Kent and South London	0	23	119	33	7	182	0.0	12.6	65.4	18.1	3.8
Lincolnshire and Northamptonshire	0	20	147	58	10	235	0.0	8.5	62.6	24.7	4.3
Northumberland Durham and Tees	4	101	213	70	8	396	1.0	25.5	53.8	17.7	2.0
Shropshire Herefordshire Worcestershire and Gloucestershire	0	22	174	86	8	290	0.0	7.6	60.0	29.7	2.8
Solent and South Downs	0	35	126	37	10	208	0.0	16.8	60.6	17.8	4.8
Staffordshire Warwickshire and West Midlands	0	35	113	62	15	225	0.0	15.6	50.2	27.6	6.7
Wessex	0	55	230	60	11	356	0.0	15.4	64.6	16.9	3.1
West Thames	0	16	165	80	13	274	0.0	5.8	60.2	29.2	4.7
Yorkshire	0	82	414	55	15	566	0.0	14.5	73.1	9.7	2.7
England	8	738	2971	822	140	4679	0.2	15.8	63.5	17.6	3.0

Source: Environment Agency Catchment Planning System

1. 2017 data for England has not been published. From 2016, England moved to a triennial reporting system. The 2019 classification data is expected to be published in May 2020.

	Bathing Water Classification				Total Designated Bathing Waters	Percentage of Total			
	Excellent	Good	Sufficient	Poor		Excellent	Good	Sufficient	Poor
Year: 2017									
Wales	80	18	5	1	104	76.9	17.3	4.8	1.0
Scotland	25	34	16	10	85	29.4	40.0	18.8	11.8
Northern Ireland	12	5	5	1	23	52.2	21.7	21.7	4.3

Source: Natural Resources Wales, Scottish Environment Protection Agency, Department of Agriculture, Environment and Rural Affairs (DAERA)

10.17 Estimated abstractions from all surface and groundwaters by purpose and source: 2005 to 2017
England

million cubic metres

Purpose Source	2005	2006	2007	2008	2009	2010	2011	2012	2013	2014	2015	2016	2017
Public water supply											b		
tidal waters	0	0	0	0	6	21	13	16	12	8	7	9	12
non-tidal surface waters	3,831	3,778	3,626	3,624	3,462	3,545	3,431	3,490	3,482	3,377	3,450	3,606	3,660
groundwaters	1,819	1,747	1,707	1,724	1,730	1,715	1,730	1,672	1,684	1,661	1,651	1,666	1,660
all sources except tidal	5,649	5,525	5,333	5,348	5,192	5,260	5,161	5,161	5,167	5,038	5,101	5,273	5,320
all surface and groundwaters	5,649	5,525	5,333	5,348	5,198	5,281	5,175	5,177	5,178	5,046	5,109	5,282	5,332
Spray irrigation (1)													
tidal waters	0	0	0	0	0	0	0	0	0	0	0	0	0
non-tidal surface waters	45	52	32	31	44	56	62	30	50	50	50	46	46
groundwaters	35	45	26	25	40	46	55	20	48	38	44	38	41
all sources except tidal	80	97	58	56	84	102	116	50	97	88	94	84	87
all surface and groundwaters	80	97	58	56	84	102	117	50	97	88	94	84	87
Agriculture (excl. spray irrigation)													
tidal waters	0	0	0	0	0	0	0	0	0	0	0	0	0
non-tidal surface waters	6	3	11	3	2	2	2	2	3	3	2	1	1
groundwaters	14	14	14	9	13	22	22	23	23	23	23	25	21
all sources except tidal	21	16	25	13	15	24	25	25	24	26	25	26	22
all surface and groundwaters	21	16	25	13	15	24	25	25	25	26	25	26	22
Electricity supply industry (2)													
tidal waters	7,292	7,942	8,083	6,677	6,803	6,518	6,806	6,906	5,968	4,514	4,398	4,977	4,872
non-tidal surface waters	1,715	1,073	1,072	1,662	1,508	1,781	1,424	2,040	1,799	2,140	2,474	2,642	3,252
groundwaters	10	13	13	10	7	7	9	8	5	4	4	8	7
all sources except tidal	1,725	1,086	1,085	1,672	1,515	1,788	1,433	2,048	1,805	2,144	2,478	2,650	3,259
all surface and groundwaters	9,017	9,028	9,168	8,349	8,318	8,305	8,239	8,954	7,772	6,659	6,876	7,628	8,131
Other industry													
tidal waters	783	1,018	887	717	822	926	1,065	888	955	955	943	980	1,023
non-tidal surface waters	1,100	933	586	753	874	573	434	813	563	679	630	686	607
groundwaters	271	241	233	205	188	189	184	169	163	168	171	176	170
all sources except tidal	1,371	1,174	819	958	1,061	762	618	983	726	847	801	862	777
all surface and groundwaters	2,154	2,192	1,706	1,675	1,884	1,688	1,683	1,871	1,681	1,802	1,744	1,842	1,799
Fish farming, cress growing, amenity ponds													
tidal waters	0	0	0	0	0	0	0	0	0	0	0	0	0
non-tidal surface waters	1,115	1,105	1,013	850	768	871	679	813	785	794	718	608	769
groundwaters	141	152	151	133	143	135	129	113	119	128	123	113	119
all sources except tidal	1,256	1,257	1,164	983	911	1,007	808	926	902	922	841	722	888
all surface and groundwaters	1,256	1,257	1,164	983	911	1,007	808	926	903	922	841	722	888
Private water supply													
tidal waters	0	0	0	0	0	0	0	0	0	0	0	0	0
non-tidal surface waters	1	3	1	1	1	0	1	0	1	2	1	1	1
groundwaters	9	9	10	8	9	8	9	8	8	8	9	9	9
all sources except tidal	9	12	11	8	10	9	9	9	9	10	10	9	10
all surface and groundwaters	9	12	11	8	10	9	9	9	9	10	10	9	10
Other													
tidal waters	0	0	0	0	0	0	0	0	0	0	0	0	0
non-tidal surface waters	8	13	25	13	12	10	9	10	12	10	13	18	14
groundwaters	14	17	14	14	14	10	13	16	22	18	16	21	18
all sources except tidal	22	30	39	27	25	20	22	27	33	28	29	40	32
all surface and groundwaters	22	30	39	27	25	20	22	27	33	28	29	40	32
Total											b		
tidal waters	8,075	8,961	8,971	7,394	7,631	7,465	7,885	7,810	6,935	5,477	5,349	5,967	5,907
non-tidal surface waters	7,821	6,959	6,366	6,936	6,670	6,838	6,042	7,199	6,693	7,056	7,338	7,609	8,350
groundwaters	2,313	2,238	2,167	2,129	2,143	2,133	2,152	2,029	2,070	2,047	2,042	2,057	2,044
all sources except tidal	10,133	9,197	8,533	9,065	8,814	8,971	8,193	9,228	8,764	9,103	9,379	9,666	10,395
all surface and groundwaters	18,208	18,158	17,504	16,458	16,445	16,437	16,078	17,038	15,699	14,580	14,728	15,633	16,302

Notes:

1. Includes small amounts of non-agricultural spray irrigation

2. The 'Electricity supply industry' category includes hydropower licences

b. Indicates a break in the series where information concerning abstractions in the country of England and the Dee/Wye regional charge areas (formerly the Wales regional charge area) has been amalgamated into the 'North West' and 'Midlands' regional charge areas respectively.

Source: Environment Agency

Contact: enviro.statistics@defra.gsi.gov.uk

10.18 Estimates of remaining recoverable oil and gas reserves and resources, 2005 to 2016[7]

Oil (million tonnes)

	Reserves						Range of undiscovered resources		Range of total reserves and resources		Expected level of reserves[3]			
	Proven K7MI	Probable K7MJ	Proven plus probable K7MK	Possible K7ML	Contingent Resources[5] -	Maximum K7MM	Lower K7MN	Upper K7MO	Lower[1] K7MP	Upper[2] K7MQ	Opening stocks K7MR	Extraction[4] K7MS	Other volume changes[6] K7MT	Closing stocks K7MU
2005	516	300	816	451	-	1267	396	1581	1663	2848	816	-85	85	816
2006	479	298	777	478	-	1255	438	1637	1693	2892	816	-77	38	777
2007	452	328	780	399	-	1179	379	1577	1558	2756	777	-77	80	780
2008	408	361	769	360	-	1129	454	1561	1583	2690	780	-72	61	769
2009	378	390	768	343	-	1111	397	1477	1508	2588	769	-68	67	768
2010	374	377	751	342	-	1093	475	1374	1568	2467	768	-63	46	751
2011	413	374	788	319	-	1106	422	1321	1528	2427	751	-52	89	788
2012	405	405	811	253	-	1064	455	1342	1519	2406	788	-45	68	811
2013	404	342	746	338	-	1084	172	755	1257	1840	811	-41	-24	746
2014	374	342	716	344	-	1060	156	755	1216	1816	746	-40	10	716
2015	349	217	566	161	216	943	156	755	1099	1698	716	-45	-105	566
2016	337	178	515	177	211	903	156	755	1059	1658	566	-47	-4	515

Gas (billion cubic metres)

	Reserves						Range of undiscovered resources		Range of total reserves and resources		Expected level of reserves[3]			
	Proven K7MV	Probable K7MW	Proven plus probable K7MX	Possible K7MY	Contingent Resources[5] -	Maximum K7MZ	Lower K7N2	Upper K7N3	Lower[1] K7N4	Upper[2] K7N5	Opening stocks K7N6	Extraction[4] K7N7	Other volume changes[6] K7N8	Closing stocks K7N9
2005	481	247	728	278	-	1006	226	1035	1232	2041	827	-86	-13	728
2006	412	272	684	283	-	967	301	1049	1268	2016	728	-78	34	684
2007	343	304	647	293	-	940	280	1039	1220	1979	684	-70	33	647
2008	292	309	601	306	-	907	319	1043	1226	1950	647	-68	22	601
2009	256	308	564	276	-	840	300	949	1140	1789	601	-56	19	564
2010	253	267	520	261	-	781	363	1021	1144	1802	564	-53	9	520
2011	246	246	493	216	-	709	353	977	1062	1686	520	-42	15	493
2012	244	217	461	238	-	699	369	1010	1068	1709	493	-35	3	461
2013	241	211	452	198	-	650	133	581	783	1231	461	-33	24	452
2014	205	201	407	187	-	594	120	581	713	1174	452	-34	-11	407
2015	207	126	333	113	62	508	120	581	628	1089	407	-36	-38	333
2016	176	121	297	88	84	903	156	755	1059	1658	297	-37	1	297

Source: Department for Business, Energy and Industrial Strategy (BEIS)and Oil and Gas Authority

Notes:

All data refer to end of year. Components may not sum to totals due to rounding

1. The lower end of the range of total reserves and resources has been calculated as the sum of maximum and the lower end of the range of undiscovered resources. The lower range of total reserves and resources have decreased due to a new "Yet to Find" (YTF) reporting method for the years 2013 to 2016.

2. The upper end of the range of total reserves and resources has been calculated as the sum of maximum and the upper end of the range of undiscovered resources. The upper range of total reserves and resources have decreased due to a new "Yet to Find" (YTF)

3. Expected reserves are the sum of proven and probable reserves.

4. The negative of extraction is shown here for the purposes of the calculation only. Of itself, extraction should be considered as a positive value.

5. Data for contingent resources is only available from 2015.

6. "Other volume changes" is calculated (on unrounded figures) as the closing stock minus the opening stock minus the extraction figure.

7. Data sourced from the Oil and Gas Authority has been converted to the appropriate units using the following conversion factors:

1 tonne of crude oil = 7.5 barrels of oil equivalent. 1 cubic metre of gas = 35.315 cubic feet of gas. 1 cubic foot of gas = 1/5,800 barrels of oil equivalent.

For further details see https://www.ogauthority.co.uk/data-centre/data-downloads-and-publications/reserves-and-resources/

10.19a Local Authority Collected Waste Generation from 2004/05 to 2016/17

England *Thousand tonnes*

Household waste from:	2004/05	2005/06	2006/07r	2007/08	2008/09r	2009/10	2010/11	2011/12	2012/13	2013/14	2014/15	2015/16	2016/17
Regular household collection	15,470	14,616	14,050	13,046	12,076	11,432	11,048	10,586	10,317	10,308	10,392	10,532	10,497
Other household sources	1,205	1,314	1,173	1,073	1,026	1,070	1,047	997	1,027	1,099	1,058	1,142	1,099
Civic amenity sites	3,198	2,726	2,576	2,434	2,086	1,765	1,635	1,470	1,477	1,568	1,597	1,700	1,728
Household recycling	5,785	6,796	7,976	8,735	9,146	9,398	9,724	9,846	9,759	9,980	10,117	10,075	10,329
Total household	**25,658**	**25,454**	**25,775**	**25,287**	**24,334**	**23,666**	**23,454**	**22,899**	**22,580**	**22,967**	**23,169**	**23,449**	**23,653**
Non household sources (excl. recycling)	2,795	2,289	2,408	2,250	2,063	1,999	1,882	1,654	1,558	1,600	1,617	1,585	1,634
Non household recycling	1,167	1,003	961	969	936	877	864	866	817	950	950	998	923
Total LA collected waste	**29,619**	**28,745**	**29,144**	**28,506**	**27,334**	**26,541**	**26,200**	**25,419**	**24,955**	**25,518**	**25,737**	**26,032**	**26,210**

Source: Department for Environment, Food & Rural Affairs

There has been a revision to 2008/09 to include asbestos in "Other household sources" instead of "Non household sources".

10.19b Waste managed (tonnes) by management method and year, 2012-13 onwards

Area: Wales

	2013-14	2014-15	2015-16	2016-17
Total Municipal Waste Collected/Generated	1,557,229	1,543,357.32	1,592,177.68	1,589,794.99
Total Waste Reused/Recycled/Composted (Statutory Target) (1)	429,863	430,264.88	482,303.10	541,444.93
Non-Household Waste Reused/Recycled (2)	274,265	281,762.43	285,800.99	297,376.84
Non-Household Waste Composted (3)	128,310	141,705.77	176,763.10	163,312.12
Household Waste Reused/Recycled (4)	13,653	14,346.36	13,391.66	12,320.87
Household Waste Composted (5)	846,091	868,079.46	958,258.84	1,014,454.76
Waste sent for other recovery (6)	4,532	4,923.87	3,937.44	1,010.14
Other recovery: Recycling (7)	.	1,488.15	781.57	855.89
Other recovery: Composting (8)	.	3,435.72	3,155.87	154.25
Waste Incinerated with Energy Recovery	89,907	182,961.26	301,905.74	389,599.30
Waste Incinerated without Energy Recovery	341	245.08	655.93	495.59
Waste Landfilled	587,390	453,497.37	288,820.05	150,984.28
Percentage of Waste Reused/Recycled/Composted (Statutory Target) (9)	27.60	27.88	30.29	34.06
Percentage of Non-Household Waste Reused/Recycled (10)	17.61	18.26	17.95	18.71
Percentage of Non-Household Waste Composted (11)	8.24	9.18	11.10	10.27
Percentage of Household Waste Reused/Recycled (12)	0.88	0.93	0.84	0.77
Percentage of Household Waste Composted (13)	54.33	56.25	60.19	63.81

Source: WasteDataFlow, Natural Resources Wales

1 Total waste reused/recycled/composted as defined by the Statutory Local Authority Recovery Target (LART)
2 Non-Household waste sent to be reused/recycled as defined by the Statutory Local Authority Recovery Target (LART)
3 Non-Household waste sent to be composted as defined by the Statutory Local Authority Recovery Target (LART)
4 Household waste sent to be reused/recycled as defined by the Statutory Local Authority Recovery Target (LART)
5 Household waste sent to be composted as defined by the Statutory Local Authority Recovery Target (LART)
6 Other waste sent for recycling and/or composting that is not included in the statutory target definition
7 Other waste sent for recycling that is not included in the statutory target definition
8 Other waste sent for composting that is not included in the statutory target definition
9 Total waste reused/recycled/composted (as defined by the Statutory Local Authority Recovery Target, LART), as a percentage of total municipal waste collected/generated
10 Non-Household waste sent to be reused/recycled (as defined by the Statutory Local Authority Recovery Target, LART), as a percentage of total municipal waste collected/generated
11 Non-Household waste sent to be composted (as defined by the Statutory Local Authority Recovery Target, LART), as a percentage of total municipal waste collected/generated
12 Household waste sent to be reused/recycled (as defined by the Statutory Local Authority Recovery Target, LART), as a percentage of total municipal waste collected/generated
13 Household waste sent to be composted (as defined by the Statutory Local Authority Recovery Target, LART), as a percentage of total municipal waste collected/generated
. The data item is not applicable.

10.19c Household waste - Summary data 2017, Scotland

Local Authority	2017								2016
	Generated (tonnes)	Recycled (tonnes)	Percentage Recycled (%)	Other diversion from landfill* (tonnes)	Percentage Other diversion from Landfill (%)	Landfilled (tonnes)	Percentage Landfilled (%)	Carbon Impact (TCO2e)	Percentage Recycled (%)
Aberdeen City	87,787	38,568	43.9	17,594	20.0	31,624	36.0	198,580	39.0
Aberdeenshire	127,632	55,714	43.7	1,997	1.6	69,921	54.8	325,283	43.5
Angus	56,278	31,072	55.2	11,840	21.0	13,367	23.8	115,900	56.7
Argyll and Bute	50,437	19,536	38.7	7,323	14.5	23,578	46.7	127,000	33.9
City of Edinburgh	200,720	82,277	41.0	5,206	2.6	113,237	56.4	507,553	44.6
Clackmannanshire	27,201	16,196	59.5	2,867	10.5	8,138	29.9	55,349	56.5
Dumfries and Galloway	76,289	21,245	27.8	23,318	30.6	31,725	41.6	220,022	22.1
Dundee City	64,297	22,844	35.5	37,486	58.3	4,017	6.2	154,653	33.6
East Ayrshire	55,842	29,547	52.9	5,353	9.6	20,941	37.5	119,536	53.3
East Dunbartonshire	56,445	27,043	47.9	6,295	11.2	23,107	40.9	130,755	48.5
East Lothian	50,612	26,885	53.1	227	0.4	23,499	46.4	112,099	51.8
East Renfrewshire	47,564	31,937	67.1	1,480	3.1	14,146	29.7	95,976	60.8
Falkirk	74,651	41,728	55.9	8,504	11.4	24,419	32.7	149,870	51.3
Fife	178,478	97,546	54.7	12,146	6.8	68,786	38.5	349,263	54.7
Glasgow City	224,525	59,876	26.7	13,706	6.1	150,943	67.2	680,071	25.2
Highland	130,190	56,704	43.6	1,493	1.1	71,966	55.3	321,198	44.5
Inverclyde	27,565	15,771	57.2	1,528	5.5	10,265	37.2	56,720	53.4
Midlothian	42,725	22,049	51.6	1,819	4.3	18,857	44.1	98,306	51.4
Moray	51,090	29,517	57.8	0	0.0	21,573	42.2	97,536	59.1
Na h-Eileanan Siar	14,453	3,461	23.9	1,639	11.3	9,353	64.7	38,652	24.1
North Ayrshire	62,954	35,125	55.8	1,778	2.8	26,052	41.4	137,512	55.3
North Lanarkshire	156,813	62,037	39.6	7,765	5.0	87,011	55.5	406,737	41.1
Orkney Islands†	10,798	1,980	18.3	4,855	45.0	3,325	30.8	29,073	19.4
Perth and Kinross	74,789	41,578	55.6	4,715	6.3	28,496	38.1	141,124	54.7
Renfrewshire	83,526	39,923	47.8	19,871	23.8	23,641	28.3	195,536	48.5
Scottish Borders	53,471	21,324	39.9	1,555	2.9	30,593	57.2	134,667	39.0
Shetland Islands	9,754	778	8.0	6,798	69.7	2,178	22.3	31,975	7.9
South Ayrshire	56,772	31,349	55.2	5,295	9.3	20,128	35.5	126,084	49.9
South Lanarkshire	151,740	71,753	47.3	0	0.0	79,987	52.7	370,821	53.0
Stirling	40,774	22,470	55.1	6,187	15.2	12,117	29.7	84,605	54.7
West Dunbartonshire	43,551	20,735	47.6	3,654	8.4	19,162	44.0	104,858	48.5
West Lothian	71,051	43,556	61.3	6,688	9.4	20,807	29.3	145,935	48.5
Total Scotland	**2,460,772**	**1,122,124**	**45.6**	**230,983**	**9.4**	**1,106,959**	**45.0**	**5,863,249**	**45.0**

Source: Scottish Environment Protection Agency (SEPA)

10.19c Household waste - Summary data 2017, Scotland

* Other waste diverted from landfill is the fate of waste material not reused, recycled or landfilled. It includes household waste treated by incineration, mechanical biological and heat treatment. It also includes composted wastes that do not reach the quality standards set by PAS 100/110. It also includes any Incinerator Bottom Ash and Metals recycled as a result of treatment, and excludes the residue from incineration which is landfilled. It also includes any weight loss during the treatment process. It does not include temporary storage of treated waste pending a recycling or disposal market.

† Includes treated waste sent to interim storage pending a recycling or disposal market. More waste will be recorded as generated than managed.

Please note local authorities report the management of waste in the same period when that waste is collected to avoid discrepancies with the total waste generated and managed. The figures are accurate at the time of publication. Data may be updated in accordance with SEPA's revision policy.

10.19d-i: Local authority collected (LAC) municipal waste sent for preparing for reuse, dry recycling, composting, energy recovery and landfill by district council and waste management group

Northern Ireland, 2016/17

Units: Tonnes
KPI (j)

Authority				LAC municipal waste						
	preparing for reuse	dry recycling	composting	preparing for reuse, dry recycling and composting	energy recovery rate (mixed residual LACMW)	energy recovery rate (specific streams e.g. wood)	recovery rate	landfill rate	un-classified	arisings
Antrim & Newtownabbey	12	25,283	21,484	46,779	5,000	6,063	11,063	33,315	474	91,631
Ards & North Down	111	20,181	25,575	45,867	355	5,013	5,368	43,636	78	94,949
Armagh City, Banbridge & Craigavon	34	26,129	24,935	51,098	29,082	2,617	31,699	21,365	181	104,342
Belfast	735	39,274	21,170	61,180	32,097	2,587	34,683	75,973	399	172,235
Causeway Coast & Glens	124	21,758	11,997	33,879	11,517	2,852	14,369	31,421	88	79,758
Derry City & Strabane	98	24,705	6,943	31,746	19,830	3,348	23,177	19,397	161	74,481
Fermanagh & Omagh	53	16,434	8,373	24,860	448	1,238	1,686	27,135	197	53,878
Lisburn & Castlereagh	111	14,316	17,086	31,514	287	2,481	2,767	39,380	314	73,976
Mid & East Antrim	66	17,516	16,602	34,183	3,378	2,228	5,606	35,342	57	75,188
Mid Ulster	0	19,364	21,832	41,196	7,706	2,219	9,925	31,505	207	82,833
Newry, Mourne & Down	19	19,060	12,828	31,907	39,189	2,501	41,690	9,015	111	82,723
arc21	1,053	135,631	114,745	251,429	80,305	20,873	101,178	236,661	1,434	590,702
NWRWMG	222	46,463	18,940	65,626	31,347	6,199	37,547	50,818	249	154,239
Northern Ireland	1,362	244,022	188,825	434,209	148,888	33,146	182,034	367,484	2,268	985,994

Source: NIEA

Notes: The tonnage of waste sent for recycling includes recycling from both clean/source segregated collection sources (as shown in Table 8) and recycling from residual waste processes.

Unclassified waste is calculated as a residual amount of municipal waste after municipal waste sent for preparing for reuse, for dry recycling, composting, energy recovery and to landfill have been accounted for.

It is not extracted directly from the WasteDataFlow system. The majority of the total unclassified tonnage can be attributed to moisture and/or gaseous losses.

Small negative tonnages can arise in the unclassified column if more waste is sent for treatment in the year than was actually collected as is more likely at councils operating transfer stations.

10.19d-ii: Percentage of local authority collected (LAC) municipal waste sent for preparing for reuse, dry recycling, composting, energy recovery and landfill by district council and waste management group

Northern Ireland, 2016/17

KPI (e2)

Units: Percentages
KPI(f)

Authority				LAC municipal waste					
	preparing for reuse rate	dry recycling rate	composting rate	preparing for reuse, dry recycling and composting rate	energy recovery rate (mixed residual LACMW)	energy recovery rate (specific streams e.g. wood)	recovery rate	landfill rate	un-classified
Antrim & Newtownabbey	0.0	27.6	23.4	51.1	5.5	6.6	12.1	36.4	0.5
Ards & North Down	0.1	21.3	26.9	48.3	0.4	5.3	5.7	46.0	0.1
Armagh City, Banbridge & Craigavon	0.0	25.0	23.9	49.0	27.9	2.5	30.4	20.5	0.2
Belfast	0.4	22.8	12.3	35.5	18.6	1.5	20.1	44.1	0.2
Causeway Coast & Glens	0.2	27.3	15.0	42.5	14.4	3.6	18.0	39.4	0.1
Derry City & Strabane	0.1	33.2	9.3	42.6	26.6	4.5	31.1	26.0	0.2
Fermanagh & Omagh	0.1	30.5	15.5	46.1	0.8	2.3	3.1	50.4	0.4
Lisburn & Castlereagh	0.2	19.4	23.1	42.6	0.4	3.4	3.7	53.2	0.4
Mid & East Antrim	0.1	23.3	22.1	45.5	4.5	3.0	7.5	47.0	0.1
Mid Ulster	0.0	23.4	26.4	49.7	9.3	2.7	12.0	38.0	0.3
Newry, Mourne & Down	0.0	23.0	15.5	38.6	47.4	3.0	50.4	10.9	0.1
arc21	0.2	23.0	19.4	42.6	13.6	3.5	17.1	40.1	0.2
NWRWMG	0.1	30.1	12.3	42.5	20.3	4.0	24.3	32.9	0.2
Northern Ireland	0.1	24.7	19.2	44.0	15.1	3.4	18.5	37.3	0.2

Source: NIEA

Notes: Rates calculated by dividing total tonnage of LAC municipal waste sent in each category by total LAC municipal waste arisings.
Unclassified waste is calculated as a residual amount of municipal waste after municipal waste sent for preparing for reuse, for dry recycling, composting, energy recovery and to landfill have been accounted for.

It is not extracted directly from the WasteDataFlow system. The majority of the total unclassified tonnage can be attributed to moisture and/or gaseous losses.

Small negative tonnages can arise in the unclassified column if more waste is sent for treatment in the year than was actually collected as is more likely at councils operating transfer stations.

10.19d-iii: Household waste sent for preparing for reuse, dry recycling, composting and landfill by district council and waste management group

Northern Ireland, 2016/17

Units: Tonnes

Authority	Household waste preparing for reuse	Household waste dry recycling	Household waste composting	Household waste preparing for reuse, dry recycling and composting	Household waste landfilled	Household waste arisings
Antrim & Newtownabbey	12	15,700	21,473	37,185	30,091	78,219
Ards & North Down	111	15,680	25,533	41,324	37,357	84,001
Armagh City, Banbridge & Craigavon	34	22,242	24,184	46,460	19,746	95,148
Belfast	735	35,720	20,600	57,055	59,808	144,915
Causeway Coast & Glens	124	17,978	11,997	30,099	28,285	71,189
Derry City & Strabane	98	20,978	6,938	28,014	18,669	69,090
Fermanagh & Omagh	53	14,354	8,373	22,780	25,714	50,283
Lisburn & Castlereagh	111	10,395	17,086	27,592	36,542	67,173
Mid & East Antrim	66	13,106	16,602	29,773	30,909	65,698
Mid Ulster	0	16,297	21,832	38,129	26,702	73,947
Newry, Mourne & Down	19	17,845	12,768	30,632	8,083	76,302
arc21	1,053	108,446	114,062	223,562	202,790	516,308
NWRWMG	222	38,956	18,936	58,113	46,955	140,279
Northern Ireland	1,362	200,295	187,387	389,045	321,906	875,965

Source: NIEA

Note: The tonnages of waste sent for preparing for reuse, for dry recycling, composting and landfill may not always equal the waste arisings because the recycling measures were defined to capture outputs from recycling processes which exclude energy recovery.

10.19d-iv: Percentage of household waste sent for preparing for reuse, dry recycling, composting and landfill by district council and waste management group

Northern Ireland, 2016/17

Units: Percentages

Authority	Household waste preparing for reuse rate	Household waste dry recycling rate	Household waste composting rate	KPI (a2) Household waste preparing for reuse, dry recycling and composting rate	KPI (b) Household waste landfill rate
Antrim & Newtownabbey	0.0	20.1	27.5	47.5	38.5
Ards & North Down	0.1	18.7	30.4	49.2	44.5
Armagh City, Banbridge & Craigavon	0.0	23.4	25.4	48.8	20.8
Belfast	0.5	24.6	14.2	39.4	41.3
Causeway Coast & Glens	0.2	25.3	16.9	42.3	39.7
Derry City & Strabane	0.1	30.4	10.0	40.5	27.0
Fermanagh & Omagh	0.1	28.5	16.7	45.3	51.1
Lisburn & Castlereagh	0.2	15.5	25.4	41.1	54.4
Mid & East Antrim	0.1	19.9	25.3	45.3	47.0
Mid Ulster	0.0	22.0	29.5	51.6	36.1
Newry, Mourne & Down	0.0	23.4	16.7	40.1	10.6
arc21	0.2	21.0	22.1	43.3	39.3
NWRWMG	0.2	27.8	13.5	41.4	33.5
Northern Ireland	0.2	22.9	21.4	44.4	36.7

Source: NIEA

Notes: Rates calculated by dividing total tonnage of household waste sent in each category by total household waste arisings.
The percentages of waste sent for preparing for reuse, for dry recycling, composting and landfill may not equal 100% because the recycling measures were defined to capture outputs from recycling processes which exclude energy recovery.

10.19d-v: Household waste per capita and per household by district council and waste management group

Northern Ireland, 2016/17

Units: Kilogrammes per capita and tonnes per household

Authority	Household waste arisings	Population (2015)	KPI (p) Household waste arisings (kg per capita)	Housing stock (at Apr 2016)	KPI (h) Household waste arisings (tonnes per household)
Antrim & Newtownabbey	78,219	141,032	555	56,215	1.391
Ards & North Down	84,001	159,593	526	66,911	1.255
Armagh City, Banbridge & Craigavon	95,148	210,260	453	78,674	1.209
Belfast	144,915	339,579	427	146,575	0.989
Causeway Coast & Glens	71,189	143,525	496	56,651	1.257
Derry City & Strabane	69,090	150,142	460	58,463	1.182
Fermanagh & Omagh	50,283	115,799	434	44,607	1.127
Lisburn & Castlereagh	67,173	141,181	476	56,570	1.187
Mid & East Antrim	65,698	137,821	477	56,260	1.168
Mid Ulster	73,947	145,389	509	50,619	1.461
Newry, Mourne & Down	76,302	177,816	429	64,842	1.177
arc21	516,308	1,097,022	471	447,373	1.154
NWRWMG	140,279	293,667	478	115,114	1.219
Northern Ireland	875,965	1,862,137	470	736,387	1.190

Source: NIEA, NISRA, LPS

Notes: The population figures are NISRA mid-year population estimates for 2015.
The number of occupied households is estimated from the total housing stock adjusted for vacant properties using the 2011 Census.

10.20a Waste arisings from households (Million tonnes) UK, 2010 – 2017

Waste from households arisings (million tonnes)	2010	2011	2012	2013	2014	2015	2016	2017
UK	27.0	26.8	26.4	25.9	26.8	26.7	27.3	26.9
England	22.1	22.2	22.0	21.6	22.4	22.2	22.8	22.4
Scotland	2.6	2.5	2.4	2.3	2.3	2.4	2.4	2.3
Wales	1.3	1.3	1.3	1.3	1.3	1.3	1.3	1.3
NI	0.8	0.8	0.8	0.8	0.8	0.8	0.8	0.8

Source: Waste Data Flow, Department for Environment, Food & Rural Affairs

1. The 'waste from households' calculation was first published by Defra in May 2014. It was introduced for statistical purposes to provide a harmonised UK indicator to be reported against the Waste Framework Directive (2008/98/EC). It is calculated on a calendar year basis by each of the four UK countries using almost identical methodologies.

2. The waste from household measure is a narrower measure than the 'household waste' measure which was previously used in England. Waste from households excludes waste not considered to have come directly from households, such as recycling from street bins, parks and grounds.

3. Waste arising from households in the UK decreased by 1.5 per cent between 2016 and 2017 to 26.9 million tonnes.

10.20b Waste from households, England, 2011 to 2017
(Waste Prevention Metric)

	2011	2012	2013	2014	2015	2016	2017
Total waste generated from households (Million tonnes)	22.2	22	21.6	22.4	22.2	22.8	22.4
Waste generated (kg per person)	421	412	402	413	406	412	403

Source: Department for Environment, Food & Rural Affairs, Office for National Statistics

1. Total waste from households amounted to 22.4 million tonnes in 2017, an decrease of 1.8 per cent on 2016.
2. A breakdown of the previous measure of household waste covering national, regional and local authorities can be downloaded on the gov.uk website.

Notes: Waste from households' includes waste from: Regular household collection, Civic amenity sites, 'Bulky waste' 'Other household waste'. It does not include street cleaning/sweeping, gully emptying, separately collected healthcare waste, asbestos waste. 'Waste from households' is a narrower measure than 'municipal waste' and 'council collected waste'.

10.20c Waste from households, UK and country split, 2010 -2017

thousand tonnes and % rate

Year	Measure	UK	England	NI	Scotland	Wales
2010	Arisings ('000 tonnes)	26,954	22,131	829	2,649	1,344
	Of which recycled (excl. IBAm) ('000 tonnes)	10,878	9,112	314	861	591
	Recycling rate (excl. IBAm)	**40.4%**	**41.2%**	**37.8%**	**32.5%**	**44.0%**
2011	Arisings ('000 tonnes)	26,792	22,170	810	2,482	1,329
	Of which recycled (excl. IBAm) ('000 tonnes)	11,492	9,596	324	921	651
	Recycling rate (excl. IBAm)	**42.9%**	**43.3%**	**40.0%**	**37.1%**	**49.0%**
2012	Arisings ('000 tonnes)	26,428	21,956	783	2,383	1,306
	Of which recycled (excl. IBAm) ('000 tonnes)	11,594	9,684	319	911	681
	Recycling rate (excl. IBAm)	**43.9%**	**44.1%**	**40.7%**	**38.2%**	**52.1%**
2013	Arisings ('000 tonnes)	25,929	21,564	781	2,310	1,274
	Of which recycled (excl. IBAm) ('000 tonnes)	11,433	9,523	324	916	669
	Recycling rate (excl. IBAm)	**44.1%**	**44.2%**	**41.5%**	**39.6%**	**52.5%**
2014	Arisings ('000 tonnes)	26,795	22,355	806	2,348	1,285
	Of which recycled (excl. IBAm) ('000 tonnes)	12,035	10,025	344	962	704
	Recycling rate (excl. IBAm)	**44.9%**	**44.8%**	**42.6%**	**41.0%**	**54.8%**
2015	Arisings ('000 tonnes)	26,675	22,225	818	2,354	1,278
	Of which recycled (excl. IBAm) ('000 tonnes)	11,795	9,752	344	989	709
	Recycling rate (excl. IBAm)	**44.2%**	**43.9%**	**42.1%**	**42.0%**	**55.5%**
	Of which recycled (incl. IBAm)('000 tonnes)	11,898	9,849	z	991	713
	Recycling rate (incl. IBAm)	**44.6%**	**44.3%**	**z**	**42.1%**	**55.8%**
2016	Arisings ('000 tonnes)	27,300	22,770	845	2,378	1,307
	Of which recycled (excl. IBAm) ('000 tonnes)	12,198	10,074	366	1,017	741
	Recycling rate (excl. IBAm)	**44.7%**	**44.2%**	**43.3%**	**42.8%**	**56.7%**
	Of which recycled (incl. IBAm)('000 tonnes)	12,351	10,217	z	1,020	749
	Recycling rate (incl. IBAm)	**45.2%**	**44.9%**	**z**	**42.9%**	**57.3%**
2017	Arisings ('000 tonnes)	26,897	22,437	843	2,345	1,271
	Of which recycled (excl. IBAm) ('000 tonnes)	12,093	9,959	390	1,018	726
	Recycling rate (excl. IBAm)	**45.0%**	**44.4%**	**46.3%**	**43.4%**	**57.1%**
	Of which recycled (incl. IBAm)('000 tonnes)	12,282	10,139	z	1,019	733
	Recycling rate (incl. IBAm)	**45.7%**	**45.2%**	**z**	**43.5%**	**57.6%**

z = Not applicable - see Notes for users

Recycling rate = 'Recycled' as a percentage of 'Arisings'

IBAm = Incineration bottom ash metal

Source: Waste Data Flow, Department for Environment, Food & Rural Affairs

10.20c Waste from households, UK and country split, 2010 -2017

Notes for users:

1) UK estimates for 'Waste from households' have been calculated in accordance with the Waste Framework Directive.

2) 'Waste from households' includes waste from:

   Regular household collection

   Civic amenity sites

   'Bulky waste'

   'Other household waste'

3) 'Waste from households' excludes waste from:

   Street cleaning/sweeping

   Gully emptying

   Separately collected healthcare waste

   Soil, Rubble, Plasterboard & Asbestos wastes

4) Whilst the general approach is consistent across UK countries, aggregation method and the wording of some questions completed by Local Authorities varies.

5) Users should be aware that individual UK countries other than England publish household recycling estimates using alternative measures and as such may differ from the estimates published here.

6) Local Authorities in England may also use an alternative measure to 'Waste from Households'.

7) a) The NI waste from households data previously reported in Dec 2015 used the England WfH calculation for the years 2013 and 2014.

   b) A new WfH calculation specific to NI has been used for 2015 which now correctly excludes certain Construction & Demolition wastes from the Recycled tonnage.

   c) In order to provide a uniform comparison ALL previous years for NI have been recalculated.

8) England figure for 2015 revised in February 2018 due to methodological changes.

9) Incineration bottom ash (IBA) metals are now included within the recycling rate calculations, though the date from which these have been included varies for each country:

   England and Scotland have both included IBA metals for data from 2015 onwards, when Q100 was introduced. England data only includes IBA metals from April 2015, when Q100 came into full use by all local authorities.

   For Wales, Q100 was introduced in 2012. Wales figures for 2012-2014 have been revised in this release to remove IBA metals. These are included for 2015 and 2016 in line with the other UK countries.

   NI did not have any incinerators that burnt municipal waste collected by local authorities in this period, so no WfH IBA metal was counted as recycled.

10) Minor revisions were made to historical figures for UK, England, NI and Scotland in February 2019.

10.20d Packaging waste and recycling / recovery, split by material, UK

	Packaging waste arising (thousand tonnes)		Total recovered / recycled (thousand tonnes)		Achieved recovery / recycling rate (%)		EU target recovery / recycling rate (%)	
	2015	2016	2015	2016	2015	2016	2015	2016
Metal	736	736	440	506	59.8%	68.7%	50.0%	50.0%
of which: Aluminium	177	177	76	90	42.9%	50.8%	z	z
of which: Steel	559	559	364	416	65.1%	74.4%	z	z
Paper and cardboard	4,749	4,749	3,667	3,892	77.2%	81.9%	60.0%	60.0%
Glass	2,399	2,399	1,577	1,609	65.7%	67.1%	60.0%	60.0%
Plastic	2,260	2,260	891	1,015	39.4%	44.9%	22.5%	22.5%
Wood	1,310	1,310	375	405	28.6%	30.9%	15.0%	15.0%
Other materials	23	23	0	0	0.0%	0.0%	z	z
Total (for recycling)	**11,476**	**11,476**	**6,950**	**7,427**	**60.6%**	**64.7%**	**55.0%**	**55.0%**
Energy from Waste	z	z	476	767	4.1%	6.7%	z	z
Total (for recycling and recovery)	**11,476**	**11,476**	**7,427**	**8,194**	**64.7%**	**71.4%**	**60.0%**	**60.0%**

Source: Defra Statistics

z = Not applicable

1) These statistics have been calculated to fulfil a reporting requirement to Eurostat at UK level in relation to the EC Packaging and Packaging Waste Directive (94/62/EC). Calendar year figures are submitted to Eurostat in the June of the year after next and so the latest figures for 2016 were submitted in June 2018.

2) Figures are compiled from the National Packaging Waste Database (NPWD) and industry reports.

3) Only includes obligated packaging producers (handle 50 tonnes of packaging materials or packaging and have a turnover more than £2 million a year)

4) "Recovery' in these tables refer specifically to waste used for energy recovery.

5) The recovery figure includes waste sent to facilities that did not meet the R1 recovery energy efficiency thresholds, but were considered eligible based on the Packaging Directive, which states "'energy recovery' shall mean the use of combustible packaging waste as a means to generate energy through direct incineration with or without other waste but with recovery of the heat".

6) In 2013, 72.7% of packaging waste was either recycled or recovered. The main driver for the drop after 2013 was the adoption of a new paper and cardboard 'placed on market' estimate published in a bespoke industry report. This represents a significant increase in the estimated level of waste arisings compared to that assumed for 2012 and 2013.

7) The packaging arisings estimates have remained unchanged since 2014 because in the absence of reliable total packaging waste arisings figures, they are based on estimates of packaging placed on the market. Since 2014, our research suggested that there would 0% growth in sales, or if there were any growth it would be off-set by minimisation/prevention activity, and so the arisings figures have been held flat.

10.21 CIEH Survey of Local Authority Noise Enforcement Activity 2018-19
England - Data represents 143 (45%) of local authorities (LAs)

Area	Data collection		Noise Complaints Received		Notices Served by LAs		Noise-related prosecutions	LA Staffing Levels
	Number of LAs	% of LAs	Total	Average number *(per 10,000 people)*	Total	Average number *(per 10,000 people)*	Total	No. FTE professionals allocated to dealing with noise complaints *(per 10,000 people)*
England	**143**	**45**	**143,054**	**61**	**2543**	**1**	**101**	**0.2**
Region								
Greater London	8	24	39,200	183	779	4	18	0.2
East of England	17	22	10,785	49	131	1	13	0.3
South East	33	49	20,832	51	418	1	5	0.3
South West	16	53	11,161	35	262	1	7	0.1
West Midlands	20	67	20,102	49	398	1	26	0.2
East Midlands	24	60	12,647	45	141	1	1	0.3
Yorkshire	10	48	19,759	75	276	1	23	0.1
North West	11	28	5,703	35	80	0.5	3	0.3
North East	4	33	2,865	53	58	1	5	0.2

Sector breakdown of Noise Complaints	Data collection	Noise Complaints Received	
	Number of LAs[1]	Total	Average number *(by local authority)*
Residential	112	69,369	619
Cmmericialor leisure	109	12,907	118
Construction	99	5,027	51
Industrial	106	2,854	27
Other[2]	98	9,259	94

Source: Chartered Institute of Environmental Health (CIEH)

1. The number of the local authorities represented in each figure is included as not all local authorities record or use the same sector categories

2. Other sources of noise complaints recorded by local authorities include noise in the street, vehicles, machinery and equipment; dogs, agriculture, alarms, military, traffic, aircrafts and railways.

Notes:

CIEH did not collect data for the years 2016/17 and 2017/2018 as they worked to imporve the methd of data collection.

Compared with the last time CIEH collected noise data in 2015/16, the 2018/19 data shows a 9% increase in the number of noise complaints in the 65 local authorities which participated in the survey in both years.

Data from Wales had not been published at the time this edition of AOS was printed. It will be included in the next edition of AOS.

10.22: Government revenue from environmental taxes in the UK, 2006 to 2017

£ million

Environmental tax	2006	2007	2008	2009	2010	2011	2012	2013	2014	2015	2016	2017
Energy taxes	**24,859**	**26,035**	**26,503**	**27,731**	**29,114**	**29,336**	**29,795**	**31,130**	**32,519**	**34,010**	**35,255**	**36,345**
Tax on Hydrocarbon oils[1]	23,448	24,512	24,790	25,894	27,013	26,923	26,703	26,697	27,094	27,415	27,989	27,973
Climate Change Levy[2]	711	690	717	693	666	675	624	1,098	1,506	1,752	1,881	1,878
Fossil Fuel Levy[3]												
Gas Levy[4]												
Hydro-Benefit[5]												
Renewable Energy Obligations[6]	700	833	996	1,099	1,243	1,423	1,842	2,391	2,931	3,691	4,479	5,236
Contracts for Difference[7]												483
Emissions Trading Scheme (EU-ETS)[8]				45	192	315	278	339	418	493	386	335
Carbon Reduction Commitment[9]							348	605	570	659	520	440
Transport taxes	**6,233**	**7,667**	**7,853**	**8,099**	**8,910**	**9,597**	**10,108**	**10,534**	**10,854**	**10,869**	**11,043**	**11,244**
Air Passenger Duty[10]	961	1,883	1,876	1,800	2,094	2,605	2,766	2,960	3,154	3,119	3,150	3,398
Rail Franchise Premia[11]	125	244	285	496	792	993	1,275	1,275	1,501	1,611	1,656	1,380
Vehicle Registration Tax[12]	137	141	134	122	123	120	125	138	151	169	171	162
Northern Ireland Driver Vehicle Agency[13]		15	19	18	16	16	16	16	16	16	16	16
Motor vehicle duties paid by businesses[14]	1,245	1,288	1,259	1,266	1,279	1,539	1,679	1,789	1,865	2,076	2,205	2,214
Motor vehicle duty paid by households[14]	3,765	4,096	4,265	4,364	4,561	4,281	4,201	4,312	4,110	3,823	3,784	4,012
Boat Licenses[15]												
Air Travel Operators Tax[16]			15	33	45	43	46	44	57	55	61	62
Pollution/Resources taxes	**1,145**	**1,236**	**1,308**	**1,137**	**1,375**	**1,403**	**1,379**	**1,494**	**1,506**	**1,403**	**1,450**	**1,300**
Landfill Tax[17]	804	877	954	842	1,065	1,090	1,094	1,191	1,143	1,028	1,024	904
Fishing Licenses[18]	20	20	20	20	20	23	21	21	21	21	21	21
Aggregates levy[19]	321	339	334	275	290	290	264	282	342	354	405	375
Total environmental taxes	**32,237**	**34,938**	**35,664**	**36,967**	**39,399**	**40,336**	**41,282**	**43,158**	**44,879**	**46,282**	**47,748**	**48,889**
Total taxes and social contributions	491,125	517,443	521,265	491,267	529,135	556,105	559,632	579,174	598,911	622,564	658,019	696,370
As a percentage of total taxes and social contributions (%)	**6.6**	**6.8**	**6.8**	**7.5**	**7.5**	**7.3**	**7.4**	**7.5**	**7.5**	**7.4**	**7.3**	**7.0**
Gross Domestic Product (GDP)[20]	1,465,902	1,541,442	1,579,796	1,537,213	1,587,466	1,644,546	1,694,417	1,761,347	1,844,295	1,895,839	1,969,524	2,049,629
As a percentage of GDP (%)	**2.2**	**2.3**	**2.3**	**2.4**	**2.5**	**2.5**	**2.4**	**2.5**	**2.4**	**2.4**	**2.4**	**2.4**

Source: Office for National Statistics

10.22: Government revenue from environmental taxes in the UK, 2006 to 2017

Notes

1. Tax on hydrocarbon oils is also known as fuel duty and include taxes on unleaded petrol (including super unleaded), leaded petrol, lead replacement petrol, ultra low sulphur petrol, diesel and ultra low sulphur diesel.

2. Climate Change Levy is a tax on energy delivered to non-domestic users and was introduced in 2001. From 2013 it includes the Carbon Price Floor, which taxes fossil fuel used to generate electricity.

3. Fossil Fuel Levy was introduced in 1990 and is a tax paid by suppliers of electricity from non-renewable energy sources, it was set at 0% following the introduction of the Climate Change Levy.

4. The Gas Levy was introduced in 1981. It was a tax on gas suppliers who arranged gas supply contracts. The tax was removed in the Finance Act 1998. Data are included to 1998.

5. Hydro benefit is a tax on energy suppliers designed to protect domestic consumers from the high costs of distributing electricity in the North of Scotland, it ceased in 2005 and was replaced with alternative schemes to help fund this.

6. Renewable Energy Obligation tax was introduced in 2002 for Great Britain and 2005 in Northern Ireland. It requires suppliers of electricity to generate a certain proportion of electricity from renewable sources.

7. The Contracts for Difference scheme was introduced in 2014 and was included within Renewable Energy Obligations data in previous versions of this table. The scheme is designed to incentivise investment in renewable energy generation. The basic principle being that generators are offered a contract with a known strike price for renewable electricity sold, if the market price for electricity is below this strike price, the generator is paid the difference from government, if the market price for electricity is above the strike price, the generator pays back the difference to government.

8. The European Union emissions trading scheme was introduced in 2005. It is designed to help limit greenhouse gas emissions from heavy energy using installations by setting a cap on allowable greenhouse gas emissions from such installations. Companies with such installations receive or buy emissions allowances which they can trade with one another as needed. A limit on the total number of allowances available ensures they have a value. Each year a company must surrender enough emissions allowances to cover its emissions, otherwise fines are imposed. If a company reduces its emissions, it can keep spare allowances to cover its future needs or sell them to other companies. Data available from 2009.

9. The Carbon Reduction Commitment was introduced in 2010 and is designed to improve energy efficiency and cut carbon dioxide emissions in private and public sector organisations that are high energy users. Data is available from 2012.

10. Air Passenger Duty was introduced in 1994 and is charged on all passenger flights from UK airports.

11. Rail Franchise Premia refers to the premium paid by train companies to UK government to provide specified train services. The franchising system began in the 1990's as part of the privatisation of British Rail.

12. Vehicle registration tax includes revenue from tax on vehicle registration in the United Kingdom but excludes Northern Ireland from 2007 onwards.

13. The Northern Ireland Driver Vehicle Agency was set up in 2007, data refers to revenue from tax on vehicle registration in Northern Ireland.

14. Motor Vehicle duties paid by businesses and households are also known as vehicle excise duty, this is payable annually by owners of most types of vehicles.

15. Boat licences refer to an annual charge required by owners of boats who use or keep their boats on inland waterways in the UK. Data is available for 1997 to 2000.

16. Data is available from 2008.

17. Landfill tax was introduced in 1996 and is payable by landfill site operators dependant on the amount and type of waste they send to landfill.

18. Fishing licences are required to fish for certain species of fish in various locations across the United Kingdom.

19. Aggregates levy was introduced in 2002 and is a tax on sand, gravel or rock that has been dug from the ground, dredged from the sea or imported into the UK, it is generally payable by the quarrying industry but can also apply when aggregate is removed in the course of infrastructure projects.

20. Gross Domestic Product at current prices: Office for National Statistics YBHA EDP4 series.

: indicates data is not available.
All data are presented in current prices i.e. not adjusted for inflation.
Totals may not sum due to rounding.

Housing

Chapter 11

Housing

Permanent dwellings (Table 11.1, 11.3)

Local housing authorities include: the Commission for the New Towns and New Towns Development Corporations; Communities Scotland; and the Northern Ireland Housing Executive. The figures shown for housing associations include dwellings provided by housing associations other than the Communities Scotland and the Northern Ireland Housing Executive and include those provided or authorised by government departments for the families of police, prison staff, the Armed Forces and certain other services.

New dwellings completed by dwelling type and number of bedrooms, Wales (Table 11.4b)

In Wales, new house building is undertaken by the private sector, Registered Social Landlords (RSLs) and local authorities. The information presented here is based on the reports of local authority building inspectors and the National House Building Council (NHBC). It does not include information from private approved inspectors. The exclusion of this information means that there is currently an undercount in the number of dwellings completed though this is estimated to be quite small.

The information shows the number of new dwellings completed in Wales and is collected in order to assess the level of new house building across Wales during the period. The data is used to help monitor trends in both the overall level of Welsh housing stock and the changes in its tenure distribution over time. Data is also used by the Welsh Government and local authorities to assess levels of housing supply across Wales and as an indication as to whether housing need is being met.

Figures on housing completions are from records kept for building control purposes. It is sometimes difficult for data providers to identify whether a dwelling is being built for RSLs or for a private developer. This may lead to an understatement of RSL completions recorded in these tables, and a corresponding overstatement of private enterprise figures. This problem is more likely to occur with starts than completions.

New house building funded with capital grant funding includes funding via Social Housing Grant, Recycled Social Housing Grant and Strategic Capital Investment Fund.

Mortgage arrears and repossessions (Table 11.6)

Properties taken into possession include those voluntarily surrendered. The UKF arrears figures are for the end of the year. Changes in the mortgage rate have the effect of changing monthly mortgage repayments and hence the number of months in arrears which a given amount represents. Arrears figures are for both homeowners and buy to let mortgages except for bottom row (of all 3+ month arrears cases) which is exclusively for homeowners.

Households in Temporary Accommodation under homelessness provisions (Tables 11.7, 11.8, 11.9)

Comprises households in accommodation arranged by local authorities pending enquiries or after being accepted as owed a main homeless duty under the 1996 Act (includes residual cases awaiting re-housing under the 1985 Act). Excludes "homeless at home" cases. The data shown for Wales includes "homeless at home" cases.

11.1a Dwelling stock: by tenure[1], England[2][3]

Thousands of dwellings

	Owner Occupied	Rented Privately or with a job or business	Rented from Private Registered Providers	Rented from Local Authorities	Other public sector dwellings	All Dwellings
31 March [3]						
2000	14,600	2,089	1,273	3,012	101	**21,075**
2001	14,735	2,133	1,424	2,812	103	**21,207**
2002	14,846	2,197	1,492	2,706	112	**21,354**
2003 [4]	14,752	2,549	1,651	2,457	104	**21,513**
2004 [4]	14,986	2,578	1,702	2,335	83	**21,684**
2005 [4]	15,100	2,720	1,802	2,166	82	**21,870**
2006 [4]	15,052	2,987	1,865	2,087	82	**22,073**
2007 [4]	15,093	3,182	1,951	1,987	75	**22,288**
2008 [4]	15,067	3,443	2,056	1,870	74	**22,511**
2009 [4]	14,968	3,705	2,128	1,820	74	**22,694**
2010 [4]	14,895	3,912	2,180	1,786	66	**22,839**
2011 [4]	14,827	4,105	2,255	1,726	63	**22,976**
2012 [4] [P]	14,754	4,286	2,304	1,693	75	**23,111**
2013 [4] [P]	14,685	4,465	2,331	1,682	73	**23,236**
2014 [4] [P]	14,674	4,623	2,343	1,669	64	**23,372**
2015 [4] [P]	14,684	4,773	2,387	1,643	55	**23,543**
2016 [4] [P]	14,801	4,832	2,430	1,612	57	**23,733**
2017 [4] [P R]	15,050	4,798	2,444	1,602	56	**23,950**

Source: Department for Communities and Local Government

Contact: 0303 44 41864

E-Mail: housing.statistics@communities.gsi.gov.uk

1. For detailed definitions of all tenures, see Definitions of housing terms in Housing Statistics home page.
2. Figures for census years are based on census output.
3. Series from 1992 to 2001 for England has been adjusted so that the 2001 total dwelling estimate matches the 2001 Census.
   Series from 2002 to 2011 for England has been adjusted so that the 2011 total dwelling estimate matches the 2011 Census.
   Estimates from 2002 are based on local authority and Private Registered Provider (housing association) dwelling counts,
   the Labour Force Survey and, from 2003, the English Housing Survey. Estimates may not be strictly comparable between periods.
4. From 2003 the figures for owner-occupied and the private rental sector for England have been produced using a new
   improved methodology as detailed in the dwelling stock release. Previous to this vacancy was not accounted for.
5. Shared ownership dwellings are currently included in owner-occupied dwellings.

6. Private Registered Provider here refers to registered providers of social housing (previously known as Housing
Associations or Registered Social Landlords). These figures include all self-contained units only. As such this figure does not match Live Table
100 which include self contained units and bedspaces.

R- Revised from previous publication
P - Provisional

Data for earlier years are less reliable and definitions may not be consistent throughout the series Stock estimates are expressed to the nearest thousand but should not be regarded as accurate to the last digit. Components may not sum to totals due to rounding.

11.1b Dwelling stock: by tenure[1,2], Wales

Thousands of dwellings

	Owner Occupied	Rented Privately or with a job or business	Rented from Housing Associations	Rented from Local Authorities	All Dwellings
31 March					
2000 [5]	914	106	54	193	**1,267**
2001 [5]	941	90	55	188	**1,275**
2002 [5]	957	89	57	183	**1,285**
2003 [5]	966	97	57	176	**1,296**
2004 [5]	979	104	64	160	**1,307**
2005 [5]	989	109	65	156	**1,319**
2006 [5]	997	113	66	154	**1,331**
2007 [5]	1,002	122	67	153	**1,343**
2008 [5]	1,001	135	89	130	**1,355**
2009 [5]	989	157	107	113	**1,366**
2010 [5]	984	170	110	111	**1,375**
2011 [5]	981	180	134	89	**1,384**
2012 [5]	977	189	135	88	**1,389**
2013 [5]	983	188	135	88	**1,394**
2014 [5]	981	196	135	88	**1,400**
2015 [5]	974	208	136	88	**1,406**
2016 [5] [R]	986	202	137	87	**1,413**
2017 [5]	990	203	139	87	**1,419**

Source: Welsh Assembly Government

Contact: 0303 44 41864

E-Mail: housing.statistics@communities.gsi.gov.uk

1. For detailed definitions of all tenures, see Definitions of housing terms in Housing Statistics home page.

2. Owner-occupied tenure includes owner-occupied, intermediate and other.

3. April data for census years are based on census output.

4. Data for years 1969 to 1990 sourced from Department of Environment publications:

   Housing and Construction Statistics, 1967-1979, 1977-1987,1980-1990 and 1990-1997.

5. The tenure split between owner-occupied and privately rented dwellings has been calculated from 1997 onwards using

   information from the Labour Force Survey. These figures were revised in January 2011 following a re-weighting of the

   Labour Force Survey data.

R - Revised

P - Provisional

These data are produced and published separately by the Welsh Assembly Government, and although the figures in this table are correct at the time of its latest update they may be superseded before the next update. Data for earlier years are less reliable and definitions may not be consistent throughout the series. Stock estimates are expressed to the nearest thousand but should not be regarded as accurate to the last digit. Components may not sum to totals due to rounding.

11.1c Dwelling stock: by tenure[1,2,3,4,5], Scotland

Thousands of dwellings

	Owner Occupied	Rented Privately or with a job or business	Rented from Housing Associations	Rented from Local Authorities	All Dwellings
31 December[4]					
2000	1,472	155	137	557	**2,322**
31 March[4]					
2001	1,439	181	139	553	**2,312**
2002	1,477	179	143	531	**2,329**
2003	1,505	188	238	416	**2,347**
2004	1,513	213	251	389	**2,367**
2005	1,536	225	251	374	**2,387**
2006	1,559	234	251	362	**2,406**
2007	1,562	259	261	346	**2,428**
2008	1,592	259	269	330	**2,451**
2009	1,590	285	268	326	**2,469**
2010	1,584	303	272	323	**2,482**
2011	1,580	320	275	320	**2,495**
2012	1,545	366	277	319	**2,508**
2013	1,537	389	277	318	**2,521**
2014	1,545	394	277	318	**2,534**
2015	1,552	402	278	317	**2,549**
2016	1,558	414	278	317	**2,567**
2017	1,579	413	279	315	**2,585**

Source: Scottish Government

Contact: 0303 44 41864

E-Mail: housing.statistics@communities.gsi.gov.uk

1. For detailed definitions of all tenures, see Definitions of housing terms in Housing Statistics home page
2. April data for census years are based on census output
3. Data for years 1969 to 1989 sourced from Department of Environment publications:
   Housing and Construction Statistics, 1967-1979, 1977-1987,1980-1990 and 1990-1997.
4. Estimates from 1990 onwards are based on the Census (1991, 2001, 2011), council tax records and exemptions,
   social sector stock counts, and private tenure splits from the Scottish Household Survey and are not strictly comparable
   with earlier figures. These are not all collected for the same timescales and may not be exact counts, even rounded to
   the nearest thousand.
5. In order to include vacant private sector dwellings in table 107, estimates for vacant private sector stock in the Scottish
   statistics have been apportioned according to the % of occupied dwellings for the private sector tenures.

R- Revised from previous publication
P- Provisional

These data are produced and published separately by the Scottish Government, and although the figures in this table are correct at the time of its latest update they may be superseded before the next update. Data for earlier years are less reliable and definitions may not be consistent throughout the series. Stock estimates are expressed to the nearest thousand but should not be regarded as accurate to the last digit. Components may not sum to totals due to rounding.

11.1d Household Tenure 2008-09 to 2017-18[1,2,3,4,5]

All households   *Percentages*

Tenure	2008-09	2009-10	2010-11	2011-12	2012-13	2013-14	2014-15	2015-16	2016-17	2017-18
Owned outright	36	36	35	35	36	38	37	37	39	40
Owned with mortgage[2]	33	34	33	31	30	29	31	28	27	29
Rented- NIHE[3]	14	13	12	14	12	12	11	13	13	12
Rented other[4]	16	16	19	19	20	19	20	21	20	18
rented from housing association	..	..	..	..	*4*	*3*	*4*	*4*	*4*	*4*
rented privately	..	..	..	..	*16*	*16*	*16*	*17*	*16*	*14*
Rent free[5]	1	2	1	1	1	1	1	1	1	1
Bases=100%	2,474	2,761	2,718	2,778	2,710	2,736	2,521	2,494	2,532	4,359

Source: Continuous Household Survey

1. See Appendix 1: Data Sources - Supply at https://www.communities-ni.gov.uk/sites/default/files/publications/communities/ni-housing-stats-16-17-appendix1.docx

2. Includes properties being purchased through the co-ownership scheme.

3. NIHE - Northern Ireland Housing Executive

4. Includes properties which are rented from a housing association, rented privately.

5. Includes squatting and rent free

11.2 Dwelling stock: by tenure[1], Great Britain

Thousands of dwellings

	Owner Occupied	Rented Privately or with a job or business	Rented from Housing Associations	Rented from Local Authorities	Other public sector dwellings	All Dwellings
1 April [2]						
1981	12,020	2,354	454	6,127	..	20,954
31 December [3]						
1981	11,936	2,288	468	6,388	..	21,077
1982	12,357	2,201	480	6,196	..	21,233
1983	12,740	2,121	497	6,060	..	21,419
1984	13,099	2,040	516	5,959	..	21,615
1985	13,440	1,963	537	5,863	..	21,803
1986	13,790	1,885	551	5,776	..	22,002
1987	13,968	2,139	586	5,599	..	22,293
1988	14,424	2,077	614	5,412	..	22,527
1989	14,832	2,069	652	5,190	..	22,743
1990	15,099	2,123	702	5,015	..	22,940
31 March [3,4]						
1991	15,155	1,990	701	4,966	167	22,979
1992	15,367	2,058	733	4,879	151	23,190
1993	15,523	2,120	811	4,759	153	23,366
1994	15,695	2,184	884	4,634	150	23,546
1995	15,869	2,255	976	4,496	145	23,739
1996	16,013	2,332	1,078	4,369	141	23,931
1997	16,215	2,361	1,132	4,273	132	24,113
1998	16,461	2,367	1,205	4,140	121	24,296
1999	16,734	2,324	1,319	3,983	110	24,469
2000	16,949	2,350	1,458	3,788	101	24,645
2001	17,115	2,404	1,618	3,553	103	24,794
2002	17,280	2,465	1,692	3,420	112	24,968
2003 [5]	17,223	2,834	1,946	3,049	104	25,156
2004 [5]	17,478	2,895	2,017	2,884	83	25,358
2005 [5]	17,625	3,054	2,118	2,696	82	25,576
2006 [5]	17,608	3,334	2,182	2,603	82	25,810
2007 [5]	17,657	3,563	2,279	2,486	75	26,059
2008 [5]	17,660	3,837	2,414	2,330	74	26,317
2009 [5]	17,547	4,147	2,503	2,259	74	26,529
2010 [5]	17,463	4,385	2,562	2,220	66	26,696
2011 [5]	17,388	4,605	2,664	2,135	63	26,855
2012 [5]	17,276	4,841	2,716	2,100	75	27,008
2013 [5]	17,205	5,042	2,743	2,088	73	27,151
2014 [5]	17,200	5,213	2,755	2,075	64	27,306
2015 [5]	17,210	5,383	2,801	2,048	55	27,498
2016 [5]	17,345	5,448	2,845	2,016	57	27,713

Source: Ministry of Housing, Communities and Local Government
Welsh Assembly Government, Scottish Government
Department for Social Development (Northern Ireland)
Contact: 0303 44 41864
E-Mail: housing.statistics@Communities.gsi.gov.uk

1. For detailed definitions of all tenures, see Definitions of housing terms in Housing Statistics home page. 'Other public sector dwellings' figures are currently only available for England.
2. Figures for census years are based on census output
3. Data for years 1969 to 1990 sourced from Department of Environment publications:
   Housing and Construction Statistics, 1967-1979, 1977-1987,1980-1990 and 1990-1997.
   Great Britain totals from 2002 are derived by summing country totals at 31st March.
   For 1991 to 2001 Scotland stock levels from the year before is added into the UK total.
4. Series from 1992 to 2011 for England has been adjusted so that the 2001 and 2011 total dwelling estimate matches the 2001 and 2011 Census.
   Estimates from 2002 are based on local authority and housing association dwelling counts, the Labour Force Survey and, from 2003, the English Housing Survey. Estimates may not be strictly comparable between periods.
5. From 2003 the figures for owner-occupied and the private rental sector for England have been produced using a new improved methodology as detailed in the dwelling stock release. Previous to this vacancy was not accounted for.

R- Revised from previous publication
P- Provisional

Data for earlier years are less reliable and definitions may not be consistent throughout the series. Stock estimates are expressed to the nearest thousand but should not be regarded as accurate to the last digit. Components may not sum to totals due to rounding.

11.3 House building: permanent dwellings completed, by tenure[1] and country[2, 3]

Number of dwellings

Financial Year		United Kingdom	England	Wales	Scotland	Northern Ireland
All Dwellings						
1995-96		197,710	154,600	9,170	24,690	9,250
1996-97		185,940	146,250	10,090	20,700	8,910
1997-98		190,760	149,560	8,430	22,590	10,180
1998-99		178,290	140,260	7,740	20,660	9,640
1999-00		184,010	141,800	8,710	23,110	10,400
2000-01		175,370	133,260	8,330	22,110	11,670
2001-02		174,200	129,870	8,270	22,570	13,490
2002-03		183,210	137,740	8,310	22,750	14,420
2003-04		190,590	143,960	8,300	23,820	14,510
2004-05		205,400	155,890	8,490	26,470	14,540
2005-06 [4]		210,320	163,400	8,250	24,960	13,710
2006-07 [4]		215,220	167,680	9,330	24,280	13,930
2007-08		215,860	170,610	8,660	25,790	10,800
2008-09		178,560	140,990	7,120	21,020	9,430
2009-10		151,230	119,910	6,170	17,130	8,020
2010-11		136,040	107,870	5,510	16,450	6,210
2011-12		145,830	118,510	5,580	16,040	5,720
2012-13		133,050	107,980	5,450	14,100	5,530
2013-14		138,620	112,330	5,840	15,140	5,320
2014-15	R	153,060	124,640	6,170	16,740	5,500
2015-16	R	169,240	139,710	6,900	16,860	5,770
2016-17	R	178,860	147,520	6,830	17,200	6,460
2017-18	R	192,040	160,980	6,660	17,720	7,100
2018-19	P	..	169,070	5,780	..	7,810
Private Enterprise						
1995-96		156,540	123,620	6,880	19,200	6,850
1996-97		153,450	121,170	7,520	17,490	7,270
1997-98		160,680	127,840	6,490	17,980	8,370
1998-99		154,560	121,190	6,440	18,780	8,140
1999-00		160,520	124,470	7,860	19,070	9,120
2000-01		152,740	116,640	7,390	18,200	10,510
2001-02		153,580	115,700	7,490	18,310	12,070
2002-03		164,300	124,460	7,520	18,940	13,390
2003-04		172,370	130,100	7,860	20,460	13,950
2004-05		183,710	139,130	7,990	22,450	14,150
2005-06 [4]		185,840	144,940	7,880	20,260	12,760
2006-07 [4]		188,560	145,680	8,990	21,040	12,850
2007-08		187,280	147,170	8,320	21,660	10,140
2008-09		145,300	113,800	6,430	16,110	8,960
2009-10		116,420	93,030	5,290	11,140	6,960
2010-11		103,900	83,180	4,510	10,720	5,480
2011-12		108,840	89,120	4,750	10,150	4,830
2012-13		103,230	84,550	4,710	9,890	4,070
2013-14		110,090	89,630	5,160	11,090	4,200
2014-15	R	118,700	96,270	5,330	12,560	4,540
2015-16		135,470	111,350	5,650	13,430	5,050
2016-17		145,670	120,450	5,590	13,430	5,360
2017-18		155,960	131,770	5,470	13,130	5,880
2018-19	P	..	138,660	4,490	..	6,870
Housing Associations						
1995-96		38,170	30,230	2,130	4,780	1,040
1996-97		30,950	24,630	2,550	2,960	810
1997-98		28,550	21,400	1,940	4,490	730
1998-99		22,870	18,890	1,270	1,750	960
1999-00		23,170	17,270	850	3,960	1,090
2000-01		22,250	16,430	900	3,800	1,110
2001-02		20,400	14,100	710	4,200	1,390
2002-03		18,610	13,080	780	3,720	1,030

11.3 House building: permanent dwellings completed, by tenure[1] and country[2, 3]

Number of dwellings

Financial Year		United Kingdom	England	Wales	Scotland	Northern Ireland
2003-04		18,020	13,670	420	3,370	560
2004-05		21,550	16,660	480	4,020	390
2005-06 [2]		24,160	18,160	350	4,700	950
2006-07 [2]		26,400	21,750	350	3,230	1,080
2007-08		28,330	23,220	340	4,100	660
2008-09		32,430	26,690	690	4,580	470
2009-10		34,030	26,520	880	5,580	1,060
2010-11		30,380	23,550	990	5,110	740
2011-12		33,950	27,460	830	4,780	890
2012-13		27,500	22,060	740	3,240	1,450
2013-14		26,480	21,790	670	2,910	1,110
2014-15		31,880	27,020	840	3,060	960
2015-16		30,770	26,470	1,250	2,320	730
2016-17		30,330	25,230	1,240	2,750	1,100
2017-18		32,570	27,210	1,120	3,130	1,210
2018-19	P	..	27,860	1,230	..	940

Local Authorities

Financial Year		United Kingdom	England	Wales	Scotland	Northern Ireland
1995-96		3,010	760	160	720	1,360
1996-97		1,540	450	20	240	820
1997-98		1,520	320	-	110	1,080
1998-99		870	180	30	120	540
1999-00		320	60	-	70	190
2000-01		380	180	50	110	50
2001-02		230	60	70	70	30
2002-03		300	200	10	90	-
2003-04		210	190	20	-	-
2004-05		130	100	30	-	-
2005-06 [4]		320	300	20	-	-
2006-07 [4]		260	250	-	10	-
2007-08		250	220	10	30	-
2008-09		830	490	-	340	-
2009-10		780	370	-	410	-
2010-11		1,760	1,140	-	610	-
2011-12		3,080	1,960	-	1,110	-
2012-13		2,330	1,360	-	960	-
2013-14		2,060	910	10	1,140	-
2014-15	R	2,480	1,360	-	1,120	-
2015-16	R	3,000	1,900	-	1,100	-
2016-17	R	2,870	1,830	-	1,020	-
2017-18	R	3,510	2,000	80	1,460	-
2018-19	P	..	2,560	60	..	-

1. For detailed definitions of all tenures see definitions of housing terms on Housing Statistics home page
2. Northern Ireland data prior to 2005 is sourced from the Department of Communities, which use different definitions and adjust their data. Further information can be viewed at:
   https://www.communities-ni.gov.uk/publications/review-new-dwelling-starts-and-completions
3. These figures are for new build dwellings only. The Department also publishes an annual release entitled 'Housing Supply: net additional dwellings, England' which is the primary and most comprehensive measure of housing supply in England.
4. Figures from October 2005 to March 2007 in England are missing a small number of starts and completions that were inspected by independent approved inspectors. These data are included from June 2007.

Totals may not equal the sum of component parts due to rounding to the nearest 10

This table has been discontinued as a result of the reorganisation of tables. From late October 2019, this data will be published by the Office for National Statistics.

- Less than 5 dwellings
P Figure provisional and subject to revision
R Revised from previous release
.. Not available

Source:
P2 returns from local authorities
National House-Building Council (NHBC)
Approved inspector data returns
Welsh Assembly Government
Scottish Government
Department of Finance and Personnel (DFPNI)
District Council Building Control (NI)

Contact: Anthony Myers
Telephone: 0303 444 2246
Email: housing.statistics@communities.gov.uk

11.4a Housebuilding: permanent dwellings completed, by house and flat, number of bedroom and tenure[1], England

Percentage of all dwellings

Financial Year	2004 /05 [2]	2005 /06 [2]	2006 /07[2]	2007 /08[2]	2008 /09[2]	2009 /10[2]	2010 /11[2]	2011 /12[2]	2012 /13[2]	2013 /14[2]	2014 /15[2]	2015 /16[2]	2016 /17[2]
Private Enterprise													
Houses													
1 bedroom	1	-	0	-	1	0	1	0	1	1	1	0	0
2 bedrooms	8	8	7	7	9	9	11	10	10	11	12	10	10
3 bedrooms	30	25	24	23	25	25	28	32	36	33	38	38	38
4 or more bedrooms	29	23	24	23	23	28	32	29	33	32	33	31	33
All	67	56	56	53	58	63	72	71	79	76	83	80	81
Flats													
1 bedroom	7	10	10	8	10	10	4	6	4	6	4	6	5
2 bedrooms	25	32	33	38	31	26	23	22	17	18	12	14	14
3 bedrooms	1	1	1	1	1	1	1	1	0	0	0	0	0
4 or more bedrooms	-	-	-	-	0	0	0	0	0	0	0	0	0
All	33	44	44	47	42	37	28	29	21	24	17	20	19
Houses and flats													
1 bedroom	7	11	10	8	11	10	5	7	5	7	5	6	6
2 bedrooms	33	40	40	45	40	35	34	32	26	29	25	24	23
3 bedrooms	30	26	25	23	26	26	29	32	36	33	38	38	38
4 or more bedrooms	29	23	24	24	23	29	32	29	33	32	33	31	33
All	100	100	100	100	100	100	100	100	100	100	100	100	100
Housing Associations													
Houses													
1 bedroom	3	1	1	1	1	1	1	1	1	2	1	2	2
2 bedrooms	23	23	21	19	22	18	22	27	24	27	27	28	28
3 bedrooms	33	23	23	19	17	19	22	25	32	34	32	29	25
4 or more bedrooms	8	3	2	2	3	4	6	7	6	7	4	4	4
All	67	50	46	40	43	42	51	60	63	71	64	63	59
Flats													
1 bedroom	13	21	20	20	21	24	15	10	11	10	10	11	14
2 bedrooms	17	28	33	37	35	34	34	30	27	19	26	26	27
3 bedrooms	3	1	1	-	1	0	1	0	0	0	0	0	0
4 or more bedrooms	-	-	-	2	0	0	0	0	0	0	0	0	0
All	33	50	54	60	57	58	49	40	37	29	36	37	41
Houses and flats													
1 bedroom	16	21	21	21	22	25	16	10	11	12	11	13	16
2 bedrooms	40	52	54	55	57	52	56	57	51	46	53	53	55
3 bedrooms	36	24	24	19	18	20	22	26	32	34	32	29	25
4 or more bedrooms	8	3	2	4	3	4	6	7	6	7	4	4	4
All	100	100	100	100	100	100	100	100	100	100	100	100	100
All tenures													
Houses													
1 bedroom	1	-	-	-	1	1	1	0	1	1	1	1	1
2 bedrooms	10	10	9	9	13	12	14	13	12	13	15	14	14
3 bedrooms	30	25	24	22	23	23	27	30	35	33	36	36	35
4 or more bedroom	26	20	21	20	18	21	25	24	28	28	27	26	27
All	67	55	54	51	54	57	67	68	76	75	79	77	77
Flats													
1 bedroom	7	12	11	10	13	14	7	7	5	7	5	7	7
2 bedrooms	24	32	33	38	32	29	26	24	18	18	15	17	16
3 bedrooms	1	1	1	1	1	0	1	1	0	0	0	0	0
4 or more bedroom	-	-	-	1	0	0	0	0	0	0	0	0	0
All	33	45	46	49	46	43	33	32	24	25	21	23	23
Houses and flats													
1 bedroom	8	12	12	10	14	15	8	8	6	8	6	7	8
2 bedrooms	34	41	42	47	45	40	40	38	31	32	30	30	30
3 bedrooms	31	26	25	23	23	24	27	31	36	33	37	36	36
4 or more bedroom	27	21	21	20	18	21	25	24	28	28	27	26	27
All	100	100	100	100	100	100	100	100	100	100	100	100	100

Source: Department for Communities and Local Government
Telephone: 0303 444 1291
E-Mail: housing.statistics@communities.gsi.gov.uk

1. For detailed definitions of all tenures, see Definitions of housing terms in Housing Statistics home page
2. Figures for 2001/02 onwards are based on NHBC data only, so there is some degree of variability owing to partial coverage.
3. The England worksheet and charts have been corrected following initial publication.
4. Financial Year relates to April - April

Contact:
Contact: Rosie McGarrity
Telephone: 0303 444 6770

11.4b New dwellings completed by dwelling type and number of bedrooms, Wales

| | Flats | | | | | Houses | | | | | Total |
| | Number of bedrooms | | | | | Number of bedrooms | | | | | Houses |
	1	2	3	4 or more	Total	1	2	3	4 or more	Total	and flats
2002/2003	431	754	102	0	1,287	30	773	2,847	3,373	7,023	8,310
2003/2004	407	966	111	6	1,490	51	688	2,992	3,075	6,806	8,296
2004/2005	555	972	94	17	1,638	57	770	2,888	3,139	6,854	8,492
2005/2006	649	1,585	96	1	2,331	72	645	2,755	2,446	5,918	8,249
2006/2007	960	1,807	125	9	2,901	29	819	2,983	2,602	6,433	9,334
2007/2008	843	1,906	154	25	2,928	59	728	2,667	2,282	5,736	8,664
2008/2009	1,048	1,494	88	20	2,650	59	863	2,033	1,516	4,471	7,12
2009/2010	594	1,235	68	1	1,898	64	785	2,164	1,263	4,276	6,174
2010/2011	513	767	55	5	1,340	48	813	2,079	1,225	4,165	5,505
2011/2012	528	786	15	6	1,335	85	772	1,975	1,408	4,240	5,575
2012/2013	292	574	42	2	910	99	744	2,101	1,597	4,541	5,451
2013/2014	575	659	31	4	1,269	70	743	2,148	1,613	4,574	5,843
2014/2015	540	580	13	11	1,144	46	836	2,331	1,813	5,026	6,170
2015/2016	783	699	37	7	1,526	47	865	2,646	1,816	5,374	6,900
2016/2017	647	790	48	5	1,490	60	1,016	2,512	1,755	5,343	6,833

Source: New house building data collection, Welsh Government

.	The data item is not applicable.
Bedrooms	Breakdown by number of bedrooms within property.
Dwelling Type	Breakdown by Houses and Flats.
Period	Data on starts and completions of new permanent dwellings is only available at Wales level prior to 1996-97. From 1996-97, the information is available at a local authority level. Figures on housing starts are from records kept for building control purposes. It is often difficult for data providers to identify whether a dwelling is being built for RSLs or for a private developer and this may lead to an understatement of RSL

Contact email: stats.housing@gov.wales
Web link: http://gov.wales/statistics-and-research/new-house-building/?lang=en

In Wales, new house building is undertaken by the private sector, Registered Social Landlords (RSLs) and local authorities. The information presented here is based on the reports of local authority building inspectors and the National House Building Council (NHBC). It does not include information from private approved inspectors. The exclusion of this information means that there is currently an undercount in the number of dwellings completed though this is estimated to be quite small.

The information shows the number of new dwellings completed in Wales and is collected in order to assess the level of new house building across Wales during the period. The data is used to help monitor trends in both the overall level of Welsh housing stock and the changes in its tenure distribution over time. Data is also used by the Welsh Government and local authorities to assess levels of housing supply across Wales and as an indication as to whether housing need is being met.

Figures on housing completions are from records kept for building control purposes. It is sometimes difficult for data providers to identify whether a dwelling is being built for RSLs or for a private developer. This may lead to an understatement of RSL completions recorded in these tables, and a corresponding overstatement of private enterprise figures. This problem is more likely to occur with starts than completions.

New house building funded with capital grant funding includes funding via Social Housing Grant, Recycled Social Housing Grant and Strategic Capital Investment Fund.

11.5 Mortgage possession workload in the county courts of England and Wales, 2005 - 2017

Year (calendar)	Quarter	Claims Issued	Orders			Warrants[1]	Repossessions by county court bailiffs	Properties taken into possession[2] in
			Outright	Suspended	Total			
2005		114,733	32,757	38,211	70,968	48,513	12,794	14,500
2006		131,248	46,288	44,895	91,183	66,060	20,960	21,000
2007		137,725	58,250	49,259	107,509	73,890	23,831	25,900
2008		142,741	70,804	61,994	132,798	89,748	35,792	40,000
2009		93,533	44,856	38,039	82,895	77,461	32,457	48,300
2010		75,431	32,940	29,235	62,175	63,532	23,612	38,100
2011		73,181	30,190	29,697	59,887	65,371	25,463	37,100
2012		59,877	24,129	23,935	48,064	59,040	19,728	34,000
2013		53,659	20,718	19,585	40,303	52,305	15,692	28,900
2014		41,151	16,120	13,519	29,639	41,900	11,976	20,900
2015		19,852	7,984	6,031	14,015	23,220	5,592	10,200
2016		18,456	7,274	4,481	11,755	17,627	4,754	7,700
2017 (r)		19,836	8,270	4,710	12,980	16,336	4,386	7,400
2013	Q1	14,375	5,674	5,260	10,934	13,580	4,474	8,000
	Q2	12,881	5,187	5,059	10,246	13,529	4,087	7,600
	Q3	14,256	4,974	4,723	9,697	13,039	3,733	7,200
	Q4	12,147	4,883	4,543	9,426	12,157	3,398	6,100
2014	Q1	12,706	4,648	4,277	8,925	12,391	3,709	6,400
	Q2	10,773	4,400	3,539	7,939	11,121	3,028	5,400
	Q3	9,731	3,940	3,201	7,141	10,067	2,805	5,000
	Q4	7,941	3,132	2,502	5,634	8,321	2,434	4,100
2015	Q1	5,643	2,298	1,926	4,224	6,343	1,658	3,000
	Q2	4,849	1,951	1,475	3,426	5,646	1,363	2,500
	Q3	5,012	2,055	1,385	3,440	6,255	1,423	2,500
	Q4	4,348	1,680	1,245	2,925	4,976	1,148	2,200
2016	Q1	4,739	1,736	1,281	3,017	4,848	1,355	2,100
	Q2	4,430	1,901	1,200	3,101	4,848	1,183	1,900
	Q3	4,485	1,746	939	2,685	4,353	1,212	1,900
	Q4	4,802	1,891	1,061	2,952	3,578	1,004	1,800
2017	Q1	5,545	2,210	1,284	3,494	4,439	1,128	1,900
	Q2	5,186	2,127	1,205	3,332	4,183	1,068	1,800
	Q3	4,759	2,120	1,246	3,366	4,196	1,197	1,900
	Q4 (r)	4,346	1,813	975	2,788	3,518	993	1,800

Source: HM Courts and Tribunals Service CaseMan, Possession Claim On-Line (PCOL) and Council of Mortgage Lenders (CML)

Notes:

[1] Multiple warrants may be issued per claim

[2] Council of Mortgage Lenders (CML) statistics for the latest quarter are unavailable prior to this bulletin being published. This figure relates to all repossessions made in the United Kingdom whereas all other statistics in this bulletin relate to England and Wales. Please see the CML website http://www.cml.org.uk/ for more information about these statistics.

[3] Data relating to 1999 onwards are sourced from county court administrative systems and exclude duplicate observations. Data prior to 1999 are sourced from manual counts made by court staff.

.. = data not available

(p) = provisional

(r) = revised

11.6 Mortgage arrears and repossessions

Year	2005	2006	2007	2008	2009	2010	2011	2012	2013	2014	2015	2016	2017
Number of mortgages													
at year end (000s)	11,608	11,746	11,852	11,667	11,504	11,478	11,384	11,284	11,186	11,147	11,111	11,064	10,980
of which homeowners	10,909	10,910	10,827	10,498	10,257	10,169	9,996	9,835	9,658	9,491	9,330	9,208	9,100
Repossessions during year	14,500	21,000	25,900	40,000	48,900	38,500	37,300	33,900	28,900	20,900	10,200	7,700	7,300
of which homeowners	–	19,900	23,900	37,000	44,100	33,900	31,200	27,000	23,300	15,900	7,200	5,300	4,700
Cases in mortgage arrears													
12+ months arrears	15,000	15,700	15,300	29,500	69,500	63,700	54,400	48,500	41,100	30,700	30,500	28,900	26,900
+ 6 - 12 months arrears	38,600	34,900	40,500	72,000	93,900	80,500	72,200	69,900	60,700	45,100	38,600	31,300	26,900
+ 3 - 6 months arrears	69,400	64,900	71,700	117,400	112,400	103,300	99,000	97,200	86,600	68,800	55,100	42,700	35,600
= All 3+ months arrears	122,900	115,600	127,500	219,000	275,800	247,500	225,600	215,700	188,300	144,500	124,300	102,800	89,400
of which homeowners	118,400	110,800	120,000	192,000	250,700	225,600	206,600	199,200	174,200	133,200	113,900	94,200	82,200

UK Finance, Compendium of Housing Finance Statistics and Housing Finance website

Notes:
1. Properties taken into possession include those voluntarily surrendered. The UKF arrears figures are for the end of the year. Changes in the mortgage rate have the effect of changing monthly mortgage repayments and hence the number of months in arrears which a given amount represents.
2. Arrears figures are for both homeowners and buy to let mortgages except for bottom row (of all 3+ month arrears cases) which is exclusively for homeowners. For arrears and possesions related to buy to let mortgages see Table 55.
3. For intervening years before 2000, and the Janet Ford figures for 3-5 months arrears for the years from 1985 to 1994 see earlier editions of the Review.

11.7 Households in temporary accommodation (1) by type of accommodation, and cases where duty owed but no accommodation has been secured at the end of each quarter[7] England, 2003 to 2017

number

		Total number of households in TA[1,2,4]	Total number of households in TA with children	Total number of children in TA	Bed and breakfast hotels (including shared annexes)					Nightly paid, privately managed accommodation, self-contained	
					Total number of households[6]	Total with children	Total with children and resident more than 6 weeks	Total with children and resident more than 6 weeks and pending review / appeal	Total with 16/17-year-old main applicant	Total number of households	Total with children
2003	Q1	89,040	61,510	..	12,440	5,230	2,910	..	..	4,110	3,070
	Q2	91,870	65,040	..	11,380	3,940	2,120	..	..	5,020	3,770
	Q3	94,440	67,260	..	10,310	3,200	1,590	..	..	4,690	3,460
	Q4	94,610	67,540	..	8,420	1,730	940	..	..	3,250	2,350
2004	Q1	97,680	70,580	..	7,090	820	30	..	..	3,260	2,350
	Q2	99,530	71,640	121,590	7,240	1,100	60	..	..	4,140	2,970
	Q3	101,300	72,510	122,530	7,450	1,420	180	..	..	4,270	2,920
	Q4	101,030	72,800	124,630	6,450	820	100	..	..	4,180	2,760
2005	Q1	101,070	72,670	125,860	6,780	1,180	110	..	..	4,190	2,850
	Q2	100,970	72,810	124,900	6,290	1,300	130	50	..	4,520	2,970
	Q3	101,020	74,180	127,990	6,100	1,470	150	40	..	4,520	3,140
	Q4	98,730	72,920	127,620	4,950	820	140	30	..	4,450	2,950
2006	Q1	96,370	71,560	127,650	5,150	1,020	110	30	..	4,500	2,990
	Q2	93,910	69,790	130,470	4,890	1,050	100	50	..	4,570	3,020
	Q3	93,090	69,500	129,340	4,900	1,100	120	40	..	4,870	3,190
	Q4	89,510	65,770	122,080	4,210	650	110	40	..	4,900	3,310
2007	Q1	87,120	65,210	125,430	4,310	980	80	30	..	5,140	3,470
	Q2	84,900	64,020	117,340	4,070	940	100	30	670	5,410	3,760
	Q3	82,750	62,830	117,090	4,090	900	130	30	690	5,650	3,990
	Q4	79,500	59,990	112,260	3,530	700	120	20	550	5,780	4,080
2008	Q1	77,510	59,230	110,360	3,840	1,030	160	30	560	6,210	4,450
	Q2	74,690	57,210	107,050	3,440	1,030	180	30	420	5,980	4,220
	Q3	72,130	55,850	104,640	3,230	940	160	30	400	5,680	4,120
	Q4	67,480	52,290	98,880	2,560	520	100	20	330	5,230	3,720
2009	Q1	64,000	49,030	92,590	2,450	470	70	10	340	4,980	3,470
	Q2	60,230	45,940	87,030	2,150	510	80	20	310	4,570	3,120
	Q3	56,920	43,400	82,780	2,050	510	130	20	230	4,180	2,870
	Q4	53,370	40,560	77,990	1,880	400	120	10	170	3,620	2,450
2010	Q1	51,310	39,200	74,610	2,050	630	100	10	180	3,380	2,320
	Q2	50,400	37,940	72,590	2,410	740	160	10	190	3,340	2,220
	Q3	49,680	37,620	71,460	2,660	930	140	10	210	3,520	2,380
	Q4	48,010	36,230	69,050	2,310	660	150	10	140	3,590	2,430
2011	Q1	48,240	36,640	69,660	2,750	1,030	200	10	160	3,920	2,730
	Q2	48,330	35,950	68,770	3,120	1,210	240	20	150	4,200	2,920
	Q3	49,100	36,680	69,850	3,370	1,340	310	40	140	4,350	3,100
	Q4	48,920	36,600	69,460	3,170	1,310	450	60	100	4,460	3,330
2012	Q1	50,430	37,190	70,090	3,960	1,660	480	60	150	4,860	3,670
	Q2	51,630	39,470	73,890	4,230	1,900	680	70	140	5,370	4,100
	Q3	52,960	40,090	75,460	4,120	1,920	870	100	120	5,880	4,580
	Q4	53,140	40,830	76,740	3,820	1,600	770	100	80	6,190	4,900
2013	Q1	55,320	40,450	76,040	4,510	1,970	760	50	100	7,000	5,450
	Q2	55,840	42,800	79,030	4,350	2,090	740	70	80	7,630	6,060
	Q3	57,410	42,210	78,770	4,610	2,110	800	80	70	8,400	6,420
	Q4	56,940	43,750	80,970	3,920	1,560	500	60	70	8,660	7,330
2014	Q1	58,410	44,770	83,370	4,370	1,900	440	40	60	9,340	7,960
	Q2	59,570	45,940	87,890	4,590	2,130	610	50	70	10,100	8,630
	Q3	60,900	47,460	91,090	4,680	2,140	470	60	70	11,750	10,100
	Q4	61,930	48,460	93,980	4,540	2,040	780	60	40	12,540	10,770
2015	Q1	64,710	51,210	98,620	5,270	2,560	920	50	50	13,620	11,770
	Q2	66,980	52,550	102,090	5,630	2,660	880	40	40	14,870	12,840
	Q3	68,560	53,480	103,440	5,910	3,000	1,050	100	30	15,760	13,280
	Q4	69,140	54,240	106,240	5,120	2,270	910	100	40	16,210	13,990
2016	Q1	71,670	56,430	111,060	5,960	2,890	950	60	60	17,060	14,640
	Q2	73,050	58,140	114,810	6,490	3,370	1,240	40	40	17,530	15,120
	Q3	74,750	59,380	117,510	6,680	3,450	1,450	100	30	18,420	15,740
	Q4	75,740	60,240	118,930	5,990	2,790	1,260	40	20	19,080	16,340
2017	Q1	77,220	60,980	120,520	6,580	3,010	1,300	60	30	19,570	16,620
	Q2	78,540	60,810	120,660	6,520	2,640	1,200	30	20	20,280	16,890
	Q3	79,830	61,980	122,830	6,470	2,700	1,130	250	20	20,650	17,190
	Q4	79,720	61,620	122,400	5,780	2,050	880	40	20	20,800	17,190

11.7 Households in temporary accommodation (1) by type of accommodation, and cases where duty owed but no accommodation has been secured at the end of each quarter7 England, 2003 to 2017

number

		Hostels (including reception centres, emergency units and refuges)		Private sector accommodation leased by your authority or leased or managed by a registered provider		Local authority or Housing association (LA/HA) stock		Any other type of temporary accommodation (including private landlord and not known)[3]		In TA in another local authority district	Duty owed, no accommodation secured[5]	
		Total number of households	Total with children	Total number of households	Total with children	Total number of households	Total with children	Total number of households	Total with children		Total number of households	Total with children
2003	Q1	10,050	6,040	28,370	23,240	28,250	19,690	5,810	4,240	11,470	10,580	..
	Q2	10,420	6,360	31,460	26,630	27,590	19,900	5,990	4,450	10,880	13,310	..
	Q3	10,800	6,450	35,140	29,950	27,560	19,650	5,950	4,560	9,880	15,370	..
	Q4	10,370	6,060	38,730	33,340	27,480	19,450	6,370	4,600	10,440	17,500	..
2004	Q1	10,790	6,280	42,390	36,390	27,890	20,120	6,270	4,620	9,850	15,870	..
	Q2	10,570	6,090	42,630	35,900	27,960	20,070	6,990	5,510	9,480	17,030	..
	Q3	10,380	5,960	43,720	36,650	28,220	20,070	7,260	5,500	10,030	17,100	..
	Q4	10,070	5,650	46,140	38,750	27,730	19,730	6,460	5,080	8,610	16,100	..
2005	Q1	10,280	5,830	46,530	39,170	26,630	18,610	6,670	5,030	11,660	15,290	
	Q2	9,870	5,440	46,990	39,600	27,430	19,070	5,870	4,420	11,790	16,020	10,470
	Q3	10,020	5,410	48,860	41,500	25,030	17,620	6,500	5,060	13,430	15,140	10,270
	Q4	9,230	4,990	49,910	42,310	24,220	17,110	5,970	4,750	10,700	11,570	7,840
2006	Q1	9,010	4,960	49,660	41,960	22,350	16,080	5,700	4,560	11,080	11,010	7,210
	Q2	8,940	4,820	49,320	41,740	20,790	14,880	5,400	4,270	11,590	10,210	6,920
	Q3	8,460	4,460	49,700	41,980	20,180	14,830	4,980	3,950	11,620	9,720	6,550
	Q4	7,850	3,950	48,850	40,130	18,840	13,930	4,870	3,810	9,950	8,470	5,740
2007	Q1	7,640	4,030	45,600	38,600	18,040	13,510	6,400	4,620	10,130	8,780	5,910
	Q2	7,230	3,890	44,610	37,920	17,240	12,970	6,350	4,540	10,490	9,150	6,280
	Q3	7,180	3,850	43,430	37,100	16,490	12,410	5,920	4,590	11,130	9,540	6,300
	Q4	6,620	3,490	41,730	35,380	15,910	11,780	5,930	4,560	10,820	8,080	5,510
2008	Q1	6,450	3,580	40,480	34,610	14,740	11,080	5,790	4,480	10,200	7,470	5,180
	Q2	6,020	3,350	41,130	34,900	14,030	10,660	4,090	3,050	8,720	7,890	5,630
	Q3	5,800	3,190	39,990	34,390	13,420	10,240	4,020	2,970	7,620	6,740	4,520
	Q4	5,250	2,830	38,790	33,560	11,930	8,990	3,720	2,680	7,360	6,070	4,030
2009	Q1	5,170	2,740	37,450	32,050	10,480	7,800	3,480	2,500	7,960	5,560	3,740
	Q2	4,710	2,430	35,920	30,620	9,520	6,970	3,360	2,290	7,880	4,560	2,940
	Q3	4,480	2,330	34,130	29,090	8,780	6,390	3,310	2,210	6,550	4,350	2,770
	Q4	4,150	2,150	32,430	27,540	8,180	5,950	3,100	2,070	5,780	4,150	2,540
2010	Q1	4,240	2,270	30,920	26,310	7,790	5,790	2,940	1,900	5,430	3,710	2,320
	Q2	4,320	2,380	29,820	25,210	7,650	5,570	2,860	1,830	5,630	3,780	2,510
	Q3	4,360	2,440	28,740	24,570	7,610	5,480	2,800	1,810	5,880	4,100	2,700
	Q4	4,160	2,270	27,730	23,620	7,430	5,440	2,790	1,810	5,810	4,410	2,970
2011	Q1	4,250	2,330	26,960	23,170	7,490	5,500	2,870	1,890	6,300	4,770	3,270
	Q2	4,370	2,340	26,240	22,170	7,570	5,460	2,850	1,860	6,290	4,770	3,420
	Q3	4,380	2,370	26,380	22,250	7,890	5,810	2,740	1,810	6,850	5,110	3,590
	Q4	4,310	2,380	26,080	21,800	7,990	5,840	2,910	1,950	7,350	5,490	3,550
2012	Q1	4,360	2,350	26,040	21,490	8,270	6,000	2,940	2,030	7,870	5,400	3,860
	Q2	4,350	2,610	25,960	22,240	8,600	6,470	3,130	2,160	8,170	5,500	3,830
	Q3	4,390	2,690	26,290	21,880	8,930	6,730	3,360	2,280	8,520	5,560	4,080
	Q4	4,280	2,610	26,310	22,490	9,090	6,830	3,440	2,400	9,270	5,690	3,950
2013	Q1	4,480	2,710	26,260	21,020	9,270	7,020	3,800	2,290	9,130	5,930	4,270
	Q2	4,590	2,890	24,780	21,210	10,060	7,860	4,440	2,690	11,280	5,510	3,870
	Q3	4,700	2,960	25,660	21,460	9,810	7,290	4,230	1,970	11,860	5,010	3,520
	Q4	4,710	2,950	25,460	21,450	9,560	7,280	4,620	3,190	12,190	4,930	3,460
2014	Q1	4,880	3,010	25,270	21,150	9,880	7,530	4,680	3,220	12,910	5,620	3,960
	Q2	4,980	3,100	24,800	20,730	10,120	7,870	4,990	3,490	14,130	5,310	3,880
	Q3	5,010	3,110	23,290	19,550	10,070	7,790	6,110	4,780	15,460	6,120	4,420
	Q4	5,090	3,360	23,460	19,790	10,530	8,110	5,780	4,400	15,990	5,820	4,290
2015	Q1	5,040	3,360	23,990	20,410	10,920	8,540	5,860	4,570	16,810	6,900	5,060
	Q2	5,180	3,510	23,820	20,410	11,500	8,750	5,980	4,390	17,640	6,370	4,640
	Q3	5,310	3,510	23,520	19,800	12,210	9,600	5,860	4,290	18,600	6,610	4,830
	Q4	5,360	3,670	25,580	21,130	12,480	9,790	4,400	3,400	18,670	7,490	4,800
2016	Q1	5,570	3,820	24,420	20,500	13,130	10,230	5,530	4,350	19,880	6,790	4,980
	Q2	5,490	3,860	24,990	21,170	13,540	10,830	5,020	3,780	20,650	7,020	5,280
	Q3	5,690	3,940	24,300	20,600	13,610	10,880	6,060	4,780	21,400	7,590	5,600
	Q4	5,700	4,080	24,150	20,440	14,040	11,150	6,790	5,430	21,920	8,610	6,540
2017	Q1	5,740	3,990	24,510	20,740	14,370	11,520	6,460	5,110	21,950	8,230	6,090
	Q2	5,710	3,740	23,050	19,310	15,010	11,910	7,960	6,350	22,160	8,870	6,550
	Q3	5,570	3,680	23,390	19,650	15,560	12,320	8,170	6,480	22,730	9,540	7,010
	Q4	5,440	3,510	23,430	19,710	15,750	12,450	8,520	6,770	22,410	10,020	7,380

Source: Department for Communities and Local Government; MHCLG P1E Homelessness returns (quarterly)
email: homelessnessstats@communities.gov.uk

11.7 Households in temporary accommodation (1) by type of accommodation, and cases where duty owed but no accommodation has been secured at the end of each quarter7 England, 2003 to 2017

Notes:

1 Q2 2018 was the first to utilise the H-CLIC case level collection system. The data is now also processed by MHCLG using case level information to produce tables part of this release. Previous quarters were solely based on the P1E collection where local authorities aggregated data themselves. The breakdowns have been processed by MHCLG so that 2018 Q2 can be compared to previous quarters. It has become apparent that there were minor inconsistencies in how local authorities previously applied this methodology which may account for some of the difference in totals from Q1 to Q2 2018.

2 Households in accommodation arranged by local authorities pending enquiries or after being accepted as homeless under the 1996 Act (includes residual cases awaiting re-housing under the 1985 Act) and as amended by the Homelessness Reduction Act 2018.

3 Other accommodation includes accommodation that has been leased directly by the household from a private landlord where this arrangement is temporary, supported lodgings, mobile homes such as caravans and unknown types of temporary accomodation. The Homelessness (Suitability of Accommodation) (England) Order 2003 came into force on 1 April 2004. This prohibits the use of B&B accommodation for families except in an emergency and even then for no longer than six weeks.

4 Households in TA are as reported at the end of each quarter. Q1 represents March 31st, Q2 June 30th, Q3 September 30th, Q4 December 31st.

5 Households owed a main duty but either (a) remain in accommodation from which accepted as homeless or (b) making own arrangements for temporary accommodation ("homeless at home")

6 The large rise in B&B placements from Q1 to Q2 2018 can be attributed to the impact of the HRA in April 2018 and improvements in local authorities' reporting of this information through H-CLIC. This includes correctly classifying shared annexes, which were previously being reported as self-contained by some local authorities.

Figures include imputations for missing values.
R Revised data
Q1 to Q4 refers to calendar year quarters.
Total figures are presented rounded to the nearest 10 households.
Totals may not equal the sum of components because of rounding.

11.8 Homeless Households in temporary accomodation at the end of the period - as at 31 March each year, Wales

		2006-07 Annual	2007-08 Annual	2008-09 Annual	2009-10 Annual	2010-11 Annual	2011-12 Annual	2012-13 Annual	2013-14 Annual	2014-15 Annual	2015-16 Annual	2016-17 Annual
Total accommodated at the end of quarter		3150	2880	2815	2490	2640	2770	2525	2295	2050	1875	2013
Total accommodated at the end of quarter	Private sector accommodation (1)	710	900	1070	1050	1080	1065	1010	910	855	801	786
	Public sector accommodation (2)	575	445	415	390	435	380	380	440	440	411	432
	Hostels and women's refuges	405	475	510	400	415	485	505	510	475	498	549
	Bed and breakfast	380	280	255	235	240	310	300	185	195	108	189
	Other	230	190	105	*	*	5	10	*	*	*	3
	Homeless at home	855	585	465	415	470	525	320	245	85	60	54
	Accommodation type unknown	.	.	.	.	.	.	.	.	.	.	.

Source: Welsh Government, Homelessness data collection

Contact: stats.housing@wales.gsi.gov.uk

1. Private sector accommodation includes private sector accommodation leased by the local authority, RSLs and directly with a private sector landlord

2. Public sector accommodation includes within local authority stock and RSL stock on assured shorthold tenancies

. The data item is not applicable.

* The data item is disclosive or not sufficiently robust for publication

11.9 Households in temporary accommodation by accommodation type, Scotland

All households		Social sector accommodation[1]		Hostel		Bed & Breakfast		Other[2]		Total	
		Number	%	Number	%	Number	%	Number	%	Number	%
2004	as at 31 March	3,537	55	1,586	25	1,190	18	132	2	6,445	100
	as at 30 June	3,754	56	1,514	23	1,273	19	105	2	6,646	100
	as at 30	3,894	56	1,590	23	1,331	19	110	2	6,925	100
	as at 31	4,071	59	1,521	22	1,243	18	117	2	6,952	100
2005	as at 31 March	4,136	57	1,490	20	1,516	21	159	2	7,301	100
	as at 30 June	4,324	59	1,340	18	1,413	19	264	4	7,341	100
	as at 30	4,606	60	1,320	17	1,424	19	333	4	7,683	100
	as at 31	4,525	60	1,295	17	1,323	18	356	5	7,499	100
2006	as at 31 March	4,747	59	1,328	17	1,494	19	416	5	7,985	100
	as at 30 June	4,732	60	1,342	17	1,362	17	452	6	7,888	100
	as at 30	4,880	60	1,301	16	1,491	18	439	5	8,111	100
	as at 31	4,981	62	1,235	15	1,391	17	482	6	8,089	100
2007	as at 31 March	5,164	60	1,242	14	1,528	18	643	7	8,577	100
	as at 30 June	5,075	60	1,170	14	1,588	19	690	8	8,523	100
	as at 30	5,104	61	1,134	13	1,492	18	671	8	8,401	100
	as at 31	5,460	63	1,104	13	1,348	16	721	8	8,633	100
2008	as at 31 March[3]	6,134	64	1,079	11	1,609	17	713	7	9,535	100
	as at 30 June	6,079	62	1,064	11	1,791	18	815	8	9,749	100
	as at 30	6,131	62	1,058	11	1,780	18	848	9	9,817	100
	as at 31	5,931	62	1,019	11	1,662	17	924	10	9,536	100
2009	as at 31 March	6,355	63	994	10	1,748	17	956	10	10,053	100
	as at 30 June	6,294	62	1,186	12	1,654	16	1,072	11	10,206	100
	as at 30	6,438	62	1,221	12	1,584	15	1,100	11	10,343	100
	as at 31	6,378	62	1,234	12	1,515	15	1,151	11	10,278	100
2010	as at 31 March	6,775	63	1,217	11	1,765	16	972	9	10,729	100
	as at 30 June	6,938	62	1,267	11	1,940	17	958	9	11,103	100
	as at 30	7,124	63	1,369	12	1,673	15	1,098	10	11,264	100
	as at 31	7,272	66	1,339	12	1,418	13	1,066	10	11,095	100
2011	as at 31 March	7,215	64	1,371	12	1,544	14	1,124	10	11,254	100
	as at 30 June	7,443	67	1,349	12	1,414	13	953	9	11,159	100
	as at 30	7,382	67	1,329	12	1,433	13	916	8	11,060	100
	as at 31	7,102	66	1,310	12	1,232	12	1,041	10	10,685	100
2012	as at 31 March	7,093	66	1,333	12	1,281	12	1,043	10	10,750	100
	as at 30 June	7,106	68	1,190	11	1,205	12	965	9	10,466	100
	as at 30	7,146	68	1,333	13	1,090	10	977	9	10,546	100
	as at 31	6,920	67	1,292	13	1,063	10	977	10	10,252	100
2013	as at 31 March	7,061	67	1,290	12	1,170	11	950	9	10,471	100
	as at 30 June	6,965	66	1,451	14	1,104	11	972	9	10,492	100
	as at 30	6,877	67	1,503	15	1,001	10	887	9	10,268	100
	as at 31	6,687	67	1,496	15	932	9	848	9	9,963	100
2014	as at 31 March	6,405	62	1,813	18	1,125	11	938	9	10,281	100
	as at 30 June	6,310	62	1,830	18	1,073	11	981	10	10,194	100
	as at 30	6,509	63	1,714	17	1,070	10	1,035	10	10,328	100
	as at 31	6,365	62	1,766	17	1,030	10	1,057	10	10,218	100
2015	as at 31 March	6,558	62	1,741	16	1,085	10	1,183	11	10,567	100
	as at 30 June	6,600	63	1,743	17	1,059	10	1,061	10	10,463	100
	as at 30	6,542	62	1,749	17	1,156	11	1,026	10	10,473	100
	as at 31	6,610	63	1,733	17	1,025	10	1,072	10	10,440	100
2016	as at 31 March	6,679	63	1,728	16	1,052	10	1,084	10	10,543	100
	as at 30 June	6,704	62	1,716	16	1,069	10	1,269	12	10,758	100
	as at 30	6,562	61	1,792	17	1,071	10	1,335	12	10,760	100
	as at 31	6,665	63	1,691	16	1,012	9	1,293	12	10,661	100
2017	as at 31 March	6,637	61	1,738	16	1,113	10	1,385	13	10,873	100

Source: Scottish Government

11.9 Households in temporary accommodation by accommodation type, Scotland

Notes

All figures in the table are rounded to 5 for disclosure purposes

1. Includes Glasgow Housing Association stock from 2003, and all other housing associations from June 2005 onward.

2. The category 'other' includes mainly private landlords. Prior to June 1999 the figures may also include an unknown number of local authority-owned chalets or mobile homes.

3. From 31 March 2008 there is a break in comparability in numbers in temporary accommodation in Glasgow-see notes page.

Background to discontinuity in Glasgow data from 31 March 2008

From 31 March 2008 there is a break in comparability in the information on numbers of homeless applicants in temporary accommodation in Glasgow. The number of homeless households in temporary accommodation in Glasgow includes asylum applications given indefinite leave to remain in the United Kingdom and who are in temporary accommodation. From 31 March 2008 there was an significant increase in such households as a consequence of the 'legacy' case reviews undertaken by the Home Office. This introduces a discontinuity in the statistics for both Glasgow and for Scotland in the totals for all households and households with children. To bridge the discontinuity Glasgow have provided figures on the numbers of such households included at the end of each quarter from 31 March 2008.

Banking, insurance

Chapter 12

Banking, insurance

Industrial analysis of monetary financial institutions deposits and lending (Tables 12.4 and 12.5)

These data collate information from UK MFIs on deposits from and lending to UK residents other than MFIs and are separated into 18 broad industrial categories, based upon the SIC classification system. Until Q3 2007, the analysis of lending covered loans, advances (including under reverse repos), finance leasing, acceptances and facilities (all in sterling and other currencies) provided by reporting MFIs to their UK resident non-MFI customers, as well as MFI holdings of sterling and euro commercial paper issued by these resident customers. Following a review of statistical data collected, acceptances and holdings of sterling and euro commercial paper are no longer collected at the industry level detail with effect from Q4 2007 data. Total lending therefore reflects loans and advances (including under reverse repos) only, from Q4 2007 data.

Consumer credit (Excluding Student Loans) (Table 12.12)

Following an ONS review in August 1997, data for 'other specialist lenders' were improved and revised back to January 1995. Total outstanding consumer credit was revised upwards by £2.6bn. Flows were break adjusted. Monthly data are available for lending by retailers from January 1997 but are not available for lending by insurance companies. The missing monthly data have been interpolated from quarterly data.

Within total consumer credit (excluding student loans) outstanding, credit card lending had been underestimated and 'other' consumer credit overestimated prior to January 1999 as a result of a longstanding inconsistency. The credit card element had previously covered sterling credit card lending to the UK household sector by only UK banks and building societies. Credit card lending by other specialist lenders and retailers (where they finance lending themselves) could not be separately identified and so was included within the 'other' consumer credit component.

From January 1999 onwards this inconsistency has been corrected, as credit card lending by other specialist lenders can be separately identified. As a result, data from January 1999 onwards for credit card lending and for 'other' consumer credit are not directly comparable with those for earlier periods. The change affects all three measures of credit card lending (gross, net and amounts outstanding), with an equal offsetting change to 'other' consumer credit. In non-seasonally adjusted terms, gross credit card lending was on average around £800 million per month higher since January 1999, whilst the amount outstanding of credit card debt was boosted by £4.8 billion in January 1999. The changes to net credit card lending are much smaller in absolute terms, with no discernible change to trend.

From November 2006, the Bank of England ceased to update the separate data on consumer credit provided by other specialist lenders, retailers, and insurance companies, previously contained in Table A5.6 of Monetary and Financial Statistics. The final month for which separate data are available on the Bank's Statistical Interactive Database is November 2006. The three categories have been merged into "other consumer credit lenders".

Prior to January 2008, building societies' lending was unsecured lending to individuals including sterling bridging loans (prior to October 1998 this was class 3 lending to individuals). Building societies gross lending through overdrafts is no longer included from January 2008.

http://www.bankofengland.co.uk/boeapps/iadb/newintermed.asp

12.1a The Bank of England's Annual Balance Sheet 1958-2017

This sheet gives consistent annual series for the Bank's liabilities and assets from 1958 to 1966, which were first published in the June 1967 Quarterly Bulletin and extended to 2006 using subsequent Quarterly Bulletins. An extension to 2014 is made using the new reporting method introduced in 2006. The dates in each year to which the figures relate are: last Saturday in February; 1958-1966 - Published Bank Returns, last Wednesday in February: 1966-2006 - Published Bank Returns: third Wednesday in February (the dates on which the London Clearing Banks compiled their monthly figures); 2007-2017 - Published Bank Returns: third Wednesday in February. See links at the bottom of the table for more information on the data.

Assets, £

	Total	Government debt	Government securities	Other securities	Coin and bullion	Notes in the Bank
1958	2,327,196,165	11,015,100	2,222,675,614	49,533,147	5,727,986	38,244,318
1959	2,396,154,287	11,015,100	2,309,195,369	27,721,013	3,926,655	44,296,150
1960	2,509,611,487	11,015,100	2,412,801,707	44,040,161	3,309,613	38,444,906
1961	2,766,327,702	11,015,100	2,655,953,220	52,627,117	2,331,877	44,400,388
1962	2,920,525,859	11,015,100	2,819,589,302	65,236,067	1,468,171	23,217,219
1963	2,714,385,221	11,015,100	2,585,258,886	75,731,629	1,467,915	40,911,691
1964	2,831,080,950	11,015,100	2,738,352,963	58,944,838	1,451,304	21,316,745
1965	3,042,943,845	11,015,100	2,879,556,182	101,959,343	1,469,009	48,944,211
1966	3,302,225,248	11,015,100	3,165,623,387	102,383,926	1,460,584	21,742,251

3rd Wednesday in February

	Total	Government debt	Government securities	Other securities	Coin and bullion	
1966	3,294,800,000	11,015,100	3,165,884,900	100,500,000	17,400,000	
1967	3,532,700,000	11,015,100	3,326,484,900	153,800,000	41,400,000	
1968	3,742,400,000	11,015,100	3,513,084,900	179,700,000	38,600,000	
1969	3,870,700,000	11,015,100	3,724,084,900	116,900,000	18,700,000	
1970	3,887,000,000	11,015,100	3,655,984,900	198,000,000	22,000,000	
1971	4,668,000,000	11,015,100	4,166,984,900	446,000,000	44,000,000	
1972	4,242,000,000	11,015,100	3,598,984,900	594,000,000	38,000,000	
1973	5,484,000,000	11,015,100	4,572,984,900	866,000,000	34,000,000	
1974	6,564,000,000	11,015,100	5,208,984,900	1,320,000,000	24,000,000	
1975	6,945,000,000	11,015,100	6,037,984,900	877,000,000	19,000,000	
1976	7,677,000,000	11,015,100	6,508,984,900	1,138,000,000	19,000,000	
1977	8,156,000,000	11,015,100	6,600,984,900	1,537,000,000	7,000,000	
1978	10,004,000,000	11,015,100	8,541,984,900	1,428,000,000	23,000,000	
1979	10,193,000,000	11,015,100	8,491,984,900	1,683,000,000	7,000,000	
1980	11,065,000,000	11,015,100	8,760,984,900	2,269,000,000	24,000,000	
1981	12,112,000,000	11,015,100	7,336,984,900	4,749,000,000	15,000,000	
1982	12,701,000,000	11,015,100	4,765,984,900	7,919,000,000	5,000,000	
1983	15,961,000,000	11,015,100	4,369,984,900	11,565,000,000	15,000,000	
1984	14,315,000,000	11,015,100	2,109,984,900	12,184,000,000	10,000,000	
1985	19,603,000,000	11,015,100	2,893,984,900	16,693,000,000	5,000,000	
1986	19,219,000,000	11,015,100	3,227,984,900	15,972,000,000	8,000,000	
1987	17,256,000,000	11,015,100	2,150,984,900	15,087,000,000	7,000,000	
1988	16,889,000,000	11,015,100	2,713,984,900	14,157,000,000	7,000,000	
1989	17,350,000,000	11,015,100	6,114,984,900	11,214,000,000	10,000,000	
1990	18,679,000,000	11,015,100	11,681,984,900	6,976,000,000	10,000,000	
1991	19,694,000,000	11,015,100	9,952,984,900	9,726,000,000	4,000,000	
1992	20,280,000,000	11,015,100	8,448,984,900	11,808,000,000	12,000,000	
1993	24,569,000,000	11,015,100	7,976,984,900	16,574,000,000	7,000,000	
1994	26,351,000,000	11,015,100	6,954,984,900	19,381,000,000	4,000,000	
1995	24,195,000,000	0	14,311,000,000	9,876,000,000	8,000,000	
1996	25,181,000,000	0	13,552,000,000	11,624,000,000	5,000,000	
1997	26,917,000,000	0	10,309,000,000	16,596,000,000	12,000,000	
1998	32,681,000,000	0	6,174,000,000	26,503,000,000	4,000,000	
1999	101,956,000,000	0	6,936,000,000	95,008,000,000	12,000,000	
2000	80,756,000,000	0	7,255,000,000	73,489,000,000	12,000,000	
2001	37,850,000,000	0	15,061,000,000	22,782,000,000	7,000,000	
2002	41,352,000,000	0	15,298,000,000	26,050,000,000	4,000,000	
2003	46,712,000,000	0	15,735,000,000	30,970,000,000	7,000,000	
2004	48,565,000,000	0	16,426,000,000	32,132,000,000	7,000,000	
2005	55,714,000,000	0	15,193,000,000	40,512,000,000	9,000,000	
2006	61,799,000,000	0	15,538,000,000	46,254,000,000	7,000,000	

12.1a The Bank of England's Annual Balance Sheet 1958-2017

This sheet gives consistent annual series for the Bank's liabilities and assets from 1958 to 1966, which were first published in the June 1967 Quarterly Bulletin and extended to 2006 using subsequent Quarterly Bulletins. An extension to 2014 is made using the new reporting method introduced in 2006. The dates in each year to which the figures relate are: last Saturday in February; 1958-1966 - Published Bank Returns, last Wednesday in February: 1966-2006 - Published Bank Returns: third Wednesday in February (the dates on which the London Clearing Banks compiled their monthly figures); 2007-2017 - Published Bank Returns: third Wednesday in February. See links at the bottom of the table for more information on the data.

Assets, £

	£ Short-term repo operations with BoE	£ long-term operations with BoE	Central Bank bonds and other securities acquired via market transactions	£ Ways and means advances to HM government	Other assets including loan to the Asset Purchase facility[2]
3rd Wednesday in February, Q1 observation used from 2015[1]					
2007	76,991,000,000	30,110,000,000	15,000,000,000	6,727,000,000	13,370,000,000	18,511,000,000
2008	97,616,000,000	6,609,000,000	31,999,000,000	7,618,000,000	7,370,000,000	51,638,000,000
2009	168,404,000,000	0	126,261,000,000	11,873,000,000	9,392,000,000	32,751,000,000
2010	247,280,000,000	0	20,607,000,000	12,691,000,000	370,000,000	226,303,000,000
2011	241,642,000,000	0	14,102,000,000	13,273,000,000	370,000,000	227,169,000,000
2012	311,459,000,000	0	5,545,000,000	13,892,000,000	370,000,000	305,545,000,000
2013	404,228,000,000	0	2,085,000,000	13,248,000,000	370,000,000	401,773,000,000
2014	403,089,000,000	0	1,165,000,000	16,312,000,000	370,000,000	401,554,000,000
2015	410,775,000,000	0	5,180,000,000	-	370,000,000	405,225,000,000
2016	423,246,000,000	0	18,657,000,000	-	370,000,000	404,219,000,000
2017	539,184,000,000	0	11,320,000,000	-	370,000,000	527,493,000,000

Source: Bank of England

1. Following Bank of England money market reform on 18 May 2006 the Bank of England 'Bank Return' was changed. This series forms part of the new Bank Return, with data starting on 24 May 2006. More information on changes made to the Bank's monetary policy operations and their impact on published data can be found in `The implications of money market reform for data published in Monetary and Financial Statistics' in the June 2006 issue of Bank of England: Monetary and Financial Statistics. From 2015 onwards the observations on the total balance sheet are only available quarterly in the consolidated balance sheet and with a 5 quarter lag so the nearest proxies are used for February observations. Thomas Ryland (email: ryland.thomas@bankofengland.co.uk.)

2. Other assets is the sum of loan to the Asset Purchase Facility (RPQZ4TM), other assets (RPQZ6MW), all foreign currency reserve assets (RPQZ4TN), other foreign currency assets (RPQZ6MX) and denominated bond holdings (RPQZ4TL)

3. Other liabilities is the sum of capital & reserves (equity) from monetary financial institutions (RPQZ6MS), other foreign currency liabilities (RPQZ6MT) and other liabilities (RPQZ6MO).

For more information on the data please use the following links:

Bank of England Consolidated balance sheet: https://www.bankofengland.co.uk/weekly-report/balance-sheet-and-weekly-report

Original 1967 Quarterly Bulletin article: http://www.bankofengland.co.uk/archive/Documents/historicpubs/qb/1967/qb67q2159163.pdf

Bank of England Statistical Abstracts: http://www.bankofengland.co.uk/archive/Pages/digitalcontent/historicpubs/statisticalabst.aspx

The implications of money market reform for data published in Monetary and Financial Statistics' in the June 2006 issue of Bank of England: Monetary and Financial Statistics.: http://www.bankofengland.co.uk/statistics/Documents/ms/articles/artjun06.pdf

Changes to the Bank's weekly reporting regime - Quarterly Bulletin article, 2014Q3: http://www.bankofengland.co.uk/publications/Documents/quarterlybulletin/2014/qb300614.pdf

Other data

Weekly data on the Bank of England's balance sheet 1844-2006: http://www.bankofengland.co.uk/research/Documents/onebank/balanceweekly_final.xlsx

12.1b The Bank of England's Annual Balance Sheet 1958-2017

This sheet gives consistent annual series for the Bank's liabilities and assets from 1958 to 1966, which were first published in the June 1967 Quarterly Bulletin and extended to 2006 using subsequent Quarterly Bulletins. An extension to 2014 is made using the new reporting method introduced in 2006. The dates in each year to which the figures relate are: last Saturday in February; 1958-1966 - Published Bank Returns, last Wednesday in February: 1966-2006 - Published Bank Returns: third Wednesday in February (the dates on which the London Clearing Banks compiled their monthly figures); 2007-2017 - Published Bank Returns: third Wednesday in February. See links at the bottom of the table for more information on the data.

Liabilities, £

	Notes In circulation	Notes in the Bank	Capital	Rest	Deposits	o/w Public deposits	o/w Special deposits	o/w Bankers deposits	Other accounts	Total
1958	1,962,114,543	38,244,318	14,553,000	3,900,136	308,384,168					2,327,196,165
1959	2,006,062,831	44,296,150	14,553,000	3,900,441	327,341,865					2,396,154,287
1960	2,111,915,274	38,444,906	14,553,000	3,898,192	340,800,115					2,509,611,487
1961	2,205,960,632	44,400,388	14,553,000	3,880,391	497,533,291		155,100,000			2,766,327,702
1962	2,302,141,403	23,217,219	14,553,000	3,917,764	576,696,473		241,400,000			2,920,525,859
1963	2,309,448,249	40,911,691	14,553,000	3,894,802	345,577,479					2,714,385,221
1964	2,429,044,275	21,316,745	14,553,000	3,896,193	362,270,737					2,831,080,950
1965	2,601,417,528	48,944,211	14,553,000	3,881,812	374,147,294					3,042,943,845
1966	2,778,619,008	21,742,251	14,553,000	3,882,025	483,428,964					3,302,225,248
3rd Wednesday in February										
1966	2,783,800,000	16,600,000	14,553,000	3,747,000	476,100,000	12,600,000	97,500,000	274,800,000	91,200,000	3,294,800,000
1967	2,859,800,000	40,600,000	14,553,000	3,747,000	614,000,000	16,100,000	201,400,000	282,400,000	114,100,000	3,532,700,000
1968	3,012,500,000	38,000,000	14,553,000	3,747,000	673,600,000	14,400,000	218,200,000	308,400,000	132,600,000	3,742,400,000
1969	3,132,500,000	18,000,000	14,553,000	3,747,000	702,000,000	18,500,000	231,000,000	308,200,000	144,300,000	3,870,800,000
1970	3,231,000,000	20,000,000	14,553,000	0	621,447,000	15,000,000	220,000,000	248,000,000	138,000,000	3,887,000,000
1971	3,658,000,000	42,000,000	14,553,000	0	953,447,000	15,000,000	398,000,000	314,000,000	226,000,000	4,668,000,000
1972	3,663,000,000	37,000,000	14,553,000	0	527,447,000	15,000,000	0	178,000,000	334,000,000	4,242,000,000
1973	4,166,000,000	34,000,000	14,553,000	0	1,269,447,000	22,000,000	714,000,000	230,000,000	303,000,000	5,484,000,000
1974	4,552,000,000	23,000,000	14,553,000	0	1,974,447,000	28,000,000	1,368,000,000	266,000,000	312,000,000	6,564,000,000
1975	5,306,000,000	19,000,000	14,553,000	0	1,604,447,000	21,000,000	935,000,000	275,000,000	373,000,000	6,944,000,000
1976	5,981,000,000	19,000,000	14,553,000	0	1,662,447,000	20,000,000	980,000,000	269,000,000	394,000,000	7,677,000,000
1977	6,694,000,000	6,000,000	14,553,000	0	1,441,447,000	18,000,000	711,000,000	273,000,000	439,000,000	8,156,000,000
1978	7,652,000,000	23,000,000	14,553,000	0	2,315,447,000	25,000,000	1,229,000,000	386,000,000	675,000,000	10,005,000,000
1979	8,843,000,000	7,000,000	14,553,000	0	1,328,447,000	25,000,000	255,000,000	404,000,000	644,000,000	10,193,000,000
1980	9,651,000,000	24,000,000	14,553,000	0	1,376,447,000	26,000,000	104,000,000	579,000,000	667,000,000	11,066,000,000
1981	10,160,000,000	15,000,000	14,553,000	0	1,921,447,000	32,000,000	0	602,000,000	1,288,000,000	12,111,000,000
1982	10,570,000,000	5,000,000	14,553,000	0	2,111,447,000	39,000,000	0	518,000,000	1,554,000,000	12,701,000,000
1983	10,910,000,000	15,000,000	14,553,000	0	5,021,447,000	2,286,000,000	0	537,000,000	2,199,000,000	15,961,000,000
1984	11,400,000,000	10,000,000	14,553,000	0	2,890,447,000	742,000,000	0	778,000,000	1,371,000,000	14,315,000,000
1985	11,976,000,000	4,000,000	14,553,000	0	7,608,447,000	5,267,000,000	0	672,000,000	1,670,000,000	19,603,000,000
1986	11,962,000,000	8,000,000	14,553,000	0	7,234,447,000	4,885,000,000	0	823,000,000	1,527,000,000	19,219,000,000
1987	12,503,000,000	7,000,000	14,553,000	0	4,731,447,000	2,158,000,000	0	953,000,000	1,621,000,000	17,256,000,000
1988	13,253,000,000	7,000,000	14,553,000	0	3,614,447,000	104,000,000	0	1,057,000,000	2,452,000,000	16,889,000,000
1989	14,140,000,000	10,000,000	14,553,000	0	3,185,447,000	112,000,000	0	1,319,000,000	1,755,000,000	17,350,000,000
1990	14,951,000,000	9,000,000	14,553,000	0	3,704,447,000	54,000,000	0	1,736,000,000	1,915,000,000	18,679,000,000

12.1b The Bank of England's Annual Balance Sheet 1958-2017

This sheet gives consistent annual series for the Bank's liabilities and assets from 1958 to 1966, which were first published in the June 1967 Quarterly Bulletin and extended to 2006 using subsequent Quarterly Bulletins. An extension to 2014 is made using the new reporting method introduced in 2006. The dates in each year to which the figures relate are: last Saturday in February; 1958-1966 - Published Bank Returns, last Wednesday in February: 1966-2006 - Published Bank Returns: third Wednesday in February (the dates on which the London Clearing Banks compiled their monthly figures): 2007-2017 - Published Bank Returns: third Wednesday in February. See links at the bottom of the table for more information on the data.

Liabilities, £

	Notes In circulation	Notes in the Bank	Capital	Rest	Deposits	o/w Public deposits	o/w Special deposits	o/w Bankers deposits	Other accounts	Total
1991	15,266,000,000	4,000,000	14,553,000	0	4,410,447,000	44,000,000	0	1,729,000,000	2,637,000,000	19,695,000,000
1992	15,436,000,000	12,000,000	14,553,000	0	4,815,447,000	119,000,000	0	1,417,000,000	3,278,000,000	20,278,000,000
1993	16,163,000,000	7,000,000	14,553,000	0	8,384,447,000	2,948,000,000	0	1,535,000,000	3,902,000,000	24,569,000,000
1994	17,096,000,000	4,000,000	14,553,000	0	9,236,447,000	1,129,000,000	0	1,628,000,000	6,479,000,000	26,351,000,000
1995	18,062,000,000	8,000,000	14,553,000	0	6,110,447,000	1,202,000,000	0	1,812,000,000	3,096,000,000	24,195,000,000
1996	19,405,000,000	5,000,000	14,553,000	0	5,756,447,000	873,000,000	0	1,842,000,000	3,042,000,000	25,181,000,000
1997	20,578,000,000	12,000,000	14,553,000	0	6,312,447,000	969,000,000	0	1,934,000,000	3,409,000,000	26,917,000,000
1998	22,026,000,000	4,000,000	14,553,000	0	10,636,447,000	1,103,000,000	0	2,716,000,000	6,817,000,000	32,681,000,000
1999	23,098,000,000	11,000,000	14,553,000	0	78,831,447,000	153,000,000	0	1,339,000,000	77,339,000,000	101,955,000,000
2000	24,918,000,000	12,000,000	14,553,000	0	55,811,447,000	217,000,000	0	1,360,000,000	54,235,000,000	80,756,000,000
2001	26,983,000,000	7,000,000	14,553,000	0	10,846,447,000	409,000,000	0	1,402,000,000	9,036,000,000	37,851,000,000
2002	29,046,000,000	4,000,000	14,553,000	0	12,288,447,000	372,000,000	0	1,656,000,000	10,261,000,000	41,353,000,000
2003	31,033,000,000	7,000,000	14,553,000	0	15,656,447,000	469,000,000	0	1,756,000,000	13,431,000,000	46,711,000,000
2004	33,013,000,000	7,000,000	14,553,000	0	15,530,447,000	650,000,000	0	1,929,000,000	12,951,000,000	48,565,000,000
2005	35,021,000,000	9,000,000	14,553,000	0	20,670,447,000	561,000,000	0	2,008,000,000	18,102,000,000	55,715,000,000
2006	36,793,000,000	7,000,000	14,553,000	0	24,983,447,000	803,000,000	0	2,857,000,000	21,324,000,000	61,798,000,000

Liabilities, £

	Notes In circulation	£ Short-term repo operations with BoE	FC public securities issued	Cash ratio deposits	£ reserve balances	Other liabilities[3]	Total
3rd Wednesday in February, Q1 observation used from 2015[1]							
2007	38,214,000,000	0	3,314,000,000	2,568,000,000	17,716,000,000	15,180,000,000	76,992,000,000
2008	40,933,000,000	0	3,327,000,000	2,936,000,000	23,824,000,000	26,595,000,000	97,615,000,000
2009	44,494,000,000	43,940,000,000	2,893,000,000	2,427,000,000	33,700,000,000	40,951,000,000	168,405,000,000
2010	49,486,000,000	0	3,881,000,000	2,574,000,000	155,165,000,000	36,174,000,000	247,280,000,000
2011	51,542,000,000	0	3,783,000,000	2,444,000,000	138,015,000,000	45,857,000,000	241,641,000,000
2012	54,568,000,000	0	3,870,000,000	2,386,000,000	188,975,000,000	61,660,000,000	311,459,000,000
2013	57,242,000,000	0	3,943,000,000	2,479,000,000	279,727,000,000	60,836,000,000	404,227,000,000
2014	59,391,000,000	0	3,597,000,000	4,078,000,000	305,951,000,000	30,072,000,000	403,089,000,000
2015	64,639,000,000	0	4,060,000,000	4,098,000,000	304,207,000,000	33,771,000,000	410,775,000,000
2016	69,776,000,000	0	4,194,000,000	4,136,000,000	312,245,000,000	32,895,000,000	423,246,000,000
2017	74,045,000,000	0	4,793,000,000	4,424,000,000	423,226,000,000	32,697,000,000	539,184,000,000

Source: Bank of England

12.1b The Bank of England's Annual Balance Sheet 1958-2017

This sheet gives consistent annual series for the Bank's liabilities and assets from 1958 to 1966, which were first published in the June 1967 Quarterly Bulletin and extended to 2006 using subsequent Quarterly Bulletins. An extension to 2014 is made using the new reporting method introduced in 2006. The dates in each year to which the figures relate are: last Saturday in February; 1958-1966 - Published Bank Returns, last Wednesday in February: 1966-2006 - Published Bank Returns: third Wednesday in February (the dates on which the London Clearing Banks compiled their monthly figures); 2007-2017 - Published Bank Returns: third Wednesday in February. See links at the bottom of the table for more information on the data.

1. Following Bank of England money market reform on 18 May 2006 the Bank of England 'Bank Return' was changed. This series forms part of the new Bank Return, with data starting on 24 May 2006. More information on changes made to the Bank's monetary policy operations and their impact on published data can be found in 'The implications of money market reform for data published in Monetary and Financial Statistics' in the June 2006 issue of Bank of England: Monetary and Financial Statistics. From 2015 onwards the observations on the total balance sheet are only available quarterly in the consolidated balance sheet and with a 5 quarter lag so the nearest proxies are used for February observations. Thomas Ryland (email: ryland.thomas@bankofengland.co.uk.)

2. Other assets is the sum of loan to the Asset Purchase Facility (RPQZ4TM), other assets (RPQZ6MW), all foreign currency reserve assets (RPQZ4TN), other foreign currency assets (RPQZ6MX) and denominated bond holdings (RPQZ4TL)

3. Other liabilities is the sum of capital & reserves (equity) from monetary financial institutions (RPQZ6MS), other foreign currency liabilities (RPQZ6MT) and other liabilities (RPQZ6MO).

For more information on the data please use the following links:

Original 1967 Quarterly Bulletin article: http://www.bankofengland.co.uk/archive/Documents/historicpubs/qb/1967/qb67q2159163.pdf

Bank of England Statistical Abstracts: http://www.bankofengland.co.uk/archive/Pages/digitalcontent/historicpubs/statisticalabst.aspx

The implications of money market reform for data published in Monetary and Financial Statistics' in the June 2006 issue of Bank of England: Monetary and Financial Statistics.: http://www.bankofengland.co.uk/statistics/Documents/ms/articles/artjun06.pdf

Changes to the Bank's weekly reporting regime - Quarterly Bulletin article, 2014Q3: http://www.bankofengland.co.uk/publications/Documents/quarterlybulletin/2014/qb300614.pdf

Other data

Weekly data on the Bank of England's balance sheet 1844-2006: http://www.bankofengland.co.uk/research/Documents/onebank/balanceweekly_final.xlsx

12.2a Annual clearing volumes and values

Clearing volumes thousands	Annual volumes		
	2017	2018	Change 2018 on 2017 %
Bacs			
Direct Credits	2,120,519	2,088,189	-32,330 -2%
Direct Debits	4,226,661	4,355,840	129,179 3%
Total BACS	**6,347,180**	**6,444,029**	**96,849 2%**
CHAPS Clearing Company			
Retail and Commercial (MT 103)	32,396	38,280	5,884 18%
Wholesale Financial (MT202)	9,256	10,242	986 11%
Total CHAPS	**41,652**	**48,522**	**6,870 16%**
Faster Payments			
Standing Order Payments	372,866	394,431	21,565 6%
Single Immediate Payments	1,081,096	1,384,711	303,615 28%
Forward Dated Payments	199,315	260,982	61,667 31%
Return Payments	2,530	2,676	146 6%
Total Faster Payments	**1,655,807**	**2,042,800**	**386,993 23%**
Cheque and Credit Clearing Company (C&CCC)			
Paper Clearings			
Cheques	292,961	234,935	-58,026 -20%
Credits	17,261	13,276	-3,984 -23%
Euro debits[2]	55	31	-24 -44%
Total Paper	**310,276**	**248,242**	**-62,035 -20%**
ICS[3]			
Cheques (RTPs)	16	19,482	19,466 -
Credits (ITPs)	0	301	301 -
Total ICS	**16**	**19,783**	**19,767 -**
Total Cheque and Credit	**310,293**	**268,025**	**-42,268 -14%**
Belfast Bankers' Clearing Company Ltd (BBCCL)[1]			
Cheques	9,739	7,586	-2,153 -22%
Credits	180	133	-47 -26%
Total BBCCL	**9,918**	**7,719**	**-2,199 -22%**
Total all Clearing Companies	**8,364,849**	**8,811,095**	**446,246 5%**

Clearing values £ millions	Annual values		
	2017	2018	Change 2018 on 2017 %
Bacs			
Direct Credits	3,619,384	3,631,398	12,014 -
Direct Debits	1,304,647	1,327,269	22,622 2%
Total BACS	**4,924,030**	**4,958,667**	**34,636 1%**
CHAPS Clearing Company			
Retail and Commercial (MT 103)	22,083,495	23,277,342	1,193,847 5%
Wholesale Financial (MT202)	61,998,982	60,236,600	-1,762,383 -3%
Total CHAPS	**84,082,477**	**83,513,941**	**-568,536 -1%**
Faster Payments			
Standing Order Payments	122,314	130,372	8,057 7%
Single Immediate Payments	874,684	1,045,707	171,023 20%
Forward Dated Payments	401,476	531,714	130,239 32%
Return Payments	1,217	1,285	68 6%
Total Faster Payments	**1,399,691**	**1,709,078**	**309,387 22%**
Cheque and Credit Clearing Company (C&CCC)			
Paper Clearings			
Cheques	356,438	299,734	-56,705 -16%
Credits	11,733	7,530	-4,203 -36%
Euro debits	1,139	940	-199 -17%
Total Paper	**369,310**	**308,203**	**-61,107 -17%**
ICS[3]			
Cheques (RTPs)	12	26,663	26,651 -
Credits (ITPs)	0	1,025	1,025 -
Total ICS	**12**	**27,688**	**27,676 -**
Total Cheque and Credit	**369,322**	**335,891**	**-33,430 -9%**
Belfast Bankers' Clearing Company Ltd (BBCCL)[1]			
Cheques	16,664	13,441	-3,223 -19%
Credits	1,256	1,011	-245 -20%
Total BBCCL	**17,920**	**14,452**	**-3,468 -19%**
Total all Clearing Companies	**90,793,441**	**90,532,029**	**-261,411 -**

Source: Cheque and Credit Clearing Company Ltd

12.2a Annual clearing volumes and values

1. The Belfast Bankers' Clearing Company was formed in May 2007 as a means of formalising existing rules and standards for sterling paper clearings and Euro debit clearing in Northern Ireland, and has 4 members. The role of the Company is to maintain the integrity of the clearing arrangements and ensure the system is efficient and effective.

(a) Totals, averages and percentages are calculated using unrounded data. The value of euro debit clearings are shown as £ sterling equivalent.

(b) 252 days were used to calculate the average daily statistics in 2017, with 253 in 2016. In terms of Faster Standing Orders and 365 days were used for the other payment types. Payments, 252 days were used for Inter-branch clearing volumes (i.e. items cleared between branches of the same bank) are shown separately. These data are less comprehensive due to changes in agency arrangements and individual member processing policies, for example, the increased use of electronic processing methods. It is likely, therefore, that a proportion of inter-branch transactions are not included in these totals.

2. The C&CCC ceased operation of the euro debit clearing, which processed cheques drawn in euros on UK banks, on 10 September 2018.

3. ICS is the Image Clearing System that enables digital images of cheques and credits to be exchanged between participant banks and building societies across the whole of the UK for clearing and settlement. It is gradually replacing the Paper Clearing System. It was launched on 30 October 2017 and because very small volumes were processed through the system in 2017 it is not appropriate to show a % increase in volumes between 2017 and 2018. Request to Pay (RTP) is the message type for cheques and Instruction to Pay (ITP) is the message type for credits.

12.2b Bacs volumes and values

Annual volumes thousands

	Bacs Direct Credits		Bacs Euro Credits		Direct Debits		Total volumes	
	Number	% Change	Number	% Change	Number	% Change	Number	% Change
2003	1,341,945	14.4	54	45.0	2,429,915	6.2	4,060,357	8.7
2004	1,710,673	27.5	84	55.5	2,589,934	6.6	4,602,570	13.4
2005	2,093,859	22.4	12	48.6	2,722,245	5.1	5,134,250	11.6
2006	2,171,586	3.7	157	25.9	2,857,761	5.0	5,361,749	4.4
2007	2,233,106	2.8	181	15.8	2,296,474	3.7	5,544,109	3.4
2008	2,254,875	1.0	176	-3.2	3,076,857	3.8	5,655,751	2.0
2009	2,289,813	1.5	193	10.0	3,149,153	2.3	5,638,919	-0.3
2010	2,292,942	0.1	143	26.0	3,229,338	2.5	5,672,730	0.6
2011	2,270,987	-1.0			3,322,360	2.9	5,716,999	0.8
2012	2,182,667	-3.9			3,416,651	2.8	5,616,392	-1.8
2013	2,151,718	-1.0			3,524,905	3.0	5,695,028	1.0
2014	2,150,557	-			3,671,997	4.0	5,841,232	3.0
2015	2,171,697	-			3,908,346	6.0	6,080,043	4.0
2016	2,146,835	-1.0			4,071,911	4.0	6,218,746	2.0
2017	2,120,519	-1.0			4,226,661	4.0	6,347,180	2.0
2018	2,088,189	-2.0			4,355,840	3.0	6,444,029	2.0

Average values £ millions

	Bacs Direct Credits [a]		Bacs Euro Credits		Direct Debits		Total volumes	
	Number	% Change	Number	% Change	Number	% Change	Number	% Change
2003	1,910,251	8.3	1,924	53.9	662,192	7.3	2,574,367	8.1
2004	2,131,031	11.6	2,040	6.0	750,381	13.3	2,883,452	12.0
2005	2,350,644	10.3	2,524	23.7	797,039	6.2	3,150,207	9.3
2006	2,581,682	9.8	2,819	11.7	844,832	6.0	3,429,333	8.9
2007	2,808,349	8.8	3,965	40.6	883,592	4.6	3,695,906	7.8
2008	3,006,159	7.0	4,806	21.2	935,356	5.9	3,946,321	6.8
2009	2,969,711	-1.2	5,255	9.3	885,708	-5.3	3,860,674	-2.2
2010	3,111,218	4.8	3,033	-42.3	948,137	7.0	4,062,388	5.2
2011	3,318,536	6.7			1,044,677	10.2	4,363,214	7.4
2012	3,036,714	-8.5			1,075,507	3.0	4,112,222	-5.8
2013	3,103,579	2.0			1,115,065	4.0	4,218,644	3.0
2014	3,253,279	5.0			1,167,266	5.0	4,420,546	5.0
2015	3,374,815	4.0			1,215,396	4.0	4,590,211	4.0
2016	3,541,313	4.0			1,262,235	4.0	4,803,548	5.0
2017	3,619,384	3.0			1,304,647	3.0	4,924,030	3.0
2018	3,631,398	-			1,327,269	2.0	4,958,667	1.0

Sources: UK Payments Administration Ltd, Cheque and Credit Clearing Company Ltd

(a) Values represent standing orders and Bacs Direct Credits combined.

The Euro Debit Credit Service ceased operation on 29 October 2010.

Bacs has been maintaining the integrity of payment related services since 1968, with responsibility for the schemes behind the clearing and settlement of UK automated payment methods, Direct Debit and Bacs Direct Credit. A membership company limited by Guarantee, Bacs currently has 19 members from the UK, Europe, and the US. Bacs also provides managed services for third parties, such as the Cash ISA Transfer Service, and the company owns and manages the Current Account Switch Service which has 46 direct participants.

12.2c CHAPS volumes and values

Annual volumes thousands

	CHAPS Sterling [a]		CHAPS Euro						Total volumes [a]	
			Domestic (a)		Target					
					Transmitted to		Received from			
	Number	% Change	Number	% Change	Number	% Change	Number	% Change	Number	% Change
2003	27,215	6.5	1,399	13.1	2,904	19.2	1,685	4.7	33,202	7.6
2004	28,322	4.1	1,378	-1.5	3,314	14.1	1,849	9.7	34,862	5.0
2005	29,686	4.8	1,484	7.7	3,597	8.6	1,988	7.5	36,756	5.4
2006	33,030	11.3	1,461	-1.6	4,115	14.4	2,080	4.6	40,686	10.7
2007	35,588	7.7	1,455	-0.3	4,263	3.6	2,229	7.1	43,535	7.0
2008	34,606	-2.8	220	-84.9	379	-91.1	593	-73.4	35,797	-17.8
2009	31,926	-7.7							31,926	-10.8
2010	32,169	0.8							32,169	0.8
2011	34,024	5.8							34,024	5.8
2012	33,936	-0.3							33,936	-0.3
2013	34,976	3.0							34,976	3.0
2014	36,521	4.0							36,521	4.0
2015	37,548	3.0							37,548	3.0
2016	38,964	4.0							38,964	4.0
2017	41,652	7.0							41,652	7.0
2018	48,522	16.0							48,522	16.0

Annual Values £ millions

	CHAPS Sterling [a]		CHAPS Euro (sterling equivalent)						Total values [a]	
			Domestic (a)		Target					
					Transmitted to		Received from			
	Number	% Change	Number	% Change	Number	% Change	Number	% Change	Number	% Change
2003	51,613,456	-0.5	5,114,198	22.7	15,924,879	21.5	15,923,974	21.5	88,576,506	7.7
2004	52,347,525	1.4	4,509,924	-11.8	17,238,602	8.2	17,238,737	8.3	91,334,788	3.1
2005	52,671,592	0.6	6,069,146	34.6	19,180,194	11.3	19,179,273	11.3	97,100,206	6.3
2006	59,437,370	12.8	7,365,558	21.4	21,415,027	11.7	21,419,195	11.7	109,637,149	12.9
2007	69,352,322	16.7	6,781,942	-7.9	25,266,729	18.0	25,268,856	18.0	126,669,848	15.5
2008	73,625,908	6.2	574,610	-91.5	4,413,570	-82.5	4,403,022	-82.6	83,017,110	-34.5
2009	64,616,956	-12.2							64,616,956	-22.2
2010	61,587,609	-4.7							61,587,609	-4.7
2011	63,876,772	3.7							63,876,772	3.7
2012	71,716,857	12.3							71,716,857	12.3
2013	70,138,927	-2.0							70,138,927	-2.0
2014	67,959,491	-3.0							67,959,491	-3.0
2015	68,411,178	1.0							68,411,178	1.0
2016	75,573,628	10.0							75,573,628	10.0
2017	84,082,477	11.0							84,082,477	11.0
2018	83,513,941	-1.0							83,513,941	-1.0

Sources: UK Payments Administration Ltd, Cheque and Credit Clearing Company Ltd

(a) NewCHAPS was launched on 27 August 2001 and since this date CHAPS Sterling and CHAPS Euro Domestic figures include all CHAPS traffic.

CHAPS is the UK's same day high value payment system for both wholesale financial and retail payments. Payments of any value are settled individually and irrevocably in central bank funds, and transferred over SWIFT. CHAPS is focused on systemically important, high value and time-critical payments. Currently 26 domestic and international financial institutions are Direct Participants, with several future joiners planned. These serve over 5,000 Indirect Participants. Financial institutions such as banks, central counterparties and authorised non-bank payment service providers are eligible to join CHAPS. Most of CHAPS' value is accounted for by wholesale financial transactions and international sterling flows.

12.2d Faster Payments volumes and values

Annual volumes thousands

	Standing Order Payments		Single Immedicate Payments		Forward Dated Payments		Return Payments		Total Volumes (a)	
	Number	% change	Number	% change	Number	% change	Number	% change	Number	% change
2003										
2004										
2005										
2006										
2007										
2008	37,574	*	36,325	*	8,708	*	182	*	82,789	*
2009	156,865	*	109,337	*	27,912	*	673	*	294,787	*
2010	203,055	29.4	181,195	65.7	40,632	45.6	880	30.6	425,761	44.4
2011	235,654	16.1	237,718	31.2	50,861	25.2	1,092	24.2	525,325	23.4
2012	299,630	27.1	379,844	59.8	129,829	155.3	1,788	63.6	811,090	54.4
2013	312,995	4	502,025	32	150,381	16.0	2,228	25.0	967,629	19.0
2014	329,858	5	609,879	21	159,153	6.0	2,040	-8.0	1,100,930	14.0
2015	343,642	4	730,675	20	170,339	7.0	2,378	17.0	1,247,035	13.0
2016	357,411	4.0	882,226	21	183,675	8.0	2,780	17.0	1,426,093	14.0
2017	372,866	4.0	1,081,096	23	199,315	9.0	2,530	-9.0	1,655,807	16.0
2018	394,431	6.0	1,384,711	28	250,982	31.0	2,676	6.0	2,042,800	23.0

Annual values £ millions

	Standing Order Payments		Single Immedicate Payments		Forward Dated Payments		Return Payments		Total Volumes (a)	
	Number	% change	Number	% change	Number	% change	Number	% change	Number	% change
2003										
2004										
2005										
2006										
2007										
2008	3,341	*	22,015	*	7,473	*	42	*	32,871	*
2009	26,527	*	59,015	*	20,544	*	136	*	106,223	*
2010	38,963	46.9	92,847	57.3	32,193	56.7	208	52.9	164,211	54.6
2011	50,206	28.9	135,101	45.5	49,492	53.5	308	47.9	235,044	43.1
2012	80,009	59.4	328,685	143.3	208,252	321.3	965	213	617,911	162.9
2013	88,885	11.0	423,571	29.0	257,794	24.0	1,111	15.0	771,361	25.0
2014	97,121	9.0	517,641	22.0	288,034	12.0	999	-10.0	903,794	17.0
2015	107,202	10.0	619,303	20.0	313,099	9.0	1,114	12.0	1,040,717	15.0
2016	113,558	6.0	728,642	18.0	345,630	10.0	1,178	6.0	1,189,008	14.0
2017	122,314	8.0	874,684	20.0	401,476	16.0	1,217	3.0	1,399,691	18.0
2018	130,372	7.0	1,045,707	20.0	531,714	32.0	1,285	6.0	1,709,078	22.0

Sources: UK Payments Administration Ltd, Cheque and Credit Clearing Company Ltd

(a) The UK Faster Payments Service was launched on 27 May 2008

The Faster Payment Service, operated by Faster Payments Scheme Ltd, enables internet, mobile and telephone banking payments as well as standing order payments to move from account to account, normally within seconds, 24 hours a day, 365 days a year. As at the end of 2017 there were 18 direct settling participants, with a number more intending to join in 2018. Given its scale and reach, it is considered one of the most advanced real-time payment services in the world.

12.2e Inter-bank volumes and values

Annual volumes thousands

	Exchanged in Great Britain							
	Cheques		Credits		Euro Debits		Total Volumes	
	Number	% change	Number	% change	Number	% change	Number	% change
2003	1,519,117	-6.3	140,792	-6.4	759	3.6	1,519,876	-6.3
2004	1,423,742	-6.3	132,899	-5.6	724	-4.6	1,424,465	-6.3
2005	1,325,762	-6.9	123,280	-7.2	637	-12.0	1,326,399	-6.9
2006	1,237,401	-6.7	108,309	-12.1	586	-7.9	1,237,987	-6.7
2007	1,124,869	-9.1	96,90	-11.1	531	-9.5	1,125,400	-9.1
2008	1,007,379	-23.5	86,442	-10.2	445	-16.1	1,007,824	-10.4
2009	875,533	-11.4	73,686	-14.8	351	-21.2	875,884	-13.1
2010	775,643	-12.1	61,662	-16.3	279	-20.5	775,922	-11.4
2011	682,082	-12.1	53,934	-12.5	223	-20.1	6,822,305	-12.1
2012	597,076	-12.5	46,927	-12.9	165	-25.7	597,241	-12.5
2013	525,295	-12.0	40,569	-14.0	131	-21.0	525,426	-12.0
2014	464,191	-12.0	34,962	-14.0	108	-17.0	464,299	-12.0
2015	404,134	-13.0	28,049	-20.0	89	-18.0	432,273	-13.0
2016	344,621	-15.0	21,857	22.0	80	-10.0	366,558	-15.0
2017	292,961	-15.0	17,261	-21.0	55	-25.0	310,276	-15.0
2018	234,935	-20.0	13,276	-23.0	31	-44.0	248,242	-20.0

Annual values £ millions

	Exchanged in Great Britain							
	Cheques		Credits		Euro Debits		Total Volumes	
	Number	% change	Number	% change	Number	% change	Number	% change
2003	1,240,685	-3.2	74,366	-7.6	3,898	13.8	1,244,583	-3.1
2004	1,210,057	-2.5	68,261	-8.2	3,456	-11.4	1,213,513	-2.5
2005	1,152,256	-4.8	61,844	-9.4	3,206	-7.2	1,155,642	-4.8
2006	1,171,062	1.6	59,309	-4.1	3,111	-3.0	1,174,174	1.6
2007	1,156,684	-1.2	57,347	-3.3	2,970	-4.6	1,159,653	-1.2
2008	1,075,694	-7.0	51,641	-9.9	2,980	0.4	1,078,674	-7.0
2009	870,591	-19.1	41,624	-19.4	2,993	0.4	873,584	-19.0
2010	761,081	-12.6	32,312	-22.4	1,767	-41	762,848	-12.7
2011	675,706	-11.2	27,990	-13.4	1,559	-11.8	667,226	-11.2
2012	601,256	-11	23,802	-15	1,188	-23.9	602,444	-11.0
2013	535,513	-11.0	21,109	-11.0	1,166	-2.0	536,679	-11.0
2014	498,729	-7.0	19,659	-7.0	1,289	11.0	500,018	-7.0
2015	454,838	-9.0	17,216	-12.0	1,208	-6.0	473,261	-8.8
2016	400,158	-12.0	15,022	-13.0	1,274	5.0	416,455	-12.0
2017	356,438	-11.0	11,733	-22.0	1,139	-11.0	369,310	-11.0
2018	299,734	-16.0	7,530	-36.0	940	-17.0	308,203	-17.0

Source: Cheque and Credit Clearing Company Ltd

The Cheque and Credit Clearing Company has managed the cheque clearing system in England and Wales since 1985 and in Scotland since 1996. As well as clearing cheques, the system processes bankers' drafts, postal orders, warrants, government payable orders and travellers' cheques. The company also manages the systems for the clearing of paper bank giro credits and euro cheques (drawn on GB banks). There are 11 participants in the cheque clearing system and the credit clearing system and 10 in the euro debit clearing system. The Image Clearing System (ICS) was launched on 30th October with very low volumes going through the system. There are 17 Participants in ICS. More information on the ICS Participants can be found on our website: www.chequeandcredit.co.uk.

12.3a Monetary financial institutions' (excluding central bank) balance sheet

£ millions

Amounts outstanding of sterling liabilities

Not seasonally adjusted

	Notes outstanding and cash loaded cards	Sight deposits					Time deposits							Sale and repurchase agreements				
		UK MFIs	of which intragroup banks	UK public sector	Other UK residents	Non-residents	UK MFIs	of which intragroup banks	UK public sector	Other UK residents	of which SAYE	of which cash ISAs	Non-residents	UK MFIs	of which intragroup banks	UK public sector	Other UK residents	Non-residents
RPM	B3LM	B3GL	B8ZC	B3MM	B3NM	B3OM	B3HL	B8ZE	B3PM	B3QM	B3RM	B3SM	B3TM	B3IL	B8ZA	B3UM	B3VM	B3WM
2017 Jan	7,696	111,503	94,478	15,241	1,316,897	174,390	199,656	159,778	16,586	642,895	911	268,988	179,714	43,286	17,314	1,765	122,405	84,315
Feb	7,704	120,382	104,905	14,312	1,319,172	172,824	220,016	171,176	15,772	639,394	904	268,217	179,606	41,658	16,843	3,815	123,232	85,681
Mar	7,718	138,906	123,951	13,916	1,353,802	177,016	247,893	186,401	14,051	634,702	926	268,855	173,634	41,275	14,102	2,613	123,066	95,220
Apr	7,804	140,045	123,758	15,609	1,345,004	173,275	249,930	185,931	14,052	645,780	931	272,693	176,180	40,387	14,752	2,108	138,104	87,075
May	7,721	145,488	128,334	15,310	1,353,523	179,365	263,949	195,214	14,261	639,384	946	272,963	176,003	43,006	13,953	4,551	129,669	90,796
Jun	7,695	149,562	133,677	15,413	1,374,429	182,993	292,279	212,168	14,852	641,232	943	273,545	176,983	38,401	12,167	4,945	121,744	99,990
Jul	7,853	161,110	144,328	19,190	1,363,667	179,581	293,178	209,608	15,883	638,708	915	272,697	169,628	39,522	16,176	8,121	130,035	91,255
Aug	8,051	162,469	146,447	19,068	1,375,962	181,096	305,692	218,069	16,100	640,168	924	272,005	173,867	35,032	13,285	8,943	136,861	92,958
Sep	8,106	163,962	149,673	19,847	1,395,331	171,892	314,775	224,425	16,093	635,763	1,090	271,864	172,002	33,501	12,532	1,847	140,839	107,841
Oct	8,153	139,367	125,612	19,899	1,392,222	169,642	317,746	221,798	15,465	641,736	1,100	271,156	161,953	34,544	10,285	5,101	154,876	98,028
Nov	8,107	121,486	107,770	19,549	1,404,113	172,043	279,268	182,207	15,180	643,224	1,063	270,465	155,791	37,197	10,787	5,937	150,148	103,667
Dec	8,279	106,670	93,247	19,805	1,409,530	181,934	293,013	185,383	14,371	641,684	1,049	269,991	157,201	43,590	12,904	3,500	132,701	108,887
2018 Jan	7,953	72,628	58,912	20,111	1,400,497 (b)	168,640	263,002	151,790	13,711	641,673	1,064	269,097	152,982	40,994	11,620	6,335	152,224	110,989
Feb	7,845	77,347	63,165	19,350	1,401,377	167,832	279,717	149,048	11,928	641,703	1,099	269,072	155,176	42,226	11,324	8,280	150,560	104,075
Mar	7,949	61,344	47,219	20,548	1,417,816	175,494	278,043	146,756	10,812	633,259	1,109	271,687	162,426	43,594	9,896	2,506	121,804	109,256
Apr	8,302	53,277	40,472	17,027	1,419,902	174,154	289,873	158,433	11,546	634,912	1,113	276,657	161,897	46,752	15,427	2,439	125,517	103,453
May	7,841	36,911	25,121	16,164	1,428,657	190,998	276,158	144,077	12,926	638,845	1,122	277,409	157,981	45,128	13,810	2,459	122,521	108,046
Jun	7,585	57,129	43,856	16,007	1,440,248	187,794	272,865	141,532	13,493	631,202	1,128	277,112	157,833	53,237	25,032	1,848	123,286	102,922
Jul	7,459	59,683	43,336	16,820	1,437,342	200,389	275,381	143,200	14,491	632,742	1,111	276,915	155,753	29,641	8,827	1,268	128,391	104,057
Aug	7,606	45,035	31,288	17,019	1,442,010	187,758	276,862	145,356	15,267	633,126	1,137	276,605	154,732	32,484	7,550	2,068	131,211	108,603
Sep	7,689	46,780	31,639	17,118	1,450,449	199,206	272,473	141,689	15,285	631,227	1,157	276,851	158,882	28,612 (d)	8,433	618	130,954 (e)	103,909
Oct	7,620	83,264 (a)	68,653 (a)	17,188	1,452,292	185,148	232,682 (c)	101,214 (c)	14,913	631,621	1,168	276,953	160,765	30,530	7,814	304	143,269	98,314
Nov	7,614	81,037	66,956	15,836	1,459,649	176,950	225,854	99,047	15,192	632,009	1,078	276,725	163,206	33,949	10,841	130	143,621	95,866
Dec	7,840	92,485	78,362	17,008	1,462,638	177,117	228,782	102,148	14,315	642,687	1,066	276,881	153,677	38,085	9,600	49	140,364	106,385
2019 Jan	7,542	72,483	57,860	16,751	1,444,512	176,307	223,219	96,741	15,141	641,001	1,077	277,032	148,605	38,689	10,479	200	146,440	105,745
Feb	7,473	87,914	73,839	16,282	1,439,299	177,103	223,037	96,462	14,289	646,569	1,101	277,608	154,417	40,449	9,673	1,959	153,688	96,772
Mar	7,582	137,685	122,622	15,990	1,456,653	176,684	224,599	97,879	13,734	651,632	1,109	281,396	153,274	42,160	7,821	768	131,577	113,300
Apr	7,549	133,911	119,714	16,795	1,451,966	173,579	227,184	99,653	14,674	655,527	1,119	288,185	146,296	41,899	8,923	868	147,823	108,718

12.3a Monetary financial institutions' (excluding central bank) balance sheet

£ millions

Not seasonally adjusted

Amounts outstanding of sterling liabilities

RPM	Acceptances granted B3XM	CDs and Commercial paper B2TL	CDs and other paper issued — Bonds with maturity of up to and incl. 5 years B6OI	CDs and other paper issued — Bonds with maturity of greater than 5 years B2TM	Total B3YM	Total sterling deposits B3ZM	Sterling items in suspense and transmission B3GN	Net deri-vatives B3HN	Accrued amounts payable B3IN	Sterling capital and other internal funds B3JN	Total sterling liabilities B3KN
2017 Jan	186	119,299	24,223	34,520	178,042	3,086,880	42,929	-12,810	21,392	415,255	3,561,342
Feb	187	124,491	24,053	34,543	183,087	3,119,136	44,997	21,903	21,907	426,770	3,642,417
Mar	252	134,753	23,161	34,305	192,219	3,208,566	48,528	-3,748	19,133	407,664	3,687,861
Apr	258	137,516	20,958	34,136	192,610	3,220,418	48,361	-38,604	18,570	409,362	3,665,911
May	264	134,143	23,527	33,460	191,130	3,246,700	43,690	-17,707	19,237	431,746	3,731,386
Jun	263	128,125	24,027	33,460	185,612	3,298,698	47,097	-25,745	19,210	460,976	3,807,931
Jul	217	127,960	23,779	33,560	185,298	3,295,393	41,454	-9,258	19,022	434,939	3,789,403
Aug	207	128,659	23,505	33,629	185,794	3,334,217	34,662	-2,291	18,520	443,925	3,837,085
Sep	145	127,302	22,424	33,076	182,802	3,356,641	37,043	-46,716	18,043	465,953	3,839,070
Oct	148	120,678	22,844	33,442	176,964	3,327,691	35,432	-29,840	18,754	455,427	3,815,617
Nov	146	126,435	22,780	33,921	183,136	3,290,885	45,860	-56,444	19,626	460,447	3,768,482
Dec	152	127,913	23,411	33,943	185,267	3,298,302	15,149	-61,678	21,902	474,854	3,756,809
2018 Jan	144	130,653	26,786	33,437	190,876	3,234,806 (b)	36,069	-146,781 (f)	21,088	455,839	3,608,972 (b)(f)
Feb	170	129,102	25,025	34,312	188,439	3,248,179	33,490	2,475 (g)	21,797	436,799	3,750,586 (g)
Mar	178	133,067	28,199	32,877	194,143	3,231,223	36,093	-24,358 (h)	22,316	440,814	3,714,037 (h)
Apr	261	131,111	29,292	30,536	190,939	3,231,949	34,784	70,082	19,623	452,008	3,816,748
May	338	132,646	29,752	30,465	192,863	3,229,995	42,769	59,312	23,507	446,017	3,809,441
Jun	397	128,223	29,304	30,477	188,004	3,246,265	33,049	4,137 (i)	19,887	442,767 (l)	3,753,689
Jul	326	125,026	28,953	30,727	184,706	3,240,990	39,448	-3,168 (j)	18,423	399,622 (m)	3,702,773
Aug	302	120,161	28,585	30,382	179,128	3,225,604	34,312	-26,906 (j)	19,131	435,298 (m)	3,695,046
Sep	278	114,930	31,278	30,010	176,219	3,232,009	38,870	-26,102 (j)	21,649	435,936 (n)	3,710,051
Oct	292	112,708	31,948	29,125	173,781	3,224,362	36,986	31,054 (k)	19,612	410,092 (m)	3,729,725
Nov	321	105,225	32,597	29,004	166,826	3,210,448	34,845	30,422	20,529	409,987	3,713,846
Dec	309	105,989	32,558	28,920	167,466	3,241,368	19,557	4,947	19,385	439,949	3,733,045
2019 Jan	296	108,600	33,519	29,083	171,203	3,200,592	35,439	4,424	18,467	436,540	3,703,004
Feb	304	108,004	35,373	29,410	172,787	3,224,868	35,721	-18,397	18,574	445,756	3,713,995
Mar	308	111,687	36,348	30,370	178,404	3,296,770	34,785	24,247	21,075	441,914	3,826,373
Apr	271	109,054	36,770	30,344	176,167	3,295,676	31,859	29,704	18,116	430,200	3,813,105

Source: Bank of England

12.3a Monetary financial institutions' (excluding central bank) balance sheet

£ millions

Notes to table

Movements in amounts outstanding can reflect breaks in data series as well as underlying flows. For changes data, users are recommended to refer directly to the appropriate series or data tables. Further explanation can be found at: https://www.bankofengland.co.uk/statistics/details/further-details-about-changes-flows-growth-rates-data.

(a) Due to improvements in reporting at one institution, the amounts outstanding increased by £38bn. This effect has been adjusted out of the flows for October 2018.

(b) Due to improvements in reporting at one institution, the amounts outstanding increased by £5bn. This effect has been adjusted out of the flows for January 2018.

(d) Due to improvements in reporting at one institution, the amounts outstanding decreased by £4bn. This effect has been adjusted out of the flows for September 2018.

(f) Due to improvements in reporting by one institution, amounts outstanding decreased by £66bn.

(h) Due to improvements in reporting by one institution, amounts outstanding increased by £9bn.

(i) Due to improvements in reporting at one institution, the amounts outstanding increased by £5bn.

(j) Due to improvements in reporting at one institution, the amounts outstanding increased by £10bn.

(k) Due to improvements in reporting at one institution, the amounts outstanding increased by £10bn.

(l) Due to improvements in reporting at one institution, the amounts outstanding decreased by £5bn.

(m) Due to improvements in reporting at one institution, the amounts outstanding decreased by £10bn.

(n) Due to improvements in reporting at one institution, the amounts outstanding decreased by £10bn.

Explanatory notes can be found here:
https://www.bankofengland.co.uk/statistics/details/further-details-about-monetary-financial-institutions-excluding-central-bank-balance-sheet-data

Copyright guidance and the related UK Open Government Licence can be viewed here: www.bankofengland.co.uk/legal

12.3b Monetary financial institutions' (excluding central bank) balance sheet

£ millions

Not seasonally adjusted

Amounts outstanding of foreign currency liabilities (including euro)

RPM	Sight and time deposits					Acceptances granted	Sale and repurchase agreements				
	UK MFIs	of which intragroup banks	UK public sector	Other UK residents	Non-residents		UK MFIs	of which intragroup banks	UK public sector	Other UK residents	Non-residents
	B3JL	B8ZM	B2UP	B2UV	B3NN	B3KQ	B3KL	B8ZK	B3PN	B3QN	B3RN
2017 Jan	217,530	153,305	2,689	336,977	1,817,696	3,288	78,312	35,890	1,317	151,236	628,041
Feb	206,752	144,109	2,633	336,848	1,842,486	3,789	76,870	35,761	339	181,024	653,853
Mar	197,174	136,947	2,383	328,699	1,813,204	4,008	82,344	39,816	70	172,708	648,760
Apr	192,379	133,353	2,674	329,056	1,783,247	3,580	77,778	38,108	1,340	171,046	632,021
May	194,621	135,194	3,053	340,680	1,775,830	3,183	80,168	38,516	1,092	170,582	672,362
Jun	156,845	99,947	3,130	339,052	1,789,057	3,327	74,407	35,467	255	169,837	654,366
Jul	164,065	104,084	3,222	338,603	1,798,441	2,792	75,876	34,249	2,659	183,889	688,821
Aug	172,338	108,733	3,154	346,532	1,795,818	2,541	83,413	40,431	1,855	190,514	698,269
Sep	162,434	105,987	3,402	330,970	1,728,963	2,953	80,247	39,181	2,443	181,224	733,277
Oct	160,441	104,307	3,572	331,512	1,754,574	2,785	78,816	34,941	2,221	194,611	761,236
Nov	156,376	101,797	3,273	333,947	1,743,717	2,950	83,263	39,834	2,463	204,451	757,869
Dec	159,519	103,492	3,242	322,158	1,750,941	3,472	75,320	34,218	2,730	187,517	806,778
2018 Jan	179,405	121,511	3,408	321,105	1,726,383	2,857	73,935	31,506	1,926	208,415	756,581
Feb	178,394	123,977	3,457	320,597	1,748,101	3,286	82,567	36,109	1,417	216,374	808,525
Mar	189,435	135,474	3,137	314,447	1,721,439	3,221	79,540	33,311	1,850	199,036	812,382
Apr	196,618	142,505	3,636	323,308	1,755,649	3,725	78,802	34,779	3,487	202,024	765,265
May	210,110	151,769	3,476	332,182	1,791,940	4,334	87,313	41,654	934	217,423	790,770
Jun	211,322	155,995	3,652	337,612	1,754,931	3,991	92,885	48,696	1,240	202,809	774,279
Jul	200,593	147,595	3,646	328,816	1,729,333	4,035	86,492	45,701	2,873	210,643	753,050
Aug	209,784	155,353	2,942	334,489	1,741,323	4,525	80,019	43,636	2,220	213,321	743,028
Sep	193,195	142,172	3,215	335,584	1,734,067	5,000	84,256	46,975	2,946	211,520	753,972
Oct	186,128	129,836	2,996	356,075	1,780,088	4,538	82,619	43,494	2,526	219,837	770,036
Nov	186,851	130,916	2,821	359,218	1,788,608	4,201	90,506	47,192	2,426	242,684	815,697
Dec	176,396	120,416	2,727	337,400	1,791,900	4,084	85,145	44,525	869	238,741	850,519
2019 Jan	169,909	110,521	2,774	323,453	1,761,861	3,929	85,848	45,221	849	232,874	796,631
Feb	161,265	104,926	2,520	314,798	1,780,768	3,414	87,842	47,258	1,675	207,906	804,382
Mar	150,163	93,380	2,797	323,107	1,799,664	3,619	67,063	23,313	2,913	198,992	832,740
Apr	153,195	97,625	3,246	327,230	1,803,208	3,842	69,691	26,474	1,547	201,374	838,481

12.3b Monetary financial institutions' (excluding central bank) balance sheet

£ millions

Not seasonally adjusted

Amounts outstanding of foreign currency liabilities (including euro)

RPM	CDs and Commercial paper	CDs and other paper issued			Total foreign currency deposits	Items in suspense and transmission	Net derivatives	Accrued amounts payable	Capital and other internal funds	Total foreign currency liabilities	Total liabilities
		Bonds with maturity of up to and incl. 5 years	Bonds with maturity of greater than 5 years	Total							
	B2TP	B6OK	B2TQ	B3SN	B3TN	B3UN	B3VN	B3WN	B3XN	B3YN	B3ZN
2017 Jan	113,498	207,223	247,708	568,429	3,805,515	231,738	-31,677	14,500	205,652	4,225,728	7,787,070
Feb	124,665	214,881	256,400	595,945	3,900,539	242,484	-67,864	16,901	224,225	4,316,285	7,958,706
Mar	125,439	208,200	252,862	586,502	3,835,852	205,851	-41,321	16,370	207,097	4,223,848	7,911,709
Apr	121,172	215,507	251,092	587,771	3,780,891	237,729	-3,803	17,474	190,064	4,222,355	7,888,650
May	127,906	209,566	257,713	595,185	3,836,757	230,037	-21,819	18,545	159,769	4,223,290	7,954,676
Jun	127,362	208,509	257,017	592,983	3,783,260	206,994	-26,309	17,117	174,215	4,155,276	7,963,207
Jul	131,238	204,754	260,514	596,507	3,854,874	187,327	-33,778	17,849	189,381	4,215,653	8,005,056
Aug	136,229	208,773	275,514	620,517	3,914,952	190,012	-47,255	17,447	171,594	4,246,752	8,083,837
Sep	131,124	202,427	259,948	593,500	3,819,412	207,849	-9,192	16,299	151,743	4,186,109	8,025,180
Oct	138,995	188,455	257,545	584,995	3,874,764	212,241	-21,741	16,535	159,591	4,241,390	8,057,007
Nov	140,662	206,840	253,559	601,061	3,889,368	253,712	9,997	16,751	159,740	4,329,568	8,098,050
Dec	138,958	202,216	254,763	595,937	3,907,615	98,055	21,463	16,374	158,920	4,202,426	7,959,235
2018 Jan	136,983	196,191	253,195	586,369	3,860,382	230,435	110,385 (o)	18,220	158,007	4,377,429 (r)	7,986,401 (u)
Feb	140,752	197,576	259,152	597,481	3,960,200	242,862	-55,396 (p)	18,332	185,185	4,331,183 (s)	8,081,769
Mar	140,118	225,402	255,360	620,880	3,945,367	184,818	-28,236 (q)	25,323	164,533	4,291,805 (t)	8,005,842
Apr	147,302	243,070	255,101	645,473	3,977,986	187,182	-131,581	19,006	174,469	4,227,061	8,043,809
May	152,697	251,430	257,017	661,145	4,099,628	244,570	-112,631	19,317	163,351	4,414,234	8,223,675
Jun	154,312	254,551	257,540	666,403	4,049,124	169,131	-54,166	17,946	164,498	4,346,533	8,100,222
Jul	156,115	252,487	260,738	669,340	3,988,820	182,902	-43,478	17,203	176,417	4,321,865	8,024,639
Aug	159,217	254,979	263,840	678,036	4,009,687	171,531	-16,932	17,771	155,605	4,337,662	8,032,708
Sep	152,365	256,486	263,568	672,420	3,996,174	190,052	-25,483	18,204	145,892	4,324,839	8,034,890
Oct	157,609	257,474	261,464	676,546	4,081,389	202,162	-85,334	18,941	156,822	4,373,980	8,103,706
Nov	157,357	252,623	263,137	673,117	4,166,129	206,333	-84,113	18,586	150,420	4,457,355	8,171,201
Dec	153,066	245,571	270,355	668,991	4,156,773	88,686	-53,655	17,706	122,309	4,331,818	8,064,864
2019 Jan	156,839	238,092	265,107	660,038	4,038,167	215,672	-50,104	18,711	117,670	4,340,116	8,043,119
Feb	156,325	189,648	260,736	606,709	3,971,278	209,205	-26,489	18,295	110,579	4,282,868	7,996,864
Mar	157,624	196,705	286,326	620,655	4,001,715	176,076	-74,929	18,572	115,037	4,236,472	8,062,844
Apr	156,518	191,845	271,244	619,608	4,021,422	183,570	-78,384	17,263	126,255	4,270,126	8,083,231

Source: Bank of England

12.3b Monetary financial institutions' (excluding central bank) balance sheet

£ millions

Not seasonally adjusted

Notes to table

Movements in amounts outstanding can reflect breaks in data series as well as underlying flows. For changes data, users are recommended to refer directly to the appropriate series or data tables.
Further explanation can be found at: www.bankofengland.co.uk/statistics/Pages/iadb/notesiadb/Changes_flows_growth_rates.aspx

(o) Due to improvements in reporting by one institution, amounts outstanding increased by £288bn.
(p) Due to improvements in reporting by one institution, amounts outstanding increased by £29bn.
(q) Due to improvements in reporting by one institution, amounts outstanding decreased by £14bn.
(r) Due to improvements in reporting by one institution, amounts outstanding increased by £66bn.
(s) Due to improvements in reporting by one institution, amounts outstanding decreased by £8bn.
(t) Due to improvements in reporting by one institution, amounts outstanding decreased by £9bn.
(u) Due to improvements in reporting at one institution, the amounts outstanding increased by £5bn. This effect has been adjusted out of the flows for January 2018.

Explanatory notes can be found here:
https://www.bankofengland.co.uk/statistics/details/further-details-about-monetary-financial-institutions-excluding-central-bank-balance-sheet-data

Copyright guidance and the related UK Open Government Licence can be viewed here: www.bankofengland.co.uk/legal

12.3 c Monetary financial institutions' (excluding central bank) balance sheet

£ millions

Not seasonally adjusted

Amounts outstanding of sterling assets

	Notes coin	With UK central bank		UK MFIs	of which intragroup banks	Loans			Acceptances granted				Bills			
		Cash ratio deposits	Other			UK MFIs CDs	UK MFIs commercial paper	Non-residents	UK MFIs	UK public sector	Other UK residents	Non-residents	Treasury bills	UK MFIs bills	Other UK residents	Non-residents
RPM	B3UO	B3VO	B3WO	B3NL	B8ZI	B3OL	B3PL	B3XO	B3QL	B3YO	B3ZO	B3GP	B3HP	B3RL	B3IP	B3JP
2017 Jan	10,365	4,424	375,664	273,300	250,356	3,478	27	136,718	1	-	146	40	5,357	269	126	619
Feb	10,186	4,424	393,306	295,742	273,393	3,219	21	127,280	-	-	145	41	4,131	265	188	426
Mar	10,263	4,424	409,678	329,741	310,074	3,139	21	127,998	1	-	142	110	3,751	265	138	595
Apr	10,246	4,426	415,419	329,307	308,686	2,837	21	134,581	1	-	147	110	2,167	294	183	666
May	10,743	4,428	413,198	344,516	322,877	2,840	22	132,701	-	-	154	110	2,637	284	233	511
Jun	10,437	4,581	424,012	365,168	345,633	3,039	26	128,208	2	-	159	102	2,570	315	155	477
Jul	10,326	4,584	427,044	373,115	351,969	2,992	22	122,580	-	-	117	100	3,292	-	484	366
Aug	10,909	4,583	419,209	379,093	358,816	2,932	18	134,334	-	-	115	92	3,283	-	377	416
Sep	10,439	4,584	423,784	392,276	373,536	2,892	91	123,248	-	-	117	28	4,045	-	381	530
Oct	10,818	4,584	436,432	360,424	342,590	2,676	59	126,726	-	-	121	27	5,287	-	414	615
Nov	11,039	4,586	445,652	304,479	286,570	2,740	16	122,255	-	-	118	27	4,820	-	434	271
Dec	11,920	4,708	450,708	295,719	277,942	2,771	15	122,170	-	-	115	37	4,862	-	440	250
2018 Jan	10,365	4,708	451,929	228,023	210,856	2,822	15	126,364	-	-	109	34	3,867	-	396	926
Feb	10,338	4,708	474,461	229,203	212,334	2,585	24	118,059	-	-	128	41	5,014	1	430	938
Mar	10,426	4,708	464,665	211,670	195,334	2,758	23	111,415	-	-	130	48	5,704	1	391	951
Apr	10,672	4,711	479,065	216,324	200,314	2,852	9	109,208	-	-	212	48	7,954	1	382	328
May	10,321	4,723	472,828	185,885	170,374	2,982	9	118,734	-	-	290	49	8,955	1	382	362
Jun	9,887	7,647	471,326	206,143	185,755	2,772	9	110,724	-	-	349	48	11,069	1	393	430
Jul	9,935	7,647	469,598	206,696	187,230	2,925	69	115,837	-	-	308	18	10,978	1	389	1,024
Aug	10,318	7,647	474,412	194,174	178,271	2,882	9	110,935	-	-	284	18	12,180	1	500	296
Sep	9,818	7,647	472,875	191,269	174,816	2,797	9	114,340	-	-	259	18	9,669	1	362	484
Oct	9,978	7,648	466,823	187,784	171,146	3,026	18	112,653	-	-	273	19	6,974	1	381	512
Nov	10,080	7,653	464,414	183,495	166,875	3,136	9	108,910	-	-	298	22	8,755	1	381	340
Dec	11,378	7,884	461,169	198,257	181,598	3,032	8	107,416	-	-	285	23	7,159 (y)	2	399	241
2019 Jan	10,068	7,884	461,776	172,677	155,434	3,152	10	107,553	-	-	272	25	7,762 (z)	1	425	303
Feb	9,753	7,884	463,773	187,912	170,928	3,322	29	102,764	-	-	280	24	8,030	1	418	264
Mar	9,709	7,884	449,759	239,350	221,429	2,859	12	107,519	-	-	282	26	8,684	1	496	309
Apr	10,056	7,879	460,863	238,224	220,676	2,895	12	109,758	1	-	242	28	7,271	1	464	258

12.3 c Monetary financial institutions' (excluding central bank) balance sheet

£ millions

Amounts outstanding of sterling assets

	UK MFIs	Sales and repurchase agreements				Advances			Investments					Items in suspense and collection	Accrued amounts receivable	Other assets	Total sterling assets
		of which intragroup banks	UK public sector	Other UK residents	Non-residents	UK public sector	Other UK residents	Non-residents	UK government bonds	Other UK public sector	UK MFIs	Other UK residents	Non-residents				
RPM	B3SL	B2ZG	B3KP	B3LP	B3MP	B2UK	B3OP	B3PP	B3QP	B3RP	B3TL	B3SP	B3TP	B3UP	B3VP	B3WP	B3XP
2017 Jan	34,699	17,364	2,388	182,328	61,003	8,460	1,891,296	102,614	144,060	548	29,133	223,121	57,269	39,719	13,789	30,795	3,631,756
Feb	32,510	17,098	530	180,222	65,973	8,344	1,893,382	106,713	143,570	590	27,934	220,789	56,822	46,371	13,965	30,451	3,667,539
Mar	29,889	14,177	1,372	194,529	73,256	8,312	1,899,967	106,385	145,809	484	28,183	219,078	54,960	52,979	13,013	30,552	3,749,034
Apr	27,148	14,827	422	194,147	71,776	8,359	1,907,448	109,180	147,043	435	27,899	223,042	55,334	48,866	13,163	30,484	3,765,152
May	28,819	14,028	508	199,668	72,318	7,903	1,908,994	106,687	148,848	453	28,446	222,940	55,372	48,801	13,735	30,606	3,786,476
Jun	26,587	12,242	498	201,751	77,865	8,201	1,916,592	108,916	149,239	497	39,077	243,291	55,067	51,057	12,722	29,437	3,860,046
Jul	30,040	16,251	511	199,359	66,692	42,597	1,886,661	109,216	148,655	514	38,345	246,045	56,171	44,851	13,010	29,588	3,857,279
Aug	24,758	13,360	1,475	204,449	68,630	42,903	1,896,508	109,442	147,509	518	37,706	245,307	54,160	37,670	14,443	29,669	3,870,504
Sep	26,072	12,607	517	220,854	75,635	43,130	1,911,609	109,537	142,711	460	36,924	239,695	54,345	40,691	12,253	26,324	3,903,173
Oct	28,186	10,860	388	215,970	68,389	43,032	1,908,355	112,245	145,949	495	37,749	238,845	46,560	37,246	12,501	26,483	3,870,579
Nov	32,642	10,862	593	219,426	72,679	42,663	1,917,128	113,364	136,681	525	37,557	234,563	46,582	49,726	13,404	26,653	3,840,623
Dec	37,750	12,979	793	222,053	72,517	42,763	1,917,130	112,323	133,823	543	38,391	235,168	45,254	20,241	13,740	26,964	3,813,169
2018 Jan	36,791	11,695	534	218,834	77,298	42,299	1,921,352	109,302	138,385	510	38,883	228,608	48,437	38,702	14,790	27,588	3,771,870
Feb	38,106	11,399	1,413	213,837	77,549	42,293	1,923,115	112,639	140,668	485	38,153	226,369	48,660	36,375	15,116	27,817	3,788,525
Mar	40,324	9,971	4,902	204,223	85,068	41,534	1,937,419	112,709	139,395	498	39,507	222,730	47,656	38,774	15,458	26,414	3,769,502
Apr	43,400	15,507	619	196,716	85,228	8,696 (ac)	1,974,416 (ad)	111,178	138,606	498	39,689	222,442	47,700	39,393	13,790	26,546	3,780,684
May	42,486	14,728	558	189,944	89,661	8,326	1,974,815	110,361	144,093	503	40,284	222,652	48,888	44,517	14,389	25,985	3,762,983
Jun	50,748	25,124	549	195,724	96,436	8,130	1,997,504	95,470	141,321	476	40,713	228,727	51,636	34,477	15,529	26,380	3,800,621
Jul	27,567	8,913	414	203,810	92,591	8,351	2,005,192	95,681	134,717	492	40,983	193,861	54,115	36,959	14,834	26,194	3,761,185
Aug	29,617	7,972	770	200,920	89,774	8,323	2,011,062	95,517	135,143	474	41,086	193,598	55,217	31,376	15,933	26,211	3,748,677
Sep	23,329 (aa)	8,508	266	211,925 (ab)	95,849	8,175	2,026,484	96,562	131,756	485	39,744	195,807	58,064	43,586	15,256	26,210	3,783,048
Oct	26,286	7,890	298	216,496	93,677	8,169	2,021,918	96,620	125,389	469	40,581	193,733	62,253	40,231	14,884	26,465	3,763,557
Nov	28,582	10,919	188	229,674	100,831	7,654	2,027,963	99,095	123,978	463	40,996	187,072	62,236	35,581	15,599	26,652	3,774,058
Dec	30,602	9,675	1,298	241,191	115,534	7,312	2,026,320	95,889	120,032	481	40,793	183,354	60,748	21,977	14,627	27,087	3,784,499
2019 Jan	30,083	10,556	232	234,298	105,449	7,714	2,037,370	96,939	125,690	328	41,801	183,022	66,361	37,869	15,220	27,480	3,781,761
Feb	30,336	9,752	539	226,911	107,932	7,768	2,041,197	93,211	126,936	329	42,530	182,934	68,911	36,747	15,693	27,491	3,793,920
Mar	29,913	7,900	177	235,143	126,427	7,516	2,051,520	94,940	126,051	365	43,819	185,952	69,047	38,343	15,252	27,693	3,879,048
Apr	29,373	8,985	177	233,846	115,819	7,936	2,043,011	93,842	121,081	368	43,676	183,082	69,202	31,669	18,154	27,866	3,857,052

Source: Bank of England

12.3 c Monetary financial institutions' (excluding central bank) balance sheet

£ millions

Not seasonally adjusted

Notes to table

Movements in amounts outstanding can reflect breaks in data series as well as underlying flows. For changes data, users are recommended to refer directly to the appropriate series or data tables.
Further explanation can be found at: https://www.bankofengland.co.uk/statistics/details/further-details-about-changes-flows-growth-rates-data.

(y)   Due to changes in the reporting population, amounts outstanding decreased by £1bn. This has been adjusted out of flows for December 2018.

(z)   Due to improvements in reporting at one institution, amounts outstanding decreased by £0.3bn. This has been adjusted out of the flows for January 2019.

(aa)   Due to improvements in reporting at one institution, the amounts outstanding decreased by £5bn. This effect has been adjusted out of the flows for September 2018.

(ab)   Due to improvements in reporting at one institution, the amounts outstanding increased by £5bn. This effect has been adjusted out of the flows for September 2018.

(ac)   In order to bring reporting in line with the National Accounts, English housing associations were reclassified from public corporations to PNFCs with effect from April 2018 data.
        The amounts outstanding decreased by £34bn. This effect has been adjusted out of the flows for April 2018.

(ad)   In order to bring reporting in line with the National Accounts, English housing associations were reclassified from public corporations to PNFCs with effect from April 2018 data.
        The amounts outstanding increased by £34bn. This effect has been adjusted out of the flows for April 2018.

Explanatory notes can be found here:
https://www.bankofengland.co.uk/statistics/details/further-details-about-monetary-financial-institutions-excluding-central-bank-balance-sheet-data

Copyright guidance and the related UK Open Government Licence can be viewed here: www.bankofengland.co.uk/legal

12.3d Monetary financial institutions' (excluding central bank) balance sheet

£ millions

Amounts outstanding of foreign currency assets (including euro)

| RPM | Loans and advances | | | | | | Sale and repurchase agreements | | | | | Acceptances granted | Total bills |
| | UK MFIs | of which intragroup banks | UK MFIs' CDs etc. | UK public sector | Other UK residents | Non-residents | UK MFIs | of which intragroup banks | UK public sector | Other UK residents | Non-residents | | |
	B3UL	B8ZQ	B3VL	B2UN	B3ZP	B2UH	B3WL	B8ZO	B3HQ	B3IQ	B3JQ	B3KQ	B3LQ
2017 Jan	206,911	154,446	1,201	7	248,512	1,978,013	74,364	35,740	1,989	137,091	687,479	3,288	39,168
Feb	199,983	144,529	1,112	35	254,679	2,028,597	75,379	35,565	1,072	159,167	698,801	3,789	37,152
Mar	194,249	137,113	1,037	11	245,880	1,970,277	81,217	39,816	197	154,816	705,335	4,008	36,559
Apr	187,594	133,417	896	24	247,590	1,935,696	75,737	38,118	3,253	158,797	678,921	3,580	32,843
May	188,201	135,662	817	58	243,432	1,947,331	79,461	38,388	2,861	164,829	692,111	3,183	35,796
Jun	151,243	100,452	797	58	238,187	1,954,968	72,436	35,467	1,017	164,375	687,493	3,327	38,006
Jul	156,411	104,436	909	122	238,871	1,987,300	73,391	34,263	3,154	165,094	723,094	2,792	39,348
Aug	161,433	109,158	1,106	110	242,104	2,043,328	80,817	40,407	2,356	169,270	731,457	2,541	39,073
Sep	158,005	106,543	1,031	150	228,535	1,942,944	76,914	39,182	2,635	157,215	766,670	2,953	38,875
Oct	155,334	104,453	840	108	230,376	1,954,418	72,491	34,946	2,839	166,615	810,156	2,785	35,281
Nov	150,058	101,895	919	168	224,766	1,953,149	78,774	39,832	2,680	176,826	827,889	2,950	35,405
Dec	151,671	103,814	948	113	217,015	1,992,422	71,822	34,219	3,208	160,861	844,890	3,472	36,008
2018 Jan	169,980	121,581	750	266	227,380	1,924,383	69,967	31,516	1,678	183,451	795,020	2,857	36,391
Feb	169,908	123,652	527	60	219,994	1,959,861	76,960	36,112	1,409	191,978	862,381	3,286	34,532
Mar	181,022	135,312	615	62	215,564	1,983,357	77,177	33,326	1,894	175,965	860,787	3,221	33,072
Apr	189,087	142,024	646	136	230,083	2,003,479	78,184	34,780	3,276	166,654	835,305	3,725	32,395
May	202,842	152,503	571	116	227,290	2,071,007	86,080	41,597	2,085	192,959	871,162	4,334	30,704
Jun	205,227	157,124	628	88	226,742	2,028,663	90,549	48,667	2,226	171,861	847,030	3,991	29,679
Jul	194,048	147,420	722	60	223,452	2,020,607	85,573	45,701	3,680	171,845	821,755	4,035	29,546
Aug	203,898	155,300	686	82	229,038	2,054,631	80,608	43,539	3,895	175,783	816,254	4,525	30,054
Sep	188,451	142,074	605	93	218,674	2,016,848	83,965	47,024	3,449	177,961	818,291	5,000	33,172
Oct	178,405	129,492	441	95	214,309	2,057,090	80,385	43,540	3,292	183,032	872,806	4,538	31,368
Nov	179,751	130,052	352	80	216,157	2,023,845	85,368	47,230	3,622	206,914	918,129	4,201	32,106
Dec	171,068	120,476	262	94	213,190	1,987,610	80,126	44,521	1,795	201,492	1,006,462	4,084	29,638
2019 Jan	162,803	109,448	369	82	215,520	1,918,785	83,609	45,228	2,676	198,589	910,832	3,929	32,625
Feb	154,723	104,986	437	50	209,789	1,882,465	87,204	47,305	3,524	166,479	931,497	3,414	32,567
Mar	144,364	93,836	522	38	213,344	1,893,176	67,200	23,301	3,932	158,028	953,989	3,619	32,345
Apr	147,809	98,137	622	33	218,474	1,878,037	68,832	26,439	2,368	162,915	964,287	3,842	32,414

12.3d Monetary financial institutions' (excluding central bank) balance sheet

£ millions

Amounts outstanding of foreign currency assets (including euro)

Not seasonally adjusted

RPM	UK government bonds B3MQ	Other UK public sector B3NQ	Investments UK MFIs B3XL	Other UK residents B3OQ	Non-residents B3PQ	Items in suspense and collection B3QQ	Accrued amounts receivable B3RQ	Other assets B3SQ	Total foreign currency assets B3TQ	Total assets B3UQ	Holdings of own sterling acceptances B3IM	Holdings of own foreign currency acceptances B3JM
2017 Jan	426	-	13,113	12,548	459,369	239,807	9,653	42,375	4,155,313	7,787,070	193	920
Feb	401	-	15,007	13,465	484,150	264,093	12,637	41,647	4,291,167	7,958,706	184	1,113
Mar	473	-	13,240	12,917	461,730	228,757	11,733	40,240	4,162,675	7,911,709	158	1,436
Apr	438	-	12,953	11,541	452,033	268,152	12,883	40,569	4,123,499	7,888,650	160	1,096
May	484	-	13,289	11,478	473,080	259,937	12,909	38,942	4,168,199	7,954,676	161	1,403
Jun	464	-	13,969	10,629	472,273	243,682	12,750	37,486	4,103,162	7,963,207	146	1,238
Jul	461	-	13,681	11,751	457,269	219,783	12,833	42,121	4,147,777	8,005,056	143	1,045
Aug	353	-	13,263	11,751	445,670	209,232	16,673	42,794	4,213,333	8,083,837	281	1,277
Sep	344	-	12,431	10,985	459,476	211,908	12,040	38,895	4,122,007	8,025,180	160	1,051
Oct	245	-	11,442	11,000	468,036	211,219	11,765	41,475	4,186,427	8,057,007	171	1,004
Nov	245	-	12,068	11,076	475,164	254,701	12,228	38,363	4,257,427	8,098,050	170	1,061
Dec	229	-	13,193	13,113	480,204	99,909	13,195	43,795	4,146,067	7,959,235	151	1,398
2018 Jan	232	-	11,935	12,197	491,593	233,801	12,718	39,933	4,214,532	7,986,401	150	1,260
Feb	331	-	12,445	11,197	461,088	233,902	12,356	41,029	4,293,243	8,081,769	151	1,637
Mar	313	-	12,305	10,527	434,754	190,043	13,116	42,545	4,236,340	8,005,842	156	2,108
Apr	321	-	12,370	10,603	447,331	193,120	14,411	41,999	4,263,125	8,043,809	154	2,294
May	183	-	14,132	11,904	426,688	255,846	19,053	43,737	4,460,693	8,223,675	154	2,210
Jun	189	-	13,989	11,743	426,767	181,253	17,018	41,959	4,299,601	8,100,222	160	3,189
Jul	86	-	14,101	11,596	434,134	189,645	17,411	41,157	4,263,453	8,024,639	145	2,865
Aug	83	-	14,034	11,585	421,076	178,491	17,464	41,845	4,284,031	8,032,708	166	2,892
Sep	119	-	12,417	11,861	432,351	191,166	13,851	43,569	4,251,842	8,034,890	174	2,444
Oct	120	-	12,640	11,585	427,831	206,843	12,986	42,381	4,340,148	8,103,706	156	1,804
Nov	21	-	12,746	11,990	429,882	215,163	13,444	43,370	4,397,143	8,171,201	166	2,061
Dec	122	-	12,465	10,690	412,199	91,837	12,893	44,339	4,280,365	8,064,864	225	2,473
2019 Jan	84	-	12,424	9,901	423,380	224,369	12,974	48,404	4,261,358	8,043,119	311	2,196
Feb	79	-	12,655	10,081	427,057	217,571	13,210	50,143	4,202,944	7,996,864	379	2,051
Mar	78	-	13,230	9,742	444,834	177,371	13,979	54,005	4,183,796	8,062,844	484	2,446
Apr	89	-	13,775	9,486	464,285	188,598	14,865	55,446	4,226,179	8,083,231	535	2,494

Source: Bank of England

Notes to table

Movements in amounts outstanding can reflect breaks in data series as well as underlying flows. For changes data, users are recommended to refer directly to the appropriate series or data tables.
Further explanation can be found at: https://www.bankofengland.co.uk/statistics/details/further-details-about-changes-flows-growth-rates-data.

Explanatory notes can be found here:
https://www.bankofengland.co.uk/statistics/details/further-details-about-monetary-financial-institutions-excluding-central-bank-balance-sheet-data

Copyright guidance and the related UK Open Government Licence can be viewed here: www.bankofengland.co.uk/legal

12.3e Monetary financial institutions' (excluding central bank) balance sheet

£ millions

Not seasonally adjusted

Changes in sterling liabilities

RPM	Notes outstanding & cash loaded cards	Sight deposits UK MFIs	of which intragroup banks	UK public sector	Other UK residents	Non-residents	Time deposits UK MFIs	of which intragroup banks	UK public sector	Other UK residents	of which SAYE	of which cash ISAs	Non-residents
	B4JJ	B4GA	B8ZD	B4CF	B4BH	B4DD	B4HA	B8ZF	B4DF	B4CH	B4FH	B4DH	B4ED
2017 Jan	-156	6,589	5,214	211	-16,328	2,172	7,960	-5,462	-370	-9,377	12	-795	-4,286
Feb	8	9,369	10,427	-907	1,763	-1,565	20,360	11,398	-814	-6,711	-7	-771	-108
Mar	14	18,524	19,046	-388	30,254	3,243	27,877	15,225	-1,721	-4,309	21	638	-5,971
Apr	86	1,139	-194	1,693	-9,026	-1,488	2,037	-470	1	11,247	5	3,838	4,143
May	-83	5,442	4,576	-299	8,882	6,091	14,019	9,283	209	-6,946	15	271	-177
Jun	-26	4,074	5,343	103	21,906	3,627	28,330	16,954	592	848	-3	-419	980
Jul	157	11,548	10,651	1,209	-8,195	-3,409	899	-2,559	746	-2,262	-28	-848	-7,333
Aug	199	1,759	2,519	-122	13,081	1,490	12,513	8,460	218	1,438	9	-692	4,262
Sep	55	1,493	3,227	113	13,252	-9,287	9,084	6,356	-7	-10,293	24	-141	-1,843
Oct	48	-24,594	-24,061	57	-3,428	-2,144	2,971	-2,627	-626	6,042	10	-709	-10,112
Nov	-46	-17,882	-17,842	-350	11,897	2,400	-37,626	-39,591	-285	509	-37	-691	-6,162
Dec	172	-14,816	-14,523	256	5,097	9,891	18,780	8,210	-810	-1,540	-14	-474	1,409
2018 Jan	-326	-33,582	-34,076	207	-14,402	-12,673	-30,172	-33,753	-500	3,829	15	-725	-4,679
Feb	-107	4,720	4,253	-761	880	-809	16,715	-2,742	-1,784	30	35	-25	2,194
Mar	104	-16,004	-15,946	1,198	16,439	7,662	-1,673	-2,292	-1,115	-8,444	10	2,615	7,248
Apr	353	-8,044	-6,746	-535	-936	-1,299	12,523	11,677	1,175	2,655	3	4,908	-527
May	-461	-16,366	-15,351	-863	8,756	16,843	-13,715	-14,355	1,380	3,932	10	752	-3,916
Jun	-256	20,218	18,735	-157	11,591	-3,204	-3,293	-2,545	567	-7,642	5	-297	-991
Jul	-126	2,553	-520	812	-2,906	12,596	2,516	1,668	999	1,540	-17	-197	-2,080
Aug	147	-15,101	-12,956	199	4,676	-12,746	1,481	2,155	776	384	25	-310	-1,021
Sep	83	1,278	351	114	9,421	11,448	-4,389	-3,667	18	-1,899	20	246	4,150
Oct	-69	-2,006	-1,203	70	2,191	-13,935	-1,764	-2,448	-373	200	12	102	1,851
Nov	-6	-1,820	-1,697	-1,352	6,949	-8,198	-6,817	-2,167	279	418	-90	-170	2,765
Dec	226	11,088	11,409	1,198	2,965	-3,261	2,927	3,101	-871	10,987	-13	156	-5,018
2019 Jan	-298	-20,002	-20,502	-628	-17,989	-810	-5,564	-5,408	707	-1,757	12	154	-5,072
Feb	-69	15,431	15,979	-469	-5,213	796	-182	-279	-852	5,568	23	575	4,811
Mar	108	49,776	48,788	-292	17,939	-382	1,562	1,417	-519	4,520	8	3,788	199
Apr	-32	-2,805	-1,889	726	-5,024	-3,789	2,635	1,774	880	5,265	10	6,789	-2,332

12.3e Monetary financial institutions' (excluding central bank) balance sheet

£ millions

Not seasonally adjusted

Changes in sterling liabilities

RPM	UK MFIs	of which intragroup banks	UK public sector	Other UK residents	Non-residents	Acceptances granted	CDs and Commercial paper	Bonds with maturity of and up to incl. 5 years	Bonds with maturity of greater than 5 years	Total	Total sterling deposits	Sterling items in suspense and transmission	Net derivatives	Accrued amounts payable	Sterling capital and other internal funds	Total sterling liabilities
	B4IA	B8ZB	B4EF	B4EH	B4FD	B4BK	B2SU	B6OH	B2TK	B4EJ	B4FJ	B4AK	B4GJ	B4CJ	B4DJ	B4JJ
2017 Jan	2,436	4,486	-411	26,351	364	5	-851	-725	-657	-2,233	13,083	22,337	-24,227	-31	30,126	41,132
Feb	-1,629	-471	2,049	-1,031	1,366	1	5,192	-170	23	5,045	27,188	2,118	35,703	1,193	1,644	67,855
Mar	-383	-2,740	-1,202	-166	9,538	66	10,262	-892	-238	9,133	84,494	3,531	-25,650	-2,775	-15,579	44,035
Apr	-1,532	649	-505	15,039	-7,500	5	2,763	-2,203	-170	391	15,644	-167	-34,850	-541	1,666	-18,161
May	2,619	-799	2,444	-8,436	3,721	6	-3,373	2,569	-676	-1,480	26,094	-4,671	20,891	644	20,290	63,164
Jun	-4,606	-1,785	394	-7,925	9,194	-1	-6,018	500	1	-5,518	51,998	3,407	-9,538	-28	7,233	53,047
Jul	946	4,008	3,176	8,231	-9,175	-46	-165	-249	99	-315	-3,979	-5,643	16,487	-188	-26,210	-19,376
Aug	-4,490	-2,891	823	6,826	1,703	-10	700	-273	69	496	39,987	-6,792	6,967	-502	7,008	46,867
Sep	-3,749	-753	-7,096	-1,166	10,891	-62	-1,357	-1,081	-553	-2,992	-1,661	3,663	-43,530	-799	27,538	-14,736
Oct	1,043	-2,247	3,254	14,037	-9,813	3	-6,624	420	366	-5,838	-29,151	-1,589	16,876	750	-8,569	-21,636
Nov	2,654	502	836	-4,728	5,640	-3	5,757	-63	479	6,173	-36,926	10,428	-26,204	873	8,814	-43,061
Dec	6,393	2,117	-2,437	-17,447	5,219	6	1,478	630	22	2,131	12,132	-30,711	-5,233	2,277	11,233	-10,131
2018 Jan	-2,596	-1,285	2,835	20,453	1,172	-8	2,740	3,375	-506	5,609	-64,507	20,920	-85,104	-814	-7,230	-137,061
Feb	1,232	-296	1,945	-1,664	-6,913	26	-1,550	-1,761	875	-2,437	13,374	-2,579	149,256	710	-16,107	144,547
Mar	1,367	-1,427	-5,775	-28,756	5,181	9	3,965	3,174	-1,435	5,704	-16,959	2,603	-26,833	519	2,790	-37,777
Apr	3,158	5,531	-66	2,640	-5,804	83	-1,868	1,104	-2,343	-3,107	1,915	-1,034	90,689	-2,496	16,637	106,064
May	-1,624	-1,617	20	-2,996	4,594	78	1,535	460	-70	1,924	-1,954	7,985	-10,770	3,885	-12,732	-14,047
Jun	8,109	11,222	-611	765	-5,125	59	-4,422	-448	11	-4,859	15,426	-9,720	-55,175	-3,620	-6,430	-59,775
Jul	-23,595	-16,205	-579	5,105	1,135	-71	-3,197	-351	250	-3,298	-5,275	6,399	-7,305	-1,464	-41,625	-49,396
Aug	2,388	-1,277	283	3,337	4,546	-24	-4,434	-368	-345	-5,147	-15,969	-5,136	-23,738	708	34,119	-9,867
Sep	162	883	-102	-5,172	-4,694	-25	-5,231	2,694	-372	-2,909	7,402	4,947	803	2,519	4,238	19,991
Oct	1,918	-618	-314	12,314	-5,595	14	-2,222	669	-885	-2,438	-7,865	-1,884	57,156	-2,036	-24,976	20,326
Nov	3,419	3,026	-174	353	-2,447	29	-7,483	649	-121	-6,955	-13,553	-2,140	-7,132	918	2,056	-19,857
Dec	4,136	-1,241	-81	-3,258	10,519	-12	764	-39	-84	640	31,960	-15,148	-25,476	-1,155	31,783	22,190
2019 Jan	604	880	151	6,077	-640	-13	2,612	962	163	3,737	-41,198	15,882	-522	-918	-5,301	-32,356
Feb	1,760	-807	1,759	7,248	-8,973	8	-596	1,854	327	1,584	23,275	281	-22,820	108	13,239	14,015
Mar	1,711	-1,852	-1,191	-22,111	16,528	4	3,682	975	103	4,760	72,504	-381	42,743	2,538	-11,561	105,951
Apr	-1,190	1,102	100	16,246	-4,582	-37	-2,633	422	-27	-2,237	3,854	-2,925	5,457	-2,948	-4,381	-976

Liabilities under sale and repurchase agreements — column grouping over UK MFIs / intragroup banks / UK public sector / Other UK residents / Non-residents.

CDs and other paper issued — column grouping over CDs and Commercial paper / Bonds with maturity of and up to incl. 5 years / Bonds with maturity of greater than 5 years / Total.

Source: Bank of England

Notes to table

Explanatory notes can be found here:
https://www.bankofengland.co.uk/statistics/details/further-details-about-monetary-financial-institutions-excluding-central-bank-balance-sheet-data

Copyright guidance and the related UK Open Government Licence can be viewed here: www.bankofengland.co.uk/legal

12.3f Monetary financial institutions' (excluding central bank) balance sheet

£ millions

Not seasonally adjusted

Changes in foreign currency liabilities (including euro)

RPM	Sight and time deposits					Acceptances granted	Sale and repurchase agreements				
	UK MFIs	of which intragroup banks	UK public sector	Other UK residents	Non-residents		UK MFIs	of which intragroup banks	UK public sector	Other UK residents	Non-residents
	B4GB	B8ZN	B2VP	B2VV	B2VJ	B4HM	B4HB	B8ZL	B4GG	B4GI	B4HE
2017 Jan	4,959	5,615	213	8,957	987	-675	3,849	2,982	796	12,516	26,574
Feb	-12,411	-9,523	-74	-943	16,425	472	-1,554	-41	-975	25,197	22,265
Mar	-10,131	-6,995	-245	-7,369	-24,977	233	5,631	4,095	-272	-8,694	-3,197
Apr	-615	-404	358	9,071	13,871	-335	-2,760	-993	1,279	1,956	-1
May	-929	-473	353	7,556	-32,000	-415	762	-629	-289	-4,714	31,387
Jun	-38,229	-35,646	86	-983	14,846	150	-5,956	-3,250	-845	-1,165	-17,062
Jul	6,603	3,752	114	1,085	11,371	-516	-1,842	-1,594	1,576	4,864	1,402
Aug	4,377	1,983	-142	501	-38,273	-300	5,397	5,155	-877	1,236	-7,890
Sep	-3,644	1,590	371	-430	19,997	488	-605	445	670	-949	23,847
Oct	-2,206	-1,817	149	-2,745	21,482	-184	-1,603	-4,279	-221	13,652	25,492
Nov	-3,421	-2,065	-259	5,809	3,437	197	4,934	5,021	234	10,278	3,110
Dec	2,554	1,292	-39	-12,636	1,905	523	-8,323	-5,821	250	-17,855	46,716
2018 Jan	25,289	21,064	296	10,900	27,126	-503	690	-2,015	-778	25,872	-23,337
Feb	-4,066	151	-36	-8,014	-16,585	363	7,206	4,097	-525	4,309	34,071
Mar	13,095	13,054	-269	-1,515	-3,161	-23	-1,992	-2,398	448	-14,876	15,503
Apr	6,141	6,277	449	6,818	16,955	477	-1,219	1,341	1,607	1,140	-54,327
May	11,093	7,519	-239	1,860	2,215	526	7,432	6,560	-2,582	13,007	9,795
Jun	-44	3,268	150	3,298	-48,107	-367	5,417	6,729	298	-16,069	-22,191
Jul	-12,057	-9,451	-30	-10,928	-36,325	23	-7,017	-3,315	1,618	6,294	-26,322
Aug	7,301	6,124	-734	3,240	-107	462	-7,037	-2,332	-663	1,440	-15,590
Sep	-15,303	-12,526	285	2,659	609	491	4,684	3,620	734	-573	14,914
Oct	-9,367	-13,189	-246	17,430	28,975	-526	-2,186	-3,628	-423	7,337	8,146
Nov	2,504	933	-177	776	6,709	-338	7,754	3,641	-102	22,483	44,567
Dec	-11,222	-11,027	-108	-22,342	16,211	-123	-6,022	-3,120	-1,577	-5,899	30,966
2019 Jan	-2,424	-6,916	116	-5,113	14,321	-60	3,011	1,871	2	564	-31,628
Feb	-6,501	-4,024	-221	-2,372	43,450	-484	3,417	2,902	857	-24,024	19,680
Mar	-12,619	-12,634	244	4,033	-2,038	153	-20,949	-24,286	1,213	-11,514	15,748
Apr	3,270	4,422	440	4,912	5,146	225	2,731	3,208	-1,364	2,785	6,958

12.3f Monetary financial institutions' (excluding central bank) balance sheet

£ millions

Changes in foreign currency liabilities (including euro)

RPM	CDs and other paper issued				Total foreign currency deposits	Items in suspense and transmission	Net derivatives	Accrued amounts payable	Capital and other internal funds	Total foreign currency liabilities	Total liabilities
	CDs and Commercial paper	Bonds with maturity of up to and incl. 5 years	Bonds with maturity of greater than 5 years	Total							
	B2TN	B6OJ	B2TO	B4CM	B4DM	B4GM	B4EM	B4AM	B4BM	B4FM	B4JM
2017 Jan	1,966	1,404	2,218	5,589	63,764	124,391	27,279	620	-23,761	192,292	233,424
Feb	10,850	6,608	8,403	25,860	74,263	11,241	-34,557	5,186	-1,059	55,074	122,929
Mar	990	-6,285	-3,403	-8,698	-57,718	-37,128	26,530	-551	-12,259	-81,126	-37,091
Apr	-1,318	12,947	3,732	15,361	38,184	33,661	37,829	1,218	-13,334	97,558	79,397
May	4,436	-8,703	1,181	-3,086	-1,375	-11,576	-16,993	786	-40,408	-69,567	-6,403
Jun	-766	-575	-1,064	-2,404	-51,560	-23,893	-4,047	-1,497	16,260	-64,738	-11,691
Jul	3,503	-3,186	1,754	2,071	26,728	-21,352	-5,790	574	9,499	9,659	-9,717
Aug	1,481	-1,030	-2,335	-1,884	-37,855	-210	-10,211	-1,201	2,170	-47,308	-441
Sep	491	2,186	-4,038	-1,361	38,384	22,112	30,642	-597	-3,679	86,863	72,127
Oct	7,673	-14,969	-2,565	-9,860	43,956	4,836	-12,583	258	4,015	40,481	18,845
Nov	2,507	20,580	-3,105	19,982	44,301	41,104	31,195	191	694	117,484	74,423
Dec	-2,338	-5,214	95	-7,457	5,637	-155,978	11,632	-423	-5,413	-144,545	-154,675
2018 Jan	2,053	3,315	2,456	7,823	73,379	133,552	86,448	1,934	4,740	300,054	162,993
Feb	1,089	-1,192	-1,234	-1,337	15,386	-8,785	-162,794	40	29,937	-126,216	18,331
Mar	1,232	30,891	-536	31,588	38,799	-37,049	26,717	7,065	-14,229	21,303	-16,474
Apr	6,349	13,081	240	19,670	-2,288	2,423	-100,243	-6,302	9,395	-97,015	9,049
May	2,874	2,198	-1,682	3,390	46,497	57,316	18,944	309	-14,100	108,965	94,918
Jun	416	1,517	-1,176	757	-76,859	-76,185	57,574	-1,431	1,981	-94,921	-154,696
Jul	655	-3,631	1,390	-1,586	-86,331	12,991	10,266	-807	9,851	-54,031	-103,427
Aug	2,076	-490	1,405	2,991	-8,697	-11,826	26,417	540	-22,608	-16,175	-26,042
Sep	-6,176	2,885	1,224	-2,067	6,434	19,099	-8,275	471	-4,052	13,675	33,666
Oct	3,794	-2,318	-3,568	-2,093	47,047	12,383	-59,701	768	23,194	23,690	44,016
Nov	-424	-4,428	751	-4,101	80,075	4,050	7,715	-362	2,276	93,753	73,897
Dec	-5,024	-8,156	5,329	-7,850	-7,965	-117,645	30,121	-834	-19,004	-115,328	-93,137
2019 Jan	8,068	-904	1,958	9,122	-12,090	128,917	3,903	1,204	-17,047	104,887	72,531
Feb	1,785	-46,134	-132	-44,482	-10,679	-5,575	24,348	-285	-8,859	-1,051	12,964
Mar	-923	4,030	3,149	6,256	-19,472	-23,478	-48,572	327	-1,347	-92,542	13,409
Apr	-898	-4,480	5,475	97	25,202	7,673	-3,375	-1,294	11,374	39,579	38,604

Source: Bank of England

Notes to table

Explanatory notes can be found here: http://www.bankofengland.co.uk/mfsd/iadb/notesiadb/MFIs_exICB.htm

Copyright guidance and the related UK Open Government Licence can be viewed here: www.bankofengland.co.uk/legal

12.3g Monetary financial institutions' (excluding central bank) balance sheet

£ millions

Not seasonally adjusted

Changes in sterling assets

RPM	Notes and coin	With UK central bank		Loans					Acceptances granted				Bills			
		Cash ratio deposits	Other	UK MFIs	of which intragroup banks	UK MFIs' CDs etc	UK MFIs commercial paper	Non-residents	UK MFIs	UK public sector	Other UK residents	Non-residents	Treasury bills	UK MFIs bills	Other UK	Non-residents
	B4II	B3YR	B3ZR	B4DC	B8ZJ	B4JB	B4BC	B4BD	B4EC	B4FF	B3TR	B4GD	B4BA	B4IB	B4HG	B4IC
2017 Jan	-1,502	-	18,977	-1,969	-3,402	200	2	9,152	-	-	5	1	-2,273	-41	-27	255
Feb	-179	-	17,641	22,441	23,037	-259	-6	-9,438	-	-	-1	1	-1,227	-4	62	-193
Mar	77	-	16,359	35,248	36,682	-80	-	948	-	-	-3	69	-379	-	-50	169
Apr	-17	2	5,742	-433	-1,388	-302	1	10,331	-	-	5	-	-1,584	29	46	70
May	497	2	-2,221	15,209	14,192	3	-	-1,880	-	-	7	-	470	-10	50	-155
Jun	-306	154	10,814	20,632	22,749	199	4	-4,494	2	-	5	-8	-72	31	-78	-34
Jul	-111	3	3,033	7,951	6,336	-47	-4	-5,628	-2	-	-42	-2	722	-	14	-111
Aug	583	-1	-7,836	5,978	6,848	-61	-4	11,754	-	-	-2	-8	-9	-	-107	50
Sep	-469	-	4,576	12,655	14,717	-40	73	-11,086	-	-	2	-64	762	-	4	114
Oct	378	4	12,648	-31,853	-30,946	-216	-32	3,478	-	-	4	-1	1,243	-	33	85
Nov	222	1	9,220	-55,945	-56,020	64	-43	-4,471	-	-	-2	-	-467	-	20	-344
Dec	881	122	5,056	-3,700	-3,593	32	-	30	-	-	-4	10	42	-	6	-20
2018 Jan	-1,555	-	1,227	-69,796	-69,106	50	-	3,794	-	-	-5	-3	-994	-	-44	676
Feb	-27	-	22,533	1,180	1,478	-237	9	-8,305	-	-	19	7	1,147	-	34	12
Mar	88	3	-9,797	-17,536	-17,000	173	-1	-6,643	-	-	2	7	689	-	-39	13
Apr	246	3	14,401	4,721	4,980	94	-14	-2,207	-	-	82	-	2,250	-	-18	-575
May	-351	13	-6,238	-30,435	-29,939	130	-	9,526	-	-	77	-	1,001	-	-	34
Jun	-434	2,924	-1,502	16,258	15,381	-105	-	-8,115	-	-	60	-1	2,114	-	11	68
Jul	48	-	-1,729	4,552	1,475	191	60	5,113	-	-	-41	-30	-91	-	-4	594
Aug	383	-	4,814	-14,165	-10,598	-5	-60	-4,902	-	-	-24	-	1,202	-	112	-728
Sep	-500	-	-1,537	-3,013	-3,455	-47	-	3,405	-	-	-25	-	-2,510	-	-138	188
Oct	160	1	-6,052	-3,663	-3,670	229	9	-1,625	-	-	14	-	-2,696	-	19	28
Nov	102	5	-2,387	-4,106	-3,911	110	-9	-3,742	-	-	25	4	1,781	-	-	-172
Dec	1,298	230	-3,244	15,128	14,733	-104	-	59	-	-	-13	1	-442	-	18	-99
2019 Jan	-1,310	-1	607	-25,579	-26,164	121	1	137	-	-	-14	1	893	-	26	62
Feb	-315	-	1,997	15,235	15,495	170	20	-4,788	-	-	8	-1	268	-	-7	-39
Mar	-44	-	-14,014	51,436	50,490	-463	-17	5,054	-	-	2	2	653	-	78	45
Apr	347	1	13,006	-106	266	36	-	2,239	1	-	-40	2	-1,583	-	-32	-51

12.3g Monetary financial institutions' (excluding central bank) balance sheet

£ millions

Changes in sterling assets

Not seasonally adjusted

RPM	Claims under sale and repurchase agreements					Advances			Investments					Items in suspense and collection	Accrued amounts receivable	Other assets	Total sterling assets
	UK MFIs	of which intragroup banks	UK public sector	Other UK residents	Non-residents	UK public sector	Other UK residents	Non-residents	UK government bonds	Other UK public sector	UK MFIs	Other UK residents	Non-residents				
	B4FA	B8ZH	B4BF	B4AH	B4CD	B2VK	B2VQ	B4HC	B4CA	B4IE	B4CC	B4IG	B4JC	B4BJ	B4HI	B4JI	B4AJ
2017 Jan	4,673	4,461	2,388	6,372	-6,527	-54	5,020	-2,379	-3,029	-27	-22,212	29,208	3,215	21,991	-18	-71	61,330
Feb	-1,428	-266	-2,095	-3,805	4,288	-117	2,275	4,134	-3,887	43	-1,415	-4,787	-1,879	2,574	162	-344	22,558
Mar	-2,621	-2,921	842	14,306	7,283	-32	4,863	-1,241	1,754	-106	291	-1,943	-1,414	6,607	-952	97	80,090
Apr	-2,741	649	-950	-381	-1,480	48	7,292	1,193	885	-49	-230	5,472	597	-4,112	150	-61	19,522
May	1,671	-799	86	5,521	542	94	1,460	-2,467	1,095	18	300	-1,648	125	-65	572	122	19,398
Jun	-2,232	-1,785	-10	2,083	5,547	89	7,816	2,254	3,316	44	519	3,847	-122	2,256	-1,014	-1,170	50,071
Jul	3,385	4,008	13	-2,456	-11,761	-98	4,929	316	-609	-1	-428	1,726	1,431	-6,206	288	150	-3,543
Aug	-5,282	-2,891	964	5,090	1,938	236	11,598	15	-2,981	4	-645	-1,307	-3,739	-4,265	439	81	12,481
Sep	-903	-753	-958	11,261	3,013	-463	2,713	107	-1,003	-58	-602	-2,608	2,329	2,103	-2,307	-3,132	16,019
Oct	2,114	-2,247	-129	-4,884	-7,246	-98	-2,310	2,709	2,846	35	783	1,101	-8,453	-3,443	267	160	-30,776
Nov	4,456	502	205	3,455	4,290	-369	9,287	1,119	-8,958	30	-162	-1,725	684	12,480	902	170	-25,883
Dec	5,107	2,117	177	-1,842	-1,059	123	5,125	-142	-4,783	18	692	-1,118	-1,828	-29,485	336	313	-25,912
2018 Jan	-889	-1,285	-260	-3,130	4,631	-464	6,150	-3,312	7,620	-33	858	303	4,517	18,461	1,050	624	-30,523
Feb	1,315	-296	880	-4,997	251	-6	2,433	3,342	2,808	-24	-536	-559	-39	-2,326	325	350	19,589
Mar	2,219	-1,427	3,489	-9,614	7,519	-759	13,325	75	-4,152	13	1,446	-1,857	-247	2,398	343	-1,403	-20,251
Apr	3,764	5,536	-4,283	-7,507	160	503	4,363	-1,801	678	-5	264	338	-241	1,060	-1,785	132	14,626
May	-914	-779	-61	-6,772	4,433	-293	1,307	-350	3,748	6	701	-4,566	-600	5,123	599	-561	-24,441
Jun	8,263	10,396	-9	5,780	6,775	-119	23,750	-14,818	-1,964	-27	491	801	2,552	-10,040	506	395	33,615
Jul	-23,182	-16,210	-135	8,086	-3,844	221	7,980	216	-5,672	15	209	-33,996	1,921	2,482	-695	-186	-37,916
Aug	2,050	-942	357	-2,890	-2,817	-28	6,319	-159	979	-17	166	-1,252	485	-5,583	1,099	17	-14,648
Sep	-1,153	536	-504	5,870	6,075	-148	15,412	1,001	-1,310	10	-1,265	4,300	3,499	12,318	-569	-1	39,357
Oct	2,957	-618	32	4,571	-2,172	-6	-4,115	-134	-7,561	-16	847	-1,013	4,821	-3,356	-372	255	-18,839
Nov	2,296	3,029	-110	13,177	7,154	-514	6,531	2,481	685	-6	614	-955	44	-4,650	626	189	19,174
Dec	2,020	-1,244	1,109	11,517	14,703	-342	1,457	-1,099	-6,656	18	-166	-3,688	-2,165	-13,446	-965	437	15,566
2019 Jan	-519	881	-1,065	-6,893	-10,085	-7	11,392	1,051	4,445	-17	758	-305	5,562	15,892	595	392	-3,861
Feb	253	-804	307	-7,386	2,483	54	6,160	-3,823	3,045	1	721	-33	1,487	-1,122	474	12	15,181
Mar	-423	-1,852	-362	8,293	18,494	-252	11,175	2,716	-4,956	36	1,201	763	-1,648	177	-387	203	77,763
Apr	-1,470	1,085	-	-1,297	-10,608	421	-5,854	-1,095	-2,692	3	-258	-496	392	-4,406	2,902	172	-10,465

Source: Bank of England

Notes to table

Explanatory notes can be found here:
https://www.bankofengland.co.uk/statistics/details/further-details-about-monetary-financial-institutions-excluding-central-bank-balance-sheet-data

12.3h Monetary financial institutions' (excluding central bank) balance sheet

£ millions

Not seasonally adjusted

Changes in foreign currency assets (including euro)

| RPM | Loans and advances | | | | | | Claims under sale and repurchase agreements | | | | | Acceptances granted | Total Bills |
| | UK MFIs | of which intragroup banks | UK MFIs' CDs etc. | UK public sector | Other UK residents | Non-residents | UK MFIs | of which intragroup banks | UK public sector | Other UK residents | Non-residents | | |
	B4EB	B8ZR	B4AF	B2VN	B2VT	B2VH	B4FB	B8ZP	B4EG	B4EI	B4FE	B4HM	B4GL
2017 Jan	5,246	6,372	54	-6	9,613	-23,703	4,417	2,846	199	15,213	24,494	-675	3,224
Feb	-7,356	-10,232	-90	28	4,837	41,081	924	-94	-922	17,464	6,937	472	-2,244
Mar	-5,454	-7,270	-74	-24	-7,618	-55,084	5,981	4,291	-882	-4,336	8,385	233	-531
Apr	-2,586	-550	-124	13	8,127	12,602	-3,703	-966	3,086	7,245	-8,239	-335	-2,807
May	-2,467	-83	-85	33	-7,271	-19,194	2,089	-768	-436	2,042	3,871	-415	2,480
Jun	-37,411	-35,608	-19	-	-4,820	8,531	-7,270	-3,117	-1,847	-911	-3,166	150	2,440
Jul	4,587	3,569	116	63	355	32,421	-2,421	-1,575	2,129	-2,778	-3,808	-516	1,417
Aug	1,412	2,071	188	-16	-1,252	12,570	6,171	5,108	-879	-581	-10,731	-300	-1,202
Sep	2,384	1,682	-57	45	-2,554	-653	-1,466	482	377	-4,760	25,803	488	1,491
Oct	-2,908	-2,237	-195	-43	948	6,756	-4,599	-4,282	216	9,656	41,037	-184	-3,777
Nov	-4,654	-2,124	83	61	-3,890	12,696	6,717	5,021	-165	10,504	25,092	197	438
Dec	1,069	1,519	29	-55	-8,418	32,479	-7,299	-5,815	511	-16,751	14,356	523	563
2018 Jan	23,735	20,933	-186	158	16,823	-9,794	146	-1,970	-1,497	26,729	-22,477	-503	1,516
Feb	-2,991	-218	-229	-210	-12,515	-5,744	6,061	4,059	-287	4,848	47,969	363	-2,520
Mar	13,765	13,214	90	3	-1,405	49,519	1,218	-2,366	500	-13,856	10,655	-23	-1,018
Apr	6,795	5,687	29	-23	14,777	4,994	799	1,282	1,361	-10,062	-33,249	477	-1,317
May	11,027	8,282	-79	-24	-8,161	35,093	6,821	6,461	-1,219	24,293	18,117	526	-2,261
Jun	1,091	3,567	56	-28	-1,930	-56,226	3,881	6,771	125	-22,291	-29,658	-367	-1,139
Jul	-12,120	-10,418	93	-29	-4,692	-21,110	-4,789	-3,296	1,431	-1,498	-31,203	23	-239
Aug	8,731	6,981	-37	22	3,968	18,755	-5,532	-2,449	200	2,933	-11,804	462	259
Sep	-14,599	-12,547	-32	11	-9,164	-26,182	4,112	3,771	-429	2,897	6,738	491	3,316
Oct	-11,016	-13,467	-164	1	-6,800	23,310	-3,218	-3,636	-181	4,226	45,386	-526	-2,286
Nov	1,572	425	-89	-15	1,267	-35,027	4,873	3,628	323	23,528	44,205	-338	720
Dec	-9,998	-10,364	-75	13	-2,018	-23,503	-5,864	-3,157	-1,847	-7,081	89,413	-123	-2,457
2019 Jan	-4,294	-8,045	-9	-10	7,959	-19,700	5,702	1,900	947	2,472	-69,966	-60	3,704
Feb	-6,020	-2,883	69	-32	-5,088	-8,494	5,020	2,942	894	-29,323	34,988	-484	355
Mar	-11,827	-12,299	83	-12	616	-9,631	-20,313	-24,379	366	-10,240	8,600	153	-341
Apr	3,682	4,486	100	-6	5,525	-10,989	1,735	3,184	-1,560	5,176	11,763	225	113

12.3h Monetary financial institutions' (excluding central bank) balance sheet

£ millions

Not seasonally adjusted

Changes in foreign currency assets (including euro)

RPM	UK government bonds	Other UK public sector	Investments UK MFIs	Other UK residents	Non-residents	Items in suspense and collection	Accrued amounts receivable	Other assets	Total foreign currency assets	Total assets	Holdings of own sterling acceptances	Holdings of own foreign currency acceptances
	B4EA	B4CG	B4AG	B4CI	B4DE	B4JL	B4FL	B4HL	B4IL	B4IM	B3VR	B3XR
2017 Jan	-30	-	-697	-752	8,164	129,083	356	-2,080	172,122	233,452	4	-73
Feb	-29	-	1,879	217	16,374	18,922	2,936	-1,070	100,362	122,920	-9	183
Mar	74	-	-1,750	125	-19,641	-34,487	-893	-1,203	-117,178	-37,088	-26	333
Apr	-21	-	-48	-830	-68	44,768	1,439	1,377	59,895	79,417	3	-303
May	45	-	146	8	8,389	-13,077	-188	-1,774	-25,802	-6,404	1	305
Jun	-17	-	692	-848	807	-16,641	-161	-1,268	-61,759	-11,688	-15	-160
Jul	2	-	-227	260	-18,032	-24,879	106	5,036	-6,168	-9,711	-3	-179
Aug	-117	-	-712	686	-25,478	10,705	-3,227	-169	-12,932	-452	138	209
Sep	4	-	-382	-213	31,002	10,806	-3,766	-2,418	56,131	72,150	-121	-181
Oct	-103	-	-1,015	-341	3,112	-1,003	-314	2,373	49,618	18,842	11	-57
Nov	4	-	694	183	9,565	44,924	556	-2,695	100,309	74,426	-1	75
Dec	-16	-	1,108	1,407	1,067	-155,719	940	5,445	-128,764	-154,675	-20	338
2018 Jan	14	-	-444	-842	24,210	138,269	-68	-2,275	193,514	162,990	-	-74
Feb	92	-	285	-387	-31,258	-4,134	-638	42	-1,253	18,335	1	24
Mar	-13	-	5	14	-17,458	-41,250	936	2,095	3,777	-16,474	5	500
Apr	3	-	52	-74	9,696	1,301	1,171	-2,309	-5,577	9,049	-2	146
May	-146	-	1,517	678	-31,763	59,853	4,352	735	119,357	94,915	-	-159
Jun	5	-	-228	-411	-1,298	-75,808	-2,151	-1,935	-188,314	-154,699	6	957
Jul	-104	-	35	-145	2,275	7,220	276	-940	-65,514	-103,430	-15	-342
Aug	-3	-	-164	292	-18,065	-12,392	-57	1,036	-11,397	-26,045	20	3
Sep	36	-	-1,610	544	16,205	13,540	-3,535	1,971	-5,689	33,668	9	-438
Oct	-1	-	77	-557	2,768	14,326	-995	-1,504	62,846	44,007	-19	-673
Nov	-99	-	93	394	3,794	8,106	442	973	54,721	73,895	10	252
Dec	102	-	-315	-1,682	-19,859	-123,828	-562	983	-108,702	-93,136	59	418
2019 Jan	-35	-	256	-1,095	8,346	136,722	421	5,046	76,408	72,547	86	-212
Feb	-4	-	371	-250	6,758	-3,765	416	2,375	-2,215	12,966	68	-122
Mar	-3	-	429	149	6,197	-32,491	661	3,241	-64,362	13,402	105	356
Apr	11	-	563	-66	18,822	11,518	909	1,548	49,070	38,604	51	50

Source: Bank of England

Notes to table

Explanatory notes can be found here:
https://www.bankofengland.co.uk/statistics/details/further-details-about-monetary-financial-institutions-excluding-central-bank-balance-sheet-data

Copyright guidance and the related UK Open Government Licence can be viewed here: www.bankofengland.co.uk/legal

12.4a Industrial analysis of monetary financial institutions' lending to UK residents

£ millions

Not seasonally adjusted

Amounts outstanding of lending in sterling

RPM	Agriculture, hunting and forestry	Fishing	Mining and quarrying	Manufacturing									Electricity, gas and water supply			
				Food, beverages and tobacco	Textiles, wearing apparel and leather	Pulp, paper, and printing	Chemicals, pharmaceuticals, rubber and plastics	Non-metallic mineral products and metals	Machinery, equipment and transport equipment	Electrical, medical and optical equipment	Other manufacturing	Total	Electricity, gas steam and air conditioning	Water collection and sewerage	Waste management related services and remediation activities	Total
	TBUC	TBUD	TBUE	TBUG	TBUH	TBUI	TBUJ	TBUK	TBUL	TBUM	TBUN	TBUF	TBUO	TBUP	B3F9	B3FN
2017 Jan	17,698	286	1,812	4,542	721	1,862	4,055	3,218	5,443	1,420	2,487	23,750	9,005	3,442	2,299	14,746
Feb	17,814	285	1,895	4,453	723	1,839	3,570	3,201	5,681	1,541	2,505	23,513	9,207	3,175	2,350	14,733
Mar	17,934	285	1,833	4,297	744	1,865	3,352	3,336	5,814	1,737	2,486	23,631	10,936	2,972	2,230	16,138
Apr	18,048	282	1,849	4,630	733	1,787	3,292	3,340	5,874	1,627	2,615	23,899	11,054	2,982	2,245	16,281
May	18,206	281	1,818	5,697	734	1,948	3,431	3,262	5,601	1,568	2,680	24,920	10,936	2,980	2,261	16,177
Jun	18,351	275	1,765	5,620	742	1,945	3,209	3,235	5,425	1,606	2,711	24,492	11,175	3,011	2,266	16,452
Jul	18,443	272	1,802	8,703	742	1,664	3,252	3,077	5,260	1,450	2,684	26,834	11,385	2,947	2,234	16,566
Aug	18,379	276	1,840	8,690	751	1,707	3,173	3,242	5,070	1,544	2,648	26,825	11,665	2,906	2,199	16,770
Sep	18,402	283	2,082	8,292	793	1,804	3,290	3,698	5,851	2,065	2,731	28,522	10,891	3,177	2,237	16,305
Oct	18,341	282	1,897	8,044	799	1,854	3,434	3,495	5,837	2,331	2,693	28,487	10,950	3,204	2,201	16,355
Nov	18,454	289	1,850	7,962	751	1,741	3,401	3,548	6,029	2,136	2,799	28,366	11,085	3,256	2,210	16,551
Dec	18,095	293	1,792	8,604	686	1,785	3,453	3,473	6,010	1,908	2,741	28,660	11,705	3,145	2,261	17,112
2018 Jan	18,151	289	1,843	8,367	697	1,739	3,277	3,524	5,958	1,969	2,827	28,359	11,591	3,346	2,260	17,196
Feb	18,182	290	1,800	8,361	693	1,732	3,341	3,679	6,335	1,744	2,945	28,831	11,637	3,499	2,306	17,442
Mar	18,320	290	1,806	8,493	699	1,596	3,291	3,866	6,932	1,729	2,696	29,302	11,294	3,273	2,334	16,902
Apr	18,354	280	2,127	8,426	730	1,806	3,426	4,023	7,100	1,617	2,784	29,911	11,545	3,334	2,302	17,181
May	18,549	282	1,952	8,736	724	1,577	3,424	4,139	7,280	1,757	2,781	30,417	11,469	3,372	2,322	17,162
Jun	18,663	284	1,994	8,633	723	1,749	5,878	3,948	7,201	1,745	2,747	32,625	11,670	3,436	2,375	17,481
Jul	18,786	285	2,065	8,741	696	1,795	5,933	3,938	7,096	1,774	2,674	32,647	11,675	3,422	2,469	17,566
Aug	18,854	295	1,989	8,754	637	1,775	5,946	3,923	7,310	1,882	2,689	32,914	11,923	3,527	2,401	17,852
Sep	19,065	314	1,984	8,505	691	1,818	5,935	3,992	6,959	1,950	2,656	32,506	11,942	3,574	2,300	17,817
Oct	18,852	319	2,113	8,583	691	1,848	6,103	3,813	7,119	1,906	2,669	32,732	11,975	3,461	2,317	17,753
Nov	19,041	324	1,873	8,543	665	1,889	6,199	3,781	7,751	2,055	2,687	33,569	11,254	3,697	2,316	17,268
Dec	18,612	336	1,797	9,632	658	1,712	5,937	3,674	7,717	1,901	2,646	33,877	12,333	3,502	2,352	18,186
2019 Jan	18,782	334	1,890	8,999	666	1,603	5,730	3,778	8,371	1,896	2,725	33,769	12,649	3,777	2,460	18,885
Feb	18,808	337	1,987	9,134	637	1,543	5,790	3,900	8,439	1,950	2,798	34,191	12,428	3,927	2,461	18,816
Mar	18,782	322	2,077	9,152	631	1,442	5,658	3,856	7,618	1,978	2,707	33,042	11,859	4,176	2,433	18,468
Apr	18,872	328	2,135	9,412	698	1,541	5,920	3,998	7,727	2,000	2,800	34,097	12,059	3,940	2,362	18,362

12.4a Industrial analysis of monetary financial institutions' lending to UK residents

£ millions

Not seasonally adjusted

Amounts outstanding of lending in sterling

| | | Construction | | | | | Wholesale and retail trade | | | | | Transport, storage and communication | | |
| RPM | Development of buildings | Construction of commercial buildings | Construction of domestic buildings | Civil Engineering | Other construction activities | Total | Wholesale and retail trade and repair of motor vehicles and motorcycles | Wholesale trade, excluding motor vehicles and motor cycles | Retail trade excluding motor vehicles and motor cycles | Total | Accommodation and food service activities | Transportation and storage | Information and communication | Total |
	B7EB	B3I6	B3LX	B4PK	B4PX	TBUQ	TBUS	TBUT	TBUU	TBUR	TBUV	B5PK	B5PR	TBUW
2017 Jan	15,159	4,260	4,974	3,209	5,703	33,304	11,620	10,641	17,019	39,280	23,786	11,044	8,108	19,152
Feb	15,237	4,302	5,285	3,303	5,755	33,881	12,873	10,679	16,738	40,289	24,236	11,241	7,760	19,001
Mar	15,323	4,248	5,475	3,681	5,674	34,400	12,086	10,957	17,049	40,092	24,163	11,526	7,435	18,961
Apr	15,375	4,224	5,552	3,326	5,701	34,178	11,827	10,533	17,010	39,371	24,287	11,477	7,651	19,128
May	15,353	4,201	5,680	3,432	5,586	34,252	11,794	10,348	17,328	39,470	24,022	11,471	7,966	19,437
Jun	15,273	4,094	5,122	3,357	5,638	33,483	12,280	9,826	17,371	39,478	24,153	11,354	8,198	19,553
Jul	15,443	4,115	4,946	3,406	5,743	33,653	12,065	10,062	17,248	39,374	24,197	11,083	8,736	19,819
Aug	15,260	4,237	4,988	3,042	5,821	33,348	12,475	10,099	17,535	40,109	24,377	10,991	8,773	19,764
Sep	15,343	4,291	5,129	3,163	5,878	33,804	12,174	10,771	18,520	41,465	24,287	10,959	9,548	20,506
Oct	15,245	4,294	5,110	3,086	5,752	33,487	11,856	10,879	18,489	41,223	24,629	11,432	9,630	21,062
Nov	15,213	4,419	5,425	3,136	5,639	33,832	11,773	11,134	17,920	40,827	24,784	11,073	9,340	20,413
Dec	14,865	4,240	4,982	2,875	5,593	32,555	11,851	11,068	17,957	40,876	25,210	11,606	9,081	20,687
2018 Jan	14,919	4,110	5,297	2,894	5,671	32,892	11,906	11,129	18,160	41,195	24,967	11,692	8,902	20,594
Feb	15,044	4,121	5,492	3,245	5,722	33,624	12,362	11,262	19,037	42,662	25,013	11,823	8,602	20,426
Mar	15,210	3,982	5,594	2,856	5,686	33,328	11,736	11,233	19,701	42,670	24,868	11,945	8,685	20,631
Apr	15,320	4,017	5,458	2,830	5,694	33,319	11,597	11,345	19,383	42,325	24,698	12,310	8,551	20,861
May	15,193	3,911	5,585	2,925	5,646	33,260	11,467	11,267	18,980	41,714	24,731	12,194	8,602	20,797
Jun	15,012	3,770	5,348	2,820	5,634	32,584	11,316	10,908	19,559	41,783	24,921	11,582	8,936	20,518
Jul	14,412	3,874	5,237	3,028	5,700	32,250	11,159	11,089	19,384	41,632	25,778	12,041	8,714	20,755
Aug	14,206	3,987	5,176	2,940	5,613	31,922	10,971	11,086	19,326	41,383	25,346	12,141	8,610	20,751
Sep	14,280	3,893	5,308	2,808	5,653	31,942	10,862	11,340	20,082	42,284	25,512	12,469	8,433	20,902
Oct	14,211	3,934	5,436	2,806	5,714	32,101	10,776	11,569	19,859	42,204	25,240	12,376	8,416	20,792
Nov	14,091	3,926	5,548	2,864	5,714	32,143	10,993	11,421	19,740	42,155	25,329	12,568	9,282	21,851
Dec	13,965	3,650	5,054	2,595	5,588	30,852	11,615	11,153	19,701	42,469	25,690	12,178	9,056	21,233
2019 Jan	14,172	3,718	5,358	2,628	5,621	31,497	11,841	11,687	19,979	43,508	25,929	12,677	9,162	21,839
Feb	14,258	3,726	5,501	2,682	5,645	31,812	12,338	11,535	20,326	44,199	25,680	12,564	8,816	21,379
Mar	14,323	3,702	5,597	2,670	5,509	31,802	12,191	11,208	20,903	44,302	25,541	12,357	9,112	21,469
Apr	13,952	3,702	5,920	2,479	5,486	31,538	9,180	11,539	19,837	40,555	25,430	12,344	9,517	21,861

12.4a Industrial analysis of monetary financial institutions' lending to UK residents

£ millions

Amounts outstanding of lending in sterling

| | Real estate, professional services and support activities | | | | Public administration and defence | Education | Human health and social work | Recreational, personal and community service activities | | | Financial intermediation (excluding insurance and pension funds) | | | | |
| | Buying, selling and renting of real estate | Professional, scientific and technical activities | Administrative and support services | Total | | | | Recreational, cultural and sporting activities | Personal and community service activities | Total | Financial leasing corporations | Non-bank credit grantors excluding credit unions and SPVs | Credit unions | Factoring corporations | Mortgage and housing credit corporations excluding SPVs |
RPM	TBUY	B6PD	B6PO	TBUX	TBVD	TBVE	TBVF	TBVH	TBVG	B6PT	TBVJ	TBVK	TBVL	TBVM	TBVN
2017 Jan	134,203	16,154	24,081	174,439	10,078	10,351	19,976	5,311	3,297	8,609	24,880	23,495	1	5,257	48,768
Feb	134,408	16,014	24,270	174,691	8,122	10,321	20,005	5,265	3,206	8,470	24,752	19,841	1	5,579	43,605
Mar	134,749	15,821	24,320	174,890	8,988	10,330	20,183	5,517	3,358	8,874	24,700	20,329	1	5,825	47,639
Apr	135,473	15,706	25,383	176,562	8,284	10,364	19,867	5,617	3,431	9,049	24,322	20,090	1	5,879	47,634
May	135,671	15,489	25,460	176,620	7,700	10,212	20,002	5,678	3,415	9,093	24,927	19,851	1	5,794	46,339
Jun	135,435	15,680	25,723	176,837	7,729	10,185	20,081	5,504	3,444	8,948	25,749	22,746	1	6,217	46,276
Jul	134,246	15,518	26,218	175,981	7,935	10,240	20,034	5,483	3,351	8,834	25,927	23,155	1	6,157	46,465
Aug	133,369	15,823	25,738	174,930	8,470	10,271	20,292	5,426	3,345	8,772	26,331	24,055	2	6,190	46,482
Sep	134,939	16,515	26,292	177,746	8,274	10,348	19,800	5,522	3,326	8,848	27,367	23,539	2	6,451	46,829
Oct	133,858	15,597	26,888	176,343	8,153	10,169	20,061	5,603	3,227	8,830	26,587	23,664	2	6,254	44,934
Nov	134,309	15,764	26,587	176,660	8,095	10,233	20,230	5,594	3,193	8,787	26,827	23,472	2	6,436	45,070
Dec	134,668	15,878	25,329	175,875	8,418	10,254	19,707	5,627	3,249	8,876	27,157	23,877	2	6,247	44,870
2018 Jan	134,841	15,652	26,085	176,578	8,023	10,077	19,870	5,494	3,139	8,632	27,547	24,733	2	5,765	45,284
Feb	134,773	15,843	25,869	176,485	8,935	9,966	20,095	5,486	3,198	8,684	31,081 (b)	24,745	2	9,929 (c)	45,214
Mar	135,319	15,372	25,787	176,479	12,216	10,068	19,944	5,244	3,195	8,438	31,964	25,144	2	9,837	45,528
Apr	135,555	14,853	25,945	176,352	8,361	9,980	20,059	4,964	3,181	8,145	32,549	25,638	2	9,565	46,396
May	136,721	14,608	25,664	176,993	7,798	9,970	20,125	4,930	3,159	8,089	34,139	25,932	2	9,261	47,647
Jun	135,900	15,551	25,981	177,433	7,873	10,006	19,917	5,053	3,126	8,179	34,882	26,523	2	9,540	46,227
Jul	136,263	15,208	26,824 (a)	178,295 (a)	7,829	10,156	19,975	5,025	3,196	8,222	38,917	28,132	2	9,264	46,707
Aug	137,885	15,383	26,819	180,087	8,310	10,063	20,029	4,979	3,124	8,104	38,937	28,545	2	9,275	47,714
Sep	139,191	15,664	26,556	181,410	7,636	10,094	19,973	4,854	3,139	7,992	39,130	29,588	2	9,816	45,321
Oct	139,291	15,596	26,739	181,626	7,792	9,791	20,216	4,892	3,188	8,080	38,338	29,357	2	9,836	45,720
Nov	139,610	15,944	28,196	183,750	7,214	9,909	20,089	4,864	3,266	8,130	39,080	29,414	2	9,836	46,420
Dec	140,273	15,880	27,401	183,554	7,960	9,855	19,829	4,636	3,228	7,864	42,333	29,829	2	9,632	46,571
2019 Jan	139,584	16,736	27,461	183,781	7,251	9,631	20,004	4,714	3,159	7,873	42,600	30,939	2	9,232	48,105
Feb	139,980	16,732	27,622	184,335	7,480	9,544	20,074	4,825	3,263	8,088	43,141	30,004	2	9,560	48,454
Mar	141,200	16,643	28,317	186,160	6,855	9,615	19,916	4,815	3,285	8,100	43,218	31,425	5	9,792	49,036
Apr	141,595	16,444	27,703	185,743	7,375	9,600	19,988	4,837	3,300	8,136	43,869	30,980	4	8,733	49,704

12.4a Industrial analysis of monetary financial institutions' lending to UK residents

£ millions

Not seasonally adjusted

Amounts outstanding of lending in sterling

RPM	Investment and unit trusts excluding money market mutual funds	Money market mutual funds	Bank holding companies	Financial intermediation (excluding insurance and pension funds) Venture and development capital	Securities dealers	SPVs related to securitisation	Other financial intermediaries	of which intragroup activity	Total	Insurance companies & pension funds Insurance companies	Pension funds	Total
	TBVO	TBVP	TBVQ	ZO8X	TBVR	B6PY	TBVS	B3U8	TBVI	B3V6	B3W4	TBVT
2017 Jan	9,703	119	18,859		21,908	9,771	37,996	29,801	200,758	8,460	15,742	24,202
Feb	11,085	101	18,296		22,155	9,503	38,103	29,814	193,022	8,162	16,744	24,906
Mar	9,597	119	13,376		20,107	9,291	38,267	31,501	189,250	8,713	25,179	33,892
Apr	9,735	92	13,509		21,308	10,417	38,386	31,906	191,372	9,898	27,311	37,209
May	9,440	92	14,502		22,246	9,895	37,688	31,533	190,774	9,092	27,494	36,586
Jun	10,219	102	14,146		21,681	7,876	39,251	33,345	194,264	8,358	26,652	35,010
Jul	10,599	120	14,292		22,749	8,458	39,968	32,812	197,892	7,726	24,682	32,409
Aug	10,528	149	14,152		22,081	7,663	39,104	31,792	196,736	8,844	24,580	33,424
Sep	11,513	130	14,371		22,450	7,836	38,912	31,993	199,400	7,676	26,118	33,795
Oct	11,490	123	11,027		24,525	8,712	37,867	30,940	195,184	7,232	24,656	31,888
Nov	11,629	111	11,751		24,345	8,705	37,560	30,363	195,908	7,674	25,778	33,452
Dec	9,470	107	11,122		24,342	8,323	42,059	34,562	197,575	7,900	25,286	33,186
2018 Jan	9,828	112	11,115		25,588	7,055	43,065	34,389	200,093	7,384	24,961	32,346
Feb	10,111	109	11,050		23,526	7,012	40,598 (e)	31,845 (e)	203,377	7,106	25,507	32,613
Mar	10,180	101	11,216		26,414	7,330	45,589	36,619	213,306	7,921	25,151	33,072
Apr	9,874	126	11,257		24,717	7,101	47,471	38,772	214,697	7,991	24,772	32,763
May	11,467	158	8,667		23,207	7,701	39,678	30,403	207,859	7,779	27,702	35,481
Jun	11,686	163	8,738		24,928	8,086	51,938	42,623	222,713	7,985	27,941	35,926
Jul	11,841	165	8,706		21,005	7,890	50,403	41,545	223,031	7,403	28,229	35,632
Aug	11,755	319	9,781		18,610	8,016	50,394	41,590	223,348	7,195	28,065	35,261
Sep	13,658	377	16,406		21,433	8,213	50,798	41,553	234,743	6,981	29,935	36,916
Oct	14,063	276	9,473		18,332	8,319	50,930	40,453	224,646	6,770	30,207	36,977
Nov	13,751	267	10,981		19,268	7,792	47,802	36,376	224,612	5,974	30,370	36,345
Dec	14,009	269	8,050	#	19,173	6,948	48,056	37,170	224,872	6,275	31,768	38,044
2019 Jan	14,141	268	7,401	1,371 (d)	16,877	7,180	53,296 (f)	41,937	231,413	6,276	31,172	37,449
Feb	13,776	204	7,604	1,412	15,875	7,210	52,227 (g)	42,773	229,470 (g)	5,605	30,919	36,524
Mar	13,876	264	8,150	1,600	15,055	7,607	52,357	42,272	232,385	6,865	31,947	38,812
Apr	13,578	271	8,345	1,626	18,003	10,049	47,256	37,340	232,417	5,672	30,388	36,060

12.4a Industrial analysis of monetary financial institutions' lending to UK residents

£ millions

Not seasonally adjusted

Amounts outstanding of lending in sterling

RPM	Fund management activities	Activities auxiliary to financial intermediation			Total	Total financial and non-financial businesses	Individuals and individual trusts			Total UK residents
		Central clearing counterparties	Other auxiliary activities	Total			Lending secured on dwellings inc. bridging finance	Other loans and advances	Total	
	TBVU	B7FX	B8FD	TBVV	B8FJ	Z949	TBVX	TBVY	TBVW	TBUA
2017 Jan	113,025	63,583	3,017	66,600	179,624	801,852	1,152,375	130,246	1,282,620	2,084,472
Feb	113,115	65,067	4,066	69,133	182,248	797,433	1,154,709	130,336	1,285,045	2,082,478
Mar	116,626	70,283	3,434	73,718	190,344	814,189	1,159,064	130,926	1,289,991	2,104,180
Apr	111,909	73,265	3,301	76,566	188,475	818,504	1,160,471	131,401	1,291,872	2,110,376
May	116,118	71,979	3,038	75,017	191,134	820,704	1,164,418	131,950	1,296,369	2,117,073
Jun	116,430	73,111	3,054	76,164	192,595	823,650	1,171,109	132,283	1,303,392	2,127,042
Jul	112,026	72,991	2,941	75,932	187,959	822,244	1,174,102	132,783	1,306,884	2,129,129
Aug	117,100	78,205	3,088	81,293	198,393	832,976	1,179,193	133,166	1,312,359	2,145,334
Sep	121,009	85,733	3,080	88,813	209,822	853,687	1,189,317	133,106	1,322,423	2,176,109
Oct	121,206	82,169	2,670	84,838	206,044	842,435	1,192,030	133,281	1,325,311	2,167,746
Nov	121,210	86,137	3,215	89,352	210,562	849,293	1,196,335	134,181	1,330,516	2,179,809
Dec	131,312	75,896	3,081	78,977	210,288	849,461	1,199,156	134,123	1,333,279	2,182,739
2018 Jan	123,534	81,605	2,576	84,181	207,715	848,821	1,200,878	133,319	1,334,197	2,183,018
Feb	115,693	78,139	2,532	80,671	196,364	844,787	1,201,886	133,985	1,335,871	2,180,658
Mar	119,057	64,636	2,638	67,273	186,330	847,968	1,206,815	133,294	1,340,109	2,188,078
Apr	111,475	65,009	2,641	67,649	179,125	838,539	1,207,971	133,938	1,341,908	2,180,447
May	105,860	64,895	2,552	67,447	173,307	828,486	1,210,850	134,306	1,345,156	2,173,642
Jun	107,898	68,956	2,010	70,967	178,865	851,763	1,215,269	134,875	1,350,143	2,201,906
Jul	109,796	77,647	2,009	79,656	189,451	864,357 (a)	1,218,203	135,206	1,353,409 (a)	2,217,767 (a)
Aug	109,001	75,482	2,091	77,573	186,574	863,080	1,222,433	135,562	1,357,994	2,221,075
Sep	111,568	81,316 (i)	2,253	83,569 (i)	195,137 (i)	886,227 (i)	1,224,137	136,485	1,360,622	2,246,849 (i)
Oct	116,607	79,379	2,687	82,066	198,673	879,908	1,230,984 (j)	135,989	1,366,973 (j)	2,246,881 (j)
Nov	120,577	87,811	2,083	89,894	210,471	894,073	1,234,826	136,581	1,371,406	2,265,479
Dec	123,402	91,873	1,770	93,643	217,045	902,074	1,237,320	136,726	1,374,046	2,276,120
2019 Jan	114,883	93,462	1,732	95,194	210,077	903,910	1,239,494	136,090	1,375,584	2,279,494
Feb	115,037 (h)	90,330	1,710	92,041	207,078 (h)	899,800 (g)(h)	1,241,261 (k)	135,256 (l)	1,376,516 (k)(l)	2,276,317 (g)(h)(k)(l)
Mar	117,827	97,665	1,722	99,388	217,215	914,862	1,244,549	134,846	1,379,395	2,294,257
Apr	116,304	92,184	1,662	93,846	210,150	902,646	1,246,696	135,529	1,382,225	2,284,872

Source: Bank of England

12.4a Industrial analysis of monetary financial institutions' lending to UK residents

£ millions

Notes to table

Movements in amounts outstanding can reflect breaks in data series as well as underlying flows. For changes data, users are recommended to refer directly to the appropriate series or data tables. https://www.bankofengland.co.uk/statistics/details/further-details-about-changes-flows-growth-rates-data.

(a) Due to improvements in reporting at one institution, the amounts outstanding increased by £1bn. This effect has been adjusted out of the flows for July 2018.

(b) Due to improvements in reporting at one institution, the amounts outstanding increased by £3bn. This effect has been adjusted out of the flows for February 2018.

(c) Due to improvements in reporting at one institution, the amounts outstanding increased by £4bn. This effect has been adjusted out of the flows for February 2018.

(d) Previously, this series formed part of other financial intermediaries.

(e) Due to improvements in reporting at one institution, the amounts outstanding decreased by £6bn. This effect has been adjusted out of the flows for February 2018.

(f) From January 2019 onwards, this series no longer includes venture and development capital. As result, this series decreased by £1bn. This effect has been adjusted out the flows for January 2019

(g) Due to improvements in reporting at one institution, the amounts outstanding decreased by £2bn. This effect has been adjusted out of the flows for February 2019.

(h) Due to improvements in reporting at one institution, the amounts outstanding increased by £2bn. This effect has been adjusted out of the flows for February 2019.

(i) Due to improvements in reporting at one institution, the amounts outstanding increased by £5bn. This effect has been adjusted out of the flows for September 2018.

(j) Due to improvements in reporting at one institution, the amounts outstanding increased by £2bn. This effect has been adjusted out of the flows for October 2018.

(k) Due to improvements in reporting at one institution, the amounts outstanding increased by £1bn. This effect has been adjusted out of the flows for February 2019.

(l) Due to improvements in reporting at one institution, the amounts outstanding decreased by £1bn. This effect has been adjusted out of flows for February 2019.

Explanatory notes can be found here: http://www.bankofengland.co.uk/statistics/Pages/iadb/notesiadb/industrial.aspx

Copyright guidance and the related UK Open Government Licence can be viewed here: www.bankofengland.co.uk/legal

12.4b Industrial analysis of monetary financial institutions' lending to UK residents

£ millions

Not seasonally adjusted

Amounts outstanding of lending in all currencies

RPM	Agriculture, hunting and forestry	Fishing	Mining and quarrying	Manufacturing									Electricity, gas and water supply			Total
				Food, beverages and tobacco	Textiles, wearing apparel and leather	Pulp, paper, and printing	Chemicals, pharmaceuticals, rubber and plastics	Non-metallic mineral products and metals	Machinery, equipment and transport equipment	Electrical, medical and optical equipment	Other manufacturing	Total	Electricity, gas steam and air conditioning	Water collection and sewerage	Waste management related services and remediation activities	
	TBSC	TBSD	TBSE	TBSG	TBSH	TBSI	TBSJ	TBSK	TBSL	TBSM	TBSN	TBSF	TBSO	TBSP	B3FA	B3FO
2017 Jan	17,924	309	10,251	6,655	1,179	2,463	7,193	4,685	8,018	2,850	3,233	36,276	9,580	3,443	2,462	15,486
Feb	18,030	309	9,886	6,990	1,114	2,502	6,791	4,730	8,257	2,805	3,256	36,446	9,748	3,177	2,365	15,289
Mar	18,159	308	9,613	6,448	1,134	2,483	7,198	5,206	8,042	2,868	3,416	36,796	11,516	2,973	2,323	16,813
Apr	18,295	304	7,212	6,852	1,130	2,393	7,133	4,787	8,380	2,835	3,382	36,892	11,644	2,983	2,349	16,975
May	18,472	303	7,184	8,074	1,183	2,577	7,201	4,843	8,099	2,794	3,467	38,238	11,480	2,981	2,391	16,852
Jun	18,566	297	7,057	7,723	1,261	2,599	11,930	4,834	7,843	2,779	3,486	42,454	11,749	3,012	2,373	17,134
Jul	18,653	294	6,961	15,443	1,206	2,174	11,865	4,705	8,192	2,553	3,576	49,714	11,997	2,948	2,322	17,267
Aug	18,600	298	7,244	11,078	1,254	2,290	9,687	4,909	7,600	2,692	3,496	43,006	12,259	2,907	2,286	17,452
Sep	18,628	304	8,021	10,569	1,338	2,318	9,229	4,949	7,967	3,036	3,609	43,015	11,428	3,178	2,323	16,929
Oct	18,551	303	8,592	10,389	1,355	2,316	9,174	4,652	8,493	3,342	3,456	43,178	11,518	3,205	2,257	16,981
Nov	18,655	310	7,882	10,187	1,509	2,165	9,307	4,640	8,749	3,046	3,526	43,129	11,631	3,258	2,239	17,128
Dec	18,290	314	7,791	10,850	1,277	2,212	9,575	4,739	8,181	2,790	3,419	43,043	12,307	3,146	2,273	17,727
2018 Jan	18,337	308	7,482	11,310	1,167	2,172	8,967	4,698	8,386	2,914	3,529	43,143	12,222	3,347	2,296	17,864
Feb	18,381	311	8,133	10,577	1,165	2,133	8,777	4,786	9,137	2,712	3,669	42,957	12,418	3,508	2,319	18,245
Mar	18,520	309	8,329	10,261	1,255	2,162	8,502	5,160	9,527	2,637	3,426	42,930	12,181	3,283	2,374	17,838
Apr	18,549	300	8,210	10,329	1,258	2,248	8,152	5,268	9,645	2,773	3,595	43,268	12,381	3,343	2,386	18,110
May	18,745	303	8,745	10,447	1,248	2,118	8,534	5,384	9,931	2,768	3,594	44,024	12,213	3,382	2,358	17,953
Jun	18,867	305	8,978	10,479	1,264	2,222	10,461	5,372	10,091	2,637	3,526	46,053	12,428	3,438	2,400	18,266
Jul	18,983	306	8,075	10,364	1,192	2,384	10,519	5,429	10,183	2,815	3,400	46,285	12,453	3,424	2,978	18,854
Aug	19,045	330	8,054	10,627	1,210	2,432	10,578	5,362	10,426	2,932	3,447	47,013	12,603	3,529	2,905	19,037
Sep	19,284	353	8,048	10,433	1,234	2,460	10,383	5,346	10,279	3,005	3,430	46,569	12,663	3,576	2,800	19,039
Oct	19,071	355	8,340	10,362	1,220	2,272	10,523	5,111	10,785	2,993	3,294	46,561	12,784	3,463	2,849	19,096
Nov	19,247	360	7,716	10,517	1,183	2,423	10,627	5,008	11,309	3,250	3,316	47,633	12,027	3,699	2,834	18,560
Dec	18,822	371	7,480	11,306	1,177	2,218	11,124	4,727	11,432	3,063	3,314	48,362	12,978	3,502	2,834	19,314
2019 Jan	18,964	365	6,946	10,612	1,167	2,435	13,035	4,698	11,993	3,146	3,348	50,433	13,354	3,777	3,103	20,233
Feb	18,992	367	6,698	10,748	1,098	2,300	12,344	4,926	11,766	3,317	3,340	49,840	13,041	3,927	3,116	20,085
Mar	18,962	354	7,512	10,464	1,111	2,276	10,460	4,857	11,713	3,048	3,372	47,302	12,472	4,176	3,032	19,681
Apr	19,048	360	7,332	11,107	1,185	2,320	11,304	5,083	11,750	3,217	3,723	49,689	12,855	3,940	3,011	19,806

12.4b Industrial analysis of monetary financial institutions' lending to UK residents

£ millions

Not seasonally adjusted

Amounts outstanding of lending in all currencies

RPM	Development of buildings	Construction					Wholesale and retail trade				Accommodation and food service activities	Transport, storage and communication		
		Construction of commercial buildings	Construction of domestic buildings	Civil Engineering	Other construction activities	Total	Wholesale and retail trade and repair of motor vehicles and motorcycles	Wholesale trade, excluding motor vehicles and motor cycles	Retail trade excluding motor vehicles and motor cycles	Total		Transportation and storage	Information and communication	Total
	B7EC	B3I7	B3LY	B4PL	B4PY	TBSQ	TBSS	TBST	TBSU	TBSR	TBSV	B5PL	B5PS	TBSW
2017 Jan	15,317	4,402	4,974	3,571	5,774	34,039	12,385	16,202	19,294	47,882	24,346	15,139	13,526	28,665
Feb	15,402	4,449	5,286	3,663	5,829	34,628	13,600	16,373	19,120	49,094	24,912	15,364	13,095	28,458
Mar	15,488	4,403	5,475	4,047	5,746	35,160	12,842	16,843	19,477	49,162	24,884	15,667	12,552	28,220
Apr	15,534	4,368	5,552	3,675	5,785	34,915	12,650	17,947	19,122	49,720	24,968	15,468	12,698	28,166
May	15,520	4,342	5,680	3,823	5,675	35,040	12,617	19,087	19,437	51,141	25,042	15,498	12,930	28,428
Jun	15,456	4,248	5,122	3,708	5,717	34,250	13,086	18,738	19,641	51,465	25,155	14,849	12,984	27,833
Jul	15,632	4,254	4,946	3,704	5,833	34,369	12,900	19,319	19,646	51,865	25,134	14,858	13,620	28,478
Aug	15,454	4,375	4,988	3,337	5,905	34,059	13,300	19,127	20,305	52,732	25,282	14,808	14,107	28,914
Sep	15,546	4,413	5,129	3,433	5,937	34,459	13,011	18,542	21,145	52,698	25,183	14,834	14,734	29,568
Oct	15,435	4,469	5,111	3,481	5,795	34,291	12,653	19,529	21,130	53,311	25,527	15,645	14,705	30,349
Nov	15,412	4,587	5,434	3,544	5,681	34,657	12,517	18,777	20,330	51,625	25,679	15,200	13,983	29,182
Dec	15,069	4,359	4,991	3,269	5,631	33,320	12,623	19,038	20,261	51,922	26,356	15,452	13,919	29,371
2018 Jan	15,112	4,215	5,306	3,267	5,706	33,605	12,699	18,507	19,987	51,192	26,294	15,536	13,470	29,006
Feb	15,383	4,234	5,501	3,640	5,796	34,553	13,142	18,834	21,122	53,097	26,661	15,661	13,030	28,691
Mar	15,555	4,129	5,602	3,232	5,744	34,263	12,499	19,448	21,595	53,543	26,316	15,580	12,747	28,327
Apr	15,667	4,163	5,467	3,191	5,761	34,248	12,565	20,553	21,274	54,392	26,299	15,519	12,573	28,093
May	15,538	4,064	5,593	3,295	5,698	34,188	12,442	20,363	20,607 (s)	53,413 (s)	26,742	15,219	13,140	28,359
Jun	15,489	3,889	5,356	3,159	5,682	33,575	12,291	19,661	21,168	53,119	26,518	14,435	13,488	27,923
Jul	14,888	3,998	5,245	3,343	5,755	33,229	12,104	20,726	20,611	53,441	27,195	14,922	13,175	28,097
Aug	14,547	4,101	5,185	3,248	5,672	32,753	11,913	20,699	20,560	53,172	26,794	15,020	13,224	28,244
Sep	14,610	3,995	5,315	3,089	5,723	32,733	11,634	19,509	21,607	52,750	26,391	15,254	12,794	28,048
Oct	14,537	4,036	5,478	3,089	5,805	32,945	11,581	19,597	21,387	52,566	26,086	15,339	13,133	28,472
Nov	14,410	4,035	5,590	3,128	5,775	32,938	11,792	20,060	21,353	53,204	26,083	15,648	13,845	29,493
Dec	14,266	3,767	5,095	2,813	5,602	31,544	12,411	19,825	21,182	53,419	26,654	14,987	13,094	28,081
2019 Jan	14,461	3,817	5,398	2,867	5,636	32,179	12,657	20,326	21,421	54,404	26,910	15,565	13,311	28,876
Feb	14,545	3,825	5,541	2,903	5,658	32,473	13,126	20,266	21,516	54,907	27,085	15,358	13,087	28,446
Mar	14,612	3,809	5,638	2,879	5,535	32,474	12,902	19,752	22,620	55,274	26,750	15,171	12,920	28,091
Apr	14,254	3,793	5,968	2,699	5,527	32,242	10,016	19,829	21,639	51,485	26,808	15,097	13,402	28,499

12.4b Industrial analysis of monetary financial institutions' lending to UK residents

£ millions

Not seasonally adjusted

Amounts outstanding of lending in all currencies

RPM	Real estate, professional services and support activities				Public administration and defence	Education	Human health and social work	Recreational, personal and community service activities			Financial intermediation (excluding insurance and pension funds)			
	Buying, selling and renting of real estate	Professional, scientific and technical activities	Administrative and support services	Total				Recreational, cultural and sporting activities	Personal and community service activities	Total	Financial leasing corporations	Non-bank credit grantors excluding credit unions and SPVs	Credit unions	Factoring corporations
	TBSY	B6PE	B6PP	TBSX	TBTD	TBTE	TBTF	TBTH	TBTG	B6H5	TBTJ	TBTK	TBTL	TBTM
2017 Jan	134,831	20,771	29,994	185,596	12,073	10,403	20,179	6,095	3,411	9,506	31,054	23,942	1	6,265
Feb	135,104	20,639	29,909	185,652	9,222	10,375	20,209	6,077	3,324	9,401	30,586	20,243	1	6,462
Mar	135,514	20,248	29,252	185,015	9,196	10,379	20,405	6,284	3,476	9,760	30,473	20,764	1	6,763
Apr	136,035	20,085	29,712	185,832	11,595	10,410	20,059	6,368	3,600	9,968	29,649	20,501	2	6,684
May	136,257	20,002	30,772	187,031	10,662	10,238	20,195	6,469	3,589	10,058	30,381	20,299	1	6,765
Jun	135,998	20,251	30,365	186,614	8,862	10,216	20,297	6,289	3,603	9,891	31,198	23,547	2	7,653
Jul	134,881	19,232	30,956	185,069	11,246	10,277	20,235	6,342	3,493	9,835	31,500	23,957	2	7,248
Aug	133,929	19,372	30,662	183,962	10,935	10,332	20,489	6,310	3,470	9,780	32,202	24,891	2	7,246
Sep	135,555	19,830	31,237	186,622	11,058 (bc)	10,412	19,940	6,358	3,469	9,828	32,971	24,392	2	7,351
Oct	134,417	19,071	31,764	185,252	11,100	10,233	20,186	6,478	3,302	9,781	32,684	24,485	2	7,259
Nov	134,816	19,533	31,591	185,940	10,926	10,298	20,409	6,442	3,266	9,708	32,688	24,279	2	7,283
Dec	135,140	19,499	30,078	184,718	11,724	10,310	19,902	6,689	3,347	10,035	32,699	24,754	2	7,318
2018 Jan	135,284	19,093	30,845	185,222	9,940	10,141	20,035	6,428	3,186	9,614	32,905	25,538	2	6,644
Feb	135,249	19,153	30,988	185,389	10,436	10,028	20,279	6,477	3,229	9,707	37,570 (v)	25,607	3	10,925 (w)
Mar	135,805	19,261	31,175	186,240	14,171	10,132	20,238	6,206	3,233	9,439	38,293	25,996	3	11,103
Apr	136,105	19,365	31,402	186,872	11,705	10,067	20,254	5,946	3,212	9,159	38,760	26,491	2	11,209
May	137,289	19,801	30,731 (t)	187,821 (t)	9,942	10,034	20,348	5,667	3,203	8,869	40,346	26,746	2	11,017
Jun	136,317	20,834	30,927	188,077	10,130	10,076	20,129	5,857	3,162	9,019	41,177	27,423	2	11,355
Jul	136,856	20,565	31,760 (u)	189,181 (u)	11,576	10,248	20,170	5,846	3,228	9,073	45,076	29,125	2	11,114
Aug	138,670	20,908	32,008	191,585	12,340	10,151	20,217	5,784	3,153	8,936	45,085	29,466	2	11,092
Sep	139,940	20,865	31,236	192,040	11,146	10,157	20,136	5,619	3,159	8,778	45,138	30,451	3	11,772
Oct	139,938	21,008	31,779	192,725	11,219	9,904	20,445	5,596	3,217	8,813	44,319	30,213	2	11,638
Nov	140,419	21,311	33,674	195,404	10,982	10,079	20,328	5,593	3,311	8,904	45,141	30,169	2	11,651
Dec	140,897	20,976	31,445	193,319	9,935	9,978	20,071	5,154	3,328	8,481	48,519	30,598	2	11,566
2019 Jan	140,220	21,566	32,025	193,811	10,066	9,696	20,215	5,225	3,217	8,442	48,845	31,696	2	10,956
Feb	140,574	21,349	32,275	194,198	11,103	9,607	20,360	5,346	3,295	8,641	49,241	30,739	2	10,894
Mar	141,820	21,340	32,578	195,738	10,911	9,684	20,150	5,306	3,331	8,637	49,379	32,123	5	10,995
Apr	142,294	21,014	32,195	195,503	9,811	9,641	20,219	5,338	3,336	8,675	49,995	31,679	4	9,892

12.4b Industrial analysis of monetary financial institutions' lending to UK residents

£ millions

Amounts outstanding of lending in all currencies

Not seasonally adjusted

RPM	Mortgage and housing credit corporations excluding SPVs	Financial intermediation (excluding insurance and pension funds) (contd.)									Insurance companies & pension funds			Activities auxiliary to financial intermediation
		Investment and unit trusts excluding money market mutual funds	Money market mutual funds	Bank holding companies	Venture and development capital	Securities dealers	SPVs related to securitisation	Other financial intermediaries	of which intragroup activity	Total	Insurance companies	Pension funds	Total	Fund management activities
	TBTN	TBTO	TBTP	TBTQ	ZO8Z	TBTR	B6PZ	TBTS	B3U9	TBTI	B3V7	B3W5	TBTT	TBTU
2017 Jan	49,258	14,204	180	26,225		136,033	12,015	83,163	67,279	382,340	13,499	16,980	30,478	175,239
Feb	44,089	15,477	162	25,702		136,382	11,669	95,627	72,332	386,399	13,394	17,927	31,321	183,337
Mar	48,131	13,974	191	20,854		137,486	11,425	89,378	74,340	379,440	13,785	26,393	40,178	187,607
Apr	48,084	13,651	188	20,756		142,980	12,463	88,948	73,995	383,906	15,502	28,661	44,163	185,279
May	46,811	13,720	212	21,702		140,372	12,021	88,200	74,659	380,484	14,404	28,680	43,083	187,104
Jun	46,761	14,658	188	21,310		131,558	10,042	86,645	73,758	373,560	13,462	27,507	40,968	194,784
Jul	46,956	15,072	224	21,404		128,452	10,645	87,956	73,364	373,416	12,686	25,957	38,642	186,471
Aug	46,995	14,901	274	21,241		125,492	9,307	88,294	73,245	370,847	13,969	26,006	39,975	193,671
Sep	47,303	15,785	267	21,054		120,283	9,517	87,209	73,178	366,135	13,015	27,490	40,504	188,692
Oct	45,403	15,715	250	18,104		124,077	9,533	86,576	72,930	364,087	12,179	26,099	38,278	189,450
Nov	45,547	15,918	254	19,289		130,532	9,537	87,077	73,213	372,406	12,850	27,163	40,013	191,202
Dec	45,348	13,327	151	17,881		121,278	9,401	90,102	76,781	362,260	12,868	26,510	39,378	203,306
2018 Jan	45,737	14,643	182	16,190		141,768	7,668	91,256	75,019	382,533	12,515	26,323	38,837	199,949
Feb	45,752	14,747	189	16,129		134,575	7,739	79,694 (z)	63,318 (z)	372,930	11,710	27,053	38,762	197,567
Mar	45,994	14,441	199	16,322		129,568	8,484	83,026	67,322	373,427	12,423	27,268	39,690	200,006
Apr	46,786	14,112	220	16,361		133,327	8,341	88,712	73,083	384,321	12,024	26,199	38,223	187,702
May	48,052	17,105	279	13,836		126,542	9,154	90,674	74,182	383,752	11,853	29,181	41,034	189,439
Jun	46,639	16,977	251	14,509		123,849	10,778	103,108	86,832	396,069	11,978	29,323	41,301	183,177
Jul	47,125	16,785	265	14,485		112,676	10,396	102,141	86,351	389,189	11,558	29,774	41,332	183,517
Aug	48,132	16,520	420	15,385		109,376	10,556	102,316	86,052	388,351	11,245	29,364	40,609	187,663
Sep	45,729	18,608	464	21,225		109,013	10,897	105,849	89,623	399,150	10,611	31,123	41,734	189,703
Oct	46,126	18,902	407	13,741		105,975	11,316	105,707	87,762	388,346	10,418	31,533	41,951	199,493
Nov	46,822	18,467	338	15,265		106,070	10,818	102,220	83,315	386,964	9,543	31,543	41,086	216,575
Dec	46,978	18,849	374	12,274	#	105,820	10,121	103,846	86,129	388,948	9,755	33,780	43,535	230,686
2019 Jan	48,498	18,742	369	11,774	4,105 (x)	102,227	10,383	106,801 (aa)	90,897	394,399	10,191	32,556	42,746	219,005
Feb	48,839	18,713	283	11,789	4,364	104,225	8,142 (y)	104,822 (ab)	91,169	392,051 (y)(ab)	9,144	32,556	41,699	208,138 (y)
Mar	49,424	18,994	334	12,615	4,845	103,716	8,137	106,056	91,425	396,622	10,975	33,551	44,526	211,044
Apr	50,076	19,668	448	12,537	4,958	117,652	10,578	102,779	88,122	410,266	9,552	31,527	41,079	209,654

12.4b Industrial analysis of monetary financial institutions' lending to UK residents

£ millions

Not seasonally adjusted

Amounts outstanding of lending in all currencies

| RPM | Activities auxiliary to financial intermediation (contd.) | | | Total | Total financial and non-financial businesses | Individuals and individual trusts | | | Total UK residents |
| | Central clearing counterparties | Other auxiliary activities (Other) | Total | Total | | Lending secured on dwellings inc. bridging finance | Other loans and advances | Total | |
	B3X2	B8FE	TBTV	B5H8	Z92T	TBTX	TBTY	TBTW	TBSA
2017 Jan	136,194	11,056	147,250	322,489	1,188,241	1,152,531	131,298	1,283,829	2,472,070
Feb	146,347	11,844	158,191	341,529	1,211,160	1,154,866	131,405	1,286,271	2,497,432
Mar	140,926	11,702	152,628	340,235	1,213,720	1,159,213	132,150	1,291,363	2,505,083
Apr	146,561	11,642	158,203	343,482	1,226,862	1,160,600	132,578	1,293,177	2,520,040
May	150,395	10,612	161,007	348,110	1,230,560	1,164,549	133,143	1,297,693	2,528,253
Jun	145,802	10,696	156,498	351,282	1,225,902	1,171,238	133,539	1,304,777	2,530,679
Jul	149,796	10,399	160,194	346,666	1,228,124	1,174,234	134,012	1,308,246	2,536,370
Aug	165,966	11,896	177,862	371,532	1,245,440	1,179,327	134,408	1,313,735	2,559,175
Sep	167,021	11,869	178,890	367,582	1,240,888	1,189,444	134,313	1,323,757	2,564,644
Oct	170,984	10,672	181,656	371,106	1,241,107	1,192,167	134,410	1,326,577	2,567,684
Nov	176,157	7,149	183,306	374,507	1,252,455	1,196,473	135,321	1,331,794	2,584,249
Dec	152,769	6,799	159,568	362,874	1,229,335	1,199,296	135,306	1,334,602	2,563,937
2018 Jan	170,592	6,267	176,859	376,808	1,260,362	1,201,014	134,417	1,335,431	2,595,793
Feb	174,868	6,018	180,886	378,453	1,257,014	1,202,022	135,062	1,337,085	2,594,099
Mar	150,085	6,425	156,510	356,517	1,240,229	1,206,949	134,385	1,341,334	2,581,563
Apr	151,875 (ac)	5,936	157,811 (ac)	345,513 (ac)	1,237,583 (ac)	1,208,098	134,914	1,343,012	2,580,595 (ac)
May	160,074	6,023	166,097	355,536	1,249,807	1,210,986	135,299	1,346,286	2,596,093
Jun	154,115	5,834	159,949	343,127	1,251,531	1,215,403	135,890	1,351,293	2,602,824
Jul	167,748	5,740	173,489	357,006	1,262,239 (u)	1,218,340	136,224	1,354,564 (u)	2,616,804 (u)
Aug	170,765	5,611	176,375	364,038	1,270,671	1,222,567	136,635	1,359,202	2,629,873
Sep	173,112 (ad)	6,067	179,180 (ad)	368,883 (ad)	1,285,239 (ad)	1,224,267	137,520	1,361,788 (ad)	2,647,027 (ad)
Oct	166,821	6,234	173,054	372,548	1,279,441	1,231,129 (ae)	137,040	1,368,168 (ae)	2,647,609 (ae)
Nov	188,753	5,370	194,122	410,697	1,319,677	1,234,967	137,609	1,372,576	2,692,253
Dec	173,801	4,584	178,384	409,071	1,317,387	1,237,459	137,844	1,375,304	2,692,690
2019 Jan	176,975	4,929	181,904	400,909	1,319,593	1,239,628	137,141	1,376,769	2,696,362
Feb	149,357	4,463	153,820 (y)	361,958 (y)	1,278,510 (y)(ab)	1,241,389 (af)	136,259 (ag)	1,377,648 (af)(ag)	2,656,158 (y)(ab)(af)(ag)
Mar	150,654	4,711	155,366	366,409	1,289,076	1,244,679	135,844	1,380,522	2,669,599
Apr	140,290	4,986	145,276	354,931	1,285,393	1,246,823	136,446	1,383,270	2,668,662

Source: Bank of England

12.4b Industrial analysis of monetary financial institutions' lending to UK residents

£ millions

Not seasonally adjusted

Notes to table

Movements in amounts outstanding can reflect breaks in data series as well as underlying flows. For changes data, users are recommended to refer directly to the appropriate series or data tables.
Further explanation can be found at: www.bankofengland.co.uk/statistics/Pages/iadb/notesiadb/Changes_flows_growth_rates.aspx.

(s) Due to improvements in reporting at one institution, the amounts outstanding increased by £1bn. This effect has been adjusted out of the flows for May 2018.

(t) Due to improvements in reporting at one institution, the amounts outstanding decreased by £1bn. This effect has been adjusted out of the flows for May 2018.

(u) Due to improvements in reporting at one institution, the amounts outstanding increased by £1bn. This effect has been adjusted out of the flows for July 2018.

(v) Due to improvements in reporting at one institution, the amounts outstanding increased by £3bn. This effect has been adjusted out of the flows for February 2018.

(w) Due to improvements in reporting at one institution, the amounts outstanding increased by £4bn. This effect has been adjusted out of the flows for February 2018.

(x) Previously, this series formed part of other financial intermediaries.

(y) Due to improvements in reporting at one institution, the amounts outstanding increased by £2bn. This effect has been adjusted out of the flows for February 2019.

(z) Due to improvements in reporting at one institution, the amounts outstanding decreased by £6bn. This effect has been adjusted out of the flows for February 2018.

(aa) From January 2019 onwards, this series no longer includes venture and development capital. As result, this series decreased by £4bn. This effect has been adjusted out the flows for January 2019

(ab) Due to improvements in reporting at one institution, the amounts outstanding decreased by £2bn. This effect has been adjusted out of the flows for February 2019.

(ac) Due to improvements in reporting at one institution, the amounts outstanding increased by £1bn. This effect has been adjusted out of the flows for April 2018

(ad) Due to improvements in reporting at one institution, the amounts outstanding increased by £5bn. This effect has been adjusted out of the flows for September 2018.

(ae) Due to improvements in reporting at one institution, the amounts outstanding increased by £2bn. This effect has been adjusted out of the flows for October 2018.

(af) Due to improvements in reporting at one institution, the amounts outstanding increased by £1bn. This effect has been adjusted out of the flows for February 2019.

(ag) Due to improvements in reporting at one institution, the amounts outstanding decreased by £1bn. This effect has been adjusted out of the flows for February 2019.

Explanatory notes can be found here: http://www.bankofengland.co.uk/statistics/Pages/iadb/notesiadb/industrial.aspx

Copyright guidance and the related UK Open Government Licence can be viewed here: www.bankofengland.co.uk/legal

12.4c Industrial analysis of monetary financial institutions' lending to UK residents

£ millions

Amounts outstanding of facilities granted in sterling

Not seasonally adjusted

RPM	Agriculture, hunting and forestry	Fishing	Mining and quarrying	Manufacturing									Electricity, gas and water supply			
				Food, beverages and tobacco	Textiles, wearing apparel and leather	Pulp, paper, and printing	Chemicals, pharmaceuticals, rubber and plastics	Non-metallic mineral products and metals	Machinery, equipment and transport equipment	Electrical, medical and optical equipment	Other manufacturing	Total	Electricity, gas steam and air conditioning	Water collection and sewerage	Waste management related services and remediation activities	Total
	TCCC	TCCD	TCCE	TCCG	TCCH	TCCI	TCCJ	TCCK	TCCL	TCCM	TCCN	TCCF	TCCO	TCCP	B3FE	B3FS
2017 Jan	24,722	339	3,197	9,640	1,151	2,666	7,999	5,458	11,195	3,139	3,729	44,978	18,468	7,766	2,998	29,233
Feb	24,772	339	3,240	9,623	1,135	2,687	7,972	5,426	11,399	3,156	3,740	45,137	18,334	7,428	3,054	28,816
Mar	24,883	343	3,192	9,512	1,186	2,747	8,032	5,497	11,518	3,467	3,836	45,795	20,307	7,320	2,945	30,572
Apr	24,917	333	3,120	10,434	1,180	2,731	7,971	5,471	11,705	3,164	3,843	46,498	20,273	7,320	2,885	30,478
May	25,089	332	3,128	10,744	1,198	2,856	8,131	5,379	11,215	3,093	3,906	46,521	19,853	7,282	2,921	30,057
Jun	25,238	339	3,196	10,653	1,201	3,003	7,938	5,375	10,829	3,162	3,985	46,147	20,297	7,452	2,932	30,681
Jul	25,349	341	3,359	13,910	1,182	2,661	8,008	5,260	10,607	3,149	3,977	48,755	20,370	6,999	2,857	30,226
Aug	25,378	337	3,285	13,863	1,204	2,696	8,170	5,277	10,787	3,127	3,878	49,003	20,456	7,347	2,798	30,601
Sep	25,491	329	3,480	13,750	1,231	2,844	8,263	5,789	11,385	3,627	4,080	50,968	19,728	7,455	2,848	30,031
Oct	25,515	364	3,228	13,877	1,233	2,927	7,940	5,613	11,258	3,871	3,999	50,719	19,615	7,562	2,822	29,999
Nov	25,504	385	3,217	14,092	1,215	2,863	7,490	5,640	11,418	3,793	4,070	50,582	19,494	7,696	2,863	30,052
Dec	25,413	391	3,183	13,983	1,103	2,837	7,692	5,737	11,659	3,535	3,993	50,539	20,028	7,620	2,896	30,544
2018 Jan	25,294	339	2,369	13,781	1,082	2,776	7,508	5,695	11,551	3,624	4,092	50,108	19,802	7,708	2,838	30,348
Feb	25,181	335	2,377	13,772	1,118	2,797	7,521	5,744	11,820	3,231	4,248	50,249	19,940	7,660	2,868	30,468
Mar	25,269	336	2,319	14,127	1,098	2,655	8,822	6,002	12,551	3,199	3,998	52,452	20,051	7,693	2,924	30,668
Apr	25,301	339	2,637	14,200	1,129	2,886	8,583	6,290	13,241	3,091	4,056	53,476	20,306	7,715	2,878	30,899
May	25,314	340	2,544	14,524	1,179	2,652	9,212	6,516	12,925	3,412	4,086	54,507	20,180	7,805	2,918	30,904
Jun	25,315	346	2,576	14,125	1,205	2,813	10,146	6,320	12,850	3,357	4,014	54,830	20,297	7,867	2,963	31,127
Jul	25,292	352	2,641	14,299	1,198	2,784	10,302	6,264	12,609	3,179	3,869	54,503	20,262	7,789	3,050	31,102
Aug	25,282	384	2,622	14,272	1,158	2,710	10,177	6,222	12,604	3,304	3,989	54,437	20,702	7,903	2,964	31,569
Sep	25,497	399	2,675	13,605	1,188	2,881	10,088	6,267	12,185	3,333	3,932	53,479	20,559	8,051	2,934	31,544
Oct	25,287	393	3,026	13,899	1,212	2,983	10,037	6,008	12,317	3,298	3,949	53,704	20,164	7,744	2,944	30,853
Nov	25,502	389	2,767	14,192	1,145	3,275	10,520	6,080	12,903	3,560	4,034	55,709	20,567	8,845	2,949	32,360
Dec	25,385	373	2,568	15,204	1,151	3,009	10,024	5,882	13,182	3,594	3,983	56,029	22,473	9,039	2,976	34,489
2019 Jan	25,337	389	2,572	15,231	1,112	2,723	10,097	5,965	13,756	3,582	4,050	56,516	22,796	8,800	3,072	34,668
Feb	25,266	392	2,620	15,368	1,100	2,698	10,374	5,910	13,770	3,732	4,177	57,130	22,688	8,817	3,087	34,593
Mar	25,206	380	2,689	15,433	1,127	2,592	10,417	6,169	13,199	3,756	4,170	56,862	22,276	9,425	3,274	34,975
Apr	25,177	383	2,889	15,708	1,179	2,789	10,636	6,450	13,276	3,758	4,202	57,998	22,574	9,306	3,045	34,924

12.4c Industrial analysis of monetary financial institutions' lending to UK residents

£ millions

Not seasonally adjusted

Amounts outstanding of facilities granted in sterling

| RPM | Construction | | | | | | Wholesale and retail trade | | | | Accommodation and food service activities | Transport, storage and communication | | |
| | Development of buildings | Construction of commercial buildings | Construction of domestic buildings | Civil Engineering | Other construction activities | Total | Wholesale and retail trade and repair of motor vehicles and motorcycles | Wholesale trade, excluding motor vehicles and motor cycles | Retail trade excluding motor vehicles and motor cycles | Total | | Transportation and storage | Information and communication | Total |
	B7EG	B3ID	B4PC	B4PP	B5QH	TCCQ	TCCS	TCCT	TCCU	TCCR	TCCV	B5PP	B5PW	TCCW
2017 Jan	20,060	5,194	10,237	5,235	8,113	48,840	16,217	19,606	28,130	63,953	29,128	18,967	14,156	33,123
Feb	20,344	5,211	10,330	5,339	8,111	49,335	17,079	19,993	28,053	65,124	29,522	19,722	14,007	33,729
Mar	20,299	5,148	10,461	5,897	8,042	49,847	17,007	19,697	27,994	64,698	29,632	19,739	14,415	34,153
Apr	20,396	5,095	10,498	5,497	8,058	49,544	16,644	18,809	28,443	63,896	30,192	19,676	14,400	34,076
May	20,208	5,086	10,576	5,476	7,909	49,255	16,506	18,619	28,736	63,860	29,884	19,452	14,907	34,359
Jun	20,332	4,944	10,454	5,437	7,965	49,132	17,127	18,161	28,460	63,748	29,740	19,675	15,207	34,882
Jul	20,568	5,022	10,342	5,548	8,066	49,546	16,694	18,246	28,376	63,315	29,664	19,290	15,782	35,072
Aug	20,431	5,089	10,293	5,090	8,205	49,108	16,782	18,265	28,632	63,679	29,865	19,287	15,048	34,335
Sep	20,470	5,122	10,271	5,071	8,249	49,183	16,895	18,909	29,555	65,359	30,001	19,368	15,927	35,295
Oct	20,323	5,250	10,331	4,896	8,002	48,801	16,524	18,650	30,090	65,264	30,248	19,442	15,878	35,320
Nov	20,233	5,187	10,672	5,065	7,872	49,031	16,236	18,643	29,924	64,803	30,412	19,082	15,811	34,894
Dec	20,121	5,220	10,368	4,940	7,859	48,508	16,351	18,698	29,956	65,005	30,636	19,384	15,665	35,049
2018 Jan	19,973	4,999	10,379	4,698	7,749	47,799	16,463	18,447	30,007	64,917	30,289	19,337	15,482	34,818
Feb	20,075	4,987	10,627	5,151	7,827	48,666	16,370	18,373	30,625	65,368	30,291	19,928	15,573	35,501
Mar	20,516	4,869	10,775	4,737	7,797	48,694	16,376	18,248	31,385	66,008	30,526	20,164	15,781	35,945
Apr	20,616	4,951	10,574	4,497	7,744	48,383	16,300	18,360	31,097	65,757	30,377	20,322	15,767	36,089
May	20,997	5,001	10,654	4,635	7,769	49,056	16,234	18,736	30,964	65,935	30,563	20,439	16,245	36,684
Jun	20,883	4,908	10,771	4,559	7,779	48,900	16,092	18,391	32,674	67,158	30,947	19,679	16,337	36,017
Jul	20,231	5,003	10,830	5,227	7,833	49,124	15,885	18,671	32,975	67,531	31,950	20,046	16,106	36,151
Aug	20,764	5,140	10,650	4,569	7,703	48,826	15,965	18,644	32,700	67,308	31,487	20,139	16,432	36,571
Sep	20,864	5,008	10,843	4,799	7,775	49,288	16,008	18,607	33,248	67,862	31,712	20,026	15,958	35,984
Oct	20,004	5,071	10,860	4,718	7,927	48,579	15,767	18,821	32,867	67,455	31,387	20,166	15,649	35,815
Nov	20,023	5,020	10,886	4,910	7,929	48,767	16,012	19,433	33,502	68,948	32,164	20,217	16,035	36,253
Dec	19,899	4,806	10,857	4,501	7,886	47,948	16,136	19,114	33,788	69,038	32,488	19,664	16,735	36,398
2019 Jan	19,849	4,918	10,979	4,346	7,831	47,922	16,312	19,653	33,144	69,110	32,887	20,109	17,138	37,247
Feb	20,022	4,661	10,984	4,348	7,809	47,824	16,435	19,329	33,677	69,441	32,238	20,071	16,350	36,421
Mar	19,940	4,737	11,069	4,329	7,679	47,753	16,551	18,683	34,346	69,580	32,191	19,762	16,649	36,412
Apr	19,365	4,667	11,224	4,274	7,655	47,185	12,490	18,986	32,419	63,896	32,391	20,365	16,989	37,354

12.4c Industrial analysis of monetary financial institutions' lending to UK residents

£ millions

Not seasonally adjusted

Amounts outstanding of facilities granted in sterling

RPM	Real estate, professional services and support activities				Public administration and defence	Education	Human health and social work	Recreational, personal and community service activities			Financial intermediation (excluding insurance and pension funds)			
	Buying, selling and renting of real estate	Professional, scientific and technical activities	Administrative and support services	Total				Recreational, cultural and sporting activities	Personal and community service activities	Total	Financial leasing corporations	Non-bank credit grantors excluding credit unions and SPVs	Credit unions	Factoring corporations
	TCCY	B6PH	B3S2	TCCX	TCDD	TCDE	TCDF	TCDH	TCDG	B3SR	TCDJ	TCDK	TCDL	TCDM
2017 Jan	167,262	23,591	33,285	224,138	12,404	12,865	23,336	7,678	4,295	11,973	27,443	25,028	3	5,369
Feb	166,919	23,180	32,756	222,855	10,639	12,866	23,021	7,709	4,129	11,838	27,290	21,293	3	5,730
Mar	166,799	23,360	33,133	223,292	11,918	12,904	23,237	7,714	4,328	12,042	26,406	21,844	3	5,949
Apr	167,428	23,537	33,921	224,885	11,208	12,926	23,048	7,847	4,425	12,272	25,879	21,519	3	5,984
May	168,147	23,413	34,235	225,795	10,476	12,968	23,278	7,894	4,388	12,283	26,518	20,978	2	5,903
Jun	167,956	23,546	34,910	226,411	11,361	12,839	23,349	7,709	4,433	12,142	27,360	23,959	2	6,338
Jul	166,265	23,304	35,387	224,957	11,237	12,891	23,439	7,541	4,338	11,879	27,615	24,146	2	6,299
Aug	166,542	23,379	34,882	224,803	12,151	12,938	23,502	7,444	4,333	11,776	27,958	25,210	3	6,344
Sep	168,253	24,314	35,748	228,315	11,227	13,093	23,244	7,519	4,299	11,818	29,896	24,600	3	6,531
Oct	166,761	23,834	36,105	226,700	10,892	13,107	23,385	7,584	4,213	11,797	29,075	24,748	3	6,346
Nov	166,410	23,916	35,805	226,132	10,870	13,061	23,588	7,564	4,139	11,703	30,075	24,510	3	6,595
Dec	166,840	24,226	34,924	225,990	11,069	13,105	22,914	7,760	4,222	11,982	30,290	24,947	3	6,406
2018 Jan	167,685	24,374	35,555	227,614	10,752	12,821	23,202	7,621	4,099	11,720	30,685	26,085	3	5,829
Feb	167,893	24,503	35,240	227,635	14,154	12,666	23,361	7,654	4,158	11,812	34,275 (ak)	26,224	4	9,962 (al)
Mar	169,510	23,894	35,402	228,806	15,066	12,818	23,149	7,556	4,166	11,722	35,184	26,599	4	9,880
Apr	170,664	23,923	35,706	230,293	10,977	12,756	23,180	7,258	4,129	11,386	35,665	27,163	3	9,613
May	171,672	24,055	36,604	232,331	10,334	12,898	23,239	7,192	4,159	11,351	37,554	27,178	3	9,305
Jun	171,167	25,035	37,317	233,519	10,352	12,910	23,102	7,601	4,126	11,727	38,200	27,761	3	9,581
Jul	172,500	24,532	37,724 (ai)	234,756 (ai)	10,298	12,851	22,926	7,291	4,238	11,529	41,627	29,585	3	9,310
Aug	173,340	24,550	37,368	235,257	10,769	12,935	23,165	7,277	4,165	11,442	41,711	30,021	3	9,319
Sep	174,036	24,413	36,813	235,262	9,917	13,118	23,364	7,020	4,183	11,203	41,867	31,110	4	9,856
Oct	174,784	24,372	36,894	236,051	10,101	12,831	23,522	6,968	4,293	11,261	41,090	30,849	3	9,887
Nov	175,908 (ah)	25,047	38,745	239,700 (ah)	9,524	13,058	23,344	7,157	4,914 (aj)	12,071 (aj)	41,824	31,408	3	9,901
Dec	177,429	25,178	38,488	241,096	10,237	12,733	22,989	6,603	4,453	11,056	45,023	32,081	3	9,708
2019 Jan	177,137	25,729	38,092	240,958	9,405	12,583	23,136	6,723	4,300	11,023	45,445	33,211	3	9,354
Feb	178,164	25,723	38,216	242,102	9,972	12,525	23,330	6,792	4,245	11,037	46,516	32,407	3	9,680
Mar	179,290	26,367	38,715	244,372	9,609	12,522	23,215	6,869	4,271	11,140	46,788	33,766	33	9,926
Apr	180,091	26,359	38,583	245,034	9,960	12,637	23,214	6,899	4,299	11,198	46,819	33,369	33	8,858

12.4c Industrial analysis of monetary financial institutions' lending to UK residents

£ millions

Amounts outstanding of facilities granted in sterling

RPM	Mortgage and housing credit corporations excluding SPVs	Investment and unit trusts excluding money market mutual funds	Money market mutual funds	Bank holding companies	Venture and development capital	Securities dealers	SPVs related to securitisation	Other financial intermediaries	of which intragroup activity	Total	Insurance companies	Pension funds	Total
	TCDN	TCDO	TCDP	TCDQ	ZO92	TCDR	B8EX	TCDS	B7FF	TCDI	B7FK	B3W9	TCDT
2017 Jan	49,147	12,830	164	19,114		22,851	14,397	41,794	30,276	218,141	13,440	16,552	29,992
Feb	44,004	13,854	169	18,536		23,022	14,108	41,731	30,311	209,740	12,949	17,505	30,454
Mar	47,998	12,177	168	13,632		21,018	14,052	41,590	31,987	204,836	13,861	25,911	39,772
Apr	47,980	12,394	120	13,774		22,208	15,107	41,813	32,394	206,781	15,067	28,080	43,147
May	46,680	12,342	111	14,755		23,052	13,980	41,176	32,013	205,497	14,401	28,253	42,654
Jun	46,620	13,273	118	14,469		22,690	12,194	42,699	33,869	209,723	13,134	27,411	40,546
Jul	46,794	13,878	140	14,587		23,645	12,827	43,217	33,416	213,149	12,357	25,431	37,788
Aug	46,831	13,695	164	14,474		22,875	12,122	42,578	32,356	212,253	13,151	25,331	38,482
Sep	47,183	15,263	143	14,670		23,263	12,457	42,134	32,557	216,141	14,512	26,869	41,380
Oct	45,275	15,063	142	11,318		25,273	13,564	41,551	31,505	212,357	12,046	25,404	37,450
Nov	45,534	15,067	131	12,041		25,119	11,284	40,956	30,874	211,315	12,470	26,417	38,888
Dec	45,335	12,855	124	11,385		25,199	11,277	45,209	35,069	213,031	14,757	25,979	40,736
2018 Jan	45,755	13,656	131	11,380		26,481	9,122	47,722	34,811	216,850	11,759	25,646	37,405
Feb	45,679	13,726	127	11,321		24,637	9,003	45,211 (an)	32,346 (an)	220,169	11,725	26,206	37,931
Mar	45,969	13,673	117	11,477		27,337	9,578	50,232	37,065	230,050	12,771	25,852	38,623
Apr	46,878	13,680	144	11,516		25,575	9,087	50,736	39,188	230,061	13,124	25,474	38,598
May	48,152	15,269	182	8,951		24,051	9,827	43,056	30,816	223,527	13,084	28,491	41,575
Jun	46,730	15,283	209	8,983		25,829	10,469	55,915	43,034	238,964	13,489	28,637	42,126
Jul	46,831	15,391	183	8,990		21,768	10,096	54,273	41,955	238,056	12,606	28,922	41,528
Aug	47,839	15,200	334	10,129		19,135	10,015	54,554	41,920	238,260	12,691	28,498	41,189
Sep	45,425	17,878	395	16,763		21,993	10,212	54,920	41,892	250,423	12,409	30,366	42,775
Oct	45,828	18,246	289	9,838		18,900	10,439	54,720	40,799	240,090	11,982	30,637	42,619
Nov	46,538	18,174	277	11,349		19,922	10,376	51,539	36,719	241,313	11,577	30,757	42,334
Dec	46,711	18,732	279	8,381	#	19,822	9,230	52,239	37,887	242,208	12,077	32,176	44,254
2019 Jan	48,256	18,839	288	7,729	1,767 (am)	17,463	9,400	57,078 (ao)	42,654	248,833	11,641	31,579	43,220
Feb	48,788	18,927	222	7,971	1,810	16,446	14,677	55,993 (ap)	43,386	253,441 (ap)	11,516	31,316	42,832
Mar	49,346	18,527	281	8,504	1,960	15,624	14,587	56,061	42,963	255,404	12,409	32,331	44,740
Apr	49,975	18,489	280	8,742	2,372	18,672	12,625	50,700	37,872	250,934	11,200	30,760	41,960

Financial intermediation (excluding insurance and pension funds) (contd.)

Insurance companies & pension funds

12.4c Industrial analysis of monetary financial institutions' lending to UK residents

£ millions

Not seasonally adjusted

Amounts outstanding of facilities granted in sterling

RPM	Fund management activities	Activities auxiliary to financial intermediation				Total financial and non-financial businesses	Individuals and individual trusts			Total UK residents
		Central clearing counterparties	Other		Total		Lending secured on dwellings inc. bridging finance	Other loans and advances	Total	
			Other auxiliary activities	Total						
	TCDU	B3X6	B3Y4	TCDV	B8FO	Z94D	TCDX	TCDY	TCDW	TCCA
2017 Jan	114,829	63,796	4,377	68,174	183,003	993,365	1,209,427	207,224	1,416,650	2,410,015
Feb	115,191	65,273	5,688	70,961	186,152	987,578	1,212,619	207,735	1,420,354	2,407,932
Mar	118,714	70,475	5,108	75,583	194,297	1,005,414	1,219,164	208,675	1,427,839	2,433,253
Apr	114,149	73,433	4,880	78,314	192,463	1,009,783	1,220,500	209,396	1,429,896	2,439,679
May	118,016	72,151	4,637	76,787	194,803	1,010,238	1,226,046	209,351	1,435,397	2,445,635
Jun	118,308	73,332	4,695	78,027	196,334	1,015,810	1,234,203	209,866	1,444,069	2,459,878
Jul	113,652	73,220	4,650	77,870	191,522	1,012,489	1,232,797	203,399	1,436,196	2,448,685
Aug	118,904	78,760	4,795	83,555	202,459	1,023,954	1,236,277	203,837	1,440,114	2,464,069
Sep	122,898	86,405	4,910	91,315	214,213	1,049,567	1,245,840	204,570	1,450,410	2,499,977
Oct	123,267	83,178	4,458	87,635	210,903	1,036,050	1,248,590	203,826	1,452,417	2,488,467
Nov	123,124	86,401	4,972	91,374	214,498	1,038,933	1,253,950	203,508	1,457,457	2,496,391
Dec	133,303	76,094	4,988	81,081	214,385	1,042,479	1,255,532	204,197	1,459,729	2,502,208
2018 Jan	125,626	81,888	4,496	86,383	212,010	1,038,657	1,253,853	200,648 (au)	1,454,501 (au)	2,493,158 (au)
Feb	117,945	78,366	4,568	82,934	200,879	1,037,044	1,254,310	202,402	1,456,711	2,493,755
Mar	120,643	64,835	4,658	69,493	190,136	1,042,588	1,260,315	202,544	1,462,860	2,505,447
Apr	113,580	65,207	4,527	69,734	183,313	1,033,822	1,261,985	203,142	1,465,127	2,498,949
May	108,733	65,078	4,521	69,599	178,332	1,029,433	1,265,596	215,931	1,481,527	2,510,959
Jun	110,336	69,145	4,079	73,223	183,560	1,053,475	1,271,674	217,120	1,488,794	2,542,270
Jul	111,175	77,836	4,031	81,867	193,042	1,063,632 (ai)	1,274,779	218,545	1,493,324	2,556,957 (ai)
Aug	110,450	76,134	4,227	80,361	190,811	1,062,313	1,279,036	219,596	1,498,632	2,560,946
Sep	113,033	81,564 (ar)	4,091	85,655 (ar)	198,688 (ar)	1,083,189 (ar)	1,281,681	219,557	1,501,238	2,584,428 (ar)
Oct	118,382	79,614	4,473	84,086	202,468	1,075,442	1,288,220 (as)	219,296	1,507,515 (as)	2,582,957 (as)
Nov	122,272	88,279	4,306	92,585	214,856	1,099,060 (ah)(aj)	1,291,080	219,827	1,510,908	2,609,968 (ah)(aj)
Dec	125,261	92,193	4,018	96,211	221,473	1,110,762	1,291,717	220,376	1,512,092	2,622,854
2019 Jan	116,801	93,937	3,850	97,787	214,588	1,110,395	1,292,784	220,702	1,513,486	2,623,881
Feb	116,932 (aq)	90,722	4,054	94,776	211,708 (aq)	1,112,872 (ap)(aq)	1,293,712 (at)	219,759 (av)	1,513,471 (at)(av)	2,626,343 (ap)(at)(av)
Mar	119,664	98,038	3,994	102,032	221,696	1,128,744	1,298,449	220,216	1,518,664	2,647,409
Apr	118,303	92,586	3,884	96,470	214,772	1,111,907	1,301,265	220,827	1,522,093	2,633,999

Source: Bank of England

Not seasonally adjusted

12.4c Industrial analysis of monetary financial institutions' lending to UK residents

£ millions

Notes to table

Movements in amounts outstanding can reflect breaks in data series as well as underlying flows. For changes data, users are recommended to refer directly to the appropriate series or data tables. Further explanation can be found at: www.bankofengland.co.uk/statistics/Pages/iadb/notesiadb/Changes_flows_growth_rates.aspx.

(ah) Due to improvements in reporting at one institution, the amounts outstanding increased by £1bn. This effect has been adjusted out of the flows for November 2018.

(ai) Due to improvements in reporting at one institution, the amounts outstanding increased by £1bn. This effect has been adjusted out of the flows for July 2018.

(aj) Due to improvements in reporting at one institution, the amounts outstanding increased by £1bn. This effect has been adjusted out of the flows for November 2018.

(ak) Due to improvements in reporting at one institution, the amounts outstanding increased by £3bn. This effect has been adjusted out of the flows for February 2018.

(al) Due to improvements in reporting at one institution, the amounts outstanding increased by £4bn. This effect has been adjusted out of the flows for February 2018.

(am) Previously, this series formed part of other financial intermediaries.

(an) Due to improvements in reporting at one institution, the amounts outstanding decreased by £6bn. This effect has been adjusted out of the flows for February 2018.

(ao) From January 2019 onwards, this series no longer includes venture and development capital. As result, this series decreased by £2bn. This effect has been adjusted out the flows for January 2019

(ap) Due to improvements in reporting at one institution, the amounts outstanding decreased by £2bn. This effect has been adjusted out of the flows for February 2019.

(aq) Due to improvements in reporting at one institution, the amounts outstanding increased by £2bn. This effect has been adjusted out of the flows for February 2019.

(ar) Due to improvements in reporting at one institution, the amounts outstanding increased by £5bn. This effect has been adjusted out of the flows for September 2018.

(as) Due to improvements in reporting at one institution, the amounts outstanding increased by £2bn. This effect has been adjusted out of the flows for October 2018.

(at) Due to improvements in reporting at one institution, the amounts outstanding increased by £1bn. This effect has been adjusted out of the flows for February 2019.

(au) Due to improvements in reporting at one institution, the amounts outstanding decreased by £3bn. This effect has been adjusted out of the flows for January 2018.

(av) Due to improvements in reporting at one institution, the amounts outstanding decreased by £1bn. This effect has been adjusted out of the flows for February 2019.

Explanatory notes can be found here: http://www.bankofengland.co.uk/statistics/Pages/iadb/notesiadb/industrial.aspx

Copyright guidance and the related UK Open Government Licence can be viewed here: www.bankofengland.co.uk/Pages/disclaimer.aspx.

12.4d Industrial analysis of monetary financial institutions' lending to UK residents

£ millions

Not seasonally adjusted

Amounts outstanding of facilities granted in all currencies

| RPM | Agriculture, hunting and forestry | Fishing | Mining and quarrying | Manufacturing | | | | | | | | | Electricity, gas and water supply | | | Total |
| | | | | Food, beverages and tobacco | Textiles, wearing apparel and leather | Pulp, paper, and printing | Chemicals, pharmaceuticals, rubber and plastics | Non-metallic mineral products and metals | Machinery, equipment and transport equipment | Electrical, medical and optical equipment | Other manufacturing | Total | Electricity, gas steam and air conditioning | Water collection and sewerage | Waste management related services and remediation activities | |
	TCAC	TCAD	TCAE	TCAG	TCAH	TCAI	TCAJ	TCAK	TCAL	TCAM	TCAN	TCAF	TCAO	TCAP	B3FF	B3FT
2017 Jan	25,031	370	30,930	32,304	1,851	4,669	19,292	8,731	18,583	6,147	5,678	97,255	26,783	8,477	3,315	38,575
Feb	25,066	370	30,720	33,695	1,785	4,813	32,120	8,750	18,988	6,024	5,692	111,867	26,697	8,167	3,222	38,087
Mar	25,183	373	30,176	33,313	1,831	4,840	27,766	9,258	18,818	6,139	6,051	108,016	28,612	8,063	3,223	39,898
Apr	25,258	362	27,259	36,031	1,826	4,801	28,194	8,728	19,050	6,037	5,701	110,366	28,049	8,173	3,113	39,335
May	25,451	361	28,415	36,206	1,894	4,935	30,079	8,697	19,406	5,950	5,814	112,981	27,685	8,136	3,180	39,001
Jun	25,552	368	27,826	36,270	1,974	4,584	26,269	8,745	19,097	6,013	6,342	109,295	27,730	8,290	3,167	39,187
Jul	25,663	370	27,719	29,936	1,908	4,244	25,803	8,751	19,078	5,893	5,893	101,181	27,943	8,077	3,067	39,087
Aug	25,693	366	28,271	26,991	1,993	4,337	23,379	8,814	18,678	5,651	5,749	95,592	28,406	8,246	2,999	39,651
Sep	25,799	357	28,178	25,835	2,057	4,349	22,753	8,895	18,736	6,039	6,002	94,667	27,318	8,355	3,056	38,730
Oct	25,811	395	28,288	25,970	2,002	4,407	21,835	8,766	18,795	6,097	5,843	93,715	27,186	8,402	2,995	38,583
Nov	25,811	411	27,721	25,530	2,175	4,232	22,098	8,802	19,159	6,047	5,749	93,793	27,100	8,530	3,018	38,649
Dec	25,720	417	27,659	25,735	1,975	4,173	22,397	8,915	18,875	5,822	5,666	93,558	27,695	8,473	3,036	39,205
2018 Jan	25,731	367	26,012	26,261	1,793	4,011	21,265	8,670	18,811	8,313	5,710	94,834	27,499	8,433	3,005	38,937
Feb	25,466	362	27,205	26,602	1,837	4,599	21,138	8,859	19,407	8,110	5,939	96,491	27,721	8,433	2,964	39,117
Mar	25,566	358	26,657	26,241	1,894	4,624	28,313	9,018	20,294	8,036	5,667	104,086	28,087	8,454	3,049	39,589
Apr	25,564	360	27,005	26,434	1,915	3,980	24,806	9,315	22,180 (ay)	5,588 (az)	5,741	99,959	28,321	8,433	3,085	39,839
May	25,588	362	28,105	27,089	2,032	3,990	23,541	9,488	21,365	5,793	5,840	99,139	28,084	8,487	3,065	39,636
Jun	25,619	367	28,226	27,079	2,021	4,574	23,828	9,476	21,579	5,679	5,670	99,907	28,214	8,524	3,080	39,818
Jul	25,600	374	27,482	27,575	1,924	4,742	23,408	9,599	21,165	5,882	5,532	99,827	27,844	8,359	3,628	39,831
Aug	25,566	420	27,551	27,273	1,930	4,824	23,361	9,536	21,049	6,044	5,584	99,601	28,491	8,562	3,553	40,606
Sep	25,801	439	27,710	26,496	1,889	4,929	23,085	9,549	20,780	5,929	5,485	98,141	28,412	8,846	3,565	40,823
Oct	25,594	430	29,068	26,604	1,906	4,766	23,026	8,955	21,356	5,881	5,355	97,849	28,249	8,579	3,597	40,425
Nov	25,794	425	28,370	26,167 (aw)	1,836	5,388	24,556 (ax)	9,091	21,930	6,387	5,461	100,814	28,544	9,717	3,604	41,866
Dec	25,673	408	28,073	26,234	1,816	4,838	28,065	8,670	21,596	6,435	5,422	103,076	29,582	9,900	3,592	43,074
2019 Jan	25,600	423	26,585	26,204	1,756	4,220	24,902	8,473	22,372	6,425	5,435	99,787	29,919	9,700	3,839	43,458
Feb	25,529	424	26,225	26,150	1,730	4,147	24,402	8,422	22,026	6,677	5,511	99,065	29,859	9,662	3,858	43,378
Mar	25,455	414	27,421	25,619	1,782	4,122	22,950	8,787	21,608	6,474	5,582	96,923	29,007	10,328	4,027	43,362
Apr	25,419	415	27,905	26,234	1,834	4,596	23,584	9,059	21,574	6,457	5,770	99,108	29,350	10,424	3,836	43,610

12.4d Industrial analysis of monetary financial institutions' lending to UK residents

£ millions

Amounts outstanding of facilities granted in all currencies

Not seasonally adjusted

| RPM | Construction | | | | | | Wholesale and retail trade | | | | Accommodation and food service activities | Transport, storage and communication | | |
| | Development of buildings | Construction of commercial buildings | Construction of domestic buildings | Civil Engineering | Other construction activities | Total | Wholesale and retail trade and repair of motor vehicles and motorcycles | Wholesale trade, excluding motor vehicles and motor cycles | Retail trade excluding motor vehicles and motor cycles | Total | | Transportation and storage | Information and communication | Total |
	B7EH	B3IE	B4PD	B3KK	B5PC	TCAQ	TCAS	TCAT	TCAU	TCAR	TCAV	B3NF	B5PX	TCAW
2017 Jan	20,481	5,701	10,506	6,267	8,347	51,302	17,660	29,520	35,513	82,693	32,050	27,505	31,449	58,954
Feb	20,763	5,709	10,604	6,362	8,347	51,785	18,414	29,928	35,403	83,746	32,426	28,439	31,490	59,928
Mar	20,724	5,654	10,734	6,806	8,278	52,196	18,365	30,109	35,260	83,734	32,679	28,163	31,991	60,154
Apr	20,780	5,583	10,774	6,308	8,298	51,744	17,969	30,899	35,274	84,143	33,344	27,934	31,470	59,404
May	20,475	5,571	10,855	6,367	8,150	51,418	17,893	32,119	35,614	85,626	33,115	27,841	32,312	60,153
Jun	20,638	5,508	10,743	6,288	8,233	51,411	18,587	32,310	35,553	86,450	33,339	27,812	32,009	59,821
Jul	20,898	5,548	10,631	6,576	8,330	51,982	18,066	32,469	35,767	86,303	32,630	27,888	31,402	59,291
Aug	20,763	5,560	10,579	6,003	8,491	51,395	18,180	32,511	36,203	86,894	32,865	27,860	31,710	59,570
Sep	20,947	5,458	10,553	5,999	8,506	51,463	18,263	32,091	36,978	87,332	32,986	27,799	32,058	59,858
Oct	20,652	5,598	10,663	5,802	8,252	50,967	17,922	32,825	37,473	88,220	33,278	28,310	32,856	61,166
Nov	20,571	5,498	11,011	6,008	8,127	51,214	17,609	30,851	37,722	86,182	33,476	27,999	32,312	60,311
Dec	20,468	5,553	10,702	5,904	8,114	50,741	17,727	31,957	38,167	87,851	34,008	27,875	32,938	60,812
2018 Jan	20,319	5,290	10,701	5,622	7,985	49,918	17,821	30,896	37,392 (bb)	86,108 (bb)	33,689	27,455	31,894	59,348
Feb	20,649	5,278	10,976	6,066	8,104	51,073	17,715	31,326	39,743	88,784	33,784	28,237	31,818	60,056
Mar	21,134	5,298	11,017	5,632	8,041	51,122	17,663	33,974	38,713	90,351	34,078	28,198	31,843	60,041
Apr	21,278	5,349	10,820	5,392	7,989	50,828	17,801	34,418	38,567	90,785	34,002	28,025	31,321	59,346
May	21,712	5,402	10,902	5,474	8,012	51,501	17,761	33,006	38,418 (bc)	89,185 (bc)	34,554	28,187	38,131	66,318
Jun	21,614	5,393	11,031	5,452	8,058	51,548	17,642	32,412	39,951	90,005	34,718	27,259	32,550	59,809
Jul	20,971	5,404	11,092	6,063	8,078	51,607	17,351	33,284	39,661	90,296	35,348	27,621	33,032	60,653
Aug	21,392	5,511	10,915	5,359	7,953	51,130	17,441	32,933	39,113	89,486	34,908	27,571	32,983	60,554
Sep	21,344	5,367	11,122	5,541	8,027	51,401	17,253	31,892 (ba)	39,639	88,785 (ba)	35,006	27,378	32,596	59,973
Oct	20,480	5,405	11,141	5,460	8,201	50,688	17,051	31,624	39,358	88,033	34,749	27,776	32,669	60,446
Nov	20,492	5,330	11,176	5,614	8,168	50,779	17,501	33,151	40,290	90,942	35,451	28,038	33,099	61,137
Dec	20,360	5,252	11,152	5,355	8,118	50,237	17,584	32,776	40,434	90,794	35,499	26,905	32,785	59,690
2019 Jan	20,241	5,344	11,273	5,182	8,029	50,068	17,551	32,870	39,343	89,763	36,292	27,013	32,615	59,628
Feb	20,412	5,054	11,277	4,986	8,009	49,738	17,701	32,697	39,607	90,005	35,718	26,886	32,057	58,943
Mar	20,369	5,106	11,364	4,940	7,899	49,678	17,661	32,274	40,466	90,401	35,576	26,669	32,062	58,731
Apr	19,807	5,052	11,524	4,863	7,876	49,122	13,707	32,363	38,811	84,880	35,472	27,118	32,403	59,520

12.4d Industrial analysis of monetary financial institutions' lending to UK residents

£ millions

Amounts outstanding of facilities granted in all currencies

RPM	Real estate, professional services and support activities				Public administration and defence	Education	Human health and social work	Recreational, personal and community service activities			Financial intermediation (excluding insurance and pension funds)				
	Buying, selling and renting of real estate	Professional, scientific and technical activities	Administrative and support services	Total				Recreational, cultural and sporting activities	Personal and community service activities	Total	Financial leasing corporations	Non-bank credit grantors excluding credit unions and SPVs	Credit unions	Factoring corporations	Mortgage and housing credit corporations excluding SPVs
	TCAY	B3QF	B3S3	TCAX	TCBD	TCBE	TCBF	TCBH	TCBG	B3T2	TCBJ	TCBK	TCBL	TCBM	TCBN
2017 Jan	170,347	31,621	43,648	245,616	14,412	12,969	23,746	9,309	4,713	14,022	33,744	26,251	3	6,542	49,802
Feb	170,011	31,128	42,984	244,122	11,751	12,966	23,420	9,429	4,487	13,916	33,252	22,421	3	6,747	44,654
Mar	169,639	31,391	42,640	243,670	12,142	13,001	23,643	9,247	4,710	13,957	32,330	23,046	3	7,068	48,654
Apr	170,163	31,417	42,392	243,972	14,650	13,020	23,442	9,424	4,825	14,249	31,337	22,678	3	6,969	49,035
May	170,922	31,385	43,731	246,039	13,567	13,060	23,695	9,431	4,801	14,232	32,084	22,296	2	7,017	47,572
Jun	170,598	31,441	43,892	245,931	12,669	12,933	23,767	9,286	4,817	14,103	32,918	25,413	2	7,946	47,266
Jul	169,018	30,519	44,351	243,888	14,721	12,974	23,838	9,115	4,752	13,867	33,302	25,836	2	7,567	47,442
Aug	169,567	30,649	44,385	244,600	14,781	13,054	23,907	9,076	4,743	13,819	33,929	26,630	3	7,508	47,506
Sep	171,253	31,376	45,289	247,918	14,196	13,225	23,629	9,134	4,700	13,834	35,595	25,984	3	7,610	47,819
Oct	169,651	30,691	45,267	245,608	14,028	13,228	23,710	9,239	4,453	13,692	35,273	26,057	3	7,538	45,903
Nov	169,125	31,001	45,167	245,293	13,899	13,171	23,979	9,198	4,406	13,604	36,152	25,748	3	7,610	46,170
Dec	169,877	31,422	44,325	245,624	14,574	13,211	23,377	9,629	4,460	14,090	35,962	26,265	3	7,537	45,973
2018 Jan	170,788	31,074	47,290	249,152	12,898	12,933	23,822	9,192	4,203	13,395	36,160	27,336	4	6,800	46,376
Feb	171,107	30,946	45,240	247,293	15,835	12,789	23,820	9,275	4,247	13,521	40,873 (bh)	27,454	4	11,146 (bi)	46,372
Mar	172,968	30,988	45,179	249,136	17,203	12,943	23,715	9,079	4,270	13,350	41,614	27,829	4	11,363	46,551
Apr	173,961	31,866	45,615	251,443	14,481	12,894	23,637	8,800	4,223	13,023	41,990	28,478	4	11,469	47,383
May	174,758	32,842	46,199 (be)	253,799 (be)	12,627	13,022	23,756	8,831	4,278	13,110	43,858	28,544	3	11,169	48,671
Jun	174,334	33,967	46,681	254,982	12,758	13,040	23,619	9,251	4,234	13,485	44,552	29,226	3	11,459	47,257
Jul	175,857	33,728	46,982 (bf)	256,567 (bf)	14,192	13,002	23,425	8,935	4,409	13,344	47,839	31,147	4	11,215	47,362
Aug	176,475	33,747	46,569	256,791	14,946	13,067	23,699	8,920	4,283	13,203	47,915	31,512	4	11,189	48,370
Sep	177,525	33,511	45,316	256,352	13,597	13,262	23,787	8,554	4,301	12,855	47,958	32,501	4	11,859	45,946
Oct	178,248	33,406	46,290	257,945	13,716	13,003	23,942	8,254	4,405	12,659	47,186	32,278	4	11,721	46,345
Nov	179,578 (bd)	34,075	48,272	261,924 (bd)	13,472	13,314	23,777	8,436	5,138(bg)	13,574 (bg)	48,049	32,733	4	11,760	47,050
Dec	181,127	34,045	46,372	261,544	12,372	12,963	23,441	7,663	4,675	12,338	51,291	33,372	3	11,669	47,629
2019 Jan	180,623	33,811	46,157	260,592	12,362	12,761	23,569	7,827	4,480	12,307	51,799	34,503	3	11,108	48,761
Feb	181,876	33,854	46,257	261,987	13,749	12,703	23,843	7,894	4,401	12,294	52,734	33,646	3	11,047	49,285
Mar	182,980	34,960	46,553	264,493	13,830	12,696	23,692	7,948	4,448	12,396	53,036	35,009	34	11,160	50,169
Apr	183,753	34,980	46,530	265,263	12,560	12,786	23,667	8,025	4,458	12,484	53,015	34,627	34	10,048	50,806

12.4d Industrial analysis of monetary financial institutions' lending to UK residents

Amounts outstanding of facilities granted in all currencies

Not seasonally adjusted

	Investment and unit trusts excluding money market mutual funds	Money market mutual funds	Bank holding companies	Venture and development capital	Securities dealers	SPVs related to securitisation	Other financial inter-mediaries	of which intragroup activity	Total	Insurance companies	Pension funds	Total	Fund management activities	Central clearing counterparties	Other auxiliary activities	Total
RPM	TCBO	TCBP	TCBQ	ZO92	TCBR	B7FB	TCBS	B7FG	TCBI	B7FL	B7FQ	TCBT	TCBU	B3X7	B3Y5	TCBV
2017 Jan	20,757	342	26,508		144,833	17,262	92,318	70,107	418,361	22,604	17,909	40,512	180,013	136,409	13,668	150,076
Feb	22,020	342	25,970		145,280	16,767	104,803	75,245	422,260	22,317	18,690	41,006	188,212	146,574	14,578	161,152
Mar	20,100	346	21,137		145,852	16,823	97,733	77,207	413,093	22,953	27,127	50,080	192,443	141,123	14,514	155,637
Apr	20,073	320	21,048		151,075	17,760	97,519	76,765	417,816	24,560	29,431	53,991	189,333	146,738	14,507	161,245
May	20,715	380	22,003		148,397	16,287	97,409	77,626	414,161	23,657	29,440	53,097	190,781	150,821	13,626	164,447
Jun	21,786	368	21,701		139,798	14,595	96,120	76,916	407,914	22,298	28,268	50,565	198,576	146,164	13,654	159,818
Jul	22,222	412	21,766		135,792	15,246	97,478	76,725	407,067	21,270	26,707	47,977	189,933	150,103	13,577	163,680
Aug	21,843	409	21,632		132,778	14,013	99,333	76,595	405,584	22,028	26,759	48,787	197,397	166,525	15,153	181,678
Sep	23,256	373	21,404		127,546	14,385	98,506	76,483	402,482	24,952	28,241	53,194	192,491	167,699	14,949	182,648
Oct	23,199	352	18,422		131,294	14,612	98,436	76,139	401,088	20,397	26,847	47,243	193,522	172,007	13,736	185,743
Nov	22,883	338	19,606		137,685	12,337	98,812	76,340	407,344	20,766	27,802	48,568	194,880	176,440	10,327	186,767
Dec	20,648	338	18,217		128,203	12,897	101,089	79,885	397,133	22,822	27,203	50,026	206,947	152,973	10,035	163,008
2018 Jan	22,320	351	16,528		148,798	9,873	105,256	77,993	419,803	19,774	27,008	46,781	205,200	170,881	9,397	180,277
Feb	22,191	340	16,427		141,787	9,927	92,510 (bl)	66,239 (bl)	409,032	19,379	27,752	47,132	203,319	175,100	9,269	184,369
Mar	22,041	433	16,609		136,639	10,957	96,352	70,298	410,392	20,344	27,969	48,313	205,278	150,290	9,624	159,915
Apr	21,820	395	16,834		136,970	10,544	100,394	76,247	416,283	20,227	26,901	47,128	193,952	152,077 (bo)	8,915	160,992 (bo)
May	24,613	487	14,147		129,920	11,450	101,703	77,167	414,565	20,120	30,250	50,369	196,203	160,262	9,160	169,423
Jun	24,050	481	14,781		127,495	13,267	115,607	89,912	428,178	20,653	30,058	50,711	189,518	154,319	9,090	163,409
Jul	23,629	480	14,790		115,391	12,687	115,459	89,800	420,003	20,294	30,507	50,801	187,367	167,944	8,734	176,679
Aug	23,348	643	15,759		110,991	12,803	116,693	89,504	419,228	20,289	29,837	50,126	191,875	171,462	8,746	180,208
Sep	25,801	667	21,606		111,074	13,222	120,238	93,082	430,876	19,618	31,859	51,477	193,960	173,406 (bp)	8,886	182,292 (bp)
Oct	26,439	588	14,132		108,241	13,839	119,063	91,041	419,837	18,892	32,005	50,897	204,095	167,101	9,160	176,261
Nov	26,436	566	15,660		108,433	13,789	116,180	86,569	420,661	18,654	31,971	50,625	221,381	189,273	8,645	197,918
Dec	26,508	594	12,631	#	107,694	12,752	117,913	90,032	422,056	18,883	34,229	53,112	238,870	175,322	7,962	183,284
2019 Jan	26,452	587	12,130	6,472 (bi)	104,021	12,974	117,949 (bm)	94,275	426,760	19,017	33,000	52,017	227,334	177,503	7,944	185,447
Feb	26,845	487	12,183	6,757	105,781	16,163 (bk)	116,152 (bn)	94,435	431,084 (bk)(bn)	18,595	32,990	51,585	212,543 (bk)	149,799	7,543	157,342
Mar	26,096	550	13,112	7,056	105,377	17,235	117,522	94,994	436,355	19,652	33,976	53,628	215,257	151,080	8,544	159,624
Apr	26,930	601	13,121	7,826	119,345	13,313	113,777	91,585	443,443	18,476	31,938	50,415	213,859	140,705	8,139	148,845

Financial intermediation (excluding insurance and pension funds) (contd.)

Insurance companies & pension funds

Activities auxiliary to financial intermediation

12.4d Industrial analysis of monetary financial institutions' lending to UK residents

Not seasonally adjusted

Amounts outstanding of facilities granted in all currencies

RPM	Activities auxiliary to financial intermediation (contd.) Total	Total financial and non-financial businesses	Individuals and individual trusts			Total UK residents
			Lending secured on dwellings inc. bridging finance	Other loans and advances	Total	
	B322	Z94H	TCBX	TCBY	TCBW	TCAA
2017 Jan	330,089	1,516,889	1,209,585	208,372	1,417,957	2,934,845
Feb	349,364	1,552,801	1,212,778	208,883	1,421,661	2,974,462
Mar	348,080	1,550,074	1,219,315	209,987	1,429,302	2,979,376
Apr	350,578	1,562,935	1,220,632	210,664	1,431,296	2,994,231
May	355,228	1,569,600	1,226,180	210,637	1,436,817	3,006,417
Jun	358,395	1,559,525	1,234,334	211,237	1,445,571	3,005,096
Jul	353,612	1,542,171	1,232,931	204,779	1,437,710	2,979,881
Aug	379,075	1,563,906	1,236,415	205,236	1,441,651	3,005,556
Sep	375,139	1,562,987	1,245,970	205,917	1,451,886	3,014,874
Oct	379,265	1,558,284	1,248,731	205,104	1,453,835	3,012,119
Nov	381,647	1,565,075	1,254,091	204,795	1,458,886	3,023,960
Dec	369,955	1,547,962	1,255,677	205,531	1,461,208	3,009,170
2018 Jan	385,477	1,579,005 (bb)	1,253,992	201,887 (bs)	1,455,880 (bs)	3,034,884 (bb)(bs)
Feb	387,688	1,579,448	1,254,449	203,617	1,458,066	3,037,514
Mar	365,193	1,572,092	1,260,454	203,788	1,464,241	3,036,334
Apr	354,945 (bo)	1,561,522 (bo)	1,262,115	204,273	1,466,389	3,027,911 (bo)
May	365,626	1,581,261	1,265,739	217,078	1,482,817	3,064,077
Jun	352,927	1,579,718	1,271,814	218,292	1,490,106	3,069,824
Jul	364,046 (bf)	1,586,400 (bf)	1,274,919	219,651	1,494,570 (bf)	3,080,970 (bf)
Aug	372,083	1,592,965	1,279,173	220,758	1,499,931	3,092,895
Sep	376,253 (bp)	1,606,538 (ba)(bp)	1,281,814	220,674	1,502,488	3,109,026 (ba)(bp)
Oct	380,356	1,599,634	1,288,373 (bq)	220,429	1,508,802 (bq)	3,108,436 (bq)
Nov	419,299	1,652,225 (bd)(bg)	1,291,230	220,939	1,512,168	3,164,394 (bd)(bg)
Dec	422,154	1,656,504	1,291,858	221,555	1,513,412	3,169,916
2019 Jan	412,781	1,644,753	1,292,927	221,827	1,514,755	3,159,508
Feb	369,885 (bk)	1,606,159 (bk)(bn)	1,293,850 (br)	220,835 (bt)	1,514,685 (br)(bt)	3,120,844 (bk)(bn)(br)(bt)
Mar	374,881	1,619,932	1,298,589	221,304	1,519,893	3,139,826
Apr	362,703	1,608,774	1,301,404	221,842	1,523,246	3,132,020

Source: Bank of England

12.4d Industrial analysis of monetary financial institutions' lending to UK residents

£ millions

Not seasonally adjusted

Notes to table

Movements in amounts outstanding can reflect breaks in data series as well as underlying flows. For changes data, users are recommended to refer directly to the appropriate series or data tables.

(aw) Due to improvements in reporting at one institution, the amounts outstanding decreased by £1bn. This effect has been adjusted out of the flows for November 2018.

(ax) Due to improvements in reporting at one institution, the amounts outstanding increased by £1bn. This effect has been adjusted out of the flows for November 2018.

(ay) Due to improvements in reporting at one institution, the amounts outstanding decreased by £2bn. This effect has been adjusted out of the flows for April 2018

(az) Due to improvements in reporting at one institution, the amounts outstanding increased by £2bn. This effect has been adjusted out of the flows for April 2018

(ba) Due to a change within the reporting population, the amounts outstanding decreased by £1bn. This effect has been adjusted out of the flows for September 2018.

(bb) Due to improvements in reporting at one institution, the amounts outstanding increased by £1bn. This effect has been adjusted out of the flows for January 2018.

(bc) Due to improvements in reporting at one institution, the amounts outstanding increased by £1bn. This effect has been adjusted out of the flows for May 2018.

(bd) Due to improvements in reporting at one institution, the amounts outstanding increased by £1bn. This effect has been adjusted out of the flows for November 2018.

(be) Due to improvements in reporting at one institution, the amounts outstanding decreased by £1bn. This effect has been adjusted out of the flows for May 2018.

(bf) Due to improvements in reporting at one institution, the amounts outstanding increased by £1bn. This effect has been adjusted out of the flows for July 2018.

(bg) Due to improvements in reporting at one institution, the amounts outstanding increased by £1bn. This effect has been adjusted out of the flows for November 2018.

(bh) Due to improvements in reporting at one institution, the amounts outstanding increased by £3bn. This effect has been adjusted out of the flows for February 2018.

(bi) Due to improvements in reporting at one institution, the amounts outstanding increased by £4bn. This effect has been adjusted out of the flows for February 2018.

(bj) Previously, this series formed part of other financial intermediaries.

(bk) Due to improvements in reporting at one institution, the amounts outstanding increased by £2bn. This effect has been adjusted out of the flows for February 2019.

(bl) Due to improvements in reporting at one institution, the amounts outstanding decreased by £6bn. This effect has been adjusted out of the flows for February 2018.

(bm) From January 2019 onwards, this series no longer includes venture and development capital. As result, this series decreased by £6bn. This effect has been adjusted out the flows for January 2019

(bn) Due to improvements in reporting at one institution, the amounts outstanding decreased by £2bn. This effect has been adjusted out of the flows for February 2019.

(bo) Due to improvements in reporting at one institution, the amounts outstanding increased by £1bn. This effect has been adjusted out of the flows for April 2018

(bp) Due to improvements in reporting at one institution, the amounts outstanding increased by £5bn. This effect has been adjusted out of the flows for September 2018.

(bq) Due to improvements in reporting at one institution, the amounts outstanding increased by £2bn. This effect has been adjusted out of the flows for October 2018.

(br) Due to improvements in reporting at one institution, the amounts outstanding increased by £1bn. This effect has been adjusted out of the flows for February 2019.

(bs) Due to improvements in reporting at one institution, the amounts outstanding decreased by £3bn. This effect has been adjusted out of the flows for January 2018.

(bt) Due to improvements in reporting at one institution, the amounts outstanding decreased by £1bn. This effect has been adjusted out of the flows for February 2019.

Explanatory notes can be found here: http://www.bankofengland.co.uk/statistics/Pages/iadb/notesiadb/industrial.aspx

Copyright guidance and the related UK Open Government Licence can be viewed here: www.bankofengland.co.uk/legal

12.5a Industrial analysis of monetary financial institutions' deposits from UK residents

£ millions

Not seasonally adjusted

Amounts outstanding of deposit liabilities (including under repo) in sterling

RPM	Agriculture, hunting and forestry	Fishing	Mining and quarrying	Manufacturing									Electricity, gas and water supply			Total
				Food, beverages and tobacco	Textiles, wearing apparel and leather	Pulp, paper, and printing	Chemicals, pharmaceuticals, rubber and plastics	Non-metallic mineral products and metals	Machinery, equipment and transport equipment	Electrical, medical and optical equipment	Other manufacturing	Total	Electricity, gas steam and air conditioning	Water collection and sewerage	Waste management related services and remediation activities	
	TDCB	TDCC	TDCD	TDCF	TDCG	TDCH	TDCI	TDCJ	TDCK	TDCL	TDCM	TDCE	TDCN	TDCO	BE2M	B3FI
2017 Jan	7,066	354	4,325	3,904	1,254	1,685	4,642	6,392	9,791	4,666	4,860	37,193	6,548	2,738	1,752	11,038
Feb	6,924	368	4,930	3,571	1,183	1,675	4,527	6,127	9,318	4,561	4,732	35,694	6,169	2,434	1,768	10,370
Mar	6,873	370	5,291	3,502	1,173	1,736	5,053	6,316	9,972	4,573	5,116	37,442	7,771	2,105	1,779	11,655
Apr	6,801	348	4,772	3,567	1,138	1,678	4,867	6,009	9,367	4,807	4,949	36,381	7,788	2,130	1,699	11,617
May	6,834	348	4,087	3,991	1,167	1,658	5,505	6,061	9,506	4,674	5,002	37,563	7,791	2,503	1,754	12,049
Jun	6,897	345	4,256	5,472	1,137	1,672	5,502	6,035	10,032	4,903	5,319	40,073	6,677	2,665	1,843	11,186
Jul	6,856	349	3,818	3,938	1,131	1,734	5,463	5,964	9,493	4,785	5,247	37,756	7,041	2,408	1,844	11,293
Aug	6,868	355	4,000	4,391	1,164	1,706	5,431	6,169	10,103	4,874	5,669	39,505	7,524	2,187	1,820	11,531
Sep	6,927	374	4,623	4,122	1,188	1,694	5,650	6,609	11,315	5,245	5,418	41,241	6,664	2,175	1,846	10,685
Oct	7,180	372	4,751	4,314	1,242	1,977	5,432	6,539	11,453	5,618	5,428	42,003	6,275	2,398	1,760	10,434
Nov	7,175	399	4,925	4,242	1,310	1,911	5,681	6,661	11,921	5,768	5,549	43,042	6,389	2,706	1,889	10,983
Dec	8,013	372	4,900	4,610	1,206	1,897	5,431	6,548	11,838	5,677	5,590	42,795	5,921	2,886	1,905	10,712
2018 Jan	7,727	358	4,548	3,935	1,226	1,831	5,471	6,205	11,079	5,733	5,140	40,620	5,903	2,588	2,030	10,522
Feb	7,696	363	4,406	4,013	1,206	1,781	5,581	6,008	10,583	5,520	4,786	39,479	6,414	2,864	2,030	11,308
Mar	7,709	356	3,921	3,865	1,153	1,740	5,647	6,134	11,111	5,685	4,884	40,219	7,289	2,560	1,960	11,809
Apr	7,569	345	3,847	3,842	1,180	1,765	5,184	6,224	11,076	5,592	4,746	39,608	6,193	2,510	1,963	10,667
May	7,632	346	3,896	4,228	1,182	1,615	5,331	6,301	11,605	5,512	4,760	40,536	6,562	2,770	2,044	11,377
Jun	7,600	350	3,456	4,134	1,196	1,627	5,835	6,491	11,152	5,513	4,868	40,816	6,104	2,581	2,048	10,733
Jul	7,432	347	3,395	4,458	1,171	2,701	5,176	6,508	11,646	5,302	4,943	41,904	6,486	2,644	2,151	11,281
Aug	7,333	348	2,699	4,828	1,180	1,642	5,334	6,544	11,797	5,229	5,262	41,817	7,158	2,478	2,262	11,899
Sep	7,323	340	3,173	4,395	1,172	1,591	5,926	6,581	11,258	5,067	4,885	40,877	7,437	2,521	2,215	12,173
Oct	7,527	341	3,531	4,503	1,188	1,668	5,466	6,606	11,503	5,097	5,268	41,297	7,165	2,613	2,213	11,991
Nov	7,518	363	3,906	4,366	1,248	1,766	5,576	6,643	10,936	5,192	5,560	41,288	6,408	3,021	2,204	11,633
Dec	8,444	365	4,155	4,837	1,242	1,699	5,691	6,660	11,273	4,845	5,463	41,710	6,288	3,080	2,337	11,704
2019 Jan	8,023	356	3,923	4,463	1,180	1,591	5,311	6,513	11,687	4,798	5,324	40,867	6,431	3,546	2,171	12,148
Feb	7,916	352	3,372	4,473	1,132	1,643	5,112	6,355	11,541	4,716	5,312	40,284	6,246	3,385	2,182	11,813
Mar	7,744	350	3,354	4,046	1,183	1,668	4,673	6,425	10,626	4,777	5,273	38,671	6,652	3,613	2,141	12,406
Apr	7,671	337	3,414	4,175	1,137	1,742	4,711	6,580	10,888	4,985	5,359	39,578	6,201	3,202	2,156	11,559

12.5a Industrial analysis of monetary financial institutions' deposits from UK residents

£ millions

Not seasonally adjusted

Amounts outstanding of deposit liabilities (including under repo) in sterling

| | Construction | | | | | | Wholesale and retail trade | | | | Accommodation and food service activities | Transport, storage and communication | | |
| | Development of buildings | Construction of commercial buildings | Construction of domestic buildings | Civil Engineering | Other construction activities | Total | Wholesale and retail trade and repair of motor vehicles and motorcycles | Wholesale trade, excluding motor vehicles and motor cycles | Retail trade excluding motor vehicles and motor cycles | Total | | Transportation and storage | Information and communication | Total |
RPM	B3FW	B7EK	B3LS	B3JO	B4PS	TDCP	TDCR	TDCS	TDCT	TDCQ	TDCU	B5PF	B3NI	TDCV
2017 Jan	8,842	4,508	4,401	6,205	12,729	36,685	5,732	15,569	17,825	39,126	8,378	13,192	23,506	36,698
Feb	8,412	4,498	4,630	5,990	12,402	35,932	5,915	15,701	17,734	39,350	8,489	13,874	23,920	37,794
Mar	8,947	5,037	4,803	6,560	12,813	38,161	6,442	16,715	16,701	39,858	8,771	13,763	24,455	38,218
Apr	8,899	4,834	4,838	6,665	12,846	38,082	6,232	16,345	15,913	38,490	9,032	13,887	24,332	38,219
May	8,853	5,086	4,899	6,620	13,091	38,549	6,261	16,430	17,313	40,004	9,272	14,224	24,255	38,479
Jun	9,026	5,224	5,900	7,044	13,308	40,502	6,311	16,596	17,567	40,474	9,253	13,808	24,938	38,746
Jul	9,153	5,245	5,323	6,666	13,359	39,747	6,343	15,957	16,553	38,852	9,732	13,892	25,218	39,111
Aug	9,540	5,526	5,618	6,689	13,581	40,954	6,381	16,644	17,500	40,525	10,387	14,244	25,367	39,611
Sep	9,189	5,978	5,795	6,617	13,698	41,277	6,465	16,896	17,776	41,138	10,165	14,548	26,152	40,699
Oct	9,310	6,348	6,246	6,430	13,988	42,322	6,588	17,319	16,546	40,453	10,475	14,694	26,085	40,779
Nov	9,788	6,365	5,847	6,598	13,988	42,586	6,496	17,405	19,007	42,907	10,607	16,306	27,113	43,419
Dec	9,820	6,365	6,671	6,993	14,199	44,049	6,209	17,533	21,154	44,896	9,720	14,960	26,894	41,854
2018 Jan	9,716	5,555	6,009	6,569	13,823	41,672	6,396	16,904	16,812	40,112	9,431	14,163	26,263	40,426
Feb	9,663	5,651	5,885	6,457	13,419	41,076	6,532	17,035	18,409	41,976	9,374	13,911	26,888	40,799
Mar	9,775	6,166	6,157	6,639	13,640	42,376	6,631	16,890	17,666	41,188	8,952	14,346	26,895	41,242
Apr	9,927	5,832	5,898	6,378	13,832	41,866	6,919	16,809	17,055	40,783	9,258	14,271	26,986	41,257
May	9,849	5,778	5,987	6,411	13,931	41,955	7,109	16,993	18,866	42,968	9,901	14,889	27,884	42,773
Jun	10,341	6,203	6,838	6,634	14,150	44,166	7,039	17,002	19,105	43,145	9,692	14,228	27,993	42,221
Jul	10,221	6,028	6,151	6,451	14,394	43,244	7,102	17,258	19,340	43,701	10,260	14,088	28,887	42,975
Aug	10,273	6,268	5,713	6,630	14,504	43,388	7,048	17,755	19,529	44,332	10,547	14,118	29,413	43,531
Sep	10,238	6,528	5,857	6,531	14,483	43,637	7,056	17,360	18,444	42,860	10,316	14,723	28,845	43,568
Oct	10,283	6,193	6,128	6,469	15,029	44,101	7,189	17,884	17,623	42,696	11,521 (a)	14,275	28,905	43,180
Nov	10,031	6,257	5,737	6,622	15,227	43,873	7,021	18,730	20,289	46,040	11,329	14,396	30,437	44,833
Dec	10,184	6,827	6,952	6,722	15,250	45,935	6,747	19,003	23,653	49,404	11,020	13,937	30,651	44,588
2019 Jan	9,998	6,067	5,998	6,490	14,332	42,885	6,515	18,056	18,437	43,009	9,811	14,969	31,036	46,006
Feb	10,184	5,873	5,871	6,451	14,032	42,412	6,750	17,567	20,533	44,850	9,731	15,221	31,015	46,236
Mar	10,438	6,163	5,796	6,619	14,262	43,278	6,858	17,939	19,855	44,653	9,696	15,954	30,643	46,596
Apr	10,064	5,965	5,652	6,332	14,410	42,424	6,762	18,160	19,481	44,403	10,098	14,723	30,661	45,384

12.5a Industrial analysis of monetary financial institutions' deposits from UK residents

£ millions

Not seasonally adjusted

Amounts outstanding of deposit liabilities (including under repo) in sterling

| | Real estate, professional services and support activities | | | | Public administration and defence | Education | Human health and social work | Recreational, personal and community service activities | | | Financial intermediation (excluding insurance and pension funds) | | | | |
| | Buying, selling and renting of real estate | Professional, scientific and technical activities | Administrative and support services | Total | | | | Recreational, cultural and sporting activities | Personal and community service activities | Total | Financial leasing corporations | Non-bank credit grantors excluding credit unions and SPVs | Credit unions | Factoring corporations | Mortgage and housing credit corporations excluding SPVs |
RPM	TDCX	B3ON	B6PK	TDCW	TDDB	TDDC	TDDD	TDDF	TDDE	B3S6	TDDH	TDDI	TDDJ	TDDK	TDDL
2017 Jan	43,820	73,203	26,282	143,305	28,169	20,610	20,855	12,869	18,223	31,092	7,900	3,985	1,062	894	7,457
Feb	43,619	73,357	26,636	143,612	28,169	20,836	21,123	12,930	18,190	31,120	7,869	3,620	1,063	886	3,559
Mar	46,472	76,023	27,889	150,384	25,524	19,378	21,162	12,702	18,573	31,275	7,866	3,749	1,063	765	7,621
Apr	45,350	74,500	27,667	147,517	26,379	20,807	20,965	12,214	18,534	30,748	7,700	3,779	1,084	758	7,232
May	44,955	77,578	28,582	151,116	28,349	22,886	21,239	12,479	18,826	31,305	7,885	3,899	1,082	827	6,677
Jun	48,223	81,313	29,093	158,630	29,462	21,988	21,805	12,088	18,675	30,763	8,282	3,784	1,068	981	6,446
Jul	46,626	79,984	29,967	156,577	33,993	20,799	21,570	12,750	18,617	31,367	8,204	4,504	1,069	985	6,565
Aug	46,651	80,540	29,693	156,884	35,501	20,617	21,672	13,140	18,914	32,054	8,659	4,452	1,075	971	6,233
Sep	49,686	80,701	29,785	160,172	28,906	21,831	21,446	12,744	18,916	31,659	8,221	4,243	1,061	936	6,406
Oct	47,979	80,054	30,737	158,770	31,338	22,701	21,866	12,797	19,049	31,847	8,124	4,193	1,062	777	6,064
Nov	47,981	82,059	31,565	161,605	30,532	21,554	21,926	12,676	18,877	31,553	7,760	4,485	1,034	804	6,093
Dec	51,299	77,410	30,516	159,226	28,623	20,214	21,692	13,485	18,973	32,458	8,048	4,449	1,040	302	5,926
2018 Jan	48,676	78,805	32,544 (b)	160,024 (b)	30,825	21,283	21,325	13,893	18,988	32,882	7,846	5,240	1,049	809	5,864
Feb	48,373	78,338	32,434	159,144	30,385	21,554	21,220	14,001	18,682	32,683	8,186	4,925	1,070	1,210	5,714
Mar	51,540	79,301	33,017	163,858	24,914	20,386	21,502	14,042	18,737	32,779	7,222	4,955	1,083	1,100	5,776
Apr	50,492	80,788	32,665	163,946	25,229	21,315	21,353	13,912	18,893	32,805	7,310	5,083	1,086	840	5,715
May	50,058	82,687	33,519	166,264	25,491	23,606	21,730	13,627	19,023	32,650	6,776	4,752	1,083	511	5,674
Jun	52,588	85,808	33,382	171,778	25,583	23,174	21,982	13,533	19,064	32,597	7,290	4,717	1,076	547	5,303
Jul	50,978	84,089	33,429	168,496	26,706	22,115	21,915	13,806	19,196	33,002	7,191	6,332	1,030	625	5,670
Aug	50,608	83,915	34,514	169,036	28,212	21,981	21,992	14,269	19,075	33,344	6,890	5,754	1,034	708	5,485
Sep	54,337	85,933	33,289	173,558	26,560	22,665	22,136	13,890	18,846	32,736	6,520	5,811	1,064	508	5,939
Oct	52,538	86,862	33,351	172,751	26,089	23,751	22,202	14,046	19,145	33,191	6,236	5,664	1,099	436	5,789
Nov	51,903	87,310	33,263	172,476	25,137	22,293	22,353	14,560	18,879	33,439	6,052	5,403	1,057	414	5,479
Dec	55,478	83,118	32,619	171,215	25,458	20,927	22,284	14,915	18,570	33,484	8,798	5,378	1,060	404	5,390
2019 Jan	52,430	84,378	32,211	169,019	25,898	22,704	22,326	14,888	18,780	33,667	8,968	5,482	1,080	412	5,416
Feb	51,671	81,906	31,289	164,866	26,219	23,119	22,291	14,566	18,862	33,428	8,781	5,204	1,075	463	5,283
Mar	54,904	83,849	31,645	170,399	24,283	21,826	22,605	14,857	18,820	33,677	8,644	5,708	1,074	510	5,214
Apr	52,370	82,060	32,379	166,809	26,058	22,907	22,912	14,500	18,954	33,453	8,492	5,633	1,088	526	5,512

12.5a Industrial analysis of monetary financial institutions' deposits from UK residents

£ millions

Not seasonally adjusted

Amounts outstanding of deposit liabilities (including under repo) in sterling

RPM	Investment and unit trusts excluding money market mutual funds	Money market mutual funds	Venture and development capital	Financial intermediation (excluding insurance and pension funds) (contd.)						Insurance companies & pension funds		
				Bank holding companies	Securities dealers	SPVs related to securitisation	Other financial intermediaries	of which intragroup activity	Total	Insurance companies	Pension funds	Total
	TDDM	TDDN	ZO8R	TDDO	TDDP	B3T5	TDDQ	B3U3	TDDG	B7FI	B3VH	TDDR
2017 Jan	19,541	868		17,141	31,628	82,137	57,942	36,785	230,556	25,204	26,248	51,452
Feb	18,941	1,047		17,122	32,537	77,728	59,012	37,176	223,384	24,529	26,638	51,167
Mar	20,435	398		14,602	35,114	78,135	60,350	37,442	230,099	26,749	28,541	55,289
Apr	22,208	648		15,235	32,588	76,705	59,350	37,769	227,287	23,591	27,929	51,521
May	21,354	900		14,128	31,257	75,327	60,246	38,180	223,583	23,583	29,755	53,338
Jun	21,953	522		13,940	38,089	76,904	58,365	35,683	230,334	23,647	28,202	51,850
Jul	21,332	851		13,862	34,546	77,683	57,554	35,749	227,156	22,112	27,648	49,760
Aug	21,716	1,972		14,109	34,529	75,365	57,636	36,000	226,717	23,281	27,142	50,423
Sep	22,520	936		13,592	33,967	83,171	57,150	36,239	232,202	23,897	27,905	51,802
Oct	22,516	696		10,934	36,814	81,520	54,812	34,656	227,512	24,326	27,543	51,869
Nov	23,010	697		10,446	39,392	82,012	52,110	31,167	227,843	24,152	27,127	51,279
Dec	25,577	1,085		10,251	39,267	82,754	51,673	31,685	230,372	25,216	28,988	54,204
2018 Jan	25,833	512		15,613	41,004	81,471	54,593 (d)	32,020	239,835 (d)	19,133 (f)	30,935	50,068 (f)
Feb	25,425	607		14,940	39,853	81,111	49,431	27,066	232,472	18,691	30,842	49,533
Mar	25,740	679		15,236	38,220	82,649	49,315	26,077	231,974	19,075	32,049	51,124
Apr	25,097	537		17,475	41,145	82,095	47,812	25,297	234,194	18,984	30,843	49,827
May	26,419	1,031		10,002	36,970	81,992	45,321	21,930	220,529	19,486	31,051	50,537
Jun	26,677	959		7,917	39,364	82,573	42,309	18,884	218,733	20,085	30,844	50,929
Jul	26,353	543		8,784	34,764	82,352	43,209	19,145	216,852	18,725	30,114	48,839
Aug	26,332	579		9,689	33,334	82,396	40,814	16,341	213,015	18,493	29,157	47,650
Sep	27,282	612		9,403	33,783	84,544	40,564	16,652	216,030	18,308	32,045	50,353
Oct	28,549	1,052		10,834	33,036	84,088	41,559	16,843	218,343	19,695	29,522	49,216
Nov	27,523	1,069		8,801	34,030	79,831	44,558	18,170	214,216	19,788	28,456	48,244
Dec	27,460	1,128	#	8,331	33,566	88,311	42,563	16,898	222,389	20,093	29,081	49,174
2019 Jan	27,328	1,131	2348 (c)	7,466	31,998	90,808	40,073 (e)	16,127	222,512	20,358	28,656	49,014
Feb	26,444	1,325	2151	9,179	30,760	89,287	41,514	17,014	221,464	19,584	29,921	49,505
Mar	26,934	1,324	1852	11,876	32,014	89,368	41,878	17,477	226,394	21,296	30,728	52,024
Apr	25,799	1,433	2232	10,025	34,225	89,279	39,747	15,073	223,990	20,786	31,399	52,185

12.5a Industrial analysis of monetary financial institutions' deposits from UK residents

£ millions

Not seasonally adjusted

Amounts outstanding of deposit liabilities (including under repo) in sterling

RPM	Fund management activities	Activities auxiliary to financial intermediation				Total financial and non-financial businesses	Individuals and individuals trusts	Total UK residents
		Other			Total			
		Central clearing counterparties	Other auxiliary activities	Total				
	TDDS	B7FT	B7FY	TDDT	B3Y8	Z945	TDDU	TDCA
2017 Jan	86,344	94,528	20,264	114,792	201,135	908,036	1,207,752	2,115,788
Feb	88,091	95,776	18,856	114,633	202,724	901,986	1,213,710	2,115,696
Mar	86,962	90,887	20,856	111,743	198,706	918,456	1,223,694	2,142,150
Apr	90,778	112,001	19,666	131,668	222,446	931,411	1,229,246	2,160,657
May	88,844	102,773	20,930	123,703	212,547	931,548	1,225,151	2,156,698
Jun	88,999	91,554	21,531	113,085	202,084	938,649	1,233,967	2,172,615
Jul	90,449	102,885	21,049	123,933	214,383	943,118	1,232,486	2,175,603
Aug	92,777	109,793	21,371	131,164	223,940	961,545	1,235,558	2,197,103
Sep	93,304	105,243	21,893	127,135	220,440	965,587	1,244,134	2,209,721
Oct	97,115	121,392	21,728	143,119	240,234	984,905	1,244,394	2,229,299
Nov	96,085	119,199	22,446	141,645	237,730	990,066	1,248,085	2,238,151
Dec	93,558	96,053	21,801	117,854	211,412	965,510	1,256,079	2,221,590
2018 Jan	98,445	118,292	21,508	139,799	238,245	989,901 (b)(d)(f)	1,244,650	2,234,551 (b)(d)(f)
Feb	100,662	119,648	21,281	140,929	241,591	985,060	1,248,137	2,233,197
Mar	94,794	85,018	21,777	106,795	201,588	945,897	1,260,847	2,206,744
Apr	95,664	91,421 (g)	21,250	112,671 (g)	208,335 (g)	952,205 (g)	1,259,138	2,211,343 (g)
May	105,159	90,295	22,258	112,554	217,712	959,904	1,261,668	2,221,572
Jun	96,737	89,256	23,315	112,571	209,308	956,262	1,269,821	2,226,083
Jul	94,307	104,455	24,030	128,485	222,792	965,253	1,265,801	2,231,054
Aug	98,932	103,909	23,787	127,696	226,628	967,753	1,272,949	2,240,702
Sep	95,388	102,121 (h)	23,326	125,447 (h)	220,835 (h)	969,141 (h)	1,276,511	2,245,652 (h)
Oct	99,058	108,975	23,998	132,973	232,031	983,759 (a)	1,275,828	2,259,586 (a)
Nov	98,624	111,224	23,268	134,492	233,116	982,056	1,284,382	2,266,438
Dec	97,774	103,678	23,913	127,591	225,365	987,622	1,289,438	2,277,061
2019 Jan	98,338	109,361	24,191	133,552	231,890	984,058	1,279,987	2,264,045
Feb	99,634	113,722	23,350	137,072	236,706	984,563	1,287,522	2,272,085
Mar	94,888	94,015	24,949	118,963	213,851	971,808	1,298,547	2,270,355
Apr	97,087	108,718	25,182	133,900	230,988	984,169	1,303,483	2,287,652

Source: Bank of England

12.5a Industrial analysis of monetary financial institutions' deposits from UK residents

£ millions

Not seasonally adjusted

Notes to table

Movements in amounts outstanding can reflect breaks in data series as well as underlying flows. For changes data, users are recommended to refer directly to the appropriate series or data tables. www.bankofengland.co.uk/statistics/Pages/iadb/notesiadb/Changes flows growth rates.aspx.

(a) Due to improvements in reporting at one institution, the amounts outstanding increased by £1bn. This effect has been adjusted out of the flows for October 2018.

(b) Due to improvements in reporting at one institution, the amounts outstanding increased by £1bn. This effect has been adjusted out of the flows for January 2018.

(c) Previously, this series formed part of other financial intermediaries.

(d) Due to improvements in reporting at one institution, the amounts outstanding increased by £4bn. This effect has been adjusted out of the flows for January 2018.

(e) From January 2019 onwards, this series no longer includes venture and development capital. As result, this series decreased by £2bn. This effect has been adjusted out the flows for January 2019

(f) Due to improvements in reporting at one institution, the amounts outstanding decreased by £4bn. This effect has been adjusted out of the flows for January 2018.

(g) Due to improvements in reporting at one institution, the amounts outstanding decreased by £2bn. This effect has been adjusted out of the flows for April 2018.

(h) Due to improvements in reporting at one institution, the amounts outstanding increased by £5bn. This effect has been adjusted out of the flows for September 2018.

Explanatory notes can be found here: http://www.bankofengland.co.uk/statistics/Pages/iadb/notesiadb/industrial.aspx

Copyright guidance and the related UK Open Government Licence can be viewed here: www.bankofengland.co.uk/legal

12.5b Industrial analysis of monetary financial institutions' deposits from UK residents

£ millions
Not seasonally adjusted

Amounts outstanding of deposit liabilities (including under repo) in all currencies

RPM	Agriculture, hunting and forestry	Fishing	Mining and quarrying	Manufacturing									Electricity, gas and water supply			
				Food, beverages and tobacco	Textiles, wearing apparel and leather	Pulp, paper, and printing	Chemicals, pharmaceuticals, rubber and plastics	Non-metallic mineral products and metals	Machinery, equipment and transport equipment	Electrical, medical and optical equipment	Other manufacturing	Total	Electricity, gas steam and air conditioning	Water collection and sewerage	Waste management related services and remediation activities	Total
	TDAB	TDAC	TDAD	TDAF	TDAG	TDAH	TDAI	TDAJ	TDAK	TDAL	TDAM	TDAE	TDAN	TDAO	BE2N	B3FJ
2017 Jan	7,234	380	21,176	4,979	1,720	2,046	11,444	7,699	14,188	7,685	5,974	55,735	7,747	2,758	1,985	12,490
Feb	7,089	416	22,134	4,866	1,535	2,028	10,574	7,544	14,630	7,272	5,879	54,329	7,483	2,454	2,076	12,013
Mar	7,142	421	21,866	4,798	1,652	2,139	10,872	7,472	15,210	7,457	6,287	55,887	9,067	2,129	2,066	13,262
Apr	7,074	400	18,858	4,740	1,622	2,073	10,931	7,228	14,529	7,503	5,958	54,585	8,958	2,147	1,942	13,048
May	7,086	391	19,481	5,304	1,670	1,952	12,014	7,253	15,019	7,841	6,089	57,141	9,124	2,518	1,974	13,616
Jun	7,113	389	23,140	6,928	1,703	1,984	11,547	7,159	16,123	7,739	6,224	59,406	7,748	2,723	2,044	12,515
Jul	7,076	393	21,901	5,542	1,631	2,196	11,535	7,171	15,130	7,749	6,405	57,359	8,102	2,425	2,033	12,561
Aug	7,112	402	24,136	6,222	1,660	1,995	10,783	7,521	14,161	8,300	6,805	57,447	8,636	2,211	2,008	12,855
Sep	7,189	419	23,120	5,496	1,654	1,968	11,074	7,532	15,554	8,397	6,376	58,050	7,713	2,193	2,025	11,932
Oct	7,497	415	23,134	5,679	1,721	2,309	10,887	7,533	15,776	8,685	6,415	59,005	7,195	2,413	1,972	11,579
Nov	7,462	454	23,879	5,747	1,844	2,186	11,381	7,693	16,800	8,882	6,879	61,413	7,396	2,720	2,143	12,259
Dec	8,312	420	23,469	5,884	1,643	2,296	11,322	7,583	17,011	9,225	6,718	61,683	6,831	2,902	2,107	11,840
2018 Jan	7,963	415	20,453	5,569	1,627	2,329	10,956	7,227	15,458	9,023	6,277	58,466	6,715	2,608	2,240	11,563
Feb	7,922	430	21,777	5,255	1,528	2,058	11,115	7,067	14,943	8,561	6,032	56,559	7,166	2,881	2,260	12,307
Mar	7,915	424	20,337	5,307	1,489	2,027	11,673	7,082	16,225	8,461	6,113	58,376	7,972	2,572	2,195	12,739
Apr	7,787	401	17,676	4,928	1,497	2,195	10,710	7,225	15,823	8,746	6,061	57,186	6,873	2,531	2,208	11,612
May	7,863	419	17,758	5,456	1,510	2,059	11,588	7,308	16,128	8,657	6,544	59,249	7,364	2,790	2,236	12,390
Jun	7,818	419	18,465	5,980	1,503	2,051	11,137	7,451	17,369	8,509	6,554	60,554	6,906	2,596	2,235	11,738
Jul	7,629	406	22,444	5,468	1,460	3,062	10,959	7,424	15,974	8,570	6,830	59,748	7,516	2,661	2,805	12,983
Aug	7,522	396	19,918	6,031	1,452	3,184	11,242	7,445	16,177	8,548	7,136	61,215	8,072	2,495	2,925	13,492
Sep	7,501	403	19,769	5,943	1,554	3,116	11,813	7,488	15,749	8,623	5,928	60,214	8,100	2,536	2,789	13,425
Oct	7,731	382	22,638	5,942	1,473	3,145	11,671	7,552	16,348	9,105	7,108	62,344	7,935	2,629	2,910	13,474
Nov	7,734	412	25,980	5,646	1,518	3,206	12,052	7,589	15,633	9,098	7,259	61,999	7,242	3,036	2,860	13,138
Dec	8,681	400	24,787	5,928	1,531	3,099	12,647	7,924	18,064	8,935	6,762	64,890	7,130	3,097	2,893	13,120
2019 Jan	8,253	400	21,060	5,598	1,531	1,939	11,483	7,574	17,265	8,655	6,651	60,695	7,290	3,562	3,134	13,986
Feb	8,124	390	20,183	6,263	1,474	1,907	8,921	7,394	16,782	8,840	8,855	60,438	7,008	3,401	3,172	13,581
Mar	7,951	384	18,974	5,362	1,576	1,917	9,619	7,206	16,354	9,362	8,999	60,394	7,418	3,628	3,038	14,084
Apr	7,916	374	16,654	5,487	1,436	2,009	10,700	7,477	15,980	9,082	8,934	61,103	7,089	3,219	3,056	13,364

12.5b Industrial analysis of monetary financial institutions' deposits from UK residents

£ millions

Not seasonally adjusted

Amounts outstanding of deposit liabilities (including under repo) in all currencies

	Development of buildings	Construction					Wholesale and retail trade				Accommodation and food service activities	Transport, storage and communication		
		Construction of commercial buildings	Construction of domestic buildings	Civil Engineering	Other construction activities	Total	Wholesale and retail trade and repair of motor vehicles and motorcycles	Wholesale trade, excluding motor vehicles and motor cycles	Retail trade excluding motor vehicles and motor cycles	Total		Transportation and storage	Information and communication	Total
RPM	B3FX	B7EL	B3LT	B4PG	B4PT	TDAP	TDAR	TDAS	TDAT	TDAQ	TDAU	B5PG	B3NJ	TDAV
2017 Jan	8,905	4,576	4,418	6,656	12,941	37,495	6,249	22,925	20,038	49,212	9,013	16,596	31,199	47,794
Feb	8,475	4,557	4,641	6,424	12,628	36,724	6,444	22,875	19,950	49,270	9,160	16,887	31,891	48,779
Mar	9,005	5,102	4,811	7,066	13,044	39,029	7,104	24,385	18,679	50,168	9,388	17,204	32,445	49,649
Apr	8,954	4,893	4,846	7,099	13,077	38,869	6,808	25,252	17,926	49,986	9,718	17,134	31,565	48,699
May	8,909	5,148	4,907	7,029	13,295	39,289	6,893	26,736	19,443	53,072	9,929	17,350	32,206	49,556
Jun	9,094	5,281	5,913	7,489	13,508	41,285	7,017	26,444	19,761	53,222	9,948	17,221	33,843	51,064
Jul	9,214	5,283	5,335	7,052	13,557	40,441	6,959	25,869	18,783	51,612	10,503	16,915	35,698	52,613
Aug	9,595	5,583	5,628	7,063	13,789	41,657	6,893	27,426	19,861	54,180	11,092	17,386	35,800	53,186
Sep	9,247	6,011	5,803	6,891	13,895	41,846	6,962	26,781	20,026	53,770	10,854	18,643	36,252	54,895
Oct	9,368	6,392	6,257	6,713	14,211	42,940	7,056	27,854	18,852	53,762	11,130	17,947	35,667	53,614
Nov	9,855	6,411	5,858	6,941	14,270	43,335	6,970	27,799	21,434	56,203	11,256	19,688	35,894	55,582
Dec	9,893	6,439	6,679	7,301	14,421	44,732	6,573	27,852	23,696	58,121	10,427	18,296	36,901	55,197
2018 Jan	9,780	5,614	6,017	6,971	14,059	42,442	6,825	25,971	19,400	52,196	10,142	17,253	36,537	53,790
Feb	9,777	5,705	5,901	6,924	13,685	41,991	6,910	25,901	21,191	54,001	10,051	17,026	37,477	54,503
Mar	9,885	6,219	6,172	7,072	13,865	43,212	6,994	25,836	20,256	53,086	9,558	17,141	37,148	54,290
Apr	10,013	5,889	5,931	6,771	14,051	42,655	7,306	27,055	19,622	53,983	9,938	19,225	36,832	56,057
May	9,929	5,828	6,018	6,779	14,228	42,783	7,482	26,979	21,454	55,916	10,577	18,458	38,275	56,732
Jun	10,410	6,257	6,869	6,990	14,469	44,995	7,383	26,402	21,645	55,430	10,421	17,816	38,810	56,626
Jul	10,286	6,074	6,185	6,775	14,674	43,994	7,555	25,968	22,041	55,563	10,827	17,879	39,402	57,282
Aug	10,335	6,313	5,741	6,960	14,719	44,069	7,459	26,745	21,914	56,118	11,028	17,645	40,126	57,771
Sep	10,302	6,582	5,885	6,902	14,695	44,367	7,458	26,033	21,004	54,495	10,768	18,425	39,630	58,056
Oct	10,378	6,245	6,169	6,780	15,319	44,890	7,675	26,025	20,181	53,882	12,932 (r)(s)	18,176	41,794	59,970
Nov	10,111	6,317	5,772	6,951	15,541	44,692	7,543	26,825	22,874	57,242	12,555	17,895	42,745	60,640
Dec	10,249	6,927	6,980	7,013	15,486	46,654	7,182	28,171	26,175	61,528	12,149	18,196	42,988	61,183
2019 Jan	10,064	6,130	6,029	6,796	14,550	43,569	6,949	26,374	20,969	54,292	10,204	17,808	44,091	61,899
Feb	10,246	5,935	5,902	6,730	14,250	43,063	7,279	25,378	23,007	55,663	10,141	18,007	41,516	59,522
Mar	10,500	6,270	5,827	6,907	14,536	44,039	7,417	26,154	22,202	55,772	10,108	19,323	41,994	61,316
Apr	10,134	6,057	5,681	7,818	14,622	44,312	7,323	25,914	21,992	55,229	10,574	17,531	41,367	58,898

12.5b Industrial analysis of monetary financial institutions' deposits from UK residents

£ millions

Not seasonally adjusted

Amounts outstanding of deposit liabilities (including under repo) in all currencies

RPM	Real estate, professional services and support activities				Public administration and defence	Education	Human health and social work	Recreational, personal and community service activities			Financial intermediation (excluding insurance and pension funds)				
	Buying, selling and renting of real estate	Professional, scientific and technical activities	Administrative and support services	Total				Recreational, cultural and sporting activities	Personal and community service activities	Total	Financial leasing corporations	Non-bank credit grantors excluding credit unions and SPVs	Credit unions	Factoring corporations	Mortgage and housing credit corporations excluding SPVs
	TDAX	B3TK	B6PL	TDAW	TDBB	TDBC	TDBD	TDBF	TDBE	B3S7	TDBH	TDBI	TDBJ	TDBK	TDBL
2017 Jan	44,721	82,465	32,054	159,240	31,663	21,637	21,854	14,612	19,222	33,834	8,520	4,413	1,062	944	7,648
Feb	44,497	82,290	32,192	158,979	31,018	21,835	22,189	14,608	19,202	33,810	8,278	4,075	1,063	942	4,048
Mar	47,351	84,952	33,848	166,151	27,917	20,375	22,247	14,663	19,646	34,309	8,254	4,135	1,063	870	8,122
Apr	46,207	83,711	33,482	163,400	30,330	21,769	22,029	14,116	19,565	33,681	8,025	4,120	1,084	943	7,681
May	46,779	86,894	34,777	168,450	32,415	23,819	22,273	14,249	19,799	34,048	8,219	4,447	1,083	1,034	7,139
Jun	49,448	90,210	35,323	174,981	32,746	22,920	22,810	14,389	19,733	34,122	8,583	4,093	1,069	1,323	6,914
Jul	47,744	88,012	35,634	171,390	39,696	21,702	22,602	15,121	19,734	34,855	8,524	4,981	1,070	1,406	7,014
Aug	47,781	88,811	35,928	172,520	40,371	21,599	22,683	15,279	20,036	35,315	9,148	5,033	1,076	1,277	6,635
Sep	50,782	88,625	37,061	176,467	33,481	22,823	22,352	14,844	19,955	34,800	8,684	4,739	1,062	1,110	6,777
Oct	49,031	88,614	37,738	175,383	36,871	23,812	22,787	14,954	20,019	34,972	8,957	4,616	1,063	985	6,123
Nov	49,225	90,844	38,930	178,999	36,026	22,632	22,837	14,883	19,832	34,714	8,450	4,941	1,035	992	6,153
Dec	52,524	87,071	37,675	177,270	34,264	21,359	22,701	15,794	20,049	35,842	8,506	4,865	1,042	535	5,983
2018 Jan	49,616	87,808	39,595 (u)	177,019 (u)	35,871	22,505	22,294	16,326	20,012	36,338	8,302	5,713	1,051	1,015	5,922
Feb	49,243	87,440	39,952	176,635	34,932	22,771	22,282	15,881	19,664	35,545	8,649	5,467	1,072	1,449	5,774
Mar	52,412	87,794	39,169	179,375	29,596	21,574	22,558	16,084	19,881	35,965	7,698	5,516	1,084	1,248	5,835
Apr	51,406	89,273	39,047	179,726	30,777	22,475	22,444	15,839	19,898	35,737	7,947	5,660	1,087	1,023	5,772
May	51,007	91,874 (t)	39,220 (v)	182,100	29,485	24,767	22,795	15,697	19,992	35,689	7,287	5,363	1,141	714	5,814
Jun	53,480	95,060	39,461	188,001	29,962	24,237	23,054	15,990	20,086	36,075	7,846	5,306	1,135	786	5,447
Jul	52,230	93,357	39,066	184,653	32,822	23,118	23,057	16,373	20,312	36,686	7,649	7,014	1,032	886	5,727
Aug	51,844	93,744	40,065	185,652	32,964	23,014	23,170	16,436	20,061	36,498	7,315	6,482	1,035	1,000	5,545
Sep	55,520	95,148	38,910	189,578	32,273	23,693	23,322	16,092	19,911	36,002	6,880	6,468	1,065	787	5,997
Oct	53,601	95,927	39,163	188,690	31,298	24,762	23,404	16,171	20,078	36,249	6,764	6,277	1,100	721	5,845
Nov	52,997	96,905	39,757	189,659	30,064	23,300	23,612	16,688	19,949	36,636	6,644	5,907	1,059	757	5,543
Dec	56,643	93,448	38,721	188,812	27,861	22,061	23,663	17,342	19,610	36,952	9,256	5,970	1,061	804	5,451
2019 Jan	53,439	94,206	38,855	186,500	29,147	23,770	23,616	16,860	19,781	36,641	9,435	5,945	1,081	776	5,479
Feb	52,751	91,933	38,136	182,820	30,052	24,205	23,527	16,774	19,863	36,637	9,422	5,645	1,076	565	5,342
Mar	55,918	93,706	38,618	188,242	29,647	23,023	23,810	16,947	19,822	36,770	9,135	6,225	1,076	567	5,271
Apr	53,409	92,307	39,387	185,104	30,394	24,153	24,069	16,601	19,945	36,547	9,016	6,120	1,090	603	5,587

12.5b Industrial analysis of monetary financial institutions' deposits from UK residents

£ millions

Amounts outstanding of deposit liabilities (including under repo) in all currencies

Not seasonally adjusted

RPM	Investment and unit trusts excluding money market mutual funds	Money market mutual funds	Bank holding companies	Venture and development capital	Securities dealers	SPVs related to securitisation	Other financial intermediaries	of which intragroup activity	Total	Insurance companies	Pension funds	Total
			Financial intermediation (excluding insurance and pension funds) (contd.)							Insurance companies & pension funds		
	TDBM	TDBN	TDBO	ZO8T	TDBP	B3T6	TDBQ	B3U4	TDBG	B3V2	B7FO	TDBR
2017 Jan	27,286	898	97,072		99,934	90,715	90,697	60,955	429,190	33,530	31,106	64,637
Feb	26,805	1,066	98,909		92,825	86,365	94,261	62,919	418,637	32,788	31,453	64,241
Mar	29,035	418	89,221		96,233	87,356	94,401	63,427	419,110	34,838	33,673	68,511
Apr	30,613	684	87,521		96,984	85,593	96,443	66,999	419,691	32,206	33,362	65,568
May	29,903	927	92,346		92,149	84,330	99,253	69,100	420,830	32,496	35,026	67,521
Jun	30,767	739	94,320		101,383	86,096	95,647	64,518	430,933	32,036	33,172	65,208
Jul	30,140	900	93,128		89,461	86,492	97,137	67,578	420,251	30,429	33,625	64,053
Aug	30,124	2,327	95,054		91,971	84,299	96,872	67,405	423,815	31,751	31,936	63,687
Sep	30,983	1,188	85,927		95,435	91,917	96,359	67,495	424,180	31,706	32,653	64,359
Oct	31,215	1,155	80,282		98,586	90,876	96,343	68,819	420,201	32,235	32,391	64,625
Nov	32,248	1,191	76,020		110,216	92,005	94,648	65,842	427,899	32,459	31,858	64,316
Dec	34,138	1,628	76,856		102,562	92,114	92,404	64,317	420,632	32,590	34,018	66,608
2018 Jan	35,044	947	86,118		105,781	89,919	96,559 (y)(z)	65,903	436,369 (y)(z)	26,761 (ab)	36,103	62,864
Feb	35,090	897	88,075		106,305	89,754	88,291	58,236	430,823	26,081	35,455	61,536
Mar	34,335	1,003	90,833		104,363	92,179	83,563	52,015	427,657	26,501	37,076	63,578
Apr	34,318	763	87,226		119,552	91,738	84,417	53,721	439,503	26,246	36,080	62,325
May	35,746	1,535	86,979		109,554	91,936	86,536	54,891	432,606	27,518	35,879	63,397
Jun	36,022	1,275	84,467		115,231	93,728	84,596	52,086	435,839	28,338	35,820	64,157
Jul	36,173	869	80,331		104,742	93,521	86,067	52,851	424,012	27,648	34,915	62,563
Aug	35,646	861	82,014		106,677	93,662	81,827	48,550	422,065	26,429	33,861	60,291
Sep	36,690	921	82,769		111,300	98,420	85,103	52,358	436,400	25,789	37,124	62,914
Oct	39,428	1,412	92,460		113,846	98,303	87,355	53,819	453,511	27,451	33,703	61,154
Nov	37,593	1,422	92,874		111,977	93,884	92,436	56,914	450,096	28,701	33,079	61,781
Dec	37,883	1,369	72,420	#	110,672	103,794	86,952	53,142	435,632	28,769	33,535	62,304
2019 Jan	37,212	1,538	74,871	2,951 (w)	101,905	106,185	83,547 (aa)	49,844	430,924	28,464	34,074	62,538
Feb	36,484	2,013	74,607	2,767	98,767	101,762 (x)	83,395	48,541	421,846 (x)	27,160	35,452	62,612
Mar	36,477	2,592	77,615	2,339	101,672	101,783	83,895	49,241	428,647	29,617	36,120	65,737
Apr	35,955	2,482	76,940	2,817	111,381	101,681	81,840	47,144	435,510	29,737	36,423	66,160

12.5b Industrial analysis of monetary financial institutions' deposits from UK residents

£ millions

Not seasonally adjusted

Amounts outstanding of deposit liabilities (including under repo) in all currencies

	Fund management activities	Activities auxiliary to financial intermediation			Total	Total financial and non-financial businesses	Individuals and individuals trusts	Total UK residents
		Other						
		Central clearing counterparties	Other auxiliary activities	Total				
RPM	TDBS	B7FU	B7FZ	TDBT	B5H5	Z8ZX	TDBU	TDAA
2017 Jan	183,825	174,364	31,736	206,100	389,926	1,392,510	1,215,497	2,608,007
Feb	200,346	192,734	31,129	223,862	424,208	1,414,830	1,221,710	2,636,540
Mar	203,155	172,233	33,262	205,495	408,650	1,414,081	1,231,928	2,646,009
Apr	204,287	192,878	32,382	225,260	429,547	1,427,251	1,237,522	2,664,773
May	198,177	188,175	33,056	221,231	419,409	1,438,324	1,233,782	2,672,105
Jun	202,926	163,691	33,979	197,669	400,595	1,442,397	1,242,492	2,684,890
Jul	208,199	191,754	33,872	225,626	433,825	1,462,831	1,241,146	2,703,977
Aug	210,775	207,905	34,084	241,989	452,764	1,494,822	1,244,336	2,739,158
Sep	205,683	194,028	34,866	228,894	434,577	1,475,114	1,252,645	2,727,759
Oct	217,344	214,376	34,681	249,058	466,402	1,508,130	1,253,086	2,761,216
Nov	215,687	218,677	31,867	250,544	466,231	1,525,498	1,256,786	2,782,284
Dec	209,518	178,011	31,575	209,586	419,104	1,471,982	1,265,255	2,737,237
2018 Jan	221,622 (ac)	212,190	31,496	243,686	465,308 (ac)	1,515,996 (u)(y)(z)(ab)(ac)	1,253,408	2,769,404 (u)(y)(z)(ab)(ac)
Feb	227,583	215,840	30,457	246,296	473,879	1,517,945	1,257,097	2,775,042
Mar	218,604	163,867	32,787	196,654	415,259	1,455,498	1,269,716	2,725,214
Apr	207,641	185,524 (ad)	32,155	217,679 (ad)	425,320 (ad)	1,475,603 (ad)	1,268,195	2,743,798 (ad)
May	224,020	192,737	33,532	226,270	450,290	1,504,815	1,270,772	2,775,588
Jun	207,986	181,806	34,905	216,711	424,697	1,492,490	1,278,907	2,771,397
Jul	202,755	206,635	34,944	241,579	444,334	1,502,121	1,274,910	2,777,032
Aug	211,189	210,526	34,530	245,056	456,245	1,511,426	1,282,247	2,793,673
Sep	207,997	197,829 (ae)	33,886	231,715 (ae)	439,712 (ae)	1,512,892 (ae)	1,286,024	2,798,916 (ae)
Oct	217,233	205,920	34,739	240,658	457,891	1,555,202 (r)(s)	1,285,818	2,841,020 (r)(s)
Nov	227,268	218,746 (af)	33,552	252,298 (af)	479,566 (af)	1,579,106 (af)	1,294,481	2,873,587 (af)
Dec	235,396	196,184	34,877	231,061	466,457	1,557,136	1,299,662	2,856,798
2019 Jan	222,457	209,756	34,317	244,074	466,531	1,534,024	1,289,982	2,824,006
Feb	217,131	197,581	34,115	231,696	448,826	1,501,631 (x)	1,297,354	2,798,984 (x)
Mar	211,680	173,048	35,942	208,991	420,671	1,489,570	1,308,594	2,798,164
Apr	213,506	187,059	36,477	223,537	437,042	1,507,404	1,313,646	2,821,049

Source: Bank of England

12.5b Industrial analysis of monetary financial institutions' deposits from UK residents

£ millions

Not seasonally adjusted

Notes to table

Movements in amounts outstanding can reflect breaks in data series as well as underlying flows. For changes data, users are recommended to refer directly to the appropriate series or data tables. www.bankofengland.co.uk/statistics/Pages/iadb/notesiadb/Changes_flows_growth_rates.aspx.

(r) Due to improvements in reporting at one institution, the amounts outstanding increased by £1bn. This effect has been adjusted out of the flows for October 2018.
(s) Due to improvements in reporting at one institution, the amounts outstanding increased by £1bn. This effect has been adjusted out of the flows for October 2018.
(t) Due to improvements in reporting at one institution, the amounts outstanding decreased by £1bn. This effect has been adjusted out of the flows for May 2018.
(u) Due to improvements in reporting at one institution, the amounts outstanding increased by £1bn. This effect has been adjusted out of the flows for January 2018.
(v) Due to improvements in reporting at one institution, the amounts outstanding increased by £1bn. This effect has been adjusted out of the flows for May 2018.
(w) Previously, this series formed part of other financial intermediaries.
(x) Due to improvements in reporting at one institution, the amounts outstanding increased by £2bn. This effect has been adjusted out of the flows for February 2019
(y) Due to improvements in reporting at one institution, the amounts outstanding increased by £2bn. This effect has been adjusted out of the flows for January 2018.
(z) Due to improvements in reporting at one institution, the amounts outstanding increased by £4bn. This effect has been adjusted out of the flows for January 2018.
(aa) From January 2019 onwards, this series no longer includes venture and development capital. As result, this series decreased by £3bn. This effect has been adjusted out the flows for January 2019
(ab) Due to improvements in reporting at one institution, the amounts outstanding decreased by £4bn. This effect has been adjusted out of the flows for January 2018.
(ac) Due to improvements in reporting at one institution, the amounts outstanding decreased by £2bn. This effect has been adjusted out of the flows for January 2018.
(ad) Due to improvements in reporting at one institution, the amounts outstanding decreased by £2bn. This effect has been adjusted out of the flows for April 2018.
(ae) Due to improvements in reporting at one institution, the amounts outstanding increased by £5bn. This effect has been adjusted out of the flows for September 2018.
(af) Due to improvements in reporting at one institution, the amounts outstanding increased by £1bn. This effect has been adjusted out of the flows for November 2018.

Explanatory notes can be found here: http://www.bankofengland.co.uk/statistics/Pages/iadb/notesiadb/industrial.aspx

Copyright guidance and the related UK Open Government Licence can be viewed here: www.bankofengland.co.uk/legal

12.6a Components of M4

£ millions   Seasonally adjusted

Amounts outstanding

LPM	Retail deposits and cash in M4			Wholesale deposits in M4		M4	M3 (estimate of EMU aggregate for the UK)
	Deposits	Notes and coin	Total	Deposits	of which repos		
	B3SF	VQJO	VQWK (a)	VRGP	VZZQ	AUYN	VWYZ
2017 Aug	1,592,998	74,163	1,667,162	658,856	129,715	2,338,047	2,775,103
Sep	1,596,471	73,824	1,670,295	683,740 (g)(h)	147,148 (m)	2,351,112 (g)(h)	2,786,224 (g)(h)
Oct	1,599,409	73,809	1,673,218	684,006	148,573	2,364,892	2,802,702
Nov	1,605,733	73,532	1,679,265	681,330	141,244	2,370,637	2,800,239
Dec	1,607,160	73,087	1,680,247	678,571	144,344	2,354,148	2,805,514
2018 Jan	1,620,051 (c)	73,439	1,693,490 (c)	692,077 (f)	144,519	2,386,125 (j)(k)	2,821,879 (k)(l)
Feb	1,624,374	73,237	1,697,610	677,047	140,158	2,375,380	2,823,499
Mar	1,624,252 (d)	73,018	1,697,270 (d)	664,025 (g)	128,627	2,361,186	2,791,738
Apr	1,628,983	72,515	1,701,498	663,307	126,902	2,362,016	2,793,293
May	1,634,605	73,397	1,708,002	669,895	125,399	2,377,052	2,835,188
Jun	1,642,699	73,292	1,715,990	661,419	133,269	2,374,448	2,831,453
Jul	1,647,255	73,493	1,720,748	659,459	128,330	2,383,564	2,832,069
Aug	1,652,071	73,217	1,725,288	650,908	127,955	2,377,785	2,834,937
Sep	1,652,904	73,184	1,726,088	656,912 (h)	136,335 (h)	2,383,459 (h)	2,852,754 (h)
Oct	1,659,567 (e)	73,782	1,733,349 (e)	657,377 (i)	137,693	2,392,136	2,884,182
Nov	1,666,735	73,424	1,740,159	649,132	134,964	2,389,406	2,876,191
Dec	1,671,430	74,056	1,745,486	664,803	149,782	2,409,347	2,904,439
2019 Jan	1,676,555	74,144	1,750,699	657,250	138,125	2,407,508	2,866,619
Feb	1,680,384	74,293	1,754,677	658,542	141,781	2,413,221	2,843,683
Mar	1,684,274	74,188	1,758,462	662,435	138,913	2,418,427	2,857,619
Apr	1,688,717	74,582	1,763,299	674,900	149,027	2,434,377	2,866,100

Source: Bank of England

Notes to table

Movements in amounts outstanding can reflect breaks in data series as well as underlying flows. For changes and growth rates data, users are recommended to refer directly to the appropriate series or data tables. Further details can be found at https://www.bankofengland.co.uk/statistics/details/further-details-about-changes-flows-growth-rates-data.

(a) A minor change was made to the definition of Retail M4 in October 2007. From October data onwards, non-interest-bearing bank deposits are only included in Retail M4 when reporters identify them explicitly as being taken from retail sources. There was also a change to the reporting population in October 2007. Together these led to a break in the amount outstanding of Retail M4 in October 2007. The effect of this has been removed from the flows data.

(b) Changes were introduced with effect from the publication of May 2019 data, following a review. The construction of this series now includes a seasonally adjusted intermediate other financial corporation (IOFC) component. Previously, this component was not seasonally adjusted.

(c) Due to improvements in reporting at one institution, the amounts outstanding increased by £8bn. This effect has been adjusted out of the flows for January 2018.

(d) Due to improvements in reporting at one institution, the amounts outstanding decreased by £4bn. This effect has been adjusted out of the flows for March 2018.

(e) Due to improvements in reporting at one institution, the amounts outstanding increased by £3bn.

(f) Due to improvements in reporting at one institution, the amounts outstanding decreased by £8bn. This effect has been adjusted out of the flows for January 2018.

(g) Due to improvements in reporting at one institution, the amounts outstanding increased by £4bn. This effect has been adjusted out of the flows for March 2018.

(h) Due to improvements in reporting at one institution, the amounts outstanding increased by £5bn. This effect has been adjusted out of the flows for September 2018.

(i) Due to improvements in reporting at one institution, the amounts outstanding decreased by £3bn.

(j) Due to improvements in reporting at one institution, the amounts outstanding increased by £5bn. This effect has been adjusted out of the flows for January 2018.

(k) Due to improvements in reporting at one institution, the amounts outstanding decreased by £4bn. This effect has been adjusted out of the flows for January 2018.

(l) Due to improvements in reporting at one institution, the amounts outstanding increased by £5bn. This effect has been adjusted out of the flows for January 2018.

Explanatory notes can be found here: http://www.bankofengland.co.uk/statistics/details/further-details-about-m4-data
Copyright guidance and the related UK Open Government Licence can be viewed here: www.bankofengland.co.uk/legal

12.6b Components of M4

£ millions   Not seasonally adjusted

Changes

LPM	Retail deposits and cash in M4					Wholesale deposits in M4		M4	M3 (estimate of EMU aggregate for the UK)
	Total	Deposits of which:		Notes and coin	Total	Deposits	of which repos		
		Private non-financial corporations	Household sector						
	VRLW	Z599	Z59A	VQLU	VQZA	VRLR	VWDN	AUZI	VWXK
2017 Jan	-18,116	-5,853	-10,767	-2,054	-20,170	21,120	26,351	950	18,526
Feb	3,228	-622	5,316	170	3,399	-4,261	-1,031	-862	26,341
Mar	22,311	8,180	8,157	861	23,172	7,016	-166	30,188	17,935
Apr	3,399	-1,447	7,788	825	4,224	21,617	15,039	25,841	33,602
May	2,426	3,704	-5,282	-822	1,604	-7,841	-8,436	-6,237	-1,972
Jun	12,542	6,545	8,897	631	13,173	-5,935	-7,925	7,238	3,391
Jul	-2,163	-2,620	-1,310	291	-1,873	-1,795	8,231	-3,668	5,788
Aug	5,376	3,072	590	-196	5,180	17,896	6,826	23,076	23,680
Sep	10,575	2,750	10,002	326	10,901	-4,344	-1,166	6,556	8,030
Oct	691	-949	4,108	-648	43	8,666	14,037	8,710	18,116
Nov	4,952	2,612	2,983	417	5,369	-639	-4,728	4,730	22,380
Dec	5,909	1,564	7,357	1,327	7,237	-18,366	-17,447	-11,130	-42,248
2018 Jan	-15,461	-6,557	-10,598	-2,939	-18,401	33,476	20,453	15,075	47,202
Feb	2,624	-230	3,921	-182	2,442	-7,353	-1,664	-4,911	-8,886
Mar	20,282	4,162	12,308	1,466	21,748	-33,918	-28,756	-12,170	-33,498
Apr	912	2,299	-1,109	-1,095	-184	-4,049	2,640	-4,233	6,612
May	6,343	3,995	2,969	550	6,893	9,625	-2,996	16,518	32,985
Jun	14,266	5,627	7,384	534	14,800	-12,950	765	1,850	-9,005
Jul	-2,548	-1,000	-3,748	-42	-2,590	4,591	5,105	2,001	-2,519
Aug	5,341	1,924	7,640	300	5,641	-383	3,337	5,258	10,324
Sep	6,722	1,174	3,589	129	6,851	-8,093	-5,172	-1,242	5,428
Oct	4,443	1,560	710	-399	4,044	8,118	12,314	12,161	34,956
Nov	9,586	1,521	8,530	827	10,413	-6,962	353	3,451	20,965
Dec	7,315	3,631	3,966	1,181	8,496	3,683	-3,258	12,179	-14,606
2019 Jan	-16,295	-7,387	-9,223	-2,341	-18,636	3,167	6,077	-15,470	-18,456
Feb	3,563	-3,483	8,192	158	3,721	6,073	7,248	9,794	-20,212
Mar	18,713	4,726	10,395	831	19,545	-12,017	-22,111	7,527	-3,302
Apr	1,189	-2,464	4,740	-329	859	14,468	16,246	15,327	21,571

Notes to table   Source: Bank of England

Explanatory notes can be found here: http://www.bankofengland.co.uk/statistics/details/further-details-about-m4-data

Copyright guidance and the related UK Open Government Licence can be viewed here: www.bankofengland.co.uk/legal

12.7 Counterparts to changes in M4: alternative presentation

£ millions   Not seasonally adjusted

Changes

	Public sector net cash requirement (PSNCR)	M4 private sector net purchases (-) of Central Government debt						M4 private sector net purchases of other public sector debt			
		Gilts	Sterling treasury bills	Tax instruments	National Savings	Other	Total	Local government debt (-)	Public corporation debt (-)	Other public sector purchases of M4 private sector debt (+)	Total
LPM	ABEN (g)	AVBY	VQLK	VQLG	VQLJ	VQLI	RCMD	VQLL	VQLO	VQLQ	AVBV
2017 Jan	-27,125	2,893	6,913	-18	-1,688	9,668	17,767	-105	-	353	248
Feb	-3,114	-9,636	5,500	179	-1,551	4,527	-980	369	-	-1,200	-831
Mar	22,441	-5,603	6,902	72	-599	-11,235	-10,463	576	-	-1,946	-1,370
Apr	-17,659	-6,961	12,343	-74	-884	10,208	14,632	-444	-	1,532	1,088
May	9,730	-2,098	4,862	-61	-396	-6,601	-4,293	-191	-	-387	-578
Jun	17,627	-4,800	-1,896	-79	-348	-4,818	-11,941	-64	-	623	559
Jul	-10,680	-12,027	6,629	-55	-575	5,043	-985	-163	-	185	22
Aug	2,902	-2,049	-7,580	58	-610	6,545	-3,637	-33	-	179	146
Sep	18,943	23,701	-3,558	-130	-591	-15,712	3,709	-329	-	-94	-423
Oct	-7,426	-4,277	-454	-15	-808	16,124	10,570	182	-	87	269
Nov	12,665	-544	-6,722	-65	-809	-4,494	-12,634	-119	-	-919	-1,038
Dec	20,215	-10,527	-7,134	15	-1,651	-361	-19,659	-52	-	448	396
2018 Jan	-27,771	-9,593	15,853	40	-335	9,804	15,769	221	-	133	354
Feb	582	5,950	2,127	326	-1,444	7,022	13,981	258	-	-674	-416
Mar	21,709	2,408	-3,985	208	-1,137	-24,224	-26,730	99	-	-577	-478
Apr	-8,328	-1,577	4,429	22	-1,073	3,362	5,162	-576	-	1,140	564
May	6,330	-5,023	2,582	16	-940	5,943	2,578	-152	-	184	32
Jun	12,126	-8,306	-1,632	-11	-1,090	-2,341	-13,379	10	-	772	782
Jul	-15,510	-3,045	-4,168	39	-821	881	-7,115	-470	-	80	-390
Aug	4,223	12,118	-3,026	130	-876	5,468	13,814	7	-	-494	-487
Sep	14,562	-4,759	-5,122	12	-756	-1,085	-11,709	-40	-	274	234
Oct	-3,849	-10,922	4,299	29	-1,196	7,859	69	121	-	369	490
Nov	7,611	-1,770	-1,299	18	-1,023	-938	-5,013	178	-	-324	-146
Dec	18,123	-271	-1,197	20	-1,035	-11,312	-13,795	101	-	345	446
2019 Jan	-26,061	-10,396	8,689	30	-387	18,401	16,336	-327	-	658	331
Feb	90	-13,358	-408	219	-823	11,946	-2,424	155	-	-550	-395
Mar	23,431	16,294	-9,413	19	-704	-30,551	-24,355	291	-	-788	-497
Apr	-8,014	2,762	4,057	19	-804	4,653	10,686	-34	-	706	672

£ millions   Not seasonally adjusted

Changes

	Purchases of public sector net debt (-)	External and foreign currency finance of public sector				Public sector contrib-ution	M4 lending				
		Non-residents' purchases of gilts (-)	Non-residents' purchases of £TBs (-)	Other	Total		Loans	of which reverse repos	Investments	Total	
LPM	VQLN	VQCZ	VRME	VQOC	VQDC	VWZL (h)	Z5MB	VWDP	VYAP (i)	BF37	(j)(k)
2017 Jan	18,015	7,574	815	-2,887	5,502	-3,608	11,269	6,372	29,824	41,093	
Feb	-1,811	-758	2,142	3,774	5,159	234	-1,022	-3,805	-3,716	-4,738	
Mar	-11,834	-218	1,522	-15	1,288	11,896	19,902	14,306	-1,009	18,893	
Apr	15,721	-1,949	547	630	-772	-2,711	5,714	-381	6,068	11,782	
May	-4,872	-8,384	539	-17	-7,862	-3,003	8,616	5,521	-1,683	6,933	
Jun	-11,382	-1,104	-1,213	568	-1,749	4,496	9,669	2,083	3,821	13,490	
Jul	-963	3,133	1,695	43	4,872	-6,772	2,607	-2,456	1,731	4,338	
Aug	-3,490	-6,526	-2,882	623	-8,784	-9,373	16,117	5,090	-1,293	14,824	
Sep	3,286	-3,591	-2,982	-973	-7,546	14,683	13,820	11,261	-2,627	11,193	
Oct	10,839	-1,773	-554	-1,483	-3,810	-398	-8,207	-4,884	1,025	-7,182	
Nov	-13,672	-8,649	-318	921	-8,045	-9,052	14,105	3,455	-1,744	12,362	
Dec	-19,262	-488	-3,051	1,672	-1,867	-914	3,005	-1,842	-1,142	1,862	
2018 Jan	16,123	9,299	3,531	1,077	13,907	2,259	2,353	-3,130	290	2,644	
Feb	13,565	-11,565	-152	1,197	-10,520	3,627	-2,135	-4,997	-539	-2,674	
Mar	-27,208	3,972	-1,405	536	3,103	-2,396	2,813	-9,614	-1,884	929	
Apr	5,726	2,442	2,552	310	5,305	2,703	-1,676	-7,507	334	-1,343	
May	2,610	-5,266	-919	1,114	-5,071	3,869	-6,605	-6,772	-4,549	-11,154	
Jun	-12,597	1,362	-672	1,217	1,908	1,437	29,166	5,780	800	29,966	
Jul	-7,504	17,137	-1,499	112	15,751	-7,264	14,535	8,086	-33,990	-19,455	
Aug	13,327	-14,531	170	-652	-15,012	2,538	3,203	-2,890	-1,274	1,929	
Sep	-11,475	-5,432	-388	-409	-6,229	-3,142	24,060	5,870	4,303	28,362	
Oct	559	-5,046	-667	-1,256	-6,968	-10,258	206	4,571	-1,035	-829	
Nov	-5,159	-2,418	2,803	1,002	1,387	3,838	18,245	13,177	-998	17,247	
Dec	-13,348	-11,960	-44	1,996	-10,009	-5,235	13,734	11,517	-3,766	9,968	
2019 Jan	16,667	9,548	2,175	1,353	13,076	3,682	4,418	-6,893	-331	4,087	
Feb	-2,819	8,285	-711	165	7,739	5,009	-1,809	-7,386	-66	-1,875	
Mar	-24,852	-9,370	-2,241	757	-10,854	-12,275	20,108	8,293	715	20,823	
Apr	11,358	-1,218	1,587	40	409	3,754	-7,965	-1,297	-520	-8,484	

12.7 Counterparts to changes in M4: alternative presentation

£ millions   Not seasonally adjusted

Changes

LPM	External and foreign currency flows			Total external counterparts	Net non-deposit £ liabilities (-)	M4
	Net sterling deposits from non-residents (-)	Net foreign currency liabilities (-)	Total			
	B69P	B72P	AVBW	VQLP	VWZV	AUZI
2017 Jan	5,397	-20,642	**-15,245**	-9,743	-21,290	**950**
Feb	-2,106	45,416	**43,309**	48,468	-39,667	**-862**
Mar	-4,814	-35,790	**-40,604**	-39,316	40,005	**30,188**
Apr	15,797	-37,448	**-21,651**	-22,423	38,420	**25,841**
May	-13,445	43,850	**30,405**	22,543	-40,571	**-6,237**
Jun	-11,281	3,002	**-8,278**	-10,028	-2,469	**7,238**
Jul	4,354	-15,776	**-11,422**	-6,550	10,188	**-3,668**
Aug	-4,992	34,451	**29,459**	20,675	-11,834	**23,076**
Sep	3,482	-30,739	**-27,257**	-34,803	7,938	**6,556**
Oct	11,133	9,082	**20,215**	16,405	-3,926	**8,710**
Nov	1,415	-17,146	**-15,731**	-23,776	17,152	**4,730**
Dec	-18,047	15,516	**-2,531**	-4,397	-9,547	**-11,130**
2018 Jan	21,946	-106,246	**-84,300**	-70,393	94,472	**15,075**
Feb	2,446	125,223	**127,670**	117,149	-133,533	**-4,911**
Mar	-17,588	-17,307	**-34,895**	-31,792	24,193	**-12,170**
Apr	4,584	91,547	**96,130**	101,435	-101,724	**-4,233**
May	-7,851	10,363	**2,512**	-2,559	21,291	**16,518**
Jun	508	-93,371	**-92,863**	-90,955	63,310	**1,850**
Jul	-11,479	-11,302	**-22,781**	-7,030	51,501	**2,001**
Aug	5,500	4,679	**10,179**	-4,833	-9,388	**5,258**
Sep	6,693	-19,580	**-12,887**	-19,116	-13,575	**-1,242**
Oct	8,971	38,839	**47,809**	40,842	-24,561	**12,161**
Nov	15,835	-39,127	**-23,292**	-21,905	5,658	**3,451**
Dec	8,976	6,606	**15,582**	5,573	-8,137	**12,179**
2019 Jan	-1,560	-29,076	**-30,636**	-17,560	7,397	**-15,470**
Feb	-2,151	-1,423	**-3,574**	4,165	10,234	**9,794**
Mar	5,647	27,947	**33,594**	22,740	-34,615	**7,527**
Apr	3,503	9,291	**12,794**	13,203	7,264	**15,327**

Notes to table   Source: Bank of England

(g) This estimate of the Public Sector Net Cash Requirement may differ from the headline measure published by the Office for National Statistics (ONS). The Bank of England's measure is maintained to ensure the 'Alternative Counterparts' analysis continues to balance. More details on this can be found in article www.bankofengland.co.uk/statistics/Documents/ms/articles/art2may09.pdf.

(h) Net sterling lending to the public sector includes holdings of coin. Coin is a liability of Central Government and therefore outside of the MFIs' consolidated balance sheet. Holdings of coin are a component of M4 and are therefore included here in order to reconcile the counterparts.

(i) This series includes purchases of bonds made as part of the Bank of England's Corporate Bond Purchase Scheme. Data on Central Bank holdings of securities can be found in Bankstats Table B2.2. For further information on the Bank's treatment of securities transactions in credit statistics, see: www.bankofengland.co.uk/statistics/Documents/articles/2015/2may.pdf.

(j) Please note that the compilation and descriptions of some credit series have changed from publication of April 2015 data, as described in Bankstats, April 2015, 'Changes to the treatment of loan transfers and lending to housing associations', available at www.bankofengland.co.uk/statistics/Documents/ms/articles/art1apr15.pdf.

(k) This series includes purchases of bonds made as part of the Bank of England's Corporate Bond Purchase Scheme. Data on Central Bank holdings of securities can be found in Bankstats Table B2.2. For further information on the Bank's treatment of securities transactions in credit statistics, see: www.bankofengland.co.uk/statistics/Documents/articles/2015/2may.pdf.

12.8 Selected retail banks' base rate

Daily average of 4 UK Banks' base rates, IUDAMIH [a]

Date of change	New rate	Date of change	New rate	Date of change	New rate
09-Jan-86	12.5	17-Sep-92	10.5	05-Jul-07	5.75
19-Mar-86	11.5	18-Sep-92	10	06-Dec-07	5.5
08-Apr-86	11.13	22-Sep-92	9	07-Feb-08	5.25
09-Apr-86	11	16-Oct-92	8.25	10-Apr-08	5
21-Apr-86	10.5	19-Oct-92	8	08-Oct-08	4.5
23-May-86	10.25	13-Nov-92	7	06-Nov-08	3
27-May-86	10	26-Jan-93	6	04-Dec-08	2
14-Oct-86	10.5	23-Nov-93	5.5	08-Jan-09	1.5
15-Oct-86	11	08-Feb-94	5.25	05-Feb-09	1
10-Mar-87	10.5	12-Sep-94	5.75	05-Mar-09	0.5
18-Mar-87	10.25	07-Dec-94	6.25	04-Jan-10	0.5
19-Mar-87	10	02-Feb-95	6.63	31-Dec-10	0.5
28-Apr-87	9.88	03-Feb-95	6.75	04-Jan-11	0.5
29-Apr-87	9.5	13-Dec-95	6.5	30-Dec-11	0.5
11-May-87	9	18-Jan-96	6.25	03-Jan-12	0.5
06-Aug-87	9.25	08-Mar-96	6	31-Dec-12	0.5
07-Aug-87	10	06-Jun-96	5.75	02-Jan-13	0.5
23-Oct-87	9.88	30-Oct-96	5.94	31-Dec-13	0.5
26-Oct-87	9.5	31-Oct-96	6	02-Jan-14	0.5
04-Nov-87	9.38	06-May-97	6.25	31-Dec-14	0.5
05-Nov-87	9	06-Jun-97	6.44	02-Jan-15	0.5
04-Dec-87	8.5	09-Jun-97	6.5	31-Jul-15	0.5
02-Feb-88	9	10-Jul-97	6.75		
17-Mar-88	8.63	07-Aug-97	7	04-Aug-16	0.25
18-Mar-88	8.5	06-Nov-97	7.25	02-Nov-17	0.5
11-Apr-88	8	04-Jun-98	7.5	02-Aug-18	0.75
17-May-88	7.88	08-Oct-98	7.25	11-Mar-20	0.25
18-May-88	7.5	06-Nov-98	6.75	19-Mar-20	0.10
02-Jun-88	7.75	10-Dec-98	6.25		
03-Jun-88	8	07-Jan-99	6		
06-Jun-88	8.25	04-Feb-99	5.5		
07-Jun-88	8.5	08-Apr-99	5.25		
22-Jun-88	8.88	10-Jun-99	5		
23-Jun-88	9	08-Sep-99	5.19		
28-Jun-88	9.25	10-Sep-99	5.25		
29-Jun-88	9.5	04-Nov-99	5.5		
04-Jul-88	9.88	14-Jan-00	5.75		
05-Jul-88	10	10-Feb-00	6		
18-Jul-88	10.38	08-Feb-01	5.75		
19-Jul-88	10.5	05-Apr-01	5.5		
08-Aug-88	10.88	10-May-01	5.25		
09-Aug-88	11	02-Aug-01	5		
25-Aug-88	11.75	18-Sep-01	4.75		
26-Aug-88	12	04-Oct-01	4.5		
25-Nov-88	13	08-Nov-01	4		
24-May-89	14	07-Feb-03	3.75		
05-Oct-89	15	11-Jul-03	3.5		
08-Oct-90	14	06-Nov-03	3.75		
13-Feb-91	13.5	05-Feb-04	4		
27-Feb-91	13	07-May-04	4.25		
22-Mar-91	12.5	10-Jun-04	4.5		
12-Apr-91	12	05-Aug-04	4.75		
24-May-91	11.5	04-Aug-05	4.5		
12-Jul-91	11	03-Aug-06	4.75		
04-Sep-91	10.5	09-Nov-06	5		
05-May-92	10	11-Jan-07	5.25		
16-Sep-92	12	10-May-07	5.5		

1 Data obtained from Barclays Bank, Lloyds/TSB Bank, HSBC Bank and National Westminster Bank whose rates are used to compile this series.

Source: Bank of England

2 Where all the rates did not change on the same day a spread is shown.

[a] This series was discontinued on 31-July-2015. After this date, and data shown is the Bank of England base rate. Retail banks usually match the official bank rate but they are not obliged to do so.

12.9 Average three month sterling money market rates

Date	Monthly average rate of discount, 3 month Treasury bills, Sterling [a] IUMAAJNB	Monthly average of Eligible bills discount rate, 3 month IUMAAJND	Monthly average Sterling 3 month mean interbank lending rate [b] [c] [d] IUMAAMIJ	Monthly average of Sterling certificates of deposit interest rate, 3 months, mean offer/bid [b] [e] IUMAVCDA
31-Jan-01	5.4852	5.64	5.7642	5.7338
28-Feb-01	5.4582	5.56	5.6854	5.661
31-Mar-01	5.2286	5.37	5.473	5.4443
30-Apr-01	5.1158	5.21	5.3323	5.3037
31-May-01	4.9765	5.06	5.1727	5.1479
30-Jun-01	4.991	5.08	5.189	5.1621
31-Jul-01	5.0052	5.07	5.1917	5.1673
31-Aug-01	4.7198	4.82	4.9269	4.8991
30-Sep-01	4.4297	4.57	4.6485	4.6223
31-Oct-01	4.1567	4.26	4.3594	4.3304
30-Nov-01	3.7799	3.85	3.9326	3.9084
31-Dec-01	3.8296	3.88	3.9852	3.9568
31-Jan-02	3.8321	3.91	3.978	3.957
28-Feb-02	3.868	3.92	3.9805	3.9618
31-Mar-02	3.9672	3.99	4.0609	4.0413
30-Apr-02	3.9693	4.04	4.1086	4.0802
31-May-02	3.9525	4.01	4.0795	4.0627
30-Jun-02	3.9767	4.04	4.112	4.0889
31-Jul-02	3.8408	3.94	3.9925	3.9704
31-Aug-02	3.7663	3.86	3.9174	3.9005
30-Sep-02	3.7861	3.86	3.9315	3.9057
31-Oct-02	3.7509	3.82	3.9015	3.8802
30-Nov-02	3.8033	3.84	3.9091	3.8876
31-Dec-02	3.8418	3.71	3.9488	3.929
31-Jan-03	3.7993	3.87	3.9132	3.9036
28-Feb-03	3.4993	3.65	3.6888	3.675
31-Mar-03	3.4721	3.54	3.5838	3.5729
30-Apr-03	3.4537	3.52	3.576	3.567
31-May-03	3.4366	3.52	3.5665	3.556
30-Jun-03	3.4724	3.45	3.5702	3.5602
31-Jul-03	3.3119	3.39	3.42	3.4089
31-Aug-03	3.3998	3.42	3.4518	3.4413
30-Sep-03	3.5239	3.59	3.6311	3.6216
31-Oct-03	3.6514	3.69	3.727	3.7204
30-Nov-03	3.808	3.88	3.9085	3.9038
31-Dec-03	3.8298	3.9	3.949	3.9419
31-Jan-04	3.8983	3.94	3.986	3.9771
29-Feb-04	3.9788	4.06	4.101	4.0908
31-Mar-04	4.1025	4.19	4.2328	4.2235
30-Apr-04	4.1862	4.28	4.3288	4.321
31-May-04	4.34	4.42	4.4555	4.4505
30-Jun-04	4.5793	4.68	4.7305	4.7211
31-Jul-04	4.6424	4.75	4.7859	4.7859
31-Aug-04	4.7218	4.85	4.8938	4.8864
30-Sep-04	4.6937	4.83	4.8743	4.868
31-Oct-04	4.679	4.79	4.8336	4.8319
30-Nov-04	4.6578	4.78	4.8159	4.8098
31-Dec-04	4.677	4.77	4.8057	4.7967
31-Jan-05	4.6568	4.75	4.8045	4.801
28-Feb-05	4.6858	4.78	4.822	4.8183
31-Mar-05	4.7691	4.88	4.9186	4.9069
30-Apr-05	4.7047	4.84	4.8756	4.8645
31-May-05	4.6618	4.8	4.8253	4.821
30-Jun-05	4.6175	4.76	4.7777	4.782
31-Jul-05	4.4609	4.57	4.5948	4.5967
31-Aug-05	4.4057	4.51	4.533	4.5336

12.9 Average three month sterling money market rates

Date	Monthly average rate of discount, 3 month Treasury bills, Sterling [a] IUMAAJNB	Monthly average of Eligible bills discount rate, 3 month IUMAAJND	Monthly average Sterling 3 month mean interbank lending rate [b] [c] [d] IUMAAMIJ	Monthly average of Sterling certificates of deposit interest rate, 3 months, mean offer/bid [b] [e] IUMAVCDA
30-Sep-05	4.4037	-	4.5359	4.5377
31-Oct-05	4.402	-	4.525	4.5202
30-Nov-05	4.4169	-	4.5614	4.5627
31-Dec-05	4.4287	-	4.587	4.5845
31-Jan-06	4.3906	-	4.5369	4.5398
28-Feb-06	4.3839	-	4.5203	4.519
31-Mar-06	4.3956	-	4.5263	4.5257
30-Apr-06	4.4196	-	4.5725	4.5719
31-May-06	4.5012	-	4.6519	4.6467
30-Jun-06	4.5414	-	4.6927	4.692
31-Jul-06	4.534	-	4.6838	4.681
31-Aug-06	4.7544	-	4.8968	4.8945
30-Sep-06	4.8359	-	4.9819	4.9807
31-Oct-06	4.9357	-	5.0909	5.0886
30-Nov-06	5.0095	-	5.1791	5.1782
31-Dec-06	5.0759	-	5.2453	5.2429
31-Jan-07	5.304	-	5.4545	5.4491
28-Feb-07	5.3393	-	5.5248	5.5143
31-Mar-07	5.3274	-	5.5041	5.5177
30-Apr-07	5.4329	-	5.6058	5.6939
31-May-07	5.5516	-	5.7221	5.835
30-Jun-07	5.6702	-	5.8348	5.9379
31-Jul-07	5.7742	-	5.9789	6.1118
31-Aug-07	5.7943	-	6.3389	6.3464
30-Sep-07	5.6896	-	6.5813	6.537
31-Oct-07	5.6058	-	6.2122	6.2052
30-Nov-07	5.4986	-	6.3564	6.3445
31-Dec-07	5.3034	-	6.3542	6.3489
31-Jan-08	5.1215	-	5.6071	5.6059
29-Feb-08	5.0178	-	5.6083	5.6
31-Mar-08	4.8835	-	5.8555	5.8482
30-Apr-08	4.8258	-	5.8961	5.8882
31-May-08	4.9496	-	5.7933	5.787
30-Jun-08	5.1138	-	5.9002	5.8924
31-Jul-08	5.0843	-	5.803	5.7996
31-Aug-08	4.9539	-	5.7588	5.7535
30-Sep-08	4.7425	-	5.8698	5.8573
31-Oct-08	3.6788	-	6.1772	6.1598
30-Nov-08	1.9948	-	4.3955	4.4
31-Dec-08	1.2875	-	3.2143	3.2107
31-Jan-09	0.8945	-	2.28	2.2869
28-Feb-09	0.7177	-	2.0758	2.0625
31-Mar-09	0.6035	-	1.8273	1.7864
30-Apr-09	0.625	-	1.4838	1.425
31-May-09	0.5271	-	1.3026	1.2697
30-Jun-09	0.5037	-	1.2057	1.1386
31-Jul-09	0.4397	-	1.0293	0.9185
31-Aug-09	0.3946	-	0.7963	0.69
30-Sep-09	0.3761	-	0.617	0.5136
31-Oct-09	0.4315	-	0.5636	0.5023
30-Nov-09	0.4459	-	0.5976	0.575
31-Dec-09	0.3587	-	0.6071	0.575
31-Jan-10	0.4874	-	0.6013	0.575
28-Feb-10	0.4878	-	0.5995	0.575
31-Mar-10	0.5109	-	0.6002	0.6011
30-Apr-10	0.5083	-	0.6	0.625

12.9 Average three month sterling money market rates

Date	Monthly average rate of discount, 3 month Treasury bills, Sterling [a] IUMAAJNB	Monthly average of Eligible bills discount rate, 3 month IUMAAJND	Monthly average Sterling 3 month mean interbank lending rate [b] [c] [d] IUMAAMIJ	Monthly average of Sterling certificates of deposit interest rate, 3 months, mean offer/bid [b] [e] IUMAVCDA
31-May-10	0.4976	-	0.6645	0.6776
30-Jun-10	0.4839	-	0.7045	0.7682
31-Jul-10	0.498	-	0.75	0.775
31-Aug-10	0.4945	-	0.75	0.7655
30-Sep-10	0.4966	-	0.7545	0.725
31-Oct-10	0.5061	-	0.75	0.725
30-Nov-10	0.4935	-	0.75	0.725
31-Dec-10	0.4913	-	0.755	0.7655
31-Jan-11	0.5055	-	0.7785	0.825
28-Feb-11	0.5363	-	0.775	0.825
31-Mar-11	0.5603	-	0.7811	0.8315
30-Apr-11	0.5676	-	0.8	0.875
31-May-11	0.5265	-	0.8	0.875
30-Jun-11	0.5174	-	0.8464	0.875
31-Jul-11	0.4996	-	0.86	0.875
31-Aug-11	0.4531	-	0.8873	0.9114
30-Sep-11	0.4649	-	0.9664	1.0077
31-Oct-11	0.4608	-	1.0179	1.0593
30-Nov-11	0.4387	-	1.0511	1.0968
31-Dec-11	0.2996	-	1.0878	1.15
31-Jan-12	0.3239	-	1.1086	1.175
29-Feb-12	0.3912	-	1.0971	1.156
31-Mar-12	0.4248	-	1.0752	1.1164
30-Apr-12	0.4236	-	1.0558	1.0992
31-May-12	0.3561	-	1.0143	1.0891
30-Jun-12	0.3422	-	0.9658	1.0376
31-Jul-12	0.2943	-	0.8627	0.8691
31-Aug-12	0.2398	-	0.7048	0.7405
30-Sep-12	0.2478	-	0.6418	0.6825
31-Oct-12	0.2368	-	0.535	0.575
30-Nov-12	0.2243	-	0.5034	0.575
31-Dec-12	0.249	-	0.5	0.575
31-Jan-13	0.2677	-	0.4927	0.575
28-Feb-13	0.3147	-	0.49	0.575
31-Mar-13	0.3391	-	0.4865	0.5655
30-Apr-13	0.3447	-	0.485	0.5731
31-May-13	0.3065	-	0.485	0.5512
30-Jun-13	0.3066	-	0.485	0.525
31-Jul-13	0.3124	-	0.4839	0.525
31-Aug-13	0.2812	-	0.485	0.525
30-Sep-13	0.289	-	0.4931	0.525
31-Oct-13	0.3141	-	0.495	0.525
30-Nov-13	0.2865	-	0.5017	0.5512
31-Dec-13	0.2555	-	0.5335	0.575
31-Jan-14	0.3211	-	0.5318	0.575
28-Feb-14	0.3624	-	0.5288	0.575
31-Mar-14	0.3882	-	0.53	0.575
30-Apr-14	0.3688	-	0.532	0.575
31-May-14	0.284	-	0.55	0.575
30-Jun-14	0.3586	-	0.55	0.575
31-Jul-14	0.4271	-	0.5487	0.575
31-Aug-14	0.3978	-	0.545	0.575
30-Sep-14	0.4351	-	0.5505	0.567
31-Oct-14	0.3964	-	0.5426	0.5511
30-Nov-14	0.4114	-	0.53	0.5538
31-Dec-14	0.4101	-	0.53	0.5524

12.9 Average three month sterling money market rates

Date	Monthly average rate of discount, 3 month Treasury bills, Sterling [a] IUMAAJNB	Monthly average of Eligible bills discount rate, 3 month IUMAAJND	Monthly average Sterling 3 month mean interbank lending rate [b] [c] [d] IUMAAMIJ	Monthly average of Sterling certificates of deposit interest rate, 3 months, mean offer/bid [b] [e] IUMAVCDA
31-Jan-15	0.377	-	0.53	0.5512
28-Feb-15	0.338	-	0.53	0.55
31-Mar-15	0.4295	-	0.53	0.5511
30-Apr-15	0.4322	-	0.533	0.55
31-May-15	0.4511	-	0.54	0.5532
30-Jun-15	0.467	-	0.5505	0.55
31-Jul-15	0.4895	-	0.567	0.55
31-Aug-15	0.4643	-	0.57	0.55
30-Sep-15	0.4532	-	0.57	0.55
31-Oct-15	0.48	-	0.57	0.55
30-Nov-15	0.4806	-	0.57	0.55
31-Dec-15	0.4551	-	0.57	0.5786
31-Jan-16	0.4823	-	0.566	0.6475
28-Feb-16	0.4711	-	0.565	0.65
31-Mar-16	0.4505	-	0.565	0.6476
30-Apr-16	0.4489	-	0.565	0.65
31-May-16	0.4362	-	0.565	0.65
30-Jun-16	0.4046	-	0.563	0.6386
31-Jul-16	0.3689	-	0.5414	0.4869
31-Aug-16	0.2281	-	0.468	0.367
30-Sep-16	0.2104	-	0.3627	0.3509
31-Oct-16	0.1711	-	0.3748	0.3524
30-Nov-16	0.1385	-	0.38	0.3523
31-Dec-16	0.0518	-	0.3775	0.35
31-Jan-17	0.1441	-	0.3452	0.35
28-Feb-17	0.1131	-	0.34	0.35
31-Mar-17	0.0231	-	0.3304	0.3478
30-Apr-17	0.0709	-	0.33	0.35
31-May-17	0.0592	-	0.3267	0.35
30-Jun-17	0.0759	-	0.32	0.35
31-Jul-17	-	-	0.32	0.3324
31-Aug-17	-	-	0.3123	0.2977
30-Sep-17	-	-	0.3157	0.3
31-Oct-17	-	-	0.3491	0.3
30-Nov-17	-	-	0.4364	0.4818
31-Dec-17	-	-	-	0.5
31-Jan-18	-	-	-	0.5
28-Feb-18	-	-	-	0.5
31-Mar-18	-	-	-	0.5714
30-Apr-18	-	-	-	0.70
31-May-18	-	-	-	0.6929
30-Jun-18	-	-	-	0.65

Source: Bank of England

Notes:

[a] These data have been discontinued with effect from 20th July 2017.

[b] Data provided is for general reference purposes. Every effort is made to ensure that it is up to date and accurate. The Bank of England does not warrant, or accept responsibility/liability for, the accuracy/completeness of content, or loss/damage, whether direct, indirect or consequential, which may arise from reliance on said data. We reserve the right to change information published, including timing and methods used. We give no assurance that data currently published will be in the future.

[c] No data is available because the Bank has been unable to source sufficient representative data to allow it to calculate an average rate.

[d] No data available from 5 December 2017 because the Bank has been unable to source sufficient representative data to allow it to calculate an average rate. This series has been discontinued.

[e] From 2 July 2018 the Bank ceased publishing these data. Further information can be found at https://www.bankofengland.co.uk/statistics/details/further-details-about-wholesale-sterling-certificates-of-deposit-data

12.10 Average Foreign Exchange rates

Date	Monthly average Old IMF-based Effective exchange rate index, Sterling (a) (1990 average = 100) [3] XUMAGBG	Monthly average Effective exchange rate index, Sterling (Jan 2005= 100) [1] XUMABK67	Monthly average Spot exchange rate, US$ into Sterling [2] XUMAUSS	Monthly average Spot exchange rate, Euro into Sterling [2] XUMAERS
31-Jan-01	104.43	98.2362	1.4769	1.5753
28-Feb-01	104.14	97.7179	1.4529	1.5786
31-Mar-01	105.02	98.4528	1.4454	1.5901
30-Apr-01	105.85	99.0733	1.435	1.6084
31-May-01	106.56	99.3906	1.4259	1.6304
30-Jun-01	106.77	99.2172	1.4014	1.6434
31-Jul-01	107.18	99.7551	1.4139	1.6433
31-Aug-01	105.09	98.4283	1.4365	1.5955
30-Sep-01	106.08	99.5949	1.4635	1.606
31-Oct-01	105.78	99.3536	1.4517	1.6024
30-Nov-01	106.14	99.4505	1.4358	1.6166
31-Dec-01	106.49	100.0899	1.4409	1.6151
31-Jan-02	106.9	100.26	1.4323	1.6222
28-Feb-02	107.36	100.4838	1.4231	1.6348
31-Mar-02	106.48	99.7945	1.4225	1.6224
30-Apr-02	107.13	100.5217	1.4434	1.6282
31-May-02	105.32	99.1553	1.4593	1.5914
30-Jun-02	103.58	98.0057	1.4863	1.5515
31-Jul-02	105.29	100.1562	1.5546	1.5665
31-Aug-02	105.38	100.0907	1.5377	1.5723
30-Sep-02	106.46	101.2329	1.5561	1.5861
31-Oct-02	106.67	101.4447	1.5574	1.5868
30-Nov-02	105.87	100.8857	1.5723	1.5694
31-Dec-02	105.48	100.6652	1.5863	1.5566
31-Jan-03	104.04	99.8966	1.6169	1.5222
28-Feb-03	102.42	98.4726	1.6085	1.4924
31-Mar-03	100.6	96.8021	1.5836	1.4649
30-Apr-03	99.8	95.9557	1.5747	1.4505
31-May-03	97.9	94.8862	1.623	1.403
30-Jun-03	99.64	96.6136	1.6606	1.4234
31-Jul-03	99.4	95.9155	1.6242	1.4277
31-Aug-03	99.01	95.2462	1.595	1.4286
30-Sep-03	99.24	95.5651	1.6131	1.4338
31-Oct-03	99.82	96.7185	1.6787	1.4334
30-Nov-03	100.42	97.317	1.6901	1.4426
31-Dec-03	100.29	97.8697	1.7507	1.4246
31-Jan-04	102.38	100.1608	1.8234	1.4447
29-Feb-04	104.84	102.5025	1.8673	1.4774
31-Mar-04	104.99	102.1884	1.8267	1.489
30-Apr-04	105.18	102.137	1.8005	1.5022
31-May-04	104.63	101.7871	1.7876	1.4894
30-Jun-04	105.8	103.059	1.8275	1.505
31-Jul-04	105.89	103.1662	1.8429	1.5023
31-Aug-04	105.16	102.4289	1.8216	1.4933
30-Sep-04	103.32	100.7156	1.7922	1.4676
31-Oct-04	102.17	99.8303	1.8065	1.4455
30-Nov-04	101.73	99.7587	1.8603	1.4311
31-Dec-04	103.15	101.2974	1.9275	1.4401
31-Jan-05	102.1	100	1.8764	1.4331
28-Feb-05	103.29	100.9792	1.8871	1.4499
31-Mar-05	103.21	101.0494	1.9078	1.444
30-Apr-05	104.38	102.0108	1.896	1.4652
31-May-05	103.55	100.9896	1.8538	1.4611
30-Jun-05	104.92	101.7515	1.8179	1.4952
31-Jul-05	102.14	98.8169	1.7509	1.4547
31-Aug-05	102.85	99.7011	1.7943	1.4592
30-Sep-05	103.91	100.6456	1.8081	1.4761
31-Oct-05	103.09	99.6486	1.764	1.4674
30-Nov-05	103.16	99.4195	1.7341	1.4719

12.10 Average Foreign Exchange rates

Date	Monthly average Old IMF-based Effective exchange rate index, Sterling (a) (1990 average = 100) [3] XUMAGBG	Monthly average Effective exchange rate index, Sterling (Jan 2005= 100) [1] XUMABK67	Monthly average Spot exchange rate, US$ into Sterling [2] XUMAUSS	Monthly average Spot exchange rate, Euro into Sterling [2] XUMAERS
31-Dec-05	103.32	99.5011	1.7462	1.4725
31-Jan-06	102.67	99.0364	1.7678	1.4582
28-Feb-06	102.83	98.9167	1.747	1.4637
31-Mar-06	102.09	98.3582	1.7435	1.45
30-Apr-06	101.87	98.3685	1.7685	1.4402
31-May-06	104.14	101.2652	1.8702	1.4637
30-Jun-06	n/a	100.9564	1.8428	1.456
31-Jul-06	n/a	100.9424	1.8447	1.454
31-Aug-06	n/a	102.9584	1.8944	1.4785
30-Sep-06	n/a	103.0141	1.8847	1.4811
31-Oct-06	n/a	103.1453	1.8755	1.4869
30-Nov-06	n/a	103.4686	1.9119	1.4834
31-Dec-06	n/a	104.5334	1.9633	1.486
31-Jan-07	n/a	105.5735	1.9587	1.5079
28-Feb-07	n/a	105.0313	1.9581	1.4969
31-Mar-07	n/a	103.5671	1.9471	1.4703
30-Apr-07	n/a	104.2682	1.9909	1.4713
31-May-07	n/a	103.9117	1.9836	1.4677
30-Jun-07	n/a	104.5063	1.9864	1.4805
31-Jul-07	n/a	105.2357	2.0338	1.4821
31-Aug-07	n/a	104.5357	2.0111	1.4762
30-Sep-07	n/a	103.3042	2.0185	1.4515
31-Oct-07	n/a	102.8113	2.0446	1.437
30-Nov-07	n/a	101.8978	2.0701	1.4106
31-Dec-07	n/a	99.9273	2.0185	1.3863
31-Jan-08	n/a	96.6594	1.9698	1.3383
29-Feb-08	n/a	96.1702	1.9638	1.3316
31-Mar-08	n/a	94.7859	2.0032	1.2897
30-Apr-08	n/a	93.0097	1.9817	1.258
31-May-08	n/a	93.0105	1.9641	1.2633
30-Jun-08	n/a	93.1297	1.9658	1.2636
31-Jul-08	n/a	93.2514	1.988	1.2615
31-Aug-08	n/a	91.6752	1.8889	1.2614
30-Sep-08	n/a	90.0272	1.7986	1.2531
31-Oct-08	n/a	89.6321	1.69	1.2718
30-Nov-08	n/a	83.8159	1.5338	1.2041
31-Dec-08	n/a	78.4569	1.4859	1.1043
31-Jan-09	n/a	77.2822	1.4452	1.0919
28-Feb-09	n/a	79.1632	1.4411	1.1264
31-Mar-09	n/a	77.2402	1.4174	1.0867
30-Apr-09	n/a	79.1169	1.4715	1.1157
31-May-09	n/a	80.65	1.5429	1.1295
30-Jun-09	n/a	84.1934	1.6366	1.1682
31-Jul-09	n/a	83.777	1.6366	1.1622
31-Aug-09	n/a	83.66	1.6539	1.1597
30-Sep-09	n/a	81.3425	1.6328	1.1212
31-Oct-09	n/a	79.5602	1.6199	1.0928
30-Nov-09	n/a	81.1213	1.6597	1.1126
31-Dec-09	n/a	80.4832	1.6239	1.1127
31 Jan 10	n/a	81.0263	1.6162	1.1327
28 Feb 10	n/a	80.3054	1.5615	1.1415
31 Mar 10	n/a	77.5264	1.5053	1.1092
30 Apr 10	n/a	79.3991	1.534	1.1436
31 May 10	n/a	79.256	1.4627	1.1685
30 Jun 10	n/a	81.0391	1.4761	1.2082
31 Jul 10	n/a	81.4286	1.5299	1.1959
31 Aug 10	n/a	82.6434	1.566	1.2132
30 Sep 10	n/a	81.2194	1.5578	1.1901
31 Oct 10	n/a	79.5105	1.5862	1.1412

12.10 Average Foreign Exchange rates

Date	Monthly average Old IMF-based Effective exchange rate index, Sterling (a) (1990 average = 100) [3]	Monthly average Effective exchange rate index, Sterling (Jan 2005= 100) [1]	Monthly average Spot exchange rate, US$ into Sterling [2]	Monthly average Spot exchange rate, Euro into Sterling [2]
	XUMAGBG	**XUMABK67**	**XUMAUSS**	**XUMAERS**
30 Nov 10	n/a	80.9634	1.5961	1.1701
31 Dec 10	n/a	80.4155	1.5603	1.1802
31 Jan 11	n/a	80.715	1.5795	1.1817
28 Feb 11	n/a	81.4918	1.613	1.1815
31 Mar 11	n/a	80.2508	1.6159	1.1527
30 Apr 11	n/a	79.6607	1.6345	1.1331
31 May 11	n/a	79.7731	1.6312	1.1403
30 Jun 11	n/a	78.8754	1.6214	1.1261
31 Jul 11	n/a	78.7589	1.6145	1.1308
31 Aug 11	n/a	79.6334	1.6348	1.1422
30 Sep 11	n/a	79.2594	1.5783	1.147
31 Oct 11	n/a	79.5319	1.576	1.1487
30 Nov 11	n/a	80.4724	1.5804	1.1659
31 Dec 11	n/a	80.9046	1.5585	1.1843
31 Jan 12	n/a	81.1127	1.551	1.2034
29 Feb 12	n/a	80.9911	1.5802	1.1941
31 Mar 12	n/a	81.4148	1.5823	1.1981
30 Apr 12	n/a	82.5667	1.6014	1.2161
31 May 12	n/a	83.8068	1.5905	1.244
30 Jun 12	n/a	83.0904	1.5571	1.2416
31 Jul 12	n/a	84.0006	1.5589	1.2688
31 Aug 12	n/a	84.0155	1.5719	1.2676
30 Sep 12	n/a	84.2263	1.6116	1.2524
31 Oct 12	n/a	83.6196	1.6079	1.2393
30 Nov 12	n/a	83.6808	1.5961	1.244
31 Dec 12	n/a	83.58	1.6144	1.231
31 Jan 13	n/a	82.1724	1.5957	1.2
28 Feb 13	n/a	79.7012	1.5478	1.1594
31 Mar 13	n/a	79.0951	1.5076	1.1634
30 Apr 13	n/a	80.1289	1.5316	1.175
31 May 13	n/a	80.4286	1.5285	1.1777
30 Jun 13	n/a	80.9967	1.5478	1.174
31 Jul 13	n/a	79.9698	1.5172	1.16
31 Aug 13	n/a	80.9894	1.5507	1.1649
30 Sep 13	n/a	82.7393	1.5865	1.1883
31 Oct 13	n/a	82.6676	1.6094	1.1797
30 Nov 13	n/a	83.5617	1.6104	1.1938
31 Dec 13	n/a	84.4204	1.6375	1.1947
31 Jan 14	n/a	85.4222	1.647	1.2097
28 Feb 14	n/a	85.7378	1.6567	1.2122
31 Mar 14	n/a	85.5269	1.6622	1.2021
30 Apr 14	n/a	86.1716	1.6743	1.2125
31 May 14	n/a	86.8395	1.6844	1.2267
30 Jun 14	n/a	87.6795	1.6906	1.2436
31 Jul 14	n/a	88.7387	1.7069	1.2611
31 Aug 14	n/a	87.7645	1.6709	1.2542
30 Sep 14	n/a	87.5039	1.6305	1.2639
31 Oct 14	n/a	87.3534	1.6068	1.2678
30 Nov 14	n/a	86.9783	1.578	1.2646
31 Dec 14	n/a	87.5084	1.564	1.2686
31 Jan 15	n/a	87.7587	1.5143	1.3045
28 Feb 15	n/a	90.092	1.5334	1.3503
31 Mar 15	n/a	90.457	1.4957	1.3825
30 Apr 15	n/a	90.2153	1.4967	1.3856
31 May 15	n/a	91.3582	1.547	1.3852
30 Jun 15	n/a	92.148	1.5568	1.3879
31 Jul 15	n/a	93.3535	1.556	1.4139
31 Aug 15	n/a	93.5006	1.5583	1.4004
30 Sep 15	n/a	91.8553	1.5326	1.3665

12.10 Average Foreign Exchange rates

Date	Monthly average Old IMF-based Effective exchange rate index, Sterling (a) (1990 average = 100)[3] XUMAGBG	Monthly average Effective exchange rate index, Sterling (Jan 2005= 100)[1] XUMABK67	Monthly average Spot exchange rate, US$ into Sterling[2] XUMAUSS	Monthly average Spot exchange rate, Euro into Sterling[2] XUMAERS
31 Oct 15	n/a	91.6152	1.5339	1.3657
30 Nov 15	n/a	93.4293	1.519	1.4168
31 Dec 15	n/a	91.5782	1.4983	1.3769
31 Jan 16	n/a	88.5339	1.4379	1.3257
29 Feb 16	n/a	86.7262	1.4296	1.289
31 Mar 16	n/a	85.8045	1.425	1.2809
30 Apr 16	n/a	84.9658	1.4312	1.2622
31 May 16	n/a	86.6927	1.4518	1.2846
30 Jun 16	n/a	84.9623	1.4209	1.2646
31 Jul 16	n/a	79.4225	1.3141	1.1884
31 Aug 16	n/a	78.3966	1.31	1.1687
30 Sep 16	n/a	78.7142	1.3142	1.1722
31 Oct 16	n/a	74.7325	1.2329	1.119
30 Nov 16	n/a	76.7386	1.2431	1.1533
31 Dec 16	n/a	78.3405	1.2488	1.1838
31 Jan 17	n/a	77.0503	1.2351	1.1613
28 Feb 17	n/a	77.665	1.249	1.1732
31 Mar 17	n/a	76.6249	1.2348	1.1548
30 Apr 17	n/a	78.2619	1.2652	1.1798
31 May 17	n/a	78.6497	1.2933	1.1696
30 Jun 17	n/a	77.0493	1.2813	1.1403
31 Jul 17	n/a	76.9638	1.2994	1.1281
31 Aug 17	n/a	75.4583	1.2955	1.0974
30 Sep 17	n/a	77.0986	1.3324	1.1186
31 Oct 17	n/a	77.229	1.3197	1.1227
30 Nov 17	n/a	77.5693	1.3219	1.1259
31 Dec 17	n/a	78.2454	1.3402	1.133
31 Jan 18	n/a	78.841	1.3832	1.1331
28 Feb 18	n/a	78.9052	1.3961	1.1311
31 Mar 18	n/a	79.1273	1.397	1.1328
30 Apr 18	n/a	80.1849	1.4083	1.1477
31 May 18	n/a	78.7731	1.3459	1.1397
30 Jun 18	n/a	78.441	1.3288	1.1378
31 Jul 18	n/a	78.121	1.3169	1.1269
31 Aug 18	n/a	77.3698	1.288	1.1157
30 Sep 18	n/a	78.1567	1.3062	1.1195
31 Oct 18	n/a	78.6241	1.3012	1.133
30 Nov 18	n/a	78.3167	1.2901	1.1352
31 Dec 18	n/a	76.7291	1.2661	1.1128

(a) This series was discontinued in May 2006   Source: Bank of England

(1): Please see notes and definitions for Effective exchange rate index, Sterling (Jan 2005 = 100)
https://www.bankofengland.co.uk/statistics/details/further-details-about-effective-exchange-rate-indices-data

(2): Please see notes and definitions for Spot exchange rates
https://www.bankofengland.co.uk/statistics/details/further-details-about-spot-exchange-rates-data

(3): Please see notes and definitions for Effective exchange rates
https://www.bankofengland.co.uk/statistics/details/further-details-about-effective-exchange-rate-indices-data

12.11 Average zero coupon yields

	End month level of yield from British Government Securities			Monthly average yield from British Government Securities	
	5 year Nominal Zero Coupon [c]	10 year Nominal Zero Coupon [c]	20 year Nominal Zero Coupon [c]	10 year Real Zero Coupon [a] [b]	20 year Real Zero Coupon [a] [b]
Date	IUMSNZC	IUMMNZC	IUMLNZC	IUMAMRZC	IUMALRZC
31 Jan 01	4.978	4.6718	4.3694	2.2154	1.8684
28 Feb 01	4.9824	4.7079	4.378	2.2144	1.8523
31 Mar 01	4.8425	4.7218	4.5984	2.2702	1.9601
30 Apr 01	5.126	5.065	4.8872	2.5368	2.2434
31 May 01	5.2777	5.1627	5.0197	2.5735	2.3112
30 Jun 01	5.4376	5.2329	5.0078	2.5725	2.2817
31 Jul 01	5.1702	5.0002	4.7653	2.5826	2.2629
31 Aug 01	4.9746	4.8652	4.682	2.3919	2.1494
30 Sep 01	4.8735	4.9172	4.8559	2.4992	2.3047
31 Oct 01	4.5126	4.5368	4.4576	2.5713	2.3376
30 Nov 01	4.6523	4.6479	4.5089	2.3844	2.1178
31 Dec 01	5.0788	5.0326	4.782	2.5171	2.2145
31 Jan 02	4.9044	4.8138	4.6689	2.4923	2.2395
28 Feb 02	4.9428	4.923	4.7928	2.4722	2.2767
31 Mar 02	5.3046	5.2456	5.0823	2.5074	2.3014
30 Apr 02	5.1522	5.1553	5.0257	2.4157	2.2401
31 May 02	5.2156	5.2282	5.1364	2.4142	2.2421
30 Jun 02	5.0148	4.9715	4.8164	2.3162	2.1556
31 Jul 02	4.7231	4.9201	4.8015	2.409	2.2475
31 Aug 02	4.4976	4.6394	4.4857	2.3109	2.1482
30 Sep 02	4.1825	4.412	4.3797	2.1834	2.0609
31 Oct 02	4.3315	4.5992	4.5466	2.3919	2.236
30 Nov 02	4.5062	4.6898	4.6549	2.3853	2.2838
31 Dec 02	4.1583	4.3967	4.4607	2.3083	2.2482
31 Jan 03	4.0136	4.2807	4.374	2.0797	2.0936
28 Feb 03	3.7987	4.227	4.4687	1.7905	1.9801
31 Mar 03	3.9584	4.3549	4.5931	1.8418	2.0679
30 Apr 03	4.0086	4.4202	4.6491	1.9411	2.1236
31 May 03	3.7737	4.1579	4.4323	1.7954	2.0315
30 Jun 03	3.8401	4.2698	4.6015	1.6503	1.9703
31 Jul 03	4.2291	4.6075	4.8344	1.8311	2.1601
31 Aug 03	4.4574	4.6562	4.6988	1.9459	2.141
30 Sep 03	4.3404	4.5844	4.6736	2.0418	2.1832
31 Oct 03	4.9274	5.0093	4.859	2.1484	2.2234
30 Nov 03	4.911	5.0473	4.8909	2.2143	2.2058
31 Dec 03	4.612	4.7733	4.7034	2.034	2.0784
31 Jan 04	4.7324	4.8572	4.7485	1.9359	1.9606
29 Feb 04	4.5996	4.735	4.7232	1.959	1.901
31 Mar 04	4.6154	4.7184	4.6453	1.8077	1.7684
30 Apr 04	4.8491	4.9564	4.839	1.9295	1.848
31 May 04	5.1163	5.1062	4.8917	2.0517	1.8784
30 Jun 04	5.039	5.0587	4.84	2.1008	1.8799
31 Jul 04	5.0958	5.0518	4.7933	2.0733	1.8713
31 Aug 04	4.8585	4.881	4.6547	2.0347	1.8228
30 Sep 04	4.7433	4.7878	4.6312	1.9736	1.798
31 Oct 04	4.6267	4.7002	4.5693	1.8884	1.7651
30 Nov 04	4.473	4.5606	4.4453	1.8755	1.7059
31 Dec 04	4.4287	4.4872	4.4097	1.7552	1.5992
31 Jan 05	4.4769	4.5419	4.4811	1.7474	1.5862
28 Feb 05	4.6761	4.6514	4.5249	1.7701	1.5918
31 Mar 05	4.6106	4.6382	4.5618	1.874	1.7246
30 Apr 05	4.444	4.4749	4.4264	1.7638	1.6408
31 May 05	4.2051	4.2811	4.3326	1.6959	1.5704
30 Jun 05	4.0387	4.1508	4.2073	1.6498	1.5324
31 Jul 05	4.2154	4.2884	4.3179	1.6469	1.5442
31 Aug 05	4.0641	4.1238	4.1705	1.6094	1.4946

12.11 Average zero coupon yields

Date	End month level of yield from British Government Securities			Monthly average yield from British Government Securities	
	5 year Nominal Zero Coupon [c] IUMSNZC	10 year Nominal Zero Coupon [c] IUMMNZC	20 year Nominal Zero Coupon [c] IUMLNZC	10 year Real Zero Coupon [a] [b] IUMAMRZC	20 year Real Zero Coupon [a] [b] IUMALRZC
30 Sep 05	4.1884	4.2464	4.2336	1.5097	1.4043
31 Oct 05	4.2874	4.2823	4.238	1.5665	1.3974
30 Nov 05	4.2061	4.1665	4.0854	1.541	1.2925
31 Dec 05	4.1019	4.0476	3.9626	1.4668	1.2015
31 Jan 06	4.172	4.0783	3.8631	1.2921	0.9687
28 Feb 06	4.2134	4.1192	3.9088	1.3237	0.9897
31 Mar 06	4.3946	4.3444	4.1388	1.4135	1.1011
30 Apr 06	4.6102	4.5845	4.3286	1.542	1.2818
31 May 06	4.6302	4.5389	4.2978	1.6291	1.3321
30 Jun 06	4.7356	4.6532	4.4151	1.6832	1.3862
31 Jul 06	4.6631	4.5484	4.2984	1.6457	1.3078
31 Aug 06	4.6343	4.4357	4.163	1.5515	1.2077
30 Sep 06	4.6412	4.4466	4.1548	1.4912	1.1173
31 Oct 06	4.6979	4.4204	4.0629	1.5735	1.1274
30 Nov 06	4.6926	4.4059	4.0858	1.4768	1.0326
31 Dec 06	4.9131	4.6261	4.2526	1.5656	1.116
31 Jan 07	5.1482	4.8488	4.4219	1.7909	1.2376
28 Feb 07	4.9735	4.6832	4.3078	1.8103	1.2555
31 Mar 07	5.1338	4.8558	4.4697	1.7137	1.2048
30 Apr 07	5.194	4.9249	4.5536	1.9163	1.3708
31 May 07	5.448	5.1262	4.6951	2.0546	1.4654
30 Jun 07	5.6348	5.3592	4.8891	2.2134	1.5466
31 Jul 07	5.3529	5.1163	4.6572	2.1929	1.5126
31 Aug 07	5.1209	4.9352	4.5484	1.9573	1.3084
30 Sep 07	4.9798	4.9505	4.6447	1.7847	1.2423
31 Oct 07	4.9395	4.8693	4.5916	1.7411	1.2508
30 Nov 07	4.5718	4.6382	4.5157	1.476	1.0932
31 Dec 07	4.4137	4.5176	4.3859	1.493	1.0805
31 Jan 08	4.2964	4.489	4.4193	1.3008	0.9196
29 Feb 08	4.1991	4.5271	4.503	1.329	1.0273
31 Mar 08	3.9541	4.4245	4.6048	1.0269	0.8891
30 Apr 08	4.4429	4.7338	4.7139	1.2039	1.0222
31 May 08	4.94	4.9995	4.8293	1.3551	0.9951
30 Jun 08	5.1659	5.168	4.8995	1.3521	0.8387
31 Jul 08	4.7675	4.8489	4.7434	1.2828	0.8239
31 Aug 08	4.4108	4.5304	4.5953	1.1547	0.6732
30 Sep 08	4.208	4.523	4.6924	1.2467	0.8046
31 Oct 08	3.9697	4.6995	4.9258	2.051	1.3361
30 Nov 08	3.3664	4.0107	4.5731	2.5578	1.3677
31 Dec 08	2.708	3.3892	4.066	2.1573	1.2343
31 Jan 09	2.8834	4.0768	4.7012	1.5136	1.0594
28 Feb 09	2.6227	3.8151	4.7151	1.2268	1.2058
31 Mar 09	2.4459	3.3015	4.2558	1.172	1.202
30 Apr 09	2.594	3.6361	4.5822	1.0457	1.1446
31 May 09	2.7154	3.8141	4.7127	1.0491	1.0401
30 Jun 09	2.9717	3.703	4.5708	1.0368	0.9954
31 Jul 09	3.083	3.9125	4.7431	1.1672	0.9737
31 Aug 09	2.6877	3.6313	4.1118	1.0943	0.8452
30 Sep 09	2.6807	3.7123	4.1499	1.003	0.8653
31 Oct 09	2.7662	3.7684	4.2789	0.6518	0.7198
30 Nov 09	2.6607	3.6987	4.2359	0.7194	0.6875
31 Dec 09	2.9821	4.203	4.5939	0.7586	0.7456
31 Jan 10	2.9411	4.1135	4.5279	0.8386	0.8509
28 Feb 10	2.7834	4.2373	4.7692	1.0099	1.0893
31 Mar 10	2.7852	4.1764	4.6577	0.8325	0.9807

12.11 Average zero coupon yields

Date	End month level of yield from British Government Securities			Monthly average yield from British Government Securities	
	5 year Nominal Zero Coupon [c] IUMSNZC	10 year Nominal Zero Coupon [c] IUMMNZC	20 year Nominal Zero Coupon [c] IUMLNZC	10 year Real Zero Coupon [a] [b] IUMAMRZC	20 year Real Zero Coupon [a] [b] IUMALRZC
30 Apr 10	2.7534	4.1034	4.6279	0.7105	0.9059
31 May 10	2.4101	3.7919	4.4818	0.7262	0.9368
30 Jun 10	2.2133	3.5853	4.3767	0.761	0.975
31 Jul 10	2.2281	3.5691	4.4956	0.8592	1.0467
31 Aug 10	1.7575	3.0217	3.9415	0.6655	0.9099
30 Sep 10	1.813	3.1512	4.0071	0.4585	0.7377
31 Oct 10	1.878	3.3042	4.3031	0.3731	0.7128
30 Nov 10	2.0313	3.4591	4.4002	0.4724	0.8109
31 Dec 10	2.314	3.6174	4.3314	0.6677	0.8226
31 Jan 11	2.5858	3.8827	4.5822	0.6391	0.8706
28 Feb 11	2.6508	3.8872	4.5184	0.7246	0.8861
31 Mar 11	2.6575	3.8975	4.5078	0.5509	0.8083
30 Apr 11	2.4534	3.7116	4.3435	0.5741	0.8267
31 May 11	2.2507	3.5655	4.3125	0.4157	0.8022
30 Jun 11	2.2036	3.6665	4.4964	0.2648	0.6877
31 Jul 11	1.7512	3.1755	4.2227	0.1862	0.5977
31 Aug 11	1.6065	2.9038	4.0173	-0.1472	0.4051
30 Sep 11	1.4048	2.5096	3.5397	-0.2034	0.3054
31 Oct 11	1.3719	2.5642	3.4276	-0.1199	0.3441
30 Nov 11	1.2752	2.4275	3.1294	-0.3125	0.1156
31 Dec 11	1.001	2.105	3.0454	-0.4715	-0.0284
31 Jan 12	0.9544	2.0983	3.0189	-0.5915	-0.1548
29 Feb 12	1.001	2.2582	3.2967	-0.5792	-0.0198
31 Mar 12	1.0698	2.3026	3.447	-0.5298	0.0418
30 Apr 12	1.1164	2.2225	3.3938	-0.5906	-0.0172
31 May 12	0.6748	1.6781	2.8994	-0.6885	-0.0622
30 Jun 12	0.8284	1.8651	2.9815	-0.5981	-0.0537
31 Jul 12	0.566	1.6111	2.7932	-0.6425	-0.0247
31 Aug 12	0.5849	1.6236	2.8217	-0.6936	-0.0165
30 Sep 12	0.6798	1.7105	2.9454	-0.6624	0.0823
31 Oct 12	0.8255	1.8749	3.0256	-0.5891	0.1512
30 Nov 12	0.8264	1.7969	2.9544	-0.629	0.098
31 Dec 12	0.8853	1.8876	3.023	-0.7249	-0.0074
31 Jan 13	1.098	2.1883	3.3222	-0.8252	-0.093
28 Feb 13	0.9145	2.1117	3.2498	-1.0107	-0.1884
31 Mar 13	0.7439	1.9006	3.0961	-1.2516	-0.2985
30 Apr 13	0.73	1.7922	2.9541	-1.3174	-0.4836
31 May 13	1.02	2.1634	3.2631	-1.0712	-0.348
30 Jun 13	1.3933	2.6171	3.5337	-0.6342	-0.1241
31 Jul 13	1.2511	2.5581	3.5392	-0.5751	-0.0065
31 Aug 13	1.5644	2.8293	3.6373	-0.3741	0.0945
30 Sep 13	1.5498	2.7591	3.5437	-0.2171	0.1536
31 Oct 13	1.5122	2.7045	3.4942	-0.3777	0.0186
30 Nov 13	1.6323	2.8776	3.6515	-0.2864	0.0642
31 Dec 13	1.9931	3.1571	3.7316	-0.1927	0.1004
31 Jan 14	1.7819	2.8347	3.5521	-0.1593	0.1263
28 Feb 14	1.798	2.8561	3.5484	-0.1815	0.1037
31 Mar 14	1.8714	2.8687	3.5515	-0.2982	0.0133
30 Apr 14	1.857	2.8078	3.4992	-0.3476	-0.0308
31 May 14	1.7997	2.7029	3.412	-0.3825	-0.0841
30 Jun 14	1.9945	2.812	3.4523	-0.2886	-0.0462
31 Jul 14	1.9847	2.7598	3.3233	-0.3072	-0.0794
31 Aug 14	1.6987	2.3916	2.9663	-0.4659	-0.2378
30 Sep 14	1.781	2.4682	3.0612	-0.5305	-0.2812
31 Oct 14	1.5647	2.3054	2.9766	-0.6111	-0.4436
30 Nov 14	1.3003	1.9853	2.6605	-0.7293	-0.5389

12.11 Average zero coupon yields

	End month level of yield from British Government Securities			Monthly average yield from British Government Securities	
	5 year Nominal Zero Coupon [c]	10 year Nominal Zero Coupon [c]	20 year Nominal Zero Coupon [c]	10 year Real Zero Coupon [a] [b]	20 year Real Zero Coupon [a] [b]
Date	IUMSNZC	IUMMNZC	IUMLNZC	IUMAMRZC	IUMALRZC
31 Dec 14	1.1987	1.8196	2.4808	-0.9257	-0.7589
31 Jan 15	0.9086	1.3867	2.0083	-1.0857	-0.9306
28 Feb 15	1.2811	1.8567	2.4648	-0.8858	-0.7547
31 Mar 15	1.0851	1.6605	2.2909	-0.92	-0.7803
30 Apr 15	1.3017	1.9447	2.5347	-1.1025	-0.9588
31 May 15	1.263	1.9178	2.5294	-0.9409	-0.8146
30 Jun 15	1.4567	2.1552	2.7485	-0.8422	-0.7103
31 Jul 15	1.3892	2.0074	2.5863	-0.829	-0.7394
31 Aug 15	1.3454	1.9669	2.5998	-0.8785	-0.8826
30 Sep 15	1.184	1.7919	2.5118	-0.9077	-0.8598
31 Oct 15	1.2853	1.9457	2.6502	-0.89	-0.8081
30 Nov 15	1.2399	1.8741	2.5946	-0.7969	-0.74
31 Dec 15	1.3752	2.0097	2.7092	-0.7875	-0.7594
31 Jan 16	0.9573	1.6355	2.3858	-0.8497	-0.7912
29 Feb 16	0.6909	1.4096	2.3061	-0.9706	-0.8878
31 Mar 16	0.8064	1.5041	2.3188	-1.0452	-0.9193
30 Apr 16	0.9684	1.6947	2.4476	-1.0659	-0.8943
31 May 16	0.9044	1.5474	2.2818	-1.0012	-0.8729
30 Jun 16	0.3959	1.0263	1.7905	-1.2024	-1.055
31 Jul 16	0.2752	0.7998	1.567	-1.638	-1.4033
31 Aug 16	0.2182	0.6615	1.2793	-1.9604	-1.6859
30 Sep 16	0.2591	0.7968	1.5211	-2.052	-1.8127
31 Oct 16	0.6341	1.2877	1.9205	-2.1096	-1.832
30 Nov 16	0.6747	1.4765	2.1026	-1.8772	-1.6208
31 Dec 16	0.5495	1.2955	1.9308	-1.8447	-1.551
31 Jan 17	0.7134	1.4858	2.1078	-1.9408	-1.6249
28 Feb 17	0.4196	1.1303	1.8546	-1.9647	-1.6763
31 Mar 17	0.4384	1.1343	1.8285	-1.9685	-1.6775
30 Apr 17	0.4062	1.1067	1.8271	-2.1515	-1.8587
31 May 17	0.3898	1.0669	1.7856	-2.0011	-1.7461
30 Jun 17	0.6352	1.2971	1.9618	-1.9715	-1.7026
31 Jul 17	0.5555	1.2613	1.9555	-1.7799	-1.5366
31 Aug 17	0.4195	1.0758	1.7913	-1.9318	-1.6783
30 Sep 17	0.7839	1.4009	2.0041	-1.8845	-1.6416
31 Oct 17	0.7923	1.3894	1.9978	-1.7722	-1.5743
30 Nov 17	0.8269	1.3844	1.9538	-1.8259	-1.6367
31 Dec 17	0.7444	1.258	1.8281	-1.8414	-1.6789
31 Jan 18	1.0427	1.5477	1.9947	-1.789	-1.6514
28 Feb 18	1.0755	1.5738	1.9789	-1.5862	-1.5176
31 Mar 18	1.0304	1.424	1.7608	-1.6275	-1.589
30 Apr 18	1.0485	1.5084	1.8788	-1.6234	-1.5726
31 May 18	0.8774	1.3335	1.7559	-1.6467	-1.5473
30 Jun 18	0.9686	1.391	1.7941	-1.7328	-1.617
31 Jul 18	1.0236	1.4422	1.8254	-1.7494	-1.6433
31 Aug 18	1.0148	1.4116	1.826	-1.7649	-1.6393
30 Sep 18	1.1516	1.568	1.9675	-1.6823	-1.5407
31 Oct 18	1.0135	1.4322	1.9052	-1.7263	-1.5777
30 Nov 18	0.9318	1.3685	2.0894	-1.8489	-1.5979
31 Dec 18	0.9069	1.3052	1.8683	-2.0889	-1.7894

Source: Bank of England

[a] Calculated using the Variable Roughness Penalty (VRP) model.

[b] Users should note that on 15/02/2018 these yields were revised to the beginning of 2015.

[c] Calculated by the Bank of England using the Variable Roughness Penalty (VRP) model.

Please see notes and definitions for Yields:

https://www.bankofengland.co.uk/statistics/details/further-details-about-yields-data

12.12 Consumer credit excluding student loans

£ millions   Not seasonally adjusted

Net lending - monthly changes

			Monthly changes of			
Date	MFIs sterling net consumer credit lending to individuals [a] [d] [c] LPMVVXP	of which: mutuals [a] [b] [c] LPMB3VC	Other consumer credit lenders (excluding the Student Loans Company) sterling net consumer credit lending to individuals [a] LPMB4TH	Total (excluding the Student Loans Company) sterling net consumer credit lending to individuals LPMB3PT	credit card lending to individuals LPMVZQS	other consumer credit lending to individuals LPMB4TV
31 Jan 14	-762	-	-17	-780	-825	46
28 Feb 14	4	-	-160	-155	-90	-66
31 Mar 14	787	-	1012	1799	-276	2075
30 Apr 14	528	-	-18	510	602	-92
31 May 14	501	-	230	731	210	521
30 Jun 14	840	-	143	983	352	630
31 Jul 14	751	-	346	1097	166	931
31 Aug 14	480	-	216	696	542	154
30 Sep 14	1021	-	1125	2146	143	2003
31 Oct 14	59	-	316	375	92	283
30 Nov 14	862	-	522	1384	851	533
31 Dec 14	1115	-	496	1611	1150	461
31 Jan 15	-732	-	41	-691	-920	229
28 Feb 15	47	-	-138	-91	-118	27
31 Mar 15	1397	-	1140	2537	-203	2740
30 Apr 15	590	-	400	990	823	167
31 May 15	303	-	418	721	94	627
30 Jun 15	1327	-	418	1745	552	1193
31 Jul 15	833	-	510	1343	345	997
31 Aug 15	916	-	172	1088	580	508
30 Sep 15	999	-	1455	2454	234	2220
31 Oct 15	75	-	456	532	14	518
30 Nov 15	1579	-	755	2334	1267	1067
31 Dec 15	860	-	771	1630	1130	501
31 Jan 16	-202	-	256	54	-662	715
29 Feb 16	596	-	80	676	-143	819
31 Mar 16	1260	-	1651	2911	324	2588
30 Apr 16	201	-	561	762	470	292
31 May 16	1371	-	510	1881	601	1280
30 Jun 16	1207	-	658	1864	649	1215
31 Jul 16	743	-	462	1206	420	786
31 Aug 16	1563	-	185	1748	763	985
30 Sep 16	753	-	1712	2464	368	2096
31 Oct 16	736	-	639	1375	388	987
30 Nov 16	1523	-	1134	2657	1327	1330
31 Dec 16	144	-	860	1005	1040	-35
31 Jan 17	244	-	281	525	-645	1170
28 Feb 17	278	-	40	318	168	150
31 Mar 17	838	-	1799	2637	-135	2772
30 Apr 17	294	-	707	1002	1056	-54
31 May 17	1152	-	814	1966	289	1677
30 Jun 17	449	-	1076	1524	812	712
31 Jul 17	563	-	868	1431	445	986
31 Aug 17	997	-	625	1622	575	1047
30 Sep 17	369	-	2031	2400	672	1728
31 Oct 17	371	-	836	1207	0	1207
30 Nov 17	915	-	1289	2204	1589	616
31 Dec 17	491	-	846	1337	983	354

[a] These series may be affected by securitisations and loan transfers up to December 2009.

Source: Bank of England

[b] This series will be discontinued with effect from December 2013 data. Further information can be found in Bankstats, December 2013, `Changes to publication of data for mutually owned monetary financial institutions', available at www.bankofengland.co.uk/statistics/Documents/ms/articles/art1dec13.pdf.

[c] Separate data for banks and building societies have been discontinued from January 2010 onwards. For details, see the Monetary and Financial Statistics articleÂ http://www.bankofengland.co.uk/statistics/documents/ms/articles/artjan10.pdf. (31 Jan 2010)

[d] Please note that the compilation and descriptions of some credit series have changed from publication of April 2015 data, as described in Bankstats, April 2015, 'Changes to the treatment of loan transfers and lending to housing associations', available at www.bankofengland.co.uk/statistics/Documents/ms/articles/art1apr15.pdf.

12.13a INVESTMENT TRUSTS' ASSETS AND LIABILITIES AT MARKET VALUES

£ million

		2006	2007	2008	2009	2010	2011	2012	2013	2014	2015	2016	2017
ASSETS													
UK government securities denominated in sterling	RLLT	533	715	628	585	466	681	857	487	302	244	415	309
Index-linked	AFIS	0	0	0	0	8	95	c	45	47	c	c	35
Other[1]	K5HJ	533	715	628	585	458	586	c	442	255	c	c	274
UK government securities denominated in foreign currency	CBPP	0	0	0	0	0	0	0	0	0	0	0	0
UK local authority investments[2]	AHBR	0	0	0	0	0	0	0	0	0	0	c	c
Other UK public sector investments[3]	AHBS	0	0	0	0	0	0	0	0	0	0	0	0
UK ordinary shares Quoted[4]	AHBM	21,843	21,848	13,428	16,297	17,547	17,329	18,655	18,907	19,384	20,551	22,918	24,704
Unquoted	AHBQ	1,027	1,186	938	1,349	1,336	1,437	1,363	1,323	1,616	2,224	3,940	2,503
Overseas ordinary shares	AHCC	21,659	25,795	18,385	23,865	29,341	29,083	30,758	31,859	31,491	36,880	47,519	55,850
Other corporate securities[5] UK	CBGZ	1,071	1,259	813	665	560	529	470	498	350	427	529	514
Overseas	CBHA	741	1,038	623	939	1,304	1,344	1,129	1,217	775	698	531	927
UK authorised unit trust units	AHBT	24	0	28	33	42	76	53	86	119	331	394	341
Overseas government, provincial and municipal securities	AHBY	4	151	410	256	410	254	251	118	107	c	c	c
UK existing buildings, property, land and new construction work	CBHB	252	154	142	141	197	1,522	1,538	158	202	216	379	1,631
Other longer-term assets not elsewhere classified[6]	AMSE	2,898	3,462	3,684	3,824	4,618	5,439	5,744	6,240	8,168	c	8,665	10,710
LONGER-TERM ASSETS	**AHBD**	**50,052**	**55,608**	**39,079**	**47,954**	**55,821**	**57,694**	**60,818**	**60,893**	**62,514**	**68,830**	**85,474**	**98,035**
Short-term assets	CBGX	2,138	3,303	3,397	2,522	2,174	2,893	3,015	2,848	3,377	3,194	4,743	4,221
TOTAL ASSETS	**CBGW**	**52,190**	**58,911**	**42,476**	**50,476**	**57,995**	**60,587**	**63,833**	**63,741**	**65,891**	**72,024**	**90,217**	**102,256**
LIABILITIES													
Borrowing from UK and overseas banks[7]	CBHD	2,900	2,708	2,553	2,216	2,359	2,897	2,845	3,652	3,373	3,189	3,598	4,335
Other UK borrowing[8]	CBHG	952	988	1,012	1,062	1,123	914	865	989	1,099	1,432	1,631	1,844
Other overseas borrowing[9]	CBHI	2	0	225	185	186	126	117	11	c	162	213	226
Issued share and loan capital[10]	CBHK	5,492	5,659	3,834	3,627	3,361	4,395	4,345	3,741	3,237	3,133	4,468	4,411
TOTAL LIABILITIES	**CBHO**	**9,346**	**9,355**	**7,624**	**7,090**	**7,029**	**8,332**	**8,172**	**8,393**	**7,810**	**7,916**	**9,910**	**10,816**
NET ASSETS	**CBHM**	**42,844**	**49,556**	**34,852**	**43,386**	**50,966**	**52,255**	**55,661**	**55,348**	**58,081**	**64,108**	**80,307**	**91,440**

c Suppressed to avoid the disclosure of confidential

Source: Office for National Statistics

1 Includes securities of: 0 up to 15 years maturity; over 15 years maturity and undated maturity. Excludes treasury bills and index-linked securities.

2 Includes local authority securities; negotiable bonds; loans and mortgages.

3 Includes public corporation loans and mortgages and other public sector investments not elsewhere classified.

4 Includes investment trust securities.

5 Includes corporate bonds and preference shares.

6 Includes UK unauthorised unit trust units; UK open-ended investment companies; overseas mutual fund investments; other UK fixed assets; overseas fixed assets; overseas direct investment and other UK and overseas assets not elsewhere classified.

7 Sterling and foreign currency. Includes foreign currency liabilities on back-to-back loans and overdrafts.

8 Includes sterling and foreign currency borrowing from building societies; issue of securities (other than ordinary shares); issue of sterling commercial paper and other borrowing not elsewhere classified (such as borrowing from parent, subsidiary and associate companies and other related concerns).

9 Includes borrowing from related companies and other borrowing not elsewhere classified.

10 Quoted and unquoted. Includes ordinary shares; preference shares; deferred stocks; bonds; debentures and loan stocks.

12.13b UNIT TRUSTS AND PROPERTY UNIT TRUSTS' BALANCE SHEET
ASSETS AND LIABILITIES AT MARKET VALUES

£ million

		2006	2007	2008	2009	2010	2011	2012	2013	2014	2015	2016	2017
ASSETS													
UK government securities denominated in sterling[1]	CBHT	31,603	32,120	33,466	29,331	33,306	37,116	36,033	37,275	44,970	41,824	49,352	53,940
Ordinary shares[2]													
UK	RLIB	185,637	195,009	143,550	167,401	204,616	179,940	201,822	234,866	241,783	249,355	268,812	298,228
Overseas	RLIC	127,409	142,211	113,667	150,863	200,028	187,714	215,705	265,209	288,079	298,515	361,606	432,472
Other corporate securities[3]													
UK	CBHU	29,876	30,626	30,174	36,646	52,786	52,735	63,546	60,803	66,686	64,615	65,598	72,913
Overseas	CBHV	25,617	30,029	30,442	43,301	49,098	48,388	62,870	67,946	78,566	88,047	94,286	103,602
Overseas government, provincial and municipal securities	CBHW	3,532	3,880	5,754	5,810	9,251	14,797	19,604	19,698	23,184	26,171	38,918	53,043
UK existing buildings, property, land and new construction work	RLIE	12,781	12,480	8,518	7,248	9,484	13,220	13,498	13,576	15,706	17,432	24,207	20,628
Other longer-term assets not elsewhere classified[4]	CBHX	29,765	38,707	42,670	49,417	60,029	66,640	81,849	105,015	110,193	118,554	146,158	182,019
Short-term assets	CBHS	20,653	27,969	35,224	37,263	48,382	45,001	53,734	62,021	56,686	65,130	78,978	95,297
of which:													
Derivative contracts with UK and overseas counterparties which have a positive (asset) value5	KUU5	4,552	7,271	13,151	17,132	25,944	18,168	25,644	27,923	13,320	12,550	18,266	26,167
TOTAL ASSETS	CBHR	466,873	513,031	443,465	527,280	666,980	645,551	748,661	866,409	925,853	969,643	1,127,915	1,312,142

12.13b UNIT TRUSTS AND PROPERTY UNIT TRUSTS' BALANCE SHEET ASSETS AND LIABILITIES AT MARKET VALUES

£ million

		2006	2007	2008	2009	2010	2011	2012	2013	2014	2015	2016	2017
LIABILITIES													
Borrowing from UK and overseas banks[6]	RLLF	2,544	2,603	3,486	1,541	991	1,992	3,430	3,154	2,834	2,072	c	c
Other UK borrowing[7]	RLLH	16	36	80	26	74	350	83	34	109	c	c	c
Other overseas borrowing[8]	RLLI	0	0	0	54	0	0	0	0	0	c	0	0
Derivative contracts with UK and overseas counterparties which have a negative (liability) value5	KUU6	3,701	6,382	11,890	15,294	23,676	17,464	22,832	26,255	12,122	14,265	20,559	21,976
Other creditors, provisions and liabilities not elsewhere classified	KUU7	3,370	2,562	3,442	5,524	7,228	14,087	5,460	1,740	4,424	5,880	2,540	9,504
Liability attributable to unit and share holders	RLLG	457,242	501,448	424,567	504,841	635,011	611,658	716,856	835,226	906,364	947,406	1,101,587	1,278,369
TOTAL LIABILITIES	**RLLE**	**466,873**	**513,031**	**443,465**	**527,280**	**666,980**	**645,551**	**748,661**	**866,409**	**925,853**	**969,643**	**1,127,915**	**1,312,142**

Source: Office for National Statistics

c Suppressed to avoid the disclosure of confidential data.
Components may not sum to totals due to rounding.

1 Includes securities of: 0 up to 15 years maturity; over 15 years maturity; undated maturity and index-linked. Excludes treasury bills.

2 Quoted and unquoted. Includes investment trust securities.

3 Includes corporate bonds and preference shares.

4 UK and overseas. Includes UK government securities denominated in foreign currency; local authority and public corporation securities; mutual fund investments; other UK fixed assets; overseas fixed assets; direct investment and other assets not elsewhere classified.

5 Includes credit default products; employee stock options; other options; other swaps; futures; forwards and other derivative contracts not elsewhere classified.

6 Sterling and foreign currency. Includes foreign currency liabilities on back-to-back loans and overdrafts.

7 Includes sterling and foreign currency borrowing from building societies; issue of securities (other than ordinary shares); issue of sterling commercial paper and other borrowing not elsewhere classified (such as borrowing from parent, subsidiary and associate companies and other related concerns).

8 Includes borrowing from related companies and other borrowing not elsewhere classified.

12.14 SELF-ADMINISTERED PENSION FUNDS[1] BALANCE SHEET ASSETS AND LIABILITIES AT MARKET VALUES

£ million

		2005	2006	2007	2008	2009	2010	2011	2012	2013	2014	2015	2017
ASSETS													
UK government securities denominated in sterling	AHVK	94,325	104,910	113,617	98,577	108,871	122,007	167,372	199,400	227,528	260,553	289,169	431,657
Index-linked[2]	AHWC	48,821	53,858	64,797	58,564	70,317	83,016	110,808	142,565	159,843	178,465	195,291	288,326
Other	J8Y5	45,504	51,052	48,820	40,013	38,554	38,991	56,564	56,835	67,685	82,088	93,878	143,331
UK government securities denominated in foreign currency	RYEX	45	27	42	4	126	89	1,287	5,695	275	351	32	165
UK local authority investments[3]	AHVO	4	2	5	0	2	274	113	263	291	284	276	2,225
Other UK public sector investments[4]	JE5J	846	1,118	1,351	1,259	2,347	2,911	1,703	1,611	3,862	582	224	2,210
UK PUBLIC SECTOR SECURITIES	RYHC	95,220	106,057	115,015	99,840	111,346	125,281	170,475	206,969	231,956	261,770	289,701	436,257
UK corporate bonds[5]	JX62	47,912	54,626	57,306	55,741	64,351	59,469	61,586	64,508	64,581	76,987	74,441	82,484
Sterling	GQFT	47,061	53,190	55,550	53,765	62,073	57,086	59,322	61,896	61,938	73,601	71,297	79,677
Foreign currency	GQFU	851	1,436	1,756	1,976	2,278	2,383	2,264	2,612	2,643	3,386	3,144	2,807
UK ordinary shares[6]	AHVP	199,199	208,473	152,048	110,571	116,710	121,882	108,631	111,913	109,777	95,981	84,403	77,111
UK preference shares[6]	J8YF	153	276	235	775	833	16	11	29	51	77	53	422
Overseas corporate securities	JRS8	203,562	224,514	215,068	175,747	214,230	222,751	211,562	226,274	232,277	259,644	245,442	257,633
Bonds	RLPF	19,891	30,867	44,387	47,612	57,920	60,826	61,129	60,424	61,655	70,977	75,726	75,772
Ordinary shares	AHVR	183,060	192,978	169,598	127,525	155,577	161,043	149,638	165,141	170,007	187,923	169,255	180,919
Preference shares	RLPC	611	669	1,083	610	733	882	795	709	615	744	461	942
Mutual fund investments	JRS9	192,033	215,218	254,936	211,724	286,493	369,955	385,337	448,583	476,623	449,664	499,047	622,147
UK	J8Y7	150,316	168,635	194,833	148,798	203,523	266,058	264,652	293,071	302,024	283,997	317,215	399,566
Unit trust units[7]	JX63	105,429	115,872	147,536	110,558	148,642	178,027	172,120	208,155	212,566	191,218	207,703	238,028
Other[8]	JX64	44,887	52,763	47,297	38,240	54,881	88,031	92,532	84,916	89,458	92,779	109,512	161,538
Overseas	JE4P	41,717	46,583	60,103	62,926	82,970	103,897	120,685	155,512	174,599	165,667	181,832	222,581

12.14 SELF-ADMINISTERED PENSION FUNDS[1] BALANCE SHEET ASSETS AND LIABILITIES AT MARKET VALUES

£ million

		2005	2006	2007	2008	2009	2010	2011	2012	2013	2014	2015	2017
CORPORATE SECURITIES	RYHN	642,859	703,107	679,593	554,558	682,617	774,073	767,127	851,307	883,309	882,353	903,386	1,039,797
Overseas government, provincial and municipal securities	AHVT	19,037	21,776	22,434	21,527	16,900	17,335	21,424	23,768	23,462	22,643	22,306	26,994
Loans	JRT4	9	42	417	518	1,768	2,343	1,656	1,313	324	94	601	1,001
UK[9]	JE5E	6	6	12	0	116	81	77	170	c	c	c	520
Overseas[10]	AHVZ	3	36	405	518	1,652	2,262	1,579	1,143	c	c	c	481
Fixed assets[11]	JRT5	31,742	34,608	30,466	22,892	24,957	30,159	32,991	30,705	33,695	37,734	40,630	45,265
UK	JE5F	31,616	34,435	30,306	22,818	24,718	28,991	32,178	30,372	33,064	36,937	39,655	44,411
Overseas	GOLB	126	173	160	74	239	1,168	813	333	631	797	975	854
Investment in insurance managed funds, insurance policies and annuities	RYHS	86,336	92,387	103,610	86,541	70,318	80,613	93,823	99,896	110,723	122,194	134,665	158,012
Other longer-term assets not elsewhere classified[12]	J8YA	21,423	31,234	41,455	45,301	51,100	69,419	81,562	90,893	109,368	84,875	82,323	116,088
OTHER LONGER-TERM ASSETS	RYHW	158,547	180,047	198,382	176,779	165,043	199,869	231,456	246,575	277,572	267,540	280,525	347,360
LONGER-TERM ASSETS	RYHX	896,626	989,211	992,990	831,177	959,006	1,099,223	1,169,058	1,304,851	1,392,837	1,411,663	1,473,612	1,823,414

Source: Office for National Statistics

c Suppressed to avoid the disclosure of confidential data. Components may not sum to totals due to rounding.

1 Combined public and private sector. Data from the pension funds surveys are of lower quality than equivalent data from other institutional groups because of the difficulties in constructing a suitable sampling frame of pension funds.

2 Includes securities of: 0 up to 15 years maturity; over 15 years maturity and undated maturity. Excludes treasury bills and index-linked securities.

3 Includes local authority securities; negotiable bonds; loans and mortgages.

4 Includes public corporation loans and mortgages and other public sector investments not elsewhere classified.

5 Issued by: banks; building societies and other corporates.

6 Quoted and unquoted.

7 Authorised and unauthorised.

8 Includes property unit trusts; investment trust securities; open-ended investment companies; hedge funds and other mutual fund investments not elsewhere classified.

9 Includes sterling asset backed loans; loans to individuals secured on dwellings; other loans to individuals (including policy loans); loans to businesses and other loans not elsewhere classified. Excludes loans to UK associate companies; bank term deposits and building society investments.

10 Includes loans to parent companies; subsidiaries; associates and other loans not elsewhere classified. Excludes loans categorised as direct investment; loans covered by Export Credit Guarantee Department (ECGD), specific bank guarantees or ECGD buyer credit guarantees.

11 UK and overseas. Includes existing buildings; property; land; new construction work; vehicles; machinery and equipment; valuables and intangibles. Includes the capital value of assets bought on hire purchase or acquired (as lessee) under a finance leasing arrangement and assets acquired for hiring, renting and operating leasing purposes. Excludes the capital value of assets acquired but leased out to others under finance leasing arrangements.

12 UK and overseas. Includes certificates of tax deposit; insurance policies; annuities and loans covered by Export Credit Guarantee Department (ECGD), specific bank guarantees or ECGD buyer credit guarantees. Excludes pre-payments and debtors.

12.14 SELF-ADMINISTERED PENSION FUNDS[1] BALANCE SHEET ASSETS AND LIABILITIES AT MARKET VALUES

continued

£ million

		2006	2007	2008	2009	2010	2011	2012	2013	2014	2015	2016	2017
ASSETS													
Cash	GNOR	0	0	0	0	0	0	0	0	0	0	0	0
Balances with banks and building societies in the UK	JX5Q	22,823	25,974	20,133	23,054	23,736	23,854	22,929	22,701	25,881	22,433	21,274	28,922
Sterling	JX5S	20,211	22,615	16,802	19,099	20,043	20,159	18,901	18,193	21,841	17,525	15,757	22,620
Foreign currency	JX5U	2,612	3,359	3,331	3,955	3,693	3,695	4,028	4,508	4,040	4,908	5,517	6,302
Balances with overseas banks	GNOW	525	807	351	134	197	72	157	349	c	137	86	316
Other liquid deposits[13]	GNOX	3,891	5,392	4,300	12,431	9,631	11,956	15,265	20,505	15,084	13,133	16,182	18,039
Certificates of deposit issued by banks and building societies in the UK[14]	IX8H	4,932	8,156	6,480	2,269	2,472	6,498	2,069	1,288	455	522	804	935
Money market instruments issued by HM Treasury[15]	IX9J	22	304	548	1,709	1,109	1,734	4,116	1,167	1,698	1,828	3,767	781
UK local authority debt	AHVF	221	205	323	210	116	157	232	130	c	408	396	1,119
Commercial paper issued by UK companies[16]	GQFR	401	857	367	664	1,135	993	1,254	334	286	551	1,278	1,653
Other UK money market instruments[17]	GOZR	1,174	1,893	1,959	2,822	3,416	3,936	2,716	4,655	3,928	4,406	2,397	5,145
Money market instruments issued by non-resident businesses	GOZS	449	1,279	911	963	2,067	3,709	2,752	2,752	6,099	4,149	7,293	6,114
Other short-term assets not elsewhere classified[18]	JX5W	1,898	1,304	3,474	3,113	3,644	2,311	1,904	1,788	2,592	1,462	1,452	4,277
Balances outstanding from stockbrokers and securities dealers[19]	RYIL	24,812	28,297	30,536	11,281	7,145	8,241	6,213	5,989	5,379	7,740	9,767	9,626
Income accrued on investments and rents	RYIM	2,303	2,744	2,576	2,714	2,761	2,930	2,872	3,169	2,651	2,921	3,274	2,731
Amounts outstanding from HM Revenue and Customs[19]	RYIN	16	15	23	32	37	53	67	54	65	88	82	118
Other debtors and assets not elsewhere classified	RYIO	35,224	22,454	24,565	21,464	13,159	8,902	11,387	7,543	5,933	4,398	4,757	5,281
Derivative contracts with UK counterparties which have a positive (asset) value[20]	JRO3	24,357	29,789	35,194	61,862	99,420	170,170	186,184	191,511	222,296	225,060	252,227	232,064
Derivative contracts with overseas counterparties which have a positive (asset) value[20]	GOJU	5,995	8,652	5,835	20,534	19,803	29,445	38,324	49,910	79,679	87,428	74,623	78,450
TOTAL ASSETS	RYIR	1,118,254	1,131,112	968,752	1,124,262	1,289,071	1,444,019	1,603,292	1,706,682	1,784,104	1,850,276	2,119,834	2,218,985

12.14 SELF-ADMINISTERED PENSION FUNDS[1] BALANCE SHEET ASSETS AND LIABILITIES AT MARKET VALUES

continued

£ million

Source: Office for National Statistics

c Suppressed to avoid the disclosure of confidential data. Components may not sum to totals due to rounding.

1 Combined public and private sector. Data from the pension funds surveys are of lower quality than equivalent data from other institutional groups because of the difficulties in constructing a suitable sampling frame of pension funds.

13 Includes money market funds; liquidity funds and cash liquidity funds.

14 Sterling and foreign currency.

15 Includes treasury bills. Excludes UK government securities.

16 Sterling and foreign currency commercial paper issued by: banks; building societies; other financial institutions and other issuing companies.

17 Includes floating rate notes maturing within one year of issue.

18 UK and overseas. Excludes derivative contracts.

19 Gross value.

20 Includes credit default products; employee stock options; other options; other swaps; futures; forwards and other derivative contracts not elsewhere classified.

12.14 SELF-ADMINISTERED PENSION FUNDS[1] BALANCE SHEET ASSETS AND LIABILITIES AT MARKET VALUES

continued

£ million

		2006	2007	2008	2009	2010	2011	2012	2013	2014	2015	2016	2017
LIABILITIES													
Borrowing[21]	GQED	14,661	16,180	4,461	3,859	2,830	3,361	10,603	16,050	27,135	39,141	45,947	59,682
Balances owed to stockbrokers and securities dealers[19]	RYIS	25,954	33,965	37,912	13,707	7,312	9,688	7,816	8,267	9,869	12,346	18,251	18,997
Pensions due but not paid[22]	RYIT	285	220	167	280	271	300	5,152	2,672	3,693	3,783	5,283	7,651
Derivative contracts with UK counterparties which have a negative (liability) value[20]	JRP9	17,231	26,187	27,533	37,689	78,322	147,217	149,493	161,962	181,946	196,477	199,739	185,539
Derivative contracts with overseas counterparties which have a negative (liability) value[20]	GKGR	7,036	11,275	6,335	41,110	37,345	42,712	60,725	73,848	109,217	114,944	118,657	124,930
Other creditors, provisions and liabilities not elsewhere classified	RYIU	36,208	18,327	20,577	21,603	18,889	9,905	9,011	13,236	3,006	6,866	17,627	14,148
Market value of pension funds[23]	AHVA	1,016,879	1,024,958	871,767	1,006,014	1,144,102	1,230,836	1,360,492	1,430,647	1,449,238	1,476,719	1,714,330	1,808,038
TOTAL LIABILITIES	**RYIR**	**1,118,254**	**1,131,112**	**968,752**	**1,124,262**	**1,289,071**	**1,444,019**	**1,603,292**	**1,706,682**	**1,784,104**	**1,850,276**	**2,119,834**	**2,218,985**

Source: Office for National Statistics

c Suppressed to avoid the disclosure of confidential data. Components may not sum to totals due to rounding.

1 Combined public and private sector. Data from the pension funds surveys are of lower quality than equivalent data from other institutional groups because of the difficulties in constructing a suitable sampling frame of pension funds.

19 Gross value.

20 Includes credit default products; employee stock options; other options; other swaps; futures; forwards and other derivative contracts not elsewhere classified.

21 UK and overseas. Includes from a UK perspective: sterling and foreign currency borrowing from UK banks and building societies; borrowing arising from the issue of floating rate notes and preference shares; foreign currency liabilities on back-to-back loans; overdrafts and other borrowing not elsewhere classified. Includes from an overseas perspective: borrowing from banks and other borrowing not elsewhere classified.

22 Excludes any estimated future liabilities.

23 Net value as found in statement of net assets.

12.15a INSURANCE COMPANIES' BALANCE SHEET: LONG-TERM BUSINESS ASSETS AND LIABILITIES AT MARKET VALUES

£ million

		2006	2007	2008	2009	2010	2011	2012	2013	2014	2015	2016	2017
ASSETS													
UK government securities denominated in sterling	AHNJ	161,641	158,694	166,879	167,247	182,506	205,223	187,145	170,950	179,602	163,502	175,735	161,826
Index-linked[1]	AHQI	41,104	45,902	54,387	57,157	66,115	78,354	72,499	71,234	81,181	72,840	80,206	75,154
Other[1]	J5HZ	120,537	112,792	112,492	110,090	116,391	126,869	114,646	99,716	98,421	90,662	95,529	86,672
UK government securities denominated in foreign currency	RGBV	12	0	0	0	0	0	0	0	0	c	0	0
UK local authority investments[2]	AHNN	1,614	998	776	655	768	813	770	3,168	8,390	c	10,477	5,216
Other UK public sector investments[3]	RGCS	651	634	872	1,461	2,189	2,207	2,197	3,684	2,496	2,129	1,130	8,344
UK PUBLIC SECTOR SECURITIES	RYEK	163,918	160,326	168,527	169,363	185,463	208,243	190,112	177,802	190,488	173,188	187,342	175,386
UK corporate bonds[4]	IFLF	159,073	159,273	159,789	160,532	158,987	163,348	178,627	167,021	177,937	176,016	176,182	169,631
Sterling	IFLG	150,907	153,694	154,525	155,184	156,477	160,842	174,682	165,453	176,486	175,269	174,804	168,500
Foreign currency	IFLH	8,166	5,579	5,264	5,348	2,510	2,506	3,945	1,568	1,451	747	1,378	1,131
UK ordinary shares[5]	IFLI	299,717	293,655	188,430	209,992	207,971	177,480	162,949	170,841	160,294	145,908	143,479	137,092
UK preference shares[5]	RLOL	1,231	983	724	624	648	536	375	340	271	230	196	253
Overseas corporate securities	IFLJ	194,997	234,388	219,957	266,280	278,930	262,823	303,565	315,453	331,122	339,248	391,047	402,523
Bonds	RLOP	57,774	69,696	85,077	111,866	112,338	115,744	127,629	119,631	122,217	119,028	129,841	127,044
Ordinary shares	AHNQ	136,652	163,249	134,211	153,500	165,216	145,932	173,982	194,250	207,643	219,253	259,793	274,010
Preference shares	RLOM	571	1,443	669	914	1,376	1,147	1,954	1,572	1,262	967	1,413	1,469
Mutual fund investments[6]	IFLK	192,516	222,081	186,950	230,247	259,457	265,750	328,657	360,502	390,694	449,493	506,913	592,361
CORPORATE SECURITIES	RYEO	847,534	910,380	755,850	867,675	905,993	869,937	974,173	1,014,157	1,060,318	1,110,895	1,217,817	1,301,860
Overseas government, provincial and municipal securities	AHNS	21,078	25,787	29,053	24,601	25,363	23,727	26,928	30,013	34,234	36,854	33,332	36,727
Other longer-term assets not elsewhere classified[7]	JX8D	92,326	99,316	76,139	67,531	74,900	81,916	81,039	74,737	80,064	78,488	95,239	98,629
OTHER LONGER-TERM ASSETS	RYER	113,404	125,103	105,192	92,132	100,263	105,643	107,967	104,750	114,298	115,342	128,571	135,356
LONGER-TERM ASSETS	RYES	1,124,856	1,195,809	1,029,569	1,129,170	1,191,719	1,183,823	1,272,252	1,296,709	1,365,104	1,399,425	1,533,730	1,612,602

Source: Office for National Statistics

c Suppressed to avoid the disclosure of confidential data. Components may not sum to totals due to rounding.
1 Includes securities of: 0 up to 15 years maturity; over 15 years maturity and undated maturity. Excludes treasury bills and index-linked securities.
2 Includes local authority securities; negotiable bonds; loans and mortgages.
3 Includes public corporation loans and mortgages and other public sector investments not elsewhere classified.
4 Issued by: banks; building societies and other corporates.
5 Quoted and unquoted.
6 UK and overseas. Includes authorised and unauthorised unit trust units; investment trust securities; open-ended investment companies; hedge funds and other mutual fund investments not elsewhere classified.
7 UK and overseas. Includes loans; fixed assets and other longer-term assets not elsewhere classified.

12.15a INSURANCE COMPANIES' BALANCE SHEET: LONG-TERM BUSINESS ASSETS AND LIABILITIES AT MARKET VALUES

continued

£ million

		2006	2007	2008	2009	2010	2011	2012	2013	2014	2015	2016	2017
ASSETS													
Cash	HLGW	0	0	0	0	0	0	0	0	0	0	0	0
Balances with banks and building societies in the UK	JX2C	34,454	48,619	51,139	44,483	34,026	34,998	35,424	26,785	26,456	25,901	26,985	25,204
Sterling	JX3R	32,653	45,948	47,456	42,097	32,387	32,698	33,805	24,616	23,842	22,896	24,941	22,891
Foreign currency	JX3T	1,801	2,671	3,683	2,386	1,639	2,300	1,619	2,169	2,614	3,005	2,044	2,313
Balances with overseas banks	HLHB	1,022	1,484	2,816	3,758	2,766	1,726	1,381	3,143	3,884	1,703	1,562	1,945
Other liquid deposits[8]	HLHC	333	678	1,638	12,385	14,076	15,891	17,257	38,843	39,791	46,570	48,885	58,267
Certificates of deposit issued by banks and building societies in the UK[9]	AHND	17,256	15,177	11,401	8,177	8,622	6,158	5,769	5,318	6,080	6,483	4,680	4,444
Money market instruments issued by HM Treasury[10]	RGBM	685	785	179	1,344	509	1,392	1,081	2,688	1,261	1,312	1,310	2,986
UK local authority debt	AHNF	0	0	0	0	0	0	0	1	c	c	0	0
Commercial paper issued by UK companies[11]	JF77	2,714	2,723	2,268	1,806	646	891	496	648	458	696	271	45
Other UK money market instruments[12]	HLHL	3,538	3,071	3,159	2,959	3,208	5,501	3,540	2,055	2,212	2,358	c	5,515
Money market instruments issued by non-resident businesses	HLHM	405	859	689	1,755	2,956	1,530	2,068	907	1,047	c	c	1,717
Other short-term assets not elsewhere classified[13]	JX2E	10,749	20,965	19,555	10,076	5,232	6,159	6,008	5,582	c	3,407	3,762	3,690
Balances due from stockbrokers and securities dealers[14]	RGBU	-345	140	-399	-1,159	-834	-1,245	-1,145	-1,228	-1,604	-737	-3,675	-4,191

12.15a INSURANCE COMPANIES' BALANCE SHEET: LONG-TERM BUSINESS ASSETS AND LIABILITIES AT MARKET VALUES

continued

£ million

		2006	2007	2008	2009	2010	2011	2012	2013	2014	2015	2016	2017
SHORT-TERM ASSETS (excluding derivatives)	RYEW	70,811	94,501	92,445	85,584	71,207	73,001	71,879	84,742	82,966	89,292	91,087	99,622
Derivative contracts with UK and overseas counterparties which have a positive (asset) value[15]	IFKX	5,937	5,205	17,892	10,317	12,399	24,060	20,805	18,144	39,790	30,148	53,829	51,446
Agents' balances and outstanding premiums in respect of direct insurance and facultative reinsurance contracts[16]													
UK	RYPA	669	738	535	467	515	554	620	769	652	c	432	481
Overseas	RYPB	-136	-177	0	1	3	3	0	0	0	c	0	0
Reinsurance, coinsurance and treaty balances[17]													
UK	RYPC	3,462	-413	-1,000	3,817	-838	-1,147	-1,778	2,109	7,699	-1,078	c	c
Overseas	RYPD	1,105	836	-7,355	-8,232	-6,020	-6,035	-6,425	-5,590	-5,562	-5,077	c	c
Outstanding interest, dividends and rents[18]	RYPH	7,330	8,690	9,256	9,158	8,591	8,511	8,716	7,762	7,216	6,186	6,276	5,813
Other debtors and assets not elsewhere classified[19]	RYPF	44,923	51,373	37,342	33,162	30,366	28,806	27,717	28,595	21,833	29,100	18,685	18,611
Direct investment for non-insurance subsidiary and associate companies in the UK[20]	RYET	13,016	9,186	11,484	10,129	5,097	7,249	8,880	8,125	5,434	6,368	9,375	9,685
Direct investment for UK insurance subsidiary, associate and holding companies[20]	RYEU	6,114	7,578	7,890	7,234	6,720	7,152	4,275	3,865	6,873	2,057	1,455	1,428
Direct investment for overseas subsidiaries, associates, branches and agencies[20]	RYEV	3,341	3,832	5,011	4,141	4,131	3,340	3,470	4,296	3,542	2,762	2,988	3,705
TOTAL ASSETS	RKBI	1,281,428	1,377,158	1,203,069	1,284,948	1,323,890	1,329,317	1,410,411	1,449,526	1,535,547	1,559,670	1,717,351	1,803,167

Source: Office for National Statistics

c Suppressed to avoid the disclosure of confidential data. Components may not sum to totals due to rounding.
8 Includes money market funds; liquidity funds and cash liquidity funds.
9 Sterling and foreign currency.
10 Includes treasury bills. Excludes UK government securities.
11 Sterling and foreign currency commercial paper issued by: banks; building societies; other financial institutions and other issuing companies.
12 Includes floating rate notes maturing within one year of issue.
13 UK and overseas. Excludes derivative contracts.
14 Net of balances owed. Includes amounts due on securities bought and sold for future settlement.
15 Includes credit default products; employee stock options; other options; other swaps; futures; forwards and other derivative contracts not elsewhere classified.
16 Net of insurance liabilities.
17 Net of reinsurance bought and sold.
18 Net value.
19 Includes deferred acquisition costs.
20 Net asset value of attributable companies.

12.15a INSURANCE COMPANIES' BALANCE SHEET: LONG-TERM BUSINESS ASSETS AND LIABILITIES AT MARKET VALUES

continued

£ million

		2006	2007	2008	2009	2010	2011	2012	2013	2014	2015	2016	2017
LIABILITIES													
Borrowing	AHNI	14,369	12,426	12,467	10,866	12,542	12,084	10,660	10,112	9,578	8,983	10,892	11,965
Banks and building societies in the UK[9]	JX2G	2,867	3,801	5,089	3,305	2,099	1,980	961	1,011	461	776	1,288	1,235
Other UK[21]	ICXU	9,537	6,699	4,832	5,460	8,112	8,730	9,075	8,368	8,340	7,692	9,295	10,123
Overseas[22]	RGDD	1,965	1,926	2,546	2,101	2,331	1,374	624	733	777	515	309	607
Long-term business insurance and investment contract liabilities[23]	RKDC	1,125,221	1,205,183	1,069,993	1,153,944	1,178,823	1,171,844	1,241,671	1,277,415	1,346,456	1,380,449	1,523,568	1,612,763
Claims admitted but not paid[23]	RKBM	3,513	3,848	3,426	4,637	2,946	3,072	3,387	3,608	3,649	3,594	2,896	3,211
Provisions for taxation and dividends payable[24]	KVE9	8,120	7,489	60	1,781	2,973	1,993	2,400	2,919	4,041	2,520	4,716	6,211
Other creditors, provisions and liabilities not elsewhere classified[25]	RYPL	33,192	39,527	52,849	47,803	53,826	66,466	69,971	68,289	89,296	79,994	97,778	91,036
Excess of total assets over liabilities in respect of: long-term business; minority interests in UK subsidiary companies; shareholders' capital and reserves and any other reserves[26]	A4YP	97,013	108,685	64,274	65,917	72,780	73,858	82,322	87,183	82,527	84,130	77,501	77,981
TOTAL LIABILITIES	RKBI	1,281,428	1,377,158	1,203,069	1,284,948	1,323,890	1,329,317	1,410,411	1,449,526	1,535,547	1,559,670	1,717,351	1,803,167

Source: Office for National Statistics

c   Suppressed to avoid the disclosure of confidential data. Components may not sum to totals due to rounding.
9   Sterling and foreign currency.
21 Includes issue of securities (other than ordinary shares); issue of sterling commercial paper and other borrowing not elsewhere classified.
22 Includes borrowing from banks; related companies and other borrowing not elsewhere classified.
23 Net of reinsurers share.
24 UK and overseas. Includes deferred tax net of amounts receivable.
25 UK and overseas. Includes derivative contracts which have a negative (liability) value.
26 Includes unallocated divisible surplus and the 'net worth' of UK branches of overseas companies, including profit and loss account balances.

12.15b INSURANCE COMPANIES' BALANCE SHEET: GENERAL BUSINESS ASSETS AND LIABILITIES AT MARKET VALUES

£ million

		2006	2007	2008	2009	2010	2011	2012	2013	2014	2015	2016	2017
ASSETS													
UK government securities denominated in sterling	AHMJ	19,296	16,026	18,441	16,969	13,631	13,384	11,774	12,855	14,712	13,308	14,561	12,000
Index-linked	AHMZ	603	297	1,622	1,991	1,731	1,724	2,150	2,355	3,009	2,724	3,022	2,932
Other[1]	J8EX	18,693	15,729	16,819	14,978	11,900	11,660	9,624	10,500	11,703	10,584	11,539	9,068
UK government securities denominated in foreign currency	RYMQ	72	69	3	0	0	c	13	18	0	c	0	c
UK local authority investments[2]	AHMN	0	3	0	0	0	c	0	0	0	c	c	c
Other UK public sector investments[3]	RYMU	0	10	0	65	249	c	252	209	446	184	c	94
UK PUBLIC SECTOR SECURITIES	RYMV	**19,368**	**16,108**	**18,444**	**17,034**	**13,880**	**13,907**	**12,039**	**13,082**	**15,158**	**13,521**	**14,608**	**12,117**
UK corporate bonds[4]	IFVV	11,797	13,397	12,060	13,550	10,487	10,817	12,127	12,355	12,462	11,530	12,617	13,091
Sterling	IFVW	11,057	12,647	11,009	12,288	9,544	10,217	11,264	11,610	11,716	10,671	11,817	12,248
Foreign currency	IFVX	740	750	1,051	1,262	943	600	863	745	746	859	800	843
UK ordinary shares[5]	IFVY	10,138	9,006	10,501	9,963	9,525	8,435	8,732	7,846	3,500	3,402	3,092	2,739
UK preference shares[5]	RLOT	9	44	29	26	26	25	21	21	25	20	20	24
Overseas corporate securities	IFVZ	18,636	14,773	20,258	23,002	17,594	18,636	24,380	24,638	27,748	28,637	30,822	29,299
Bonds	RLOX	16,128	12,872	18,081	21,017	15,847	16,939	21,872	21,834	24,276	25,307	27,200	25,555
Ordinary shares	AHMQ	2,507	1,901	2,175	1,976	1,741	1,692	2,501	c	3,462	c	c	3,729
Preference shares	RLOU	1	0	2	9	6	5	7	c	10	c	c	15
Mutual fund investments[6]	IFWA	1,256	1,780	1,911	3,498	4,878	5,802	4,236	4,686	4,951	4,610	3,384	3,989
CORPORATE SECURITIES	RYNF	**41,836**	**39,000**	**44,759**	**50,039**	**42,510**	**43,715**	**49,496**	**49,546**	**48,686**	**48,199**	**49,935**	**49,142**
Overseas government, provincial and municipal securities	AHMS	8,035	4,869	8,505	7,204	5,126	5,524	9,355	9,584	8,373	9,079	8,519	8,836
Other longer-term assets not elsewhere classified[7]	JX8E	6,746	7,059	8,269	7,338	9,158	9,938	10,148	9,606	10,518	10,686	9,067	7,853
OTHER LONGER-TERM ASSETS	RYNO	**14,781**	**11,928**	**16,774**	**14,542**	**14,284**	**15,462**	**19,503**	**19,190**	**18,891**	**19,765**	**17,586**	**16,689**
LONGER-TERM ASSETS	RYNP	**75,985**	**67,036**	**79,977**	**81,615**	**70,674**	**73,084**	**81,038**	**81,818**	**82,735**	**81,485**	**82,129**	**77,948**

Source: Office for National Statistics

c Suppressed to avoid the disclosure of confidential data. Components may not sum to totals due to rounding.
1 Includes securities of: 0 up to 15 years maturity; over 15 years maturity and undated maturity. Excludes treasury bills and index-linked securities.
2 Includes local authority securities; negotiable bonds; loans and mortgages.
3 Includes public corporation loans and mortgages and other public sector investments not elsewhere classified.
4 Issued by: banks; building societies and other corporates.
5 Quoted and unquoted.
6 UK and overseas. Includes authorised and unauthorised unit trust units; investment trust securities; open-ended investment companies; hedge funds and other mutual fund investments not elsewhere classified.
7 UK and overseas. Includes loans; fixed assets and other longer-term assets not elsewhere classified.

12.15b INSURANCE COMPANIES' BALANCE SHEET: GENERAL BUSINESS ASSETS AND LIABILITIES AT MARKET VALUES

continued

£ million

		2006	2007	2008	2009	2010	2011	2012	2013	2014	2015	2016	2017
ASSETS													
Cash	HLMN	0	0	0	0	0	0	0	0	0	0	0	0
Balances with banks and building societies in the UK	JX3H	9,327	8,949	9,871	10,227	7,961	7,619	7,032	6,970	5,161	4,744	4,899	5,796
Sterling	JX43	6,501	6,907	8,170	8,763	6,730	6,532	5,988	5,583	3,974	3,480	3,363	3,908
Foreign currency	JX45	2,826	2,042	1,701	1,464	1,231	1,087	1,044	1,387	1,187	1,264	1,536	1,888
Balances with overseas banks	HLMS	1,031	1,015	1,250	1,138	981	1,150	1,841	1,369	617	976	565	731
Other liquid deposits[8]	HLMT	396	375	869	2,476	3,326	2,916	3,753	3,021	3,390	4,271	4,526	4,951
Certificates of deposit issued by banks and building societies in the UK[9]	IX8K	7,708	8,264	7,775	2,002	558	843	1,223	1,403	2,557	2,372	1,330	1,662
Money market instruments issued by HM Treasury[10]	ICWI	379	342	708	979	327	237	1,675	1,214	1,659	1,899	1,586	1,553
UK local authority debt	AHMF	0	0	0	0	0	0	0	0	0	0	0	0
Commercial paper issued by UK companies[11]	JF75	732	2,682	1,640	1,630	275	281	509	461	498	178	92	108
Other UK money market instruments[12]	HLNC	68	90	278	346	359	360	253	340	324	266	c	281
Money market instruments issued by non-resident businesses	HLND	3,516	3,261	2,277	2,321	545	745	449	1,120	570	562	c	556
Other short-term assets not elsewhere classified[13]	JX2I	1,733	612	1,576	718	725	933	1,064	781	1,413	797	963	564
Balances due from stockbrokers and securities dealers[14]	RYMA	-66	25	891	52	90	26	-1	37	-48	-34	-45	-18

12.15b INSURANCE COMPANIES' BALANCE SHEET: GENERAL BUSINESS ASSETS AND LIABILITIES AT MARKET VALUES

continued

£ million

		2006	2007	2008	2009	2010	2011	2012	2013	2014	2015	2016	2017
SHORT-TERM ASSETS (excluding derivatives)	RYME	24,824	25,615	27,135	21,889	15,147	15,110	17,798	16,716	16,141	16,031	14,792	16,184
Derivative contracts with UK and overseas counterparties which have a positive (asset) value[15]	IFVJ	45	208	685	455	906	629	3,032	4,099	3,608	238	604	501
Agents' balances and outstanding premiums in respect of direct insurance and facultative reinsurance contracts[16]													
UK	RYMF	9,060	8,216	9,022	8,171	8,470	8,222	9,584	9,089	8,829	10,212	10,453	10,047
Overseas	RYMG	480	575	55	73	-15	-200	711	868	702	750	858	855
Reinsurance, coinsurance and treaty balances[17]													
UK	RYMH	1,290	1,164	1,092	1,730	-165	-258	-182	-233	-921	-2,926	1,698	2,801
Overseas	RYMI	-48	-23	537	491	375	921	-4	217	244	-262	-1,344	-555
Outstanding interest, dividends and rents[18]	RYPN	1,067	1,255	1,108	958	765	726	724	881	581	657	436	466
Other debtors and assets not elsewhere classified[19]	RKAC	15,814	19,096	17,189	10,785	11,064	10,683	14,281	12,905	11,407	10,179	11,642	13,947
Direct investment for non-insurance subsidiary and associate companies in the UK[20]	RYNR	20,111	21,954	21,259	21,181	18,206	18,627	19,876	20,510	7,674	8,297	5,683	5,103
Direct investment for UK insurance subsidiary, associate and holding companies[20]	RYNS	4,745	6,936	7,669	7,159	6,252	6,795	6,365	6,886	c	5,152	5,753	5,414
Direct investment for overseas subsidiaries, associates, branches and agencies[20]	RYNT	9,657	9,445	10,815	12,124	9,868	9,073	8,004	7,147	c	7,737	8,586	8,170
TOTAL ASSETS	RKBY	163,030	161,477	176,543	166,631	141,547	143,412	161,227	160,903	142,025	137,550	141,290	140,881

Source: Office for National Statistics

c Suppressed to avoid the disclosure of confidential data. Components may not sum to totals due to rounding.
8 Includes money market funds; liquidity funds and cash liquidity funds.
9 Sterling and foreign currency.
10 Includes treasury bills. Excludes UK government securities.
11 Sterling and foreign currency commercial paper issued by: banks; building societies; other financial institutions and other issuing companies.
12 Includes floating rate notes maturing within one year of issue.
13 UK and overseas. Excludes derivative contracts.
14 Net of balances owed. Includes amounts due on securities bought and sold for future settlement.
15 Includes credit default products; employee stock options; other options; other swaps; futures; forwards and other derivative contracts not elsewhere classified.
16 Net of insurance liabilities.
17 Net of reinsurance bought and sold.
18 Net value.
19 Includes deferred acquisition costs.
20 Net asset value of attributable companies.

12.15b INSURANCE COMPANIES' BALANCE SHEET: GENERAL BUSINESS ASSETS AND LIABILITIES AT MARKET VALUES

continued

£ million

		2006	2007	2008	2009	2010	2011	2012	2013	2014	2015	2016	2017
LIABILITIES													
Borrowing	AHMI	19,052	18,597	20,214	15,602	14,431	14,281	12,460	12,133	7,081	5,816	4,402	4,537
Banks and building societies in the UK[9]	JX3J	3,148	675	343	744	621	606	819	794	734	393	461	688
Other UK[21]	IFHX	10,445	10,885	13,179	9,965	9,835	8,991	8,401	8,792	5,200	4,062	2,505	2,495
Overseas[22]	RYMD	5,459	7,037	6,692	4,893	3,975	4,684	3,240	2,547	1,147	1,361	1,436	1,354
General business technical reserves[23]	RKCT	77,221	71,146	76,980	70,947	58,484	57,539	61,771	59,738	59,744	56,639	57,950	56,166
Provisions for taxation and dividends payable[24]	KVF2	2,656	2,486	858	345	40	7	240	424	595	487	450	493
Other creditors, provisions and liabilities not elsewhere classified[25]	RYPR	16,226	22,069	21,149	19,976	18,367	19,362	25,757	24,989	16,985	16,601	18,720	21,118
Excess of total assets over liabilities in respect of minority interests in UK subsidiary companies, reserves[26]	A8SI	47,875	47,179	57,342	59,761	50,225	52,223	60,999	63,619	57,620	58,007	59,768	58,567
TOTAL LIABILITIES	**RKBY**	**163,030**	**161,477**	**176,543**	**166,631**	**141,547**	**143,412**	**161,227**	**160,903**	**142,025**	**137,550**	**141,290**	**140,881**

Source: Office for National Statistics

c Suppressed to avoid the disclosure of confidential data. Components may not sum to totals due to rounding.
9 Sterling and foreign currency.
21 Includes issue of securities (other than ordinary shares); issue of sterling commercial paper and other borrowing not elsewhere classified.
22 Includes borrowing from banks; related companies and other borrowing not elsewhere classified.
23 Net of reinsurers share.
24 UK and overseas. Includes deferred tax net of amounts receivable.
25 UK and overseas. Includes derivative contracts which have a negative (liability) value.
26 Includes the 'net worth' of UK branches of overseas companies, including profit and loss account balances.

12.16a Individual insolvencies

England and Wales, not seasonally adjusted

	Total individual insolvencies	Bankruptcies[1]	Debt relief orders[2]	Individual voluntary arrangements[3]
2008	106,544	67,428	z	39,116
2009	134,142	74,670	11,831	47,641
2010	135,045	59,173	25,179	50,693
2011	119,943	41,876	29,009	49,058
2012	109,640	31,787	31,179	46,674
2013	100,998	24,571	27,546	48,881
2014	99,223	20,345	26,688	52,190
2015	80,404	15,845	24,175	40,384
2016	90,657	15,044	26,196	49,417
2017	99,219	15,105	24,894	59,220

z Not applicable   Source: Office for National Statistics

[1] Figures from 2011 Q2 onwards based on the date the bankruptcy order was granted by the court or the Adjudicator. From 6 April 2016, the process for people making themselves bankrupt moved online and out of the courts.

[2] Debt Relief Orders (DROs) came into effect on 6 April 2009 as an alternative route into individual insolvency. In April 2011 a change was introduced to the legislation to allow those who have built up value in a pension scheme to apply for debt relief under these provisions.

[3] Includes Deeds of Arrangement.

12.16b Individual insolvencies

Scotland, not seasonally adjusted

	Total individual insolvencies	Sequestrations (of which LILA/MAP)[1,2]		Protected trust deeds
2008	19,991	12,449	(7,133)	7,542
2009	23,541	14,415	(8,774)	9,126
2010	20,344	11,906	(6,801)	8,438
2011	19,650	11,128	(4,812)	8,522
2012	18,402	9,630	(3,886)	8,772
2013	14,250	7,189	(2,728)	7,061
2014	11,622	6,747	(2,533)	4,875
2015	8,785	4,477	(1,509)	4,308
2016	9,779	4,474	(1,795)	5,305
2017	10,585 r	4,691	(1,884)	5,894

Source: Accountant in Bankruptcy.

[1] On 1 April 2008, Part 1 of the Bankruptcy and Diligence etc. (Scotland) Act 2007 came into force making significant changes to some aspects of sequestration (bankruptcy), debt relief and debt enforcement in Scotland. This included the introduction of the new route into bankruptcy for people with low income and low assets (LILA). Of the number or sequestrations, individuals who meet LILA criteria are shown in brackets.

[2] On 1 April 2015, part of the Bankruptcy and Debt Advice (Scotland) Act came into force making significant changes to some aspects of sequestration (bankruptcy). This included the introduction of the Minimal Asset Process (MAP), which replaced the LILA route into sequestration; mandatory debt advice for people seeking statutory debt relief; a new online process for applying for sequestration; and an additional year for people to make contributions to repaying their debts (increasing from three years to four, in line with protected trust deeds).

12.16c Individual insolvencies

Northern Ireland, not seasonally adjusted

	Total individual insolvencies	Bankruptcies	Debt relief orders[1]	Individual voluntary arrangements
2008	1,638	1,079	z	559
2009	1,958	1,236	z	722
2010	2,323	1,321	z	1,002
2011	2,839	1,615	112	1,112
2012	3,189	1,452	506	1,231
2013	3,373	1,347	593	1,433
2014	3,395	1,367	536	1,492
2015	2,690	1,071	472	1,147
2016	2,582	997	366	1,219
2017	2,878	812	529	1,537

Source: Department for Enterprise, Trade and Investment, Northern Ireland.

z Not applicable

[1] Debt relief orders came into effect on 30 June 2011.

12.17a Company insolvencies[1,2]

England and Wales, not seasonally adjusted

	Total new company insolvencies[2,3]	Compulsory liquidations[3,4]	New creditors' voluntary liquidations[3]	Administrations[5,6,7]	Company voluntary arrangements[8]	Receivership appointments[9]
2008	21,072	5,494	9,995	4,808	586	189
2009	24,036	5,643	13,491	4,077	713	112
2010	19,603	4,792	11,346	2,652	757	56
2011	20,382	5,003	12,015	2,559	758	47
2012	19,485	4,261	11,999	2,361	829	35
2013	17,709	3,632	11,487	2,004	569	17
2014	16,293 r	3,755	10,356 r	1,601	559	22
2015	14,588 r	2,889	9,904 r	1,412	372	11
2016	16,420	2,930	11,794	1,346	345	5
2017	17,317	2,806	12,886	1,316	307	2

Sources: Insolvency Service (compulsory liquidations only); Companies House (all other insolvency types).

r Revised

[1] Data from 2000Q1 are not consistent with earlier data because of a change to the methodology. This does not affect compulsory liquidations.

[2] Excludes creditors' voluntary liquidations following administration (See Table 2).

[3] Includes partnership winding-up orders.

[4] Figures from 2011 Q2 onwards based on the date the winding-up order was granted by the court.

[5] Releases prior to 2012 Q4 showed administrations separately as "Administrator Appointments" and "In Administration - Enterprise Act".

[6] The figure for Q4 2006 includes 844 separate, limited companies created and managed by "Safe Solutions Accountancy Limited" for which Grant Thornton was appointed administrator.

[7] The figure for Q3 2008 includes 728 separate managed service companies.

[8] The figure for Q2 2012 includes 104 new CVAs recorded under "Health and Social Work" in June reflecting the fact that on 20 June 2012 156 companies in the Southern Cross Healthcare Group had CVAs approved.

[9] Data before 2000 Q1 include Law of Property Act and fixed charge receiverships, which are not insolvencies but which cannot be identified separately to insolvent receiverships under the previous methodology.

12.17b Company insolvencies[2]

Scotland, not seasonally adjusted

	Total new company insolvencies[2]	Compulsory liquidations[3]	New creditors' voluntary liquidations[2]	Administrations	Company voluntary arrangements	Receivership appointments
2008	893	561	104	190	4	34
2009	1,023	548	175	253	8	39
2010	1,316	779	252	229	7	49
2011	1,477	929	282	215	12	39
2012	1,429	954	246	173	25	31
2013	885	467	259	125	16	18
2014	974	662	210	87	13	2
2015	954	580	255	107	5	7
2016	1,036 r	588 r	285	145	13	5
2017	856	482	281	85	3	5

Source: Companies House.

r **Revised**

[1] Data from 2000 Q1 are not consistent with earlier data because of a change to the methodology.

[2] Data before 2000 Q1 includes creditors' voluntary liquidations following administration as under the previous methodology it is not possible to separate these CVLs out.

[3] Includes provisional liquidations.

12.17c Company insolvencies[1,2]

Northern Ireland, not seasonally adjusted

	Total new company insolvencies[4]	Compulsory liquidations[3]	New creditors' voluntary liquidations[4,5]	Administrations[4]	Company voluntary arrangements[4]	Administrative receiverships[4]
2008	:	158	51	:	:	:
2009	:	164	83	:	:	:
2010	544	250	121	109	56	8
2011	445	208	120	68	30	19
2012	530	252	134	84	42	18
2013	367	178	98	48	37	6
2014	379	221	90	28	37	3
2015	379	225	78	36	37	3
2016	393	243	88	31	27	4
2017	310	200	69	13	27	1

Source: Department for Enterprise, Trade and Investment, Northern Ireland; Companies House

r **Revised**

[1] Includes partnerships.

[2] Data from 2009 Q4 are not consistent with earlier data because of a change to the methodology. This does not affect compulsory liquidations. Data for Northern Ireland prior to 2010 are not available under the new methodology.

[3] Source: Department for Enterprise, Trade and Investment, Northern Ireland.

[4] Source: Companies House.

[5] Data before 2009 Q4 includes creditors' voluntary liquidations following administration as under the previous methodology it is not possible to separate these CVLs out.

12.18a Monetary financial institutions' consolidated balance sheet

£ millions

Not seasonally adjusted

Amounts outstanding of liabilities

	Currency, deposits and money market instruments						Financial derivatives (net)		Other securities issued		Other liabilities		Total liabilities/ assets
	Private sector		Public sector		Non-residents								
	Sterling	Foreign currency	Sterling	Foreign currency	Sterling	Foreign currency	Sterling	Foreign currency	Sterling	Foreign currency	Sterling	Foreign currency	
LPM	VYAX	VYAY	VYAZ	VYBA	VYBB	VYBC	VWKM	VWKN	VWKO	VWKP	VWKQ	VWKR	VYBF
2017 Jan	2,247,924	561,878	39,543	4,246	495,978	2,707,400	23,929	-31,287	34,520	247,708	441,077	220,183	6,993,100
Feb	2,249,401	589,970	42,028	3,278	496,227	2,769,649	72,207	-68,084	34,543	256,400	453,121	241,162	7,139,900
Mar	2,283,910	569,228	36,220	2,673	508,240	2,728,068	47,413	-41,079	34,305	252,862	431,358	223,483	7,076,683
Apr	2,310,127	569,250	37,310	4,276	498,933	2,679,311	9,730	-3,770	34,136	251,092	432,910	207,560	7,030,864
May	2,303,924	579,559	42,127	4,599	508,772	2,715,474	32,377	-21,323	33,460	257,713	455,412	178,343	7,090,438
Jun	2,311,207	575,621	41,095	3,697	523,016	2,706,952	15,614	-26,022	33,460	257,113	484,509	191,364	7,117,627
Jul	2,305,404	590,314	50,849	6,180	503,241	2,753,945	32,960	-32,891	33,560	260,514	458,513	207,268	7,169,858
Aug	2,329,544	610,594	51,500	5,421	517,404	2,774,660	48,142	-46,930	33,629	275,514	467,126	189,089	7,255,695
Sep	2,351,608	584,931	42,430	6,125	512,452	2,738,423	-8,080	-9,111	33,076	259,948	488,408	168,062	7,168,271
Oct	2,360,636	597,048	46,780	6,052	491,520	2,793,114	5,635	-21,809	33,442	257,545	478,701	176,152	7,224,816
Nov	2,366,129	614,867	46,282	6,124	491,471	2,791,404	-20,012	10,139	33,921	253,559	484,618	176,519	7,255,023
Dec	2,355,197	583,265	42,850	6,362	506,578	2,842,107	-19,614	21,677	33,943	254,763	501,156	175,336	7,303,622
2018 Jan	2,370,201 (s)	601,667	46,269	5,620	495,388	2,759,451	-114,129 (u)	110,738 (x)	33,437	253,195	481,715	176,268	7,219,820 (s)
Feb	2,365,413	609,029	47,361	5,213	488,154	2,837,147	35,651 (v)	-55,394 (y)	34,312	259,152	463,003	203,569	7,292,611
Mar	2,353,292	593,945	39,863	5,208	506,269	2,834,277	16,571 (w)	-28,235 (z)	32,877	255,360	467,133	189,886	7,266,445
Apr	2,353,661	609,910	41,565	7,361	496,828	2,838,997	103,695	-131,823	30,536	255,101	475,975	193,514	7,275,319
May	2,370,230	635,962	42,005	5,053	517,537	2,908,914	100,147	-113,136	30,465	257,017	473,538	182,717	7,410,450
Jun	2,372,159	628,224	41,117	5,308	504,213	2,858,939	42,497	-54,542	30,477	257,540	466,625	182,507	7,335,064
Jul	2,374,138	627,958	43,277	6,946	519,619	2,813,445	31,091	-43,651	30,727	260,738	422,454	193,687	7,280,426
Aug	2,378,423	636,901	44,236	5,694	505,563	2,820,350	7,948	-17,024	30,382	263,840	458,952	173,450	7,308,715
Sep	2,381,320 (t)	637,026	42,288	6,609	513,657	2,818,909	2,400	-25,390	30,010	263,568	461,914	164,124	7,296,434
Oct	2,393,342	667,308	42,910	5,977	505,472	2,886,968	60,195	-85,543	29,125	261,464	434,710	175,794	7,377,722
Nov	2,397,194	691,365	41,524	5,733	494,721	2,935,180	55,626	-84,301	29,004	263,137	434,994	169,050	7,433,228
Dec	2,409,258	663,605	40,321	4,123	496,154	2,966,592	37,729	-53,722	28,920	270,355	463,489	140,081	7,466,905
2019 Jan	2,393,650	640,581	41,304	4,120	494,396	2,875,890	40,662	-50,143	29,083	265,107	459,831	136,526	7,331,007
Feb	2,403,510	592,514	40,203	4,738	492,647	2,868,275	14,198	-26,606	29,410	260,736	468,544	128,934	7,277,102
Mar	2,409,414	594,373	35,718	6,231	511,015	2,925,738	69,368	-75,104	30,370	266,326	468,205	133,642	7,375,295
Apr	2,426,025	598,735	39,807	5,300	495,061	2,929,183	65,313	-78,562	30,344	271,244	454,223	143,558	7,380,229

Source: Bank of England

12.18a Monetary financial institutions' consolidated balance sheet

£ millions

Not seasonally adjusted

Notes to table

Movements in amounts outstanding can reflect breaks in data series as well as underlying flows. For changes data, users are recommended to refer directly to the appropriate series or data tables.

Further explanation can be found at: www.bankofengland.co.uk/statistics/Pages/iadb/notesiadb/Changes_flows_growth_rates.aspx.

(s)   Due to improvements in reporting at one institution, the amounts outstanding increased by £5bn. This effect has been adjusted out of the flows for January 2018.

(t)   Due to improvements in reporting at one institution, the amounts outstanding increased by £5bn. This effect has been adjusted out of the flows for September 2018.

(u)   Due to improvements in reporting by one institution, amounts outstanding decreased by £66bn.

(v)   Due to improvements in reporting by one institution, amounts outstanding increased by £8bn.

(w)   Due to improvements in reporting by one institution, amounts outstanding increased by £9bn.

(x)   Due to improvements in reporting by one institution, amounts outstanding increased by £66bn.

(y)   Due to improvements in reporting by one institution, amounts outstanding decreased by £8bn.

(z)   Due to improvements in reporting by one institution, amounts outstanding decreased by £9bn.

Explanatory notes can be found here: www.bankofengland.co.uk/statistics/details/further-details-about-monetary-financial-institutions-consolidated-balance-sheet-contributions-data

Copyright guidance and the related UK Open Government Licence can be viewed here: www.bankofengland.co.uk/legal

12.18b Monetary financial institutions' consolidated balance sheet

£ millions

Not seasonally adjusted

Amounts outstanding of assets

	Loans						Securities (other than financial derivatives)						Other assets	
	Private sector		Public sector		Non-residents		Private sector		Public sector		Non-residents			
	Sterling	Foreign currency	Sterling	Foreign currency	Sterling	Foreign currency	Sterling	Foreign currency	Sterling	Foreign currency	Sterling	Foreign currency	Sterling	Foreign currency
LPM	VYBG	VYBH	VYBI	VYBJ	VYBK	VYBL	VYBM (aa)	VYBN	VYBO	VYBP	VYBQ	VYBR	VYBS	VYBT
2017 Jan	2,072,809	385,805	16,586	1,995	301,412	2,737,797	226,409	12,548	610,527	426	60,028	465,869	63,446	37,442
Feb	2,074,570	418,126	13,390	1,107	301,333	2,802,631	225,227	13,465	633,323	401	60,196	491,557	62,327	42,249
Mar	2,095,761	407,141	13,791	223	308,916	2,748,652	224,450	12,917	636,643	473	58,695	468,360	61,818	38,843
Apr	2,101,362	412,545	11,313	3,277	316,688	2,685,603	228,983	11,541	639,479	438	59,472	458,598	61,901	39,662
May	2,110,113	413,646	11,446	2,919	312,815	2,712,171	228,928	11,478	643,009	484	59,555	479,762	62,411	41,700
Jun	2,119,587	409,425	11,640	1,076	315,894	2,717,630	249,148	10,629	631,862	464	59,173	478,937	70,800	41,363
Jul	2,087,663	409,178	46,699	3,321	299,097	2,790,918	251,920	11,143	632,471	461	60,286	464,194	71,290	41,216
Aug	2,101,200	413,100	48,099	2,466	312,541	2,852,236	251,272	11,751	634,274	353	58,220	452,901	72,130	45,153
Sep	2,133,311	388,078	48,169	2,785	308,698	2,781,709	245,499	10,985	620,717	344	58,314	465,867	66,220	37,576
Oct	2,124,113	397,236	49,177	2,947	307,574	2,830,557	244,574	11,000	625,337	245	50,507	474,657	67,387	39,505
Nov	2,137,998	402,302	48,553	2,848	308,466	2,845,752	240,248	11,076	617,247	245	50,518	481,661	68,479	39,630
Dec	2,140,390	379,784	49,002	3,321	307,161	2,904,640	240,906	13,113	617,145	229	49,218	486,992	69,871	41,848
2018 Jan	2,141,415 (ab)(ac)	411,349	47,007	1,944	314,115	2,784,071	234,265	12,197	612,169	232	52,351	498,204	71,390	39,112
Feb	2,138,611	413,915	48,970	1,469	309,397	2,889,874	231,966	11,197	615,169	331	52,535	467,978	71,847	39,352
Mar	2,142,405	392,632	52,381	1,956	310,062	2,911,742	228,327	10,527	611,675	313	51,565	440,830	71,500	40,530
Apr	2,173,840 (ad)	399,340	17,525 (ah)	3,412	306,319	2,906,898	228,008	10,603	615,716	321	51,567	453,436	69,549	38,784
May	2,166,326	423,893	18,098	2,201	319,450	3,011,150	228,240	11,904	628,618	183	52,750	432,968	70,345	44,322
Jun	2,194,430	401,666	20,018	2,314	303,258	2,943,572	234,264	11,743	621,048	189	55,461	432,924	71,888	42,291
Jul	2,208,675	396,756	20,026	3,740	305,310	2,908,689	199,390	11,596	612,757	86	57,923	440,307	71,806	43,365
Aug	2,211,609	407,649	21,552	3,977	296,702	2,938,967	199,124	11,585	614,418	83	59,054	427,237	72,951	43,805
Sep	2,240,563 (ae)	398,795	18,389	3,543	307,247	2,904,127	201,257	11,861	602,051	119	61,860	438,450	69,704	38,466
Oct	2,240,242	397,902	15,807	3,387	303,277	2,996,808	199,165	11,585	600,424	120	66,031	433,989	71,462	37,521
Nov	2,258,046	423,819	16,876	3,703	308,790	3,007,569	192,328	11,990	595,960	21	65,949	436,044	72,248	39,884
Dec	2,268,444	415,344	16,057	1,889	318,621	3,054,906	188,586	10,690	596,875	122	64,412	418,435	71,268	41,257
2019 Jan	2,272,491	415,114	15,967	2,758	309,767	2,898,704	188,319	9,901	605,732	84	70,083	429,405	72,828	39,852
Feb	2,268,284 (af)(ag)	377,438	16,595	3,574	303,615	2,886,634	188,199	10,081	603,686	79	72,628	433,863	73,187	39,238
Mar	2,288,117	372,858	16,784	3,970	329,203	2,925,198	191,290	9,742	603,007	78	72,863	451,080	73,510	37,596
Apr	2,276,194	383,464	15,630	2,401	319,783	2,920,527	188,375	9,486	603,334	89	72,965	470,109	76,645	41,227

Source: Bank of England

12.18b Monetary financial institutions' consolidated balance sheet

£ millions

Not seasonally adjusted

Notes to table

Movements in amounts outstanding can reflect breaks in data series as well as underlying flows. For changes data, users are recommended to refer directly to the appropriate series or data tables. Further explanation can be found at: www.bankofengland.co.uk/statistics/Pages/iadb/notesiadb/Changes_flows_growth_rates.aspx.

(aa) This series includes purchases of bonds made as part of the Bank of England's Corporate Bond Purchase Scheme. Data on Central Bank holdings of securities can be found in Bankstats Table B2.2. For further information on the Bank's treatment of securities transactions in credit statistics, see: www.bankofengland.co.uk/statistics/Documents/articles/2015/2may.pdf.

(ab) Due to improvements in reporting at one institution, the amounts outstanding decreased by £2bn. This effect has been adjusted out of the flows for January 2018.

(ac) Due to improvements in reporting at one institution, the amounts outstanding decreased by £2bn. This effect has been adjusted out of the flows for January 2018.

(ad) In order to bring reporting in line with the National Accounts, English housing associations were reclassified from public corporations to PNFCs with effect from April 2018 data. The amounts outstanding increased by £34bn. This effect has been adjusted out of the flows for April 2018.

(ae) Due to improvements in reporting at one institution, the amounts outstanding increased by £5bn. This effect has been adjusted out of the flows for September 2018.

(af) Due to improvements in reporting at one institution, the amounts outstanding decreased by £2bn. This effect has been adjusted out of the flows for February 2019.

(ag) Due to improvements in reporting at one institution, the amounts outstanding decreased by £1bn. This effect has been adjusted out of the flows for February 2019.

(ah) In order to bring reporting in line with the National Accounts, English housing associations were reclassified from public corporations to PNFCs with effect from April 2018 data. The amounts outstanding decreased by £34bn. This effect has been adjusted out of the flows for April 2018.

Explanatory notes can be found here: www.bankofengland.co.uk/statistics/details/further-details-about-monetary-financial-institutions-consolidated-balance-sheet-contributions-data

Copyright guidance and the related UK Open Government Licence can be viewed here: www.bankofengland.co.uk/legal

12.18c Monetary financial institutions' consolidated balance sheet

£ millions

Not seasonally adjusted

Changes in liabilities

Currency, deposits and money market instruments

LPM	Private sector Sterling	Private sector Foreign currency	Public sector Sterling	Public sector Foreign currency	Non-residents Sterling	Non-residents Foreign currency	Financial derivatives (net) Sterling	Financial derivatives (net) Foreign currency	Other securities issued Sterling	Other securities issued Foreign currency	Other liabilities Sterling	Other liabilities Foreign currency	Total liabilities/ assets
	VYAA	VYAB	VYAC	VYAD	VYAE	VYAF	VWKG	VWKH	VWKI	VWKJ	VWKK	VWKL	VYAI
2017 Jan	966	21,178	-818	974	-1,118	27,151	-33,848	28,087	-657	2,218	38,100	-23,247	58,985
Feb	-827	23,541	2,202	-983	87	50,832	49,268	-35,163	23	8,403	-10,789	4,158	90,751
Mar	30,056	-20,702	-5,800	-603	10,873	-35,602	-24,794	26,998	-238	-3,403	-15,479	-12,853	-51,547
Apr	25,835	13,653	1,089	1,682	-4,733	17,413	-37,683	37,675	-170	3,732	-451	-12,293	45,750
May	-6,225	1,189	4,818	244	9,752	-1,082	22,647	-16,608	-676	1,181	19,009	-39,948	-5,699
Jun	7,239	-3,638	-1,032	-904	14,191	-5,933	-18,263	-4,273	1	-1,064	18,923	14,647	19,894
Jul	-3,628	7,080	6,902	1,671	-20,235	15,807	17,346	-5,230	99	1,754	-27,124	9,751	4,193
Aug	23,107	-587	651	-916	14,243	-48,125	15,182	-10,866	69	-2,335	-3,549	843	-12,282
Sep	6,565	283	-9,690	926	-8,987	49,588	-55,327	30,527	-553	-4,038	42,169	-4,245	47,216
Oct	8,764	8,919	4,357	-93	-20,758	47,430	13,715	-12,725	366	-2,565	-8,980	4,357	42,788
Nov	4,790	22,060	-498	104	81	20,857	-25,246	31,398	479	-3,105	8,705	637	60,261
Dec	-11,149	-33,505	-3,432	210	14,995	42,414	398	11,689	22	95	10,456	-5,913	26,281
2018 Jan	15,112	37,555	3,418	-581	-11,460	5,157	-94,515	86,612	-506	2,456	2,304	6,299	51,849
Feb	-4,886	-4,479	1,092	-512	-7,258	17,953	149,780	-163,162	875	-1,234	-16,471	30,340	2,037
Mar	-12,148	-7,477	-7,498	63	18,105	35,528	-19,080	26,736	-1,435	-536	-3,969	-7,229	21,061
Apr	-4,233	11,299	5,106	2,080	-9,115	-22,868	83,373	-100,488	-2,343	240	18,642	3,264	-15,043
May	16,513	15,262	440	-2,416	20,695	14,324	-3,549	18,678	-70	-1,682	-16,838	-13,410	47,947
Jun	1,844	-12,061	-888	218	-14,222	-69,037	-57,650	57,683	11	-1,176	-4,736	547	-99,467
Jul	1,992	-4,781	2,109	1,594	15,398	-63,602	-11,406	10,449	250	1,390	-40,445	8,973	-78,078
Aug	5,248	4,204	443	-1,294	-13,714	-13,738	-23,143	26,488	-345	1,405	34,109	-22,011	-2,349
Sep	-1,260	3,385	-585	938	7,463	11,950	-5,548	-8,078	-372	1,224	16,393	-3,605	21,905
Oct	12,116	24,754	622	-660	-7,991	39,473	57,795	-60,001	-885	-3,568	-30,561	24,303	55,396
Nov	3,410	21,918	-1,387	-248	-10,352	45,789	-11,068	7,732	-121	751	6,294	1,950	64,668
Dec	12,146	-30,382	-1,172	-1,649	2,357	39,298	-17,897	30,221	-84	5,329	25,229	-19,888	43,509
2019 Jan	-15,467	-5,846	494	101	-1,663	-16,085	2,933	4,004	163	1,958	-9,051	-15,931	-54,388
Feb	9,795	-40,410	-1,101	692	-2,593	31,833	-26,464	24,310	327	-132	16,254	-9,173	3,340
Mar	7,484	-5,681	-4,451	1,432	19,532	20,389	55,269	-48,648	103	3,149	-20,449	-1,054	27,074
Apr	15,321	5,664	3,952	-936	-12,279	6,694	-4,055	-3,375	-27	5,475	-184	10,070	26,319

Notes to table

Source: Bank of England

Explanatory notes can be found here: www.bankofengland.co.uk/statistics/details/further-details-about-monetary-financial-institutions-consolidated-balance-sheet-contributions-data

Copyright guidance and the related UK Open Government Licence can be viewed here: www.bankofengland.co.uk/legal

12.18d Monetary financial institutions' consolidated balance sheet

£ millions

Not seasonally adjusted

Changes in assets

| | Loans | | | | | | Securities (other than financial derivatives) | | | | | | Other assets | |
| | Private sector | | Public sector | | Non-residents | | Private sector | | Public sector | | Non-residents | | | |
LPM	Sterling VYAJ	Foreign currency VYAK	Sterling VYAL	Foreign currency VYAM	Sterling VYAN	Foreign currency VYAO	Sterling VYAP (ai)	Foreign currency VYAQ	Sterling VYAR	Foreign currency VYAS	Sterling VYAT	Foreign currency VYAU	Sterling VYAV	Foreign currency VYAW
2017 Jan	11,269	24,384	44	193	505	5,416	29,824	-752	-4,454	-30	3,774	7,857	-17,696	-1,351
Feb	-1,022	25,761	-3,433	-893	-685	49,420	-3,716	217	5,904	-29	-1,334	17,138	-1,165	4,589
Mar	19,902	-9,348	400	-891	7,084	-48,214	-1,009	125	5,562	74	-1,026	-20,379	-506	-3,321
Apr	5,714	15,575	-2,477	3,084	10,048	4,851	6,068	-830	850	-21	1,016	6	117	1,748
May	8,616	-6,169	684	-403	-3,869	-14,568	-1,683	8	1,142	45	176	8,325	409	1,589
Jun	9,669	-4,213	-21	-1,847	3,094	7,402	3,821	-848	3,485	-17	-184	808	-1,808	552
Jul	2,607	-4,009	565	2,236	-17,346	34,374	1,731	260	-395	2	1,466	-17,821	508	15
Aug	16,117	-1,107	1,331	-940	13,175	3,054	-1,293	686	-10,021	-117	-3,924	-25,377	-132	-3,734
Sep	13,820	-6,292	-651	423	-7,907	23,682	-2,627	-213	5,653	4	2,402	30,417	-5,774	-5,720
Oct	-8,207	8,412	1,008	173	-1,100	41,187	1,025	-341	3,006	-103	-8,524	3,271	1,176	1,805
Nov	14,105	7,292	-625	-105	766	37,508	-1,744	183	-8,866	4	730	9,469	1,089	454
Dec	3,005	-24,035	450	456	-1,207	49,239	-1,142	1,407	-4,815	-16	-1,844	1,303	1,329	2,154
2018 Jan	2,353	42,972	-1,995	-1,340	5,818	-32,813	290	-842	7,709	14	4,668	24,192	1,755	-932
Feb	-2,135	-6,148	1,963	-497	-4,713	42,690	-539	-387	2,780	92	-99	-30,999	650	-622
Mar	2,813	-15,595	3,411	503	670	61,328	-1,884	14	-13,284	-13	-153	-18,155	-291	1,696
Apr	-1,676	5,643	-1,497	1,338	-4,223	-28,120	334	-74	9,305	3	-308	9,693	-2,053	-3,407
May	-6,605	16,646	650	-1,243	13,599	52,140	-4,549	678	3,654	-146	-754	-31,728	834	4,773
Jun	29,166	-24,629	1,996	96	-16,224	-88,549	800	-411	-1,453	5	2,511	-1,431	936	-2,277
Jul	14,535	-7,504	9	1,402	2,057	-53,985	-33,990	-145	-5,173	-104	1,862	2,226	-101	830
Aug	3,203	8,392	1,525	223	-8,681	8,179	-1,274	292	1,445	-3	467	-18,144	1,234	794
Sep	24,060	-6,928	-3,162	-418	10,651	-18,078	4,303	544	-583	36	3,505	16,216	-3,102	-5,137
Oct	206	-3,722	-2,582	-179	-3,866	65,926	-1,035	-557	-7,099	-1	4,846	2,914	1,788	-1,242
Nov	18,245	24,508	1,069	307	5,500	7,515	-998	394	1,341	-99	-17	3,821	763	2,319
Dec	13,734	-9,416	334	-1,834	13,589	60,787	-3,766	-1,682	-6,773	102	-2,256	-19,820	-889	1,399
2019 Jan	4,418	11,069	-209	937	-8,840	-79,434	-331	-1,095	4,388	-35	5,617	8,121	1,442	-438
Feb	-1,809	-33,952	627	862	-6,219	31,441	-66	-250	3,282	4	1,474	7,611	351	-8
Mar	20,108	-9,600	189	354	26,834	3,209	715	149	-16,959	-3	-1,655	5,429	308	-2,005
Apr	-7,965	11,306	-1,324	-1,565	-9,129	1,109	-520	-66	9,023	11	353	18,392	2,998	3,695

Source: Bank of England

Notes to table

(ai) This series includes purchases of bonds made as part of the Bank of England's Corporate Bond Purchase Scheme. Data on Central Bank holdings of securities can be found in Bankstats Table B2.2. For further information on the Bank's treatment of securities transactions in credit statistics, see: www.bankofengland.co.uk/statistics/Documents/articles/2015/2may.pdf.

Explanatory notes can be found here: www.bankofengland.co.uk/statistics/details/further-details-about-monetary-financial-institutions-consolidated-balance-sheet-contributions-data

Copyright guidance and the related UK Open Government Licence can be viewed here: www.bankofengland.co.uk/legal

12.19 Selected interest rates, exchange rates and security prices

	Monthly average of 4 UK Banks' base rates	Monthly average rate of discount, 3 month Treasury bills, Sterling	Monthly average yield from British Government Securities, 20 year Nominal Par Yield	Monthly average Spot exchange rate, US$ into Sterling
	IUMAAMIH [b] [c]	IUMAAJNB [a]	IUMALNPY [d] [e]	XUMAUSS
31-Jan-10	0.5	0.4874	4.4161	1.6162
28-Feb-10	0.5	0.4878	4.5158	1.5615
31-Mar-10	0.5	0.5109	4.5698	1.5053
30-Apr-10	0.5	0.5083	4.558	1.534
31-May-10	0.5	0.4976	4.3143	1.4627
30-Jun-10	0.5	0.4839	4.2079	1.4761
31-Jul-10	0.5	0.498	4.1754	1.5299
31-Aug-10	0.5	0.4945	3.9824	1.566
30-Sep-10	0.5	0.4966	3.8954	1.5578
31-Oct-10	0.5	0.5061	3.8876	1.5862
30-Nov-10	0.5	0.4935	4.1149	1.5961
31-Dec-10	0.5	0.4913	4.2568	1.5603
31-Jan-11	0.5	0.5055	4.3204	1.5795
28-Feb-11	0.5	0.5363	4.405	1.613
31-Mar-11	0.5	0.5603	4.298	1.6159
30-Apr-11	0.5	0.5676	4.2933	1.6345
31-May-11	0.5	0.5265	4.1371	1.6312
30-Jun-11	0.5	0.5174	4.1277	1.6214
31-Jul-11	0.5	0.4996	4.0974	1.6145
31-Aug-11	0.5	0.4531	3.7132	1.6348
30-Sep-11	0.5	0.4649	3.4205	1.5783
31-Oct-11	0.5	0.4608	3.2581	1.576
30-Nov-11	0.5	0.4387	3.0313	1.5804
31-Dec-11	0.5	0.2996	2.9803	1.5585
31-Jan-12	0.5	0.3239	2.906	1.551
29-Feb-12	0.5	0.3912	3.089	1.5802
31-Mar-12	0.5	0.4248	3.1744	1.5823
30-Apr-12	0.5	0.4236	3.1259	1.6014
31-May-12	0.5	0.3561	2.9073	1.5905
30-Jun-12	0.5	0.3422	2.725	1.5571
31-Jul-12	0.5	0.2943	2.6453	1.5589
31-Aug-12	0.5	0.2398	2.65	1.5719
30-Sep-12	0.5	0.2478	2.7497	1.6116
31-Oct-12	0.5	0.2368	2.7969	1.6079
30-Nov-12	0.5	0.2243	2.7846	1.5961
31-Dec-12	0.5	0.249	2.8391	1.6144
31-Jan-13	0.5	0.2677	3.0069	1.5957
28-Feb-13	0.5	0.3147	3.1035	1.5478
31-Mar-13	0.5	0.3391	2.9654	1.5076
30-Apr-13	0.5	0.3447	2.776	1.5316
31-May-13	0.5	0.3065	2.9096	1.5285
30-Jun-13	0.5	0.3066	3.1584	1.5478
31-Jul-13	0.5	0.3124	3.2719	1.5172
31-Aug-13	0.5	0.2812	3.4083	1.5507
30-Sep-13	0.5	0.289	3.4662	1.5865
31-Oct-13	0.5	0.3141	3.3198	1.6094
30-Nov-13	0.5	0.2865	3.3978	1.6104
31-Dec-13	0.5	0.2555	3.497	1.6375

12.19 Selected interest rates, exchange rates and security prices

	Monthly average of 4 UK Banks' base rates IUMAAMIH [b] [c]	Monthly average rate of discount, 3 month Treasury bills, Sterling IUMAAJNB [a]	Monthly average yield from British Government Securities, 20 year Nominal Par Yield IUMALNPY [d] [e]	Monthly average Spot exchange rate, US$ into Sterling XUMAUSS
31-Jan-14	0.5	0.3211	3.4287	1.647
28-Feb-14	0.5	0.3624	3.3668	1.6567
31-Mar-14	0.5	0.3882	3.3469	1.6622
30-Apr-14	0.5	0.3688	3.3127	1.6743
31-May-14	0.5	0.284	3.2387	1.6844
30-Jun-14	0.5	0.3586	3.3055	1.6906
31-Jul-14	0.5	0.4271	3.228	1.7069
31-Aug-14	0.5	0.3978	2.9986	1.6709
30-Sep-14	0.5	0.4351	2.978	1.6305
31-Oct-14	0.5	0.3964	2.7955	1.6068
30-Nov-14	0.5	0.4114	2.7235	1.578
31-Dec-14	0.5	0.4101	2.4744	1.564
31-Jan-15	0.5	0.377	2.1112	1.5143
28-Feb-15	0.5	0.338	2.2503	1.5334
31-Mar-15	0.5	0.4295	2.2853	1.4957
30-Apr-15	0.5	0.4322	2.2061	1.4967
31-May-15	0.5	0.4511	2.4791	1.547
30-Jun-15	0.5	0.467	2.6379	1.5568
31-Jul-15	0.5	0.4895	2.5968	1.556
31-Aug-15	-	0.4643	2.4307	1.5583
30-Sep-15	-	0.4532	2.4175	1.5326
31-Oct-15	-	0.48	2.4421	1.5339
30-Nov-15	-	0.4806	2.5375	1.519
31-Dec-15	-	0.4551	2.4749	1.4983
31-Jan-16	-	0.4823	2.386	1.4379
29-Feb-16	-	0.4711	2.1972	1.4296
31-Mar-16	-	0.4505	2.222	1.425
30-Apr-16	-	0.4489	2.2451	1.4312
31-May-16	-	0.4362	2.1985	1.4518
30-Jun-16	-	0.4046	1.9487	1.4209
31-Jul-16	-	0.3689	1.5432	1.3141
31-Aug-16	-	0.2281	1.2661	1.31
30-Sep-16	-	0.2104	1.3562	1.3142
31-Oct-16	-	0.1711	1.655	1.2329
30-Nov-16	-	0.1385	1.9506	1.2431
31-Dec-16	-	0.0518	1.9985	1.2488
31-Jan-17	-	0.1441	1.9869	1.2351
28-Feb-17	-	0.1131	1.9125	1.249
31-Mar-17	-	0.0231	1.8129	1.2348
30-Apr-17	-	0.0709	1.6923	1.2652
31-May-17	-	0.0592	1.7489	1.2933
30-Jun-17	-	0.0759	1.7134	1.2813
31-Jul-17	-	-	1.88	1.2994
31-Aug-17	-	-	1.7476	1.2955
30-Sep-17	-	-	1.823	1.3324
31-Oct-17	-	-	1.9186	1.3197
30-Nov-17	-	-	1.8509	1.3219
31-Dec-17	-	-	1.7831	1.3402

Source: Bank of England

[a] These data have been discontinued with effect from 20th July 2017.
[b] Data obtained from Barclays Bank, Lloyds Bank, HSBC, and National Westminster Bank whose rates are used to compile this series.
   Where all the rates did not change on the same day a spread is shown.
[c] This series will end on 31-July-2015
[d] Calculated using the Variable Roughness Penalty (VRP) model.
[e] The monthly average figure is calculated using the available daily observations within each month.

12.20 Mergers and Acquisitions in the UK by other UK Companies:
Category of Expenditure

£ million

		Expenditure [2]				Percentage of Expenditure [2]		
		Cash						
	Total	Independent Companies	Subsidiaries	Issues of Ordinary Shares[2]	Issues of Fixed Interest Securities[2]	Cash	Issues of Ordinary Shares	Issues of Fixed Interest Securities
	DUCM	DWVW	DWVX	AIHD	AIHE	DWVY	DWVZ	DWWA
Annual								
2006	28,511	14,573	8,131	5,472	335	..	..	2
2007	26,778	13,671	6,507	4,909	1,691	76	18	6
2008	36,469	31,333	2,851	1,910	375	94	5	1
2009	12,195	2,937	709	8,435	114	30	69	1
2010[1]	12,605	6,175	4,520	1,560	350	85	12	3
2011	8,089	4,432	2,667	719	271	87	10	4
2012	3,413	1,937	789	419	268	82	10	8
2013	7,665	3,690	3,475	353	147	92	6	2
2014	8,032	3,249	1,947	2,782	51	65	35	−
2015	6,920	3,365	1,871	1,418	265	74	22	4
2016	24,688	5,493	5,308	13,471	418	43	55	2
2017	18,783	8,036	3,097	7,286	364	59	39	2
Quarterly								
2011 Q1	1,500	552	651	240	57	80	16	4
2011 Q2	3,346	2,355	704	204	83	92	6	2
2011 Q3	1,452	828	462	75	87	89	5	6
2011 Q4	1,791	697	850	200	44	87	11	2
2012 Q1	1,070	518	199	323	30	67	30	3
2012 Q2	1,041	575	269	54	143	81	5	14
2012 Q3	610	409	100	8	93	84	1	15
2012 Q4	692	435	221	34	2	95	5	−
2013 Q1	2,825	567	2,216	26	16	98	1	1
2013 Q2	2,438	1,992	316	80	50	95	3	2
2013 Q3	1,166	587	332	230	17	79	20	1
2013 Q4	1,236	544	611	17	64	94	1	5
2014 Q1	1,613	896	103	612	2	62	38	−
2014 Q2	1,625	478	1,051	50	45	94	3	3
2014 Q3	3,152	476	656	2,019	−	36	64	−
2014 Q4	1,642	1,399	137	101	4	94	6	−
2015 Q1	1,755	1,075	314	281	84	79	16	5
2015 Q2	2,739	854	789	1,019	77	60	37	3
2015 Q3	1,195	666	464	64	2	95	5	−
2015 Q4	1,231	770	304	54	102	88	4	8
2016 Q1	11,871	630	3,603	7,617	22	36	64	−
2016 Q2	6,780	1,983	388	4,237	172	35	62	3
2016 Q3	3,575	1,610	943	875	148	71	24	5
2016 Q4	2,462	1,270	374	742	76	67	30	3
2017 Q1	3,295	2,615	419	138	122	92	4	4
2017 Q2	4,855	..	..	136	92	..	3	2
2017 Q3	5,379	..	..	..	..	..	..	..
2017 Q4	5,254	1,922	390	..	..	44	..	..
2018 Q1	7,176	1,936	1,957	..	..	54	..	..

† indicates earliest revision, if any
− indicates data is zero or less than £0.5m
Disclosive data indicated by ..
2 A new method for compiling M&A statistics was introduced from Q1 2018, and as a consequence there is a discontinuity in the number of transactions reported.

Source: Mergers and Acquisitions Surveys, Office for National Statistics

Service industry

Chapter 13

Service industry

Annual Business Inquiry (Tables 13.1, 13.3 and 13.4)

The Annual Business Inquiry (ABI) estimates cover all UK businesses registered for Value Added Tax (VAT) and/or Pay As You Earn (PAYE). The businesses are classified to the 2007 Standard Industrial Classification (SIC(2007)) headings listed in the tables. The ABI obtains details on these businesses from the Office for National Statistics (ONS) Inter-Departmental Business Register (IDBR).

As with all its statistical inquiries, ONS is concerned to minimise the form-filling burden of individual contributors and as such the ABI is a sample inquiry. The sample was designed as a stratified random sample of about 66,600 businesses; the inquiry population is stratified by SIC(2007) and employment using the information from the register.

The inquiry results are grossed up to the total population so that they relate to all active UK businesses on the IDBR for the sectors covered.

The results meet a wide range of needs for government, economic analysts and the business community at large. In official statistics the inquiry is an important source for the national accounts and input-output tables, and also provides weights for the indices of production and producer prices. Additionally, inquiry results enable the UK to meet statistical requirements of the European Union. Data from 1995 and 1996 were calculated on a different basis from those for 1997 and later years. In order to provide a link between the two data series, the 1995 and 1996 data were subsequently reworked to provide estimates on a consistent basis.

Revised ABI results down to SIC(2007) 4 digit class level for 1995–2007, giving both analysis and tabular detail, are available from the ONS website at: www.statistics.gov.uk, with further extracts and bespoke analyses available on request. This service replaces existing publications.

From 2008 to 2014 the Total net capital expenditure, Total capital expenditure - acquisitions, and Total capital expenditure - disposals published values included a small element of Not yet in production (NYIP). From 2015 onwards NYIP is not estimated for and is no longer included in the published values.

Retail trade: index numbers of value and volume (Table 13.2)

The main purpose of the Retail Sales Inquiry (RSI) is to provide up-to-date information on short period movements in the level of retail sales. In principle, the RSI covers the retail activity of every business classified in the retail sector (Division 52 of the 2007 Standard Industrial Classification (SIC(2007)) in Great Britain. A business will be classified to the retail sector if its main activity is one of the individual 4 digit SIC categories within Division 52. The retail activity of a business is then defined by its retail turnover, that is the sale of all retail goods (note that petrol, for example, is not a retail good).

The RSI is compiled from the information returned to the statutory inquiries into the distribution and services sector. The inquiry is addressed to a stratified sample of 5,000 businesses classified to the retail sector, the stratification being by 'type of store' (the individual 4 digit SIC categories within Division 52) and by size. The sample structure is designed to ensure that the inquiry estimates are as accurate as possible. In terms of the selection, this means that:

• each of the individual 4 digit SIC categories are represented – their coverage depending upon the relative size of the category and the variability of the data
• within each 4 digit SIC category the larger retailers tend to be fully enumerated with decreasing proportions of medium and smaller retailers

The structure of the inquiry is updated periodically by reference to the more comprehensive results of the Annual Business Inquiry (ABI). The monthly inquiry also incorporates a rotation element for the smallest retailers. This helps to spread the burden more fairly, as well as improving the representativeness between successive benchmarks.

Service industry

13.1a Retail Trade, except of motor vehicles and motorcycles

Standard Industrial Classification (Revised 2007) - Divison	Description	Year	Number of enterprises	Total turnover	Approximate gross value added at basic prices (aGVA)	Total purchases of goods, materials and services	Total employment		costs	Total capital expenditure			Total stocks and work in progress		
							point in time[1]	average during the year[1]		net[2]	acquisitions[2]	disposals[2]	value at end of year	value at beginning of year	increase during year
			Number	£ million	£ million	£ million	Thousand	Thousand	£ million	£ million	£ million	£ million	£ million	£ million	£ million
47	Retail trade, except of motor vehicles and motorcycles	2008	194,677	311,745	65,123	246,237	3,054	3,106	38,466	9,069	10,495	1,426	26,110	25,519	591
		2009	187,890	319,318	69,924	249,058	3,073	3,139	38,914	7,784	9,113	1,330	26,539	25,759	781
		2010	187,230	332,131	71,500	261,676	3,016	3,040	39,816	6,149	9,258	3,108	28,743	26,425	2,317
		2011	189,119	342,147	70,871	271,203	3,032	3,060	41,592	9,051	10,879	1,828	30,356	28,964	1,392
		2012	187,616	349,327	73,876	275,052	3,029	3,062	42,184	9,477	10,789	1,312	30,021	28,720	1,301
		2013	189,828	359,088	81,368	278,326	3,052	3,084	43,441	9,510	10,627	1,117	31,791	29,863	1,929
		2014	186,737	371,522	83,728	288,613	3,130	3,154	44,996	11,228	12,543	1,314	33,571	32,461	1,110
		2015	186,922	375,289	85,967	290,764	3,143	3,191	47,233	11,145	12,265	1,120	34,774	32,952	1,822
		2016	195,501	389,106	85,051	304,566	3,111	3,159	47,803	13,216	14,149	934	37,482	36,354	1,128
		2017	207,608	406,446	87,872	320,626	3,159	3,169	50,404	12,442	13,392	950	37,635	35,636	1,999

Source: Annual Business Survey (ABS)

The sum of constituent items in tables may not always agree exactly with the totals shown due to rounding.

1. Total employment - point in time and Total employment - average during the year are from the Business Register and Employment Survey (BRES). Caution should be taken when combining financial data from the ABS with employment data from BRES due to differences in methodology. More information can be found in the ABS Technical Report.

2. From 2008 to 2014 the Total net capital expenditure, Total capital expenditure - acquisitions, and Total capital expenditure - disposals published values included a small element of Not yet in production (NYIP). From 2015 onwards NYIP is not estimated for and is no longer included in the published values.

13.1b Retail trade, except of motor vehicles and motor-cycles

£ million (Inclusive of VAT)

	2009	2010	2011	2012	2013	2014	2015	2016
TOTAL TURNOVER	350,269	367,446	385,692	393,423	404,534	418,185	422,119	438,171
RETAIL TURNOVER	331,256	347,144	363,280	368,176	378,917	387,983	394,339	405,464
Fruit (including fresh, chilled, dried, frozen, canned and processed)	6,469	6,821	7,662	6,846	7,132	7,351	8,503	8,642
Vegetables (including fresh, chilled, dried, frozen canned and processed)	9,949	10,533	10,677	12,380	13,219	13,273	12,675	12,982
Meat (including fresh, chilled, smoked, frozen, canned and processed)	17,855	17,909	19,541	20,254	21,230	21,794	21,778	21,096
Fish, crustaceans and molluscs (including fresh, chilled, smoked, frozen, canned and processed)	2,989	2,978	3,295	3,453	3,458	3,486	3,692	3,623
Bakery products and cereals (including rice and pasta products)	16,719	17,875	18,818	20,078	20,352	20,372	22,187	20,757
Sugar, jam, honey, chocolate and confectionery (including ice-cream)	8,618	8,844	8,688	9,162	9,829	10,233	9,725	9,848
Alcoholic drink	15,609	16,618	17,445	18,184	19,219	19,605	18,979	19,443
Non-alcoholic beverages (including tea, coffee, fruit drinks and vegetable drinks)	7,756	8,018	9,176	9,161	9,652	9,860	9,984	10,070
Tobacco (excluding smokers requisites e.g. pipes, lighters etc)	11,411	11,369	11,838	11,885	11,405	11,607	11,696	11,595
Milk, cheese and eggs (including yoghurts and cream)	11,562	10,918	11,219	11,441	11,579	11,767	11,766	11,216
Oils and fats (including butter and margarine)	1,440	1,444	1,613	1,611	1,695	1,826	1,698	2,015
Food products not elsewhere classified (including sauces, herbs spices, soups)	5,149	4,427	4,528	3,890	3,858	3,971	4,352	3,955
Pharmaceutical products	3,636	3,946	4,390	4,725	4,916	5,359	5,060	4,826
National Health Receipts	11,792	12,067	11,405	11,680	11,100	12,262	11,875	11,218
Other medical products and therapeutic appliances and equipment	3,213	4,210	3,638	4,067	3,913	3,712	5,130	4,631
Other appliances, articles and products for personal care	13,148	14,109	15,107	15,702	16,776	17,172	17,917	17,920
Other articles of clothing, accessories for making clothing	2,505	3,395	3,781	4,051	4,317	4,321	5,337	5,109
Garments	35,556	36,706	37,976	38,730	41,466	43,253	43,691	45,662
Footwear (excluding sports shoes)	7,685	7,458	7,720	8,720	8,765	9,213	10,270	9,710
Travel goods and other personal effects not elsewhere classified	1,722	2,352	2,518	2,506	3,183	3,081	3,197	2,697
Household textiles (including furnishing fabrics, curtains etc)	4,535	4,862	4,188	4,158	4,731	4,786	5,265	5,069
Household and personal appliances whether electric or not	6,736	6,917	7,278	7,367	6,768	8,351	8,205	8,461
Glassware, tableware and household utensils (including non-electric)	2,824	2,754	3,369	3,668	3,899	3,622	3,573	3,828
Furniture and furnishings	13,226	12,313	13,475	12,395	13,606	14,492	16,773	17,002
Audio and visual equipment (including radios, televisions and video recorders	6,016	6,245	5,353	5,821	4,916	4,124	3,913	4,422
Recording material for pictures and sound (including audio and video tapes, blank and pre recorded records etc)	3,052	2,787	2,547	2,668	2,560	2,034	2,031	2,284
Information processing equipment (including printers, software, calculators and typewriters)	4,384	4,530	5,425	4,862	5,475	5,058	5,871	5,640
Decorating and DIY supplies	6,820	7,319	7,528	5,580	6,500	6,952	6,883	7,421
Tools and equipment for house and garden	2,964	3,088	3,566	3,887	4,235	3,916	4,476	4,761
Books	2,623	2,648	2,579	3,084	2,623	2,636	3,001	2,888
Newspapers and periodicals	3,405	3,947	3,675	3,410	2,906	3,022	2,890	2,709
Stationery and drawing materials and miscellaneous printed matter	4,423	4,025	4,432	4,397	4,620	4,304	4,372	4,892
Carpets and other floor coverings (excluding bathroom mats, rush and door mats)	2,840	2,824	3,159	3,144	2,806	3,114	3,494	3,161
Photographic and cinematographic equipment and optical instruments	1,854	1,625	1,643	1,298	736	809	869	955
Telephone and telefax equipment (including mobile phones)	3,249	3,143	3,665	4,437	4,481	3,277	3,483	6,131
Jewellery, silverware and plate; watches and clocks	5,693	6,045	6,235	6,626	6,340	7,449	7,659	10,519
Works of art and antiques (including furniture, floor coverings and jewellery)	1,347	1,756	1,493	1,254	1,807	2,374	3,455	1,813
Equipment and accessories for sport, camping, recreation and musical instruments	4,416	4,864	5,118	4,944	4,355	5,350	4,491	5,555
Spare parts and accessories for all types of vehicle and sales of bicycles	924	1,148	983	950	1,244	1,520	1,675	1,343
Games, toys, hobbies (including video game software, video game computers that plug into the tv, video-games cassettes and CD-ROM'S)	7,484	7,514	7,492	7,154	7,018	7,855	8,147	8,310
Other goods not elsewhere classified (including sale of new postage stamps and sales of liquid and solid fuels)	5,007	5,896	5,929	5,788	6,689	7,570	6,532	9,703
Non-durable household goods (including household cleaning, maintenance products) & paper products and other non-durable household goods	5,280	4,727	4,923	4,936	5,596	6,199	5,753	5,782
Natural or artificial plants and flowers	2,973	3,609	4,023	3,501	3,772	3,523	4,895	4,437
Pets and related products (including pet food)	3,884	3,896	4,166	4,154	4,664	4,697	4,697	5,789
Petrol, diesel, lubricating oil and other petroleum products	30,515	36,664	40,001	40,171	39,507	37,432	32,425	35,574

Source: Annual Business Survey (ABS)

The sum of constituent items in tables may not always agree exactly with the totals shown due to rounding.

13.2 Retail trade: index numbers of value and volume of sales[1]

Great Britain

Non-seasonally adjusted

			2008	2009	2010	2011	2012	2013	2014	2015	2016	2017
		Sales in 2017 £ thousand							Weekly average (2016=100)			
Value												
All retailing	J5AH	405,780,949	81.3	81.9	84.1	88.5	90.5	92.9	95.7	96.7	100	104.7
Large	J5AI	319,261,807	78.8	80.1	83.9	88.3	90.7	93.4	95.7	97.3	100	104.3
Small	J5AJ	86,519,128	90.8	88.7	85.1	89.2	89.9	91.1	95.6	94.2	100	106
All retailing excluding automotive fuel	J43S	366,086,256	79.7	81.3	83.3	86.3	88.6	91.5	95	96.8	100	104.3
Predominantly food stores	EAFS	158,112,834	82.1	86.5	87.9	91.7	94.3	97.1	98.2	98.3	100	102.4
Predominantly non-food stores	EAFT	168,934,659	85.5	84.1	86.2	87.5	88.9	90.5	95.3	97.6	100	103.5
Non specialised predominantly non-food stores	EAGE	34,819,554	69	70.2	75.1	78.1	82.9	86.6	91.2	95	100	101.9
Textile, clothing, footwear and leather	EAFU	48,610,646	83.2	84.2	88.7	92	93.4	95.7	99.1	102.2	100	106.3
Household goods stores	EAFV	33,210,825	104	98.8	94.9	92.8	92	89.1	94	99	100	101.6
Other specialised non-food stores	EAFW	52,293,634	86.6	83.9	85.8	86.4	86.8	89.3	95.5	94.2	100	103.3
Non-store retailing	J596	39,038,763	40.7	43.3	47.9	55.1	60.4	70.5	78.6	85.7	100	117.6
Automotive fuel	J43H	39,694,693	96.1	87.9	92	109.6	109.3	106	102.1	96	100	107.7
Volume												
All retailing	J5DD	405,780,949	87.7	88.0	87.1	87.3	87.6	88.6	91.7	95.4	100.0	101.7
All retailing excluding automotive fuel	J448	366,086,256	86.5	87.0	87.0	86.7	87.3	88.8	92.2	95.5	100.0	101.9
Predominantly food stores	EAGW	158,112,834	96.7	97.8	96.0	94.8	94.5	94.1	94.7	96.6	100.0	99.9
Predominantly non-food stores	EAGX	168,934,659	86.4	85.8	86.6	85.8	86.4	87.9	93.1	96.7	100.0	101.3
Non specialised predominantly non-food stores	EAHI	34,819,554	71.9	72.3	75.7	76.4	80.3	83.7	88.6	93.8	100.0	100.1
Textile, clothing, footwear and leather	EAGY	48,610,646	82.6	88.1	92.8	93.8	94.2	95.3	98.6	102.0	100.0	103.7
Household goods stores	EAGZ	33,210,825	105.2	99.0	92.8	88.6	87.0	85.0	90.5	97.7	100.0	99.4
Other specialised non-food stores	EAHA	52,293,634	87.2	84.0	84.4	83.2	83.3	86.0	92.9	93.3	100.0	101.2
Non-store retailing	J5CL	39,038,763	41.2	44.3	48.0	53.9	58.7	68.3	76.1	84.9	100.0	114.7
Automotive fuel	J43V	39,694,693	98.3	96.7	87.3	91.1	89.0	86.8	87.5	94.0	100.0	99.9

1 See chapter text.

Source: Office for National Statistics

Please note that the indices have been re-referenced so the value of 100 is in 2016

13.3 Wholesale and retail trade and repair of motor vehicles and motorcycles

Standard Industrial Classification (Revised 2007) - Divison	Description	Year	Number of enterprises	Total turnover	Approximate gross value added at basic prices (aGVA)	Total purchases of goods, materials and services	Total employment		costs	Total capital expenditure			Total stocks and work in progress		
							point in time[1]	average during the year[1]		net[2]	acquis-itions[2]	disposals[2]	value at end of year	value at beginning of year	increase during year
			Number	£ million	£ million	£ million	Thousand	Thousand	£ million	£ million	£ million	£ million	£ million	£ million	£ million
45	Wholesale and retail trade and repair of motor vehicles and motorcycles	2008	67,683	135,669	20,957	115,372	532	535	10,925	1,044	2,110	1,066	16,136	15,402	735
		2009	66,372	125,764	18,348	105,457	500	539	10,445	885	1,669	784	13,624	15,545	-1,921
		2010	66,239	132,077	22,150	111,243	511	510	10,439	1,035	1,747	711	15,008	13,640	1,368
		2011	67,298	137,364	23,419	115,292	523	535	10,658	1,292	2,067	775	16,451	15,107	1,344
		2012	67,274	142,461	22,347	120,649	510	522	10,615	1,174	1,927	753	17,067	16,490	577
		2013	67,754	152,510	24,236	129,877	525	537	11,570	1,530	2,296	766	18,139	16,525	1,613
		2014	68,108	168,247	26,857	143,563	547	521	11,895	2,155	2,873	718	20,692	18,606	2,086
		2015	70,517	180,214	30,983	152,154	559	573	12,361	2,444	3,379	935	23,139	20,322	2,817
		2016	74,674	192,893	28,086	166,376	576	590	13,121	2,198	3,082	884	24,042	22,576	1,466
		2017	75,910	196,998	27,526	171,715	563	555	13,537	2,246	3,170	924	26,269	24,190	2,079

Source: Annual Business Survey (ABS)

The sum of constituent items in tables may not always agree exactly with the totals shown due to rounding.

1. Total employment - point in time and Total employment - average during the year are from the Business Register and Employment Survey (BRES). Caution should be taken when combining financial data from the ABS with employment data from BRES due to differences in methodology. More information can be found in the ABS Technical Report.

2. From 2008 to 2014 the Total net capital expenditure, Total capital expenditure - acquisitions, and Total capital expenditure - disposals published values included a small element of Not yet in production (NYIP). From 2015 onwards NYIP is not estimated for and is no longer included in the published values.

13.4 Accommodation and food service activities

Standard Industrial Classification (Revised 2007) Section, Division, Group, Class	Description	Year	Number of enterprises	Total turnover	Approximate gross value added at basic prices (aGVA)	Total purchases of goods, materials and services	Total employment		costs
							point in time[1]	average during the year[1]	
			Number	£ million	£ million	£ million	Thousand	Thousand	£ million
I	Accommodation and food service activities	2011	130,336	72,322	34,826	37,563	1,937	1,869	19,987
		2012	128,831	74,355	37,764	37,272	1,970	1,910	21,132
		2013	131,322	77,033	38,357	38,625	2,029	1,970	21,611
		2014	131,979	81,937	42,842	39,433	2,108	2,044	22,743
		2015	135,740	85,526	45,214	40,376	2,182	2,134	24,832
		2016	149,063	92,785	48,564	44,178	2,308	2,252	26,466
		2017	152,846	96,777	50,418	46,551	2,325	2,290	28,764
55	Accommodation	2011	15,159	18,566	9,838	8,775	429	407	5,331
		2012	15,205	19,822	11,149	8,677	426	409	5,752
		2013	15,365	20,222	11,353	8,860	433	416	5,646
		2014	15,513	22,646	13,964	8,866	440	419	5,857
		2015	15,842	23,132	13,943	9,292	455	444	6,286
		2016	16,808	25,272	15,447	9,916	484	473	6,890
		2017	17,351	26,206	15,792	10,540	468	448	7,281
55.1	Hotels and similar accommodation	2011	9,575	14,150	7,619	6,532	358	342	4,424
		2012	9,449	15,186	8,572	6,609	351	339	4,823
		2013	9,364	15,764	8,925	6,852	356	344	4,703
		2014	9,300	17,391	10,715	6,784	357	347	4,758
		2015	9,312	17,535	10,657	6,912	369	366	5,133
		2016	9,583	18,946	11,759	7,228	390	387	5,567
		2017	9,744	19,335	11,911	7,441	376	368	5,918
55.2	Holiday and other short stay accommodation	2011	3,296	1,476	726	774	33	30	351
		2012	3,247	1,497	913	590	33	31	344
		2013	3,411	1,562	969	601	32	30	343
		2014	3,582	1,864	1,276	598	36	31	417
		2015	3,821	2,103	1,356	780	36	33	430
		2016	4,289	1,971	1,289	693	40	36	452
		2017	4,536	2,388	1,357	1,043	39	34	521
55.3	Camping grounds, recreational vehicle parks and trailer parks	2011	1,775	2,739	1,385	1,378	34	31	507
		2012	1,796	2,771	1,415	1,360	37	34	516
		2013	1,818	2,630	1,292	1,310	39	37	524
		2014	1,854	2,950	1,659	1,319	*	*	582
		2015	1,915	2,867	1,607	1,294	42	39	568
		2016	2,039	3,606	1,955	1,687	46	42	708
		2017	2,086	3,859	2,126	1,822	44	39	724
55.9	Other accommodation	2011	513	200	108	92	4	4	49
		2012	713	368	250	118	5	4	70
		2013	772	265	168	98	6	5	76
		2014	777	440	314	166	*	*	101
		2015	794	627	324	305	8	7	155
		2016	897	749	445	308	8	7	163
		2017	985	624	397	234	8	7	118
56	Food and beverage service	2011	115,177	53,756	24,988	28,787	1,507	1,462	14,656
		2012	113,626	54,533	26,615	28,594	1,544	1,501	15,380
		2013	115,957	56,811	27,003	29,765	1,596	1,553	15,964
		2014	116,466	59,291	28,878	30,567	1,668	1,625	16,886
		2015	119,898	62,395	31,270	31,084	1,727	1,690	18,546
		2016	132,255	67,512	33,117	34,262	1,824	1,780	19,576
		2017	135,495	70,572	34,626	36,011	1,857	1,842	21,484
56.1	Restaurants and mobile food service activities	2011	63,712	25,690	12,301	13,447	767	765	6,886
		2012	64,460	25,934	13,367	13,332	782	783	7,187
		2013	67,144	28,740	14,328	14,454	859	860	8,067
		2014	68,876	29,571	14,975	14,789	885	834	8,531
		2015	72,793	32,326	16,562	15,924	937	901	9,601
		2016	83,634	35,336	18,015	17,398	1,006	967	10,497
		2017	86,630	37,652	18,918	18,899	1,021	998	11,509

13.4 Accommodation and food service activities

Standard Industrial Classification (Revised 2007) Section, Division, Group, Class	Description	Year	Number of enterprises	Total turnover	Approximate gross value added at basic prices (aGVA)	Total purchases of goods, materials and services	Total employment		costs
							point in time[1]	average during the year[1]	
			Number	£ million	£ million	£ million	Thousand	Thousand	£ million
56.2	Event catering and other food service activities	2011	7,967	8,553	4,418	4,138	231	212	3,204
		2012	7,651	8,925	4,430	4,521	262	249	3,688
		2013	8,003	8,269	3,928	4,343	233	222	3,381
		2014	8,221	9,122	4,644	4,474	255	260	3,682
		2015	8,753	9,439	5,082	4,357	253	256	4,004
		2016	10,501	10,222	5,367	4,873	265	264	4,027
		2017	10,957	10,546	5,802	4,841	272	269	4,318
56.21	Event catering activities	2011	6,267	7,837	4,119	3,715	213	196	2,993
		2012	5,514	7,844	3,879	3,968	241	228	3,411
		2013	5,885	3,196	1,394	1,809	86	82	967
		2014	6,149	3,465	1,534	1,934	96	99	1,004
		2015	6,906	3,458	1,684	1,784	88	89	1,061
		2016	8,542	5,362	2,782	2,600	135	136	1,924
		2017	8,656	5,316	2,880	2,507	142	140	2,025
56.29	Other food service activities	2011	1,700	716	299	422	18	17	211
		2012	2,137	1,081	551	552	21	20	278
		2013	2,118	5,073	2,534	2,534	147	140	2,414
		2014	2,072	5,657	3,110	2,540	159	162	2,678
		2015	1,847	5,981	3,398	2,573	165	167	2,943
		2016	1,959	4,860	2,585	2,274	130	128	2,102
		2017	2,301	5,230	2,922	2,334	130	129	2,293
56.3	Beverage serving activities	2011	43,498	19,513	8,269	11,203	509	484	4,566
		2012	41,515	19,675	8,817	10,742	501	469	4,504
		2013	40,810	19,802	8,748	10,968	503	471	4,516
		2014	39,369	20,599	9,259	11,303	527	531	4,674
		2015	38,352	20,630	9,626	10,803	538	533	4,941
		2016	38,120	21,954	9,735	11,990	554	548	5,052
		2017	37,908	22,374	9,906	12,271	564	575	5,656

Source: Annual Business Survey (ABS)

The following symbols and abbreviations are used throughout the ABS releases;

*   Information suppressed to avoid disclosure

..   not available

-    nil or less than half the level of rounding

The sum of constituent items in tables may not always agree exactly with the totals shown due to rounding.

1. Total employment - point in time and Total employment - average during the year are from the Business Register and Employment Survey (BRES). Caution should be taken when combining financial data from the ABS with employment data from BRES due to differences in methodology. More information can be found in the ABS Technical Report.

2. From 2008 to 2014 the Total net capital expenditure, Total capital expenditure - acquisitions, and Total capital expenditure - disposals published values included a small element of Not yet in production (NYIP). From 2015 onwards NYIP is not estimated for and is no longer included in the published values.

13.4 Accommodation and food service activities

Standard Industrial Classification (Revised 2007) Section, Division, Group, Class	Description	Year	Total capital expenditure			Total stocks and work in progress		
			net [2]	acquisitions [2]	disposals [2]	value at end of year	value at beginning of year	increase during year
			£ million	£ million	£ million	£ million	£ million	£ million
I	Accommodation and food service activities	2011	3,311	4,339	1,028	1,484	1,368	117
		2012	3,930	4,704	774	1,716	1,713	3
		2013	3,946	4,837	891	1,544	1,523	21
		2014	5,629	6,888	1,259	1,585	1,494	92
		2015	7,212	8,046	835	1,587	1,460	126
		2016	7,791	8,671	880	2,084	1,927	157
		2017	7,520	8,174	654	1,891	1,729	162
55	Accommodation	2011	1,416	1,641	225	506	443	63
		2012	1,898	2,050	152	469	452	17
		2013	1,836	2,045	209	397	413	-16
		2014	1,656	2,275	619	472	438	34
		2015	2,869	3,167	298	431	385	46
		2016	3,382	3,625	244	702	641	61
		2017	2,579	2,834	254	639	553	86
55.1	Hotels and similar accommodation	2011	946	1,137	191	201	176	25
		2012	1,282	1,397	115	185	172	13
		2013	1,173	1,335	162	207	189	18
		2014	1,159	1,728	569	195	190	5
		2015	2,034	2,283	250	184	187	-3
		2016	2,538	2,725	187	248	219	29
		2017	1,849	2,046	197	228	226	2
55.2	Holiday and other short stay accommodation	2011	169	184	15	59	45	15
		2012	254	269	16	64	58	6
		2013	195	216	21	37	34	3
		2014	223	234	11	54	48	6
		2015	372	388	16	83	59	24
		2016	219	240	21	61	58	3
		2017	211	219	9	55	48	7
55.3	Camping grounds, recreational vehicle parks and trailer parks	2011	238	257	18	245	222	23
		2012	258	279	21	217	219	-3
		2013	254	278	24	152	190	-38
		2014	237	269	32	220	198	22
		2015	359	390	31	163	138	25
		2016	593	625	33	390	361	29
		2017	504	551	47	355	277	77
55.9	Other accommodation	2011	62	63	1	1	1	-
		2012	104	105	1	3	3	-
		2013	214	216	2	1	-	-
		2014	36	44	8	3	2	1
		2015	104	105	1	1	1	-
		2016	33	36	3	2	2	-
		2017	16	18	2	1	1	-
56	Food and beverage service	2011	1,895	2,698	803	978	925	54
		2012	2,032	2,654	622	1,247	1,261	-13
		2013	2,110	2,792	682	1,147	1,110	37
		2014	3,973	4,613	640	1,113	1,056	58
		2015	4,343	4,880	537	1,155	1,075	80
		2016	4,409	5,046	637	1,382	1,286	96
		2017	4,941	5,340	399	1,252	1,177	75
56.1	Restaurants and mobile food service activities	2011	1,203	1,327	123	412	379	33
		2012	1,380	1,512	132	562	532	30
		2013	1,449	1,645	196	480	454	26
		2014	2,427	2,600	173	451	418	34
		2015	2,555	2,687	133	593	524	69
		2016	2,508	2,721	213	695	660	35
		2017	2,981	3,106	125	621	592	29

13.4 Accommodation and food service activities

Standard Industrial Classification (Revised 2007) Section, Division, Group, Class	Description	Year	Total capital expenditure			Total stocks and work in progress		
			net [2]	acquisitions [2]	disposals [2]	value at end of year	value at beginning of year	increase during year
			£ million	£ million	£ million	£ million	£ million	£ million
56.2	Event catering and other food service activities	2011	175	188	14	145	140	5
		2012	127	135	8	158	154	4
		2013	110	117	7	176	167	9
		2014	125	158	32	179	180	-1
		2015	148	168	20	160	150	10
		2016	263	297	34	184	169	15
		2017	266	281	15	178	163	16
56.21	Event catering activities	2011	116	129	13	123	124	-1
		2012	113	120	7	133	131	2
		2013	48	52	5	77	68	9
		2014	73	95	22	75	72	3
		2015	106	114	7	73	65	9
		2016	163	186	23	102	90	12
		2017	188	201	13	93	82	11
56.29	Other food service activities	2011	59	60	1	22	16	6
		2012	14	15	1	24	23	1
		2013	62	65	3	100	99	1
		2014	52	63	11	103	107	-4
		2015	42	54	13	86	85	1
		2016	100	111	11	82	79	3
		2017	79	80	1	86	81	5
56.3	Beverage serving activities	2011	517	1,183	666	421	405	16
		2012	524	1,007	483	527	575	-47
		2013	551	1,030	479	491	488	2
		2014	1,420	1,855	434	483	458	25
		2015	1,640	2,024	384	403	402	1
		2016	1,638	2,027	389	503	457	46
		2017	1,694	1,953	259	452	422	30

Source: Annual Business Survey (ABS)

The following symbols and abbreviations are used throughout the ABS releases;
* Information suppressed to avoid disclosure
.. not available
- nil or less than half the level of rounding

The sum of constituent items in tables may not always agree exactly with the totals shown due to rounding.

1. Total employment - point in time and Total employment - average during the year are from the Business Register and Employment Survey (BRES). Caution should be taken when combining financial data from the ABS with employment data from BRES due to differences in methodology. More information can be found in the ABS Technical Report.

2. From 2008 to 2014 the Total net capital expenditure, Total capital expenditure - acquisitions, and Total capital expenditure - disposals published values included a small element of Not yet in production (NYIP). From 2015 onwards NYIP is not estimated for and is no longer included in the published values.

Sources:

This index of sources gives the titles of official publications or other sources containing statistics allied to those in the tables of this Annual Abstract. These publications provide more detailed analyses than are shown in the Annual Abstract. This index includes publications to which reference should be made for short–term (monthly or quarterly) series.

Table number	Government department or other organisation

Chapter 1: Area

1.1	ONS Geography Codes; Office for National Statistics - Standard Area Measurement for UK Local Authority Districts (SAM 2017)

Chapter 2: Parliamentary elections

2.1	British Electoral Facts 1832-2012; Plymouth University for the Electoral Commission
2.2a	Chronology of British Parliamentary By-elections 1833-1987; British Electoral Facts 1832-2006;
	House of Commons Library, RP10/50 By-election results 2005-10; SN05833 By-elections since 2010 General Election
	CBP 7417 By-elections since the 2015 General Election; CBP-8280 By-elections since the 2017 General Election
2.2b	British Parliamentary Election Results; House of Commons Library By-election results
2.3a	House of Commons Library Briefing Paper CBP7594, National Assembly for Wales Elections: 2016
2.3b	British Electoral Facts 1832-2006; House of Commons Library Research and Briefing Papers
2.4	British Electoral Facts 1832-2006; Electoral Office for Northern Ireland
2.5	House of Commons Library briefing paper CBP7639

Chapter 3 International Development

3.1	Department for International Development
3.2	Department for International Development
3.3	Department for International Development

Chapter 4 Labour Market

4.1	Labour Force Survey, Office for National Statistics
4.2	Labour Force Survey, Office for National Statistics
4.3	Labour Force Survey, Office for National Statistics
4.4	Labour Force Survey, Office for National Statistics
4.5a	Eurostat, OECD, National Statistical Offices. Labour market statistics
4.5b	Labour Disputes Survey, Office for National Statistics
4.6	Annual Civil Service Employment Survey, Office for National Statistics
4.7	Labour Force Survey, Office for National Statistics
4.8	Labour Force Survey, Office for National Statistics
4.9	Office for National Statistics
4.10	Labour Market Statistics,Office for National Statistics; Nomisweb
4.11a	Annual Survey of Hours and Earnings, Office for National Statistics
4.11b	Annual Survey of Hours and Earnings, Office for National Statistics
4.12a	Annual Survey of Hours and Earnings, Office for National Statistics
4.12b	Annual Survey of Hours and Earnings, Office for National Statistics
4.13	Office for National Statistics
4.14a	Monthly wages and salaries survey
4.14b	Monthly wages and salaries survey
4.14c	Monthly wages and salaries survey
4.14d	Monthly wages and salaries survey
4.15a	Annual Survey of Hours and Earnings, Office for National Statistics
4.15b	Annual Survey of Hours and Earnings, Office for National Statistics
4.16	Annual Survey of Hours and Earnings, Office for National Statistics
4.17	Certification Officer Annual Report 2017/18

Chapter 5 Social Protection

5.1	HM Revenue and Customs; Department for Work and Pensions
5.2	HM Revenue and Customs; National Insurance Fund Account Great Britain
5.3	HM Revenue and Customs
5.4	Department for Work and Pensions
5.5	Department for Work and Pensions
5.6	Department for Work and Pensions
5.7	Stat-Xplore, Department for Work and Pensions
5.8	Stat-Xplore, Department for Work and Pensions
5.9a	HM Revenue and Customs
5.9b	HM Revenue and Customs
5.10	HM Revenue and Customs
5.11	HM Revenue and Customs

Sources

Table number	Government department or other organisation
5.12	Stat-Xplore, Department for Work and Pensions
5.13	Stat-Xplore, Department for Work and Pensions
5.14	Stat-Xplore, Department for Work and Pensions
5.15a	War Pensions Computer System
5.15b	War Pensions Computer System
5.16	Stat-Xplore, Department for Work and Pensions
5.17	Stat-Xplore, Department for Work and Pensions
5.18	Stat-Xplore, Department for Work and Pensions
5.19	Stat-Xplore, Department for Work and Pensions

Chapter 6 External Trade

6.1	Office for National Statistics
6.2	Office for National Statistics
6.3	Office for National Statistics
6.4	Office for National Statistics
6.5	Office for National Statistics
6.6	Office for National Statistics
6.7	Office for National Statistics
6.8	Office for National Statistics
6.9	Office for National Statistics
6.10	Office for National Statistics
6.11	Office for National Statistics
6.12	Office for National Statistics
6.13	Office for National Statistics
6.14	Office for National Statistics
6.15	Office for National Statistics
6.16	Office for National Statistics
6.17	Office for National Statistics
6.18	Office for National Statistics

Chapter 7 Research and development

7.1a	Office for National Statistics
7.1b	Office for National Statistics
7.2	Office for National Statistics
7.3	Office for National Statistics
7.4	Office for National Statistics
7.5a	Office for National Statistics
7.5b	Office for National Statistics

Chapter 8 Personal income, expendiure & wealth

8.1	Survey of Personal Incomes, HM Revenue and Customs
8.2	Office for National Statistics
8.3	Office for National Statistics
8.4	Office for National Statistics
8.5	Office for National Statistics

Chapter 9 Lifestyles

9.1	Department for Culture, Media and Sport
9.2	National Readership Surveys Ltd; PAMCo
9.3	Department for Culture, Media and Sport
9.4	Taking Part Survey, Department for Digital, Culture, Media and Sport
9.5	British Film Institute
9.6	comScore, BFI RSU analysis
9.7	International Passenger Survey, Office for National Statistics
9.8	International Passenger Survey, Office for National Statistics
9.9	International Passenger Survey, Office for National Statistics
9.10	Visit England
9.11	Visit England
9.12	Gambling Commission
9.13	Office for National Statistics
9.14a	Taking Part Survey, Department for Culture, Media and Sport
9.14b	Taking Part Survey, Department for Culture, Media and Sport
9.14c	Taking Part Survey, Department for Culture, Media and Sport
9.15	Taking Part Survey, Department for Culture, Media and Sport
9.16	Community Life Survey, Cabinet Office
9.17	International Passenger Survey, Office for National Statistics
9.18	Office for National Statistics
9.19	RAJAR/Ipsos MORI/RSMB

Table number	Government department or other organisation

Chapter 10 Environment

10.1	Ricardo Energy & Environment, Office for National Statistics
10.2	Ricardo Energy & Environment, Office for National Statistics
10.3	Department for Business, Energy & Industrial Strategy
10.4	Ricardo Energy and Environment, Department for Business, Energy & Industrial Strategy, Office for National Statistics
10.5	Department for Business, Energy & Industrial Strategy
10.6	Department for Business, Energy & Industrial Strategy
10.7	Department for Business, Energy & Industrial Strategy
10.8	Department for Environment, Food and Rural Affairs; Food and Agriculture Organization of the United Nations; Eurostat; Kentish Cobnuts Association; British Geological Survey; Office for National Statistics;
10.9	Met Office; National Hydrological Monitoring Programme, Centre for Ecology and Hydrology
10.10	Met Office
10.11	Environment Agency
10.12	Water Classification Hub, Scottish Environment Protection Agency
10.13	Water PLCs; Environment Agency; National Hydrological Monitoring Programme, Centre for Ecology and Hydrology
10.14	Office of Water Services (OFWAT)
10.15	Environment Agency
10.16	Environment Agency; Natural Resources Wales, Scottish Environment Protection Agency, Department of Agriculture, Environment and Rural Affairs (DAERA)
10.17	Environment Agency
10.18	Department for Business, Energy and Industrial Strategy (BEIS); Oil and Gas Authority
10.19a	Department for Environment, Food and Rural Affairs
10.19b	WasteDataFlow, Natural Resources Wales
10.19c	Scottish Environment Protection Agency
10.19d	Northern Ireland Environment Agency, NISRA, LPS
10.20a	Department for Environment Food and Rural affairs
10.20b	Department for Environment Food and Rural affairs; Office for National Statistics
10.20c	WasteDataFlow, Department for Environment, Food and Rural Affairs
10.20d	Department for Environment Food and Rural affairs
10.21	The Chartered Institute of Environmental Health
10.22	Office for National Statistics

Chapter 11 Housing

11.1a	Department for Communities and Local Government
11.1b	Welsh Assembly Government
11.1c	Scottish Government
11.1d	Continuous Household Survey
11.2	Ministry of Housing, Communities and Local Government; Welsh Assembly Government, Scottish Government; Department for Social Development (NI)
11.3	P2 returns from local authorities; National House Building Council (NHBC); Approved inspector data returns; Welsh Assembly Government; Scottish Government; Department of Finance and Personnel (DFPNI); District Council Building Control (NI)
11.4a	Department for Communities and Local Government
11.4b	New house building data collection, Welsh Government
11.5	HM Courts and Tribunals Service CaseMan; Possession Claim OnLine (PCOL); Council of Mortgage Lenders (CML)
11.6	UK Finance, Compendium of Housing Finance Statistics and Housing Finance website
11.7	Department for Communities and Local Government ; P1E Homelessness returns (quarterly)
11.8	Welsh Government, Homelessness data collection
11.9	Scottish Government

Chapter 12 Banking and Finance

12.1a	Bank of England
12.1b	Bank of England
12.2a	Cheque and Credit Clearing Company Ltd
12.2b	UK Payments Administration Ltd, Cheque and Credit Clearing Company Ltd
12.2c	UK Payments Administration Ltd, Cheque and Credit Clearing Company Ltd
12.2d	UK Payments Administration Ltd, Cheque and Credit Clearing Company Ltd
12.2e	Cheque and Credit Clearing Company Ltd
12.3a	Bank of England
12.3b	Bank of England
12.3c	Bank of England
12.3d	Bank of England
12.3e	Bank of England
12.3f	Bank of England
12.3g	Bank of England
12.3h	Bank of England
12.4a	Bank of England
12.4b	Bank of England
12.4c	Bank of England
12.4d	Bank of England
12.5a	Bank of England

Sources

Table number	Government department or other organisation
12.5b	Bank of England
12.6a	Bank of England
12.6b	Bank of England
12.7	Bank of England
12.8	Bank of England
12.9	Bank of England
12.10	Bank of England
12.11	Bank of England
12.12	Bank of England
12.13a	Office for National Statistics
12.13b	Office for National Statistics
12.14	Office for National Statistics
12.15a	Office for National Statistics
12.15b	Office for National Statistics
12.16a	Office for National Statistics
12.16b	Accountant in Bankruptcy (AiB)
12.16c	Department for Enterprise, Trade and Investment, Northern Ireland (DETINI)
12.17a	Insolvency Service; Companies House
12.17b	Companies House
12.17c	Department for Enterprise, Trade and Investment Northern Ireland (DETINI); Companies House
12.18a	Bank of England
12.18b	Bank of England
12.18c	Bank of England
12.18d	Bank of England
12.19	Bank of England
12.20	Mergers and Acquisitions Surveys, Office for National Statistics

Chapter 13 Services

13.1a	Annual Business Survey (ABS), Office for National Statistics
13.1b	Annual Business Survey (ABS), Office for National Statistics
13.2	Office for National Statistics
13.3	Annual Business Survey (ABS), Office for National Statistics
13.4	Annual Business Survey (ABS), Office for National Statistics